WORLDWIDE PRACTICAL PETROLEUM RESERVOIR ENGINEERING METHODS

WORLDWIDE PRACTICAL PETROLEUM RESERVOIR ENGINEERING METHODS

H.C. "Slip" Slider

PennWell Books
PennWell Publishing Company
Tulsa, Oklahoma

PennWell Publishing Company
1421 South Sheridan Road/P. O. Box 1260
Tulsa, Oklahoma 74101

Library of Congress Cataloging in Publication Data

Slider, H.C.
Worldwide practical petroleum reservoir engineering methods.

Includes bibliographical references and index. 1. Oil reservoir engineering.
2. Oil fields—Production methods. I. Title.
TN871.S557 1983 0000 622′.3382 0000 83-4067
ISBN 0-87814-234-7

Printed in the United States of America

1 2 3 4 5 87 86 85 84 83

Dedicated to Jennie, who continues to be an understanding wife—most of the time.

Contents

Dedication v
Preface xi
Preface to Practical Petroleum Reservoir Engineering Methods xv
List of Problems xix

1 Acquiring Reservoir Engineering Data 1
Reservoir data requiring predrilling planning 2
Reservoir data obtainable early in the life of a well 4
Obtaining routine reservoir data 5

2 Reservoir Fluid Flow Fundamentals 11
Characteristics of the Darcy equation 11
Characteristics of various flow regimes 26
Steady-state flow 33
General problems in fluid flow calculations 46
Correcting for static pressure differences 55
Additional problems 61
Notes 66
Additional references 67

3 Unsteady-State and Pseudosteady-State Flow 69
Physical description 69
Radial diffusivity equation 72
Constant-terminal-rate solution 79
Pseudosteady-state flow 93
Constant-terminal-pressure solution 103
Effective compressibility 115
Superposition 118
Linear unsteady-state flow 133
Spherical unsteady-state flow 133
Additional problems 135
Notes 140
Additional references 141

4 Well-Pressure Behavior Analysis 143
Productivity-index tests 143
Constant-rate drawdown tests 154
Reservoir limit tests 159
Pressure-buildup analysis 168
Interpreting drillstem-test data 220
Type-curve matching 229

Pulse-test interpretation 233
Additional problems 245
Notes 252
Additional references 254

5 Gas Reservoir Engineering 257
Natural gas properties 259
Material balance 273
Fluid flow in gas reservoirs 298
Equipment capacity limitations on deliverability 344
Predicting reservoir performance 351
Predicting the behavior of microdarcy reservoirs 362
Determining the size of a gas reservoir 366
Additional problems 371
Notes 380
Additional references 380

6 Fluid Distribution and Frontal Displacement 383
Initial saturation distribution in a reservoir 384
Reservoir saturation distribution during displacement 405
Interface tilt during displacement 430
Additional problems 440
Notes 443
Additional references 444

7 Material Balance 445
A general material-balance equation 447
Pressure-volume-temperature (PVT) relationships 453
Behavior of produced GOR 457
Material balance with gas liberation 462
Pore-volume changes in material balance 464
Modifications of the general material-balance equation 468
General difficulties in applying material-balance equations 470
Predicting gas-drive behavior 474
Predicting water-drive reservoir behavior 486
Increasing primary recovery 500
Additional problems 507
Notes 511

8 Decline-Curve Analysis 513
Decline-rate definition 514
Constant-percentage decline 517
Hyperbolic decline 523
Harmonic decline 540
Additional problems 541
Notes 542
Additional references 543

9 Waterflooding and Its Variations 545
The waterflood displacement mechanism 546
Predicting total flood recovery 551
Predicting rate versus time performance 580
Predicting waterflood performance by analogy or from pilot-flood results 598
Waterflooding variations 608

Additional problems 617
Notes 623
Additional references 623

10 Enhanced Oil Recovery 625

Steam and hot-water injection 629
In situ combustion 640
Micellar and surfactant floods 647
Miscible displacement 651
Polymer flooding 659
Feasibility analysis of EOR processes 661
Notes 664
Additional references 665

11 Computer Modeling of a Reservoir 667

Preparing a reservoir modeling study 668
An incompressible flow model 669
The general reservoir model 672
History matching 677
Notes 679

Appendix A Reservoir Engineering Symbols **681**
Appendix B Empirical Reservoir Engineering Data **691**
Appendix C Problem Solutions **733**
Index 821

Preface

Reservoir engineering in the U.S. emphasizes the problems associated with solution-gas-drive reservoirs. In fact, there are very few reservoirs in the U.S. currently producing under primary production that approximate steady-state conditions. Many years ago when only five or six days of capacity production were permitted per month, there were many water-drive reservoirs in the U.S. that produced under steady-state conditions. However, as we approached capacity production rates, most of these water-drive reservoirs became dominated by solution-gas drive because the production rates under this drive exceeded the water encroachment capabilities of the reservoirs. Thus, today most of the reservoir engineering techniques taught in the U.S. emphasize solution-gas-drive problems.

Most non-U.S. production is from reservoirs that produce under steady-state conditions, but these reservoirs are operated by U.S. personnel trained in U.S. reservoir engineering methods that emphasize nonsteady-state conditions. Thus, it is common for pseudosteady-state and other methods to be misapplied in non-U.S. areas. For example, the Horner method is based on infinite-acting equations but is routinely applied to wells that are in steady state at the time of shutin; the Matthews, Brons, and Hazebroek method of determining average pressures, devised for pseudosteady-state reservoirs, is used to determine average pressures in steady-state reservoirs; and reservoir computer models utilize only one outside cell for the water drive.

In addition to the misapplication of nonsteady-state methods to steady-state reservoirs, other conditions arise overseas that are not common to U.S. operations. Commingling production from different reservoirs is often done. Production at very low drawdowns (20–30 psia) also is common. Fresh water may be produced from depths of as much as 5,000 ft, and hemispherical flow is frequently encountered. Also, wells may produce at 5,000–10,000 b/d with almost no gas in solution (less than 60 scf/stb). The reservoirs may be so permeable that the producing gas-oil ratio from a solution-gas-drive reservoir can be controlled simply by raising or lowering the tubing. Wells may be completed with an open hole because the explosives needed for perforating require military supervision. Natural reservoir temperatures at 5,000 ft may be 300°F. In general, producing rates are in a completely different range than those in the U.S. where a 1,000-b/d well is unusual, while such a well overseas would represent a marginal economic situation. There, 50,000-b/d wells are not unusual, and 100,000-b/d wells are frequently encountered.

Practical Petroleum Reservoir Engineering Methods was first published in 1976, based on the author's 25 years of experience in the U.S. oil patch with limited foreign experience. Since that time, the author has gained extensive overseas experience through consulting and teaching in Japan, Germany, Colombia, Canada, Indonesia, Saudi Arabia, Nigeria, Angola, and England. From 1979–80 he spent eight months in overseas reservoir engineering assignments while on a professional leave from his duties at Ohio State University. This experience convinced him that the current emphasis on reservoir engineering education and reservoir engineering methods does not adequately emphasize non-U.S. reservoir conditions. Consequently, *Practical Petroleum Reservoir Engineering Methods* has been expanded extensively so it will more properly emphasize operations outside the U.S. Therefore, engineers using this book can be better prepared to solve reservoir engineering problems worldwide.

In addition, *Practical Petroleum Reservoir Engineering Methods* was written as a self-teaching book for practicing reservoir engineers. However, the book was used as a text in the petroleum engineering departments of several universities. Therefore, to facilitate its further use in formal education, the author has included additional problems following most chapters that can be used in teaching. Generally, one or more problems similar to the example problems in the text, which have complete solutions in appendix C, have been included at the end of each chapter without solutions. These problems have been designed for use as homework assignments.

The author also recognized that several subjects had been omitted from the first edition that were necessary to the general knowledge of the practicing reservoir engineer. Therefore, three new chapters are included. Chapter 1, "Acquiring Reservoir Engineering Data," has been added because of the dire need for reservoir data and the general problems associated with obtaining such information. It tells the engineer specifically what data should be obtained and when it should be collected. A chapter has also been added on enhanced oil recovery so the engineer can learn to recognize reservoirs with EOR possibilities. This chapter provides methods to eliminate all but the most promising possibilities by showing ways to obtain the most optimistic predictions of the behavior of proposed EOR projects. Promising prospects should then be referred to EOR specialists in the company or in a consulting firm.

A chapter also has been added on computer modeling of a reservoir. Here, the objective is to teach the engineer the fundamentals of computer modeling so he can intelligently and effectively use a reservoir model. No effort is made to teach the engineer to design and program a reservoir model since this is a highly specialized task.

In addition, the book has been expanded to include the following subjects:

- Waterflood predictions when a slug is used
- Calculation of drive indices
- Computer calculations of gas deviation factors
- Capillary-pressure/residual-oil relationships
- Applications of Q_{tD} functions to the infinite-acting prediction of gas-well behavior
- Analytical prediction of combination water- and solution-gas-drive reservoir behavior
- Rate-versus-time predictions for gas reservoirs
- Vogel IPR data
- Use of an apparent well radius to account for well damage with constant-pressure methods
- Type-curve methods
- Pulse-test analysis
- Real-gas functions
- Spherical and linear infinite-acting flow and their applications
- Analysis of microdarcy gas reservoirs
- Gentry and Fetkovich methods of decline-curve analysis

The biggest single objective of this book is still *understanding*.

Preface to Practical Petroleum Reservoir Engineering Methods

Petroleum reservoir engineering deals with the problems of maximizing producing rate and ultimate recovery of oil and gas reservoirs. We face a constant threat of an energy crisis through gasoline shortage, curtailment of natural gas deliveries, shortage of home heating oil, manufacturing plant shut downs due to fuel shortage, and periodic electrical service interruptions. But few realize that all of these problems could be pushed many years into the future by some major breakthrough in reservoir engineering.

It is amazing, with the ever-increasing energy demand, that so few persons, even scientists, are aware of the inefficiency in oil and gas production. Most people believe that oil and gas are produced from underground caverns or holes and that once a well quits producing all of the oil and gas has been removed from that particular hole.

Not so. Practically all oil and gas is produced from the almost microscopic-size pores that exist in underground rocks. These rocks have a texture almost like concrete.

Imagine a block of concrete ½ mile on each side and 30 ft thick that is saturated with oil. Now, imagine drilling a 6-in. hole in the center of this oil-saturated block from which you remove the oil as it flows into the hole. It does not require much imagination to recognize that very little of the oil would flow into the hole, that flow would be at a very low rate, and that when the flow stopped we would have produced a very small fraction of the oil from the cement block. This is the problem of reservoir engineering.

Most commercial oil reservoirs contain oil that has gas in solution in the oil when production is initiated. As the pressure in the oil declines, the gas will come out of solution just like the carbon dioxide bubbles out of soda pop when the bottle is opened. This expanding gas helps push the oil toward the well. Nevertheless when the oil quits flowing only about 15% of the oil in the reservoir may have been produced. It is seldom that as much as 25% of the oil can be produced from such a reservoir.

Some oil and gas reservoirs may have water moving into them from one or more sides. This water helps to move the oil toward the well, but the displacement is very inefficient. A reservoir such as this may produce as much as 50% of its oil before the produced fluid is 100% water.

The job of the reservoir engineer is to produce oil and gas reservoirs in such a way that the recovery is maximized and the rate at which the petroleum is produced is maximized. Control of the amount of gas or water produced with oil, proper placement of wells, use of the proper distance between wells, injection of water or other fluids into the reservoir, and many other means are used to help maximize the recovery.

Nevertheless, with the present state of the art today, it has been estimated that when all of the reservoirs that have produced oil in the U.S. are abandoned only about one-third of the oil that we know existed in the reservoirs initially will have been produced. In other words, twice as much oil as has been produced will remain unproducible in reservoirs. We know where this oil is; in most cases wells already exist, but we do not know how to produce it.

This book provides the basic fundamentals and many of the useful practical reservoir engineering methods that can be used today. Some engineer or scientist who learns his reservoir engineering from this book may be the one to discover the key to unlock economically the capillary, interfacial tension and other forces that hold the petroleum in our reservoirs today. Such a breakthrough would fulfill the energy demands in this country for many years to come. In the meantime, practicing reservoir engineers throughout the world will carefully control the production of the existing reservoirs to coax from them as much oil and gas as possible.

This book is written for engineers and students who wish to learn and understand the methods of reservoir engineering. It is also aimed at the practicing petroleum engineer who unknowingly misuses reservoir engineering techniques such as the Horner plot, Stiles prediction of flood behavior, gas-well testing procedures set forth by state control agencies, productivity index tests, and others because he does not understand them. When misapplied, these techniques yield more wrong answers than right.

Most practicing engineers would prefer to do their engineering work on a "cookbook" basis that simply requires filling out forms. These engineers may not find this book to their liking because the author believes such an approach to reservoir engineering is impossible and often dangerous due to the wrong answers obtained.

On the other hand the engineers who are intrigued by theory, mathematical methods, and computer analysis but have very little knowledge of what is required of the practicing engineer may find that this book is too practical and contains too many rules of thumb for their taste in engineering. However, it is believed that they too will profit from this book by gaining a better understanding of the practical and physical limitations on the reservoir engineering methods.

The author's objective is to provide the simplest possible method for obtaining accurate answers to practical problems. Unfortunately these methods cannot generally be applied by simply following a "cookbook," filling out a form, or supplying data to a canned computer program. On the other hand the author has tried to avoid the presentation of mathematically intriguing or scientifically interesting ideas that have little possibility for practical application.

Several practical but unpolished computer programs dealing with reservoir engineering techniques will be presented or discussed in this book. The author feels that the digital computer is a godsend for reservoir engineering since so many of the practical problems must be solved by trial and error. Nevertheless, a computer program is pretty much of a "cookbook" solution to a problem, and consequently, only those computer programs that are completely understood by the engineer should be used by him. If the engineer cannot make the same calculations by hand given enough time, he should avoid use of the program. Any engineer who answers "a computer method" to the question, "What technique did you use to make this calculation?" is grossly overestimating the ability of the computer. A computer printout has a tremendous influence in convincing managers that your analysis is correct. Except for this political advantage, the digital computer should be looked upon simply as a very fast calculator.

A considerable amount of time and effort will be spent in this book in "unlearning" many of the well-known reservoir engineering concepts. For example, most engineers feel that a Horner plot should be used in virtually all pressure-buildup analyses, when actually, there are very few cases where the Horner plot is necessary or even theoretically correct. Everyone "knows" that if a pressure builds up to an unchanging value during a shutin period that this unchanging value is the average reservoir pressure for the drainage area of this well. However, it will be shown that this simply is not true. There are many such misconceptions that are well known and widely misused in the practice of reservoir engineering, and it thus becomes necessary to discredit such ideas in order to avoid wrong answers.

Little effort has been expended in providing the book with a formal logical organization. Rather than having each subject fit beautifully under a particular concise heading, it is the author's opinion that it is much more important to have the sequence of subjects introduced in such a way that the opportunity for understanding is maximized. This, it appears, is seldom the same presentation sequence that would result from a formal, concise, logical, outline-type organization. The only thing that appears to be gained by a formal organization is the ability of the reader to find the subject material he desires. This objective will be attained by use of an extensive subject index. Thus, the sequence of subjects covered in this book conforms to the sequence the author has found most effective in teaching the material presented, and the outline organization based on the most popular criteria leaves much to be desired.

There is no effort in this book to cover everything that is known on a particular subject or even to reference all of the useful information on a particular subject. The author has simply tried to present the methods that he has found most effective in nearly 25 years of reservoir engineering practice and industry teaching. This does not mean that only the methods presented herein or referenced are good reservoir engineering methods. But it does mean that the methods presented herein are good, useful, practical reservoir engineering methods.

The author has attempted to keep the working equations presented in this text to a minimum since development of a multitude of special equations for all possible special situations adds considerable confusion to any scientific field. For example, the general material-balance equation presented herein will provide the answers for more than 95% of the problems dealing with material balance without a multitude of special equations for conditions above the bubble point, with water drive only, with gas-cap drive only, etc.

The reader will note that problems are interspersed throughout the text. It is suggested that these problems be solved as they are encountered and the solution compared with the complete solution presented in appendix C. If the text is conscientiously used in this fashion, the reader will find that a minimum of instructor guidance will be required. The text was prepared in this way so that it could be used by the many engineers who work on reservoir engineering or even petroleum engineering in general. The extensive nomenclature list, complete with specific units, will be found helpful for self-teaching.

Readers who desire a much easier (but much less effective) learning experience, can use the problems and appendix C problem solutions as example problems. However, in taking the easy mental route, he will end up with a lesser practical working knowledge of the material than his more ambitious colleague who attempts to work all of the problems before turning to the prepared solutions.

One additional note should be made concerning this book. The author has inserted many personal opinions, evaluations, conclusions, etc. He is often criticized for doing this without specifically stating that these are personal thoughts. If the reader does not find a reference or logical proof of a particular statement he can safely assume that it is a personal opinion based on the author's experience and knowledge of the subject.

List of Problems

Problem 2.1: Calculating permeability data from lab test **16**
Problem 2.2: Radial steady-state flow from a damaged well **41**
Problem 2.3: Determining average permeabilities **55**
Problem 2.4: Correcting for static pressure differences **58**
Problem 3.1: Application of the P_{tD} function **82**
Problem 3.2: Application of the Ei function **85**
Problem 3.3: Calculating the time needed to obtain a particular pressure drop in a reservoir **87**
Problem 3.4: Choosing a pressure-function solution **92**
Problem 3.5: A pseudosteady-state flow-test analysis **101**
Problem 3.6: Water disposal in an aquifer **114**
Problem 3.7: Pressures resulting from multiple wells in a reservoir **121**
Problem 3.8: Accounting for variations in producing rates **126**
Problem 3.9: Simulating boundary effects **130**
Problem 4.1: Productivity-index evaluation **151**
Problem 4.2: Using the Vogel IPR curve **152**
Problem 4.3: Constant-rate drawdown tests **156**
Problem 4.4: Determining the distance to a reservoir barrier from a drawdown test **163**
Problem 4.5: Pressure buildup from an unchanging well pressure **176**
Problem 4.6: Pressure-buildup analysis of an old well **178**
Problem 4.7: Determining the average pressure in the drainage area of a pseudosteady-state well **199**
Problem 4.8: Analysis of a two-rate pressure buildup **204**
Problem 4.9: Pressure falloff analysis **208**
Problem 4.10: Pressure buildup in a North Sea well demonstrating spherical flow **219**
Problem 4.11: Analyzing DST data **225**
Problem 4.12: Pulse-test analysis **239**
Problem 5.1: Determining gas characteristics **273**
Problem 5.2: Determining gas production for use in material balance **278**
Problem 5.3: Calculating the static bottom-hole pressure from surface pressure measurements **283**

Problem 5.4: Application of graphical gas material-balance techniques **288**
Problem 5.5: Calculating water encroachment in a gas reservoir **297**
Problem 5.6: Simultaneous solution of reservoir pressure and water encroachment **298**
Problem 5.7: Turbulence effects in gas wells **302**
Problem 5.8: Conventional gas-well back-pressure test **309**
Problem 5.9: Adjustment of a gas deliverability curve for a change in well spacing or depletion **311**
Problem 5.10: Using isochronal data **321**
Problem 5.11: Determining approximate isochronal data from conventional drawdown data **329**
Problem 5.12: Pressure drops in the producing system **350**
Problem 5.13: Determining gas-well spacing for completing a pipeline purchasing contract **355**
Problem 5.14: Predicting rate versus time behavior for a gas reservoir **362**
Problem 5.15: Predicting the pseudosteady-state flow rate for a microdarcy gas reservoir containing a massive hydraulic fracture **366**
Problem 5.16: Evaluating a pressure-buildup is a gas reservoir **370**
Problem 6.1: Determining the static saturation distribution from reservoir capillary pressure data **394**
Problem 6.2: Conversion of laboratory capillary pressure data to reservoir capillary pressure data **397**
Problem 6.3: Using the J function to average capillary pressure data **400**
Problem 6.4: Calculating a fractional flow curve **415**
Problem 6.5: Application of the Buckley-Leverett equation **419**
Problem 6.6: Evaluating the frontal position by material balance **420**
Problem 6.7: Using the Welge graphical method to calculate displacement in a waterflood **428**
Problem 7.1: Gas material balance **449**
Problem 7.2: Gas-cap expansion **449**
Problem 7.3: PVT lab data exercise **456**
Problem 7.4: Determining gas-oil ratios **461**
Problem 7.5: Material balance in a solution-gas-drive reservoir **463**
Problem 7.6: Calculating oil-zone shrinkage **463**
Problem 7.7: Determining the original stock-tank barrels of oil in the reservoir **464**

Problem 7.8: Determining PVT data empirically **472**
Problem 7.9: Predicting the behavior of a gas-drive reservoir **479**
Problem 7.10: Analyzing an undersaturated oil reservoir with a water drive **491**
Problem 7.11: Analyzing a combination solution-gas-water-drive reservoir **498**
Problem 7.12: Calculating drive indices **504**
Problem 8.1: Using constant-percentage decline to calculate the future life and rates of a well **520**
Problem 8.2: Using a plot of rate versus cumulative production during constant-percentage decline **523**
Problem 8.3: Application of the hyperbolic decline curves **533**
Problem 8.4: The change in reserves and life following a workover **540**
Problem 9.1: Saturation distribution in a waterflood **550**
Problem 9.2: Evaluating gross swept volume **560**
Problem 9.3: Calculating the total displacement efficiency and recovery from a waterflood **575**
Problem 9.4: Calculating reduced-time curves for use in a waterflood prediction **586**
Problem 9.5: Calculating oil and water production rates from reduced-time curves **588**
Problem 9.6: Calculating individual strata recovery curves **590**
Problem 9.7: Calculating individual strata injectivity curves **595**
Problem 9.8: Predicting total flood recovery by analogy **600**
Problem 9.9: Predicting the rate versus time performance by analogy **606**
Problem 9.10: Calculating injectivity curves when using a slug **616**

1

Acquiring Reservoir Engineering Data

One of the biggest problems in reservoir engineering is having reliable, accurate data to work with. Much of the data requires prior planning to obtain, and much of the data must be obtained during completion of the well or during the initial stages of production. Although most data are relatively inexpensive to obtain, its economic value is difficult to explain initially. However, later in the life of the reservoir when such data are needed to predict the reservoir performance accurately or to determine the economics of a proposed enhanced oil recovery scheme, the economic value becomes clear and the cost of the data becomes inconsequential. Then it is too late to determine some of the most important data accurately.*

It is the responsibility of the production manager to keep costs at a minimum. He has a staff of engineers that are paid to guide him so he does not do something foolish from an engineering standpoint. As indicated, much of the reservoir data must be obtained during drilling or during the initial testing of the well. However, consider the production manager's position. He has spent years and literally millions of dollars of the company's money in exploring, drilling, and completing an exploration well. Now that he is in a position to start putting oil in the tanks, some engineer wants to produce the well at a very low rate and even suggests shutting in the well for an extended period to get reservoir data that may have some vague economic value 5–10 years from now. Obviously, the manager will often forego the reservoir engineering tests.

*To ensure that the engineer realizes the importance of reservoir data acquisition and has the knowledge necessary for the proper planning of such acquisition, the first chapter is devoted to this problem. It is realized that students and many inexperienced petroleum engineers do not have sufficient background to be able to use this chapter until they have studied some of this book. It is recommended that such persons use this chapter as an appendix and come back to it as needed.

The manager's situation is easy to understand, but will our positions as responsible engineers be as easy to explain 5–10 years later when it becomes clear that we did not insist on the appropriate tests? We must put forth our best efforts to explain just what this lack of data can mean later when we cannot model the reservoir to explain where the excess water or gas is coming from or when we cannot conclusively show that a waterflood will work or when we cannot evaluate the applicability of some-as-yet undiscovered process to this particular reservoir.

At the time of drilling and completion, the engineer must be prepared with his best arguments and his most persuasive (but tactful) manner to influence the production manager. One argument is to show the possible financial gain: "If these data make it possible to recover just one-tenth of 1% of additional oil, the company will gross X additional dollars." This technique is usually impressive. If the tests are not made, the engineer should then write a letter to the file, with a copy to the manager, explaining his ideas. When he is later asked why the data were not obtained, the engineer can go to the file and indicate that the proper tests were requested.

Reservoir Data Requiring Predrilling Planning

Log data, core data, and perhaps drillstem-test (DST) data can only be obtained if plans are made to acquire the data prior to drilling the well. All of these data can be extremely important in reservoir studies.

Logging programs should be carefully designed cooperatively with the company logging engineer, the geologist, and a representative of the logging service company. The engineer should plan to determine net-pay thickness, porosity, and saturations from the well logs. A suite of logs designed for this purpose must of course consider the particular type of reservoir rock, geographical area, anticipated reservoir fluids, and drilling methods employed. Most reservoir engineers do not have the detailed knowledge of logging methods necessary to perform such a design on their own, so use all of the help that is available.

A word of caution is probably worthwhile. Geologists are not nearly as interested in porosities and saturation data as is the reservoir engineer—the geologist's interest is principally in the indicated lithology. Consequently, we should be very careful that the logging company has quantitative capabilities. In general, it is dangerous to try to save money in logging when quantitative log data are needed. Obtaining accurate logs and accurate log interpretations is difficult under the best conditions, so be leery of "cut-rate" logging services.

The need for core data is most often felt when enhanced oil recovery studies are made late in the life of a reservoir. When cores are available from storage, lab data applicable to a particular reservoir can often be obtained by using restored-state lab methods. Otherwise, analyses must be based on empirical data and analogy with a corresponding amount of uncertainty. At such times cores and core data may be invaluable.

Routine reservoir work requires a knowledge of the absolute permeability, statistical permeability variation, and position of the original water-oil and gas-oil contacts. The permeability data are available only from core analysis, and the original water-oil and gas-oil contacts are generally best evaluated from a study of core saturations. In addition, the porosity values obtained from logs are backed up by core data, and core data often clearly indicate deficiencies in the logging program. Saturation data from conventional cores are of course relatively useless in terms of the absolute values obtained (see chapter 5). Core data do not make quantitative log data unnecessary because the log data are obtained on each well and a log samples a much larger portion of the reservoir.*

Drillstem tests (DST) may be included in a drilling prognosis. The point that needs to be made is that a DST should be run in a quantitative manner that permits determining as much reservoir data as possible. This problem is discussed in detail in chapter 4, but we should perhaps emphasize a few of the important points here. A DST provides a unique opportunity to obtain some good reservoir data early in the life of a well. Nevertheless, it is surprising the number of DST that are run in such a manner that these data cannot be obtained from the test data, i.e., only the production data are measured with sufficient accuracy to be useful. Consequently, the engineer's first responsibility on a DST is to make clear to the service company that pressure data are desired of sufficient accuracy for quantitative analysis.

Also, make sure the DST is designed in such a way that every effort is made to record the initial reservoir pressure directly. Briefly, this requires an initial flow period long enough to relieve the reservoir around the wellbore of the super pressure caused by the static mud pressure but a flow period short enough to avoid having to use a Horner plot to determine the reservoir pressure. A surface readout of the bottom-hole pressures and the multiple flow-rate capabilities of most DST equipment minimize the problems associated with the direct measurement of the initial well pressure. This equipment permits us to start with a very short flow period of perhaps 5 min, observe the pressure buildup, run another short flow period, observe the buildup, and continue this procedure until we are satisfied that we have observed the initial reservoir pressure.

*The problem of how many development wells to core to obtain a good description of the reservoir is beyond the scope of this book. The most important point to establish is the need for a coring program.

When run with sufficient accuracy and care, DST pressure buildups allow us to determine the undamaged effective permeability and a measurement of the damage, e.g., the skin factor or the damage ratio. If the well completion plans include massive well treatments such as a massive hydraulic fracture, a test run before the treatment may be the only chance of determining the undamaged permeability.

If well logs or other information indicates the presence of a gas-oil or water-oil contact in the producing reservoir, care should be taken to determine the position of such contacts as accurately as possible before production is initiated. Sidewall samples should be taken to see if the contacts can be accurately defined. If these results are questionable, sidewall fluid samples should be employed to define the contacts. Remember that when production is initiated, coning may take place and in some cases the ability to define the contacts from this well may be lost forever.

Reservoir Data Obtainable Early in the Life of a Well

The undamaged effective permeability, measurement of well damage, initial gas in solution, initial reservoir pressure, distance to the closest reservoir barrier, and description of the drainage area can be most accurately determined during the initial flow period or buildup period of a well. Also, a fluid sample for laboratory analysis of the oil formation volume factor, gas in solution, reservoir oil viscosity, and gas deviation factor (PVT data) can best be obtained soon after the initial flow test.

Initial flow tests represent one of the best but most overlooked sources of reservoir data. Initial flow tests are run at a time when the conditions in the reservoir are best known. After a well has been produced for a period of time, the gas saturation, reservoir pressure, and fluid viscosities may be questionable. However, during the initial flow period, all of these values can be accurately ascertained. Also, the well is infinite acting at this time, which may be questionable later in the well life. All of these things make it possible to evaluate a drawdown or buildup test accurately to obtain the undamaged permeability, a measurement of the damage, and some description of the drainage area (see chapter 4). Care should be taken to measure the initial produced gas-oil ratio accurately. This generally represents the most precise measurement of the original gas in solution.

An initial flow test is not always easy to obtain. There may be a lot of pressure from management to get the well to peak production, and any delays such as a controlled flow test may be bypassed. In addition to this political difficulty, there are generally testing difficulties to overcome. The biggest problem is that during the drilling of the reservoir section of the well, mud filtrate, mud particles, and cuttings are forced into the reservoir, which damages the permeability around the wellbore. This damage would cause no testing difficulty, except that the debris is not stable initially and continues to move out of the formation during the initial flow period. This means that the Δp_{skin}—the additional pressure drop caused by damage around the wellbore—is being reduced continually during the initial flow period. An analysis of the pressure data under these conditions is impossible. With the well flowing at a constant rate, the bottom-hole pressure may be increasing instead of decreasing as it would if Δp_{skin} were constant. Then it is necessary to continue to produce the well and observe the bottom-hole pressure until the pressure becomes constant or is declining. Then if a constant-rate drawdown test is desired, the well should be shut in until the pressure approaches the initial pressure and the constant-rate drawdown test can be initiated.*

Once the reservoir pressure has declined to a value less than the saturation pressure—the pressure at which the first bubble of gas is liberated—it is very difficult to obtain a representative sample of the original hydrocarbons in the reservoir. This is caused by the difficulty of obtaining a sample with the correct ratio of free gas to oil. Bottom-hole sampling devices and sampling techniques are available for attempting to obtain a representative sample under these conditions, but the fact remains that it is very difficult to do.

However, a sample of the oil can be taken with a bottom-hole sampler while the well is flowing at a bottom-hole pressure greater than the bubble-point pressure with little or no difficulty. Thus, it is very important for the engineer to make certain that a PVT sample is obtained from a reservoir soon after production from that reservoir begins. Here again, there will probably be a reluctance on the part of management to approve shutting in the well or restricting production to get the flowing bottom-hole pressure as high as possible. However, this technique is required to obtain the PVT sample.

Obtaining Routine Reservoir Data

Meaningful reservoir studies must be completed so the ultimate oil and gas recovery and the production rates can be maximized. Such studies are based on material balance, which requires the average reservoir pressure and the corresponding cumulative production of oil, gas, and water by wells and reservoirs.

Probably the only set of these data that is apt to be correct without any effort on the engineer's part is the total oil production. Since oil is

*The design and analysis of a constant-rate drawdown test are discussed in chapter 4. Here, we simply want to point out that a properly run initial flow test provides useful reservoir information that may be much more difficult or impossible to obtain later in the life of the well.

always sold, an accurate measurement of the amount of oil produced is obtained. When gas is sold, the total gas production also is known accurately. However, on a worldwide basis most gas produced as a result of oil production is flared—burned in a flare. The engineer should also take careful note of when gas sales begin because gas sales often are initiated sometime during the life of the reservoir when the demand for natural gas reaches a point that makes gas sales economical.

Prior to the time when gas sales were initiated, the engineer should view the recorded gas production with considerable skepticism. It is generally accepted that in the absence of continued supervision, the field personnel responsible for measuring and recording gas production do not view this as an important task. Thus, measurements are seldom carried out with care and may even be boiler housed, i.e., recorded without actually having been measured. This is sometimes apparent from the nature of the data. For example, on one foreign offshore production platform that was handling several wells producing from a solution-gas-drive reservoir, the same produced gas-oil ratio figures were recorded for several months. Obviously, the gas production had not been measured during this period, although the total gas production was dutifully recorded.

The engineer should be continuously concerned about the accuracy of the gas production being recorded when gas is not being sold, liquefied, or used in some other manner. It is strongly recommended that constant-displacement direct-reading gas meters be used when pressures and capacities permit. It is much easier to check such readings than it is to check calculations from an orifice-meter chart.

One production figure that is generally known even less accurately than flared gas is the produced water. However, water production is just as important in material balance calculations as gas and oil production. Of course, there is never any economic value to the water produced with the oil. Consequently, the pumper—the person generally responsible for maintaining production records—worries little about the numbers recorded for water production. Generally, only a halfhearted effort is made to measure the water produced, and it is not unusual to find daily water production figures reported exactly the same year after year. Here again, the use of positive-displacement meters on tanks that are dumped periodically is strongly recommended, and the engineer should be continuously concerned about the accuracy of the water production being recorded.

Even when the total oil, gas, and water production recorded for a field is accurate, the engineer must be concerned about the manner in which the production is allocated to the individual wells in the field. The allocation of production to the different wells should be on the basis of well tests that are carefully run periodically (perhaps every month) on each well. Unless carefully supervised, the pumper may not put the wells on test each month and may use last month's test as the basis for the test reported for this month. When excess production capacity exists, the pumper may be tempted to obtain his assigned production from one well rather than from all of the wells covered by the allocation. Picture the reservoir engineer madly trying to juggle lateral permeability and porosity variations to match the pressure distribution when the production recorded for four wells was actually taken from one well. Here again, the engineer should have a continued obligation to ensure that well tests are carried out as designed, that production is taken from the well to which it is assigned, and that production is allocated to the individual wells as accurately as possible.

Even when the total production of oil, gas, and water for a field is correct and the allocation of the production by wells is accurate, the reservoir engineer still has a very difficult task in allocating production from a well to the individual reservoirs if production is commingled. In the U.S., commingling production from two or more separate reservoirs in a common tubing string is carefully controlled. In most producing states commingling production is permitted only when it is necessary from an economic point of view. However, in non-U.S. areas it is common practice to produce as many as four or five separate reservoirs through one tubing string so little is known about how much of the oil comes from each reservoir. This technique is often used although each reservoir would support a well or dual or triple completions—in effect, running two or three strings of tubing in one well—to avoid commingling.

Capacity production rates can be greatly reduced by commingling production from reservoirs of different permeabilities. Since essentially the same well pressure must exist in each of the commingled reservoirs, it is impossible to produce each reservoir at its maximum rate. Also, when reservoir studies are necessary to determine how best to produce the reservoir, there is almost no logical means of determining how much of the production has come from each reservoir. Then the total ultimate recovery must surely suffer. Therefore, when development wells are drilled, inform the manager of the difficulties that result from commingling production and design the development so each reservoir is produced individually.

If a reservoir contains an initial gas-oil or water-oil contact, one of the biggest unknowns associated with the reservoir behavior is the amount of oil left behind the advancing gas-oil or water-oil contact. Perhaps the best way to determine these values accurately is to observe the advance of the contacts as the reservoir is produced. In very permeable reservoirs produced with small pressure drawdowns, or pressure drops, the advance may be observed from the produced gas-oil ratio or the water-oil ratio of individual wells as the contact reaches the subject well. However, in most reservoirs production results in the formation of a water or gas cone that greatly complicates the interpretation of the production data to obtain the advance of the contacts (see chapter 6).

In such cases it may be worthwhile and economical to drill and equip wells for the purpose of observing the contact advance. Such wells are never produced. Most often they are equipped with plastic casing opposite the reservoir. Then special logging devices can be used to determine the advance of the contacts without the influence of the production from the well. On the surface it appears that such wells are extremely expensive. However, when millions of barrels of oil are at stake, investment in a series of such observation wells may show a considerable profit.

If material balance studies, including reservoir modeling, are to be completed on a reservoir, it is necessary to determine an average reservoir pressure that corresponds to cumulative gas, oil, and water production. Since production rates vary throughout the reservoir, the resulting reservoir pressures also vary throughout the reservoir and the determination of the volumetrically weighted-average reservoir pressure is not easy. This measurement is further complicated by the fact that wells can seldom be shut in until the well pressure quits changing. Even then this pressure may not represent the average pressure in the well's drainage area.

Regardless of the difficulties involved in interpreting the data, it is necessary to measure the shutin pressure periodically in enough of the wells of a reservoir to provide an accurate measurement of the average pressure. Generally, it is necessary to shut in the wells and observe the change in the shutin pressure with time so this pressure behavior can be interpreted to determine the average pressure in the drainage volume prior to shutin.

This problem is discussed in detail in chapter 4, but one additional word of caution should be added here. Remember that most of the methods of determining average pressures were devised for volumetric reservoirs producing under pseudosteady state. Reservoirs that produce under steady state, as do most of the non-U.S. wells, require different methods. Regardless of the methods used, make certain that the average pressure in the reservoir is determined periodically.

Table 1–1 provides the engineer with a quick reference to the data needed at various stages in planning and developing a reservoir. This list has been developed chronologically through the life of the well.

The importance of obtaining accurate reservoir data at the appropriate time in the life of a reservoir cannot be over emphasized. The engineer has only one opportunity to obtain such data as cores, logs, and initial reservoir pressures from a reservoir. Also, without proper planning the engineer will never know the amounts of oil, gas, and water produced from the individual wells and reservoirs or the corresponding reservoir pressures. Without accurate reservoir data reservoir engineering may be pure speculation.

TABLE 1–1 Reservoir Engineering Data Checklist

During development planning	a. Provide for the individual production of each reservoir by using multiple or dually completed wells or by planning to deplete one reservoir at a time and plugging back.
	b. Use positive-displacement meters for oil, gas, and water on each well to give the best possible allocation of the production to the reservoir in each well.
During drilling	a. Use a logging program capable of determining porosities and saturations.
	b. Provide for sufficient cores to give a good statistical analysis of porosities and permeabilities. Store the cores not needed for core analysis.
	c. When conventional drillstem tests are run, make sure the test procedure includes an effort to record the initial reservoir pressure directly. Also obtain pressure data suitably accurate for quantitative analysis, i.e., tell the testing company a quantitative analysis of the pressure data will be made.
	d. When the logging program indicates an initial gas-oil or water-oil contact, use sidewall or fluid samples to define the contact(s).
During initial production	a. Measure the initial reservoir pressure using a bottom-hole pressure bomb before the well is produced.
	b. Produce the well until the rate is reasonably constant or declining. Shut in the well until the shutin pressure approaches the initial reservoir pressure. Then proceed with a constant-rate drawdown and buildup.
	c. If a well is a successful exploration well or if it is suspected that it is producing from a new reservoir, take a bottom-hole fluid sample for PVT analysis as soon as possible after the initial flow test.
During the production life	a. Design a test program to permit allocating production accurately to individual wells and reservoirs.
	b. Periodically check pumping and metering equipment to ensure that measurements and allocations of production to individual wells and reservoirs are accurate.
	c. Design a pressure survey program that provides for periodic determination of the average reservoir pressure.
	d. Consider using plastic-cased observation wells (nonproducing) and appropriate logging methods to observe the advance of gas-oil or water-oil contacts in a reservoir.

2

Reservoir Fluid Flow Fundamentals

In addition to teaching the fundamental concepts of fluid flow and their applications, one of the principal objectives of this text is to integrate the treatment of reservoir fluid flow so the engineer can readily comprehend the relationship between steady-state, pseudosteady-state, and unsteady-state fluid flow and can recognize the characteristics of each. In so doing, the limitations of each flow regime will become apparent.

All practical fluid flow equations are based on two concepts: the Darcy equation and material balance. The simpler concepts of reservoir engineering are based on one of these concepts. However, the more complex concepts—and quite often the most useful ones—are based on both material balance and the Darcy equation.

Fluid flow can be characterized or categorized so many different ways that it is virtually impossible to consider all of the possibilities. Fluid flow can be classified according to the geometric configuration involved, the compressibility of the fluids, and the constancy of flow rates and pressure with respect to time. Also, flow can be single or multiphase. A particular working equation would probably apply to only one combination of each of these classifications, which makes it virtually impossible to consider all of the possible combinations. Consequently, the means of accounting for the various classifications are given, but the engineer must devise derivations for peculiar cases that may be encountered.

Characteristics of the Darcy Equation

Darcy's Law is an empirical relationship derived for the vertical flow of fluid through packed sand. Darcy would probably be shocked to note how we have stretched his empirical relationship to fit our particular needs in reservoir engineering. The continued use of this empirical equation to solve complex reservoir engineering problems seems analogous to using a horse to go to the moon. However, there appears to be no easy way to back up and start over with a more theoretical treatment of fluid flow, so we use Darcy's empirical relationship to the best of our ability.

Darcy showed that the apparent velocity of a fluid flowing through a porous media is proportional to the pressure gradient, dp/dx, and he assumed it would be proportional to the reciprocal of the viscosity, μ. He then wrote an equation for the velocity of fluid traveling through a particular porous media and added a proportionality constant characteristic of that particular media, a permeability, k:

$$v' = -(k)\frac{1}{\mu}\frac{p'}{x'} \tag{2.1}$$

The negative sign is added because if x' is measured in the direction of flow, the pressure p' will decline as x' increases. This decline results in a negative value for $(\Delta p'/\Delta x')$. Thus, the minus sign must be added to make the velocity v' positive. The prime marks on v, p, and x in Eq. 2.1 denote Darcy units. Darcy units are odd compared with modern oilfield units. If we substitute for the apparent velocity, v', the expression q'/A', we obtain the Darcy equation in the volumetric rate form:

$$q' = -\frac{kA'}{\mu}\frac{\Delta p'}{\Delta x'} \tag{2.2}$$

Where:

q' = cc/sec
A' = Gross cross-sectional area, sq/cm
dp'/dx' = Pressure gradient, atm/cm
μ = Viscosity, cp

The unit of the resulting constant, k, using this combination of units was termed the darcy by Mr. Darcy. This system of units, although close to the recommended international units being adopted, would be very difficult to use in the oil field. Consequently, no attempt is made in this text to state equations in the international (SI) units. Thus, the strange (to everyone except the oil producer) system of oilfield units is employed. If we state the rate q' as a function of reservoir barrels per day, A' as a function of square feet, and the pressure gradient dp'/dx' as a function of pounds per square inch per foot, Eq. 2.2 can be changed to the more useful form:

$$q = -\frac{1.127kA}{\mu}\frac{\Delta p}{\Delta x} \tag{2.3}$$

This form is used as the Darcy equation.*

The Darcy equation has some serious limitations that we must recognize. Darcy's experiments were run at relatively low flow rates and at low pressure drops using small heads of water as a driving pressure. Consequently, since flow rate is proportional to pressure drop in the Darcy equation, this expression does not apply to turbulent flow.

The additional pressure drop that is not predicted by the Darcy equation is referred to here as *turbulence* and may actually be caused by the continuous change in direction and pore cross section that characterizes flow through a reservoir pore. Under these conditions conventional viscous flow with parallel, nonintersecting flow paths becomes very difficult to obtain. The proportionality between the volumetric flow rate and the pressure drop for a liquid is lost at a velocity that is much less than that predicted by the Reynolds' number. Thus, the term *non-Darcy flow* has been adapted by many industry educators and researchers.[1] The term turbulence, as used in this book, is meant to include the non-Darcy effects.

Each time the Darcy equation is used in the form of Eq. 2.3 to derive some working equation, we must realize that this particular equation does not apply to turbulent flow. This does not appear to be a serious practical limitation when we apply the Darcy equation to the flow of liquid. Generally, liquid flow rates are too small for the additional pressure drop caused by turbulence to have great significance. However, the opposite is true when we consider the flow of natural gas. In this case the flow rate is possibly 100 times as great as the flow of liquid under similar conditions due to the viscosity differences. Thus, in the flow of gas, an additional pressure drop generally does exist because of turbulence, and the gas flow equations are modified in some way to account for this non-Darcy flow effect.

None of the liquid flow equations have been modified to account for turbulence, and no data have come to the author's attention that document the turbulent flow of liquids in the reservoir. Nevertheless, it seems likely that a well producing at 40,000 b/d from 200 ft of thickness, as some wells do in Saudi Arabia, Iran, and elsewhere, may encounter turbulence. It is seldom necessary to flow test or to be concerned with the capacity of such ultrahigh-capacity wells. Thus, they may be flowing under turbulent conditions, but the engineer has no

*Two difficulties may arise when using this equation for those who are familiar with other fluid-flow work. First, the rate, q, is stated in reservoir barrels per day, whereas many other equations state the rate in stock-tank barrels per day. Second, the permeability is stated in darcies rather than in millidarcies (thousandths of a darcy). This system of units is employed throughout the text unless it is specifically stated that some other units are used for a special case.

reason to analyze them. The practical significance is this: If flow data from a high-capacity well (say more than 50 b/d/ft) do not appear to fit the theory, the engineer may consider turbulence as an explanation of the anomaly.*

Absolute permeability. Undoubtedly, the Darcy work was never meant to apply to multiphase flow. It was meant to describe the flow of one fluid saturating 100% of the porous media. Under these conditions the permeability to a particular fluid is independent of the nature (viscosity) of the fluid. In other words the permeability to a 100% saturating fluid is a constant and is characteristic of the porous media known as absolute permeability. Reported differences of the permeability to different phases is believed to be a result of the reaction of the porous media to one or more of the phases. Consequently, we can think of the absolute permeability as a fixed characteristic of a particular porous media similar to the porosity of pore-size distribution.

Although the absolute permeability of a porous media is a physical property of that media, it appears to have some abstract qualities that are not exhibited by other properties such as porosity and pore-size distribution. There is an indication that permeability of a naturally occurring porous media is a function of the size of the sample. It may be that such observations are simply a matter of undetected plugging of the pores caused by cutting the samples. However, wide variations have been established in the permeabilities of sandstones and limestones that appear to be homogeneous in nature.

A very uniform sandstone sample (for example, a 6-in.-long, 3½-in.-diameter core) can be cut into as many permeability plugs as possible (normally 1 in. in length and ¾ in. in diameter), and the measured permeabilities of this homogeneous sandstone will probably vary as much as 300%. It is unclear how the effective or average permeability is related to the permeability distribution obtained in such a situation. Until some conclusive evidence is found to the contrary, it is recommended that the arithmetic average permeability be used as the characteristic permeability of a formation.

Permeability distributions are often characterized as indicated in Fig. 2–1, which is a plot of permeability versus the cumulative thickness of samples with a permeability greater than that of the subject permeability. In Fig. 2–1, note that 15 ft of the formation has a permeability greater than 110 md.

It is sometimes helpful to prepare a histogram of permeability distributions. If a geometric progression is used to define the limits of each permeability group considered, a distribution approaching normal may

*Various efforts have been made to adapt the Darcy equation to turbulent flow. One of these is discussed in some detail in chapter 5.

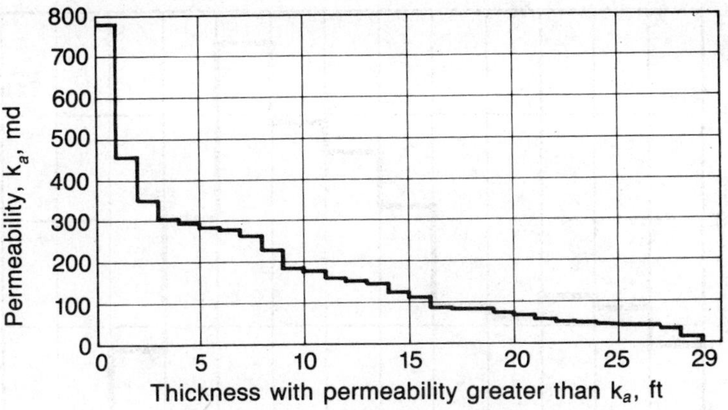

Fig. 2–1 Permeability distribution

be obtained. Actually, the shape of the histogram shown in Fig. 2–2 is typical of permeability distribution. Some engineers believe that the peak of the permeability histogram best characterizes the actual permeability behavior of a formation.

The permeability of pores of a particular size (for example, the permeability of a fracture) can be calculated by mathematical means. Also, in chapter 7 a permeability for a bundle of capillary tubes of a particular size is calculated, which gives some insight into the relationship between permeability and porosity for an idealized case.

Multiphase flow. Since all hydrocarbon-bearing reservoirs contain water that may or may not be mobile, it is necessary to extend the permeability concepts and to use permeability as a function of the flowing-phase saturation. For example, the permeability of a particular porous media to oil when the oil saturation is 50% and the water saturation is 50% may only be 45% of the permeability exhibited when the formation is 100% saturated with oil. At 60% oil saturation and 40% water saturation, the permeability based on the flow rate of the oil may be 70% of the permeability of the formation when it is 100% saturated with oil. This ratio of a permeability at a particular saturation to the permeability at 100% saturation is termed the *relative permeability,* k_r. Consequently, the effective permeability to a particular phase is the absolute permeability, k, multiplied by the relative permeability k_r. Thus:

$$k_o = kk_{ro};\ k_g = kk_{rg};\ \text{and}\ k_w = kk_{rw} \tag{2.4}$$

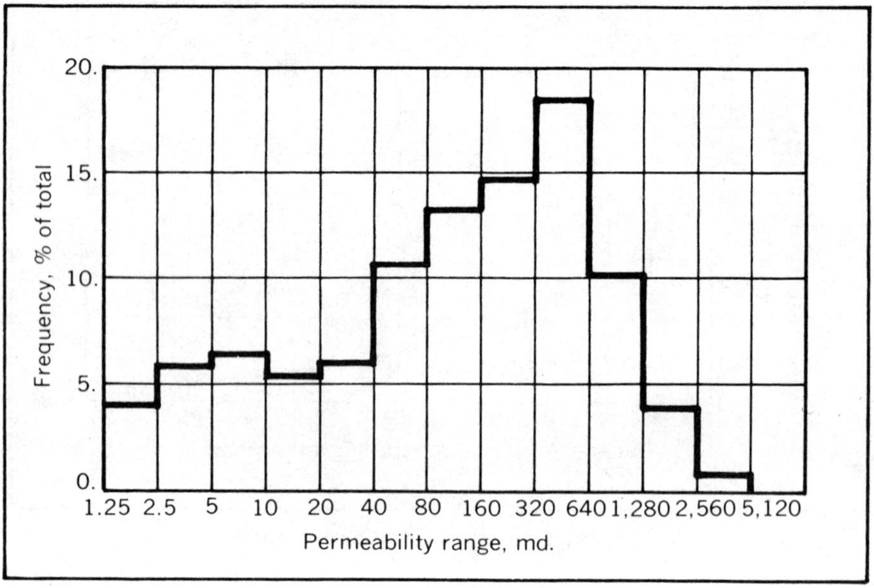

Fig. 2–2 Permeability histogram

In these relationships the subscript r refers to relative, and the subscripts o, g, and w refer to oil, gas, and water. To understand the permeability and relative permeability concepts better, work the following problem. The solution is shown in appendix C.

PROBLEM 2.1 Calculating Permeability Data from Lab Tests

The following laboratory data are given from relative permeability tests:

Cross-sectional area of core = 5 sq cm
Pressure at outlet face of core = 1 atm
Pressure at inlet face of core = 2 atm

Core length = 3 cm
Viscosity of water = 1.0 cp
Viscosity of oil = 1.25 cp

Saturation and rate data are as follows:

Saturation, %		*Flow Rates, cc/sec*	
Water	*Oil*	*Water*	*Oil*
100	0	0.50	0.00
90	10	0.30	0.00 critical
80	20	0.15	0.01
60	40	0.03	0.10
40	60	0.01	0.25
30	70	0.00 critical	0.38

What is the absolute permeability? What is the permeability to oil at a water saturation of 30%? Calculate the relative permeability values.

To obtain effective permeabilities to use in flow equations, the absolute permeability must be multiplied by the relative permeability to the flow phase of interest, according to Eq. 2.4. The relative permeability data resulting from the calculations in problem 2.1 are shown in Fig. 2–3. They exhibit typical relative permeability characteristics.

Note that at any particular saturation the relative permeabilities do not total 1.0. We might say that they mutually interfere with the

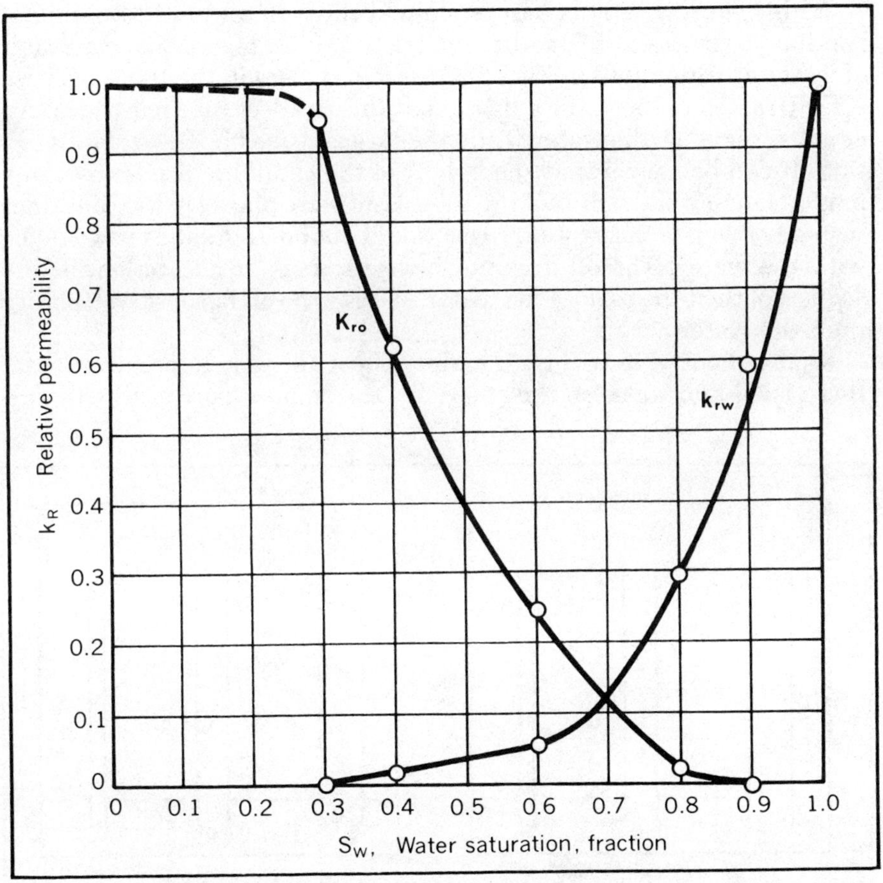

Fig. 2–3 Relative permeability versus water saturation

flow of each other. Also, both phases have an irreducible saturation at which the relative permeability to that phase is 0. As anticipated, the relative permeabilities at 100% saturation of a particular phase are 1.0. This is generally an assumed value for k_{ro} because when running relative permeability data the first point investigated is normally the 100% water saturation point. However, it is impossible to drive the water saturation below the irreducible water, in this case 0.3, without artificially cleaning the core and starting again with 100% oil saturation.

The shapes of the relative permeability curves are also characteristic of the wetting qualities of the two fluids. When water and oil are considered together, water usually is the wetting phase. Therefore, the water, or wetting phase, would occupy the smallest pores while the nonwetting, or oil phase, would occupy the largest pores.

This characteristic is best explained through an understanding of capillary attraction. If small-diameter tubes are placed in a container of water as illustrated in Fig. 2–4, the water rises in the tubes because it is attracted to the surface. Note that the smallest tube has the greatest attraction for the water while the largest tube has the least attraction. It can be shown that the height of the capillary rise is inversely proportional to the radius of the tube. If oil were placed in the container instead of water, the capillary rise would not be as great as it would be with the water. The oil does not have as great an attraction for the surface of the tube as does the water. We say the oil has less wettability than the water.

Consequently, if we have a variety of pore sizes in a reservoir and if there is oil and water in the reservoir, the smaller pore sizes will pull

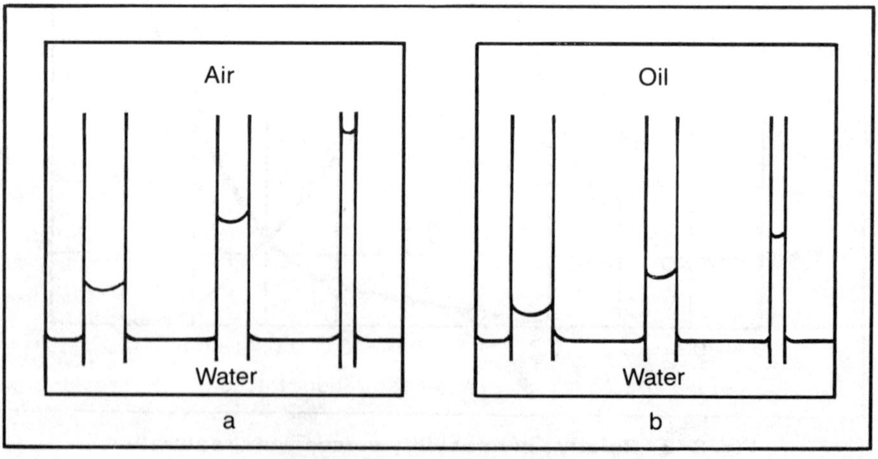

Fig. 2–4 Capillary effects

the water into them by capillary attraction, leaving the oil in the larger pores. Thus, we find that the smaller pores tend to be occupied by water while the larger pores tend to be occupied by oil.

We may think that since the pores in most reservoirs are very small—even microscopic in size—the pore sizes would be very uniform. However, even in reservoirs with the smallest pores, we find a wide variety of sizes. A typical pore-size distribution is shown in Fig. 2–5. Note that if all of the pores were the same size, they would have the same attraction for the water and there would be a random distribution of the oil and water.*

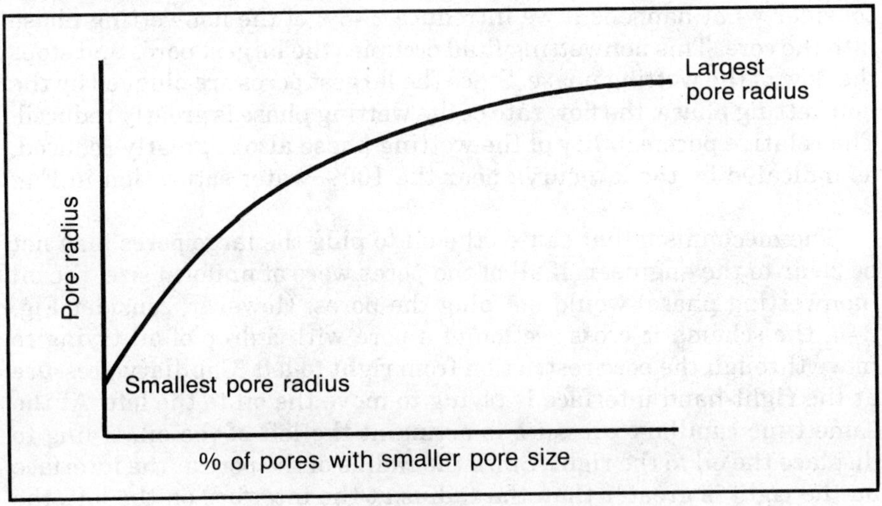

Fig. 2–5 Pore-size distribution

Because of the difference in wettability, we find that the shape of the wetting and nonwetting relative permeability curves are different. Possibly this difference is best illustrated by looking at the relative permeability to one phase at the irreducible saturation of the other phase. The relative permeability to water at an irreducible oil saturation of 10% (90% water) is about 0.6. The relative permeability to the nonwetting phase, oil, at the irreducible water saturation of 0.3 approaches 1.0. In this case it is 0.95. One practical effect of this observation is that we normally assume the effective permeability of the nonwetting phase in the presence of an irreducible saturation of the wetting phase is equal to the absolute permeability. Consequently, oil flowing in the presence of connate water or irreducible water saturation is assumed to have a permeability equal to the absolute perme-

*A quantitative treatment of capillary pressure is given in chapter 6.

ability. Similarly, gas flowing in a reservoir in the presence of irreducible water saturation is assumed to have a permeability equal to the absolute permeability. Also, since the relative permeability to the wetting phase at the nonwetting-phase irreducible saturation is much less than the relative permeability to the nonwetting phase at the irreducible wetting-phase saturation, decidedly different-shaped curves are obtained for the two phases.

To understand this concept, consider the portion of the relative permeability curves where the saturation is near 100% for the wetting phase. We know that when the wetting-phase saturation (or any saturation) is 100%, the relative permeability to this phase is 1.0. Now consider what happens if we introduce 2–3% of the nonwetting phase into the core. This nonwetting fluid occupies the largest pores and stops the flow of the wetting phase. Since the largest pores are plugged by the nonwetting phase, the flow rate of the wetting phase is greatly reduced. The relative permeability of the wetting phase also is greatly reduced, as indicated by the k_{rw} curve near the 100% water saturation in Fig. 2–3.

The mechanism that causes the oil to plug the large pores may not be clear to the engineer. If all of the pores were of uniform size, the oil (nonwetting phase) would not plug the pores. However, consider Fig. 2–6, the schematic cross section of a pore with a drop of oil trying to move through the pore restriction from right to left. Capillary pressure at the right-hand interface is trying to move the oil to the left. At the same time capillary pressure is acting at the left of the oil, trying to displace the oil to the right. Since the radius of the pore at the interface on the right is greater than the radius at the interface on the left, the capillary pressure on the right is less than the capillary pressure on the left. Thus, the applied pressure gradient must be greater than the difference in the capillary pressures or the oil will not move through the restriction and the pore is plugged.

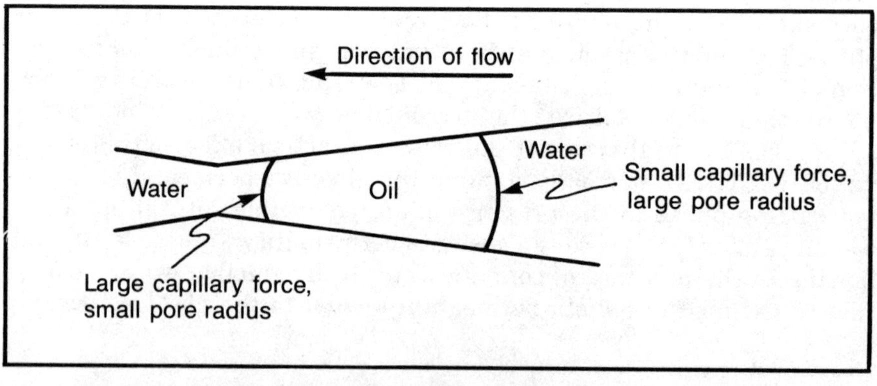

Fig. 2–6 Pores plugged by discontinuous nonwetting phase

Now consider the nonwetting-phase saturation permeability curve near the 100% nonwetting-phase saturation. At 100% saturation the relative permeability to the nonwetting phase is 1.0. Note what happens if we put 2–3% of the wetting phase in the core. This 2–3% is attracted to the smallest pores and thus occupies them. When these smallest pores are occupied by the nonwetting phase, they undoubtedly contribute little to the total flow of the nonwetting phase because they are the smallest. Consequently, when the smallest pores are occupied by the wetting phase, they cause almost no reduction in the flow rate of the nonwetting phase.

We find that the nonwetting-phase relative permeability curve has an S shape while the wetting-phase relative permeability curve is concave. These shapes are not always apparent in lab data because the only data often reported are between the irreducible nonwetting- and the irreducible wetting-phase saturations.

The same general observations apply to gas-oil relative permeability data, as shown in Fig. 2–7. These data also may be termed gas-liquid relative permeability, since they are plotted against the liquid saturation. This is typical of gas-oil relative permeability data in the presence of connate water. Since the connate or irreducible water normally occupies the smallest pores in the presence of oil and gas, it appears to make little difference whether these pores are occupied by water or oil, which would also be essentially immobile in these small pores. Most labs run gas-oil relative permeabilities without water in the core. Consequently, in applying the gas-oil relative permeability data to a reservoir, the total liquid saturation is normally used as a basis for evaluating the relative permeability to the gas and oil. We can say that the laboratory oil saturation is equal to the reservoir oil plus the irreducible water saturation.

Note that the relative permeability curve representing oil changes completely from the shape of the relative permeability curve for oil in the water-oil system. In the water-oil system, oil is normally the nonwetting phase; in the presence of gas, oil is the wetting phase. Consequently, in the presence of water only, the oil relative permeability curve takes on the shape of an S. In the presence of gas, the oil relative permeability curve takes on the shape of the wetting phase, i.e., it is concave upward. The irreducible gas saturation in the reservoir is generally very small. In fact, it is generally smaller than the irreducible gas saturation predicted by lab analysis. This factor is also called the *equilibrium gas saturation.*

It can be concluded that residual saturations in general are smaller in the reservoir than those predicted in the laboratory. That is, the irreducible water saturation, the residual oil saturation, and the equilibrium gas saturation are less in the reservoir than those predicted from laboratory relative permeability measurements. This fact is prob-

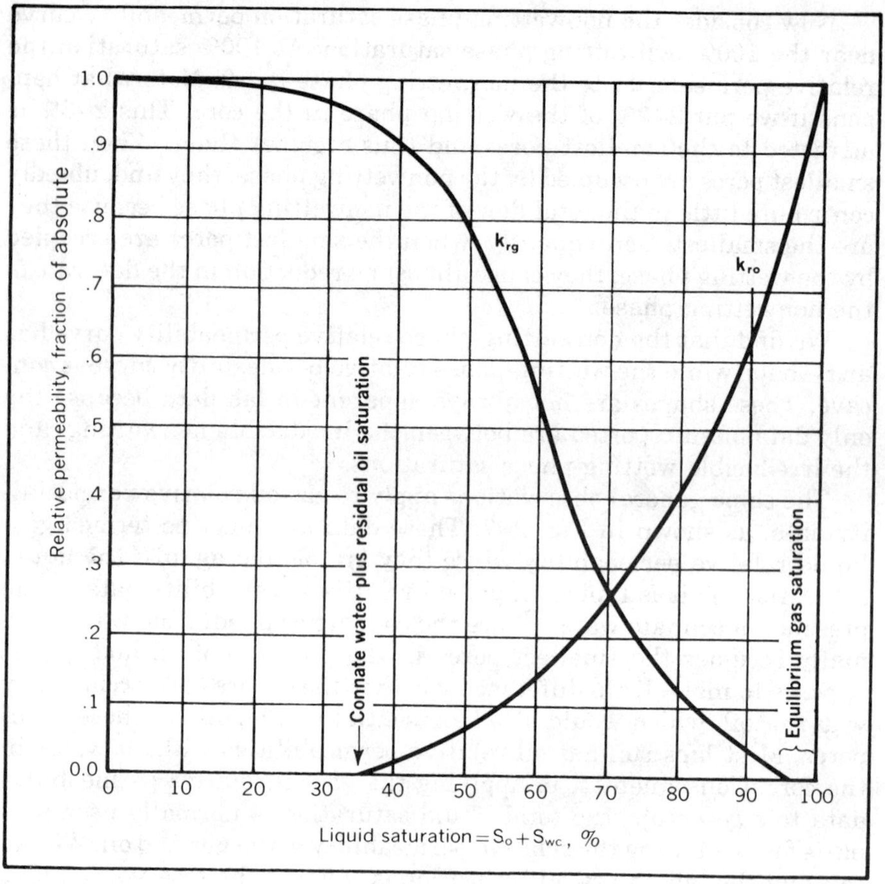

Fig. 2–7 Gas-oil relative permeability data

ably the result of the time factors involved. In the lab we are faced with using finite displacement times, but in the reservoir times actually approach infinite values for all practical purposes.

Three-phase relative permeability data have been run by several investigators (Fig. 2–8). However, these data generally have no practical significance because they have limited use. It is very difficult to find a situation where three phases flow concurrently in the reservoir.

For example, when water displaces oil and gas, the oil soon forms a bank between the displaced gas and the advancing water. Consequently, the oil displaces gas with only gas and oil flowing at a particular point in the reservoir. The water displaces oil with only oil and water flowing at other points of the reservoir. Only for very short periods is it possible for water to displace both oil and gas so all three are

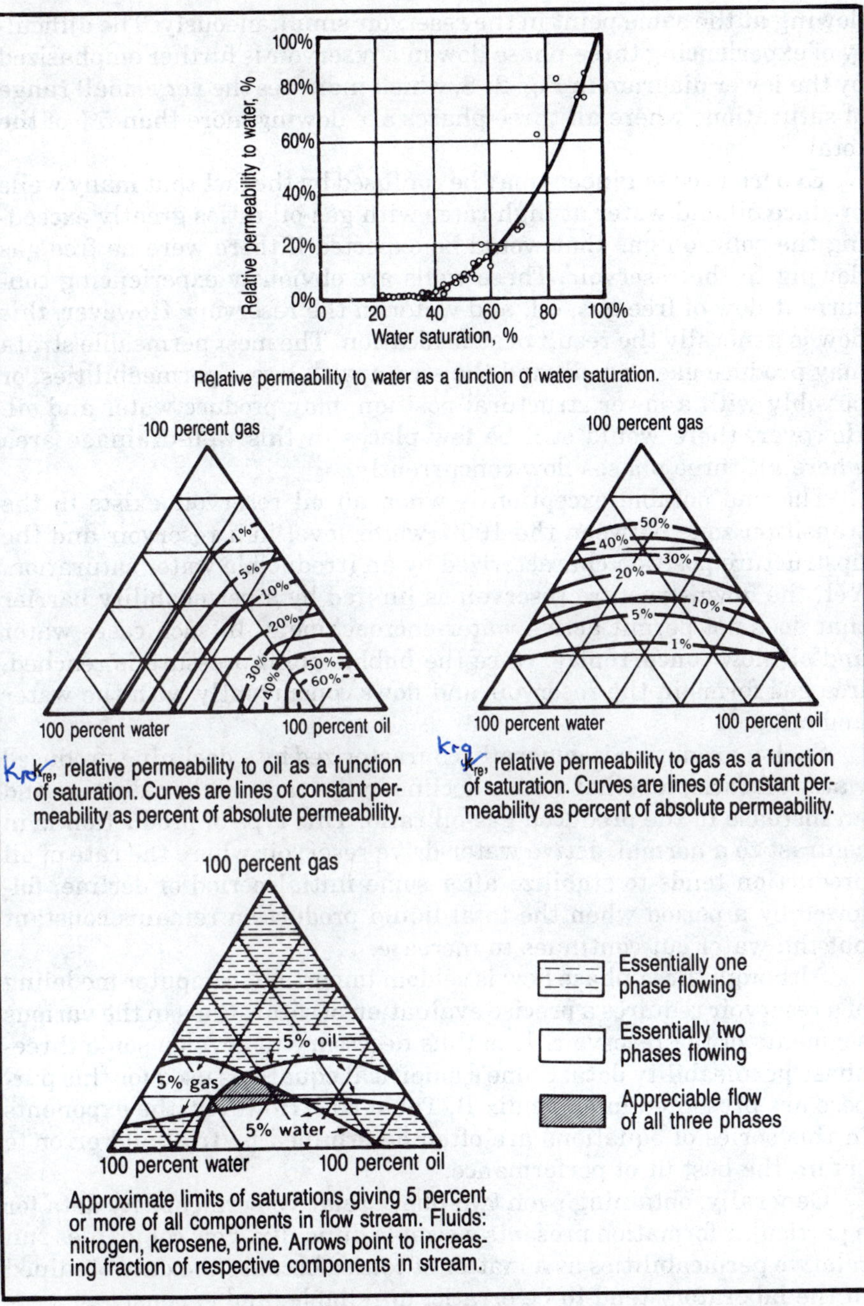

Relative permeability to water as a function of water saturation.

k_{re}, relative permeability to oil as a function of saturation. Curves are lines of constant permeability as percent of absolute permeability.

k_{re}, relative permeability to gas as a function of saturation. Curves are lines of constant permeability as percent of absolute permeability.

Approximate limits of saturations giving 5 percent or more of all components in flow stream. Fluids: nitrogen, kerosene, brine. Arrows point to increasing fraction of respective components in stream.

Fig. 2–8 Three-phase relative permeability data (after Amyx, Bass, and Whiting, *Petroleum Reservoir Engineering*, courtesy McGraw-Hill, 1960)

flowing at the same point in the reservoir simultaneously. The difficulty of experiencing three-phase flow in a reservoir is further emphasized by the lower diagram in Fig. 2–8, which indicates the very small range of saturations where all three phases are flowing more than 5% of the total.

Experienced engineers may be confused by the fact that many wells produce oil and water at high rates with gas-oil ratios greatly exceeding the solution gas that would be expected if there were no free gas flowing in the reservoir. These wells are obviously experiencing concurrent flow of free gas, oil, and water in the reservoir. However, this flow is generally the result of stratification. The most permeable strata may produce gas and oil, and the strata with lesser permeabilities, or possibly with a lower structural position, may produce water and oil. However, there would still be few places in this well-drainage area where all three phases flow concurrently.

The one notable exception is when an oil reservoir exists in the transition zone between the 100% water level in a reservoir and the upstructure position characterized by an irreducible water saturation. Yet, the downstructure reservoir is limited by a permeability barrier that does not permit active water encroachment. In such cases water and oil flow concurrently. Once the bubble-point pressure is reached, free gas forms in the reservoir and flows concurrently with the water and oil.

Such a formation is generally characterized by a declining produced water-oil ratio, together with a decline in the rate of oil production and an increase in the produced gas-oil ratio. This type of production is in contrast to a normal, active water-drive reservoir where the rate of oil production tends to stabilize after some initial period of decline, followed by a period when the total liquid production remains constant but the water cut continues to increase.

Although three-phase flow is seldom important, computer modeling of a reservoir requires a precise evaluation of saturations in the various segments of the reservoir. It is thus necessary to develop some three-phase permeability data. Stone's empirical equations used for this purpose are presented in appendix B, Table B1.[2] However, the exponents in this series of equations are often determined by trial and error to obtain the best fit of performance.

Generally, obtaining even two-phase relative permeability data for a particular formation presents a sizable difficulty. Few companies run relative permeabilities as a matter of course because the data obtained in the laboratory tend to be erratic, unreliable, and expensive.

Fig. 2–9 illustrates a portion of a suite of relative permeability curves run by a major oil company for one reservoir. The interpretation of such data to obtain one relative permeability curve that applies to a

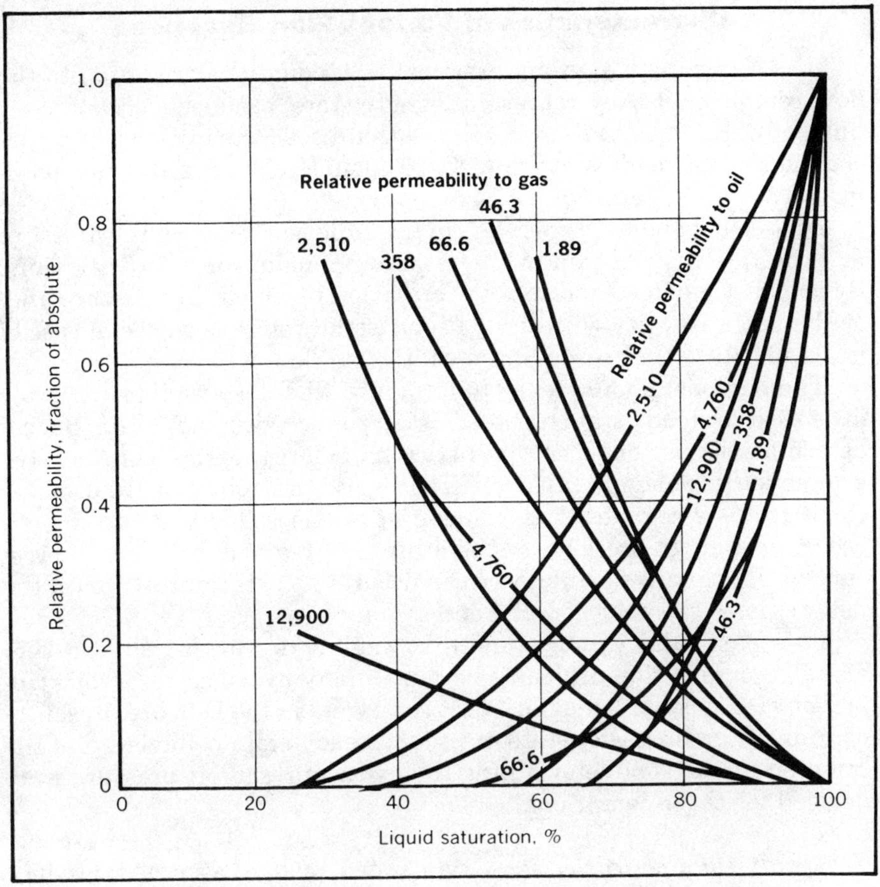

Fig. 2–9 Example family of relative permeability curves

particular formation is a major assignment. If such a normalizing task must be performed, it should be done on the basis of permeability distribution. However, there is little reason to believe that the single relative permeability curve so obtained will represent the behavior of the reservoir.

The author has never seen a situation in which the laboratory relative permeability data matched the indicated relative permeability data of a solution-gas-drive reservoir. When predicting solution-gas-drive behavior, the best results are obtained by calculating the relative permeability characteristics from the past performance of the reservoir and then extrapolating these factors to lower liquid saturations (see chapter 7). Many engineers believe that relative permeability data, as good as or better than lab-measured data, can be obtained using empirical equations such as those of Table B1 in appendix B.

Characteristics of Various Flow Regimes

It is convenient to group practical flow equations according to the flow regime that they represent: steady state, pseudosteady state, or unsteady state. Actually, we will see that pseudosteady state is a special case of unsteady-state flow. Care should be exercised when defining any of these terms.

In this text steady state refers to the situation in which the pressure and the rate distribution in the reservoir remain constant with time. By contrast, unsteady state is the situation in which the pressure and/or the flow rate vary with time. Pseudosteady state is a special case of unsteady state that resembles steady-state flow.

The engineer should learn to recognize whether a well or reservoir is nearest to steady state, unsteady state, or pseudosteady state. Such a determination is almost always necessary before the engineer can solve a reservoir problem because different working equations are used for the three flow regimes. For example, it is impossible to determine an accurate average pressure in the drainage area of a well from a pressure-buildup analysis unless we can accurately determine which of the flow regimes the subject well represents.

It is particularly important that engineers working in non-U.S. areas recognize this difficulty because many overseas reservoirs approach steady-state behavior. Most reservoirs in the U.S. are presently dominated by unsteady-state or pseudosteady-state behavior, and the common methods of determining the average reservoir pressure were devised for these conditions.

Steady-state characteristics. As noted, many reservoirs produce under steady-state flow. This flow occurs when a reservoir is producing with a strong water drive so every reservoir barrel of oil produced is replaced by a barrel of water. An idealized water-drive reservoir may look like Fig. 2–10. The pressure and rate distribution for such a system may be similar to Fig. 2–11. This pressure and rate distribution remains the same as long as the drainage area is in steady-state flow. Note that Eq. 2.3 can be solved for the pressure gradient $\Delta p/\Delta x$ at any radius:

$$\left(\frac{\Delta p}{\Delta r}\right)_r = \frac{q\mu}{1.127kA_r} \tag{2.5}$$

The minus sign of Eq. 2.3 has been dropped because the distance, r, is now measured against the direction of flow. Therefore, the pressure decreases with a decrease in radius, and $\Delta p/\Delta r$ is positive.

The cross-sectional area is subscripted with an r to indicate that it is a function of the radius. Thus, the pressure gradient is also a function

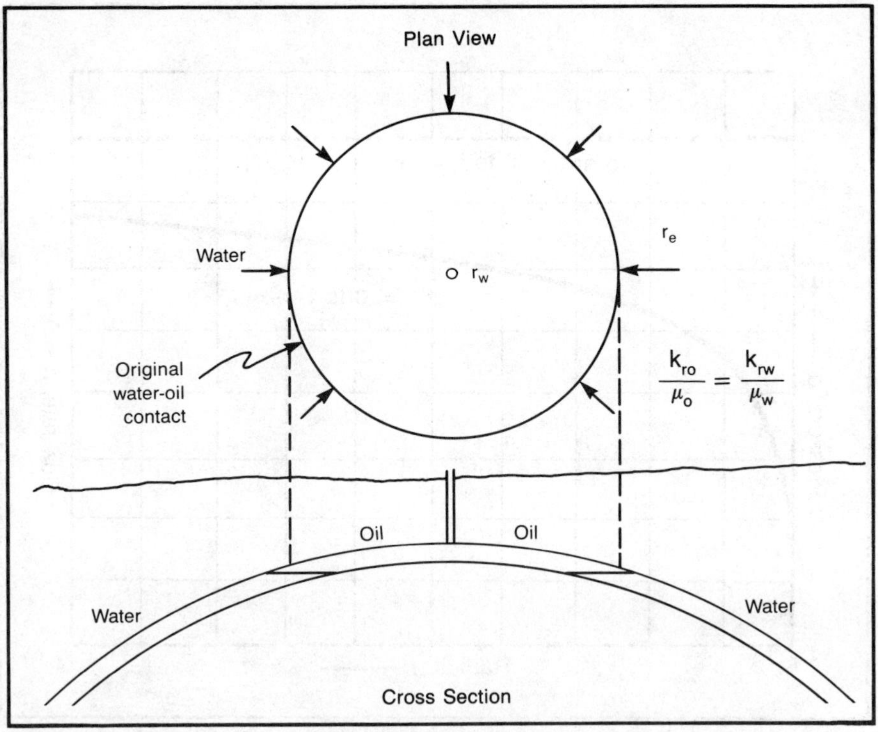

Fig. 2–10 Schematic of a water-drive reservoir

of the radius and is similarly subscripted. For a particular radius and a particular flow rate, q, the slope of the plot of pressure versus radius ($\Delta p/\Delta r$) remains constant as long as there is no change in the saturation that would change the effective permeability, k. Consequently, as long as the flow rate remains constant, the pressure distribution also remains constant.

This idea can be expanded to apply equally well for compressible fluids, such as gas, if the flow rate q is stated in mass units, such as standard cubic feet. Thus, well-pressure and flow-rate histories can be used to determine whether a well is in steady state. If the flow rate is constant and the bottom-hole pressure remains constant, there is little doubt that the drainage area of this well is in steady-state flow.

For such a situation to occur, the flow across the external drainage radius, r_e, must be equal to the flow across the well radius at r_w and the same fluid must be crossing both radii. This condition is never strictly met in a reservoir. However, a strong water drive, whereby the water-influx rate equals the producing rate, gives a pressure and rate history almost identical to the one described in Fig. 2–11. Pressure mainte-

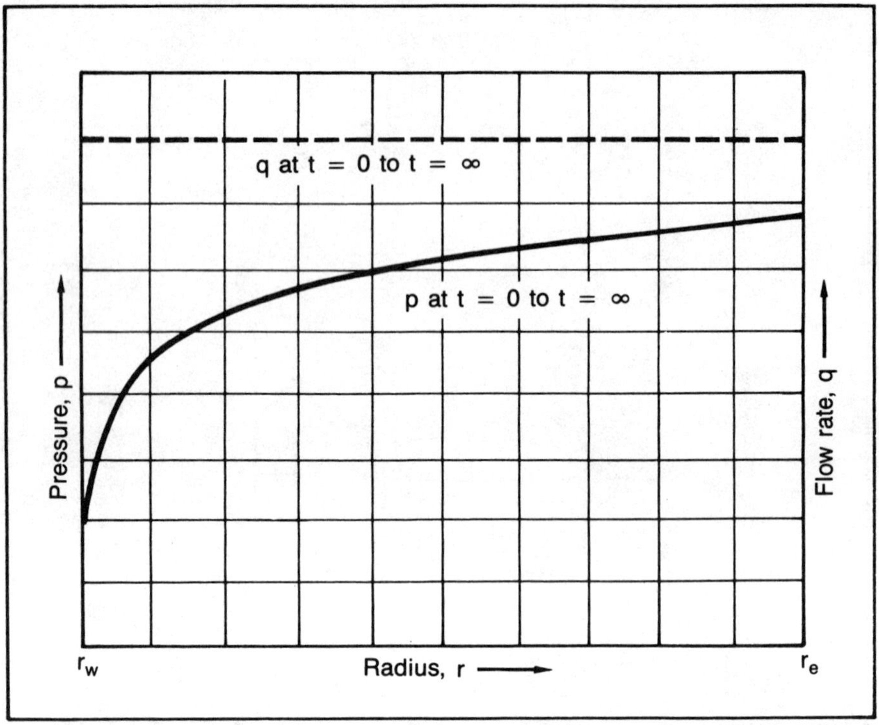

Fig. 2–11 Radial steady-state flow

nance by water injection downdip or by gas injection updip also approximates steady-state conditions, as do most pattern waterfloods after the initial stages of injection.

Steady-state equations are also useful in analyzing the conditions near the wellbore. Even in an unsteady-state system, the flow rate near the wellbore is almost constant so the conditions around the wellbore are almost constant. Thus, steady-state flow equations can be applied to this portion of the reservoir without significant error.

Unsteady-state characteristics. Fig. 2–12 shows the pressure and rate distributions for a system similar to the radial steady-state system of Fig. 2–11. However, in this case all of the production is caused by the expansion of the fluid in the reservoir. There is no water influx or other movement of fluid across r_e. The schematic reservoir model is shown in Fig. 2–13. Thus, the rate at r_e is zero, and it increases to a maximum at the well radius, r_w. In the steady-state case the flow across the outer boundary, r_e, is equal to the flow across r_w, the well radius. With flow across r_e equaling zero, the only energy causing the flow of fluid is the

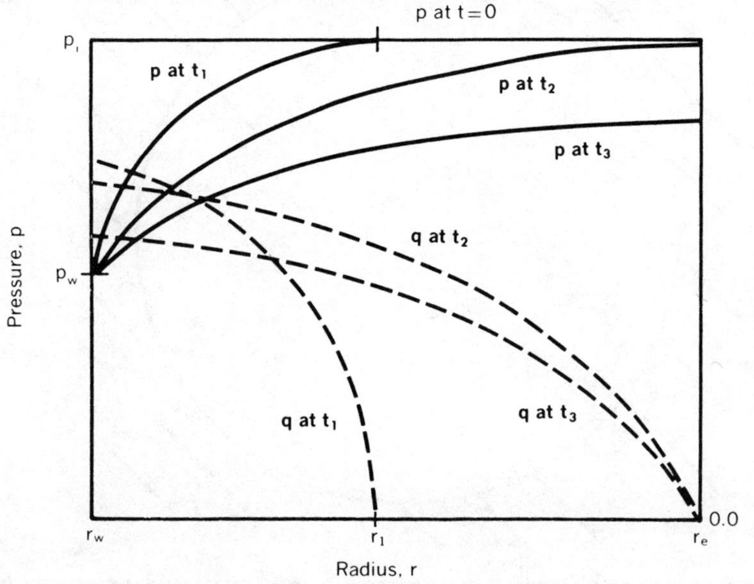

Fig. 2–12 Pressure and rate distributions for unsteady-state radial flow with constant well pressure

expansion of the fluids themselves. Initially, the pressure is uniform throughout the reservoir at p_i, which represents the zero producing time.

Examine Fig. 2–12, which shows the pressure and rate history for an unsteady-state system. The production rate is controlled so the pressure at the well is constant. This condition approximates the flow of a well against a fixed choke size or a well that is kept pumped down. After a short time of producing the well at such a rate that the well pressure remains constant, we obtain a pressure distribution shown as p at t_1. At this time only a small portion of the reservoir has been affected or has had a significant pressure drop.

Now remember that the flow taking place is caused by the expansion or compressibility of the fluid. Consequently, if no pressure drop exists in the reservoir at a particular point, or outside of that point, no flow can be taking place at that particular radius. This condition can be shown by the mathematical expression for compressibility:

$$c = (\Delta V/V)/\Delta p \tag{2.6}$$

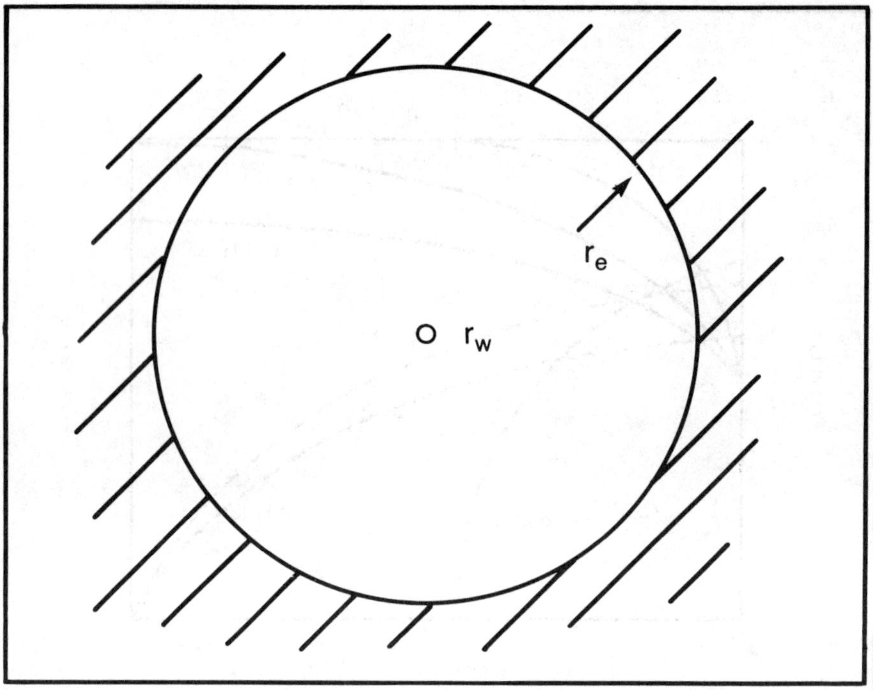

Fig. 2–13 Reservoir model for unsteady-state flow

Eq. 2.6 represents the compressibility of any material and is simply the fractional change in the volume per unit of pressure change. The expansion of fluid in the reservoir is represented by ΔV, which is equal to $cV\Delta p$. The fluid cannot expand without a drop in pressure. Thus, as shown in the plot of q at t_1, the rate at r_e is zero and it increases with a reduction in radius until the maximum rate in the reservoir is obtained at r_w. Fig. 2–12 is schematic and is not meant to be quantitative. The pressure and rate distributions at time t_1 represent an instant in time, and they move through these positions immediately as the production continues to affect more of the reservoir. That is, new areas of the reservoir experience a significant pressure drop and are subjected to flow until the entire reservoir is affected as shown by the pressure at t_2. The rate, q, at t_2 indicates that the flow rate at this time extends throughout the reservoir since all of the reservoir has been affected and has a significant pressure drop.

Note that the rate at the well has declined somewhat from time t_1 to t_2 since the same pressure drop $(p_i - p_w)$ is effective over a much larger volume of the reservoir. When the pressure in the entire reservoir has been affected, it drops throughout the reservoir as production continues. Therefore, the pressure distribution can be as shown for p at t_3 in

Fig. 2–12. The rate has declined somewhat during time t_1 to t_2 because of the increase in the radius over which flow is taking place. This rate continues to decline from t_2 to t_3 because the total pressure drop from r_e to r_w ($p_e - p_w$) is declining. Fig. 2–12 is an example of unsteady-state flow by our definition since the pressure and the rate are both changing with time, except for the one pressure that we have maintained artificially—the pressure at the well, p_w.

From time t = 0 to time t_2 when a pressure drop is finally affected throughout the entire reservoir, the pressure and rate distributions are not affected by the size of the reservoir or the position of the external drainage radius, r_e. During this time we say that the reservoir is infinite acting because the outer drainage radius, r_e, can be mathematically infinite. We will find that even in reservoir systems dominated by steady-state flow the effect of changes in well rates or pressures is governed by unsteady-state flow equations until these changes have occurred for a sufficient length of time to affect the entire reservoir so the reservoir again reaches a steady-state condition.

As expected, the mathematics governing unsteady-state flow with the pressure and rate varying both with time and with radius give a complex expression called the *radial-diffusivity equation,* which is a second-degree partial-differential equation. However, general solutions to this equation make it possible for us to insert constants into various expressions to obtain the rate behavior if constant pressure conditions prevail, as in Fig. 2–12. If a constant well rate prevails, as in Fig. 2–14, another solution allows us to analyze the pressure behavior. We can apply both the constant-rate and constant-pressure solutions to variable rates and variable pressures using a technique known as *superposition.*

Pseudosteady-state characteristics. Fig. 2–14 illustrates the pressure and rate distribution for the same unsteady-state system discussed in Fig. 2–12 and illustrated in Fig. 2–13. However, in this particular case the rate at the well, q_w, is held constant. This condition is comparable to a prorated well or one that is pumping at a constant rate. Again, at time t = 0 the pressure throughout the reservoir is uniform at p_i. Then after a short production time, t_1 at a constant rate, only a small portion of the reservoir has experienced a significant pressure drop. Consequently, the reservoir is flowing only out to a radius r_1. As production continues at the constant rate, the entire reservoir eventually experiences a significant pressure drop, shown as p at t_2 in Fig. 2–14.

Soon after the entire reservoir pressure has been affected, a rather unexpected situation arises. The change in the pressure with time at all radii in the reservoir becomes uniform. Therefore, the pressure distributions at subsequent times are parallel, as illustrated by the dis-

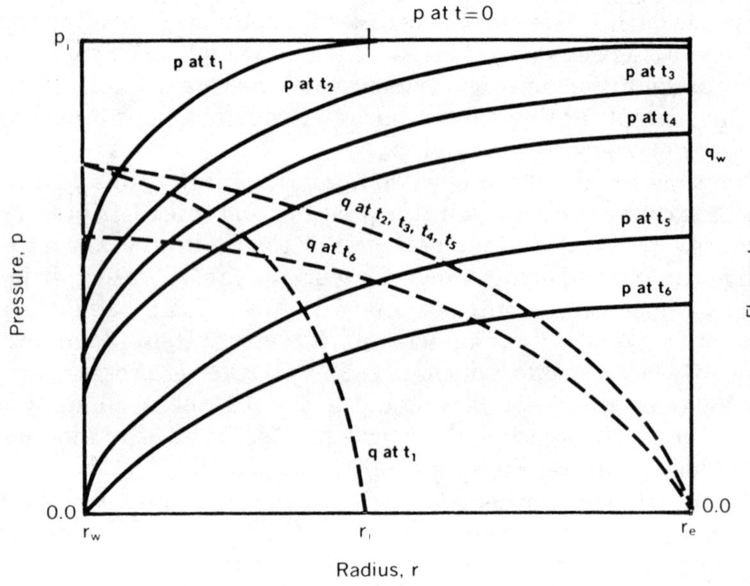

Fig. 2–14 Unsteady-state radial flow with constant producing rate—pseudosteady state, t_2 to t_5

tributions at times t_3, t_4, and t_5. Mathematically, this is equivalent to $\Delta p/\Delta t$ being constant. This situation continues with uniform changes in pressure with time at all radii and with subsequent parallel pressure distributions, until the reservoir can no longer sustain a constant flow rate at the wellbore. This point occurs when the pressure at the well, r_w, has reached its physical lower limit. Note that during the time when the change in pressure with time throughout the reservoir is constant, the rate distribution remains constant. This can be seen by examining Eq. 2.3, written for the rate of flow at a particular radius q_r:

$$q_r = \frac{1.127kA_r}{\mu}\left(\frac{\Delta p}{\Delta r}\right)_r \tag{2.7}$$

As noted for a particular radius, A_r is a constant. Also, unless some saturation change occurs in the reservoir, the permeability, k, remains constant. Note that $(\Delta p/\Delta r)$ at any particular radius represents the slope of the pressure-versus-radius plot. As long as the pressure distributions remain parallel, the slope of the plot at a particular radius and the rate at that radius will be constant.

This situation, which exists after the reservoir has been produced at a constant rate long enough to affect the entire reservoir, causing a constant change in pressure with time at all radii and resulting in parallel pressure distributions and corresponding constant rate distributions, is termed *pseudosteady-state flow*. It is easy to see how the name was obtained. Since all of the terms in the Darcy equation appear to remain constant or do remain constant, it is normal to assume that steady-state flow does exist. In fact, Craft and Hawkins refer to this phenomena as steady-state flow from a bounded reservoir.[3] Others refer to this flow regime as semisteady-state flow. However, using the definition of steady state previously cited, it can be seen that the absolute pressure is changing throughout the reservoir with time. Thus, by our definition the system would be unsteady state. Hence, the term pseudo-, or false, steady state is used for a system that appears to represent steady state but is actually unsteady state.

Pseudosteady-state flow is recognized as a special case of unsteady-state flow because simplified equations similar to those for steady-state flow can be derived to describe the behavior of a reservoir during this time. Also, a period of the reservoir production covered by pseudosteady-state flow represents a large portion of the reservoir life. Some engineers prefer to call this stabilized flow. However, care should be used in assuming that stabilized flow is synonymous with pseudosteady state, since the term *stabilized* is used in a broad sense to describe many different things.

Pseudosteady-state flow fundamentals form the basis for the interpretation of stabilized gas-well back-pressure tests, productivity index-type flow tests, and many other important reservoir engineering problems.

Steady-State Flow

As noted, steady-state flow represents the situation that exists when the pressure and rate distributions throughout the reservoir do not change with time. In order for this to occur, the mass flow rate into the reservoir must equal the mass flow rate out of the reservoir. These conditions may be closely approximated when a reservoir has a strong water drive or a large gas-cap drive, or when it is experiencing secondary recovery on a pattern basis. In many other instances the actual deviations from steady state are sufficiently minor so steady-state equations are warranted. Rates that vary over long distances in the reservoir may be nearly constant over short distances. Similarly, other reservoir conditions that may change over long periods may be substantially unchanged over short periods. Figs. 2–12 and 2–14 show that in unsteady-state flow at relatively large times, the flow rate near the wellbore is virtually constant. Thus, steady-state flow equations can be used to represent short periods of time for flow around the wellbore. By limiting the application of these equations to conditions near the wellbore for short time periods, we make the flow rate nearly constant.

Flow equations for specific geometries. The Darcy equation can be applied to specific geometries and compressibilities to obtain equations that are more readily applied than the basic Darcy equation. Some of these equations can be found in Table 2–1, which lists steady-state and pseudosteady-state flow equations. The linear flow equation for liquid

TABLE 2–1 Flow Equations

	Steady State	
Geometry	*Gas*	*Liquid*
Linear	$q_g = \dfrac{0.112Ak_{avg}(p_1^2 - p_2^2)}{T_f z \mu L}$	$q = \dfrac{1.127k_{avg}A}{\mu}\dfrac{(p_1 - p_2)}{L}$
Radial	$q_g = \dfrac{0.703k_{avg}\,h(p_2^2 - p_1^2)^n}{\mu T_f z\,\ell n\,(r_2/r_1)}$	$q = \dfrac{7.08k_{avg}\,h(p_1 - p_2)}{\mu\,\ell n\,(r_1/r_2)}$
Hemispherical	$q_g = \dfrac{0.703k_{avg}(p_2^2 - p_1^2)^n}{\mu T_f z\left(\dfrac{1}{r_1} - \dfrac{1}{r_2}\right)}$	$q = \dfrac{7.08k_{avg}(p_2 - p_1)}{\mu\left(\dfrac{1}{r_1} - \dfrac{1}{r_2}\right)}$
Five-Spot		$q = \dfrac{3.541k_{avg}\,h(p_{wi} - p_{wp})}{\mu\left(\ell n\dfrac{d}{r_w} - 0.619\right)}$
Seven-Spot		$q = \dfrac{4.721k_{avg}\,h(p_{wi} - p_{wp})}{\mu\left(\ell n\dfrac{d}{r_w} - 0.569\right)}$

Where:

- n = Turbulence constant
- q = reservoir b/d
- q_g = Mscfd
- A = ft^2
- k_{avg} = darcies
- p = psia
- p_{wi} = psia at injection well
- p_{wp} = psia at producer
- μ = cp
- z = Compressibility factor
- L = ft of length
- t_f = Formation temperature in °R = (°F + 460)
- r_e = External radius, ft
- r_w = Well radius, ft
- d = ft between input and producer

flow is probably the simplest of these equations since the compressibility of a liquid is so small that for the purposes of steady-state flow we can consider the liquid flow rate q as being constant. As noted, dp/dx in Eq. 2.3 is the slope of a plot of p versus x. This slope is a straight line for linear flow. Thus, Δx is the length of the linear system L, and Δp is the difference in pressures $p_2 - p_1$ (the pressure decreases as x increases). Then the Darcy equation becomes the linear flow equation for incompressible flow:

$$q = \frac{1.127kA}{\mu} \frac{(p_2 - p_1)}{L} \tag{2.8}$$

The radial incompressible flow equation is obtained by substituting for the cross-sectional area, A, the surface of a cylinder, $2\pi rh$, in the Darcy equation:

$$q = \frac{1.127k(2\pi rh)}{\mu} \left(\frac{\Delta p}{\Delta r}\right) \tag{2.9}$$

The finite difference form $\Delta p/\Delta r$ is used for convenience here. As noted, $\Delta p/\Delta r$ is the slope of a plot of pressure versus radius, which is the slope of a tangent to a curve at a particular radius. Thus, $\Delta p/\Delta r$ is constant only for a very small change in the radius Δr. From Eq. 2.9 the change in pressure, Δp, over a small change in radius, Δr, is:

$$\Delta p = \frac{q\mu}{7.08kh} \left(\frac{\Delta r}{r}\right) \tag{2.10}$$

Note that $q\mu/7.08kh$ is the same for all Δr when we are dealing with an incompressible fluid. Thus, we can write Eq. 2.10 for all Δr cylinders in the reservoir from the well radius, r_w, to the external radius, r_e (Fig. 2–15). If we sum all of the pressure drops from the drainage radius r_e to the well radius r_w, we would get the total change in pressure, $p_e - p_w$. Since these sums are equivalent to the corresponding integrals, we can show that the sum of all $\Delta r/r$ values is $\ell n\,(r_e/r_w)$. Thus, we obtain:

$$\sum_{p_w}^{p_e} \Delta p = \frac{q\mu}{7.08kh} \sum_{r_w}^{r_e} (\Delta r/r) \tag{2.11}$$

$$\int_{p_w}^{p_e} \Delta p = \frac{q\mu}{7.08kh} \int_{r_w}^{r_e} (\Delta r/r) \tag{2.12}$$

$$p_e - p_w = \frac{q\mu}{7.08kh} \ell n(r_e/r_w) \tag{2.13}$$

$$q = \frac{7.08kh}{\mu} \frac{(p_e - p_w)}{\ell n(r_e/r_w)} \tag{2.14}$$

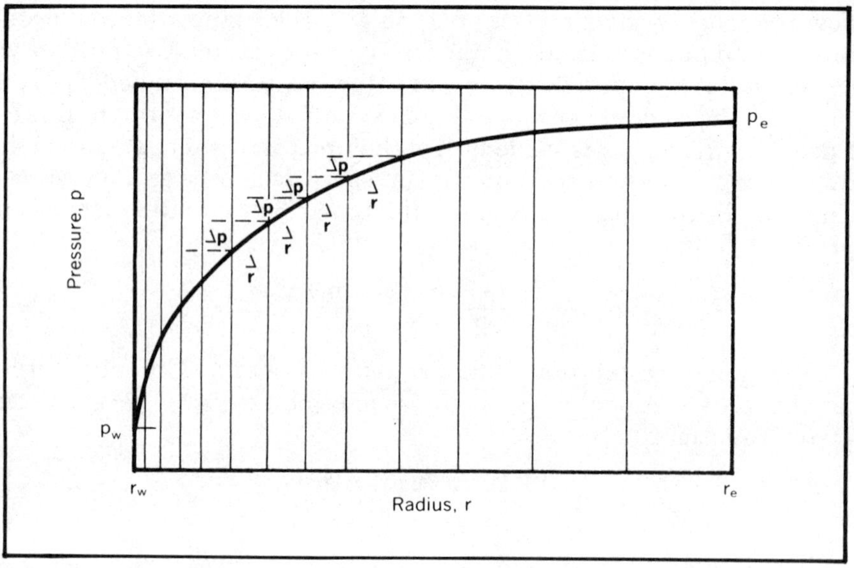

Fig. 2–15 Radial segments in radial flow

Eq. 2.14 is one of the best known and most used of the reservoir flow equations. Note that if we choose to sum the $\Delta r/r$ and Δp values from r_1 to r_2 and from p_1 to p_2, respectively, we obtain a more general equation that can be applied between any two radii:

$$q = \frac{7.08kh}{\mu} \frac{(p_1 - p_2)}{\ell n\,(r_1/r_2)} \qquad (2.15)$$

When thick reservoirs contain a bottom-water drive or a gas cap, the wells are often completed with perforations open in a very thin portion of the total thickness of the formation as far removed from the water or gas as possible. In these cases flow may be hemispherical in geometry in the vicinity of the wellbore. Fig. 2–16 illustrates hemispherical flow from a very thick reservoir. Although the flow is definitely hemispherical near the well, it tends to become more radial (cylindrical) as the distance from the well increases and as the vertical component of the flow decreases. Such well characteristics have been observed in the North Sea area and certainly are representative of many reservoirs throughout the world.

The hemispherical flow equation for incompressible fluids can be derived in much the same way that the radial flow equation was derived. In this case the cross-sectional area A is $2\pi r^2$. When this expression is substituted into the Darcy equation, the variables are separated, and the integration is performed, the hemispherical steady-state incompressible flow equation is obtained:

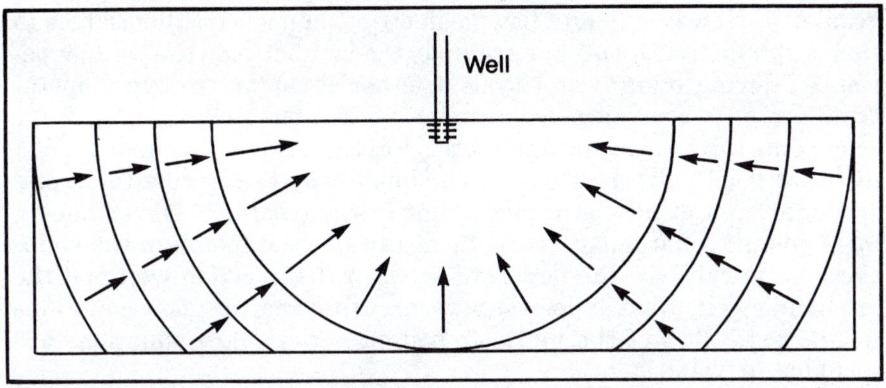

Fig. 2–16 Reservoir cross section illustrating hemispherical flow near wellbore and equipressure lines

$$q = \frac{7.08k(p_2 - p_1)}{\mu\left(\frac{1}{r_1} - \frac{1}{r_2}\right)} \qquad (2.16)$$

Eq. 2.16 is generally used in combination with a radial flow equation.* Care must be exercised in approximating the radii used in hemispherical flow. The internal radius, r_1, is not the well radius. It is a function of the well radius and the perforated thickness in a perforated completion. An approximation of r_1 for use in the hemispherical flow equation can be obtained by setting the cross-sectional area at the well equal to the hypothetical cross-sectional area at the well in hemispherical flow:

$$2\pi r_w h = 2\pi\,(r_{1\ \text{spherical}})^2 \qquad (2.17)$$

$$r_{1\ \text{spherical}} = (r_w h)^{0.5} \qquad (2.18)$$

Some wells are completed in an open hole, i.e., without casing opposite the producing formation. Although this is unusual, one of the most prolific producing areas, Saudi Arabia, completes its wells in this manner. In such cases it may be necessary to add the area of the bottom of the hole to the left-hand side of Eq. 2.17 to obtain a realistic hemispherical well radius. The external radius, r_2, is a function of the total formation thickness.

If only a small section of the formation near the center is open to production, a spherical flow equation may be advisable. In this case we simply multiply the right-hand side of Eq. 2.15 by 2.0 to obtain a spherical flow equation.

When a cross-sectional area can be simply stated, as in the cases described, it is relatively simple to account for any particular flow

*See the section on approximating complex geometries in this chapter.

geometry. However, many flow geometries do not lend themselves to such a simple treatment. For example, the flow between the wells experienced during many waterfloods or other secondary recovery operations presents a much more complex system for analysis. Fig. 2–17 represents the streamline and pressure distribution for a quadrant of a five-spot flood pattern. There is no simple way to describe the cross-sectional area at any particular point in the reservoir. Nevertheless, more complex mathematical methods can be used to obtain the equation analytically for the flow rate between the injection well and the producing well. Muskat derives several of these pattern flow equations in his text.[4] Two of the most often used pattern flow equations are included in Table 2–1.

Well damage. In deriving the flow equations, we have assumed that the permeability is uniform throughout the drainage area. Actually, this is not an accurate description of the situation. When drilling

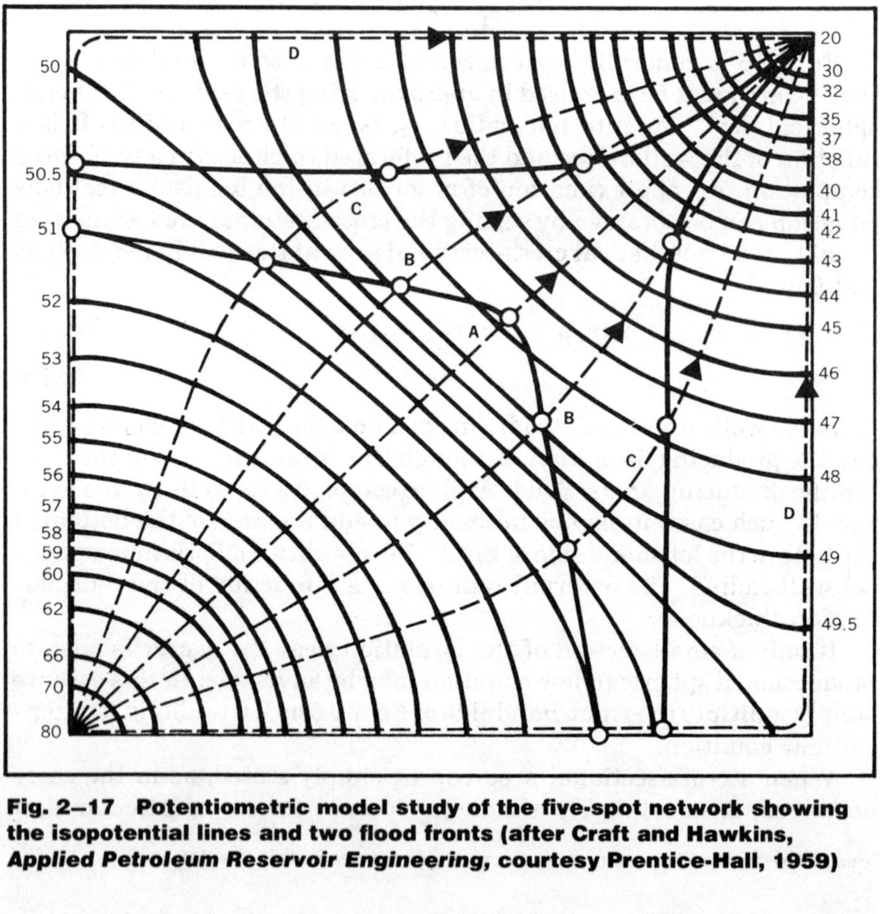

Fig. 2–17 Potentiometric model study of the five-spot network showing the isopotential lines and two flood fronts (after Craft and Hawkins, *Applied Petroleum Reservoir Engineering*, courtesy Prentice-Hall, 1959)

the well, the producing formation is always damaged, i.e., the permeability near the wellbore is reduced. During completion, wells with limited production capacities are normally treated by such methods as acidizing or fracturing, which in effect increase the permeability near the wellbore. Thus, the permeability near the wellbore is always different from the permeability away from the well where the formation has been unaffected by drilling and completion. This situation is most readily handled by including the additional pressure drop, or the reduction in pressure, in the flow equation as Δp_{skin}.

To illustrate, consider Eq. 2.15, the radial liquid steady-state flow equation. By rearranging this equation and using one of the pressures as the well pressure, we can show that the pressure at any radius r during radial steady-state flow of a liquid is:

$$p_r = p_w + \frac{0.141\, q\mu}{kh} \ell n\,(r/r_w) \tag{2.19}$$

Then, a plot of p_r versus the radius, r, can be prepared as shown by the plot in Fig. 2–18 labeled "no damage." Now assume that the permeability of the reservoir near the well was damaged during drilling. At the flow rate represented by the no-damage curve, an additional pressure drop of Δp_{skin} occurs. Then the pressure-versus-radius plot is modified to the curve shown as "with damage" in Fig. 2–18, and the well pressure that satisfies Eq. 2.14 is ($p_w + \Delta p_{skin}$). The equation then can be written as:

$$q = \frac{7.08 k_{undamaged} h(p_e - p_w - \Delta p_{skin})}{\mu \;\; \ell n\,(r_e/r_w)} \tag{2.20}$$

If the permeability around the wellbore is improved by the completion, the Δp in the affected area around the hole is reduced. Then the Δp_{skin} becomes negative in Eq. 2.20, which results in the pressure distribution shown in Fig. 2–18 as "with improvement." This figure assumes a constant producing rate.

It is often more confusing to think about the skin effect when the pressure drop remains constant as in Fig. 2–19. First, consider the damage curve relative to the no-change curve. If the permeability around the wellbore is reduced, the average permeability in the drainage area is reduced. Then for a particular pressure drop ($p_e - p_w$), the rate is reduced. If the rate is reduced, the pressure drop in the unaffected area from r_e to r_s is less than the pressure drop for no damage because the permeability in this area is unchanged but the rate is reduced. The remainder of the total pressure drop occurs in the damaged zone from r_s to r_w and is thus increased.

On the other hand when the permeability around the wellbore is increased, the average permeability in the drainage area is increased. If the total pressure drop is held constant, the rate also increases. With

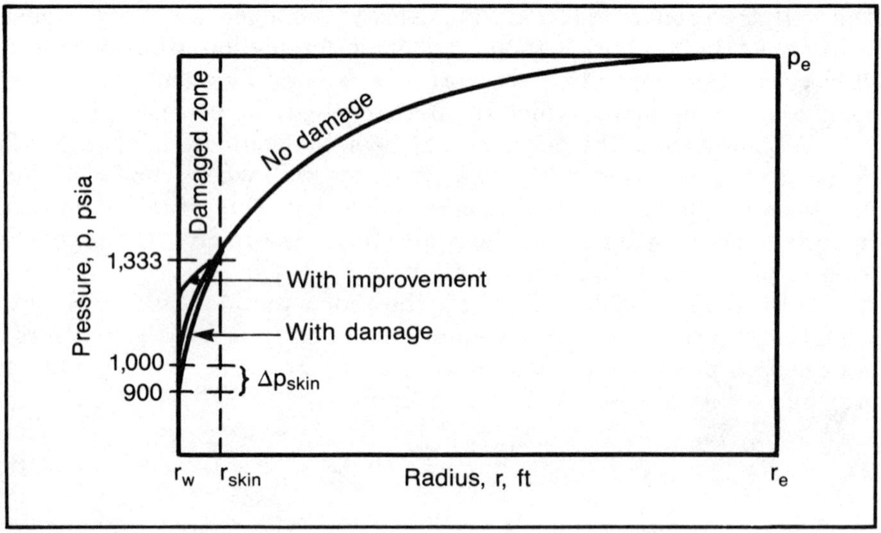

Fig. 2–18 Δp_{skin} with rate and p_e constant

this rate increase the pressure drop in the unaffected area is increased. Then the remainder of the total pressure drop in the increased permeability zone is decreased with the improved permeability.

Note that Δp_{skin} cannot be shown on Fig. 2–19 as it is in Fig. 2–18 because Δp_{skin} is defined as the additional pressure drop at the same rate. Each of the pressure plots in Fig. 2–19 represents a different rate.

To test understanding of radial flow and the Δp_{skin} concept, work problem 2.2 and check the solution against the one found in appendix C. In working this problem, remember that the rate in the equation is in reservoir barrels. Thus it is necessary to correct the rate of flow stated in stock-tank barrels per day to reservoir barrels by multiplying by the oil formation volume factor. This oil formation volume factor is the ratio between the amount of oil in the reservoir at the reservoir pressure and temperature and the amount of stock-tank oil that results from this mass of reservoir oil when the pressure and temperature are reduced to stock-tank conditions. This reduction in pressure and temperature causes liberation of gas and shrinkage of the liquid volume.

Ideally, oil formation volume factors are determined in the laboratory for a specific hydrocarbon system. However, lab data for specific reservoirs are not always available, and it is necessary to obtain estimates of these formation volume factors from empirical data. Such relationships are included in appendix B. A more complete discussion of the oil formation volume factor can be found in chapter 7.

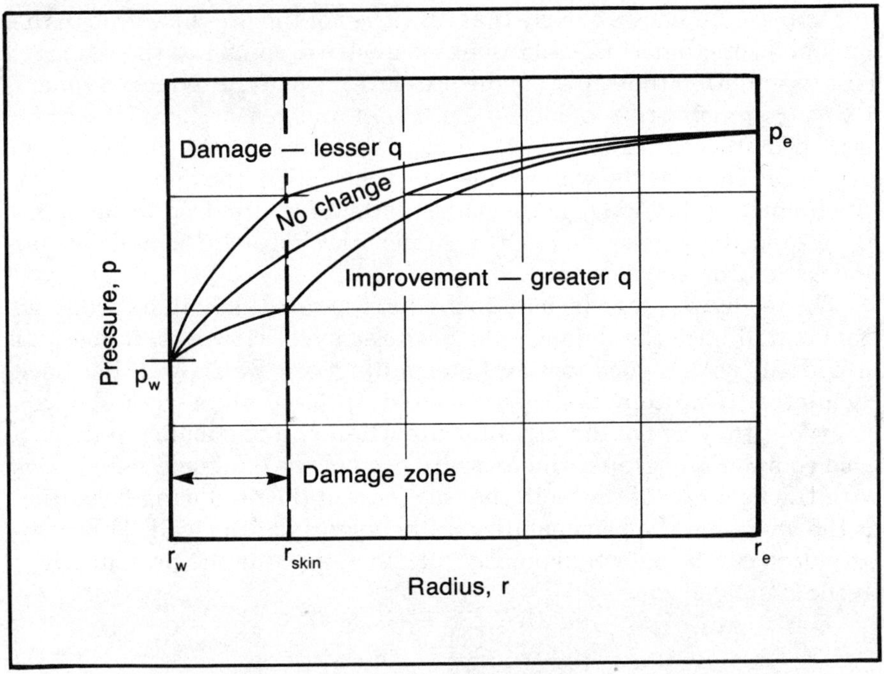

Fig. 2–19 Δp_{skin} effects with p_e and p_w constant

PROBLEM 2.2: Radial Steady-State Flow from a Damaged Well

Fig. 2–18 shows the physical significance of this problem.

A. An undersaturated water-drive reservoir has produced for several years without substantial change in the reservoir pressure. A pressure-buildup curve in well 3 indicates an undamaged reservoir permeability to oil of 10 md. A shutin well indicated a pressure at the drainage boundary of 2,000 psia for well 3. The flowing pressure in well 3 is 900 psia while it is flowing at a rate of 80 stb/d. What is Δp_{skin} if the following reservoir data are known?

Effective drainage radius = 700 ft
Well radius = 0.7 ft
Average net thickness = 6.9 ft
Reservoir oil viscosity = 0.708 cp
Oil formation volume factor = 1.25 res bbl/stb

B. What is the producing rate of this well if the pressure drop in the skin is reduced to zero by well treatments and if the same well pressure (900 psia) is maintained?

C. What is the total pressure drop in the damaged zone if the damage exists to a radius of 7.0 ft? The pressure drop from 700 ft to 7 ft can be calculated using Eq. 2.13 and a well radius of 7 ft.

Problem 2.2 shows clearly that Δp_{skin} is not the pressure drop in the damaged zone but is the additional pressure drop due to the damage. This assumption that Δp_{skin} is the pressure drop in the damaged zone is a very common error made by engineers and researchers. It can be argued mathematically that the term "skin" means the radius of the skin is the same as the well radius, i.e., the depth of the damage is zero, in which case the Δp_{skin} is the same as the pressure drop in the damaged zone. However, such an argument provides little help in our understanding the concept.

The term Δp_{skin} can be used to predict the rate that will be produced from a well once the damage has been removed. However, it does not provide us with a good tool for determining how well a well has been completed. If we compare the measured Δp_{skin} in all of the wells in a reservoir, they do not directly indicate whether a particular well has a good completion because the pressure drop in the damaged zone varies with the flow rate of the well, the thickness of the producing formation at the well, and the permeability in the vicinity of the well. This pressure drop can be shown using Eq. 2.19 to determine the pressure drop in the damaged zone:

$$(p_{rs} - p_w) = \frac{0.141q\mu}{kh} \ell n\,(r_s/r_w) \tag{2.21}$$

Consequently, the additional pressure drop, Δp_{skin}, is also proportional to the same factors. To compare the effectiveness of one completion relative to another, we can then normalize the Δp_{skin} by dividing it by the parameters that vary from well to well. For reasons that become clear in chapters 3 and 4, it is mathematically desirable to normalize Δp_{skin} by the entire group of terms in Eq. 2.21 (0.141 $q\mu/kh$). The resulting factor is the skin factor, S, introduced by van Everdingen.[5]

$$S = \frac{\Delta p_{skin}}{0.141q\mu/kh} \tag{2.22}$$

We can include the effect of damage or improvement in the permeability around the wellbore in most reservoir radial flow equations by using Δp_{skin} or the skin factor, S. We can substitute the well pressure minus Δp_{skin} for the well pressure in an equation that treats the permeability as uniform throughout, such as Eq. 2.20. Also, we can substitute a function of S for Δp_{skin} according to Eq. 2.22:

$$q = \frac{7.08k_{undamaged}\,h[p_e - p_w - (0.141q\mu S/kh)]}{\mu \ell n(r_e/r_w)} \tag{2.23}$$

Then solve for the flow rate to obtain:

$$q = \frac{7.08k_{undamaged}\,h(p_e - p_w)}{\mu[\ell n(r_e/r_w) + S]} \tag{2.24}$$

These two techniques can be used to modify all of the radial flow equations that are derived by assuming no damage or improvement around the wellbore.

Steady-state and pseudosteady-state radial flow equations when written without Δp_{skin} or S should include a permeability that is the average of the damaged permeability around the wellbore and the undamaged permeability. This method is possible because the external drainage radius, r_e, is constant for steady-state and pseudosteady-state systems. However, this is not the case for infinite-acting behavior where the external-drainage radius is increasing with time and thus the average permeability is changing with time.

We will see that the Δp_{skin} and skin factor terms are not always convenient to use. Another way of handling the damage or improvement around the wellbore is to adjust the well radius to an effective or apparent radius, r_{wa}, so correct answers are obtained from equations when the Δp_{skin} is taken as zero and the undamaged permeability is used:

$$q = \frac{7.08 k_{undamaged} h (p_e - p_w)}{\mu \ell n \dfrac{r_e}{r_{wa}}} \tag{2.25}$$

When this expression for q is equated to the right-hand side of Eq. 2.24, we find that both equations give the same flow q when:

$$r_{wa} = r_w e^{-s} \tag{2.26}$$

If there is damage around the wellbore, r_{wa} is smaller than r_w. If there is improvement, r_{wa} is larger than r_w. Eq. 2.26 can also apply to unsteady-state and pseudosteady-state flow.

Another way to express well damage is to note the ratio between the well rate and the rate that would be obtained if there were no damage. All other factors being equal, the flow rates are proportional to the effective drop ($p_e - p_w - \Delta p_{skin}$) and the damage ratio, DR, is:

$$DR = q_{undamaged}/q_{actual} \tag{2.27}$$

$$DR = (p_e - p_w)/(p_e - p_w - \Delta p_{skin}) \tag{2.28}$$

The damage ratio simply gives the ratio between the producing rate that would be obtained if damage were eliminated and the present producing rate. If the DR is less than 1.0, the permeability is improved.

Some companies use the reciprocal of the DR and call it the productivity ratio, PR:

$$PR = q_{actual}/q_{undamaged} \tag{2.29}$$

$$PR = (p_e - p_w - \Delta p_{skin})/(p_e - p_w) \tag{2.30}$$

When dealing with nontechnical persons such as managers, the engineer should consider using the damage ratio or the productivity ratio as a measure of damage since this concept is much easier to understand than the skin factor or the equivalent well radius. However, the engineer will find that the skin factor has a much greater mathematical utility.

Gas flow equations. Gas flow equations differ from liquid equations because the volumetric flow rate, q, varies with pressure caused by the compressibility of the gas. Thus, 1.0 b/d of gas at 1,000 psi does not represent the same weight or mass of gas as 1.0 b/d at 100 psi. To make the gas flow rate a constant, it is stated in terms of the equivalent gas flow rate when the gas is measured at standard conditions. This rate is equivalent to stating the gas flow rate in mass units, since the rate in standard cubic feet is directly proportionate to the pounds of gas. Using the gas equation (Eq. 2.31), the flow rate can be determined in barrels per day at the reservoir pressure and temperature as a function of the gas flow rate in thousands of standard cubic feet per day (Mscfd):

$$pV = ZnRT \tag{2.31}$$

Where:

V = Volume
Z = Gas deviation factor
n = Number of mols of gas
R = Gas constant for the particular system of units used

If psi, cubic feet, and degrees Rankine are employed, R is 10.73. Recognizing the fact that one mol of any gas at standard conditions (14.7 psia and 60°F) occupies 379 cu ft, we can write an expression for V, the volume of 1.0 scf at reservoir pressure p and reservoir temperature T:

$$(V/scf) = Z(1/379)\,(10.73)T/p \tag{2.32}$$

If we multiply this expression by the flow rate in thousands of standard cubic feet per day, q_g, and convert to barrels by dividing the expression by 5.615 cu ft/bbl, we obtain the expression:

$$q = \frac{5.04 q_g T z}{p} \tag{2.33}$$

Here, T and p are the reservoir temperature and pressure, respectively; q_g is the rate in Mscfd; and z is the gas deviation factor evaluated at the reservoir pressure and temperature.*

*The evaluation of the gas deviation factor is covered in chapter 4 and in appendix B.

Eq. 2.33 can be used to derive gas flow equations for the various geometries. For example the radial gas flow equation can be derived from Eq. 2.7 by substituting for q according to Eq. 2.33 and by substituting $2\pi rh$ for A, as in the radial incompressible flow equation:

$$\frac{5.04q_gTz}{p} = \frac{1.127k(2\pi rh)}{\mu}\frac{\Delta p}{\Delta r} \tag{2.34}$$

Rearranging Eq. 2.34, we obtain:

$$p\Delta p = \frac{5.04q_gTz\mu}{7.08kh}\frac{\Delta r}{r} \tag{2.35}$$

Eq. 2.35 then applies to any small Δr increment of the radial flow system (Fig. 2–15). By summing all of the $p\Delta p$ values and the $\Delta r/r$ values for all Δr cylinders from the external boundary, r_e, to the well radius, r_w, we obtain the radial steady-state flow equation for gas:

$$\sum_{p_w}^{p_e} p\Delta p = \frac{5.04q_gTz\mu}{7.08kh}\sum_{r_w}^{r_e}\frac{\Delta r}{r} \tag{2.36}$$

Integrating to obtain the summation terms, we obtain:

$$\int_{p_w}^{p_e} p\Delta p = \frac{5.04q_gTz\mu}{7.08kh}\int_{r_w}^{r_e}\frac{\Delta r}{r} \tag{2.37}$$

$$\frac{p_e^2}{2} - \frac{p_w^2}{2} = \frac{5.04q_gTz\mu}{7.08kh}(\ell n\ r_e - \ell n\ r_w) \tag{2.38}$$

$$q_g = \frac{0.703kh(p_e^2 - p_w^2)}{\mu zT\ \ell n(r_e/r_w)} \tag{2.39}$$

Gas flow equations are often confusing because they use different pressure and temperature bases for q and different temperature scales for the reservoir temperature, T. Eq. 2.39 uses q_g in thousands of standard cubic feet per day where the base pressure is 14.7 psia and the base temperature is 520°R (60°F). The engineer should remember that Eq. 2.39 is based on the Darcy equation, which is inaccurate for turbulent flow. Consequently, Eq. 2.39 as written does not represent a practical equation for most gas wells. However, an empirical exponent to the pressure-squared term can be used to account for turbulence and other nonideal flow conditions, such as the normal variation of temperature with gas expansion and the variation of the gas deviation factor and gas viscosity with pressure (see chapter 5). Eq. 2.39 then becomes:

$$q_g = \frac{0.703kh(p_e^2 - p_w^2)^n}{\mu zT\ \ell n(r_e/r_w)} \tag{2.40}$$

The exponent n must be evaluated by well testing. Remember, this equation—as with the radial incompressible steady-state flow equation—is written in terms of an average permeability, unless a separate term is added to account for the additional pressure drop caused by damage around the wellbore. If it is necessary or desirable to use the undamaged permeability, the additional pressure drop caused by damage can be accounted for by substituting $[\ell n\,(r_e/r_w) + S]$ for $[\ell n\,(r_e/r_w)]$. Then, $\Delta(p^2)_{skin} = q_g\mu zT\ S/0.703kh$.

Steady-state flow equations are seldom applied to gas flow because steady state seldom applies to a gas reservoir. Even when a gas reservoir has a strong water drive, the vast difference in the viscosities of the two phases generally means that the water cannot encroach at a rate even approaching the rate of gas withdrawal. In the past many gas wells in the U.S. did produce under steady-state conditions because gas production was restricted by a lack of market. However, such a situation is uncommon today. Nevertheless, we will see that the pseudosteady-state flow equation can be applied to gas reservoirs, and the equation is almost as simple as Eq. 2.40.

The linear gas flow equation can be similarly derived by substituting for q in Eq. 2.3 according to Eq. 2.33 and by recognizing that the cross-sectional area remains constant. The resulting equation is:

$$\frac{5.04q_gTz}{p} = \frac{1.127kA}{\mu}\frac{\Delta p}{\Delta x} \tag{2.41}$$

Eq. 2.41 can be rearranged so the variables are separated into $p\Delta p$ and Δx terms. Then the summation or integration of these terms from p_1 and x_1 to p_2 and x_2 results in Eq. 2.42:

$$q_g = \frac{0.112\ Ak(p_1{}^2 - p_2{}^2)}{Tz\mu L} \tag{2.42}$$

Here, L is the difference between x_1 and x_2.

Since turbulence is normally not a problem in linear flow, this equation is not used with an exponent on the p^2 term as it was in the radial flow equation (Eq. 2.40). To explain further, at the very small cross-sectional areas represented in flow into a small wellbore, the velocities become very high and turbulent flow is usually reached. However, in linear flow the cross-sectional area is constant and the velocities are not as critical in regard to turbulence. An equation similar to Eqs. 2.39 and 2.42 can also be derived for the hemispherical flow of gas. This equation is listed in Table 2–1.

General Problems in Fluid Flow Calculations

Various difficulties are associated with most types of reservoir flow calculations. Having established some knowledge of fluid flow fundamentals and the working equations for the steady-state flow regime, we can now consider several difficulties that arise in applying these steady-state flow equations as well as the flow equations for pseudo-steady state and unsteady state.

Approximating complex geometries. We have considered the simple geometries of linear, radial, and hemispherical flow, but these geometries are never encountered in ideal form. Linear flow equations are almost never applied by themselves to a reservoir problem. Linear flow in a reservoir without accompanying radial or spherical flow would imply production from a trough or a ditch in a reservoir, which of course is absurd except where massive hydraulic fracture treatments are used. Consequently, in most cases when linear flow equations are used, they are combined with radial or spherical flow equations to account for the very high pressure drops that accompany the radial or spherical flow into a well.

Radial flow equations alone do not exactly fit existing drainage areas in the U.S. because leases are relatively small and operators must protect against drainage across lease lines. The system lends itself to an overlying rectilinear well spacing. Therefore, the drainage area of a particular well is normally approximated by a square or a rectangle. Generally, such a drainage system can be readily analyzed using radial flow equations by simply calculating an effective radius from the drainage area:

$$\pi r_e^2 = (\text{acres/well})(43{,}540) \tag{2.43}$$

$$r_e = [(\text{acres/well})(43{,}560)/\pi]^{1/2} \tag{2.44}$$

This approximation seldom introduces a significant error because the deviation from radial flow is near the extremity of the system where the cross-sectional area is so large that the pressure drop is extremely small. Using 10-acre spacing and an 8-in.-diameter well, the maximum error possible is 5%. This factor is determined by comparing a well radius of 330 ft, the distance to the nearest boundary, to a well radius of 457 ft, the distance to the corner of the 10-acre rectangle. A 5% error may appear large to engineers new to reservoir engineering. However, the uncertainties that accompany reservoir description are such that an answer with a 5% error may compare favorably in accuracy with an electrical engineering error in the third significant figure.

Another means of applying a radial flow equation to an actual flow problem is to correct the actual flow rate, representing flow from a part of a circle, to the equivalent full-circle flow rate. Then apply the flow equation as required. For example, Fig. 2–20 shows a small reservoir with a well near the apex of two intersecting faults forming an angle of

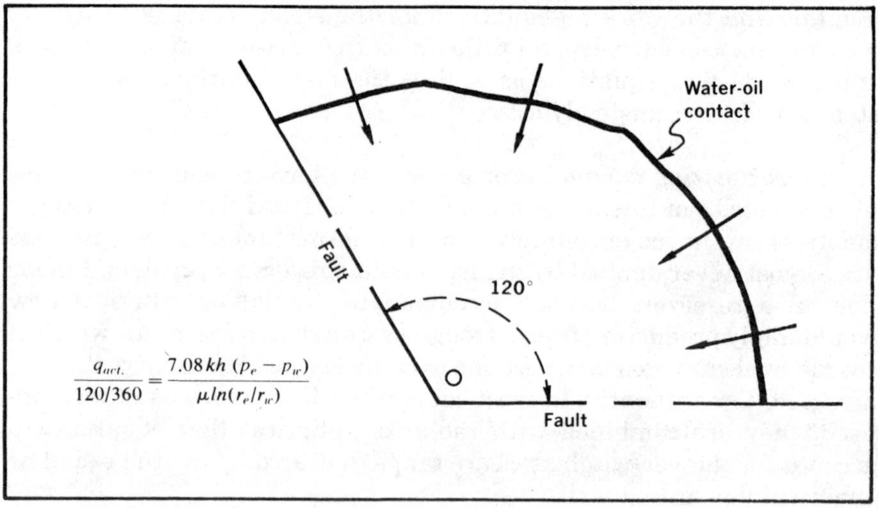

Fig. 2–20 Application of a radial flow equation to flow from a portion of a circle

120° or one-third of a full circle. If we wish to apply a radial flow equation to this situation, we can simply divide the actual q from the well by ⅓. Then the quotient is used as the q in Eqs. 2.14, 2.20, 2.24, etc.

This method is possible because in radial flow there is no flow across the radial spokes of the system. Consequently, the radial flow equations can be applied to a portion of a system as accurately as they can be applied to a full 360° radial system. The pressure distribution is the same in both cases.

Another way of handling this problem is to modify all radial flow equations by including θ in the right-hand side where θ is the fraction of a circle represented by the flow (one-third in the example case).

When drainage areas are more complex, the methods of obtaining flow equations become more complex. Suppose that a drainage area for a well is as shown in Fig. 2–21. A flow equation has not been given for this type of geometry; however, we can break the problem into two simple geometries that have the same total area. In this particular case the drainage area is modeled using a combination of linear and radial flow systems in series.

The actual dimensions of the two systems are set by ensuring that the total drainage area is equal to the total area of the model as indicated. Then with this model it can be assumed that the pressure drops in the two systems are additive. Thus, the total pressure drop from the water-oil contact to the well radius equals the pressure drop in the linear system, represented by the first term, plus the pressure drop in the radial system (half a circle), as indicated in the second term. If desired, this resulting expression can be solved for the flow rate, q, to give a close approximation of the flow characteristics of this drainage area.

In general if a complex geometry can be broken into a combination of simple geometries—flowthrough that occurs in series—we can write an expression for the pressure drop for each of the simple geometries. The total of these pressure drops represents the total system pressure drop. In performing this type of approximation, the engineer should be very careful to include some radial or spherical flow in the model if the flow considered is terminated by a well. In other words, if we use the pressure in a well as one of the terminal flow pressures, our modeling should include at least a portion of a radial or spherical flow system. There is no way to have flow into a well under linear conditions.

If the drainage area is broken up so the combination of simple geometries represents geometries with flow in parallel systems, the rates are additive. Such a situation is illustrated in Fig. 2–22. This figure is similar to the geometry described in Fig. 2–20, but the distance to the water-oil contact is not uniform. Consequently, a closer approximation of the flow characteristics of the drainage system is obtained by modeling the actual drainage geometry into a combination of two radial systems. The total area of the two radial systems is equal to the drainage area of the actual system. When the flow geometry is modeled in this way, flow takes place through parallel systems and the flow rates in the two systems are additive. This situation is indicated in the flow equation in Fig. 2–22 with the first term representing 45/360 of a radial

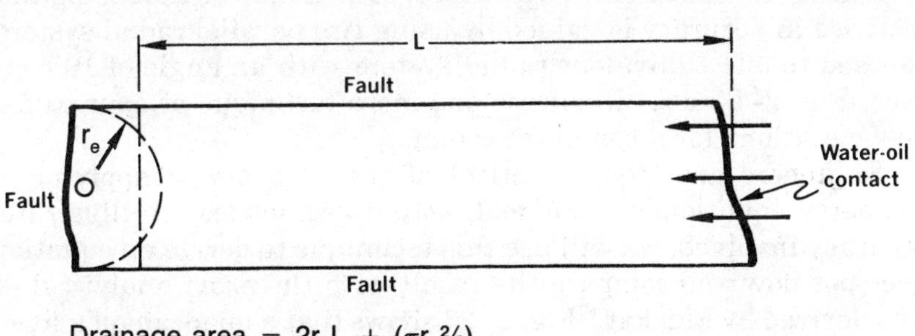

$$\text{Drainage area} = 2r_eL + (\pi r_e^2/2)$$

$$p_{we} - p_w = \frac{q\mu L}{1.127kA} + \frac{(q/.5)\mu \ell n(r_e/r_w)}{7.08kh}$$

Fig. 2–21 Approximating flow equations with two simple systems in series

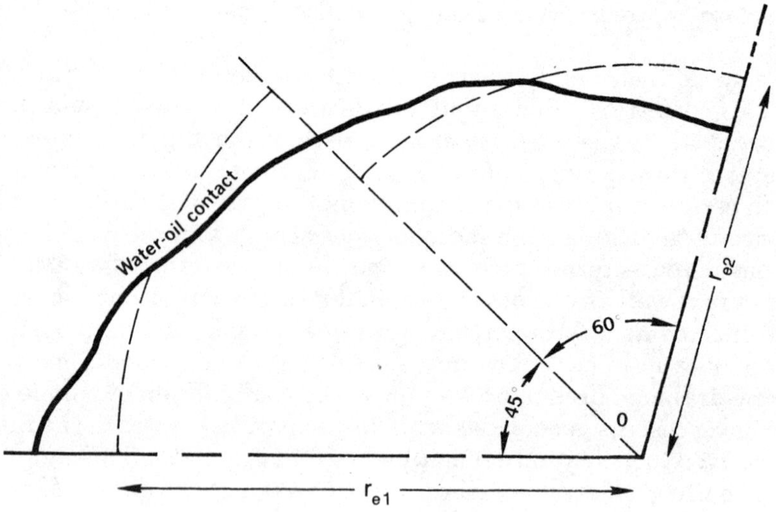

$$q = \frac{45}{360}\,\frac{7.08\,kh\,(p_{woc} - p_w)}{\mu\,\ell n\,(r_{e1}/r_w)} + \frac{60}{360}\,\frac{7.08\,kh\,(p_{woc} - p_w)}{\mu\,\ell n\,(r_{e2}/r_w)}$$

Fig. 2–22 Approximating flow equations with two simple systems in parallel

flow system and the second term representing $^{60}/_{360}$ of a radial system.

Due to the nature of radial flow, it is doubtful that a significant increase in accuracy is gained by using two parallel radial systems as opposed to one equivalent radial system with an angle of 105°. However, Fig. 2–22 does illustrate one more technique of approximating flow equations for a complex system.

Engineers are often skeptical of the accuracy of applying ideal geometry equations to nonideal, actual geometries. To illustrate the accuracy involved, we will use this technique to derive an equation for five-spot flow and compare the result with the exact analytical equation derived by Muskat.[6] Fig. 2–23 shows that a quadrant of a five-spot system can be approximated by two radial systems back to back whose total area is the same as the area of the quadrant. That is, each radial system should be equal to one-half of the quadrant area:

$$\tfrac{1}{4}\pi(r_{ep})^2 = \tfrac{1}{4}\pi(r_{ei})^2 = d^2/4 \qquad (2.45)$$

$$r_{ep} = r_{ei} = 0.57d \qquad (2.46)$$

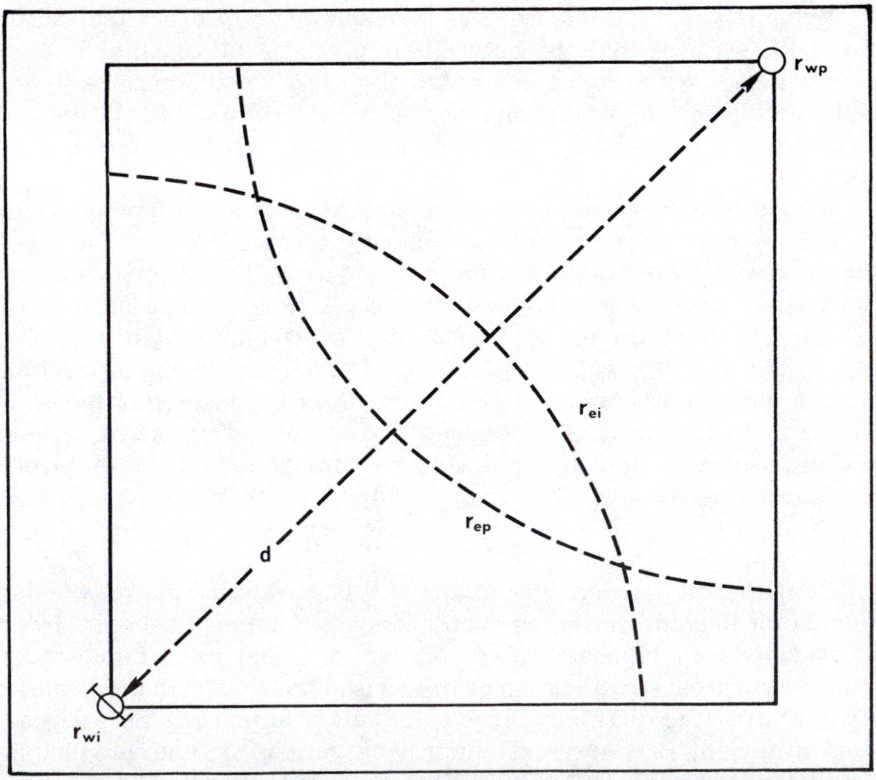

Fig. 2–23 Approximating five-spot flow with two radial systems

The total pressure drop for this quadrant, $p_{wi} - p_{wp}$, is then equal to the pressure drop on the injection side plus the pressure drop on the producing side. Using the radial incompressible liquid flow equation to denote these two pressure drops, we obtain:

$$p_{wi} - p_{wp} = \frac{q\mu \ell n(0.57d/r_{wi})}{7.08k_{avg}h} + \frac{q\mu \ell n(0.57d/r_{wp})}{7.08k_{avg}h} \quad (2.47)$$

When Eq. 2.47 is solved for q, assuming that r_{wi} equals r_{wp}, we obtain:

$$q = \frac{3.54k_{avg}h(p_{wi} - p_{wp})}{\mu[\ell n(d/r_w) - 0.571]} \quad (2.48)$$

The exact analytical equation derived by Muskat is:

$$q = \frac{3.54k_{avg}h(p_{wi} - p_{wp})}{\mu[\ell n(d/r_w) - 0.619]} \quad (2.49)$$

When the approximate Eq. 2.48 is compared to the exact analytical Eq. 2.49, we note that the only difference is in the constant in the denominator. When we consider that the log of the distance between wells divided by the well radius normally ranges from 7.0–10, we see that the error in modeling the five spot using two radial systems is less than 1%.

It should be emphasized that the approximation of the flow equation in this manner is performed to illustrate the accuracy that can be obtained with a combination of simple geometries. The engineer should not use Eq. 2.48 for calculations, since Eq. 2.49 is more accurate.

Eq. 2.49 uses the average permeability, which implies that the damage in the injection well is the same as the damage in the producing well. A more exact application is possible by writing the equation as a function of the undamaged permeability and including the Δp_{skin} for the injection well and the producing well in the pressure drop term; i.e., substitute $(p_{wi} - p_{wp} - \Delta p_{skin\ i} - \Delta p_{skin\ p})$ for $(p_{wi} - p_{wp})$ in Eq. 2.49.*

Determining average permeabilities. Permeability varies widely from point to point in the reservoir. This variation may be statistical, as observed for a homogeneous formation, or it may be systematic, as would occur from stratification or from a gradual change in permeability in a particular direction. Since sedimentary formations were deposited in ancient seas over extremely long periods of time, producing formations have different permeabilities at different vertical depths or, in the case of dipping formations, at different positions at right angles to the bedding plane. The drainage volume of a well may also have a systematic variation in permeability radially because various degrees of damage occur during drilling or treating the well. Since most flow equations simply require one permeability value, we need to determine the average permeability for such systematic variations.

Consider a vertical variation in permeability as shown in Fig. 2–24. This drawing shows three beds of different permeabilities and different thicknesses. Linear flow is depicted, and it is obvious that the flow rate through these parallel beds would be equal to the total of the flow rates through the individual beds:

$$q = q_1 + q_2 + q_3 \tag{2.50}$$

If we use the Darcy equation to write an expression for each of these flow rates including an expression for the total rate, q_{total}, based on an average permeability, k_{avg}, and the total thickness, h, we obtain:

*When the geometry is very complex, it may be necessary to use a digital computer to analyze the model. Reservoir computer-modeling techniques are discussed in chapter 11.

$$\frac{1.127k_{avg}hw\Delta p}{\mu L} = \frac{1.127k_1h_1w\Delta p}{\mu L} + \frac{1.127k_2h_2w\Delta p}{\mu L} + \frac{1.127k_3h_3w\Delta p}{\mu L} \tag{2.51}$$

Instead of writing the cross-sectional area A in each of these terms, the A is equated to hw to show that the width, w, is canceled with the other factors except permeability and thickness. Thus, Eq. 2.51 simplifies to:

$$k_{avg} = \frac{k_1h_1 + k_2h_2 + k_3h_3}{h} \tag{2.52}$$

When this approach is generalized, we can write the sum of the kh products as one term:

$$k_{avg} = \frac{\sum_{j=1}^{j=n}(k_jh_j)}{h} \tag{2.53}$$

In Eq. 2.53, n is the total number of zones considered.

If the same derivation is followed for radial flow, we find that the same equation results. Consequently, Eq. 2.53 applies to both parallel linear and radial flow.

If the variation in permeability is lateral rather than vertical, the pressure drops are added instead of the flow rates. This situation is shown in Fig. 2–25. Using the radial incompressible flow equation to evaluate Δp for each term, we find that:

$$\Delta p = \Delta p_1 + \Delta p_2 + \Delta p_3 \tag{2.54}$$

$$\frac{q\mu \ell n(r_e/r_w)}{7.08k_{avg}h} = \frac{q\mu \ell n(r_o/r_i)_1}{7.08k_1h} + \frac{q\mu \ell n(r_o/r_i)_2}{7.08k_2h} + \frac{q\mu \ell n(r_o/r_i)_3}{7.08k_3h} \tag{2.55}$$

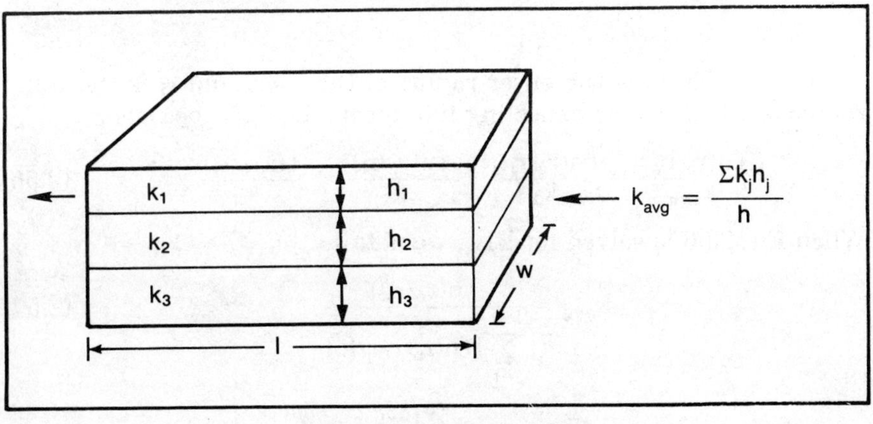

Fig. 2–24 Parallel flow

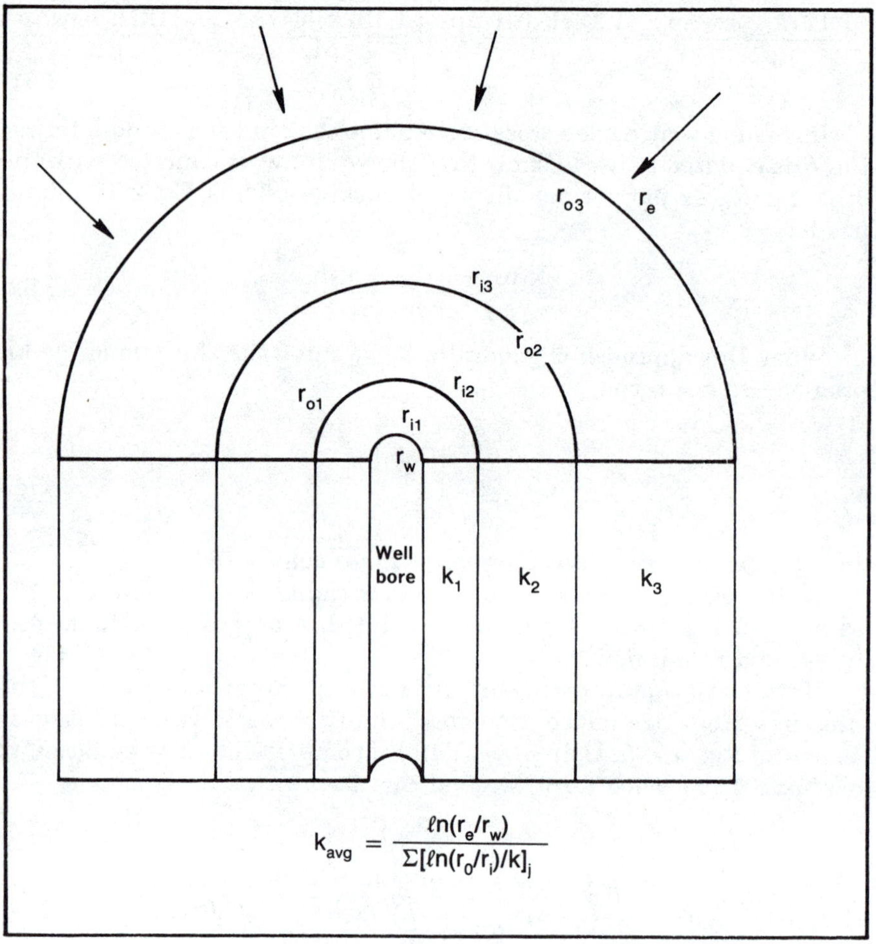

Fig. 2–25 Radial series flow

In Eq. 2.55, r_o is the outer radius of the zone, and r_i is the inner radius of the zone. By canceling like terms, Eq. 2.55 becomes:

$$\frac{\ell n(r_e/r_w)}{k_{avg}} = \frac{\ell n(r_o/r_i)_1}{k_1} + \frac{\ell n(r_o/r_i)_2}{k_2} + \frac{\ell n(r_o/r_i)_3}{k_3} \tag{2.56}$$

When Eq. 2.56 is solved for k_{avg}, we obtain:

$$k_{avg} = \frac{\ell n(r_e/r_w)}{\sum_{j=1}^{j=n} [\ell n(r_o/r_\ell)_j/k_j]} \tag{2.57}$$

There appears to be little likelihood that an equation would be needed to calculate k_{avg} for a lateral variation in permeability with linear flow. However, for completeness we will consider this equation. It can be derived in the same way that the series flow equation was derived for radial flow, except the Δp values are determined from the linear flow equation rather than the radial flow equation. When this derivation is carried out, all of the factors except the lengths and permeabilities of the individual zones are canceled and we are left with the general expression:

$$k_{avg} = \frac{L}{\sum_{j=1}^{j=n} \left(\frac{L}{k}\right)_j} \tag{2.57a}$$

To understand the permeability averaging techniques better, work problem 2.3 and check the results against the solution in appendix C.

PROBLEM 2.3: Determining Average Permeabilities

A. A stratified aquifer contains 2 ft of 200-md sand, 3 ft of 100-md sand, and 5 ft of 10-md sand. What is the steady-state water-injection rate into this 8-in.-diameter well when the well pressure is 600 psia and the reservoir pressure at a distance of 100 ft is 500 psia? Assume $\mu = 1.0$ and B_w, the water formation volume factor = 1.0 res bbl/stb of water. Note that steady-state rates can seldom be maintained in an aquifer.

B. A pressure-buildup analysis indicates an undamaged reservoir permeability of 75 md. A well in this reservoir has wellbore damage to a radius of 10 ft that reduces the permeability to this radius to 10% of the undamaged permeability, 7.5 md. What average permeability is calculated from a flow test of this well, i.e., what is the average permeability of the well drainage area?

Well radius = 1 ft
Well drainage radius = 1,000 ft

Correcting for Static Pressure Differences

All of the equations considered to this point apply only to horizontal flow if the total pressure drop is used in the equation. They can also be applied to nonhorizontal flow if the pressure drop applies only to the drop in pressure caused by flow and does not include the static pressure differences. That is, if pressures are measured at different depths in a reservoir, the static pressure difference between the two pressures must be deleted from the pressures before they are put into a flow equation.

The Darcy equation can be modified to include the static pressure differences, but the nature of the correction is such that the equation has limited use except for linear flow. As indicated, linear flow has limited applications in reservoir engineering. Nevertheless, the equation is used in discussing fluid displacement and the effects of gravity on fluid displacement, so we will indicate the basis for the Darcy equation including the static pressure drop.

The Darcy equation can be corrected to include the static pressure difference by stating the pressure drop caused by the fluid flow as a function of the total pressure drop and the static pressure difference. Fig. 2–26 indicates fluid flow updip with a bed angle, α. If the flow rate is zero, there is still a pressure difference between A and B caused by the difference in the static head of fluid, ΔD, between points A and B. Since the change in pressure with depth for fresh water is 0.433 psi/ft, the change in pressure with depth for a fluid whose specific gravity is γ is 0.433 γ. The difference in static pressure between points A and B is then 0.433 $\gamma\Delta D$. Consequently, we can determine the flowing pressure drop from the total pressure drop in a system from Eq. 2.58:

$$\Delta p_{flow} = \Delta p_{total} \pm 0.433\ \gamma\Delta D \tag{2.58}$$

This equation is written with a positive sign for flow downdip since the pressure drop in the direction of flow for a zero rate would be negative, which would give a positive absolute sign to the term. That is, ΔD in the direction of flow would be negative, which would give a positive sign to the second term. At any rate it seems simpler to write the equa-

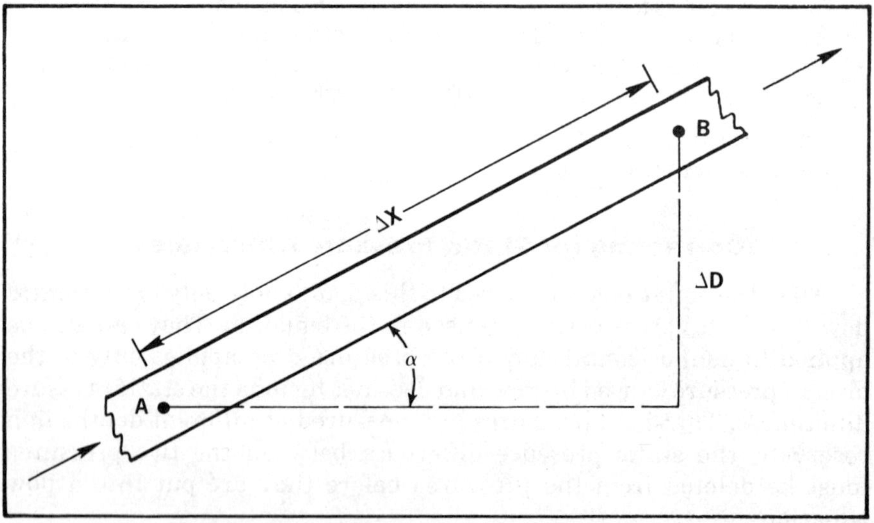

Fig. 2–26 Linear flow in a tilted reservoir

tion with plus and minus signs rather than to attempt to define the terms more specifically.

Eq. 2.58 can be used to correct total pressure differences to flowing pressure differences regardless of the prevailing flow geometry. Note also that the same result is obtained by correcting all of the pressures to some datum level in the reservoir and using the differences in those pressures as the flowing pressure difference. This technique is used by most companies for correcting data. For example, if the pressures at A and B in Fig. 2–26 are corrected to the datum level of A, the pressure at B would be corrected to the datum level at A by adding 0.433 $\gamma\Delta D$ to the pressure at B and the flowing pressure would be:

$$\Delta p_{flow} = p_A - (p_B + 0.433\ \gamma\Delta D) \tag{2.59}$$

The $p_A - p_B$ expression is the pressure drop, and Eq. 2.59 is equivalent to Eq. 2.58.

As noted, we can also modify the basic Darcy equation to include static pressure differences. When we substitute the right-hand side of Eq. 2.58 for Δp in the Darcy equation, we obtain:

$$q = \frac{1.127kA}{\mu} \frac{(\Delta p_{total} \pm 0.433\gamma\Delta D)}{\Delta x} \tag{2.60}$$

Note that $\Delta p_{total}/\Delta x$ is the total pressure gradient, and $\Delta D/\Delta x$ is the sine of the angle α (Fig. 2–26). Consequently, Eq. 2.60 can be written as:

$$q = \frac{1.127kA}{\mu} [(\Delta p/\Delta x)_{total} \pm 0.433\gamma \sin \alpha] \tag{2.61}$$

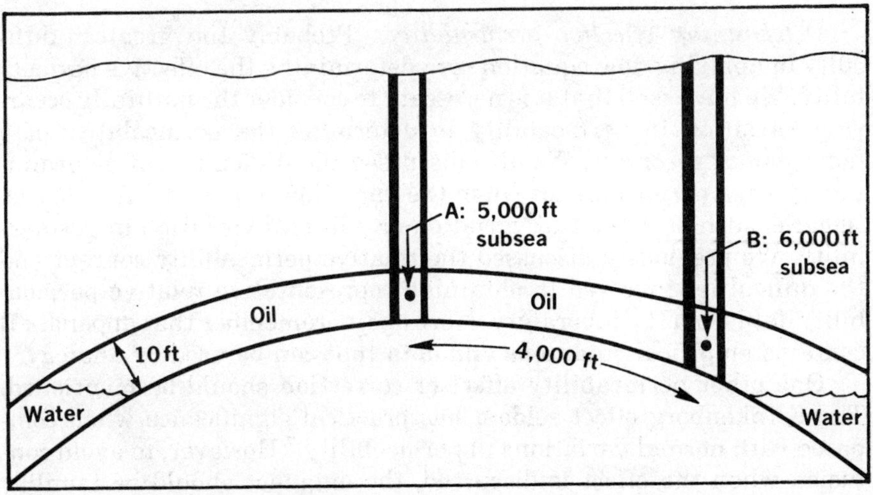

Fig. 2–27 Reservoir cross section for problem 2.4

Eq. 2.61—the Darcy equation as a function of the total pressure drop—has limited application because radial flow in a dipping formation has a different dip angle, α, along each radial ray. Consequently, for all flow geometries except linear flow, it is simpler to correct all reservoir pressures to some common datum level. Then we can use the difference in these pressures as the flowing pressure difference or use Eq. 2.58 to obtain the pressure drop caused by flow. However, Eq. 2.61 also helps identify the gravity forces that affect efficiency.

Try problem 2.4 and check the solution against that in appendix C to understand static pressure correction better.

PROBLEM 2.4: Correcting for Static Pressure Differences

Fig. 2–27 diagrams producing well A, shutin well B, and the reservoir data. Water drive activity is such that pressures do not vary with time. Assume radial flow and no well damage. The reservoir data are as follows:

Permeability to oil = 92 md
Oil reservoir viscosity = 0.708 cp
Well radius = 0.4 ft
Reservoir thickness = 10 ft
Reservoir oil specific gravity = 1.0
Saturation pressure = 2,000 psi
Flowing bottom-hole pressure at A = 2,250 psi
Shutin bottom-hole pressure at B = 3,000 psia (shutin well)

A. Find the producing rate in res b/d for the perfect dome indicated.

B. What is the producing rate if the reservoir is flat (r_e = 4,000 ft) and the same pressures prevail?

Determining effective permeability. Probably the greatest difficulty in applying flow equation is in determining the effective permeability. We have seen that it is necessary to consider the naturally occurring variation in permeability to determine the permeability of a homogenous reservoir. We also discussed the difficulties of determining average permeabilities when the variation in permeability is systematic, such as a vertical variation or a lateral variation in permeability. We previously discussed the relative permeability concept and the difficulties involved in obtaining representative relative permeability data from the laboratory. Here again, remember that appendix B contains empirical equations and data that can be used by analogy.

One other permeability effect or correction should be mentioned. The Klinkenberg effect seldom has practical significance when compared with normal variations in permeability.[7] However, to avoid confusion when the effect is discussed, the engineer should be familiar with this concept. The Darcy equation describes viscous fluid flow, and when conditions are such that viscous flow cannot take place, the

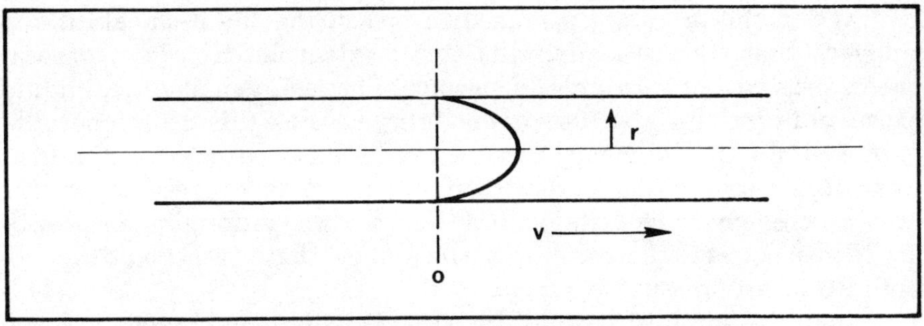

Fig. 2–28 Velocity profile for viscous flow in a tube

Darcy equation is inaccurate. Such a situation arises when gas flows at very low pressures. Remember that flow is through very small pores of the reservoir rock. When the pressure for a gas gets low enough, the distance between gas molecules becomes comparable to the size of the pores. Under these circumstances viscous flow cannot take place.

In viscous flow in a tube, we have various cylinder surfaces in the pipe moving at different velocities or shearing the fluid at different rates. A velocity profile normally exists in a tube where viscous flow is taking place, with the velocity at the surface of the tube being zero and the velocity at the center of the tube being a maximum (Fig. 2–28).

However, if the distance between molecules is comparable to the pore radius r, only one molecule at a time can flow through the small tube. There could be no velocity profile under these conditions. Since there is a variety of pore sizes in the normal petroleum reservoir, this critical condition is reached at various pressures for various pore sizes. Consequently, as the pressure gets lower, the flow deviates more from the viscous flow of the Darcy equation.

This phenomena is presented graphically in Fig. 2–29 where the permeability to gas is plotted versus the reciprocal of the arithmetic

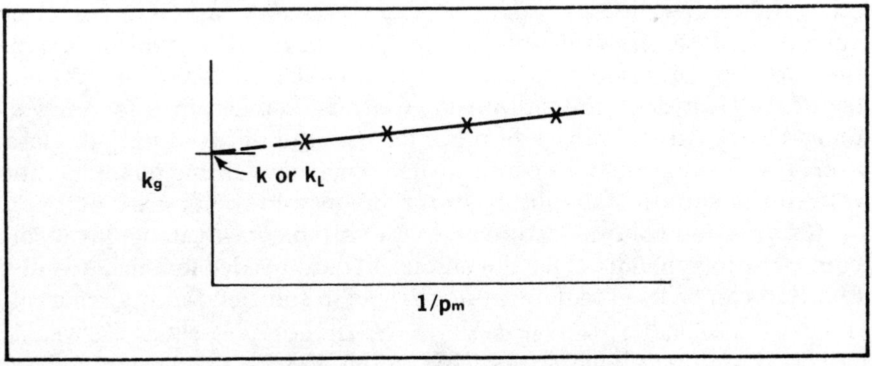

Fig. 2–29 The Klinkenberg effect

average of the pressure, p_m, used in measuring the permeability. It appears that, theoretically, this curve extrapolated to a reciprocal mean pressure of zero gives a permeability equivalent to the liquid permeability or the absolute permeability because this situation would represent an infinite mean pressure and because gas acts like liquid at very high pressures. However, the Klinkenberg effect presented in this way is somewhat misleading since there is some minimum permeability to gas, equal to the absolute permeability, that is reached before an infinite mean pressure is reached.

When the smallest pore in the porous media is in viscous flow, the entire pore structure is experiencing viscous flow. Since there is a particular pressure for a particular pore size at which viscous flow begins, it is obvious that there is a mean pressure for a particular pressure above which no change in permeability is experienced. However, when a scale for the reciprocal pressure is chosen such that the Klinkenberg effect is obvious, the difference between this minimum gas permeability and the zero point on such a scale is insignificant.

The only practical application of the Klinkenberg effect is in interpreting laboratory data. In the laboratory it is convenient to measure permeability with gas at a low pressure. Consequently, the engineer should be careful that in the laboratory permeability data are run in a gas pressure range where the Klinkenberg effect is negligible. Once this situation has been ascertained, it would appear that for all practical purposes the engineer can forget the Klinkenberg effect until he changes core-analysis laboratories.*

Determining viscosities and flow rates at reservoir conditions. The engineer is cautioned to determine whether a flow equation requires a rate stated at reservoir conditions or in stock-tank barrels. In this book all rates are stated in reservoir volumes. Consequently, stock-tank-barrel rates must be converted to reservoir volumes by multiplying the stock-tank-barrel rate by the oil formation volume factor. However, in other references the rate q may refer to stock-tank barrels or reservoir barrels. To determine the units of rate q in a particular equation, examine the flow equation to see if it contains the oil formation volume factor, B_o. If it does not contain B_o, we must convert to reservoir volumes before substituting a number for the rate in the equation. However, if the equation does contain the formation volume factor B_o, the rate in the equation should be stated in stock-tank barrels.

If formation volume factors are not available for a particular reservoir, refer to appendix B for the empirical data needed for such a study. Fig. B16 can be used to determine the gas in solution for any reservoir

*For a more quantitative treatment of this subject, refer to the Amyx, Bass, and Whiting text, *Petroleum Reservoir Engineering* (McGraw-Hill, 1960).

pressure at or below the saturation pressure. Then this gas in solution can be used in Fig. B18 to determine the oil formation volume factor at that pressure (also see chapter 7).

Many engineers introduce large errors into calculations by failing to realize that the viscosity of liquids in the reservoir is much different than the viscosity of stock-tank liquids. This is true of both water and oil. We must remember that in the reservoir the stock-tank oil has a large amount of gas in solution. This solution gas greatly reduces the viscosity of the oil. Water may also have small amounts of gas in solution, but this effect is generally very small.

However, the effect of pressure and temperature on the viscosity of both oil and water is considerable. Here again, refer to appendix B for empirical viscosity data to evaluate the amount of gas in solution in the reservoir oil at various reservoir conditions as well as the effects of the gas and the reservoir pressure and temperature on the viscosity. We should start with the evaluation of the gas in solution using Fig. B16. With this value we can find the reservoir oil viscosity using Fig. B14 in a two-step correlation. First, determine the gas-free viscosity from the top graph in Fig. B14 and use this value to determine the reservoir oil viscosity from the lower graph of Fig. B14.

Empirical gas-viscosity data are also included in appendix B. Again, the effect of pressure and temperature on gas viscosity is pronounced and is somewhat more complex than it is for liquids. At low pressures the gas viscosity decreases as the pressure decreases since the molecules become widely separated. However, at high pressures the gas reacts as a liquid, and a reduction in pressure increases the viscosity.

Two gas-viscosity correlations are included in appendix B. Fig. B13 gives an easy-to-use correlation of gas viscosity with temperature, pressure, and gas gravity, but the pressure range is limited. Figs. B19 and B20 provide a correlation that covers a much higher range of pressures and temperatures. These figures include the effects of the gas composition through the use of the reduced pressure and temperature and the effects of some of the more common nonhydrocarbons on gas viscosity.

Additional Problems

2.5 A well is producing at a rate of 748 stb/d of oil with no water at a bottom-hole pressure of 900 psia from a reservoir with a very strong water drive (assume steady state). What is the formation thickness if the lab data of problem 2.1 are applicable and a pressure drawdown analysis indicates that Δp_{skin} at this rate is 115 psia?

Wellbore diameter = 8 in.
Distance between wells = 600 ft
Estimated pressure midway between wells = 1,200 psia
Reservoir oil viscosity = 2.5 cp
Formation volume factor = 1.25 res bbl/stb

2.6 Rework problem 2.5 assuming that this is a gas reservoir instead of an oil reservoir and that it is producing 100 MMscfd. The Δp_{skin} and the additional pressure drop caused by turbulence total 200 psia. No water is produced.

Reservoir gas-deviation factor = 0.9
Reservoir gas viscosity = 0.01 cp
Reservoir temperature = 120°F

2.7 Well 1 in Fig. 2–30 is located near the fault and is offset in all other directions by similar wells at a distance of 600 ft. Find the average permeability-thickness product for this well's drainage area based on the absolute permeability. Material balance calculations indicate the current saturations to be S_o = 0.565, S_{we} = 0.300, and S_g = 0.135.

Well 1 producing rate = 750 stb/d
Oil formation volume factor = 1.20
Pressure midway between wells
(from pressure buildup analysis) = 1,000 psia
Reservoir oil viscosity = 2.5 cp
Bottom-hole pressure = 800 psia
Well radius = 0.3 ft

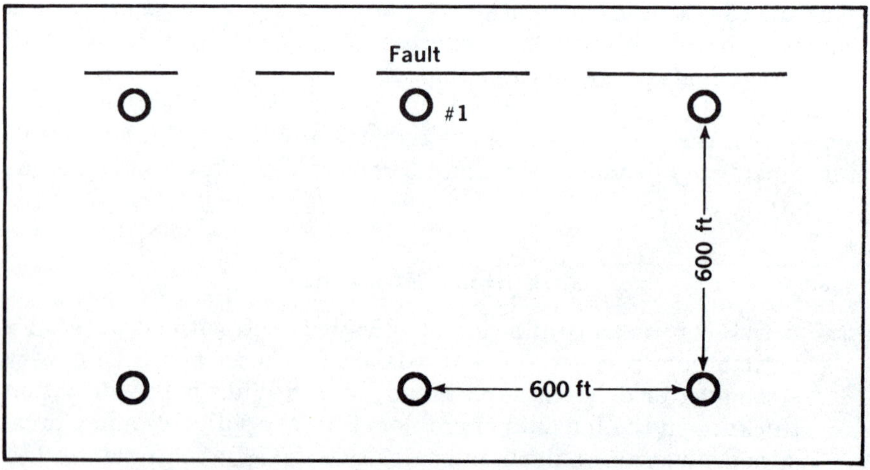

Fig. 2–30 Well pattern for problem 2.7

Relative permeability data as in Fig. 2–31. Assume no well damage and steady-state flow.

2.8 A gas-cap-reservoir discovery well produces at an initial gas-oil ratio (GOR) of 2,000 scf/stb from a reservoir whose pressure is 2,000 psia and temperature is 120°F. Assuming that the oil zone contains no free gas and the gas zone contains no oil, estimate the gas-bearing thickness in the well given the following:

Solution gas = 500 scf/stb
B_o = 1.25
k_g/k_o = 1.0 (assumes relative permeability to gas and oil are both 1.0)
μ_o = 1.2 cp
μ_g = 0.02 cp
Producing interval = 30 ft

2.9 Using the Wyllie equation from appendix B, calculate the relative permeability data for the core sample analyzed in problem 2.1. Assume the core is a well-cemented sandstone and the irreducible water saturation is 30%.

2.10 Verify the constant 1.127 of Eq. 2.3.

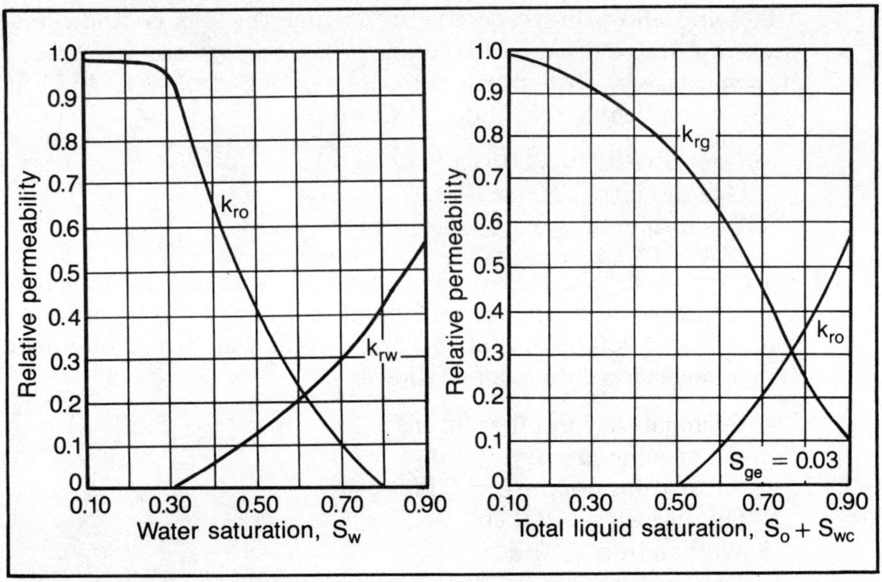

Fig. 2–31 Relative permeability data (synthetic) for problems 2.7 and 2.12

2.11 A bottom-hole heater is to be used in a well with a damaged wellbore to decrease the oil viscosity.

A. What will be the producing rate if the effect of the heater is equivalent to reducing the viscosity in a 7-ft radius around the well from 7.08 to 0.708 cp.

B. Find Δp_{skin}. Assume steady-state flow and the following reservoir data:

Pressure at the external boundary = 2,000 psia
Pressure at the well = 950 psia
Oil formation volume factor = 1.5
k_o undamaged = 100 md
k_o damaged = 10 md
Radius of damaged zone = 10 ft
Reservoir thickness = 10 ft
Drainage radius = 700 ft
Well radius = 0.7 ft

2.12 A well drains a radial reservoir consisting of two beds with undamaged permeabilities of 100 and 75 md. Near the well the formation is damaged to a depth of 33 ft, resulting in reduced permeabilities of 60 and 30 md, respectively. If the well radius is 0.33 ft and no vertical movement of fluid takes place, what is the steady-state rate of gas when the gas saturation is 30% and p_e and p_w are 1,000 and 800 psia, respectively. Assume there is no additional pressure drop caused by turbulence. Bed thicknesses are 2 and 3 ft, respectively. The external drainage radius is 3,300 ft, and Fig. 2–31 is applicable. No water is flowing.

Gas viscosity = 0.01 cp
Gas deviation factor = 0.97
Reservoir temperature = 140°F

2.13 An undersaturated water-drive reservoir is shaped as shown in Fig. 2–32, with water encroachment as indicated. The pressure along A is 3,000 psia, and the pressure at well B is 2,500 psia. Other reservoir data are as follows:

Permeability to oil = 92 md
Oil specific gravity = 1.0
Saturation pressure = 2,000 psia
Oil viscosity = 2.0 cp
Well radius = ⅓ ft
Δp_{skin} = 105 psi

Water-drive activity is such that pressures do not vary with time. Find the producing rate in res b/d.

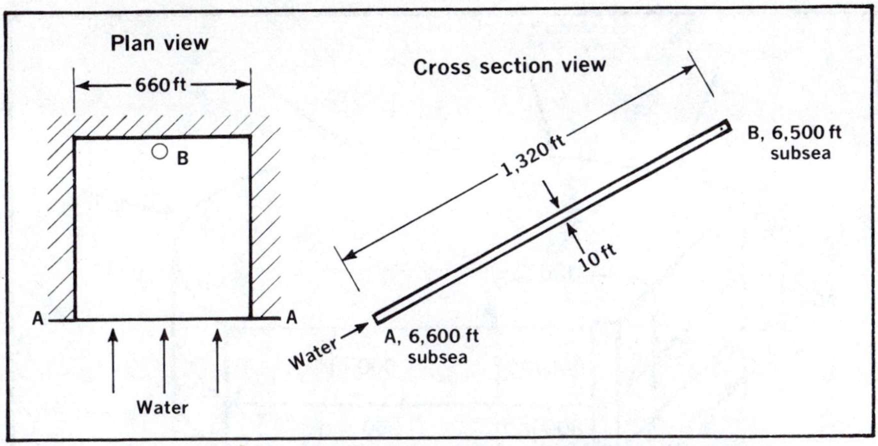

Fig. 2–32 Diagrams for problem 2.13—not to scale

2.14 A. What is the skin factor of the well in problem 2.2A?

B. Acidizing decreases the skin damage so the skin factor is 2.0. -2.0 What is the producing rate if the well pressure continues to be 900 psia and p_e is 2,000 psia? What is Δp_{skin}? Find the pressure drop from 7 to 0.7 ft and compare this with the pressure drop from 7 to 0.7 ft if there is no damage and the flow rate is the same.

2.15 A well drains a reservoir consisting of two beds with undamaged absolute permeabilities of 4,000 and 950 md (Fig. 2–33). Near the well the formation is damaged to a depth of 33 ft, resulting in reduced permeabilities of 600 and 300 md, respectively. The well radius is 0.33 ft, and no vertical movement of fluid takes place. The reservoir is in the form of a dome, with the highest point at the well and the lowest points along the external radius. The well pressure measured at a subsea depth of 2,500 ft is 1,125 psia. The pressure at the external radius measured at a subsea depth of 3,500 ft is 1,875 psia.

A. What is the average permeability of this drainage area? What is the average permeability without damage?

B. What is the steady-state flow rate of oil if viscosity is 0.71 cp, specific gravity is 1.0, and formation volume factor is 1.1? Only gas and oil are flowing (no water). The relative permeabilities are shown in Fig. 2–31, and the gas saturation is 20%. State the rate in stock-tank barrels per day.

C. What is Δp_{skin} and the skin factor?

2.16 The solution to problem 2.1 calculates the water-oil relative permeabilities of a core. The resulting data are plotted in Fig. 2–3.

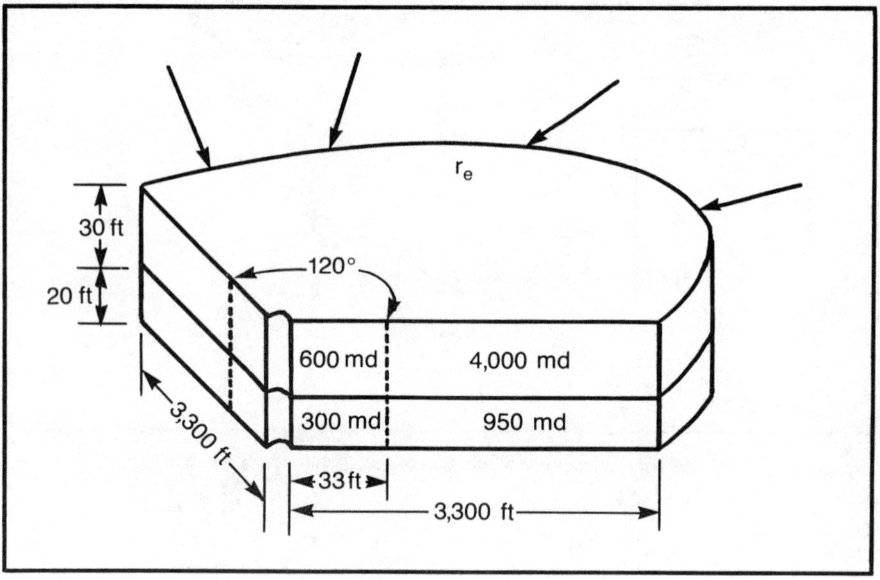

Fig. 2–33 Diagram for problem 2.15

Relative permeability measurements are to be run on another sample from the same formation with an absolute permeability of 400 md. If the following data apply to the flow tests and the relative permeabilities remain the same as those in problem 2.1, what are the flow rates at water saturations of 30, 40, 60, 80, 90, and 100% in cc/sec?

A = 10 sq cm
L = 6 cm
Water viscosity = 0.6 cp
Oil viscosity = 0.75 cp
Downstream pressure = 1.0 atm
Upstream pressure = 3.0 atm

Notes

1. D.L. Katz et al., *Handbook of Natural Gas Engineering* (New York: McGraw-Hill, 1959).

2. H.L. Stone, "Probability Model for Estimating Three-Phase Relative Permeability" (SPE of AIME, 1970), volume I, 214.

3. B.C. Craft and M.F. Hawkins, *Applied Petroleum Reservoir Engineering* (Englewood Cliffs: Prentice-Hall), volume 149, 280–283.

4. A.F. van Everdingen, "The Skin Effect and Its Influence on the Productive Capacity of A Well," *Trans.*, AIME (1953), volume 198, 171–176.

5. M. Muskat, *The Flow of Homogeneous Fluids through Porous Media* (New York: McGraw-Hill, 1937).

6. L.J. Klinkenberg, "Permeability of Porous Media to Liquids and Gases," *Drilling and Production Practice* (API, 1941), p. 200.

Additional References

Amyx, J.W.; Bass, D.M. Jr.; and Whiting, R.L. *Petroleum Reservoir Engineering*. New York: McGraw-Hill, 1960, pp. 91–93.

Conte, S.D. *Elementary Numerical Analysis*. New York: McGraw-Hill, 1965. pp. 191–197.

Sheridegger, Adrian E. *The Physics of Flow through Porous Media*. Toronto: University of Toronto Press, 1960.

3

Unsteady-State and Pseudosteady-State Flow

Unsteady-state flow is flow that occurs while the rates and/or the pressures are changing with time. Consequently, it covers all reservoir flow, except for the specific situation when the rates and pressures do not change with time.

Many engineers seem to be leery of unsteady-state flow applications probably because the material is generally presented in a mathematical form that is not completely understood. Therefore, the physical and mathematical aspects of unsteady-state methods are discussed without going into the detailed math required for rigorous derivation of some of the expressions. The guidelines and working equations of pseudosteady-state flow also are presented, since this part of unsteady-state flow embraces most of the productive life of a reservoir and lends itself to much simpler mathematical analysis. In fact, the equations governing pseudosteady-state flow are almost as simple to apply as steady-state equations.

A complete understanding of unsteady-state flow is necessary for any competent reservoir engineer. It is the backbone of such useful techniques as pressure-buildup analysis, drawdown analysis, interference tests, reservoir boundary delineation, prediction of water encroachment, prediction of disposal-well behavior, gas-well tests, and drillstem-test (DST) analysis.

Physical Description

Refer to Figs. 2–12 and 2–14, which give typical pressure and rate plots versus radius at different times for a reservoir producing under unsteady-state conditions. The reservoir represented by these diagrams is a circular drainage area with uniform permeability to the outer radius where the reservoir or well drainage ends abruptly with all production due to the expansion of the reservoir fluids.*

Fig. 2–12 represents conditions that result from producing this idealized reservoir at a rate to keep the well pressure constant. This situation is comparable to flowing a well against a constant choke size or keeping a well pumped down to a particular level. Under these conditions note from Fig. 2–12 that at some small producing time the reservoir pressure is affected significantly only to a particular radius, r_1. Since the reservoir is producing only because the reservoir fluid is expanding, the flow rate in the reservoir at any radius greater than r_1 is zero. No pressure drop has occurred to affect an expansion of the fluid and the subsequent flow. However, as flow from the well continues more of the reservoir is affected until eventually all of the reservoir has suffered a pressure drop.[1]

The time effect of the production and the fact that it takes some period of time before the entire reservoir is affected are one of the concepts with which engineers have the most trouble. Consequently, we will try to examine this concept in different ways to gain a physical understanding of the phenomena.

Consider the segmented reservoir represented in Fig. 3–1. The same pressure, p_i, exists throughout the reservoir when production is initiated. Consider what happens at the wellbore at time $t = 0$ when pressure in the well or at the internal radius of ΔV_1 is dropped to p_w by removing fluid from the well. This decrease causes a pressure drop across the sand face. Then according to the Darcy equation, flow occurs. As the fluid flows from ΔV_1 to the wellbore, the pressure drops in ΔV_1, which causes the remaining fluid to expand. This expansion provides the fluid for flow into the wellbore.

Once enough fluid has been removed from ΔV_1 to cause a significant pressure drop, a pressure difference exists between ΔV_1 and ΔV_2. According to the Darcy equation, this difference results in flow from ΔV_2 to ΔV_1. The flow of fluid from ΔV_2 causes a pressure drop in ΔV_2 and a corresponding expansion of the remaining fluid in ΔV_2, which provides the fluid for flow into ΔV_1. The flow of fluid from ΔV_2 to ΔV_1 also tends to maintain the pressure in ΔV_1.

When enough flow has taken place from ΔV_2 to cause a significant pressure drop in ΔV_2, a pressure difference exists between ΔV_3 and ΔV_2. Flow takes place from ΔV_3 to ΔV_2. This flow tends to maintain the pressure in ΔV_2 and eventually causes a pressure drop in ΔV_3 that is large enough to initiate flow from ΔV_4 to ΔV_3 because the pressure drop exists between the two segments.

*Lest the engineer be discouraged by the idealized nature of this reservoir, it should be explained that methods of applying the theoretical equations to nonideal (actual) reservoirs will be discussed and illustrated. However, it is simpler to derive and describe the behavior of simple reservoirs initially.

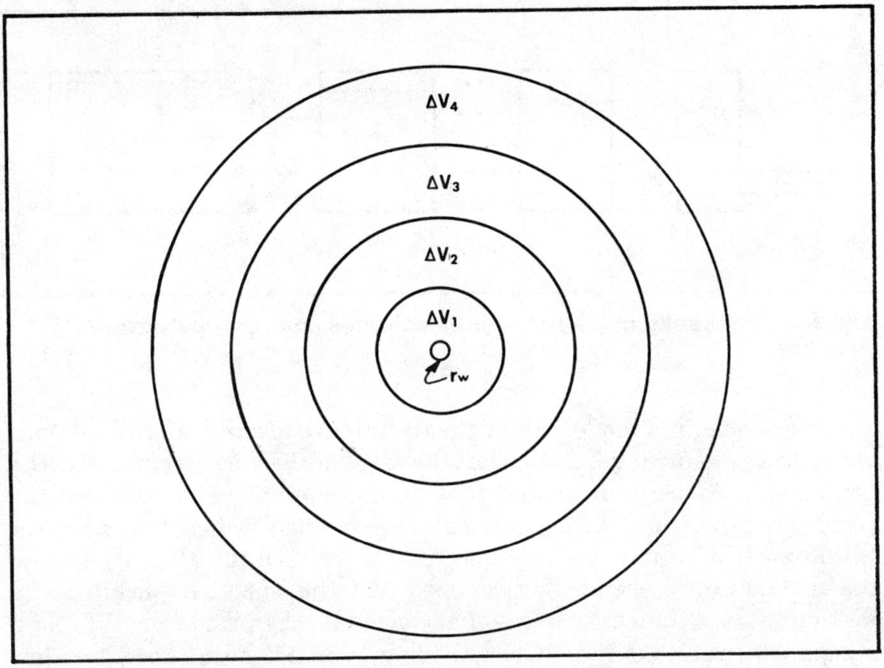

Fig. 3–1 Segmented circular reservoir for discussion of the time effect in unsteady-state flow

Physically, this process requires time so the pressure effect can move across the reservoir. Note that as the pressure effect moves out into the reservoir, it continues to have a smaller effect on the pressure of each subsequent reservoir segment as the segment radii increase. The increase in radius causes an increase in the segment size, and thus, a larger amount of fluid withdrawal is required to obtain the same pressure drop. Also, note that as the radius increases, the cross-sectional area, $2\pi rh$ or A in the Darcy equation, increases and the pressure gradient ($\Delta p/\Delta r$) decreases correspondingly.

Hawkins explained the physical concepts of this time phenomenon by using a hydraulic analog.[2] He modeled a segmented reservoir such as that in Fig. 3–1, representing the potential expansion capacity of each segment by a container with that volume. These containers are connected by pipes that are sized according to the relative resistance to flow between the various segments. Fig. 3–2 shows the schematic model of the reservoir in Fig. 3–1. Note the relative sizes of the containers representing different segments and the relative size of the pipes connecting the various segments. The only factor affecting the relative resistance to flow between the segments, or pipe size, is the cross-sectional area (A in the Darcy equation), which increases in proportion to the increase in radius.

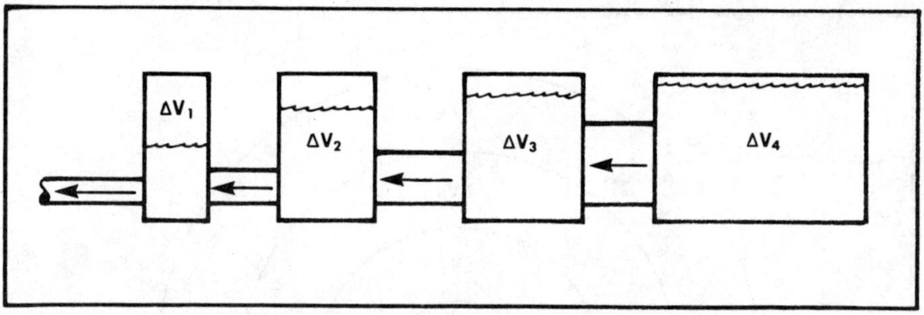

Fig. 3–2 Hydraulic analog of slightly compressible unsteady-state fluid flow

To operate this model, all of the containers are initially filled with water to the same level. Note that the level of the water represents the pressure in each segment and that as the water level is reduced the potential expansion of that particular segment is reduced. Thus, after filling each container to the same level, a valve from ΔV_1, which represents flow into the well, is opened and the model is permitted to discharge. It is then easy to see that flow will take place from ΔV_1 for a substantial period of time before the flow from ΔV_2 to ΔV_1 occurs. Flow from ΔV_3 to ΔV_2 and from ΔV_4 to ΔV_3 are similarly delayed.

Radial Diffusivity Equation

Whereas steady-state flow is governed by very simple equations, unsteady-state flow is best described by a partial-differential equation known as the *radial diffusivity equation:*

$$\frac{\partial p}{\partial t} = \eta \left[\frac{1}{r}\frac{\partial p}{\partial r} + \frac{\partial^2 p}{\partial r^2}\right] \tag{3.1}$$

The reservoir diffusivity constant η is $6.33k/\phi\mu c$ where ϕ is the porosity stated as a fraction—pore volume divided by the total volume. When written in a finite-difference form, the equation is much more meaningful to most engineers:

$$\frac{\Delta p}{\Delta t} = \eta \left[\frac{1}{r}\frac{\Delta p}{\Delta r} + \frac{\Delta(\Delta p/\Delta r)}{\Delta r}\right] \tag{3.2}$$

In this form Eq. 3.2 simply relates the slopes of three different plots at a particular time and a particular radius. These plots are pressure versus radius at different times, the slopes of the plot of pressure versus radius plotted versus the radius at different times, and pressure versus time at different radii.

We are familiar with the nature of the plot of pressure versus radius at various times from Figs. 2–11, 2–12, and 2–14. If we evaluate all of the slopes representing different combinations of radius and time and plot them versus the radius for various times, we may obtain data similar to that in Fig. 3–3. The slope of these curves then represents the term $\Delta(\Delta p/\Delta r)/\Delta r$, which is the change in the slope with respect to time. The term $\Delta p/\Delta t$ is the slope of a set of curves p versus t where each curve represents a particular radius, as illustrated in Fig. 3–4.

Note that only the pressure curve for the well radius declines from time zero. Every other curve initially has a period of time when the pressure remains at the initial pressure. A certain period of production must prevail before the pressure at a particular radius is affected. A pressure disturbance, increased or decreased, introduced at the well diffuses throughout the reservoir—hence the name diffusivity equation.

We should also note the significance of the zero value of $\Delta p/\Delta r$ in Fig. 3–3. This zero value occurs at the radius where the flow rate is zero and the initial pressure is as yet unaffected.

Since the terms of Eqs. 3.1 and 3.2 have the physical significance described, it should be obvious that the solution of the equation for a given reservoir is a series of curves relating the pressure, radius, and time that satisfy the radial diffusivity equation.

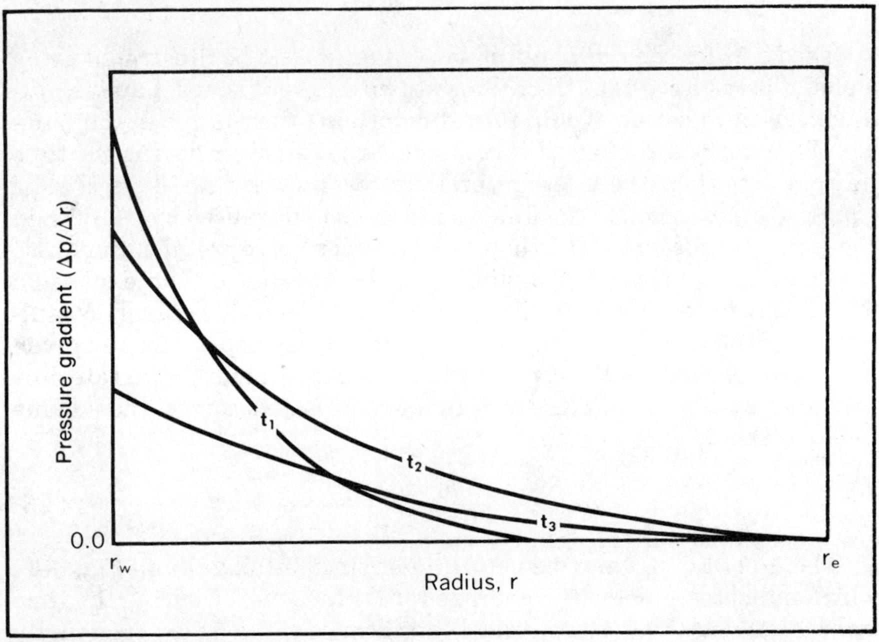

Fig. 3–3 Plot of pressure gradient versus radius for unsteady-state flow

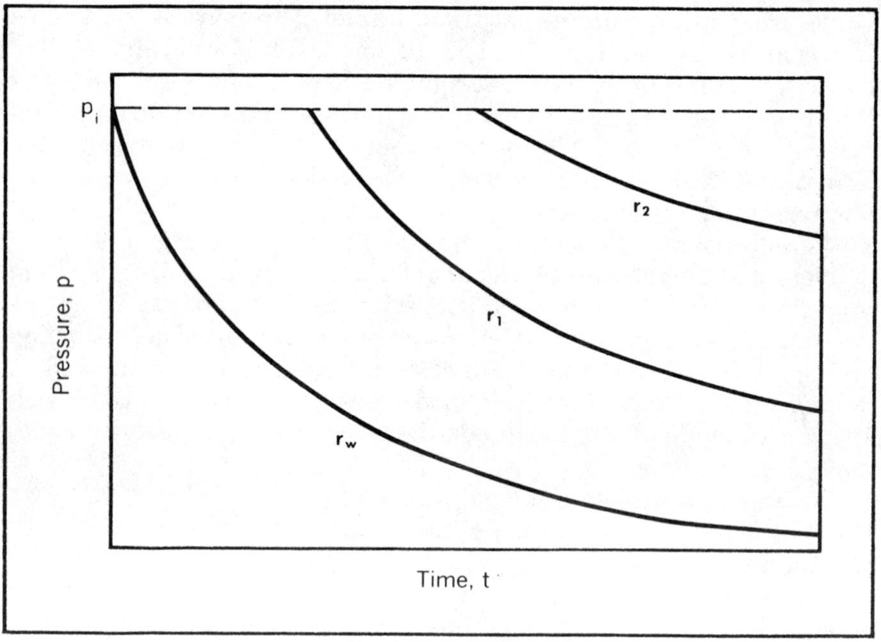

Fig. 3–4 Plot of pressure versus time for unsteady-state flow

The basis for the radial diffusivity equation can be illustrated using a plot of pressure versus the radius at two different times, t and t + Δt, as shown in Fig. 3–5. We first must determine the change in the number of barrels of fluid in a Δr increment of the reservoir during the time interval Δt when the average pressure declines by Δp. This change equals the expansion of the fluid in the increment caused by the drop in the average pressure, Δp. When the increment pore volume, $2\pi r\Delta r h\phi$, is multiplied by the compressibility, c, the amount of fluid expansion for 1.0 psi of pressure change is determined. When this change is multiplied by the average change in pressure per day (Δp/Δt) for the given increment, we obtain the contribution of the increment to the rate during this period of time. The constant 5.615 is used to correct the volume to barrels:

$$\Delta q = \frac{2\pi r\Delta r h\phi c}{5.615}\left(\frac{\Delta p}{\Delta t}\right) \qquad (3.3)$$

The Δq of Eq. 3.3 may be better understood by examining Fig. 3–6, which indicates a plot of the average rate versus radius during the time interval Δt and shows the physical significance of the Δq expression. By rearranging Eq. 3.3, we obtain the slope of the q-versus-r plot:

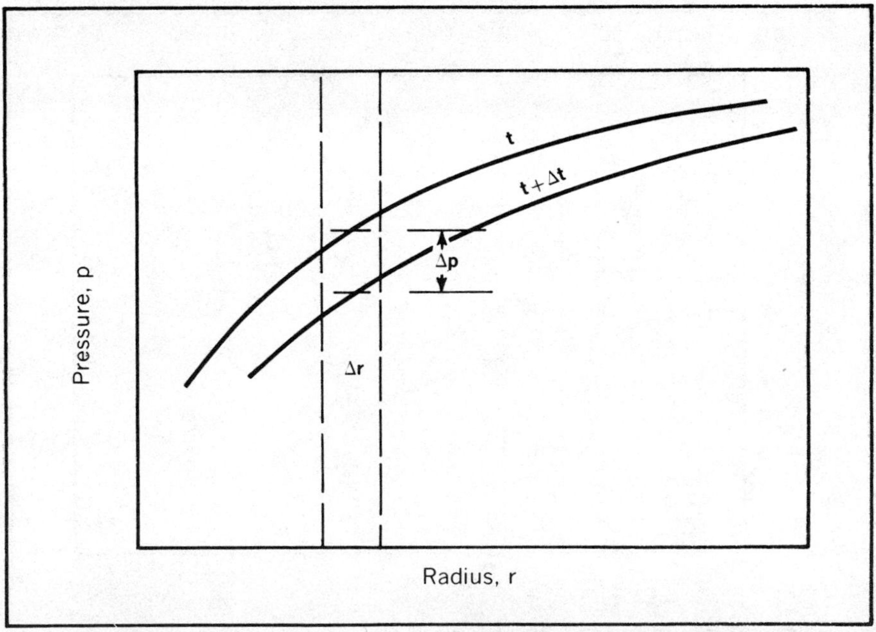

Fig. 3–5 Pressure versus radius at successive times during unsteady-state flow—for derivation of radial diffusivity equation

$$\frac{\Delta q}{\Delta r} = \frac{2\pi rh\phi c}{5.615}\left(\frac{\Delta p}{\Delta t}\right) \quad (3.4)$$

We can obtain the radial diffusivity equation by first deriving an expression for the slope of a plot of q versus r by differentiating the Darcy equation with respect to the radius. The resulting expression is equated to the right-hand side of Eq. 3.4. Then the Darcy equation written with the radial cross-sectional area, $2\pi rh$, substituted for A is:

$$q = \frac{1.127k(2\pi rh)}{\mu}\left(\frac{\Delta p}{\Delta r}\right) \quad (3.5)$$

Since the radius r and $\Delta p/\Delta r$ both vary with the rate q, the equation for the differential of a product must be used:

$$\frac{\Delta(uv)}{\Delta x} = u\frac{\Delta v}{\Delta x} + v\,\frac{\Delta u}{\Delta x} \quad (3.6)$$

Letting $u = r$ and $v = (\Delta p/\Delta r)$, we find:

$$\frac{\Delta q}{\Delta r} = \frac{1.127k2\pi h}{\mu}\left[r\frac{\Delta(\Delta p/\Delta r)}{\Delta r} + \frac{\Delta p}{\Delta r}\right] \quad (3.7)$$

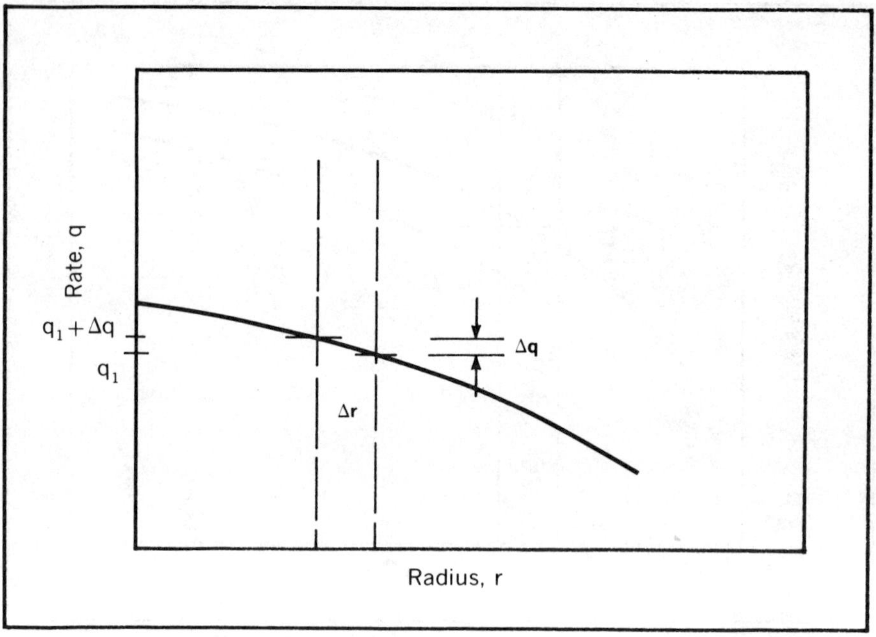

Fig. 3–6 Average rate versus radius for Fig. 3–5—for derivation of radial diffusivity equation

When the right-hand side of Eq. 3.7 is equated to the right-hand side of Eq. 3.4, the expression $2\pi h$ cancels. Rearranging the equation, we obtain:

$$\frac{\Delta p}{\Delta t} = \left(\frac{6.33k}{\phi\mu c}\right)\left[\frac{1}{r}\frac{\Delta p}{\Delta r} + \frac{\Delta(\Delta p/\Delta r)}{\Delta r}\right] \tag{3.8}$$

Eq. 3.8 is identical to Eq. 3.1 if we recognize the diffusivity constant, η as being:

$$\eta = \frac{6.33k}{\phi\mu c} \tag{3.9}$$

The constant 6.33 is the product of 1.127 and 5.615.

To most practical engineers the diffusivity equation seems to be ethereal—something dreamed up by reservoir engineers to solve their particular problems. However, it is doubtful that another equation in petroleum engineering is used more widely by other types of engineers. The radial diffusivity equation describes most forms of radial unsteady-state mass and energy transfer. When applied to the flow of heat, the pressure in Eq. 3.2 becomes the temperature and the diffusivity constant describes the heat-transfer characteristics of the material.[3] For molecular diffusion, we substitute concentrations and use molecular activity characteristics in the constant. For electricity, we use voltages and electrical characteristics. Diffusivity equations are almost universal in their applications.

Engineers who have difficulty understanding the time effect in unsteady state may find it easier to apply the diffusivity equation to another physical system, such as the flow of heat. Consider a lead cylinder with a small cylindrical hole in its center. Suppose this lead cylinder is heated to some elevated temperature and is placed in an insulating atmosphere. Then a coolant is introduced into the center hole of the cylinder so the temperature in the hole is kept constant. Physically, it seems easy to believe that there will not be an instantaneous change in the temperature throughout the cylinder and that it will take time for the temperature at the outer radius of the cylinder to be lowered. This concept is analogous to the compressible flow of liquid in an enclosed circular reservoir with the pressure maintained at a constant value at the well and the pressure at the outer boundary requiring time before it is affected by lowering the pressure at the well.

General solutions. The correct solution for applying the radial diffusivity equation to a particular reservoir is a combination of pressure, radius, and time curves that satisfies the radial diffusivity equation, Eq. 3.2. Anyone who understands the physical significance of Eq. 3.2 can eventually find a set of curves to satisfy a particular reservoir situation. For many of us, this problem requires a trial-and-error approach. Fortunately, more scientific methods are available.

Hurst and others have supplied us with general solutions that can be applied to any reservoir.[4] To illustrate the nature of these solutions, divide Eq. 3.2 by η. Then combine the η in the resulting equation with t and treat the product (ηt) as one variable:

$$\frac{\Delta p}{\Delta(\eta t)} = \frac{1}{r}\frac{\Delta p}{\Delta r} + \frac{\Delta(\Delta p/\Delta r)}{\Delta r} \tag{3.2a}$$

Based on Eq. 3.2a, a series of graphs can be developed similar to Fig. 2–11, 2–12, 2–14, 3–3, and 3–4 but with the time, t, replaced with the product (ηt). These curves are more general because they avoid the variation of permeability, porosity, viscosity, and compressibility included in the constant, η.

In a more general manner these reservoir constants have been combined with the time, pressure, and radius of Eq. 3.2 so the equation is stated in general terms that apply to any reservoir of a particular size (r_e/r_w) that is being produced at a constant rate, q:

$$\frac{\Delta p_D}{\Delta t_{Dw}} = \frac{1}{r_D}\frac{\Delta p_D}{\Delta r_D} + \frac{\Delta(\Delta p_D/\Delta r_D)}{\Delta r_D} \tag{3.10}$$

In Eq. 3.10 the reservoir constants and variables are combined as follows:

$$t_{Dw} = \frac{\eta t}{r_w^{\,2}} \tag{3.11}$$

$$\Delta p_D = \frac{\Delta p}{(0.141 q\mu/kh)} \tag{3.12}$$

$$r_D = \frac{r}{r_w} \tag{3.13}$$

If we substitute expressions for t, Δp, and r in Eq. 3.2 according to Eqs. 3.11, 3.12, and 3.13, Eq. 3.10 results. Grouping the variables and constants according to Eqs. 3.11, 3.12, and 3.13 is somewhat arbitrary. However, the radial diffusivity constant and the well radius include all of the reservoir constants that alter the time effect in a reservoir. Eq. 3.12 also contains the reservoir parameters that affect the magnitude of the pressure drop at any particular radius and time. This point can be best illustrated by examining the radial steady-state flow equation written in terms of the pressure at any radius from Eq. 2.19:

$$q = \frac{7.08kh}{\mu} \frac{(p_r - p_w)}{\ell n\,(r/r_w)} \tag{3.14}$$

When solved for the pressure drop, we obtain:

$$p_r - p_w = \frac{0.141 q\mu}{kh} \ell n\,(r/r_w) \tag{3.15}$$

Eq. 3.15 shows that the pressure drop in the reservoir is proportional to the group of terms 0.141qμ/kh, which is used to normalize the pressure drop in Eq. 3.12. Note that Eq. 3.12 applies only if the rate at the well is constant.

Based on Eq. 3.10, a relationship between Δp_D, t_{Dw}, and r_D can be obtained that applies to any reservoir of a particular size, r_e/r_w, that meets the characteristics of the particular solution. These solutions are dimensionless in character since Eqs. 3.11, 3.12, and 3.13 are dimensionless.

Many engineers are confused by dimensionless groups. If this concept is troublesome, simply forget it and assume that the group of terms has units. If the general solutions are properly used, we will never be aware of the fact that the groups are dimensionless except when the work is used to describe the theory. Generally, in this text we prefer the term "reduced" to "dimensionless" since it seems to be less confusing.

General solutions can be obtained using the various groups of reduced terms in Eqs. 3.11–3.13 rather than the corresponding time, pressure, and radius. Such a general solution can be applied to any reservoir by introducing the particular reservoir constants into the various groups. These general solutions are of two types. The constant-terminal-rate solution provides the pressure change throughout a reservoir, in which the rate at one extremity, or terminal, of the radial reservoir is constant. The constant-terminal-pressure solution provides the cumulative flow at any particular time for a reservoir in which the pressure at one of the reservoir extremities is constant.

Constant-Terminal-Rate Solution

The constant-terminal-rate solution is by far the most useful of the two general solutions to the radial diffusivity equation. It is an integral part of pressure-buildup analyses, interference tests, drawdown analyses, isochronal tests of gas wells, and many other important reservoir analysis techniques. This general type of solution can be obtained in two or three different forms using slightly different assumptions and methods of mathematical analysis.* The various solutions overlap, and all of them have particular uses and limitations. It is then very important that we understand the limitations of each. Thus, each solution is considered separately. Then a general consideration of the entire problem provides a practical means of determining when each solution should be used.

The Hurst-van Everdingen p_{tD} solution. The constant-rate solution developed by Hurst and van Everdingen analyzes a reservoir model that is perfectly radial, with the producing well in the exact center and no flow across the external radius, r_e.[5] Intially, the pressure in the model is p_i and is uniform throughout. At time zero a constant rate of production begins. It is then possible to calculate the pressure at the well at any time by using the p_{tD} functions from Fig. 3–7 or from the appropriate equations.

Fig. 3–7 indicates that the p_{tD} values are plotted versus t_{Dw}, the reduced time based on the well radius, with a different curve representing each reservoir size. The reservoir size is presented in dimensionless form as the ratio of the external radius of the reservoir and the well radius. Do not be confused by the fact that the reservoir sizes in Fig. 3–7 are so small as to be absurd for a practical situation. Hurst applied his data principally to the analysis of an aquifier that supplied the water drive for an oil reservoir. Thus, his "well" was the external boundary of the oil reservoir, and dimensionless reservoir sizes of less than 10 were practical for his application.

*William Hurst is generally given credit for devising the first constant-rate solution in general solution form.

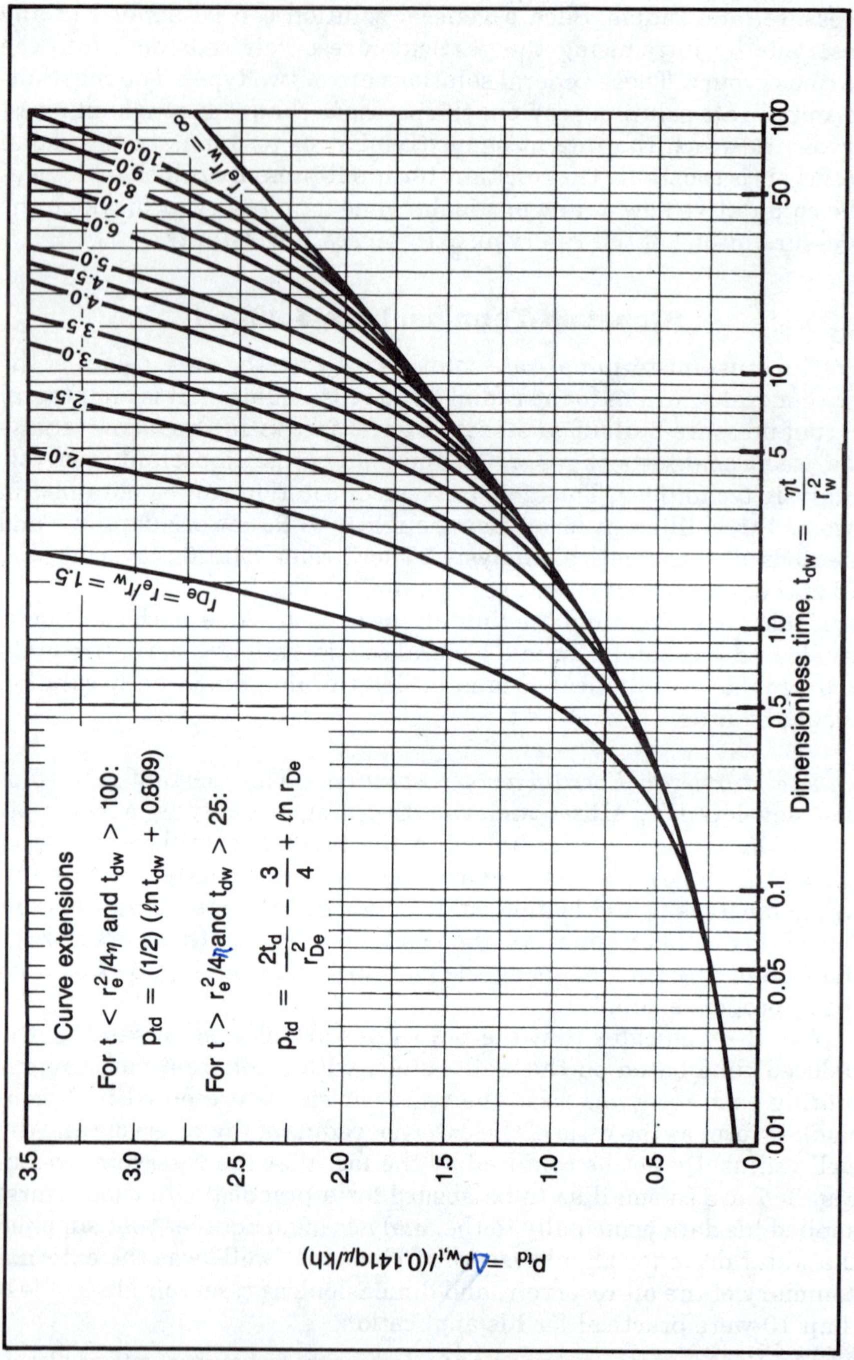

Fig. 3–7 Values for p_{td} from van Everdingen and Hurst

Note that Fig. 3–7 contains a curve with a dimensionless reservoir size of infinity. This curve has no practical physical meaning, but it has considerable mathematical significance because it represents the reservoir prior to the time when the pressure at the outer boundary has been affected. Prior to the time when the pressure at the outer boundary is affected, it makes no difference where the outer boundary is located. During this time, mathematically, the radius is infinite.

Reservoirs of different sizes behave exactly the same way—infinite acting—until the outer boundary of the smallest reservoir is affected. Consequently, in Fig. 3–7 the infinite reservoir curve forms the limit for the other curves since they all act the same until their outer boundaries are affected. Note further that the larger the r_e/r_w ratio of a particular curve is, the longer it follows the infinite curve.

The p_{tD} functions are used by first calculating the reduced time, based on the well radius, at which the pressure is desired using the appropriate real time in days. The dimensionless reservoir size, r_e/r_w, is then calculated to determine which curve of Fig. 3–7 should be used. Using the calculated t_{Dw} and the dimensionless reservoir size, the corresponding pressure function, p_{tD}, can be obtained from Fig. 3–7. This value is used in Eq. 3.16, which defines the pressure function, to calculate the pressure drop and the corresponding pressure. This equation is obtained from the general pressure function, Eq. 3.12, by substituting p_{tD} for the general pressure function, Δp_D, and the difference between the intitial reservoir pressure and the well pressure at some time, t, for Δp:

$$p_{tD} = \frac{p_i - p_{w,t}}{(0.141q\mu/kh)} \tag{3.16}$$

When Eq. 3.16 is rearranged and we account for Δp_{skin}, we obtain:

$$p_{w,t} = p_i - \frac{0.141q\mu}{kh} p_{tD} - \Delta p_{skin} \tag{3.17}$$

If a function of S is substituted for Δp_{skin} according to Eq. 2.22 and the resulting expression is simplified, we obtain:

$$p_{w,t} = p_i - \frac{0.141q\mu}{kh} [p_{tD} + S] \tag{3.17a}$$

The practical significance of the Hurst-van Everdingen solution can probably best be emphasized by a theoretical problem. In working problem 3.1, first calculate the reduced time based on the time and radius at which the pressure is desired. Remember that the time unit of the equation is days. Then find p_{tD} from Fig. 3–7. Assume the reservoir is infinite acting after only 6 min of production. The p_{tD} value obtained is then used as in Eq. 3.17 to obtain $p_{w,t}$.

PROBLEM 3.1: Application of the p_{tD} Function

The discovery well in a reservoir is put on a drillstem test and flows at a rate of 18 stb/d. What will the well pressure be after 6 min of production if the reservoir characteristics are as follows:

$k = 0.1$ md
$\phi = 0.0695$
$r_w = 0.5$ ft
$p_i = 3{,}000$ psia
$r_e = 3{,}000$ ft
$B_o = 1.39$
$p_s = 1{,}000$ psia (saturation pressure)
$\Delta p_{skin} = 0$
$c = 6.33 \times 10^{-6}$/psi
$\mu_o = 0.4$ cp
$h = 141$ ft

Problem 3.1 is not a practical problem; it merely illustrates the physical significance of the p_{tD} function. Also, the characteristics of the reservoir demonstrate a variety of ideas in subsequent parts of the problem. However, the engineer should recognize that the problem can be changed slightly to be practical. Normally, in a drillstem test we know the flow rate and the well pressure at a particular time. Then one or more of the reservoir properties can be calculated, such as the permeability or the permeability thickness product.

In applying the Hurst-van Everdingen p_{tD} functions, the engineer should notice that pressure can be calculated only at the radius where the flow rate is known. Normally, this restricts the application to calculating well pressures. Occasionally, the pressure at the outer boundary of an oil or gas reservoir can be calculated by applying the solution to the surrounding aquifier. However, such applications are unusual. The important point is that, mathematically, the p_{tD} function can be applied to calculate the pressure only at the point in the reservoir where the rate is constant. We can illustrate this fact by writing Eq. 3.17 without Δp_{skin} and by subscripting the rate with the same r that is used to subscript the pressure. This r is also used in the reduced time, t_D.

$$p_{r,t} = p_i - \frac{0.141 q_r \mu}{kh} p_{tD} \tag{3.18}$$

If we want to calculate the pressure at some point in the reservoir where the rate is unknown and is not constant, we must use a different mathematical solution.

The Ei-function solution. The Ei-function solution can be used to calculate pressure at any point in the reservoir using the flow rate at the well.[6] The solution takes the same form as the p_{tD} solution, but it is based on a different set of assumptions and a different mathematical approach.

For an infinite-acting reservoir if $\eta t/r_w^2$ is greater than 100, Eq. 3.19 applies:

$$\Delta p_D = (1/2)\left(-\text{Ei}\,\frac{-1}{4t_D}\right) \tag{3.19}$$

In Eq. 3.19 t_D is calculated using the radius at which the pressure is desired. When working with the p_{tD} function, we defined t_D as being based on the well radius. In that case the well radius is the only radius at which the pressure can be calculated because the p_{tD} application is restricted to the radius where the rate is known. However, since the Ei-function solution lets us calculate the pressure at any radius, we must always be careful to base t_D on the radius at which the pressure is desired.* The Ei-pressure function can then be used with the rate at the well to determine the pressure at any point in the reservoir. Eq. 3.20, which is similar to Eq. 3.17, uses the Ei-pressure function:

$$p_{r,t} = p_i - \frac{0.141 q_w \mu}{kh}\left[1/2\left(-\text{Ei}\,\frac{-1}{4t_D}\right)\right] - \Delta p_{skin} \tag{3.20}$$

The Δp_{skin} in Eq. 3.20 has no significance when the equation is applied to determine the pressure at any point other than the well because the well damage applies only at the wellbore. Thus, when the radius r is not equal to r_w, Eq. 3.20 becomes:

$$p_{r,t} = p_i - \frac{0.141 q_w \mu}{kh}\left[1/2\left(-\text{Ei}\,\frac{-1}{4t_D}\right)\right] \tag{3.20a}$$

Two important conditions must be met before Eqs. 3.20 and 3.20a can be used. The reservoir must be infinite acting, and $\eta t/r_w^2$ must be greater than 100. These limitations result from assumptions that must be made to obtain the Ei-function solution mathematically. The first limit, an infinite-acting reservoir, is self-explanatory. However, a surprising number of practicing engineers and researchers either are unaware of this limit or prefer to ignore it, since no mention is made of the limitation when various applications of the Ei-function solution are used or proposed.

The other limit on the Ei-function solution is not as often a practical factor, but it is also generally ignored. This limit, which states that

*Other books choose to base the dimensionless-time reference on the same radius in all cases. For example, Earlougher uses $t_D = \eta t/r_w^2$ and the argument for the exponential integral as $4(t_D r_w^2/r_e^2)$, (R.C. Earlougher Jr., 1977. *Advances in Well Test Analysis,* Dallas and New York: SPE of AIME). Others also use the dimensionless time for nonradial systems such as $t_D = \eta t/A$, where A is the area being drained by the well.

$\eta t/r_w^2$ must exceed 100, is arbitrary and it may be shortened or extended timewise depending on the accuracy desired in a particular situation. The limit arises because it is necessary to assume that production is from a point rather than from a finite radius, r_w. Consequently, a certain amount of time must transpire before the effect of the inaccuracy becomes negligible. For all practical purposes this occurs when the rate at the well is equal to the theoretical rate at a radius of zero. In chapter 2 we examined the pressure-versus-radius distribution at various times, and Figs. 2–12 and 2–14 show that as producing time increases, the slope of the rate-versus-radius curve at the wellbore becomes almost flat.

Fig. 3–8 illustrates this point by comparing the rate at the theoretical zero radius with the rate at the well radius. In this example we may assume that $\eta t/r_w^2$ becomes greater than 100 at a time of about t_2 when the rate at r_w is about equal to the rate at $r = 0.0$.

The Ei function is known as the exponential integral that occurs frequently in mathematical analyses. The exponential integral is defined as:[7]

$$Ei(-x) = -\int_x^\infty \frac{e^{-u}du}{u} \tag{3.21}$$

Eq. 3.21 can be evaluated from the series:

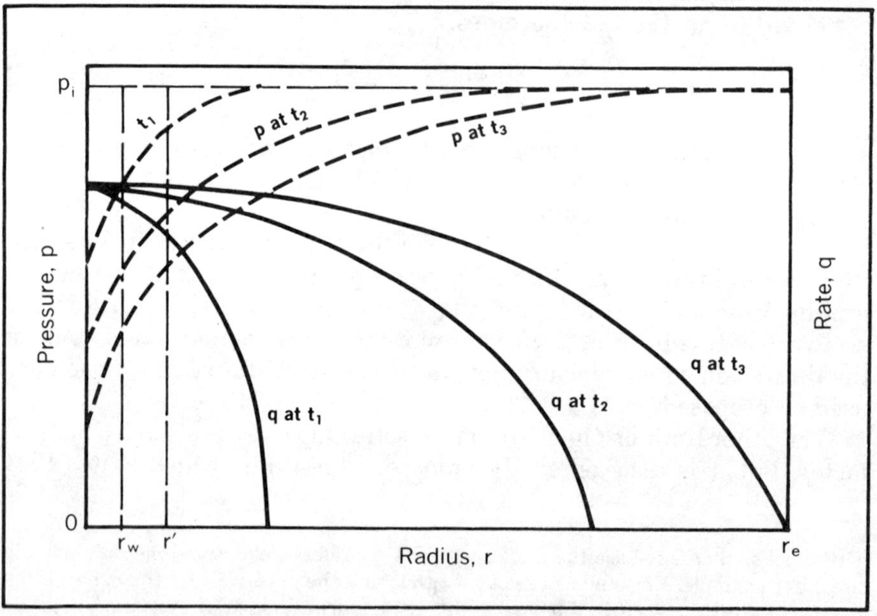

Fig. 3–8 Comparison of rates near the wellbore during unsteady-state flow with a constant rate at the well

$$Ei(-x) = \ell n\ x - \frac{x}{1!} + \frac{x^2}{2 \times 2!} - \frac{x^3}{3 \times 3!} + \ . \ . \ . \qquad (3.22)$$

Tables of the exponential integral can be found in many math tables, and most reservoir engineering books provide such graphs.[8] Since we always use the exponential as part of the pressure function as in Eq. 3.19, we simply provide a plot of the pressure function versus t_D as in Fig. 3–9. These data are often misinterpreted by the engineer. In Fig. 3–9 the Ei function includes the ½ value, and the time scale is for t_D, not $4t_D$.

When applying the Ei-function solution, the first step is to determine if the function limitations are met—infinite-acting reservoir and $\eta t/r_w^2$ greater than 100. Mistakes are often made in applying Ei functions when calculating the group of terms $\eta t/r_w^2$. Many engineers use the radius at which the pressure is being calculated rather than r_w in this group of terms. The fallacy of this practice is apparent when the reason for the test is considered. We are simply comparing conditions at the zero radius with conditions or times based on the well radius.

Since production is assumed from a zero radius or a point, the Ei-function solution is often referred to as the point-source solution. If we consider the geometry as cylindrical (three dimensional) instead of radial (two dimensional), we may think of the well being a line instead of a point. Then the function can be referred to as a line-source solution. This term is also used to describe the exponential-integral solution. The test to see if $\eta t/r_w^2$ is greater than 100 may be referred to as the test to check the validity of the point-source or line-source assumption. Be careful that this test is not confused with the calculation of t_D, in which the radius is the radius at which the pressure is desired. Thus, $\eta t/r_w^2$ may or may not be the same as t_D, which is the basis for the Ei function.

Problem 3.2 may help clarify the limitations and physical meaning of the Ei-function solution. Try to calculate all of the answers before checking the solution in appendix C.

PROBLEM 3.2: Application of the Ei Function

In problem 3.1 a DST flow period at a rate of 18 stb/d is described, and the pressure at the well after 6 min of production is calculated. What is the pressure at a radius of 5 ft at this time? What is the pressure at a radius of 5 ft after 1 hr of production? What is the pressure at 50 ft after 1 hr of production?

In problem 3.2 the application of the Ei-function solution clearly shows that for all practical purposes the pressure drop at a radius greater than r_w is insignificant for a period of time. However, it should be carefully noted that a finite pressure drop actually occurs almost

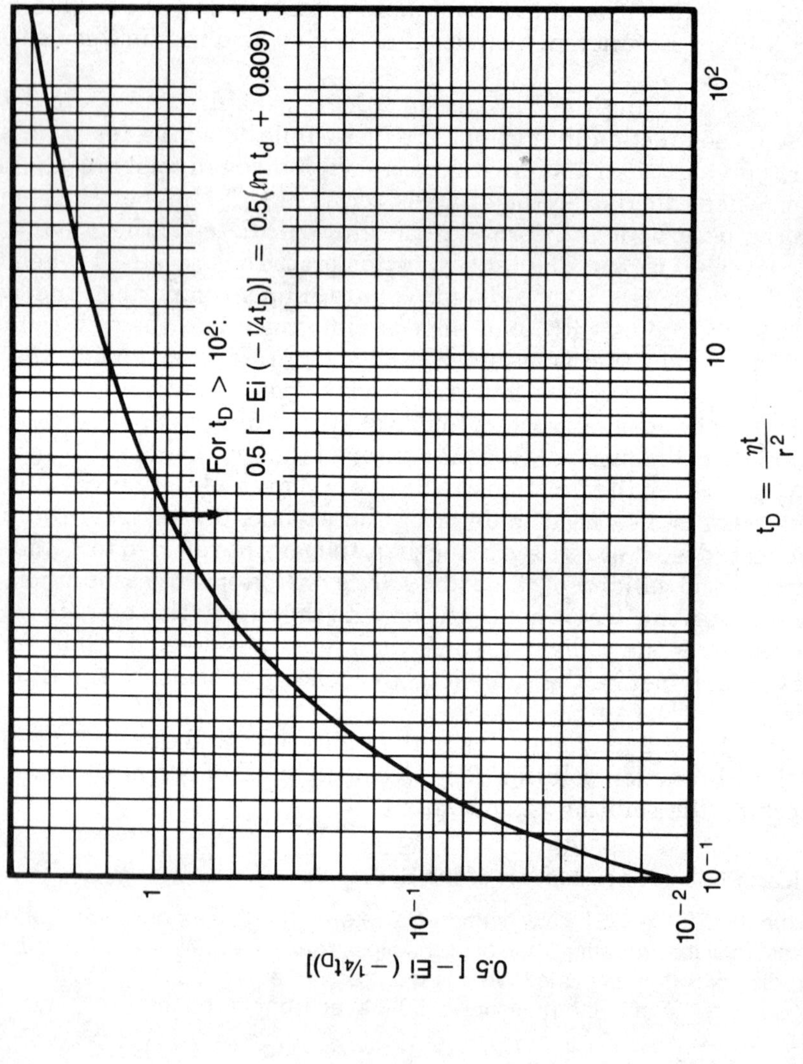

Fig. 3–9 Exponential-integral pressure function (after Earlougher, *Advances in Well Test Analysis*, Monograph No. 5, © 1977, SPE-AIME)

immediately at all parts of the reservoir. This is easily seen when we examine the nature of the exponential integral in Eq. 3.22. We find that for any finite value of $(-1/4t_D)$, there is a finite value for $(1/2)\left(-Ei\frac{-1}{4t_D}\right)$. Consequently, we recognize that a pressure drop occurs at a particular radius at a particular time.

The magnitude of "significant" seems to be a function of the measuring device used. Prior to the quartz-type pressure-measuring device, an Amerada-type instrument was used to measure downhole pressures. The Amerada bomb can be read in normal pressure ranges to an accuracy of perhaps 5–10 psi. Consequently, a smaller pressure change is considered insignificant with the Amerada bomb. The quartz-type bomb gives an accuracy of perhaps 0.01 psi under the right conditions, so the term "insignificant" takes on a different meaning.

Many reservoir engineers must determine whether two wells are producing from the same reservoir. The common approach is to put one well on production and observe the effect of this production on the other well. The engineer reasons that if the effect of the production in the observation well is noticeable, in a reasonable length of time, the wells are assumed to be in the same reservoir. Conversely, if no effect is seen in the observation well in a reasonable length of time, the wells are not in the same reservoir. Many engineers have reached an erroneous conclusion that wells are not in the same reservoir and have been subsequently proven wrong because of geological information available from further exploration or because of very similar pressure histories of the two wells. In either case the engineer could have saved himself the error and embarrassment by making a simple calculation to check the magnitude of the interference that he was trying to read from a pressure gauge.

Such a calculation involves determining the magnitude of the pressure change that can be measured and determining how long it takes to obtain such a pressure drop at the point in the reservoir being observed. In this case the normal calculating procedure is reversed. Using the assumed pressure drop, the corresponding Ei function is calculated from Eq. 3.20a. The corresponding reduced time can then be read from Fig. 3–9, and the producing time to obtain the desired pressure drop can be calculated from the definition of the reduced time. This procedure is illustrated in problem 3.3, and the solution is shown in appendix C.

PROBLEM 3.3: Calculating the Time Needed to Obtain a Particular Pressure Drop in the Reservoir

To plan an interference test for the reservoir in problem 3.1, determine how long it takes the production from well 1 to cause a pressure drop of 7.5 psi in well 2, which is 600 ft away.

Extension equations for the Ei and the p_{tD} functions. Pressure functions provided in Figs. 3–7 and 3–9 only extend to a t_D of 100. Values for reduced-time functions greater than 100 can be directly calculated by the extension equations listed in the illustrations. For the p_{tD} functions two extensions are needed to cover the reservoirs: one for when they are finite acting, the pressure at the outer boundary has been affected, and one for when they are infinite acting. In Fig. 3–7 a reservoir becomes finite acting when the producing time in days is greater than $r_e^2/4\eta$, which is an acceptable approximation.[9]

The limits set on this p_{tD} equation are again arbitrary, but they give acceptable answers based on the average engineering opinion. Consequently, for $t > r_e^2/4\eta$ and $t_{Dw} > 100$, the p_{tD} function can be calculated from Eq. 3.23:

$$p_{tD} = \frac{2(t_D + \frac{1}{4})}{r_{De}^2 - 1} - \frac{3r_{De}^4 - 4r_{De}^4 \ln r_{De} - 2r_{De}^2 - 1}{4(r_{De}^2 - 1)^2} \quad (3.23)$$

Eq. 3.23 is actually the first two terms of the general p_{tD} equation that defines all p_{tD} functions.[10] The third and last term is written to include some Bessel functions. Fortunately, when t_D becomes large enough relative to the reservoir size r_{De}, which is r_e/r_w, the Bessel function term approaches zero. The accuracy of Eq. 3.23 can be seen by examining Fig. 3–10 where the curves are based on the complete p_{tD} equation (not included in this text), and the large dots represent points calculated using Eq. 3.23 if r_D equals 10.

It is also useful to recognize that the significant portion of Eq. 3.23 can be reduced to:

$$p_{tD} = \frac{2t_D}{r_{De}^2} + \ln r_{De} - \frac{3}{4} \quad (3.24)$$

Eq. 3.24 is determined by recognizing that many numbers in Eq. 3.23 are insignificant compared with other numbers in the equation for a practical case. For example, when $t_D > 100$ and $r_{De} > 10$, the ¼ in the numerator of the first term of Eq. 3.23 is insignificant when compared to t_D of 100. Also, the 1.0 in the denominator of the first term is insignificant when r_{De}^2 equals 100. In the second term, only expressions containing r_{De}^4 are significant. Note that the denominator contains one such expression when the square of the difference is expanded. When all such comparisons are considered, Eq. 3.24 results.

If the reservoir is infinite acting and the reduced time is greater than 100, Fig. 3–7 gives an extension equation of:

$$p_{tD} = \frac{1}{2}(\ln t_D + 0.809) \quad (3.25)$$

If we again use the as-yet unproved critical time, $t = r_e^2/4\eta$ as the time when a reservoir becomes finite acting, we can use Eq. 3.25 to

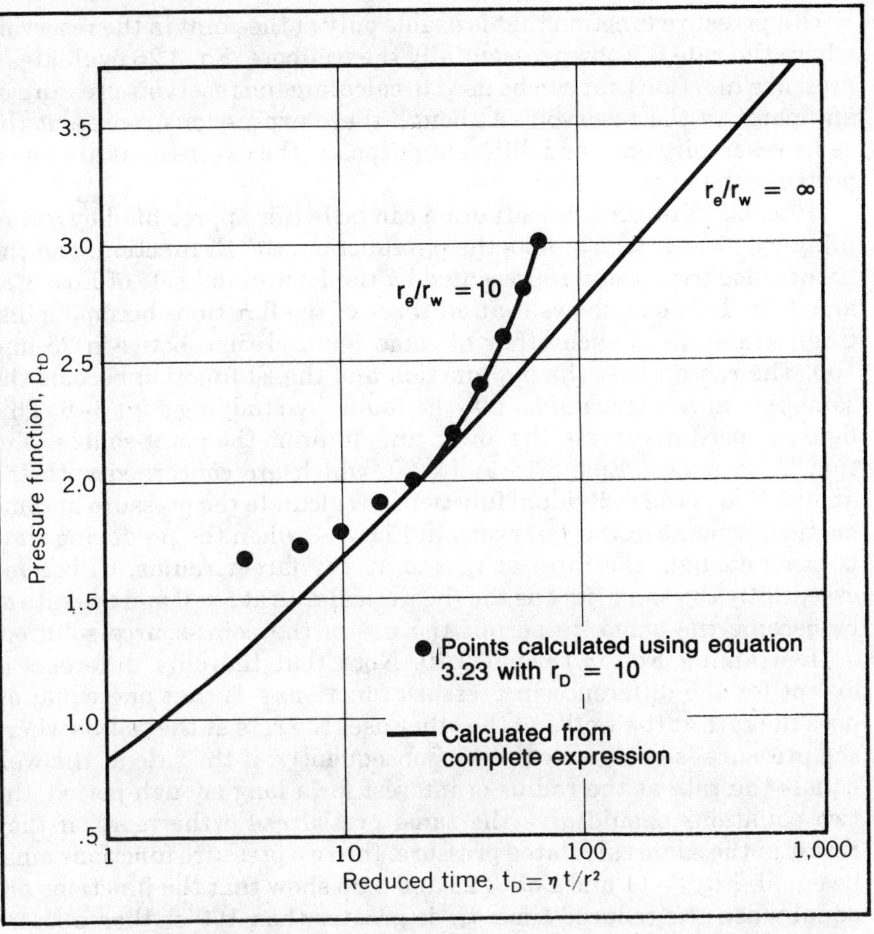

Fig. 3–10 Comparison of p_{tD} values with those calculated by Eq. 3.23

calculate the p_{tD} function, providing the time is less than this critical time and the reduced time is greater than 100. (This equation is used by practicing engineers and researchers almost completely without regard for the fact that it applies only to infinite-acting reservoirs.)

Now consider the extension equation listed in Fig. 3–9. This represents the pressure function for an infinite-acting reservoir when the reduced time, t_D, is greater than 100. The extension equation is:

$$(1/2)\left(-\text{Ei}\ \frac{-1}{4t_D}\right) = (1/2)(\ell n\ t_D + 0.809) \tag{3.26}$$

When we compare Eqs. 3.25 and 3.26, we find a rather unexpected situation. Both of these equations apply to an infinite-acting reservoir when the reduced time is greater than 100. However, Eq. 3.25 calculates a pressure function that is usable only at the point in the reservoir where the rate is known, essentially the wellbore. Eq. 3.26 evaluates a pressure function that can be used to calculate the reservoir pressure at any point in the reservoir. Although these expressions represent the same reservoir times and different purposes, the expressions are unexpectedly the same.

Possibly, this unusual situation can be better appreciated by examining Fig. 3–11, which plots the p_{tD} function, the Ei function, and the natural log expression represented by the right-hand side of Eqs. 3.25 and 3.26. This plot shows that all three of the functions become indistinguishable from each other at some reduced time between 25 and 100. The reason that the p_{tD} function and the Ei function become the same for large values of t_D may be found by studying Fig. 3–8. This figure is used to explain the lower time limit on the point-source solution. Also review Eqs. 3.18 and 3.20, which are concerned with the application of the individual functions to calculate the pressure at some particular point in the reservoir. In Fig. 3–8 when the producing time is large enough, the rates at r_w and at any larger radius, r', become essentially the same, just as the theoretical rate at $r = 0$ and the rate at r_w become the same, to permit the use of the point-source solution.

Reexamine Eqs. 3.18 and 3.20. Note that the only difference—except for the difference in pressure functions—is that one equation uses the rate at the well and the other uses the rate at the radius where the pressure is being calculated. Consequently, if the rate at the well equals the rate at the radius of interest for a long enough period, the two equations should give the same pressure. Furthermore, if they result in the same calculated pressure, the two pressure functions must be equal. Fig. 3–11 and Eqs. 3.25 and 3.26 show that the functions are equal when the reduced time, t_D, is greater than 100 or thereabouts.

Choosing the best pressure function. Actually, only two pressure-function solutions exist for the radial diffusivity equation. However, both basic solutions can be approximated by the same log equation when the reservoir is infinite acting and the reduced time is greater than 100. Therefore, it is common to think in terms of three solutions: p_{tD}, Ei, and natural log. The simplest of these is of course the log solution since it does not require graphs and since the expression can be mathematically manipulated to obtain other expressions.

Since the three pressure-function expressions have different limitations, it is often confusing to determine which should be used. The p_{tD} function can be used for infinite- and finite-acting reservoirs to find the pressure at the well or wherever the rate in the reservoir is known. The Ei solution determines the pressure at any point in the reservoir only if the reservoir is infinite acting and $\eta t/r_w^2$ is greater than 100. The log

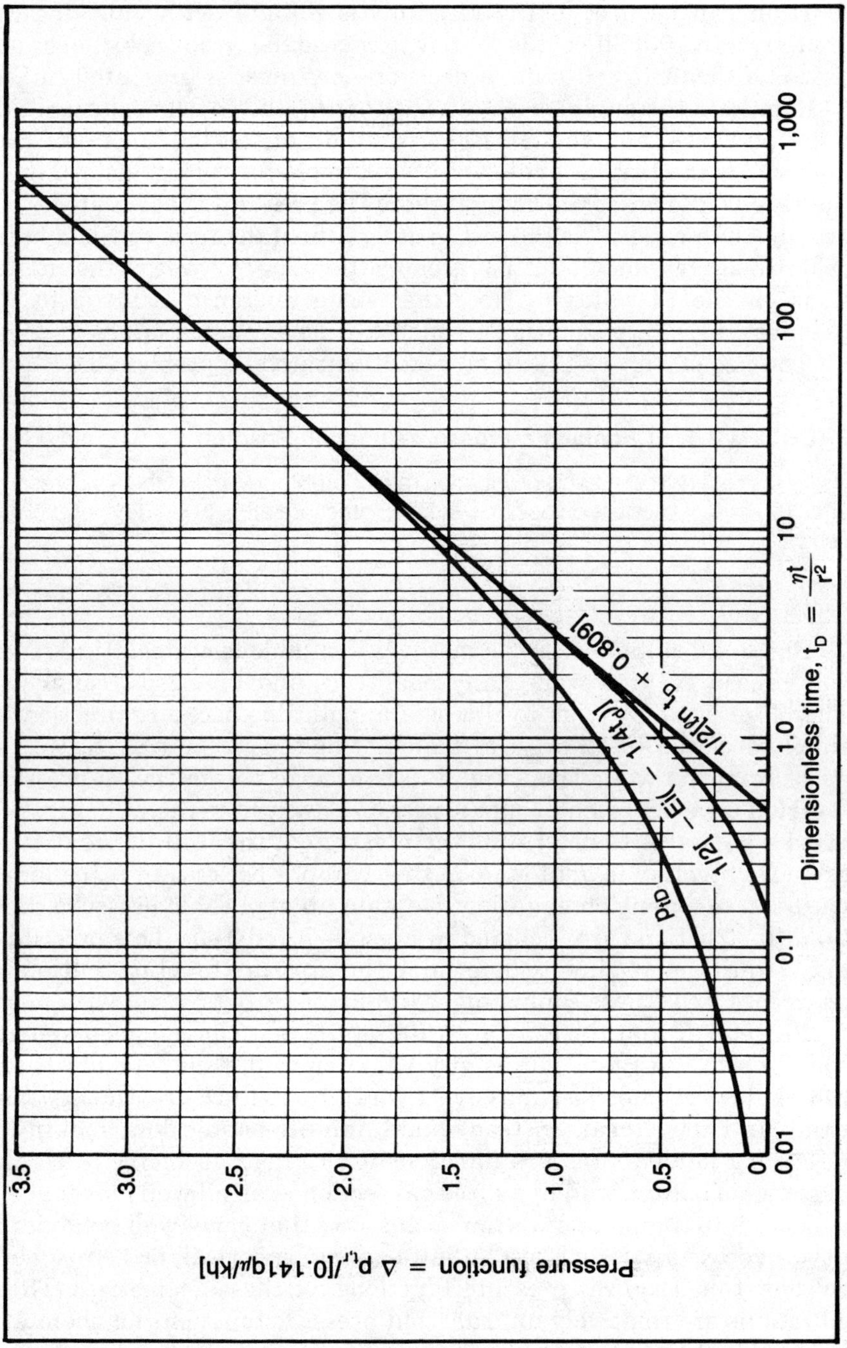

Fig. 3–11 Comparison of pressure functions for infinite reservoir

function can be used if the reservoir is infinite acting and if t_D is greater than 100. Since it is so easy to choose the wrong solution or one that is difficult to evaluate, a decision flow sheet is presented in Fig. 3–12 to help the engineer obtain the correct and easiest solution. The question "Is t_D greater than 100?" is a limit on the log function; "Is q_r known?" is the test for applicability of the p_{tD} function to determine if the rate is known at the radius where the pressure is calculated; "Is t greater than $r_e^2/4\eta$?" is the test to determine if the reservoir is infinite or finite acting; and "Is $\eta t/r_w^2$ greater than 100?" is one of the limitations on the Ei solution. Note that some reference is made to the pseudosteady-state equations, which will be discussed later.

The use of Fig. 3–12 can be best illustrated by problem 3.4.

PROBLEM 3.4: Choosing a Pressure-Function Solution

Using the reservoir data and production rate of problem 3.1, find the well pressure after 100 min of production. What is the well pressure after 1,000 days of production?

Fig. 3–12 summarizes the methods available for calculating reservoir pressures. However, many engineers find Fig. 3–12 confusing. Therefore, we may wish to choose an applicable pressure function by simply deciding whether we are calculating the pressure at the well or away from the well. Then Fig. 3–9 is used to obtain the Ei-pressure function if we are calculating the pressure away from the well. Fig. 3–7 gives a p_{tD} value to determine the pressure at the well. If the desired pressure function is not found on these graphs because t_D is too large, use a curve-extension equation to obtain an applicable pressure function. (Instructions are included in Figs. 3–7 and 3–9.) This procedure may require more time to determine the pressure function, but many engineers find it less confusing than Fig. 3–12.

Remember that the constant-rate solutions we have considered are solutions to the radial diffusivity equation and, thus, are limited to radial flow geometries. Massive hydraulic-fracture treatments may result in early linear unsteady-state, infinite-acting flow, making it necessary to have pressure functions for a linear geometry. Also, the presence of bottom water or a free gas section in an oil well may make it necessary to complete a well in such a way that early well behavior is governed by unsteady-state, infinite-acting, spherical, or hemispherical flow that requires pressure functions for these geometries. These situations are relatively unusual, but pressure functions for them will be discussed later in this chapter.

The engineer should remember that any radial flow equation or solution can be used for pie-shaped radial systems by adjusting the rate

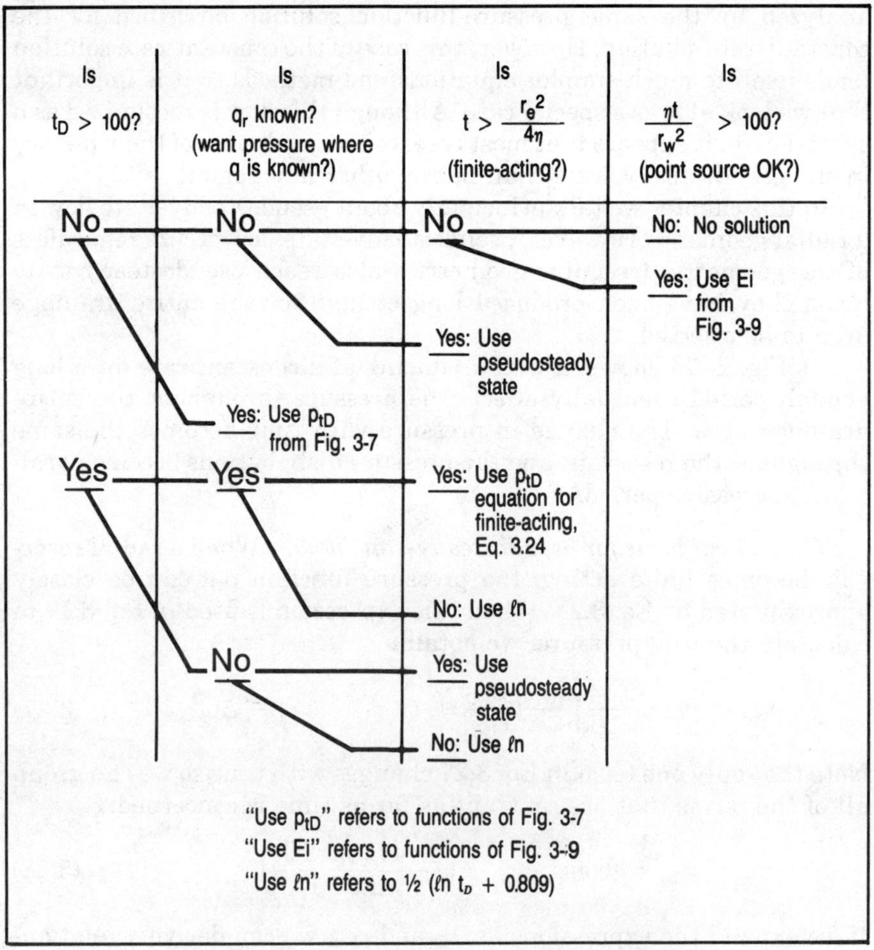

Fig. 3–12 Unsteady-state pressure functions

to that of an equivalent full-radial system. This adjustment is made by dividing the actual rate by the fraction of a circle through which flow is taking place. (This technique is illustrated in chapter 1 for a radial steady-state system.)*

Pseudosteady-State Flow

Pseudosteady-state flow is actually the finite-acting portion of the constant-rate solution to the radial diffusivity equation. Thus, it can be

*The constant-terminal-rate flow of gas is discussed in chapter 5. The same pressure function solutions are used to predict the change in the pressure-squared term, $\Delta(p^2)$, rather than the change in pressure.

analyzed by the same pressure-function solution described for the constant-rate solution. However, this case of the constant-rate solution lends itself to much simpler equations and methods so it is important that we look at it as a special case. Although this flow is recognized as a special case, it appears that most reservoirs spend more of their history in the pseudosteady state than in any other flow regime.

In this chapter we talk principally about pseudosteady-state flow in a radial geometry. However, pseudosteady-state flow occurs regardless of the geometry. Irregular geometries also reach pseudosteady state when they have been produced long enough for the entire drainage area to be affected.

As Fig. 2–14 shows, a well producing at a constant rate for a long enough period eventually affects the pressure throughout the entire drainage area. The change in pressure with time becomes the same throughout the reservoir, and the pressure distributions become parallel at successive periods.

Theoretical basis for pseudosteady-state flow. When a radial reservoir becomes finite acting, the pressure function p_{tD} can be closely approximated by Eq. 3.24. When this expression is used in Eq. 3.17 to calculate the well pressure, we obtain:

$$p_{w,t} = p_i - \frac{0.141 q_w \mu}{kh}\left(\frac{2t_D}{{r_{De}}^2} + \ell n\, r_D - 3/4\right) - \Delta p_{skin} \tag{3.27}$$

Note that only one term in Eq. 3.27 changes with time so we can group all of the terms that are constant as far as time is concerned:

$$p_w = (\text{constant}) - \left(\frac{0.141 q_w \mu}{kh}\,\frac{2t_D}{{r_{De}}^2}\right) \tag{3.28}$$

If we expand the expressions for t_D and r_{De}, we can derive a relationship between p_w and real time:

$$p_w = (\text{constant}) + \frac{0.141 q_w \mu}{kh}\,\frac{(2)(6.33)kt}{\phi \mu c {r_w}^2({r_e}^2/{r_w}^2)} \tag{3.29}$$

$$p_w = (\text{constant}) - \frac{1.79 q_w t}{\phi h c {r_e}^2} \tag{3.30}$$

$$(\Delta p_w/\Delta t) = \frac{1.79 q_w}{\phi h c {r_e}^2} \tag{3.31}$$

Eq. 3.31 is obtained from Eq. 3.30 by recognizing that a plot of p_w versus t gives a line whose slope is equivalent to the right-hand side of Eq. 3.31. Thus, the change in well pressure with time is constant, and we have pseudosteady-state flow when the p_{tD} function is governed by Eq. 3.24.

Eq. 3.31 applies only to the well pressure, but we can show that the expression is the same as the change in the average reservoir pressure needed to give pseudosteady-state flow. Under pseudosteady-state flow the pressure is declining uniformly with time at all points in the reservoir. Therefore, we can equate this change in pressure with time to the material-balance equivalent since we know that all production and flow results from the expansion of the fluid as the pressure declines. Consequently, the daily producing rate per unit of pressure drop is the reservoir pore volume, $\pi r_e^2 h\phi$ for radial flow, multiplied by the compressibility, c. When this product is multiplied by the daily pressure change, $\Delta p/\Delta t$, we obtain the production rate in barrels per day:

$$q_w = \frac{(\pi r_e^2 h\phi)c(\Delta p/\Delta t)_{pseudo}}{5.615} \qquad (3.32)$$

Eq. 3.32 can be rearranged to obtain the change in pressure with time during pseudosteady-state radial flow:

$$(\Delta p/\Delta t)_{pseudo} = \frac{1.79q}{\phi hcr_e^2} \qquad (3.33)$$

This expression for the change in the average reservoir pressure with time under pseudosteady-state flow is of course the same as the change in the well pressure with time under finite-acting behavior, according to the constant-rate solution to the radial diffusivity equation, Eq. 3.31.

Since Eqs. 3.32 and 3.33 are material-balance equations, c must represent the total effective compressibility and q the total rate of reservoir voidage including the oil, gas, and water produced. Also, note that the change in pressure with time during pseudosteady-state flow remains constant as long as the flow rate, q, and the compressibility, c, remain constant. Since both of these conditions are required for the constant-rate solution to the radial diffusivity equation and since pseudosteady-state flow is part of this solution, the change in pressure with time during pseudosteady-state flow is constant.

Eq. 3.33 is an overlooked but useful equation. It may be the only practical means of determining the effective compressibility of the reservoir when it contains substantial gas and oil saturation. Determining an average compressibility based on the saturations and the compressibility of gas, water, and the formation looks good on paper. However, it seldom appears to be correct probably because of the difficulty of accurately determining gas production data, which is generally the case when gas is not being metered for sale.

Eq. 3.33 is basically a material-balance equation rather than a flow equation. The rate must then represent the total reservoir voidage rate and must include gas if free gas is flowing in the reservoir. The total rate then is the oil rate, q_oB_o, plus the free gas rate, $(R - R_s)B_g$. In this expression R is the producing gas-oil ratio and R_s is the gas in solution in the oil. Also note that an effective radius can be used in Eq. 3.33 without sacrificing accuracy, regardless of the irregularity of the drainage area, because Eq. 3.33 is basically a material-balance equation rather than a flow equation.

Eq. 3.33 can also be written in terms of the total drainage volume in cubic feet, V_b. In this form it can be applied to any flow geometry:

$$(\Delta p/\Delta t)_{pseudo} = \frac{5.615q}{c\phi V_b} \tag{3.34}$$

Since the pressure gradient $(\Delta p/\Delta r)$ at any particular radius in the reservoir remains constant, all of the factors in the radial form of the Darcy equation (Eq. 2.7) are constant:

$$q_r = \frac{1.127kA_r}{\mu}\left(\frac{\Delta p}{\Delta r}\right)_r \tag{2.7}$$

Therefore, many engineers define this flow regime as steady state. Since the pressures are actually declining, the flow is unsteady state by the definitions of this text. However, it appears to be steady state—hence the term pseudo-, or false, steady state.*

Practical flow equations. Flow equations similar in form and simplicity to steady-state equations can be easily derived for pseudosteady-state flow. We use the same techniques employed in chapter 2 to account for geometry and compressibility. The geometry is accounted for by substituting $2\pi rh$ for the cross-sectional area, A. The compressibility of the fluid is calculated by substituting a function of the well rate, q_w, for the general flow rate, q, in the Darcy equation. Eq. 3.32 shows that the flow rate, q_w, is the pore volume multiplied by the compressibility and the change in pressure with time. The flow rate at any radius would be exactly the same, except that the volume providing the flow rate at a particular radius is caused only by the volume outside the particular radius. Thus, we can state the flow rate at any radius as:

$$q_r = \frac{(\pi r_e^2 - \pi r^2)h\phi c(\Delta p/\Delta t)_{pseudo}}{5.615} \tag{3.35}$$

If we take a ratio of Eqs. 3.35 and 3.32, we obtain the relationship between q_w, the rate at the well, and q_r, the rate at any radius:

$$\frac{q_r}{q_w} = \frac{r_e^2 - r^2}{r_e^2} = 1 - \frac{r^2}{r_e^2} \tag{3.36}$$

*The term *semisteady state* is used by some engineers to describe this flow.

$$q_r = q_w\left(1 - \frac{r^2}{r_e^2}\right) \tag{3.37}$$

If this expression is substituted for q, $2\pi rh$ is substituted for A_r in the Darcy equation, and the factors are rearranged, we obtain:

$$q_w\left(1 - \frac{r^2}{r_e^2}\right)\frac{\Delta r}{r} = \frac{7.08kh}{\mu}\Delta p \tag{3.38}$$

When Eq. 3.38 is integrated from well conditions to conditions at any radius, r:

$$q_w \int_{r_w}^{r} (\Delta r/r) - q_w \int_{r_w}^{r} (r\Delta r/r_e^2) = \frac{7.08kh}{\mu} \int_{p_w}^{p_r} \Delta p \tag{3.39}$$

$$q_w\, \ell n(r/r_w) - q_w\left(\frac{r^2}{2r_e^2} - \frac{r_w^2}{2r_e^2}\right) = \frac{7.08kh}{\mu}(p_r - p_w) \tag{3.40}$$

When Eq. 3.40 is rearranged and Δp_{skin} is accounted for, we obtain:

$$q_w = \frac{7.08kh(p_r - p_w - \Delta p_{skin})}{\mu\left[\ell n\frac{r}{r_w} - \frac{r^2}{2r_e^2} + \frac{r_w^2}{2r_e^2}\right]} \tag{3.41}$$

If we choose to account for the well damage by using the skin factor S instead of Δp_{skin} we obtain:

$$q_w = \frac{7.08kh\,(p_r - p_w)}{\mu\left[\ell n\frac{r}{r_w} - \frac{r^2}{2r_e^2} + \frac{r_w^2}{2r_e^2} + S\right]} \tag{3.41a}$$

Eqs. 3.41 and 3.41a represent the most accurate forms of the pseudosteady-state flow equation. However, the term relating the squares of the well radius and the external radius is normally insignificant when compared with the other terms in the denominator. Therefore, this term is assumed to be zero to obtain:

$$q_w = \frac{7.08kh}{\mu}\frac{(p_r - p_w - \Delta p_{skin})}{\left(\ell n\frac{r}{r_w} - \frac{r^2}{2r_e^2}\right)} \tag{3.42}$$

$$q_w = \frac{7.08kh(p_r - p_w)}{\mu\left[\ell n\frac{r}{r_w} - \frac{r^2}{2r_e^2} + S\right]} \tag{3.42a}$$

The form of the pseudosteady-state equation most often encountered is Eq. 3.42 or Eq. 3.42a, written in terms of the external radius and pressure at the external radius:

$$q_w = \frac{7.08kh}{\mu} \frac{(p_e - p_w - \Delta p_{skin})}{\left(\ell n \frac{r_e}{r_w} - ½\right)} \tag{3.43}$$

$$q_w = \frac{7.08kh(p_e - p_w)}{\mu\left[\ell n \frac{r_e}{r_w} - ½ + S\right]} \tag{3.43a}$$

Care should be taken by the engineer to avoid misapplying Eq. 3.43 or Eq. 3.43a. They can be applied only when using the pressure and radius at the external boundary of the reservoir. Since the general steady-state radial flow equation similar to Eq. 3.43 or Eq. 3.43a can be applied between any two radii, the engineer often erroneously assumes that the same is true of these pseudosteady-steady state equations.

In Fig. 3–12, which helps determine which constant-rate solution is best suited for a specific case, pseudosteady-state flow calculations are recommended for some situations. If the well pressure or external pressure is known, the pressure at the radius of interest can be determined directly by solving Eq. 3.42 or Eq. 3.42a. However, in the applications covered by Fig. 3–12, we would only know the initial well pressure. When pseudosteady state is needed, the reservoir is finite acting and the pressure is desired away from the wellbore. Under these conditions the Ei function cannot be used because the reservoir is no longer infinite acting. The p_{tD} function cannot be used because we do not know the flow rate at the radius where the pressure is desired. In such a case the pressure at the well at the desired time can be calculated using the p_{tD} function, and the pressure at the desired radius can be calculated with Eq. 3.42 or Eq. 3.42a.

Time limits on pseudosteady-state flow. According to the constant-rate solution to the radial diffusivity equation when Eq. 3.24 applies, the change in well pressure with time is equal to the change in pressure with time required for pseudosteady-state flow. If we can determine when Eq. 3.24 governs the p_{tD} function, we can find the lower limit on pseudosteady-state flow. When the constant-rate solutions were studied, we simply used a time limit of $t = r_e^2/4\eta$. Consequently, our problem is deriving the time limit that we have previously assumed.[11]

To do this, try to define all of the p_{tD} values for the entire range of producing times for a particular-sized reservoir with a particular r_{De}. If we exclude the period when the reduced time, t_D, is less than 100, we can conclude that during the early producing times when the reservoir is infinite-acting the p_{tD} values would be governed by the log solution, Eq. 3.25. Also, we would know that at large times when the reservoir is finite acting, the p_{tD} function would be governed by Eq. 3.24. This situation is illustrated by Fig. 3–13. The problem is when to switch from the infinite-acting solution to the finite solution, which is pseudosteady state. Fortunately, the two solutions come very close together, so choosing a time to switch from one to the other is not critical as far as quantitative work is concerned. The real time representing the t_D at which we choose to switch is the critical time that is often referred to as the *stabilization-time equation* or the *time equation*. Many approaches have been used to derive this equation. The method presented here seems to be better defined than any of the others.

We will choose to switch from one solution to the other when the two solid-line curves in Fig. 3–13 come closest together. This point occurs when the difference between the two is minimal. Fig. 3–14 shows a plot of the difference versus the reduced time. To find the point where the difference is a minimum, evaluate the point on the curve where the slope is zero:

$$\text{Difference} = \left(\frac{2t_{Ds}}{r_{De}^2} + \ell n\, r_{De} - 3/4\right) - 1/2(\ell n\, t_{Ds} + 0.809) \qquad (3.44)$$

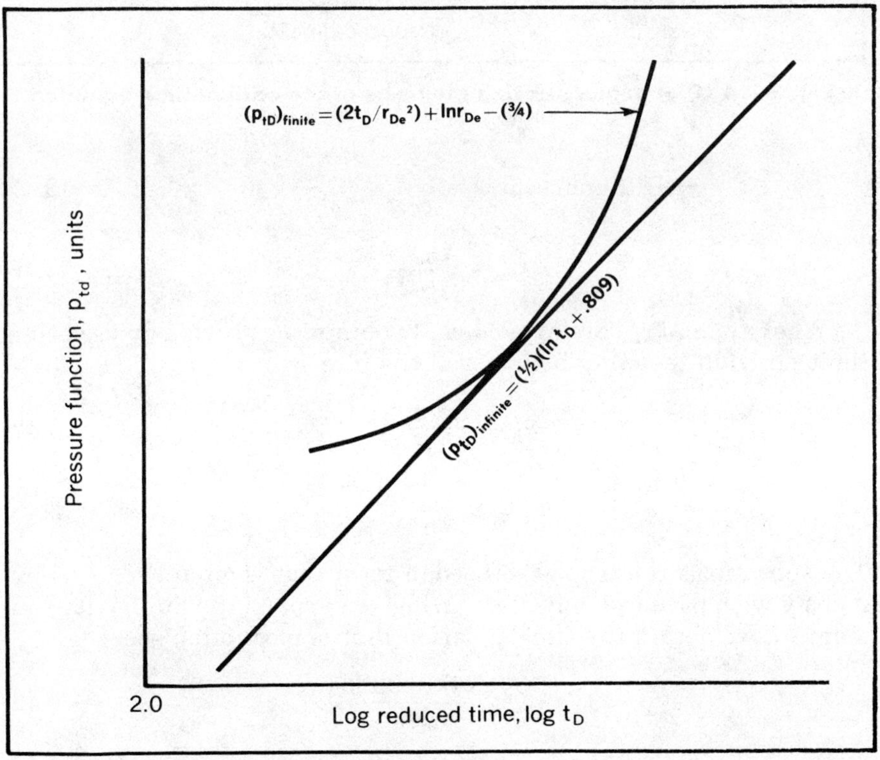

Fig. 3–13 Comparison of infinite- and finite-acting pressure functions

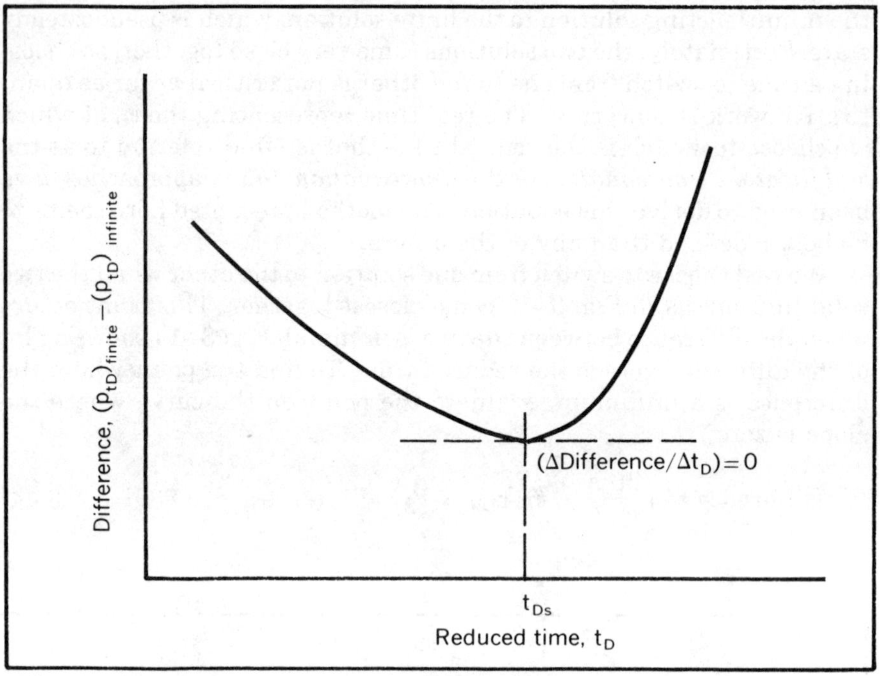

Fig. 3–14 Plot demonstrating the basis of the critical-time equation

$$(\Delta \text{Difference}/\Delta t_D) = \frac{2}{r_{De}^2} - \frac{1}{2t_{Ds}} = 0 \tag{3.45}$$

$$t_{Ds} = \frac{r_{De}^2}{4} \tag{3.46}$$

When t_{Ds} and r_{De} are expanded, we obtain the previously used time limit for infinite-acting behavior of the reservoir:

$$\frac{\eta t_s}{r_w^2} = \frac{r_e^2/r_w^2}{4} \tag{3.47}$$

$$t_s = \frac{r_e^2}{4\eta} \tag{3.48}$$

The subscript s refers to stabilized, a term that is often used synonymously with pseudosteady state. When we expand the diffusivity constant, η, we obtain the time equation that is most often seen:

$$t_s = r_e^2 \Big/ \frac{(4)\,(6.33)k}{\phi\mu c} \tag{3.49}$$

$$t_s = \frac{0.04\phi\mu c r_e^2}{k} \tag{3.50}$$

The constant, 0.04, in Eq. 3.50 does not exactly fit the reciprocal of (4) (6.33). However, the method used to derive this critical relationship is not exact. The time defined should be used simply as a guide. Since this constant, 0.04 has been in common use for several years, the author has chosen to continue its use rather than introduce another constant that would not greatly increase the accuracy of the critical time.

One confusing aspect of the time equation is that the equation does not contain q. Intuitively, we expect the time of travel of a disturbance through the reservoir to be a function of the magnitude of the disturbance. When the mathematics of the situation are studied carefully, we find that this is one more instance when our intuition leads us astray. The magnitude of the pressure drop is affected by the rate, which is part of the conversion factor from the pressure function (Eq. 3.18). However, the speed of travel of the disturbance through the reservoir is independent of the magnitude of the disturbance. An analogous situation is the time of travel of sound through the air. For example, the sound from thunder travels about 1 mile in 5 sec, regardless of the loudness of the thunder (magnitude of the disturbance). The loudness of the sound we hear is a function of the intensity of the thunder, but the time it takes the noise to reach us is independent of the intensity.

Although the time equation was derived to obtain the lower time limit on pseudosteady-state flow, it can also be used as an excellent means of determining the time required for a well that is normally in steady state to reach steady state again after a rate change. This information is often useful in analyzing well behavior in a reservoir with a strong water drive. Also, following a rate change in a well that is in pseudosteady state, the same period of time denoted in Eq. 3.50 is required to reestablish pseudosteady-state flow. The normal well producing at capacity in a solution-gas-drive reservoir continues to reach a series of pseudosteady states as the rate declines because the pressure decreases in the reservoir. The rate of decline is so small relative to the time to reach pseudosteady state that for all practical purposes the well continues to produce in pseudosteady state.

To emphasize the physical significance of some of the pseudosteady-state equations, some practice calculations will be helpful. Check the derived solution against that in appendix C.

PROBLEM 3.5: A Pseudosteady-State Flow-Test Analysis

A reservoir fluid has a viscosity of 0.5 cp and a saturation pressure of 2,500 psia. It is flowing from a radius, r_e, of 2,000 ft and a pressure of 3,500 psia—determined from a pressure-buildup test—to a well radius of 0.5 ft and a pressure of 3,200 psia. The bottom-hole pressure has been declining at a constant rate of 2 psi/day.

The producing rate is a constant 850 stb/d. The oil formation volume factor is 1.1, and the effective compressibility is 10^{-5}/psi.

A. What is the porosity thickness product?

B. What is the permeability thickness product?

C. If the bottom-hole pressure had been constant and a water drive had been evident (steady-state flow), what would be the permeability thickness product?

Note that in problem 3.5 the error introduced in calculating the capacity, kh, for the formation using a steady-state equation rather than an unsteady-state equation amounts to only about 6%, which is not overly important in reservoir calculations. This error is typical of the one introduced by assuming steady-state flow when pseudosteady-state flow actually exists. Flow around the wellbore, where most of the pressure drop occurs, is practically steady state—the rates and pressures are constant—when the reservoir is in unsteady state or pseudosteady state. Consequently, the calculations are dominated by this pressure drop in the vicinity of the wellbore.

Another confusing point connected with pseudosteady state is that, given the same pressure drop from the outer boundary to the well, a well in pseudosteady state produces at a greater rate than a well in steady state. Mathematically, this rate results from the constant, ½, in the denominator of Eq. 3.43, which is the only difference between the radial pseudosteady-state and the steady-state equations. Physically, the pseudosteady rate is greater than the steady-state rate because the average flow rate in pseudosteady state is less than the rate at the well. Since pseudosteady-state production results from fluid expansion, the rate varies from a maximum at the well to zero at the external drainage radius. In steady state the rate is the same at all points in the reservoir. Thus, the average rate is the same as the rate at the well.

Since the pressure drop is proportional to the rate, the total pressure drop necessary to obtain a particular well rate in pseudosteady state is less than the total pressure drop needed to obtain the same well rate in steady state. Consequently, the original observation that a pseudosteady-state well produces at a higher rate for a particular total pressure drop than a steady-state well with the same pressure drop is logical.

Do not be confused by this comparison. At the initial total pressure drop the pseudosteady-state well produces more than the steady-state well. However, early in the life of the pseudosteady-state well, the well pressure reaches its lower limit because pressure throughout the reservoir is declining continuously. When this lower limit occurs, the total pressure drop and the production rate begin to decline. However, in the steady-state well the total pressure drop remains essentially constant throughout the well life, and the production rate declines only when the displacing water or gas begins to be produced in the well.

Constant-Terminal-Pressure Solution

In the constant-rate solution the rate is known to be constant at some part of the reservoir, and the pressures are calculated throughout the reservoir. Conversely, in the constant-terminal-pressure solution the pressure is known to be constant at some point in the reservoir, and the cumulative flow at any particular time across the subject radius can be evaluated.[12] The constant-pressure solution is not as confusing as the constant-rate solution simply because we know less about the former. Only one constant-pressure solution is available, so we are not faced with the decision as to which solution to use.

The qualitative behavior of a radial reservoir during constant-pressure operation has been shown in Fig. 2–12. The constant-pressure solution is used most widely for calculating the water encroachment into the original oil or gas zone from the water drive of a reservoir.[13] The solution is also useful for predicting the disposal rate of water injected into an aquifer. Although the solution is not used as widely as is probably justified, it appears to be an excellent means of predicting the recovery rates from gas reservoirs prior to the time when the gas wells reach pseudosteady state.

The constant-pressure solution takes the form of a relationship between the reduced time and the reduced cumulative flow, Q.

$$Q_{tD} = \frac{Q}{1.12\phi hcr^2\Delta p} \tag{3.51}$$

Eq. 3.51 can be obtained by solving Eq. 3.12 for the rate and Eq. 3.11 for the time t.

$$q = \frac{\Delta p}{0.141(\Delta p_D)\mu/kh} \tag{3.52}$$

$$t = t_D r^2/\eta \tag{3.53}$$

Recognizing that for a constant Δp the rate, q, varies, we could define the cumulative flow at any time as:

$$Q = \sum_0^t q\Delta t \tag{3.54}$$

By substituting in Eq. 3.54 according to Eqs. 3.52 and 3.53 with η written as $6.33k/\phi\mu c$ and simplifying the expression we obtain:

$$Q = 1.12\phi hcr^2\Delta p \sum_{t_D = 0}^{t_D = t_D} \frac{1}{\Delta p_D} \Delta t_D \tag{3.55}$$

Thus, comparing Eqs. 3.51 and 3.55, we see that:

$$Q_{tD} = \sum_{t_D = 0}^{t_D = t_D} \frac{1}{\Delta p_D} \Delta t_D \tag{3.56}$$

Remember that Eq. 3.56 must be applied by superposition. Consequently, the relationship is not as simple as it may appear.

Some resulting data are plotted in Figs. 3–15 through 3–18. Unfortunately, the mathematical relationship between t_D and Q_{tD} is such that no simple relationship exists between the two over any significant range of reduced-time values as is the case with the pressure functions. This makes it necessary either to prepare numerous graphs or to list the data in tabular form as in Tables 3–1 and 3–2.

Table 3–1 is for an infinite-acting reservoir, and Table 3–2 is for a finite-acting reservoir. However, the data in Table 3–2 are limited by reservoir size. From a practical standpoint the data in Table 3–2 can be used only to predict water encroachment from an aquifer into the hydrocarbon-bearing portion of the reservoir since the dimensionless reservoir sizes only range from 1.5–10. When the constant-pressure solution is applied to the water-bearing portion of a hydrocarbon reservoir, the well radius of r_{De} is the internal radius of the aquifer or the external radius of the oil reservoir. Thus, the resulting r_{De} values are very small. The r_{De} values of Table 3–2 would be completely unrealistic if we were to calculate cumulative flow into a well.

The data of Table 3–2 represent the original published data of the Hurst-van Everdingen paper.[14] However, many major oil companies have extended these data using the equations from the subject paper. A competent programmer with a knowledge of Bessel functions can generate data for any range of dimensionless reservoir sizes needed. Without these additional data, the application of the Q_{tD} function is somewhat limited.*

The application of the constant-pressure solution is similar to that of the constant-rate data. The reduced time, t_D, is based on the radius at which the cumulative flow is desired and at which the constant pressure is known. The t_D value also incorporates the time at which the

*Fortunately, the Exxon Production Research Company permitted the use of their data in this book (Figs. 3–15 through 3–18), so the applications of the original Hurst-van Everdingen data can be greatly extended.

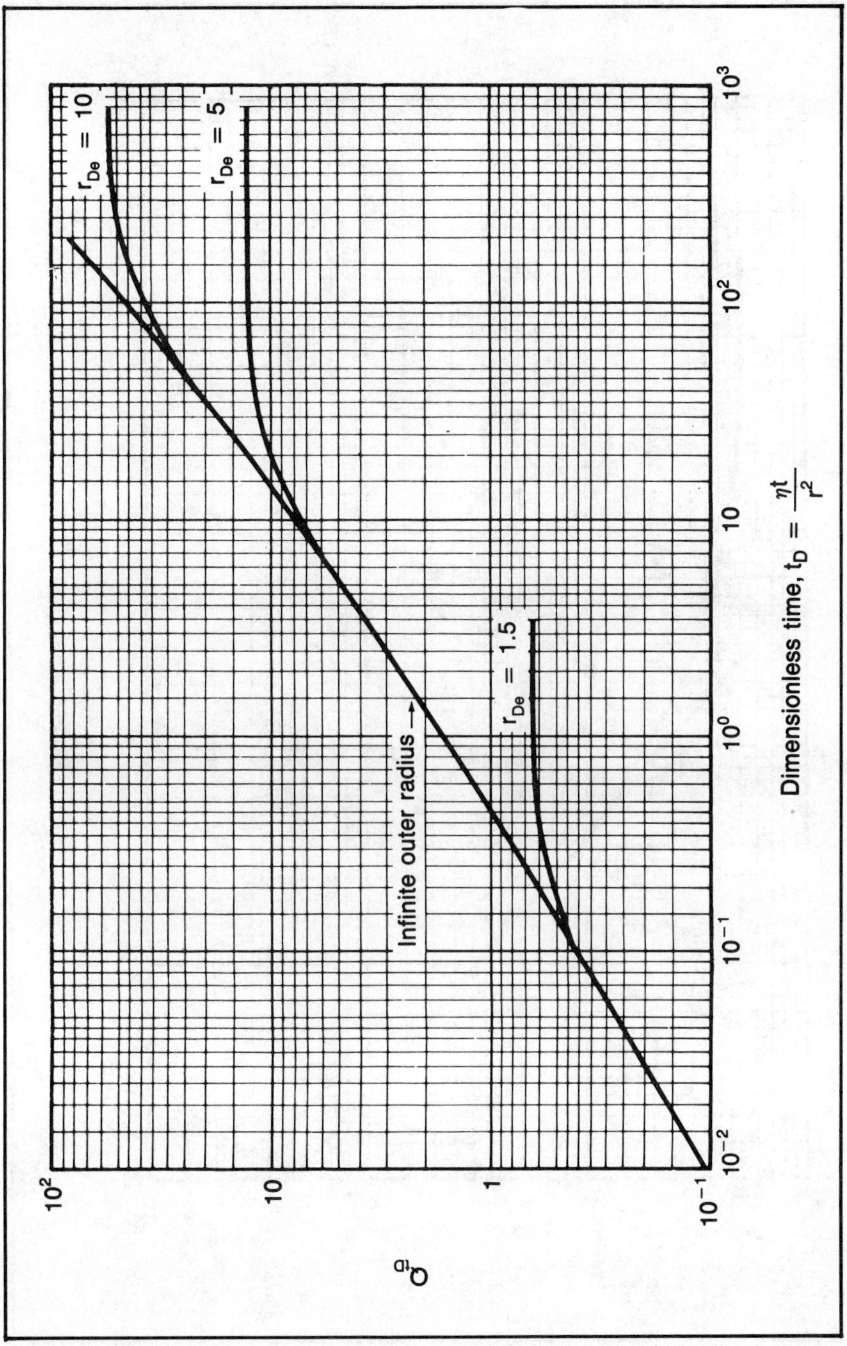

Fig. 3–15 Constant pressure functions (courtesy Exxon Production Research)

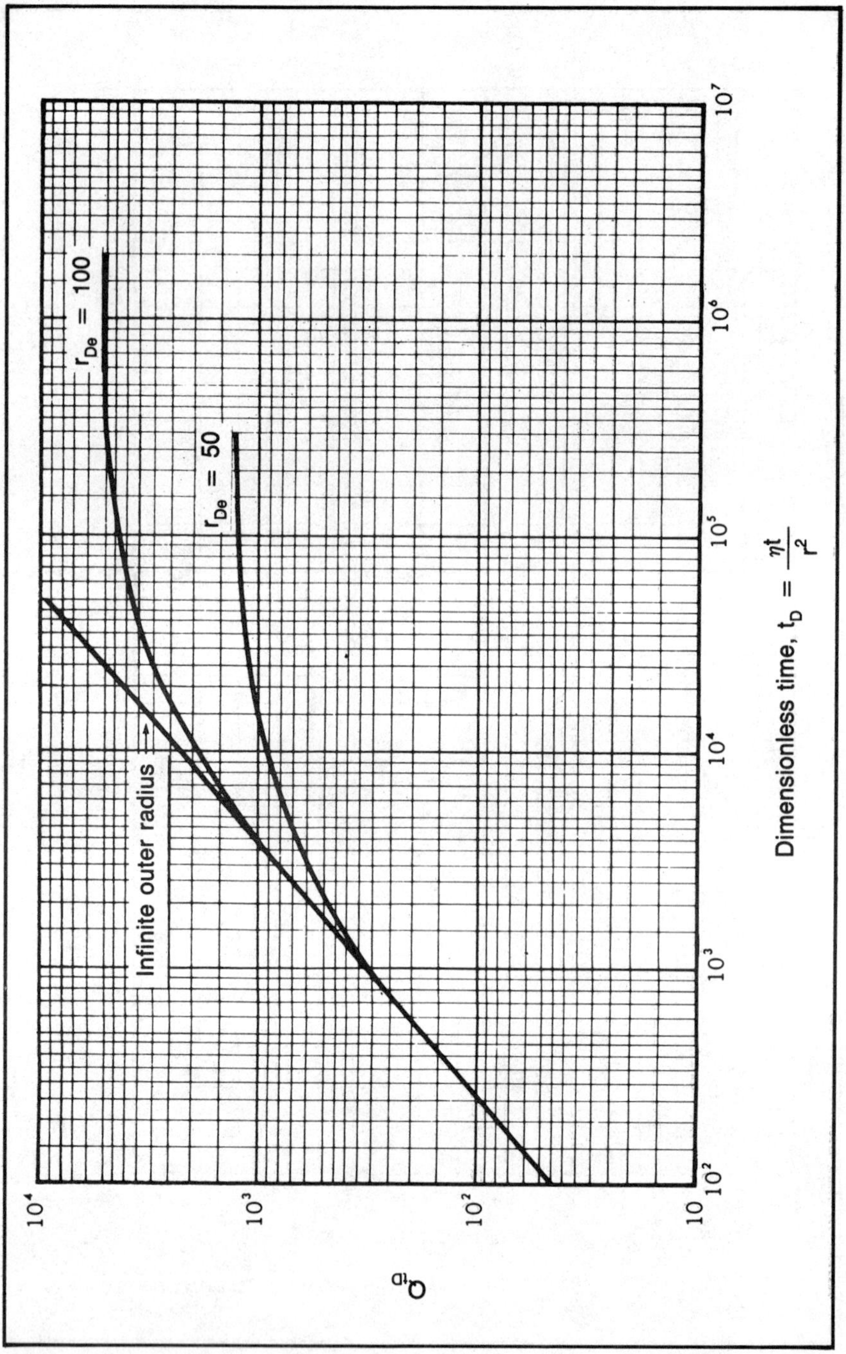

Fig. 3–16 Constant pressure functions (courtesy Exxon Production Research)

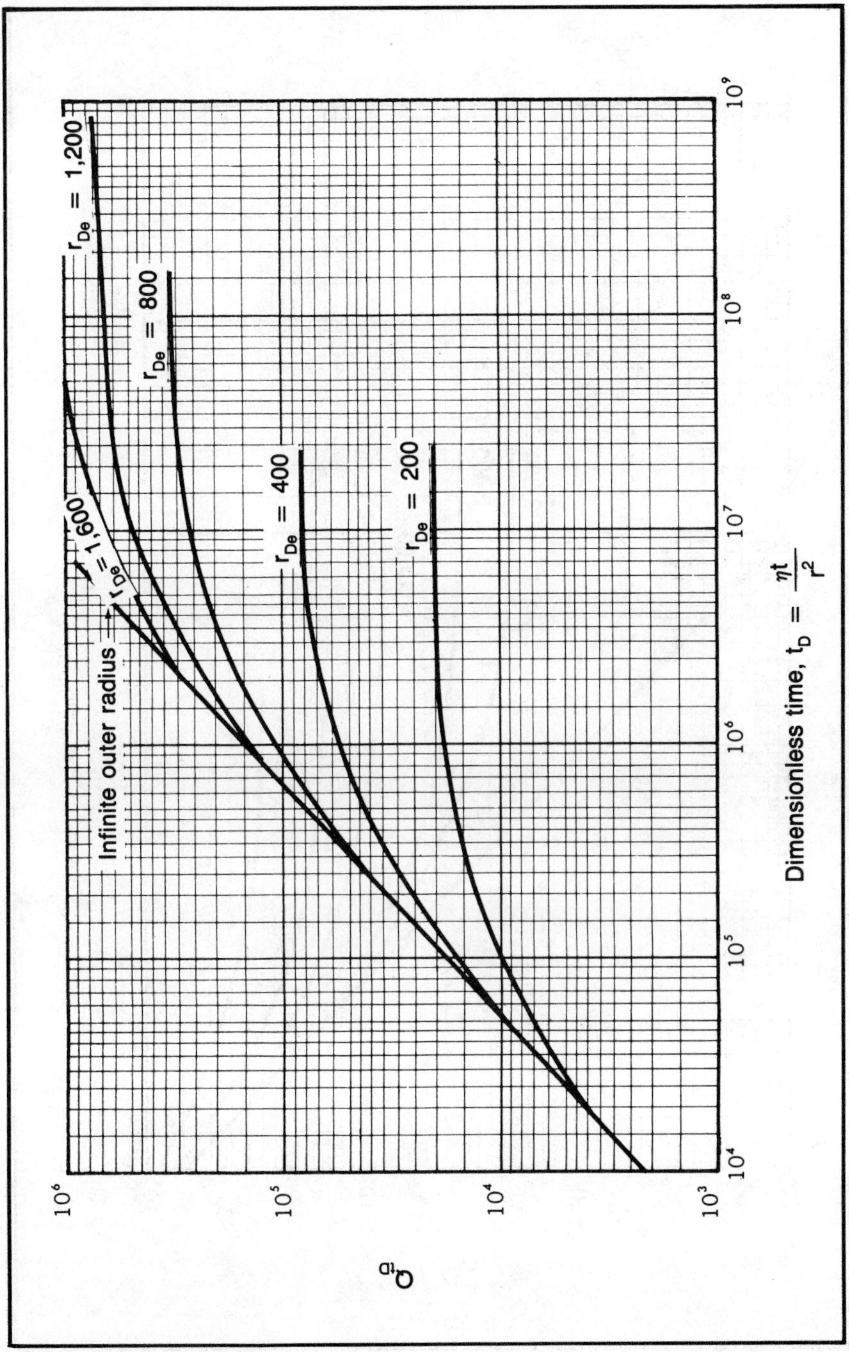

Fig. 3–17 Constant pressure functions (courtesy Exxon Production Research)

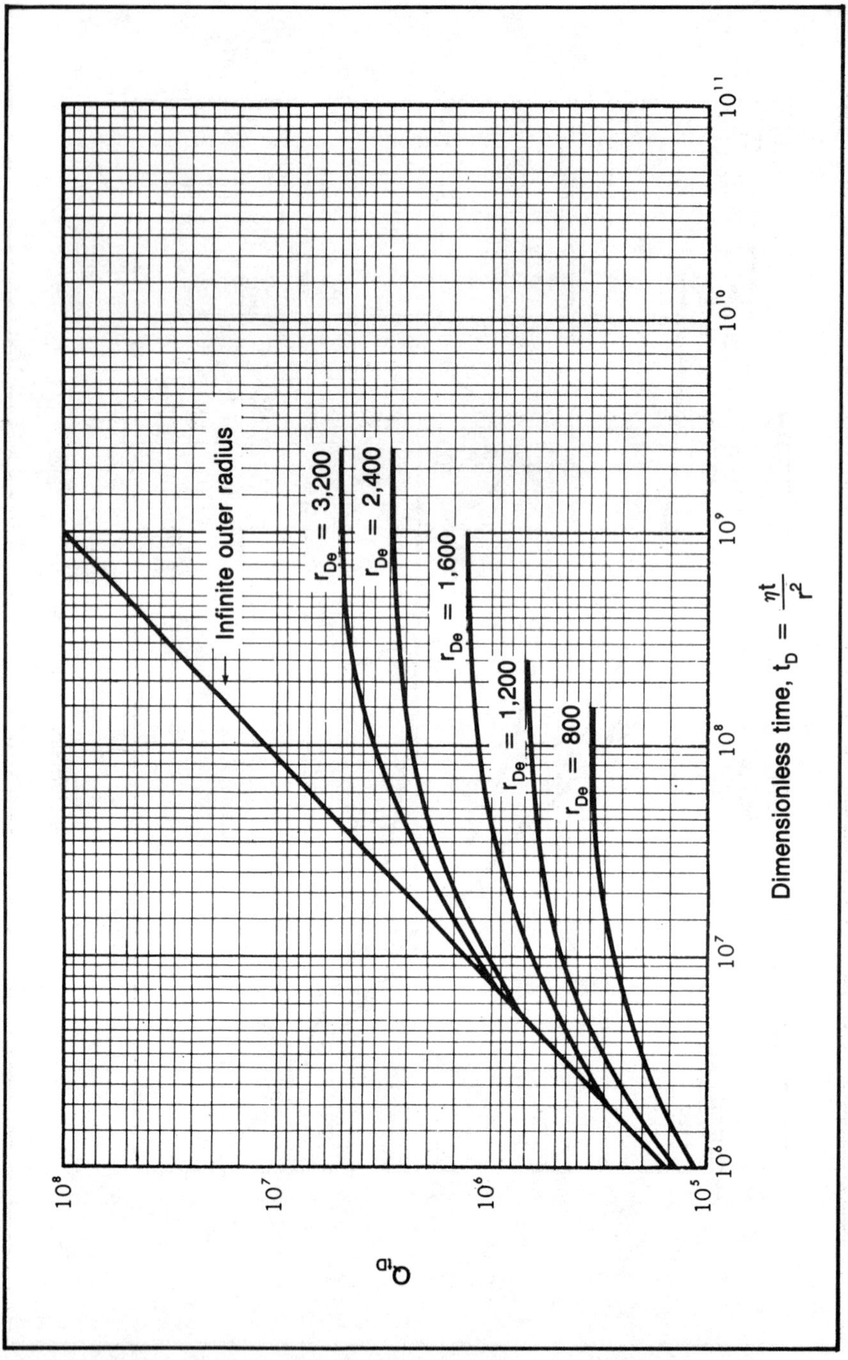

Fig. 3–18 Constant pressure functions (courtesy Exxon Production Research)

TABLE 3–1 Hurst-van Everdingen Constant-Pressure Q_{tD} Functions for Infinite-Acting Radial Reservoirs

t_D	Q_{tD}	t_D	Q_{tD}	t_D	Q_{tD}	t_D	Q_{tD}
0.00	0.000	41	21.298	96	41.735	355	121.966
0.01	0.112	42	21.701	97	42.084	360	123.403
0.05	0.278	43	22.101	98	42.433	365	124.838
0.10	0.404	44	22.500	99	42.781	370	126.270
0.15	0.520	45	22.897	100	43.129	375	127.699
0.20	0.606	46	23.291	105	44.858	380	129.126
0.25	0.689	47	23.684	110	46.574	385	130.550
0.30	0.758	48	24.076	115	48.277	390	131.972
0.40	0.898	49	24.466	120	49.968	395	133.391
0.50	1.020	50	24.855	125	51.648	400	134.808
0.60	1.140	51	25.244	130	53.317	405	136.223
0.70	1.251	52	25.633	135	54.976	410	137.635
0.80	1.359	53	26.020	140	56.625	415	139.045
0.90	1.469	54	26.406	145	58.265	420	140.453
		55	26.791	150	59.895	425	141.859
1	1.569	56	27.174	155	61.517	430	143.262
2	2.447	57	27.555	160	63.131	435	144.664
3	3.202	58	27.935	165	64.737	440	146.064
4	3.893	59	28.314	170	66.336	445	147.461
5	4.539	60	28.691	175	67.928	450	148.856
6	5.153	61	29.068	180	69.512	455	150.249
7	5.743	62	29.443	185	71.090	460	151.640
8	6.314	63	29.818	190	72.661	465	153.029
9	6.869	64	30.192	195	74.226	470	154.416
10	7.411	65	30.565	200	75.785	475	155.801
11	7.940	66	30.937	205	77.338	480	157.184
12	8.457	67	31.308	210	78.886	485	158.565
13	8.964	68	31.679	215	80.428	490	159.945
14	9.461	69	32.048	220	81.965	495	161.322
15	9.949	70	32.417	225	83.497	500	162.698
16	10.434	71	32.785	230	85.023	510	165.444
17	10.913	72	33.151	235	86.545	520	168.183
18	11.386	73	33.517	240	88.062	525	169.549
19	11.855	74	33.883	245	89.575	530	170.914
20	12.319	75	34.247	250	91.084	540	173.639
21	12.778	76	34.611	255	92.589	550	176.357
22	13.233	77	34.974	260	94.090	560	179.069
23	13.684	78	35.336	265	95.588	570	181.774
24	14.131	79	35.697	270	97.081	575	183.124
25	14.573	80	36.058	275	98.571	580	184.473
26	15.013	81	36.418	280	100.057	590	187.166
27	15.450	82	36.777	285	101.540	600	189.852
28	15.883	83	37.136	290	103.019	610	192.533
29	16.313	84	37.494	295	104.495	620	195.208
30	16.742	85	37.851	300	105.968	625	196.544
31	17.167	86	38.207	305	107.437	630	197.878
32	17.590	87	38.563	310	108.904	640	200.542
33	18.011	88	38.919	315	110.367	650	203.201
34	18.429	89	39.272	320	111.827	660	205.854
35	18.845	90	39.626	325	113.284	670	208.502
36	19.259	91	39.979	330	114.738	675	209.825
37	19.671	92	40.331	335	116.189	680	211.145
38	20.080	93	40.684	340	117.638	690	213.784
39	20.488	94	41.034	345	119.083	700	216.417
40	20.894	95	41.385	350	120.526	710	219.046

TABLE 3–1 *continued*

t_D	Q_{tD}	t_D	Q_{tD}	t_D	Q_{tD}	t_D	Q_{tD}
720	221.670	1,175	337.142	1,900	510.861	4,050	990.108
725	222.980	1,180	338.376	1,925	516.695	4,100	1,000.858
730	224.289	1,190	340.843	1,950	522.520	4,150	1,011.595
740	226.904	1,200	343.308	1,975	528.337	4,200	1,022.318
750	229.514	1,210	345.770	2,000	534.145	4,250	1,033.028
760	232.120	1,220	348.230	2,025	539.945	4,300	1,043.724
770	234.721	1,225	349.460	2,050	545.737	4,350	1,054.409
775	236.020	1,230	350.688	2,075	551.522	4,400	1,065.082
780	237.318	1,240	353.144	2,100	557.299	4,450	1,075.743
790	239.912	1,250	355.597	2,125	563.068	4,500	1,086.390
800	242.501	1,260	358.048	2,150	568.830	4,550	1,097.024
810	245.086	1,270	360.496	2,175	574.585	4,600	1,107.646
820	247.668	1,275	361.720	2,200	580.332	4,650	1,118.257
825	248.957	1,280	362.942	2,225	586.072	4,700	1,128.854
830	250.245	1,290	365.386	2,250	591.806	4,750	1,139.439
840	252.819	1,300	367.828	2,275	597.532	4,800	1,150.012
850	255.388	1,310	370.267	2,300	603.252	4,850	1,160.574
860	257.953	1,320	372.704	2,325	608.965	4,900	1,171.125
870	260.515	1,325	373.922	2,350	614.672	4,950	1,181.666
875	261.795	1,330	375.139	2,375	620.372	5,000	1,192.198
880	263.073	1,340	377.572	2,400	626.066	5,100	1,213.222
890	265.629	1,350	380.003	2,425	631.755	5,200	1,234.203
900	268.181	1,360	382.432	2,450	637.437	5,300	1,255.141
910	270.729	1,370	384.859	2,475	643.113	5,400	1,276.037
920	273.274	1,375	386.070	2,500	648.781	5,500	1,296.893
925	274.545	1,380	387.283	2,550	660.093	5,600	1,317.709
930	275.815	1,390	389.705	2,600	671.379	5,700	1,338.486
940	278.353	1,400	392.125	2,650	682.640	5,800	1,359.225
950	280.888	1,410	394.543	2,700	693.877	5,900	1,379.927
960	283.420	1,420	396.959	2,750	705.090	6,000	1,400.593
970	285.948	1,425	398.167	2,800	716.280	6,100	1,421.224
975	287.211	1,430	399.373	2,850	727.449	6,200	1,441.820
980	288.473	1,440	401.786	2,900	738.598	6,300	1,462.383
990	290.995	1,450	404.197	2,950	749.725	6,400	1,482.912
1,000	293.514	1,460	406.606	3,000	760.833	6,500	1,503.408
1,010	296.030	1,470	409.013	3,050	771.922	6,600	1,523.872
1,020	298.543	1,475	410.214	3,100	782.992	6,700	1,544.305
1,025	299.799	1,480	411.418	3,150	794.042	6,800	1,564.706
1,030	301.053	1,490	413.820	3,200	805.075	6,900	1,585.077
1,040	303.560	1,500	416.220	3,250	816.090	7,000	1,605.418
1,050	306.065	1,525	422.214	3,300	827.088	7,100	1,625.729
1,060	308.567	1,550	428.196	3,350	838.067	7,200	1,646.011
1,070	311.066	1,575	434.168	3,400	849.028	7,300	1,666.265
1,075	312.314	1,600	440.128	3,450	859.974	7,400	1,686.490
1,080	313.562	1,625	446.077	3,500	870.903	7,500	1,706.688
1,090	316.055	1,650	452.016	3,550	881.816	7,600	1,726.859
1,100	318.545	1,675	457.945	3,600	892.712	7,700	1,747.002
1,110	321.032	1,700	463.863	3,650	903.594	7,800	1,767.120
1,120	323.517	1,725	469.771	3,700	914.459	7,900	1,787.212
1,125	324.760	1,750	475.669	3,750	925.309	8,000	1,807.278
1,130	326.000	1,775	481.558	3,800	936.144	8,100	1,827.319
1,140	328.480	1,800	487.437	3,850	946.966	8,200	1,847.336
1,150	330.958	1,825	493.307	3,900	957.773	8,300	1,867.329
1,160	333.433	1,850	499.167	3,950	968.566	8,400	1,887.298
1,170	335.906	1,875	505.019	4,000	979.344	8,500	1,907.243

TABLE 3–1 *continued*

t_D	Q_{tD}	t_D	Q_{tD}
8,600	1,927.166	2.5×10^7	2.961×10^6
8,700	1,947.065	3.0×10^7	3.517×10^6
8,800	1,966.942	4.0×10^7	4.610×10^6
8,900	1,986.796	5.0×10^7	5.689×10^6
9,000	2,006.628	6.0×10^7	6.758×10^6
9,100	2,026.438	7.0×10^7	7.816×10^6
9,200	2,046.227	8.0×10^7	8.866×10^6
9,300	2,065.996	9.0×10^7	9.911×10^6
9,400	2,085.744	1.0×10^8	1.095×10^7
9,500	2,105.473	1.5×10^8	1.604×10^7
9,600	2,125.184	2.0×10^8	2.108×10^7
9,700	2,144.878	2.5×10^8	2.607×10^7
9,800	2,164.555	3.0×10^8	3.100×10^7
9,900	2,184.216	4.0×10^8	4.071×10^7
10,000	2,203.861	5.0×10^8	5.032×10^7
12,500	2,688.967	6.0×10^8	5.984×10^7
15,000	3,164.780	7.0×10^8	6.928×10^7
17,500	3,633.368	8.0×10^8	7.865×10^7
20,000	4,095.800	9.0×10^8	8.797×10^7
25,000	5,005.726	1.0×10^9	9.725×10^7
30,000	5,899.508	1.5×10^9	1.429×10^8
35,000	6,780.247	2.0×10^9	1.880×10^8
40,000	7,650.096	2.5×10^9	2.328×10^8
50,000	9,363.099	3.0×10^9	2.771×10^8
60,000	11,047.299	4.0×10^9	3.645×10^8
70,000	12,708.358	5.0×10^9	4.510×10^8
75,000	13,531.457	6.0×10^9	5.368×10^8
80,000	14,350.121	7.0×10^9	6.220×10^8
90,000	15,975.389	8.0×10^9	7.066×10^8
100,000	17,586.284	9.0×10^9	7.909×10^8
125,000	21,560.732	1.0×10^{10}	8.747×10^8
1.5×10^5	2.538×10^4	1.5×10^{10}	1.288×10^9
2.0×10^5	3.308×10^4	2.0×10^{10}	1.697×10^9
2.5×10^5	4.066×10^4	2.5×10^{10}	2.103×10^9
3.0×10^5	4.817×10^4	3.0×10^{10}	2.505×10^9
4.0×10^5	6.267×10^4	4.0×10^{10}	3.299×10^9
5.0×10^5	7.699×10^4	5.0×10^{10}	4.087×10^9
6.0×10^5	9.113×10^4	6.0×10^{10}	4.868×10^9
7.0×10^5	1.051×10^5	7.0×10^{10}	5.643×10^9
8.0×10^5	1.189×10^5	8.0×10^{10}	6.414×10^9
9.0×10^5	1.326×10^5	9.0×10^{10}	7.183×10^9
1.0×10^6	1.462×10^5	1.0×10^{11}	7.948×10^9
1.5×10^6	2.126×10^5	1.5×10^{11}	1.17×10^{10}
2.0×10^6	2.781×10^5	2.0×10^{11}	1.55×10^{10}
2.5×10^6	3.427×10^5	2.5×10^{11}	1.92×10^{10}
3.0×10^6	4.064×10^5	3.0×10^{11}	2.29×10^{10}
4.0×10^6	5.313×10^5	4.0×10^{11}	3.02×10^{10}
5.0×10^6	6.544×10^5	5.0×10^{11}	3.75×10^{10}
6.0×10^6	7.761×10^5	6.0×10^{11}	4.47×10^{10}
7.0×10^6	8.965×10^5	7.0×10^{11}	5.19×10^{10}
8.0×10^6	1.016×10^6	8.0×10^{11}	5.89×10^{10}
9.0×10^6	1.134×10^6	9.0×10^{11}	6.58×10^{10}
1.0×10^7	1.252×10^6	1.0×10^{12}	7.28×10^{10}
1.5×10^7	1.828×10^6	1.5×10^{12}	1.08×10^{11}
2.0×10^7	2.398×10^6	2.0×10^{12}	1.42×10^{11}

TABLE 3–2 Hurst-van Everdingen Constant-Pressure Q_{tD} Functions for Finite-Acting Radial Reservoirs

$r_{De} = 1.5$		$r_{De} = 2.0$		$r_{De} = 2.5$		$r_{De} = 3.0$		$r_{De} = 3.5$		$r_{De} = 4.0$		$r_{De} = 4.5$		$r_{De} = 5$		$r_{De} = 6$		$r_{De} = 7$		$r_{De} = 8$		$r_{De} = 9$		$r_{De} = 10$	
t_D	Q_{tD}	t_D	Q_{tD}	t_D	Q_{tD}	t_D	Q_{tD}	t_D	Q_{tD}	t_D	Q_{tD}	t_D	Q_{tD}	t_D	Q_{tD}	t_D	Q_{tD}	t_D	Q_{tD}	t_D	Q_{tD}	t_D	Q_{tD}	t_D	Q_{tD}
5.0×10^{-2}	0.276	5.00×10^{-2}	0.278	1.0×10^{-1}	0.408	3.0×10^{-1}	0.755	1.00	1.571	2.00	2.442	2.5	2.835	3.0	3.195	6.0	5.148	9.00	6.861	9	6.861	10	7.417	15	9.965
6.0×10^{-2}	0.304	7.50×10^{-2}	0.345	1.5×10^{-1}	0.509	4.0×10^{-1}	0.895	1.20	1.761	2.20	2.598	3.0	3.196	3.5	3.542	6.5	5.440	9.50	7.127	10	7.398	15	9.945	20	12.32
7.0×10^{-2}	0.330	1.00×10^{-1}	0.404	2.0×10^{-1}	0.599	5.0×10^{-1}	1.023	1.40	1.940	2.40	2.748	3.5	3.537	4.0	3.875	7.0	5.724	10	7.389	11	7.920	20	12.26	22	13.22
8.0×10^{-2}	0.354	1.25×10^{-1}	0.458	2.5×10^{-1}	0.681	6.0×10^{-1}	1.143	1.60	2.111	2.60	2.893	4.0	3.859	4.5	4.193	7.5	6.002	11	7.902	12	8.431	22	13.13	24	14.09
9.0×10^{-2}	0.375	1.50×10^{-1}	0.507	3.0×10^{-1}	0.758	7.0×10^{-1}	1.256	1.80	2.273	2.80	3.034	4.5	4.165	5.0	4.499	8.0	6.273	12	8.397	13	8.930	24	13.98	26	14.95
1.0×10^{-1}	0.395	1.75×10^{-1}	0.553	3.5×10^{-1}	0.829	8.0×10^{-1}	1.363	2.00	2.427	3.00	3.170	5.0	4.454	5.5	4.792	8.5	6.537	13	8.876	14	9.418	26	14.79	28	15.78
1.1×10^{-1}	0.414	2.00×10^{-1}	0.597	4.0×10^{-1}	0.897	9.0×10^{-1}	1.465	2.20	2.574	3.25	3.334	5.5	4.727	6.0	5.074	9.0	6.795	14	9.341	15	9.895	28	15.59	30	16.59
1.2×10^{-1}	0.431	2.25×10^{-1}	0.638	4.5×10^{-1}	0.962	1.00	1.563	2.40	2.715	3.50	3.493	6.0	4.986	6.5	5.345	9.5	7.047	15	9.791	16	10.361	30	16.35	32	17.38
1.3×10^{-1}	0.446	2.50×10^{-1}	0.678	5.0×10^{-1}	1.024	1.25	1.791	2.60	2.849	3.75	3.645	6.5	5.231	7.0	5.605	10.0	7.293	16	10.23	17	10.82	32	17.10	34	18.16
1.4×10^{-1}	0.461	2.75×10^{-1}	0.715	5.5×10^{-1}	1.083	1.50	1.997	2.80	2.976	4.00	3.792	7.0	5.464	7.5	5.854	10.5	7.533	17	10.65	18	11.26	34	17.82	36	18.91
1.5×10^{-1}	0.474	3.00×10^{-1}	0.751	6.0×10^{-1}	1.140	1.75	2.184	3.00	3.098	4.25	3.932	7.5	5.684	8.0	6.094	11	7.767	18	11.06	19	11.70	36	18.52	38	19.65
1.6×10^{-1}	0.486	3.25×10^{-1}	0.785	6.5×10^{-1}	1.195	2.00	2.353	3.25	3.242	4.50	4.068	8.0	5.892	8.5	6.325	12	8.220	19	11.46	20	12.13	38	19.19	40	20.37
1.7×10^{-1}	0.497	3.50×10^{-1}	0.817	7.0×10^{-1}	1.248	2.25	2.507	3.50	3.379	4.75	4.198	8.5	6.089	9.0	6.547	13	8.651	20	11.85	22	12.95	40	19.85	42	21.07
1.8×10^{-1}	0.507	3.75×10^{-1}	0.848	7.5×10^{-1}	1.299	2.50	2.646	3.75	3.507	5.00	4.323	9.0	6.276	9.5	6.760	14	9.063	22	12.58	24	13.74	42	20.48	44	21.76
1.9×10^{-1}	0.517	4.00×10^{-1}	0.877	8.0×10^{-1}	1.348	2.75	2.772	4.00	3.628	5.50	4.560	9.5	6.453	10	6.965	15	9.456	24	13.27	26	14.50	44	21.00	46	22.42
2.0×10^{-1}	0.525	4.25×10^{-1}	0.905	8.5×10^{-1}	1.395	3.00	2.886	4.25	3.742	6.00	4.779	10	6.621	11	7.350	16	9.829	26	13.92	28	15.23	46	21.69	48	23.07
2.1×10^{-1}	0.533	4.50×10^{-1}	0.932	9.0×10^{-1}	1.440	3.25	2.990	4.50	3.850	6.50	4.982	11	6.930	12	7.706	17	10.19	28	14.53	30	15.92	48	22.26	50	23.71
2.2×10^{-1}	0.541	4.75×10^{-1}	0.958	9.5×10^{-1}	1.484	3.50	3.084	4.75	3.951	7.00	5.169	12	7.208	13	8.035	18	10.53	30	15.11	34	17.22	50	22.82	52	24.33
2.3×10^{-1}	0.548	5.00×10^{-1}	0.983	1.0	1.526	3.75	3.170	5.00	4.047	7.50	5.343	13	7.457	14	8.339	19	10.85	35	16.39	38	18.41	52	23.36	54	24.94
2.4×10^{-1}	0.554	5.50×10^{-1}	1.028	1.1	1.605	4.00	3.247	5.50	4.222	8.00	5.504	14	7.680	15	8.620	20	11.16	40	17.49	40	18.97	54	23.89	56	25.53
2.5×10^{-1}	0.559	6.00×10^{-1}	1.070	1.2	1.679	4.25	3.317	6.00	4.378	8.50	5.653	15	7.880	16	8.879	22	11.74	45	18.43	45	20.26	56	24.39	58	26.11
2.6×10^{-1}	0.565	6.50×10^{-1}	1.108	1.3	1.747	4.50	3.381	6.50	4.516	9.00	5.790	16	8.060	18	9.338	24	12.26	50	19.24	50	21.42	58	24.88	60	26.67
2.8×10^{-1}	0.574	7.00×10^{-1}	1.143	1.4	1.811	4.75	3.439	7.00	4.639	9.50	5.917	18	8.365	20	9.731	25	12.50	60	20.51	55	22.46	60	25.36	65	28.02
3.0×10^{-1}	0.582	7.50×10^{-1}	1.174	1.5	1.870	5.00	3.491	7.50	4.749	10	6.035	20	8.611	22	10.07	31	13.74	70	21.43	60	23.40	65	26.48	70	29.29
3.2×10^{-1}	0.588	8.00×10^{-1}	1.203	1.6	1.924	5.50	3.581	8.00	4.846	11	6.246	22	8.809	24	10.35	35	14.40	80	22.13	70	24.98	70	27.52	75	30.49

3.4×10^{-1}	0.594	9.00×10^{-1}	1.253	1.7	1.975	6.00	3.656	8.50	4.932	12	6.425	24	8.968	26	10.59	39	14.93	90	22.63	80	26.26	75	28.48	80	31.61
3.6×10^{-1}	0.599	1.0	1.295	1.8	2.022	6.50	3.717	9.00	5.009	13	6.580	26	9.097	28	10.80	51	16.05	100	23.00	90	27.28	80	29.36	85	32.67
3.8×10^{-1}	0.603	1.1	1.330	2.0	2.106	7.00	3.767	9.50	5.078	14	6.712	28	9.200	30	10.98	60	16.56	120	23.47	100	28.11	85	30.18	90	33.66
4.0×10^{-1}	0.606	1.2	1.358	2.2	2.178	7.50	3.809	10.00	5.138	15	6.825	30	9.283	34	11.26	70	16.91	140	23.71	120	29.31	90	30.93	95	34.60
4.5×10^{-1}	0.613	1.3	1.382	2.4	2.241	8.00	3.843	11	5.241	16	6.922	34	9.404	38	11.46	80	17.14	160	23.85	140	30.08	95	31.63	100	35.48
5.0×10^{-1}	0.617	1.4	1.402	2.6	2.294	9.00	3.894	12	5.321	17	7.004	38	9.481	42	11.61	90	17.27	180	23.92	160	30.58	100	32.27	120	38.51
6.0×10^{-1}	0.621	1.6	1.432	2.8	2.340	10.00	3.928	13	5.385	18	7.076	42	9.532	46	11.71	100	17.36	200	23.96	180	30.91	120	34.39	140	40.89
7.0×10^{-1}	0.623	1.7	1.444	3.0	2.380	11.00	3.951	14	5.435	20	7.189	46	9.565	50	11.79	110	17.41	500	24.00	200	31.12	140	35.92	160	42.75
8.0×10^{-1}	0.624	1.8	1.453	3.4	2.444	12.00	3.967	15	5.476	22	7.272	50	9.586	60	11.91	120	17.45			240	31.34	160	37.04	180	44.21
		2.0	1.468	3.8	2.491	14.00	3.985	16	5.506	24	7.332	60	9.612	70	11.96	130	17.46			280	31.43	180	37.85	200	45.36
		2.5	1.487	4.2	2.525	16.00	3.993	17	5.531	26	7.377	70	9.621	80	11.98	140	17.48			320	31.47	200	38.44	240	46.95
		3.0	1.495	4.6	2.551	18.00	3.997	18	5.551	30	7.434	80	9.623	90	11.99	150	17.49			360	31.49	240	39.17	280	47.94
		4.0	1.499	5.0	2.570	20.00	3.999	20	5.579	34	7.464	90	9.624	100	12.00	160	17.49			400	31.50	280	39.56	320	48.51
		5.0	1.500	6.0	2.599	22.00	3.999	25	5.611	38	7.481	100	9.625	120	12.0	180	17.50			500	31.50	320	39.77	360	48.91
				7.0	2.613	24.00	4.000	30	5.621	42	7.490					200	17.50					360	39.88	400	49.11
				8.0	2.619			35	5.624	46	7.494					220	17.50					400	39.94	440	49.28
				9.0	2.622			40	5.625	50	7.497											440	39.97	480	49.36
				10.0	2.624																	480	39.98		

cumulative flow is desired. The dimensionless reservoir size, r_{De}, is calculated so the appropriate data can be used. Care should be exercised in using the time equation $r_e^2/4\eta$ to determine if a reservoir is infinite acting when r_{De} is small. The derivation of this expression and the corresponding t_s forms show that r_{De} must be equal to or greater than 100 for Eq. 3.24 to be accurate. Consequently, the safest procedure in using Table 3–2 is to attempt to determine Q_{tD} from the table for the appropriate r_{De}. If the smallest t_D in the table is greater than the t_D for the desired Q_{tD}, we can assume the reservoir is still infinite acting and can then evaluate Q_{tD} from Table 3–1.

When the appropriate Q_{tD} has been interpolated from Table 3–1 or Table 3–2 or when it has been read from Figs. 3–15, 3–16, 3–17, or 3–18, Q_{tD} can be converted to barrels using Eq. 3.51, which defines the reduced cumulative flow expression. It is convenient to rearrange the equation so the cumulative flow in barrels, Q, can be directly calculated. We also subscript Q and Δp to make it clear that all of the r's in the equation are the same and that we are calculating the cumulative flow at the same radius at which the pressure and the corresponding change in pressure are constant:

$$Q_r = 1.12\phi hcr^2 \, \Delta p_r \, Q_{tD} \tag{3.57}$$

When Eq. 3.57 is applied to calculate the cumulative flow at the well, either Δp_{skin} must be excluded from Δp_r or the apparent radius, r_{wa}, must be used as the radius. Remember that the apparent radius is the radius needed to give the correct answers if Δp_{skin} is set equal to zero. (In chapter 2 it is shown to be equal to $r_w e^{-S}$.) Problem 3.6 clarifies the application of Eq. 3.57. Note that c_e is the effective water compressibility, a concept that will be discussed.

PROBLEM 3.6: Water Disposal in an Aquifer

An aquifer has the following characteristics:

p_i = 1,000 psi
k_w = 10 md
h = 14.1 ft
ϕ = 0.2
r_w = 0.5 ft
μ_w = 0.5 cp
c_e = 6.33×10^{-6}/psi
r_e = 10,000 ft
S = 1.7

A. Show that, for the purposes of water injection, this aquifer is acting as an infinite reservoir after 100 days of injection.

B. If the injection pump can maintain a bottom-hole pressure of 2,200 psia, determine the disposal rate for the first 100 days.

Note that a dimensionless rate function can be used to determine the rate directly without determining cumulative values. The reduced pressure change, Δp_D of Eq. 3.12, can be used as the pressure change per reduced flow rate. By this definition we can then show that the reduced rate is as in Eq. 3.58 or Eq. 3.59:

$$q_{tD} = \frac{q}{(7.08kh\Delta p/\mu)} \tag{3.58}$$

$$q_{tD} = \frac{1}{\Delta p_D} \tag{3.59}$$

Eq. 3.58 can also be obtained by dividing the reduced volume, Q_{tD} of Eq. 3.51, by the reduced time, t_D of Eq. 3.53.

To obtain the reduced rates versus the reduced time, it is necessary to differentiate the Q_{tD}-versus-t_D curves, which directly gives the q_{tD} values. Then these values can be plotted versus the reduced time.[15,*] Some companies also use reduced-rate data in this way.

The constant-pressure solution can also be applied to the flow of gas (see chapter 5). In fact, as previously suggested it appears that much more use should be made of the constant-pressure solutions to predict the deliverability of gas reservoirs before the well reaches pseudo-steady state, which may require an extended period of time for a tight gas reservoir.

Effective Compressibility

Compressibility has previously been referred to as simply water compressibility, oil compressibility, or gas compressibility. However, the concept is more complex than this.

By referring to the derivation of the radial diffusivity equation, we see that the value required for the compressibility of a system is the amount of fluid forced from a unit of pore volume for 1.0 psi of pressure drop. If the pore volume is occupied 100% by one fluid and remains constant, we would evaluate only the compressibility of the one fluid, which would be the effective ~~permeability.~~ compressibility

However, an oil- or gas-producing reservoir always contains interstitial water, and a small change in the pore volume always accompanies a decline in reservoir pressure. This change in volume may force as much fluid from the pore volume as does the compressibility of the liquid. Consequently, we desire one value, an effective compressibility, c_e, that can be applied to the pore volume to give the amount of fluid forced from the pore volume with each 1.0 psi of pressure change. This change in volume, c_eV_p, then equals the expansion of the oil, water, and

*Plots of q_{tD} versus t_D are shown in appendix B, Fig. B21.

gas in the pore volume plus the reduction in the pore volume resulting from the decline in the reservoir pressure:

$$c_eV_p = c_oS_oV_p + c_gS_gV_p + c_wS_wV_p + \Delta V_p \qquad (3.60)$$

The change in pore volume occurs because a pressure imbalance is created in the reservoir as a result of the pressure depletion that accompanies the production from most reservoirs. Fig. 3–19 is a schematic of a microscopic view of a hydrocarbon reservoir. Normally, two different pressures are associated with a reservoir. The pressure we have been discussing up to this point is the pressure in the pores of the formation—the pressure between the sand grains—called the reservoir pressure. However, there is a second pressure associated with the reservoir rock.

If we could insert a pressure-measuring device such as a transducer into the solid material of the formation—the sand grains or carbonates—we would normally record a pressure that is different from the pressure in the pores. This pressure is caused by the weight of the formations above the reservoir that are supported almost entirely by the mechanical strength of the rock. Thus, there is a pressure in the solid material of the formation that is generally referred to as the *rock pressure*. The general distribution of sands, carbonates, shales, water, hydrocarbons, and other materials that compromise the sedimentary formations of the earth is such that, worldwide, the rock pressure is about equal to 1 psi for every foot the reservoir is beneath the surface. We say that the normal rock pressure gradient is about 1.0 psi/ft of depth. However, this pressure gradient varies widely throughout the earth.

In contrast to the normal rock pressure gradient of 1.0 psi/ft, the reservoir pressure—the pressure in the pores—is normally about 0.5 psi/ft of depth. A continuous porosity exists from any sedimentary reservoir to the surface. In terms of virtually infinite geologic time, all of these formations are permeable to the extent that the pressures between formations have been stabilized. Therefore, the initial pressure in a reservoir is about equal to the static pressure caused by a column of fluid that extends to the surface. Since most of the fluid is salt water, we find that the normal reservoir pressure is about equal to the pressure gradient of a normal salt water, or about 0.5 psi/ft.

As the reservoir pressure is depleted by production, the rock pressure or overburden pressure remains unchanged. Although the overburden is supported almost entirely by the mechanical strength of the rock, it is also supported to a small extent by the pressure in the pores. Consequently, when the reservoir pressure declines, the support of the overburden is slightly reduced and there is a slight compaction of the formation and a very small change in the thickness h. When this

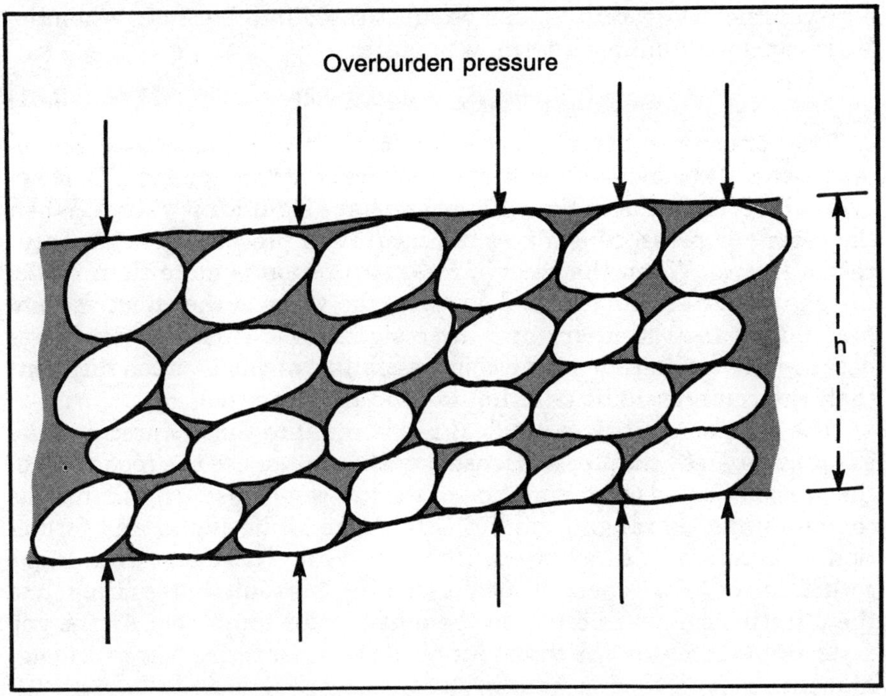

Fig. 3–19 Schematic microscopic diagram of subsurface hydrocarbon reservoir

change occurs, the bulk volume of the formation is reduced. Since the porosity is the pore volume divided by the bulk volume and since the pore volume is equal to the difference between the bulk volume and the volume of solids in the formation, the porosity is $(V_b - V_s)/V_b$ or $1 - (V_s/V_b)$. Thus, decreases in the bulk volume increase the ratio of the solids volume to the bulk volume and cause a reduction in the porosity.

Also, note the effect of the decline in pressure on the volume of solids. As shown in Fig. 3–19, most solids in the formation are surrounded by the pressure in the pores. Consequently, when the reservoir pressure declines, the compacting force on the solids is reduced and the solids volume increases slightly. From the expression for porosity, we see that an increase in the solids volume decreases the porosity. Thus, the reduction in bulk volume and the increase in solids volume tend to reduce the pore volume.

The change in the pore volume with the change in pressure can be measured in the laboratory for a particular formation and is used as the formation compressibility, $c_f = (\Delta V_p/V_p)/\Delta p$. Thus, the change in the pore volume, $\Delta V_p = c_f \Delta p\, V_p$, can be substituted into Eq. 3.60. When the V_p is canceled from each term, we obtain:

$$c_e = c_o S_o + c_g S_g + c_w S_w + c_f \quad (3.61)$$

The compressibility of oil, gas, water, and the formation—pore volume—can all be determined from empirical data in appendix B. However, all of the terms in Eq. 3.61 seldom have significant values. When the reservoir pressure is above the saturation pressure, the gas saturation is zero. When the reservoir gas saturation is more than 2–3%, the gas compressibility term dominates the value of the effective compressibility and the other terms are insignificant. This situation exists because the magnitude of the compressibility of gas is much different than the compressibilities of liquids and the formation.

The compressibility of an ideal gas is equal to the reciprocal of the gas pressure (see chapter 5). Consequently, we can use the reciprocal of the pressure as an approximation of the gas compressibility. Thus, gas compressibilities range from 10^{-3}–10^{-4}/psi, while liquid and formation compressibilities are from 10^{-5}–10^{-6}/psi. Therefore, even a gas saturation of 1.0% generally means that the gas contributes as much to the effective compressibility as the other terms combined. As the gas saturation increases, the contribution of the other terms becomes insignificant.

The formation compressibility should not be confused with the change in pore volume that often accompanies the production from a reservoir with an initial reservoir pressure similar to the rock pressure. When such a situation occurs, the overburden pressure may be totally supported by the fluids in the pore space. That is, the solids are not in sufficient contact to provide any mechanical strength to the formation. Under these conditions there will be no normal decline in reservoir pressure until enough fluid is removed so the solids are in good contact and start supporting the overburden pressure. Until such a condition is reached, production can take place without a substantial reduction in reservoir pressure. In other words, the pore-volume change is equal to the production. This phenomena usually is observed in gas reservoirs. Such reservoirs are referred to as *abnormally pressured* or *geopressured* (see chapter 5).

Superposition

If we were limited in our practical applications of the constant-rate and constant-pressure solutions to situations in which we simply had one well producing from a reservoir at a constant rate or a constant pressure, the two solutions to the radial diffusivity equation would hardly be worthwhile. However, using the concept of superposition, we can account for the effects of producing from more than one well and for the effects of rate and pressure changes. We can even extend the use of infinite-acting solutions by artificially setting up boundary conditions. The concept of superposition is not the product of a reservoir engineer. The same technique is applied to many equations governing energy mass transfer.

Accounting for the effects of multiple wells. In applying the constant-rate solution, we often wish to account for the effects of more than one well on the pressure at some point in the reservoir.[16] We can do this by simply evaluating the pressure drop caused by each producing rate as though the others did not exist. Then we add the pressure drops caused by the individual effects to obtain the total pressure drop resulting from all of the rates. In other words we simply superimpose one effect upon the other—hence the name superposition. It is undoubtedly simpler to say that we add the effects.

To illustrate the meaning of this concept consider Fig. 3–20, which shows the distance between two wells in a reservoir. If these wells have been producing for a short time and they are still infinite acting, we can determine the pressure drop caused by the production of well 2 at its well radius by assuming the other well does not exist. We can also calculate the pressure drop caused by the production from well 1 at a radius r by using the Ei-function solution, since we desire the pressure at a radius r removed from the wellbore where the rate is known. The calculation is made by assuming that well 2 does not exist. The actual pressure drop at well 2 caused by the production of both wells concurrently is the sum of the individually calculated pressure drops.

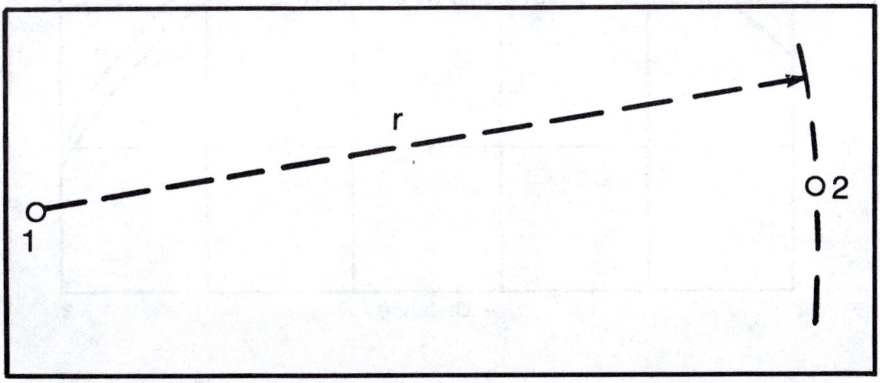

Fig. 3–20 Plan view of two wells for superposition discussion

In such an application, all producing times must be terminated at the time of the desired pressure calculation. It is not necessary that all wells start producing at the same time. In fact, it is unusual for such a coincidence to occur. By investigating the pressures at various points in the reservoir simultaneously, the complete pressure distribution can be described at any particular time. Consider Fig. 3–21. Using the Ei-pressure function, we can calculate the pressure distribution caused by one well as if the other does not exist. Then the actual pressure distribution represents the sum of the pressure drops caused by the individual wells.

We should also note that there is no limit to the number of wells that can be considered. In Fig. 3–22, if we want to calculate the pressure at point X, we simply calculate the pressure drop caused by the individual wells at the appropriate distance r and add those pressure drops to determine the total pressure drop at X. Problem 3.7 should clarify the physical significance of this discussion. The solution is shown in appendix C.

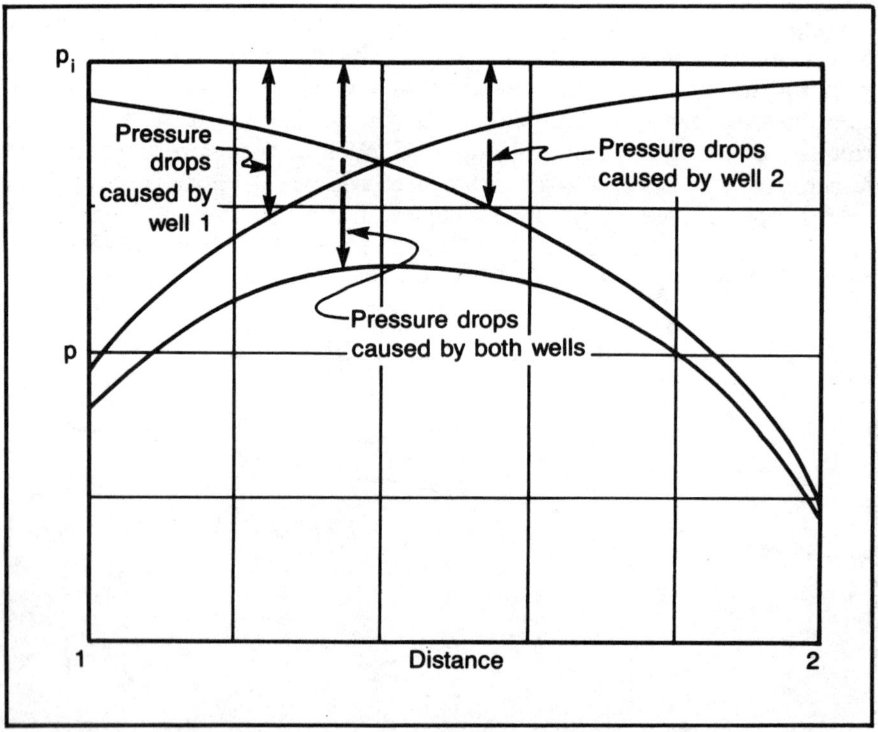

Fig. 3–21 Pressure distribution between wells

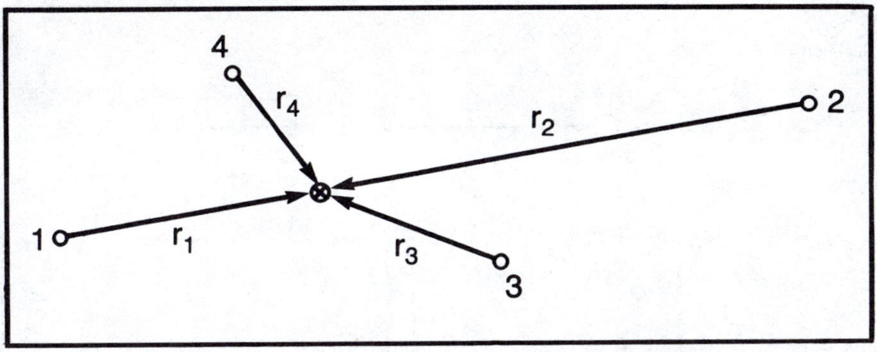

Fig. 3–22 Diagram for superposition discussion

PROBLEM 3.7: Pressure Resulting from Multiple Wells in a Reservoir

Wells 1 and 2 are drilled in an undeveloped reservoir. Well 1 is completed and produced at a constant rate of 540 stb/d for 52.5 days before well 2 is completed with initial production of 1,080 stb/d. Given the listed reservoir data, what is the pressure in well 2 after it has produced for 10 days? At this time well 1 has produced for 62.5 days, and the nearest reservoir boundary is 10,000 ft from each well. The reservoir data are as follows:

p_i = 3,000 psia
k = 100 md
h = ~~14~~ ft 14.1 ft
ϕ = 0.20
r_w = 0.5 ft
$\eta = 10^5$
p_s = 2,235 psia
μ_o = 0.5 cp
$c_e = 6 \times 10^{-5}$ 6.33×10^{-5}
B_o = 1.39 (assume constant)
Δp_{skin} = 0.0
Distance between wells = 2,500 ft

Accounting for rate-change effects. In addition to accounting for the effects of withdrawing from more than one point in the reservoir in applying the constant-rate solution, we also account for the effect of rate changes. This calculation may be best explained by referring to Fig. 3–23. In this figure two wells are shown at a distance d apart. Well 1 has produced for 10 days at a rate of 100 b/d. We know that we can calculate the pressure in either well by accounting for the effect of each well for the specified rate and time at the appropriate radius as though the other well were not there. Such a calculation can be made regard-

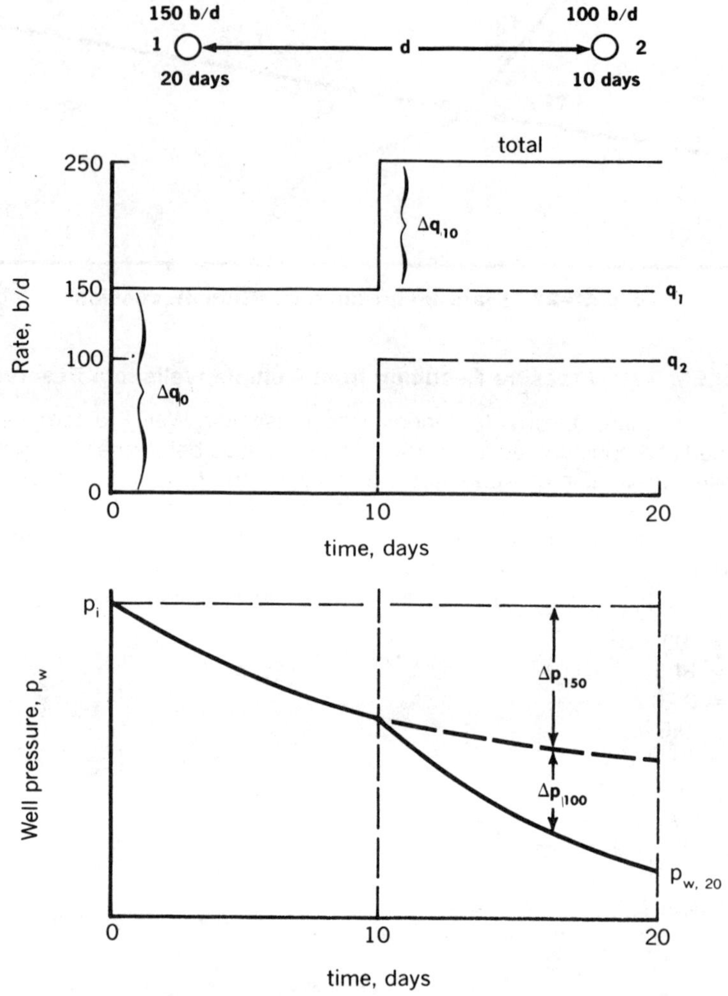

$$p_w = p_i - \Delta p_{150\ b/d} - \Delta p_{100\ b/d}$$

$$p_{w,20} = p_i - (0.141 q_1 \mu/kh)(\Delta p_D)_{20\ days} - (0.141 q_2 \mu/kh)(\Delta p_D)_{10\ days}$$

$$p_{w,20} = p_i - (0.141 \Delta q_0 \mu/kh)(\Delta p_D)_{20\ days} - (0.141 \Delta q_{10} \mu/kh)(\Delta p_D)_{10\ days}$$

Fig. 3–23 Diagram pressure drop caused by variable rate

less of the magnitude of the distance d between the wells. Consequently, we can perform such an analysis even if the two wells existed at the same point in the reservoir (d = 0).

Such an assumption, d = 0, may appear to be ridiculous. However, this idea can be used to model a rate change or a series of rate changes so we can analyze the effect of these changes. If we have two wells producing at different rates for different lengths of time at the same point in the reservoir, the effect on the reservoir pressure is the same as one well producing at a rate equal to the sum of the individual rates. In Fig. 3–22, the effect of the two wells producing as indicated with the distance between them being zero is the same as one well producing with a history equivalent to the total production of the two wells. That is, production is 150 b/d for 10 days, and the rate is increased for the next 10 days to 250 b/d.

This example provides a procedure for analyzing the effects of a rate change in a well. We must model this rate change by considering a number of wells at the same point in the reservoir having a total production equal to the one-well rate history. Mechanically, this can be done by having each well represent the rate change producing for the time remaining. In Fig. 3–23, if we have one well producing at the rate history marked "total," we can analyze it by considering the effects of a well producing at the first rate change when time equals zero (150 − 0) for the remaining time of 20 days. Then add the effect of a second well producing at the second rate change (250 − 150) for the remaining time of 10 days.

This technique can be extended to any number of rate changes, either positive or negative. Note that regardless of how many terms or rate changes are involved for a single well, each pressure-drop term contains the same group of constants, 0.141μ/kh. Thus, the mathematical expression can be written as:

insert minus sign

$$p_{w,t} = p_i \, \frac{0.141\mu}{kh} \sum_{j=1}^{j=m} \left[\Delta q_j \, (\Delta p_D)_j \right] \qquad (3.62)$$

In Eq. 3.62, the engineer must understand the meaning of Δq_j and must recognize that the pressure function Δp_D is based on the remaining length of time. Eq. 3.62 can be written in a much more explicit mathematical form, but experience shows that the more complex form is often misunderstood and misused.

The rate change, Δq, of Eq. 3.62 can be defined as $(q_{new} - q_{old})$ in order to keep the rate-change sign correct. As stated, some rate changes may be negative, which results in a pressure-increase affect.

To examine the meaning of a negative rate-change effect, refer to Fig. 3–24, which indicates a rate of 250 b/d for 10 days with a reduction in rate for the next 10 days to 150 b/d. Note that if the 250-b/d rate were continued for the entire 20 days, the resulting pressure would be the initial pressure less the pressure drop caused by 250 b/d of production for a period of 20 days. Obviously, this pressure would be too low for the actual history because the well did not produce at 250 b/d the full time. Consequently, some method must be available for reducing this pressure drop. This is accomplished by considering the effect of the negative rate change from 250 to 150 as being equal to the effect of an injection well operating at a rate of 100 b/d. That is, a negative producing rate is the same as an injection rate, and it causes a corresponding pressure increase. Thus, the actual pressure resulting from the rate history in Fig. 3–24 is the initial pressure less the reduction in pressure caused by 20 days of production at 250 b/d plus the increase in pressure from an injection well operating for 10 days at 100 b/d. This result is the same as that obtained from Eq. 3.62.

Engineers evaluating the effect of rate changes or pressure changes often want to "simplify" the calculations. One such method is to calculate the effect of a particular rate for the period it prevails and then add the effect of the next rate for the time that it prevails. When applied to Fig. 3–24, this simplification gives a well pressure equal to the initial pressure less the effect of a rate of 250 b/d for 10 days, less the effect of a rate of 150 b/d for 10 days. This type of analysis is incorrect. The production, or injection, periods must be continuous to the time of interest—the time of the desired pressure. When the times are not continuous, as in the example cited, a negative rate change is introduced when the positive rate change is discontinued—at 10 days in the example.

To correct this calculation, we must add the effect of a negative rate change of 250 b/d for 10 days, which would bring us back to where we started. Another way of explaining this problem is to recognize that there is no way of freezing or fixing a pressure distribution established in a reservoir that is other than a uniform pressure throughout. The pressure distribution does not remain fixed in the reservoir when we stop production. As soon as we discontinue production, the effect of the negative rate change immediately starts a buildup back to a uniform static pressure distribution throughout the reservoir.

Note that we can put together the effects of withdrawal from multiple points in the reservoir and the variation in rate to obtain the effect of varying the rate at more than one point in the reservoir. We can then apply Eq. 3.62 to evaluate these effects by recognizing that each j counter has both a radius and a time associated with it. A particular rate change, Δq_j, is then associated with a particular radius from the

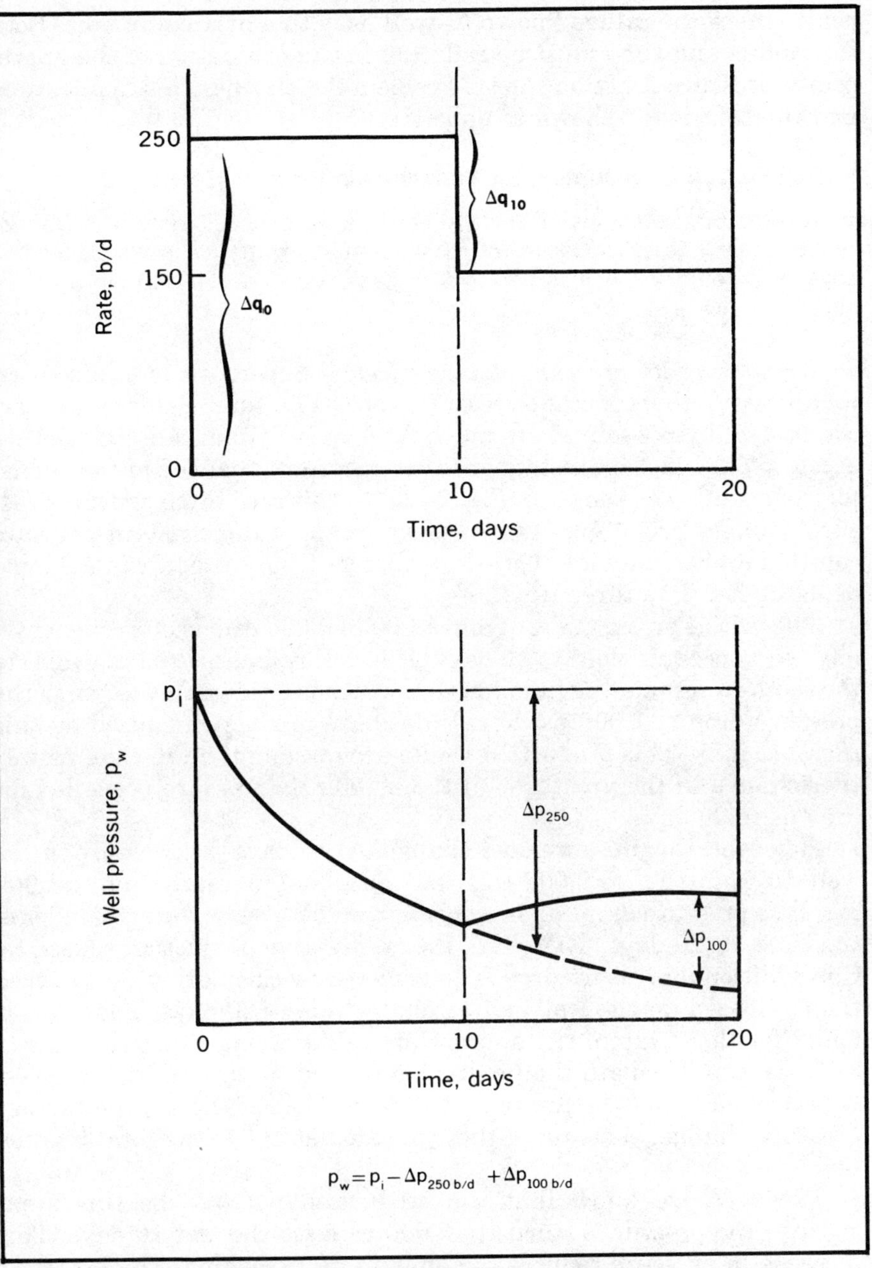

Fig. 3–24 The effect of a negative rate change on pressure

point where the rate is known as well as with a particular time. Both the radius and time must be reflected in the evaluation of the appropriate pressure function $(\Delta p_D)_j$. Problem 3.8 clarifies this application, and the solution is shown in appendix C.

PROBLEM 3.8: Accounting for Variations in Producing Rates

In problem 3.7 assume that after well 1 has produced for 62.5 days at 540 stb/d, its production rate is reduced to 180 stb/d. What is the pressure in well 2 after 72.5 days of production? At this time well 2 has produced at 1,080 stb/d for 20 days.

Accounting for pressure-change effects. Superposition is also used in applying the constant-pressure case. Pressure changes are accounted for in this solution in much the same way that rate changes are accounted for in the constant-rate case. Assume that the pressure history of a well is as shown in Fig. 3–25. In this case three distinct pressure changes occur so it is necessary to apply the constant-pressure solution to each individual pressure change. This procedure is the same as using Eq. 3.57 three times.

The initial pressure drop from 5,000 to 3,000 psia is effective for 30 days so t_D is equivalent to 30 days. The well radius is used to evaluate the reduced cumulative function Q_{tD}, which is used in Eq. 3.57 with the pressure drop of 2,000 psi to calculate the production caused by this pressure drop. This production would accumulate if there were no further change in the pressure—if the 3,000-psia pressure prevailed for the entire 30 days.

However, the pressure does change. At 10 days the pressure in the well drops further to 2,000 psia, an additional pressure drop of 1,000 psi. This pressure drop can be considered effective for the remainder of the time, or 20 days. Therefore, the cumulative production caused by this additional pressure drop is determined by calculating the reduced time t_D based on the well radius and a time of 20 days. Then Q_{tD} is found for this t_D from the appropriate table. Using this value in Eq. 3.57, we can calculate the production caused by this additional pressure drop of 1,000 psi acting for 20 days. Then if there is no further pressure change, the sum of the two calculations is the total production.

However, we know that the production will not be this great because the pressure returns to 3,000 psia for the last 10 days. This increase in pressure reduces the amount of production. Therefore, we treat this as a negative pressure drop, which results in a negative production value when applied in Eq. 3.57. The pressure increase is −1,000 psi, and the Q_{tD} is determined for a t_D that is based on a time of 10 days and the well radius. The sum of the production caused by these

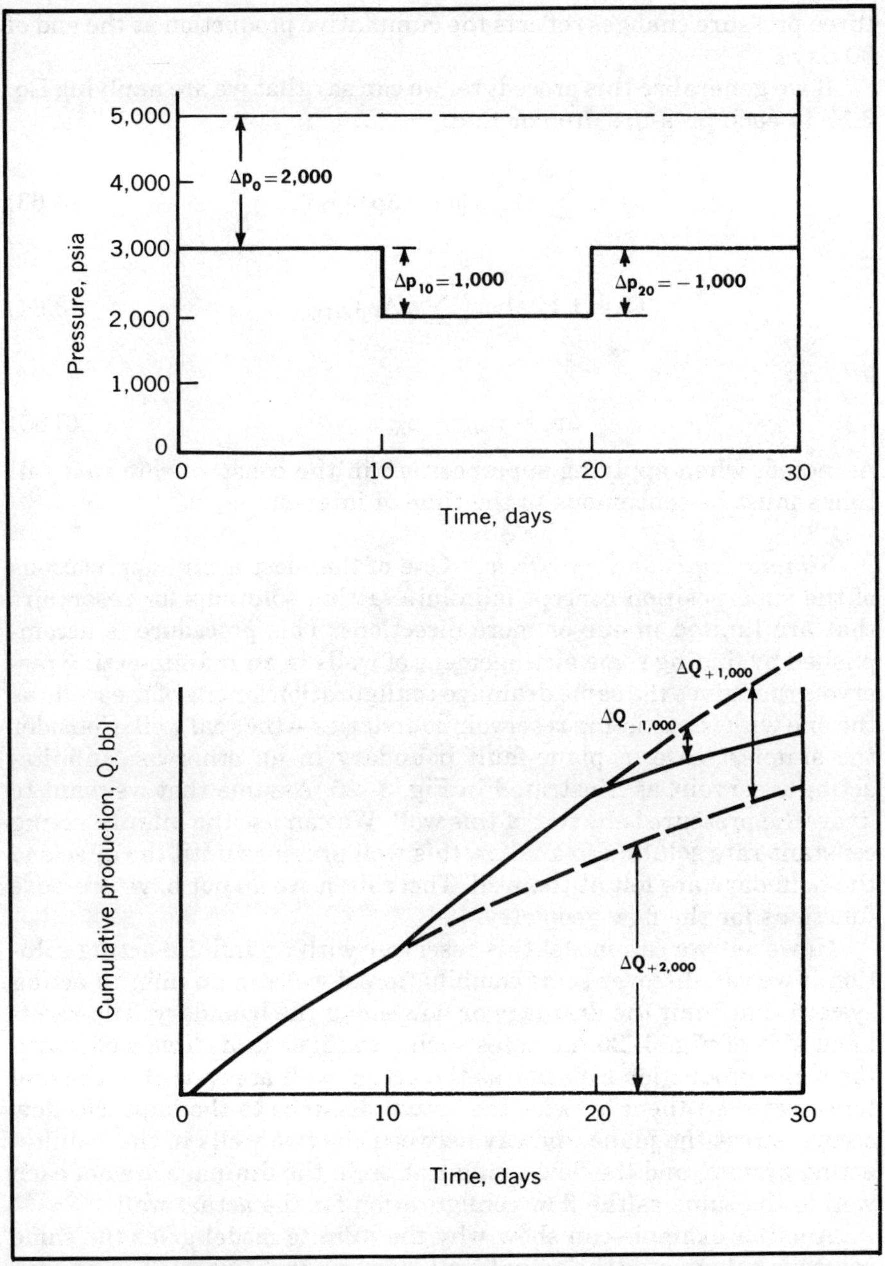

Fig. 3–25 Effect of pressure changes on cumulative production

three pressure changes reflects the cumulative production at the end of 30 days.

If we generalize this procedure, we can say that we are applying Eq. 3.57 to each pressure drop or that:

$$Q = \sum_{j=1}^{j=m} (1.12\phi hcr^2 \Delta p_j Q_{tDj}) \tag{3.63}$$

$$Q = 1.12\phi hcr^2 \sum_{j=1}^{j=m} (\Delta p_j Q_{tDj}) \tag{3.64}$$

Where:

$$\Delta p_j = p_{old} - p_{new} \tag{3.65}$$

As noted, when applying superposition in the constant-rate case, all times must be continuous to the time of interest.

Simulating boundary effects. One of the most useful applications of the superposition concept is infinite-acting solutions for reservoirs that are limited in one or more directions. This procedure is accomplished by finding some arrangement of wells in an infinite-acting reservoir that gives the same drainage configuration for one of the wells as the one with the existing reservoir boundaries—the real well. Consider the simplest case, a plane-fault boundary in an otherwise infinite-acting reservoir, as illustrated in Fig. 3–26. Assume that we want to know the pressure behavior of this well. We can use the infinite-acting constant-rate solution to analyze this well pressure until the effects of the boundary are felt at the well. Thereafter, we do not have pressure functions for the flow geometry.

However, we can model this reservoir with an infinite-acting solution if we can discover some combination of wells in an infinite-acting system that limit the drainage or flow along the boundary. The right-hand side of Fig. 3–26 indicates such a configuration. Two wells with the same production capacity as the actual well are spaced so the distance between them is twice the actual distance to the fault. No flow occurs across the plane midway between the two wells in the infinite-acting system, and the flow configuration in the drainage area of each well is the same as the flow configuration for the actual well.

Another example can show why the infinite model gives the same pressure behavior as the actual well. Assume that two wells in an infinite-acting reservoir are producing at that same rate for the same length of time. Under such conditions it is clear that any oil produced from the area to the right of the drainage boundary comes from the right-hand well and any fluid from the left of the boundary is produced

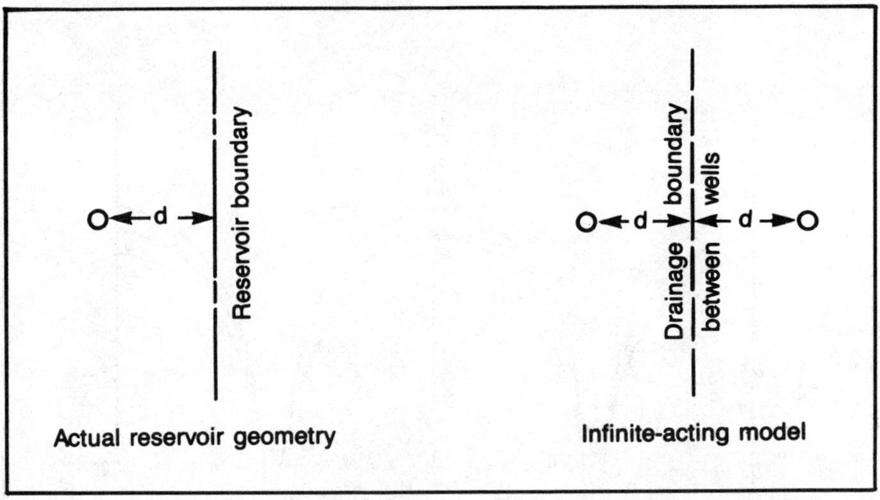

Fig. 3–26 Simulating a plane barrier in an otherwise infinite-acting reservoir

from the left-hand well. Therefore, there is no flow across the drainage boundary. Since no flow crosses this plane, a barrier can be placed in the reservoir along the plane without interfering with the production of either well. The pressure behavior with a barrier in the reservoir is the same as the pressure behavior of two wells in the infinite-acting system.

This situation is analogous to having two drains of the same size in a tub of fluid, as shown in Fig. 3–27. Drainage occurs in exactly the same way with or without a barrier at the no-flow line. Without the barrier there is pressure communication across the no-flow line, but there is no movement of fluid across this line. The same situation exists with the infinite-acting model in Fig. 3–26. Pressure communication crosses the drainage boundary, but there is no fluid movement across this line.

If we analyze the pressure behavior of the two wells in the infinite-acting system, we have the pressure behavior of the well with the reservoir boundary. This pressure behavior equals the initial reservoir pressure less the pressure drop caused by production at the well radius, less the pressure drop caused by the production from the second well at radius 2d:

$$p_w = p_i - \frac{0.141q\mu}{kh}(\Delta p_D)_w - \frac{0.141q\mu}{kh}(\Delta p_D)_{2d} \tag{3.66}$$

In Eq. 3.66, the subscripts w and 2d indicate that the pressure functions should be evaluated at reduced times based on radii of r_w and 2d,

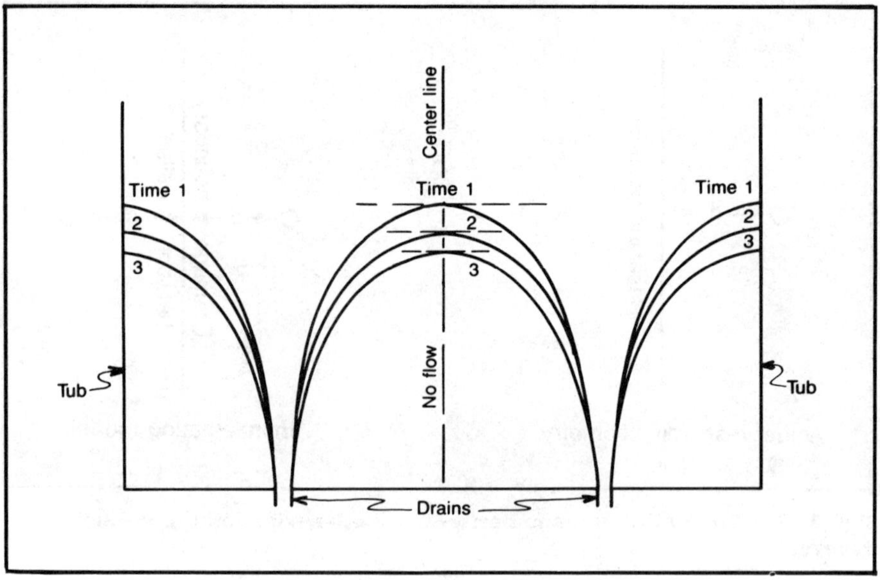

Fig. 3–27 Simulated water surface in draining tub

respectively. Artificial boundaries such as this can be set up only if the reservoir is infinite acting in all other directions since the analysis calculates pressure drops away from the wellbore. This procedure requires the Ei-function solution, which can only be used for infinite-acting behavior. Try problem 3.9 and check the solution in appendix C.

PROBLEM 3.9: Simulating Boundary Effects

Assume the reservoir in problem 3.7 contains a limiting fault 1,250 ft from well 1. What is the well pressure in this well after it has produced 52.5 days at a rate of 540 stb/d?

The additional wells that are considered in simulating a boundary are often referred to as image or ghost wells. Most boundary configurations can be simulated with the proposed technique. However, many combinations require computer facilities to be practical since a large number of ghost wells must be included.

Two boundaries at right angles to each other are simple to simulate. For example, in Fig. 3–28 one well, two boundaries, and the three image wells required to analyze the pressure behavior are shown. Note that two wells are not enough to model the two drainage boundaries. For example, consider only the real well and the two image wells, i1

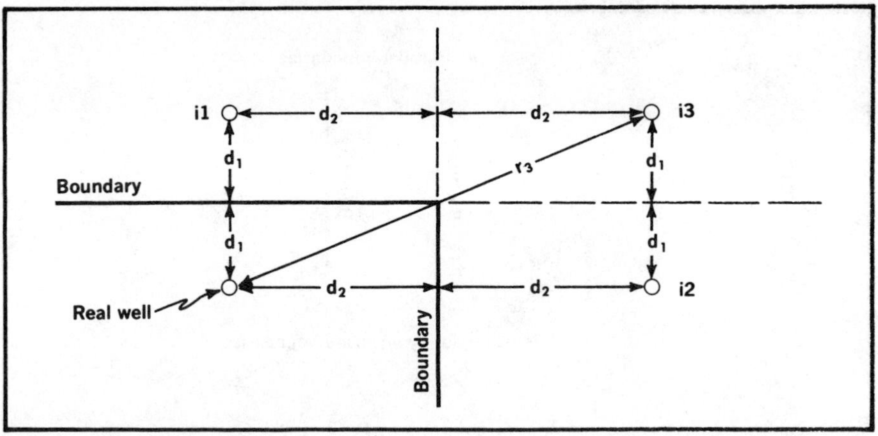

Fig. 3–28 Simulating two plane perpendicular barriers in an otherwise infinite-acting reservoir

and i2, in an infinite system. If we delineate the drainage area of each well, we find that the drainage areas do not conform to the required drainage boundaries. Consequently, four pressure-drop terms are required to determine the pressure in the real well. Note that the radius at which the effect of image well i3 is needed is the distance r_3, which is the hypotenuse of the right triangle indicated in Fig. 3–28:

$$p_w = p_i - \frac{0.141q\mu}{kh}(\Delta p_D)_w - \frac{0.141q\mu}{kh}(\Delta p_D)_{2d1} - \frac{0.141q\mu}{kh}(\Delta p_D)_{2d2} - \frac{0.141q\mu}{kh}(\Delta p_D)_{r3} \qquad (3.67)$$

Even a simple geometry can cause considerable difficulty in simulating boundaries. For example, in Fig. 3–29a the real well is between two parallel, equidistant boundaries. If we add image well i1, we stop drainage across the right boundary. Then if we add image wells i2 and i3, we stop the flow across the left-hand boundary. However, with only image wells i1, i2, and i3, we have imbalanced production across the right boundary so we must add wells i4 and i5. Then flow is imbalanced across the left boundary, and we must add wells i6 and i7. This creates imbalanced flow across the right boundary, and so the process continues.

It may appear that no solution is possible, but remember that when the image wells in the pattern are far enough removed from the real well, they do not affect the real-well pressure during the time of interest. Consequently, we can determine this limiting distance for a particular time and know that we will not improve the accuracy of the

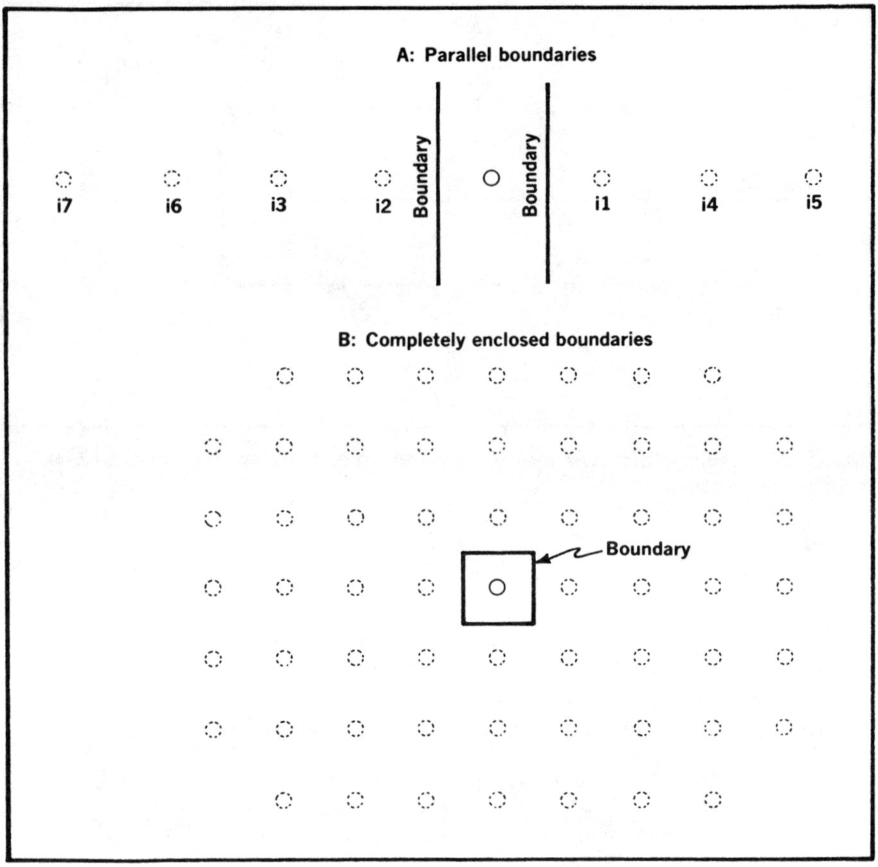

Fig. 3–29 Simulation requiring infinite ghost-well arrays

calculation for a particular time by adding image wells past this distance. This technique lends itself quite readily to computer applications.

In Fig. 3–29b it is possible to analyze a completely closed drainage area using an infinite-acting solution with a large grid of image wells. Remember that the infinite-acting model is required only when pressure functions are not available for the flow geometry being considered. Pressure functions for only the simplest geometries are included in this text, and the Q_{tD} functions that we have considered do not lend themselves to this type of boundary simulation.*

*Earlougher lists pressure functions for many geometries in appendix C of his SPE Monograph, *Advances in Well Test Analysis* (Dallas and New York: SPE of AIME, 1977).

Linear Unsteady-State Flow

Linear unsteady-state flow fundamentals are seldom needed in petroleum reservoir engineering. The author has encountered only one such type of application. Common practice is to produce low-permeability gas reservoirs using a very large artificial fracture treatment known as MHF (massive hydraulic fracture). This method is used to create a vertical fracture that is several hundred feet long. The fracture provides a combination of circumstances that leads to linear unsteady-state flow when the well is first put on production or is shut in. Thus, it may be necessary to use a linear unsteady-state flow equation to analyze such a well. Katz et al., developed an equation for infinite-acting unsteady-state flow that can be converted to the symbols of this book to provide the equation:[17]

$$p_w = p_i - \frac{1.259q}{hL}\left(\frac{\mu}{k\phi c}\right)^{0.5}(t_{days})^{0.5} \tag{3.68}$$

Where:

L = the total length of the fracture

Eq. 3.68 is applicable when t_{Dxf} is less than 0.1 where t_{Dxf} is the dimensionless time based on the length of the fracture in one direction, x_f, which is 0.5L:

$$t_{Dxf} = \eta t/x_f^2 \tag{3.69}$$

Gringgarten, Ramey, and Raghaven provide the equation in the form of a pressure function for use in the constant-rate solution, Eq. 3.17, where:[18]

$$p_{tD} = (\pi t_{Dxf})^{0.5} \tag{3.70}$$

When Eq. 3.70 is substituted for p_{tD} in Eq. 3.17, Eq. 3.68 is obtained.* Earlougher also provides pressure functions for the behavior of a fractured well when it no longer approximates linear flow.[19]

Spherical Unsteady-State Flow

Spherical or hemispherical flow often prevails for a long enough period so flow geometry characteristics may be discernible and in many cases useful. Spherical or hemispherical flow occurs when only a small portion of the reservoir thickness is perforated. This situation may result when bottom water or gas-cap gas is to be avoided. Use of a sidewall fluid sampler that involves one perforation may also result in spherical flow. Thus, an infinite-acting spherical flow equation may be

*Eq. 3.68 is applied in chapter 5.

useful. Culham presented such an equation, which when converted to the symbols used in this book gives:[20]

$$p(r, t) = p_i - \frac{0.141q\mu}{kr}\left[\left(\tfrac{1}{2}\right)\text{erfc}\left(\frac{1}{4t_D}\right)^{0.5}\right] \tag{3.71}$$

When the argument for erfc is less than 0.1 or t_D is greater than 25, erfc is governed by Eq. 3.72:

$$\text{erfc}(\tfrac{1}{4}t_D)^{0.5} = \left[1 - \frac{1}{\sqrt{\pi t_D}}\right] \tag{3.72}$$

Then when t_D is greater than 25, Eq. 3.71 becomes:

$$p(r, t) = p_i - \frac{0.141q\mu}{kr}\left[\tfrac{1}{2}\left(1 - \frac{1}{\sqrt{\pi t_D}}\right)\right] \tag{3.73}$$

Since a well is not drilled with spherical geometry, i.e., the wellbore is not spherical in shape, the radius to be used for the well in a spherical flow equation presents a problem. Moran and Finklea developed a relationship between the actual cylindrical wellbore radius and an equivalent spherical radius:[21]

$$r_{ew} = b/2\ \ell n\ (b/r_{wr}) \tag{3.74}$$

Thus, in using the spherical flow equations, Eq. 3.74 should first be used to determine the effective well radius from the actual well radius and the producing thickness, or the perforated thickness, b. When Eq. 3.74 is applied to hemispherical flow, the perforated thickness b should be replaced with 2b.

Eqs. 3.71 and 3.73 do not contain corrections for well damage or improvement. Either can be modified to include the well damage by simply adding Δp_{skin} to the right-hand side of the equation so that we obtain a modified equation such as this modification of 3.73:

$$p_{r,t} = p_1 - \frac{0.141q\mu}{kr_{ew}}\left[\tfrac{1}{2}\left(1 - \frac{1}{\sqrt{\pi t_D}}\right)\right] - \Delta p_{skin} \tag{3.75}$$

We can define the spherical skin factor, S_s, as:

$$S_s = \Delta p_{skin}/(0.141q\mu/kr_{ew}) \tag{3.76}$$

We can then write Eq. 3.75 as:

$$p_{r,t} = p_i - \frac{0.141q\mu}{kr_{ew}}\left[\tfrac{1}{2}\left(1 - \frac{1}{\sqrt{\pi t_D}}\right) + S_s\right] \tag{3.77}$$

Eq. 3.77 can then be used to evaluate either drawdown or buildup data that are dominated by spherical flow (see chapter 4).

One more point should perhaps be made before leaving the subject of infinite-acting spherical flow. Note that when a partially perforated or partially penetrating well is put on production, it first experiences radial (cylindrical) flow when only a small amount of the reservoir has been affected. Then it passes through a period dominated by spherical flow, and finally, when a large area of the reservoir has been affected, it is again dominated by radial flow behavior.

Unsteady-state and pseudosteady-state flow fundamentals provide the basis for tests such as pressure buildup, constant-rate drawdown, interference, productivity index, gas-well deliverability, isochronal, and many other well analyses. The common misapplication of well test and analysis methods in the industry undoubtedly results from a lack of understanding of the fundamentals covered. A thorough knowledge of this chapter should forestall the misapplication of the common methods of analysis and provide a basis for other methods that are not widely used in the industry. These fundamentals are used throughout our discussion of reservoir engineering.

Additional Problems

3.10 The discovery well in a reservoir is put on DST and flows at a rate of 18 stb/d. After 6 min of production, the well pressure has declined from an initial reservoir pressure of 3,380 psia to 2,878 psia. What is the net-pay thickness of the reservoir if the well is undamaged, i.e., $\Delta p_{skin} = 0$. Reservoir data are as follows:

$k_o = 0.2$ md
$\phi = 14\%$
$r_w = 0.5$ ft
$r_e = 3{,}000$ ft
$B_o = 1.39$
$p_s = 996$ psia, saturation pressure
$c_e = 6 \times 10^{-6}$/psi
$\mu_o = 0.4$ cp

3.11 A. Calculate and plot the pressure-versus-radius distribution around the well in problem 3.1 after it has produced for 1 hr at a rate of 18 stb/d.

B. In problem 3.3 we calculated the time needed to obtain a pressure drop of 7.5 psi 600 ft away. This anomaly is typical of the type we might read with confidence from an Amerada gauge. By using a quartz-type gauge, assume that we can measure a pressure disturbance of 1.0 psi. Calculate the time corresponding to this pressure drop. Reservoir data are as follows:

$k = 0.1$ md
$\phi = 0.0695$
$r_w = 0.5$ ft

p_i = 3,000 psia
r_e = 3,000 ft
B_o = 1.39
p_s = 1,000 psia
Δp_{skin} = 0
c_e = 6×10^{-6}/psi
μ_o = 0.4 cp
h = 141 ft

3.12 The discovery well in a reservoir is put on DST and flows at a rate of 1,800 stb/d. After 6 min of production, the well pressure has declined from an initial reservoir pressure of 3,300 psi to 2,878 psia. If the net pay thickness is 28 ft, what is Δp_{skin}? Reservoir data are as follows:

k_o = 100 md
ϕ = 28%
r_w = 0.5 ft
r_e = 20,000 ft
B_o = 1.39
p_s = 996 psia, saturation pressure
c = 15.8×10^{-6}/psi
μ_o = 0.8 cp

3.13 In planning an interference test in a reservoir, we conclude that, using a quartz gauge, we can measure a pressure anomaly of 0.5 psi with confidence. If the distance between wells is 5,000 ft and the rate change is 5,000 stb/d, how much time will it take to measure a 0.5-psi anomaly? A 5-psi anomaly? Reservoir data are as follows:

k = 500 md
r_w = 0.33 ft
r_e = 20,000 ft
p_s = 1,000 psia
μ_o = 0.4 cp
ϕ = 25%
p_i = 3,000 psia
B_o = 1.39
c = 18.9×10^{-6}/psi
h = 90 ft

3.14 In planning a liquid-hydrocarbon storage project in the Saudi Arabia Qatif-Sulaiy aquifer, the competence of the caprock is tested by injecting water into the aquifer at a rate of 21,000 res b/d for 3 months. Any increase in pressure is observed in a permeable zone in the caprock 130 ft above the aquifer. Calculate and plot the pressure distribution (pressure versus radius) in the aquifer after 3 months given the following. Assume infinite-acting behavior.

$h = 150$ ft
$\mu_w = 0.4$ cp
$k_w = 2$ darcies
$\phi = 25\%$
$c_e = 5 \times 10^{-6}$/psi
$r_w = 0.33$ ft
$p_i = 2{,}867$ psia
$S = +3$

3.15 Calculate and plot on graph paper the semilog and coordinate plots of well pressure versus time with a common pressure scale from 1–1,000 days for the well in problem 3.12. Identify the straight-line portion on both plots. The following data apply:

$k_o = 100$ md
$\phi = 28\%$
$r_w = 0.5$ ft
$r_e = 20{,}000$ ft
$B_o = 1.39$
$p_s = 996$ psia, saturation pressure
$c = 15.8 \times 10^{-6}$/psi
$\mu_o = 0.8$ cp
$\Delta p_{skin} = -19$ psia

3.16 A stabilized, four-point drawdown test is run on a gas well when the average pressure in the drainage area is 2,800 psia. The last flow rate of 4.5 MMscfd gave a stabilized flowing bottom-hole pressure of 2,445 psia. The well drains 2,250 acres. Assume that turbulence effects are negligible although this is not realistic. Other reservoir data are as follows:

Permeability = 74 md
Porosity = 15%
Gas viscosity = 0.021 cp
$c = 3.57 \times 10^{-4}$/psi
$z_{avg} = 0.87$
Net thickness = 4.0 ft

A. What must the bottom-hole pressure be in this well to produce at a rate of 5 MMscfd if p_e has declined to 2,000 psia and the other reservoir parameters remain unchanged?

B. How long must each rate be maintained to reach stabilized conditions during the flow test?

C. During the flow test at 4.5 MMscfd where p_i = 2,800 psia, how rapidly does the average reservoir pressure decline once pseudosteady state is reached? Assume the reservoir temperature is 110°F

3.17 In storing LPG in an aquifer, an injection pump is used to maintain a bottom-hole pressure of about 1,200 psia. Calculate the cumulative injection after 1, 2, 5, 10, 25, 50, 100, and 200 days and evaluate the average injection rate during each period. Assume an infinite-acting aquifer and the following reservoir data:

p_i = 1,000 psia
k_w and k_o = 1,517 md
h = 150 ft
ϕ = 28%
r_w = 0.33 ft
μ_w and μ_o = 0.5 cp
$c_e = 6.3 \times 10^{-6}$/psi
S = 1.5

3.18 A discovery well in a reservoir has produced at a rate of 2,500 stb/d for 50 days when an offset well 5,000 ft away in the same reservoir is put on production at a rate of 5,000 stb/d. Calculate and plot the pressure distribution between these wells when the offset has produced for 10 days; i.e., the discovery has produced for a total of 60 days. The nearest reservoir boundary to either well is 20,000 ft away, and the reservoir data are as follows:

p_i = 2,400 psia
h = 70 ft
k = 300 md
ϕ = 0.24
r_w = 0.33 ft
p_s = 2,200 psia
μ_o = 0.4 cp
$c_e = 15 \times 10^{-6}$/psi
B_o = 1.39
$S_{discovery}$ = +2.1
S_{offset} = −0.5
S = 1.5

3.19 In problem 3.8 the pressure in well 2 is calculated after it has produced for 20 days at 1,080 stb/d. Well 1 has produced for 62.5 days at 540 stb/d and then for 10 days at 180 stb/d. What is the pressure in well 1 at this time? What is the pressure at a spot in the reservoir 1,500 ft from each well? The data from problem 3.8 are applicable.

3.20 In problem 3.9 a well pressure is calculated with a fault 1,250 ft from the well. What is the well pressure if the reservoir contains a second fault that is 500 ft from the well and at right angles to the first fault? The well has produced at 540 stb/d for 52.5 days, and the following reservoir data are applicable:

p_i = 3,000 psia
k = 100 md
h = 14 ft   14.1 ft
ϕ = 0.20
r_w = 0.5 ft
η = 10^5
p_s = 2,235 psia
μ_o = 0.5 cp
c_e = 6×10^{-5}/psi 6.33×10^{-5}/psi
B_o = 1.39, assume constant
Δp_{skin} = 0.0

3.21 In problem 3.17 the liquid hydrocarbon injection into an aquifer caused by a bottom-hole pressure of 2,200 psia acting for a total cumulative injection time of 200 days is calculated. What is the total cumulative injection after 300 days if only a pressure of 1,900 psia can be maintained for the last 100 days, i.e., from 200 to 300 days?

p_i = 1,000 psia
k_w and k_o = 1,517 md
h = 150 ft
ϕ = 28%
r_w = 0.33 ft
μ_w and μ_o = 0.5 cp
c_e = 6.3×10^{-6}/psi
s = 1.5

3.22 Two wells have been drilled in a reservoir as shown in Fig. 3–30. Well 1 has been producing at a constant rate of 240 stb/d for 30 days and well 2 at 120 stb/d for 5 days. What is the flowing bottom-hole pressure in well 2 if a fault (reservoir limit) exists as indicated? Assume an infinite-acting reservoir, except for the fault limitation. Reservoir data are as follows:

r_w (both wells) = 0.25 ft
c_e = 4×10^{-4}/psi
Oil viscosity = 2 cp

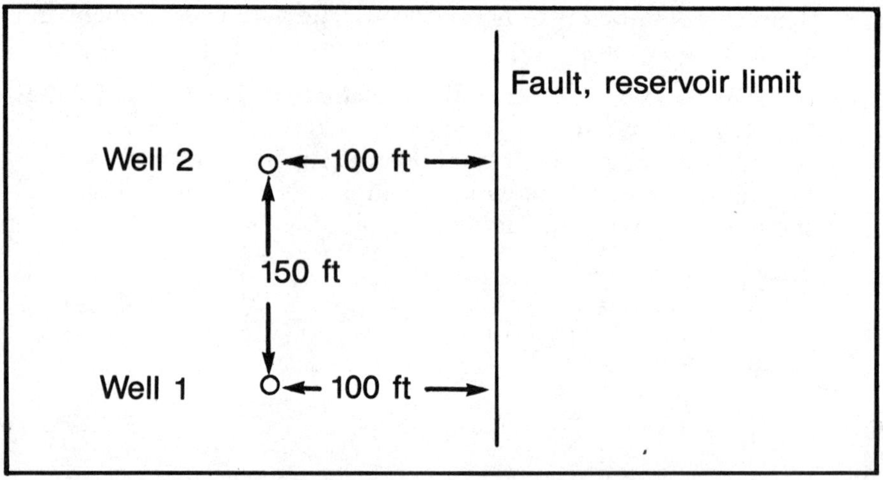

Fig. 3–30 Well spacing diagram for problem 3.22

Porosity = 0.25
Permeability = 240 md
Thickness = 12 ft
B_o = 1.2 res bbl/stb
p_i = 2,700 psia

3.23 Calculate and plot on semilog and coordinate paper the well pressure versus time from 1–1,000 days for the well in problem 3.11.

Notes

1. D.L. Katz et al., *Handbook of Natural Gas Engineering* (New York: McGraw-Hill, 1959).

2. B.C. Craft and M.F. Hawkins, *Applied Petroleum Reservoir Engineering* (Englewood Cliffs: Prentice Hall, 1959).

3. H.S. Carslaw and J.C. Jaeger, *Conduction of Heat in Solids* (Oxford: Clarendon Press, 1959).

4. A.F. van Everdingen and W. Hurst, "The Application of the Laplace Transformation to Flow Problems in Reservoirs," *Trans.,* AIME (1949), volume 186, 305.

5. van Everdingen and Hurst, 1949.

6. Thomas D. Mueller and Paul A. Witherspoon, "Pressure Interference Effects within Reservoirs and Aquifers," *JPT* (April 1965), pp. 471–474.

7. Craft and Hawkins, 1959.

8. Federal Works Agency, *Tables of Sine, Cosine, and Exponential Integrals,* WPA for the City of New York, sponsored by the U.S. National Bureau of Standards (Washington, D.C.: Superintendent of Documents, n.d.).

9. H.C. Slider, "A Simplified Method of Pressure Buildup Analysis for a Stabilized Well," *JPT* (September 1971).
10. van Everdingen and Hurst, 1949.
11. Slider, 1971.
12. van Everdingen and Hurst, 1949.
13. A.F. van Everdingen, E.H. Timmerman, and J.J. McMahon, "Application of the Material Balance Equation to a Partial Water-Drive Reservoir," *Trans.*, AIME (1953), volume 198, 51.
14. van Everdingen and Hurst, 1949.
15. M. Muskat, *The Flow of Homogeneous Fluids Through Porous Media* (New York: McGraw-Hill, 1937).
16. C.V. Theis, "The Relationship between the Lowering of Piezometric Surface and Rate and Duration of Discharge of Wells using Ground-Water Storage," *Trans.*, AGW (1935), volume II, 519.
17. Katz et al., 1959.
18. Alain C. Gringgarten, Henry J. Ramey Jr., and R. Raghavan, "Pressure Analysis for Fractured Wells," SPE paper 4051, SPE-AIME 47th Annual Fall Meeting (San Antonio: October 1972).
19. Earlougher, 1977.
20. W.E. Culham, "Pressure Buildup Equations for Spherical Flow Regime Problems," *SPEJ* (December 1974), pp. 545–555.
21. J.H. Moran and E.E. Finklea, "Theoretical Analysis for Pressure Phenomena Associated with the Wireline Formation Tester," *JPT* (June 1969), pp. 743–752.

Additional References

Brownscombe, E.R.; and Collins, F. "Pressure Distributions in Unsaturated Oil Reservoirs," *Trans.*, AIME (1950), volume 189, 371.

Fetkovich, M.J. "Decline Curve Analysis using Type Curves." *JPT* (June 1980), pp. 1065–1077.

4

Well-Pressure Behavior Analysis

One of the major problems of reservoir engineering is improving the capacity of wells to produce. This problem concerns individual wells as opposed to reservoir engineering problems that are concerned with the reservoir as a whole—controlling production to obtain maximum recovery or applying secondary recovery techniques to improve production. The individual well problem is basically concerned with fluid-flow technology, as opposed to general reservoir problems that must be concerned with material balance and fluid displacement.

Well-pressure behavior analysis is a term that includes all well-testing procedures.* In this chapter we are concerned with the interpretation of the productivity-index flow tests, the common erroneous assumptions that are made by practicing engineers concerning such tests, and the use of the Vogel curve in interpreting productivity-index test data. A different approach to transient pressure behavior, called *negative superposition,* greatly simplifies transient pressure analysis and emphasizes the shortcomings of many commonly used pressure-buildup and drawdown methods. The practical significance of pulse testing also is discussed.

This chapter indicates the most foolproof methods of determining reservoir boundaries and discusses the unlikely possibility that a transient pressure test could be used to describe a reservoir drainage shape completely. The practical use of curve-fitting techniques in solving transient pressure analysis problems also is discussed. Finally, the problem of making realistic production rate estimates from drillstem-test data is considered.

Productivity-Index Tests

Historically, the petroleum engineer has expected the production rate from a well to be proportional to the pressure drawdown, $p_e - p_w$,

*Gas-well deliverability tests are very specialized, so they are discussed separately in chapter 5.

or inversely proportional to the well pressure, p_w[1]. Furthermore, the relationship between the drawdown and the rate are often assumed to remain constant with time. These ideas have been used to predict the production rate under various operating conditions—for example, pumping a well versus flowing it, using a high-capacity submersible pump (like a Reda) versus a lower-capacity rod pump, or predicting the increased production rate of wells in a waterflood. They have also been used to predict variations in the producing rate with time or reservoir pressure depletion, especially computer reservoir models. The concepts of fluid flow covered in chapters 2 and 3 provide us with tools to investigate the validity of these ideas.

The drawdown relationship to the producing rate is known as the productivity index and is mathematically defined as:

$$J = \frac{q_{stb}}{(p_e - p_w)} \tag{4.1}$$

Where:

J = productivity index, the symbol preferred by the Society of Petroleum Engineers

Note in Eq. 4.1 that the rate is stated in stock-tank barrels, contrary to the general use in this text of reservoir barrels per day. This variation is simply a means of conforming to the general field definition for the productivity index.

As stated, the straight-line relationship between the rate and drawdown is independent of the magnitude of the rate or the reservoir depletion, which is the same as assuming that the productivity index does not change with rate or time (depletion). We will see that this assumption is true only under very special conditions.

A well producing at a constant pressure from a solution-gas-drive reservoir is in unsteady-state flow—governed by the constant-pressure solution—until the entire drainage area is affected, at which time it tends to enter pseudosteady state. The phrase "tend to enter" is used because, theoretically, the well is not in steady state until a constant rate (not pressure) is maintained for the stabilization time. We know that, during the constant-pressure unsteady-state phase of production, the rate declines somewhat as the producing time increases. However, when the entire drainage area is affected, the flow is closer to pseudosteady state than to any other flow regime for which we have working equations. Since most of the well life is spent in a flow regime approximating pseudosteady state, we should be predicting pseudosteady-state flow rates from the productivity-index data. Furthermore, it would be impossible to consider the unsteady-state constant-pressure portion of the life as being applicable to the productivity-index idea since we know that the flow rate varies during the constant-pressure unsteady-state regime.

Examine the pseudosteady-state radial equation from chapter 3:

$$q_w = \frac{7.08kh(p_e - p_w)}{\mu\left[\ell n \frac{r_e}{r_w} + ½ + S\right]} \tag{3.43a}$$

In Eq. 3.43a if we substitute $q_{stb}B_o$ for q_w and write the equation for oil flow, we can solve for the productivity index, J:

$$J = \frac{q_{stb}}{p_e - p_w} = \frac{7.08k_oh}{B_o\mu_o\left(\ell n \frac{r_e}{r_w} + ½ + S\right)} \tag{4.2}$$

Similarly, when a well produces from a strong water-drive or gas-cap-drive reservoir that tends to reach steady-state flow, its productivity index can be described by a steady-state radial flow equation similar to Eq. 4.2:

$$J = \frac{7.08k_oh}{B_o\mu_o[\ell n(r_e/r_w) + S]} \tag{4.3}$$

As noted, the steady-state and pseudosteady-state equations can be used with an average permeability, instead of an undamaged permeability, by eliminating the skin factor, S. This is possible because the drainage radius is fixed. Therefore, the weighting of the undamaged and damaged permeabilities does not vary with time as it does in infinite-acting behavior where the drainage radius increases with time. Thus, equivalent forms of Eqs. 4.2 and 4.3 can be written as:

$$J = \frac{7.08k_{avg}h}{B_o\mu_o[\ell n(r_e/r_w) - 0.5]} \tag{4.4}$$

$$J = \frac{7.08k_{avg}h}{B_o\mu_o\ell n(r_e/r_w)} \tag{4.5}$$

If the productivity index as defined in Eqs. 4.2–4.5 is to remain constant during a flow test, several conditions must be met. A particular flow rate must be maintained until the producing time at this rate exceeds the stabilization time defined as:

$$t_s = \frac{0.04\phi\mu c r_e^2}{k} \tag{3.50}$$

For this application r_e should be based on the distance to the furthest drainage boundary of the well.

In applying Eq. 3.50, care should be exercised to ensure that the effective compressibility is used, especially when substantial gas saturations exist in the reservoir (see chapter 3). The saturations normally are calculated by material balance. An equation for the calculation of the oil and gas saturations can be found in chapter 7. Actually, the stabilization time necessary to reach steady-state flow is somewhat longer than that indicated by Eq. 3.50. However, no significant error is introduced by using Eq. 3.50 for this purpose since the flow regime is nearly in complete steady state at this time.

If producing times for each rate do not exceed the stabilization time, we cannot expect to obtain an accurate productivity index. A straight-line relationship between the rate and drawdown may still be obtained because this method is insensitive or an isochronal effect occurs if about the same producing times are used for each flow rate. The isochronal effect is roughly one of affecting about the same amount of the reservoir at each rate so the final drainage radius at each rate is about the same. This provides a basis for a straight-line plot of drawdown versus rate, but this plot is not indicative of the productivity index. Using such a value results in predictions that are overly optimistic. Thus, the common practice of determining a productivity index from DST flow data is misleading and should be avoided.*

Another factor that can keep the measured productivity index from being constant during a test is the effect change in gas saturation has on the permeability to oil.[1] As mentioned, the bulk of the pressure drop in the radial flow system is near the wellbore. Consequently, in this area the largest change in gas saturation occurs. Unfortunately, this is also the area where a change in the effective permeability to oil will have the biggest effect on the average permeability of the drainage system. Since oil is the wetting phase relative to gas, small changes in the gas saturation can cause large changes in the relative permeability and in the corresponding effective permeability. This reduction in permeability can be observed even when the gas saturation is less than the equilibrium gas saturation, so no free gas flows and there is no effect on the produced gas-oil ratio.

Many engineers erroneously believe that they can ascertain a substantial change in the gas saturation around the wellbore by observing the produced gas-oil ratio. However, most produced gas originates in the reservoir, away from the area where the gas saturation increases around the wellbore. Consequently, although the permeability to gas increases with the increase in gas saturation around the wellbore, there is no gas to replace this mobile gas except gas coming out of solution as the oil passes through the low pressures near the well. Thus, there is no effect on the produced gas-oil ratio because the source of the produced gas-oil ratio—the area well removed from the well-

*The isochronal effect is discussed in detail in chapter 5.

bore—does not change pressure significantly during a well test encompassing a few days at the most.

Another difficulty that often prevents the productivity index from being constant during a test is the basic Darcy equation. Remember that the Darcy equation, which is the basis for steady-state and pseudosteady-state equations, is not valid for turbulent flow. This limit normally does not affect the applicability of equations for the flow of liquids when the flow rate is less than possibly 50 b/d/ft of sand. However, turbulence can cause a deviation in the productivity index, and the 50 b/d/ft of sand rule is just a guideline. Significant turbulence can occur at much smaller rates if the right combination of permeability, porosity, and pore-size distribution is encountered.

However, some engineers who have worked with wells in Saudi Arabia—where well rates of 40,000 b/d are not unusual—report that they have not observed turbulence effects in any wells, although producing rates may be as much as 200 b/d/ft of formation thickness. Whether this represents an exhaustive analysis or is the result of not being unduly concerned about 40,000 b/d wells is not known.

In viscous or streamline flow the pressure drop is proportional to the mass flow rate as indicated in the Darcy-based equations and the productivity index. In turbulent flow the pressure drop is roughly proportional to the square of the mass flow rate.

Insufficient producing times at each rate, a change in the gas saturation surrounding the wellbore, or turbulence may prevent productivity-index tests from indicating a constant productivity index. These factors may also cause a variation in the productivity index when it is measured at different times or stages of depletion. The sufficiency of the producing time at each rate is not particularly associated with the stage of depletion, except that the stabilization time increases almost proportionately with increases in the average gas saturation in the reservoir. Consequently, the higher the reservoir gas saturation is, the more the engineer should be concerned about the stabilization time.

Since in solution-gas-drive reservoirs the gas saturation increases as the reservoir is depleted, we can also emphasize that more consideration should be given to the stabilization time as the well life increases. However, turbulence—which seldom makes problems during oil well tests—is even less likely to cause difficulties during the later life of a well when the total production capacity may have declined considerably because the average reservoir pressure has declined.

The same change in gas saturation around the wellbore that may keep the productivity index from being a constant during testing may have an even more pronounced effect because of normal pressure depletion of the reservoir. During testing, there may be a change in saturation around the wellbore caused by pressure drawdown. During depletion of a solution-gas-drive reservoir, we are certain to have a change in the gas saturation, not only at the wellbore but throughout the reservoir. This change in gas saturation and pressure changes the mobility, k_o/μ_o, and the oil formation volume factor, B_o. Consequently, if we can predict the change in the saturation and reservoir pressure, we can predict the change in the productivity index using a ratio of the mobilities and the oil formation volume factors.[2]

$$\frac{J_1}{J_2} = \frac{(k_o/B_o\mu_o)_1}{(k_o/B_o\mu_o)_2} \tag{4.6}$$

Many engineers simply use the flow rates and corresponding well pressures to calculate the productivity index at each rate to see if there is any systematic change in the values. However, to give as much statistical strength as possible to the interpretation, it is recommended that a plot of rate versus well pressure be constructed as illustrated in Fig. 4–1. The straight-line portion of the data represents the productivity index of this well. The productivity index can be calculated by evaluating the slope of the straight-line portion of the data and recognizing that the productivity index is this slope.

Also, note that extrapolation of the curve to a q of zero yields the pressure at the external boundary, p_e. This process can be shown mathematically from the steady-state and unsteady-state equations. Physically, it simply means that at a rate of zero, the well pressure is equal to the external pressure since no flow is taking place to make the pressures different.

Eq. 4.6 can be used to predict the productivity index at any future state of reservoir depletion. Eqs. 4.2–4.5 can be used to calculate kh/μ, kh, or the permeability, k, depending on the engineer's confidence in the independent evaluation of the thickness, viscosity, and skin factor, S. If Eq. 4.4 or Eq. 4.5 is used to calculate the permeability, the average permeability can be used to determine the damage ratio, DR. In chapter 2 the damage ratio is defined as the producing rate with no damage divided by the actual producing rate, or the producing rate with damage. This ratio is proportional to the ratio of the undamaged permeability to the average permeability:

$$DR = \frac{q_{undamaged}}{q_{actual}} = \frac{(k_oh/\mu_o)_{undamaged}}{(k_oh/\mu_o)_{actual}} = \frac{(k_o)_{undamaged}}{(k_o)_{actual}} \tag{4.7}$$

Thus, if the undamaged permeability can be determined from a pressure-buildup or drawdown test and the average permeability can be determined from a productivity-index test, the damage ratio can be calculated. This procedure may sometimes be preferred to evaluating

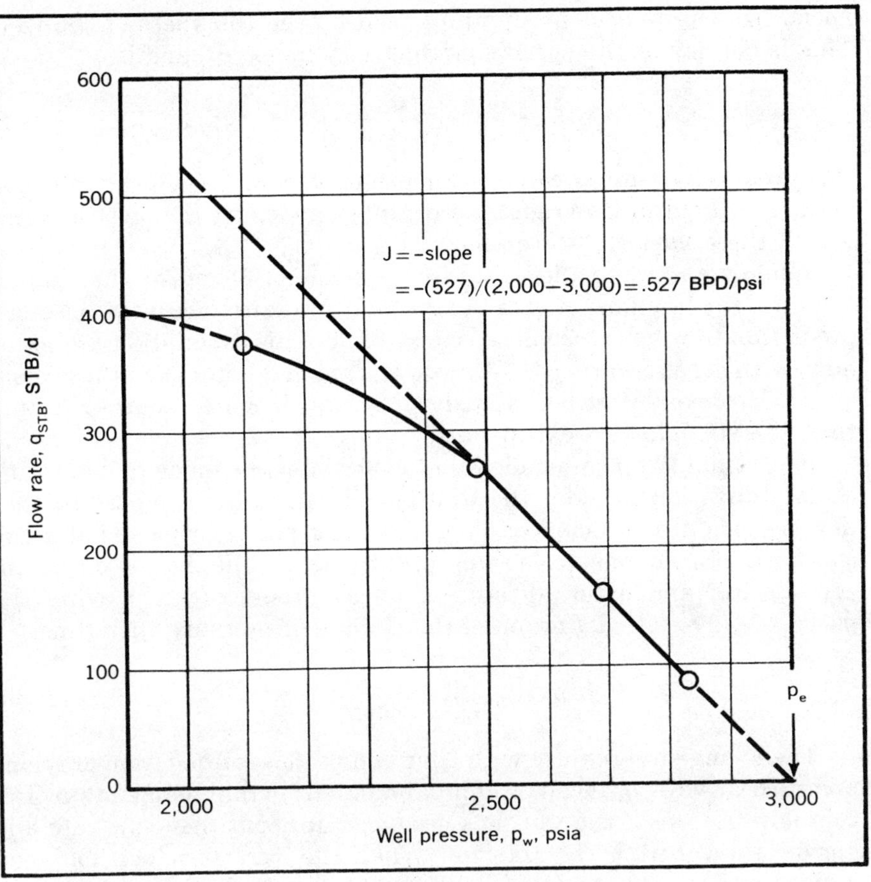

Fig. 4—1 Productivity index

the skin factor, S, since determining S from a drawdown or buildup requires a knowledge of the average compressibility.

The productivity index can be useful in determining wells that have not been properly completed or that have sustained damage such as paraffin buildup, caving, or some mechanical problem. A comparison of the productivity indices obtained at different times indicates all of these problems except the damage or improvement of the original well completion. If a measured productivity index has an unexpected decline, one of the indicated problems should be investigated. A comparison of the productivity indices run on different wells in a particular reservoir at about the same stage of depletion should also indicate the wells that may have experienced unusual difficulties or damage during completion. Since the productivity indices may vary from well to well because of the variation in thickness of the reservoir, it is helpful to normalize the indices by dividing each by the thickness of the well. This is defined as the specific productivity index, J_s, and is:

$$J_s = \frac{J}{h} = \frac{q_{stb}}{(p_e - p_w)h} \tag{4.8}$$

The productivity index can also be normalized for known variations in permeability, drainage radius, or drainage geometry in much the same way if these variations are known.

When we must predict rates in the range of flow tests that fall off the straight-line plot of rate versus well pressure, we must base our predictions on the extrapolated curved line. This calculation cannot be done with great accuracy for drawdowns much greater than those measured. However, it still obviously gives much more accurate results than the straight-line extrapolation.

It is strongly recommended that every effort be made to determine the stabilization time for the well tested rather than to assume that flow periods are sufficiently long. The most practical means of doing this for a solution-gas-drive reservoir is to determine the porosity, compressibility, and external-radius-squared product ($\phi c r_e^2$) using the pseudosteady-state definition of the change in pressure with time:

$$(\Delta p/\Delta t)_{pseudo} = \frac{1.79q}{\phi h c r_e^2} \tag{3.33}$$

The change in pressure with time can be determined from previous pressure surveys, or it can be approximated from the change in surface-tubing pressures if the well has produced at about the same rate and gas-oil ratio during the interim. When the $\phi c r_e^2$ product has been determined, it can be used in Eq. 3.50 to check the stabilization time.

A productivity-index test is often used to predict the production rate obtainable from a flowing well if that well is placed on pump, to predict the results of putting a high-capacity submersible centrifugal pump in a well that is producing large amounts of water, or to size pumps for waterflood production. In the last two cases the production rate must include the produced water.

Once the capacity of a reservoir has been determined by a productivity-index test, it must be matched with the capacity of a lift system. The simplest approach to this problem is to assume that a pump can maintain a bottom-hole pressure close to atmospheric and predict the production rate under these conditions. Another approach assumes that a flowing well has a bottom-hole pressure equal to a hydrostatic column of stock-tank oil extending from the bottom of the hole to the surface. This calculation roughly assumes that the lightening effect of gas coming out of solution in the tubing is equal to the pressure drop caused by flow through the tubing. This is a crude approximation of the bottom-hole pressure that is a function of the gas-oil ratio, the production rate, the tubing size, and numerous other factors.

Obviously, these methods leave much to be desired in the way of accuracy and do not pretend to be applicable to such methods as gas lift, plunger lift, and free pumps.*

To understand the principles of the productivity-index tests better, work the following problem and check the solution against that in appendix C.

PROBLEM 4.1: Productivity-Index Evaluation

A. The flow-test data plotted in Fig. 4–2 is obtained from a fractured well in a solution-gas-drive reservoir. This well has produced for 1 month. It drains about 40 acres and has a well radius of 4 in. The undamaged permeability to oil as determined from a DST pressure buildup is 20.5 md, and B_o is 1.25. The net thickness is 14 ft, and the reservoir oil viscosity is 1.2 cp. Plot the productivity index versus p_w. What is the damage ratio based on the straight-line portion of Fig. 4–2 data? Estimate the maximum pumping rate. What is the maximum pumping rate indicated if the test rates are all less than 270 b/d?

B. It is estimated that production of 35,000 stb of oil from this reservoir will result in a pressure at the well drainage boundary of 2,800 psia and a reservoir gas saturation of 5%. k_{ro} will be 0.5, k_{rg} will be 0.0, μ_o will be 1.1 cp, and B_o will be 1.15. Use k_{ro} in part A as 1.0. Plot the productivity index versus p_w at this time. Estimate the maximum flowing and pumping rates.

C. Productivity-index tests are planned for the well at the time described in part B. The flowing tubing-head pressure has been declining at a uniform rate of about 3 psi/d while producing at 75 stb/d. Find the product, $\phi c r_e^2$. What stabilization times are required?

The Vogel IPR curve. The difficulties of extrapolating a curve of rate versus well pressure in a solution-gas-drive reservoir are obvious from the previous discussion. A popular method of achieving this objective is the Vogel inflow performance relationship (IPR).[3] Recognizing the difficulty of the subject analysis, Vogel used a reservoir computer model to account for the change in gas saturations around the wellbore associated with the increase in drawdown. He interpreted these data to obtain a dimensionless relationship that can be used to extrapolate a rate-versus-well-pressure relationship supposedly for any well.

Vogel assumed that Fig. 4–3 applies to all solution-gas-drive wells. His study showed that the curve fit behavior accurately during the

*A comprehensive treatment of these lifting problems is beyond the scope of this book. For more information refer to *Principles of Oil Well Production*, by T.E.W. Nind (New York: McGraw-Hill, 1964) and *The Technology of Artificial Lift Methods*, by Kermit E. Brown (PennWell, 1980).

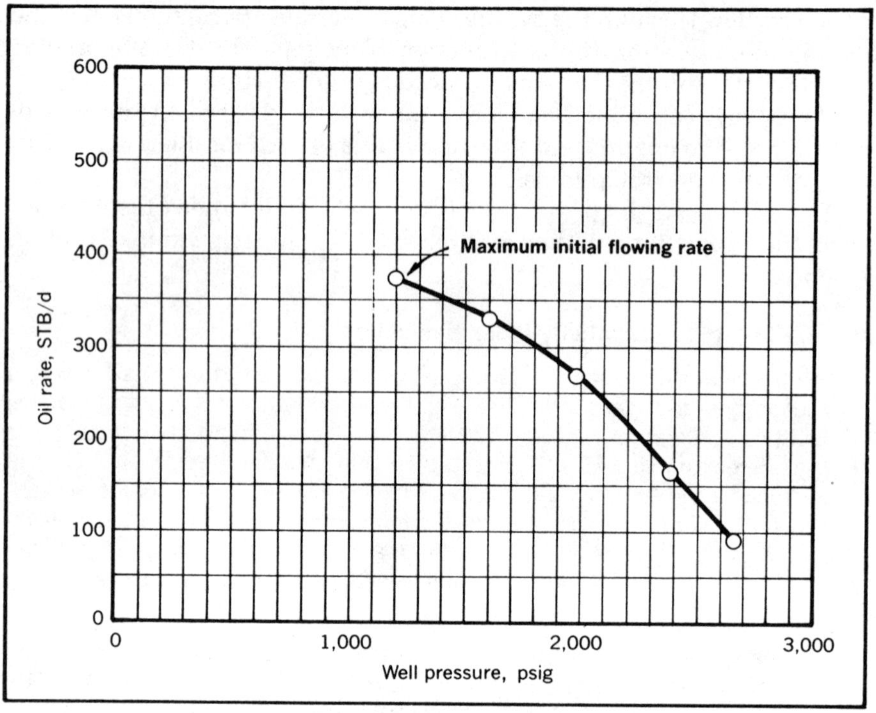

Fig. 4–2 Producing rate versus bottom-hole pressure

early depletion of a solution-gas-drive reservoir, but as pressure depletion increased, Fig. 4–3 became less accurate percentagewise. However, he noted that in terms of absolute values of barrels per day, the error for the reservoir nearing primary depletion is small because the flow rates at this time are characteristically low.

If the Vogel IPR curve applies, it is obvious that, theoretically, only one flow rate and corresponding pressure need be measured, assuming that the average reservoir pressure is known. Using the well pressure and the average reservoir pressure, Fig. 4–3 can be entered and a ratio of producing rate to maximum producing rate can be obtained. Then using the measured well rate, the maximum well rate can be calculated. Once this has been established, the rate can be determined for any assumed well rate.

Work problem 4.2 and check the solution against the one given in appendix C.

PROBLEM 4.2: Using the Vogel IPR Curve

Using the data from problem 4.1, part A, determine the maximum producing rate indicated by each data point in Fig. 4–2 (five rates). What is the best estimate of the maximum producing rate based on the IPR Vogel curve? Why? What is the producing rate if the producing bottom-hole pressure can only be lowered to 500 psig?

It seems obvious that using a multipoint productivity-index test to establish the data trend is preferable to a single-point test and the Vogel IPR curve. However, if the test data do not extend into the curved portion of the relationship, consideration should be given to using the IPR Vogel curve.

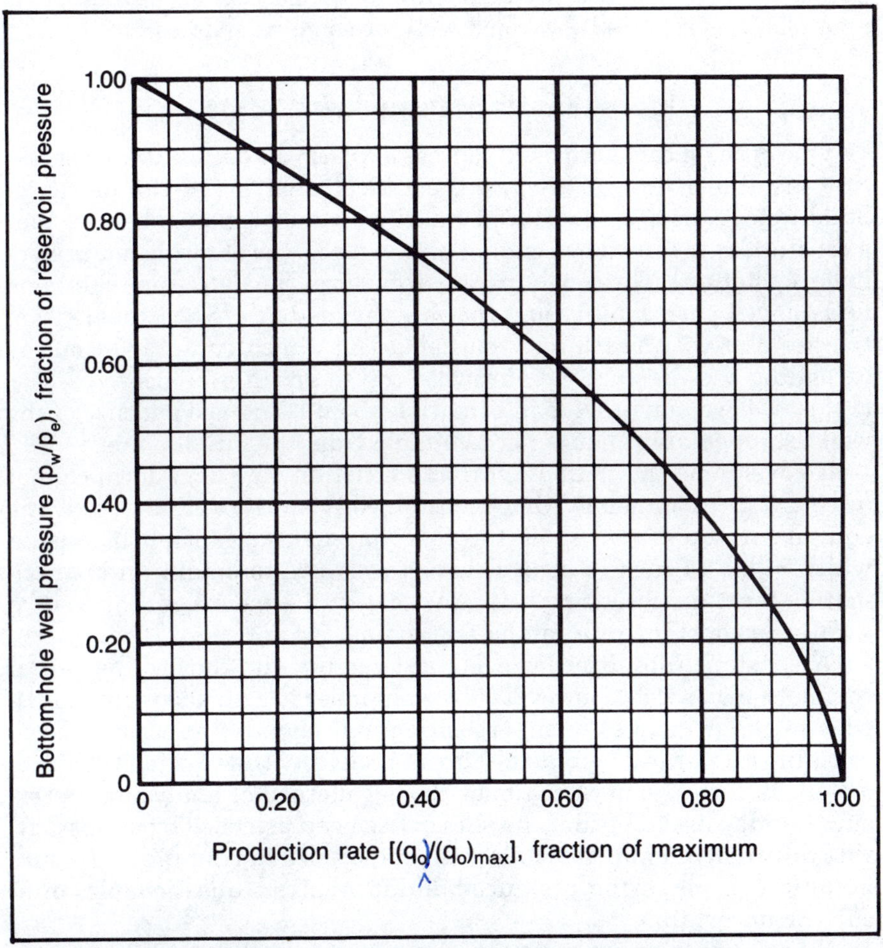

Fig. 4—3 Dimensionless inflow performance relationship for solution-gas-drive reservoirs (after Vogel, "Inflow Performance Relationships for Solution-Gas Drive Wells," courtesy *JPT*, January 1968, © SPE-AIME)

In summary, it is recommended that productivity-index data be obtained for several points that extend to a rate as high as possible. The curve extrapolation should be compared with the extrapolated values indicated by the Vogel IPR curve when the well is dominated by solution-gas drive. Good judgment also should be used in estimating the rates that prevail at higher drawdowns.

If we are running many productivity tests for a particular reservoir or a particular producing formation, we may wish to establish some empirical correction for the data. For example, we may want to base our data extrapolation on a maximum producing rate that is some fraction of the rate indicated by the straight-line portion of the data. This approach has reportedly worked well for some companies.

Constant-Rate Drawdown Tests

One of the most overlooked sources of reservoir data is the constant-rate drawdown test. Since the theoretical analysis of the pressure-buildup test has been developed and since some personnel believe that a constant-rate drawdown cannot be run, this type of test is not used to its full potential. The constant-rate drawdown can determine the same undamaged permeability and the skin factor that are obtainable from the buildup with much less mathematical difficulty. In addition, by extending the constant-rate drawdown into an interference-type test, the drawdown can provide information about the shape and size of the well drainage area that is not available from the buildup analysis.

To overcome the contention that a constant-rate drawdown cannot be run, it is recommended that the engineer visit the wellsite, study the equipment setup, and show the field foreman or roustabout pusher which valve to turn to control which pressure to obtain an approximately constant producing rate. If we can find someone willing to turn a valve, a constant rate can be approximated.

A constant-rate drawdown is most readily run during the initial completion of a well. However, if the engineer has an adequate knowledge of the principles of superposition and negative superposition, a constant-rate drawdown can be run at about any time during the life of a well. In fact, the most common Horner method of analyzing a pressure buildup assumes that the shutin has been preceded by a constant-rate drawdown. Failure to follow this procedure in running a pressure buildup can make the pressure-buildup analysis questionable, difficult, or uncertain.

When running a constant-rate drawdown test, the ideal situation is to have the drainage-area pressure uniform throughout at the start of production. Since most constant-rate drawdown surveys are performed during the initial production test on a new well, it is also important that the damage in the well be stabilized prior to initiating the test. In practical terms this means that a new well should be initially produced until it quits cleaning up. This point can generally be observed by noting when the flowing bottom-hole pressure or flowing tubing-head pressure declines in a uniform manner or when one pressure appears to be constant. This latter condition would actually mean that the well pressure is declining at a rate too small to be readily discernible. Although this initial flow period should be sufficient to stabilize or stop the change in the well damage, it should represent a cleanup period as short as possible. Then when the well is shut in prior to initiating the constant-rate drawdown, it can return to an approximately uniform pressure throughout the drainage area as quickly as possible.

This shutin following the initial cleanup flow period may get the engineer in trouble with the manager because there is political and economical pressure to get a new well on production. However, as discussed in chapter 1, we may obtain reservoir data from this initial flow test that we will never be able to measure as easily, again, if at all. The pressure to shorten or minimize the initial flow test seems to be greatest on a discovery well in a reservoir where the data obtainable from the constant-rate drawdown is most important.

Actually we will note that it is not necessary to shut in the well until it comes back to the initial pressure but only until the bottom-hole pressure versus the shutin time can be accurately extrapolated.

If we assume that the constant-rate production is initiated with the initial pressure, p_i, throughout the reservoir, the flowing bottom-hole pressure, $p_{w,t}$, is governed by the constant-rate solution to the radial diffusivity equation in chapter 3:

$$p_{w,t} = p_i - \frac{0.141q\mu}{kh}[p_{tD} + S] \tag{3.17a}$$

The p_{tD} pressure function can be used since we are considering the pressure at the well. Also, when t_D is greater than 100—which occurs in seconds or minutes in most wells—and the well is infinite acting, Eq. 3.25 applies:

$$p_{tD} = \tfrac{1}{2}(\ell n\ t_D + 0.809) \tag{3.25}$$

Of course, the well will be infinite acting as long as the producing time is less than $(r_e^2/4\eta)$. If Eq. 3.25 is substituted into equation 3.17a, we obtain:

$$p_{w,t} = p_i - \frac{0.141q\mu}{kh}[0.5(\ell n\ t_D + 0.809) + S] \tag{4.9}$$

Also:

$$\ell n\ t_D = \ell n\ (\eta t/r_w^2) = \ell n\ (\eta/r_w^2) + \ell n\ t \tag{4.10}$$

Then if we make this substitution into Eq. 4.9, we obtain:

$$p_{w,t} = p_i - \frac{0.141q\mu}{kh}\left[0.5\left(\ell n\,\frac{\eta}{{r_w}^2} + \ell n\,t + 0.809\right) + S\right] \tag{4.11}$$

Note that the only expression that changes with time in the right-hand side of Eq. 4.11 is $(0.141q\mu/kh)\ 0.5\ \ell n\ t$. If we group all of the other right-hand expressions into a constant, we obtain:

$$p_{w,t} = \text{constant} - \frac{0.141q\mu}{kh}(0.5)\ \ell n\ t \tag{4.12}$$

Eq. 4.12 shows that a plot of $p_{w,t}$ versus $\ell n\ t$ gives a straight line whose slope is the multiplier of the $\ell n\ t$. However, most engineers persist in using log paper for such plots, and most log paper is to the base 10 rather than to the base e. Therefore, it is advisable to substitute 2.303 log t for $\ell n\ t$ in Eq. 4.12 and reevaluate the slope to obtain:

$$m = 0.1625q\mu/kh \tag{4.13}$$

Therefore, by plotting the well pressure versus the log t, we obtain a straight line whose slope is as defined in Eq. 4.13. After evaluating the slope from the graph, the undamaged permeability, k, can be calculated from Eq. 4.13.

Note that the expression $(0.141q\mu/kh)$ in Eqs. 4.9 and 4.11 is equal to 0.867 m. The constant 0.867 is necessary to convert the 0.1625 in m to 0.141. If we substitute 0.867 m for its equivalent expression in Eq. 4.9, we obtain:

$$p_{w,t} = p_i - 0.867\ m\ [0.5(\ell n\ t_D + 0.809) + S] \tag{4.14}$$

Then if we read a pressure $p_{w,t}$ and its corresponding log t from the straight-line plot of $p_{w,t}$ versus log t, we can use these expressions in Eq. 4.14 to calculate the skin factor, S. Of course, the log t must be converted to $\ell n\ t_D$ before it is used in Eq. 4.14. We may prefer to convert Eq. 4.11 to a function of m for this purpose:

$$p_{w,t} = p_i - 0.867\ m\left[0.5\left(\ell n\,\frac{\eta}{{r_w}^2} + \ell n\,t + 0.809\right) + S\right] \tag{4.15}$$

Work problem 4.3 and check the solution against that in appendix C.

PROBLEM 4.3: Constant-Rate Drawdown Test

Fig. 4–4 is a plot of a flowing well pressure versus time on semilog paper. After about half of a day, the well behavior is affected by a nearby barrier in the reservoir. Using the early straight line of the plot, determine the undamaged permeability and the skin factor, S. The reservoir data are as follows:

Initial pressure = 2,800 psia
Production rate = 80 stb/d
Oil formation volume factor = 1.25 res bbl/stb
Effective porosity = 0.20
Formation thickness = 10 ft
Saturation pressure = 1,800 psia
r_w = 0.25 ft
μ = 1.2 cp
c = 10^{-5}/psi

As noted, it is easier to analyze the data from a constant-rate drawdown test if the well pressure is equivalent to the pressure throughout the reservoir when the drawdown is started. However, this condition is not always necessary and may even be undesirable if a long shutin time is required for the well pressure to approach the average pressure in the drainage area.

If the well pressure is shut in for long enough so it can be accurately extrapolated during the following flow period, the drawdown-pressure data can be interpreted as accurately as the data taken when the well pressure equals the average reservoir pressure. The extrapolated pressure is the well pressure that would have existed if the constant-rate drawdown had not been commenced. Thus, the difference between the pressure that would have existed without the constant-rate drawdown and the pressure that exists during the drawdown is the effect of the drawdown alone. A plot of these pressure differences versus the log of time can then be interpreted in a manner similar to that used for the regular constant-rate drawdown.

The slope of the plot of the pressure difference versus the log of the producing time is indicated in Eq. 4.13. The pressure difference can be used as $p_i - p_{w,t}$ in Eq. 4.14 or Eq. 4.15 to calculate the skin factor, S. For example, designating the extrapolated pressure as $p'_{w,t}$, Eq. 4.14 can be written as:

$$p'_{w,t} - p_{w,t} = 0.867\ m\ [0.5\ (\ell n\ t_D + 0.809) + S] \qquad (4.14a)$$

This negative superposition technique may greatly simplify running and interpreting a drawdown test in many cases. If the engineer exercises reasonable diligence and control over a flow test, a conventional multiple-rate flow-test analysis may be unnecessary.[4]

When discussing pressure-buildup analysis, afterflow effects must be considered. When a well is shut in for a pressure buildup, we want the rate to go to zero instantaneously at the face of the producting formation in the bottom of the well. However, since the well is shut in at the surface, the flow from the reservoir into the well continues until the gas in the wellbore is sufficiently compressed. Then the pressure in the well is large enough for flow from the reservoir to be negligible.

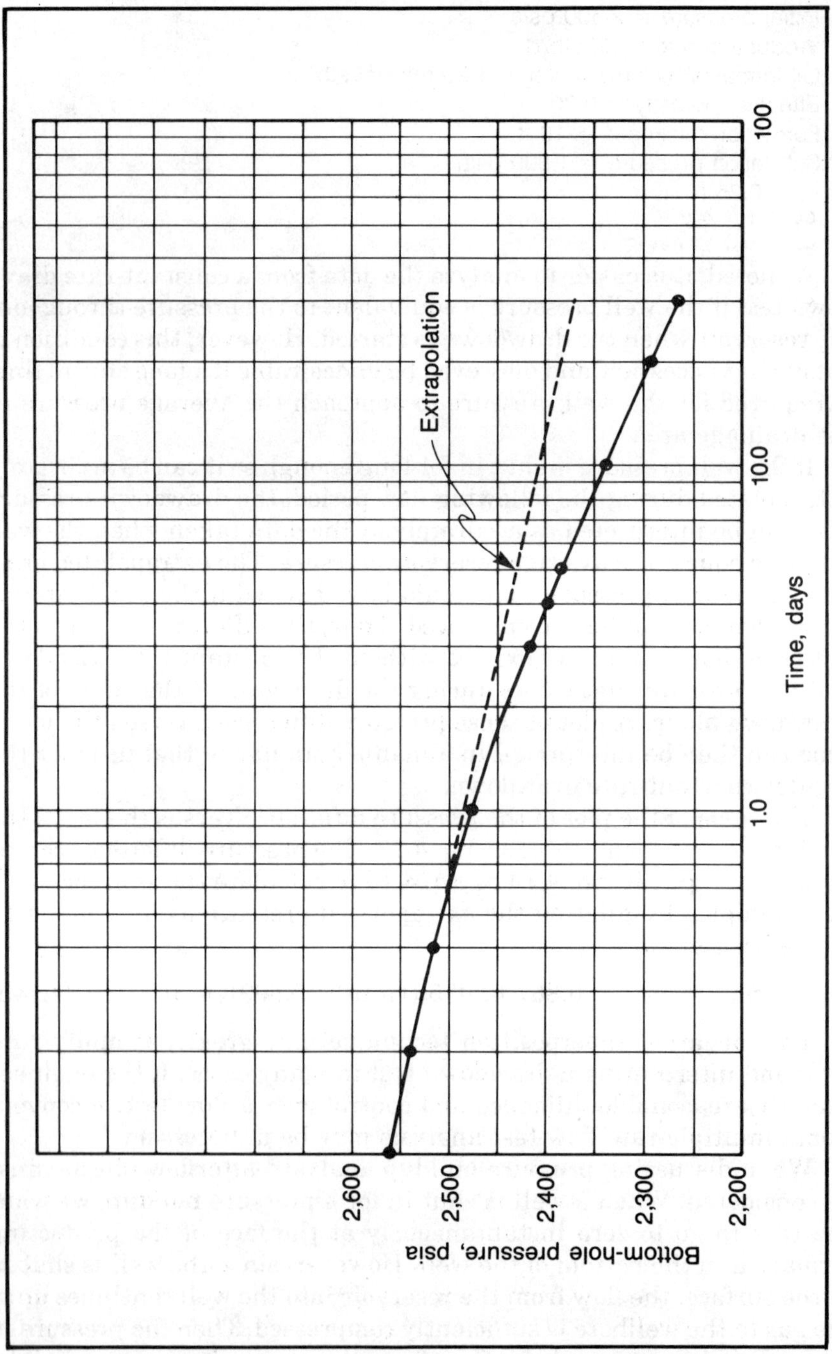

Fig. 4–4 Effect of a plane barrier on a drawdown

This afterflow effect can be extremely important, to the extent that in a very tight gas reservoir it may be impossible to run a meaningful pressure-buildup test.

Similar afterflow difficulties exist in a constant-rate drawdown test, but they are not generally as important. To demonstrate the qualitative effect, consider a very tight high-pressure gas reservoir with a large well (tubing) volume. Such a well does not produce at a high rate relative to the size of the well volume. Consequently, when production is initiated at the surface at a relatively low rate, there is no immediate significant pressure drop in the well. Therefore, the flow into the wellbore from the formation is negligible. Since a large pressure drop is necessary between the tight reservoir and the wellbore to obtain a substantial flow rate from the reservoir, some substantial time period may be needed before the flow into the wellbore is about the same as the flow from the wellbore at the surface.

With a well producing from a normal reservoir, afterflow is seldom a problem in a constant-rate drawdown. To determine when afterflow becomes insignificant, the log of the pressure difference between the reservoir and the bottom of the well can be plotted versus the log of time. While flow from the reservoir is negligible, such a plot exhibits a straight-line relationship with a slope of 1.0 and this portion of the drawdown data can be disregarded. The straight-line relationship between the bottom-hole pressure and the log of the producing time can be anticipated later. The theory behind this analysis of the afterflow will be discussed.

The upper time limit on the straight-line plot of the bottom-hole pressure versus log t is the time limit on the $0.5(\ell n\, t_D + 0.809)$ approximation of the pressure function used to develop Eq. 4.9. This limit occurs when the producing time is greater than $(r_e^2/4\eta)$, where r_e is the distance to the closest drainage boundary.

As noted, a constant-rate drawdown test can be used as an interference test. This test also can be interpreted to determine as much as possible about the drainage area shape, size, and volume.*

Reservoir Limit Tests

Reservoir boundaries cause anomalies in pressure behavior. Several methods have been devised to obtain the reservoir shape and size

*Other interpretation methods are available for analyzing the pressure behavior after the drainage boundaries have been felt at the well. However, these analytical methods rely on a simple cylindrical geometry, and the author has not found them to be practical in determining the undamaged permeability and skin factor. For a discussion of these methods, refer to *Pressure Buildup and Flow Tests in Wells,* by C.S. Matthews and D.G. Russell, SPE Monograph No. 1, (1967).

from an interpretation of pressure behavior. Many of these methods are theoretically sound but are useless from a practical standpoint.

Also, misinterpretations have been caused by an engineer's inability or carelessness in using a quantitative evaluation. Many looked-for anomalies have been too small for a conventional Amerada bottom-hole pressure gauge to record. Some quantitative calculations would have shown these erroneous conclusions. Often, the engineer has mistaken the anomaly caused by a change in gas saturation around the wellbore or an offset well for an anomaly caused by a reservoir boundary. Such problems are associated with reservoir limit or interference tests and recommended analysis methods are discussed.

The most useful test in an evaluation of reservoir limits is the drawdown test. As noted, the ideal testing method is to start producing a well in a reservoir at a constant rate with the pressure at the time production is begun being uniform throughout the reservoir. For most practical cases this situation exists only at the time a reservoir is discovered. However, the test can be run at other stages of depletion if the shutin pressure prior to the test can be accurately extrapolated.

Now consider the effect a single plane boundary would have on the flowing pressure of a well that is produced at a constant rate. A well that is a distance d from a plane reservoir boundary, as in Fig. 4–5, has the same behavior as either of the wells a distance 2d apart in an infinite-acting reservoir if both wells produce at identical rates. This relationship is true because there is no flow across the line representing all of the points equidistant from the two wells, as shown in Fig. 4–5. Therefore, the flow geometry into the wells in the infinite-acting model is exactly the same as the flow geometry into the actual well.

To analyze the pressure behavior in the infinite-acting model with both wells producing at a constant rate, we must apply the constant-rate solutions to the radial diffusivity equation, Eq. 3.12. By employing superposition, the well pressure at any particular time is equal to the initial pressure minus the pressure drop caused by the flow rate q at the well radius, minus the additional pressure drop caused by the well damage, minus the pressure drop caused by the rate q in the image well at a radius of 2d:

$$p_w = p_i - \frac{0.141 q\mu}{kh} \Delta p_{D\ real} - \Delta p_{skin} - \frac{0.141 q\mu}{kh} \Delta p_{D\ image} \tag{4.16}$$

The only way that we can evaluate the pressure drop caused by the flow rate q at a radius 2d is to use the Ei-function solution since we do not know the flow rate at the radius 2d. Do not be confused by the fact that in the infinite-acting model the real well is producing at a rate q at a distance 2d from the image well. We account for this flow in the

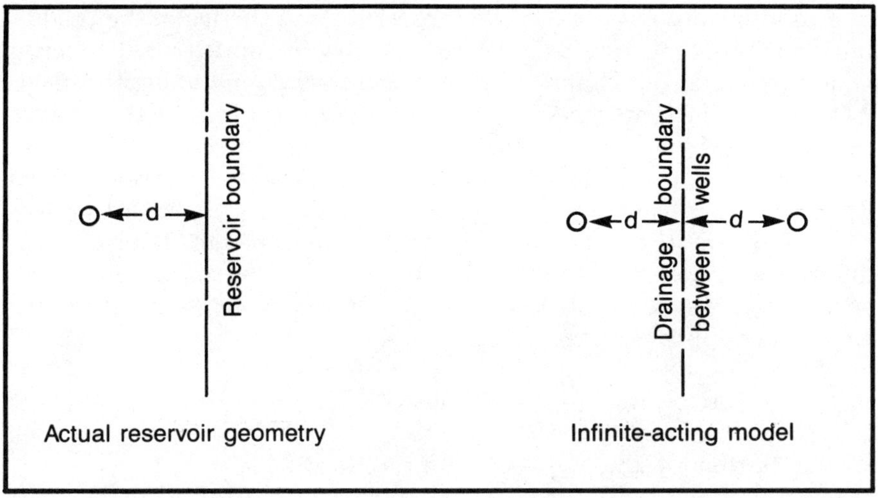

Fig. 4–5 Simulating a plane barrier in an otherwise infinite-acting reservoir

second term of Eq. 4.16, but the term has nothing to do with finding the effect of the flow from the image well at a radius of 2d.

In this case considering the last term of Eq. 4.16, we do not know the rate equivalent of the image well rate at a radius of 2d, so we cannot use the p_{tD} function to calculate the effect of the image well at the radius 2d. We must then employ the exponential pressure function, Eq. 3.19, for this purpose. Furthermore, since we can only use the exponential pressure function when the well is infinite acting, we can use the ℓn equation, Eq. 3.25, to calculate the p_{tD} function used in evaluating the first term. To use the log function it is of course necessary for the reduced-time basis for the function to be greater than 100. However, this condition normally occurs in a few minutes or seconds since the reduced time is based on the small well radius that makes the reduced time increase rapidly. Thus, using the log function for the real pressure function and the Ei function for the image pressure function, we can expand Eq. 4.16 to:

$$p_w = p_i - \frac{0.141 q\mu}{kh}\ \tfrac{1}{2}\left(\ell n\ t_{D\ real} + 0.809\right) - \Delta p_{skin} - \frac{0.141 q\mu}{kh}\ \tfrac{1}{2}\left(\text{Ei}\ \frac{-1}{4t_{D\ image}}\right) \tag{4.17}$$

An actual well pressure from a drawdown test, when plotted against the log of the producing time, has the characteristics illustrated in Fig. 4–4 if there is a plane reservoir boundary nearby. After a brief period corresponding to the time for $t_{D\ real}$ to exceed 100, there is a straight-line relationship between the well pressure and the log of the producing time. By examining Eq. 4.17, we can see why this is true. The terms p_i and Δp_{skin} do not change with time, so they do not affect the slope. Also remember that it takes some finite period of time for the production from the image well to affect the pressure at the radius 2d. That is, the function is essentially zero until the reduced time exceeds about 0.1.

Note that if we extend the early straight-line portion of this plot, the difference between this extrapolation, p_w', and the actual well pressure—after we have felt the effect of the reservoir boundary at the well—is equal to the last term of Eq. 4.17:

$$p_w' - p_w = \frac{0.141 q\mu}{kh} \, ½ \left[- \text{Ei}\left(\frac{-1}{4t_{D\ image}} \right) \right] \tag{4.18}$$

Also note that $t_{D\ image}$ is based on a radius of 2d:

$$t_{D\ image} = \frac{6.33kt}{\phi\mu c(2d)^2} \tag{4.19}$$

These equations and observations then allow us to determine the distance to a reservoir boundary. If we plot the well pressure versus the log of the producing time as in Fig. 4–4, extrapolate the early straight-line portion, and read the difference between the extrapolated straight line and the actual recorded pressure at some noted producing time, we can substitute this value for $p_w' - p_w$ in Eq. 4.18 to solve for the entire Ei function. Note that this factor includes everything in the brackets. Using the plot of the Ei function versus reduced time in Fig. 3–9, we can find a corresponding reduced time, $t_{D\ image}$. Putting this reduced time in Eq. 4.19, we can then calculate the distance to the boundary, d.

This calculating procedure can be made simpler using the early straight-line plot to evaluate $0.141q\mu/kh$. Before the well pressure is affected by the boundary, the performance is simply that of an infinite-acting constant-rate drawdown, and the slope of the pressure-versus-log of producing time plot is:

$$m = \frac{0.1625 q\mu}{kh} \tag{4.13}$$

Now we can use a function of the slope in Eq. 4.18 to obtain:

$$p_w' - p_w = 0.867 \text{ m } ½ \left(-\text{Ei} \, \frac{-1}{4t_{D\ images}} \right) \tag{4.20}$$

The slope can also be used to evaluate the mobility, k/μ, for Eq. 4.19 to calculate the distance d from the reduced time, $t_{D\ image}$. The details of this calculating procedure may be clarified by solving problem 4.4 and comparing the solution with the one in appendix C.

PROBLEM 4.4: Determining the Distance to a Reservoir Barrier from a Drawdown Test

Fig. 4–4 presents data from a constant-rate drawdown test on a discovery well drilled with a 6-in. bit. Geological evidence indicates the possibility of a nearby fault. Determine the distance to the fault using the following data:

Initial pressure = 2,800 psia
Production rate = 80 stb/d
B_o = 1.25 res bbl/stb
Effective compressibility = 10×10^{-6}/psi
Effective porosity = 0.20
Formation thickness = 10 ft
Saturation pressure = 1,800 psia

There have been many cases where drawdown data such as that in Fig. 4–4 have been erroneously interpreted to be the effect of a reservoir barrier, when the data actually resulted from another phenomenon. One of the most common effects on a drawdown that resembles the effect of a barrier is a change in gas saturation around the wellbore caused by excessive drawdown. The lower pressures around the wellbore cause a greater amount of gas to be liberated at this point in the reservoir. The increased gas saturation causes a corresponding decrease in the effective permeability to oil, and the effect is similar to that caused by a reservoir barrier. Also, interference from another producing well can create a similar effect to that of a well barrier. It may seem superfluous, but the engineer should determine whether the effect is from another well or from a barrier because this error has been repeated many times.

Since a barrier effect can be confused with other effects in the reservoir, several safeguards have been proposed to minimize the possibility of such an error. Note in Eqs. 4.16 and 4.17 that when the reduced-time base for the image well becomes greater than 100, the Ei function can be replaced by the log equation. Eq. 4.16 can then be written as:

$$p_w = p_i - \frac{0.141q\mu}{kh}\,{}^1\!/\!_2\left(\ell n\ t_{D\ real} + 0.809\right) - \Delta p_{skin}$$
$$- \frac{0.141q\mu}{kh}\,{}^1\!/\!_2\left(\ell n\ t_{D\ image} + 0.809\right) \quad (4.21)$$

Remember that the $\ell n\ t_D = \ell n(\eta/r^2) + \ell n\ t$ and the $\ell n\ (\eta/r^2)$ do not change with time although the radii for the two terms are different. In Eq. 4.21 the change in the second term with time is $(0.141q\mu/kh)0.5\ (\ell n\ t)$, which is exactly the same as the change in the value of the last term with time. Thus, the total change in well pressure with time is $2(0.141q\mu/kh)0.5\ (\ell n\ t)$ so the slope of the plot of well pressure versus the log of time is exactly double that experienced before the effect of the barrier is felt at the well.

Since the negative slope of the pressure plot versus the log of time eventually doubles, some engineers have proposed that the technique of determining the distance to a barrier be limited to those data where the slope exactly doubles. There seems to be little question that following such a procedure would keep the engineer out of trouble as far as misinterpreting the effect of a barrier is concerned. This would also mean that he would be missing many valid applications of drawdown data because in most cases the barrier must be near the well if the slope is to double in a practical length of time.

However, using the limit of t_D greater than 100 as the time when we can use the ℓn equation to determine the exponential-integral function—when the slope doubles—is ultraconservative. Fig. 3–11, which compares the ℓn, p_{tD}, and exponential-integral pressure functions, clearly shows that the ℓn equation may closely approximate the exponential-pressure function when t_D is no more than 2–3. For example, in Fig. 4–4 the slope has clearly doubled although the time of 10 days is only equivalent to a t_D of 3.0.

The recommended technique for determining if the pressure behavior observed is actually caused by a reservoir barrier or is caused by some other effect is to repeat the calculation of the distance to the barrier using different producing times and corresponding pressure differences. All of these calculations should result in the same distance to the barrier within the accuracy of the plotted data. For example, in problem 4.4 note that, at a producing time of 5 days, p_w' is 2,430 psia and the well pressure is 2,383 psia. From Eq. 4.20 we calculate an Ei-function value of 0.677, and from Fig. 3.9 the corresponding reduced time is 1.5. Putting the time of 5 days and $t_{D\ image}$ of 1.5 into Eq. 3.63, we calculate a distance to the barrier of 229 ft, which is the answer obtained in problem 3.6 using a time of 10 days. Such calculations can of course be repeated until the engineer is confident as to the nature of the pressure anomaly being analyzed.

Note that if a second barrier exists in a reservoir such that the distance to the first barrier can be established from the pressure drawdown before the second barrier begins to affect the drawdown pressure, it may be possible to determine the distance to the second barrier. For example, consider Fig. 4–6 where the distance to the nearest barrier is still d and the distance to the next closest barrier is d_2. In order to have the same drainage and flow pattern in an infinite-acting reservoir, it would be necessary to have four wells spaced as shown producing with the same rate histories.*

*A similar boundary problem is considered in chapter 3 under "Superposition." If the engineer has difficulty with this modeling of the reservoir boundaries, particularly the need for the third image well, he may restudy this section.

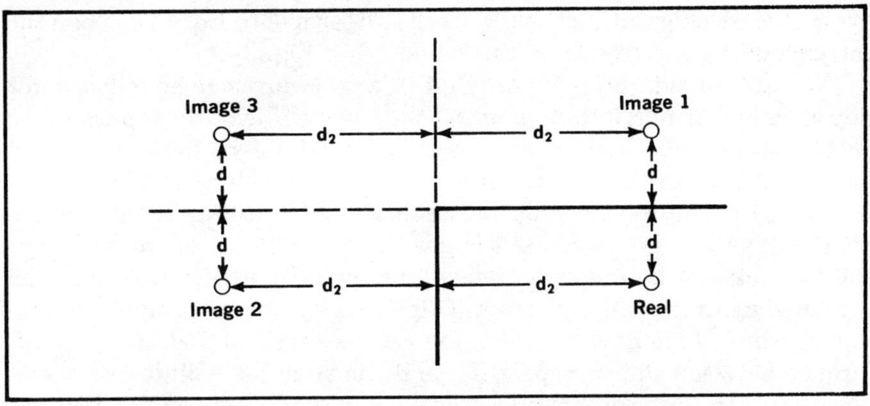

Fig. 4–6 Modeling two perpendicular boundaries

Based on the pattern in Fig. 4–6, we can write an equation for the pressure in the real well at any time, using the Ei-function equation to evaluate the pressure drops caused by the three image wells at radii of 2d, $2d_2$, and the distance of image well 3 from the real well. The distance of the latter radius is the hypotenuse of the right triangle, whose sides are 2d and $2d_2$:

$$p_w = p_i - 0.867m\ ½(\ell n\ t_{D\,real} + 0.809) - \Delta p_{skin}$$
$$- 0.867m\ ½\left(-Ei\frac{-1}{4t_{D\,image\,1}}\right) - 0.867m\ ½\left(-Ei\frac{-1}{4t_{D\,image\,2}}\right)$$
$$- 0.867m\ ½\left(-Ei\frac{-1}{4t_{D\,image\,3}}\right) \tag{4.22}$$

Note that the image-1 term becomes significant before the terms of the other image wells take on significant values. If this period is long enough, it is possible to determine the distance d to the nearest boundary as in problem 4.4. Once the distance d to the nearest boundary has been determined, the term containing the reduced time for image well 1 can be evaluated for any producing time desired. In such a case the term for image well 2 becomes significant before the term for image well 3. Thus, $p_w' - p_w$ can be evaluated for some time after the terms for image wells 2 and 3 become significant. The distance to the second closest barrier can then be calculated from Eqs. 4.19 and 4.23:

$$p_w' - p_w = 0.867m\ ½\left(-Ei\ \frac{-1}{4t_{D\,image\,1}}\right)$$
$$- 0.867m\ ½\left(-Ei\ \frac{-1}{4t_{D\,image\,2}}\right) \tag{4.23}$$

As in the previous calculation, the reduced time for image well 2 is evaluated by calculating the (½) (−Ei . . .) term with Eq. 4.23. Then the corresponding reduced time can be found in Fig. 3–9.

Careful consideration shows that this procedure can be followed for any number of barriers as long as they are sufficiently separated so each distance to a barrier can be determined before the next closest barrier affects the drawdown pressure. The procedure can be further simplified if the relationship between the distances is such that the effect of the last boundary felt is governed by the log equation before the next closest boundary is felt at the well. In such a case the well pressure again becomes a straight-line plot against the log of the producing time. This new straight line can be extrapolated, and the difference between the extrapolation and the recorded well pressure can be equated to the appropriate Ei term as in Eq. 4.20. Then the distance to the next nearest barrier can be determined as shown previously.

Determining the distance to the two closest boundaries does not give the angle between the boundaries. Also, note that it is impossible to carry the analysis further without knowing the angle between the two boundaries. If other than a right angle exists between the two, the modeling of an infinite-acting reservoir becomes more difficult. The angle between the two barriers can be determined by observing the relationship between the initial straight-line slope and the next straight-line slope. For example, in Eq. 4.22 when the reduced time for image well 3 is greater than 100 (or perhaps 3.0), all four of the pressure-drop terms are governed by the log equation. Then the slope of the well pressure versus log time is four times the initial straight-line slope when only the term for the real well is significant. This slope ratio of four is 360°/90° where the 90° is the angle between the two boundaries. If the angle between the two boundaries is 180°, only one boundary exists and the well pressure is governed by Eq. 4.21. Under these conditions the slope doubles, and the ratio of the original and final slopes is 2.0, or 360°/180°.

It can be shown that this observed relationship between the intersecting angle and the slope ratios exists for any angle between the intersecting boundaries. Turning the observation around, we find that once we have determined the ratio between the initial and final slopes (assuming that a final straight-line slope is reached), 360° can be divided by the ratio to give the angle between the two boundaries. For example, if the slope ratio is 3.0, the angle between the boundaries is 360°/3, or 120°.

The parenthetical phrase is extremely important. Nature seldom arranges the dimensions of a reservoir such that the effects of one boundary are so well separated from those of another that each will be governed by the log equation before it is confused with the next effect. It has been the author's experience that the distance to the closest boundary can be determined in most cases. Occasionally we should be able to determine the distance to the next closest boundary with a reasonable amount of confidence. Furthermore, we can seldom obtain more information from the pressure behavior with the exception of the total reservoir size.

If constant-rate production is continued until the entire reservoir has been affected, flow behavior enters the pseudosteady-state regime. Then a straight-line relationship is established between a linear plot of the well pressure versus the producing time. The slope of this plot is the change in pressure with time according to the pseudosteady-state equation:

$$(\Delta p/\Delta t)_{pseudo} = \frac{5.615q}{\phi c V_b} \tag{3.34}$$

Eq. 3.34 can be used to calculate the total reservoir volume, V_b in cubic feet. By noting the time when the pseudosteady state begins, an estimate of the distance to the furthermost point in the reservoir can be obtained. The observed time is used in the stabilization-time equation, and r_e is calculated. This r_e is the greatest distance to the drainage boundary:

$$t_s = \frac{0.04\phi\mu c r_e^{\,2}}{k} \tag{3.50}$$

This technique sounds good from a theoretical standpoint, but in practice, it leaves much to be desired because of the difficulty in determining exactly when the plot of well pressure versus time becomes straight. Since the distance desired is proportional to the square root of the time, sizable errors can result from the mechanical inability to determine this critical time accurately. It is recommended that this technique be used only in a qualitative sense with a full realization of the possible inaccuracies involved.

Many engineers use a similar technique to calculate the distance to the nearest boundary and other reservoir dimensions. They use the semilog plot of the well pressure versus time to determine when the effect of the nearest boundary is felt at the well. This time coincides with the departure of the plot from the original straight line. The subject time is then used in the stabilization-time equation, Eq. 3.50, to calculate the distance r_e to the nearest boundary.

This calculation has the same shortcomings noted for the calculation of the distance to the furthermost part of the reservoir. Also, there is the problem of accurately determining the time when the semilog plot deviates from the straight line. This is difficult enough when working with a linear plot. However, when the number desired is plotted as a log, it is much more difficult to obtain an accurate linear value. In addition, the stabilization equation is at best an approximation that gives a built-in error of unknown magnitude to this technique. Also, some independent technique such as the doubling of the slope must be used to verify that we are actually dealing with a boundary effect and not with some other phenomena. Due to these shortcomings and the availability of the method proposed in this section, it is recommended that the engineer avoid using the stabilization equation for determining reservoir dimensions, except to check other methods and to estimate the greatest distance to the drainage boundary.

One of the biggest problems associated with reservoir limit and interference tests is the failure of the engineer to determine whether the pressure anomaly desired can actually be measured with the equipment available. Many engineers need to determine if two offset wells are in the same or different reservoirs. To solve this problem, we try to observe the effect of one well on another. We may decide to double the producing rate in one well and monitor the bottom-hole pressure in the second to determine the effect of the double rate. After watching the pressure in the second well for several days, we may see no effect and may conclude that the wells are in separate reservoirs. However, subsequent development and pressure surveys may prove this conclusion erroneous.

If we had taken the time to determine qualitatively the magnitude of the pressure disturbance we were trying to observe, we would quite likely have avoided the embarrassment of this erroneous conclusion by finding that the pressure change was too small for the equipment being used. This problem was discussed in chapter 3. The procedure simply involves estimating the minimum pressure difference that can be detected with confidence and then calculating the amount of time required to obtain this pressure difference at the radius in the reservoir where the observations will be made. For details of this calculation, refer to problem 3.3.

Pressure-Buildup Analysis

Much has been written on pressure-buildup analysis, possibly more than on any other specific subject in reservoir engineering. However, because of confusion, it seems the engineer can seldom be certain that the best or even a correct method of analysis is being used. The Society of Petroleum Engineers' Monograph No. 1 is concerned mostly with pressure-buildup analysis.[5] This publication did much to organize the published methods at the time of the monograph. However, it still is the author's opinion that the monograph left a confusion of methods, the invitation to misapply some of the methods, and a dangerous set of cookbook forms that seem to imply that correct answers come from filling in forms.

Earlougher has since updated the methods in the original monograph in his 1977 SPE Monograph, *Advances in Well Test Analysis.*[6] This book gives a much more realistic approach to the problem of pressure-buildup analysis. It takes the liberty of emphasizing the practical limits on many of the published analysis methods and does not use blank forms that are to be completed. It still avoids a theoretical discussion of the inadequacies and limitations of the confusing array of published methods that apparently must be included for SPE political reasons. However, this monograph is extremely comprehensive and is referred to often in this chapter.

Probably 90% of the pressure-buildup data can be analyzed using the same method of analysis, regardless of whether the well is infinite acting, in pseudosteady state, or in steady state at the time of shutin. Most pressure-buildup methods claim to be able to determine the undamaged permeability, some measure of the well damage, and the average reservoir pressure in the drainage area at the time of shutin. The permeability can probably be evaluated with sufficient accuracy using any method, including many that are theoretically incorrect. The damage parameters are more difficult to determine accurately, and some methods for calculating the average pressure are meaningless.

The most popular methods for determining the average pressure require an independent knowledge of the drainage-area shape and size, the average porosity, the average effective compressibility, and the cumulative production—rate and time. Using these values and a function of the buildup, we can calculate the average pressure at the time of shutin. However, a careful evaluation of the data required shows that there is really no need for the buildup data since the average pressure can be directly determined by material balance using the other information. Also, methods derived for pseudosteady-state conditions are being used worldwide to determine the average pressure of wells in reservoirs that are dominated by strong water drives and are thus clearly dominated by steady-state conditions. Furthermore, the erroneous average pressures determined are in turn used to validate reservoir computer models that are then used to predict the future behavior and best operating conditions for a reservoir. Obviously, the use of improper methods in determining the average reservoir pressure can cause all kinds of difficulty for the reservoir engineer.

This inability of the reservoir engineer to evaluate average reservoir pressures accurately is probably the biggest reason for the many instances of anomalous reservoir behavior that are so often reported. Since it is difficult to determine the average reservoir pressure in a drainage area at the time of shutin, this problem is considered separately from calculations for undamaged permeability and the amount of damage.

To save the experienced engineer from shock, he is warned in advance that there is little use of the Horner method in the techniques developed and recommended in this chapter. The Horner method is theoretically correct for an infinite-acting reservoir only. Except for gas wells, the infinite-acting portion of a well life is very small. One of the few times that infinite-acting behavior characterizes flow is during drillstem tests. However, we will see that the pressure in the well during most drillstem tests is increasing at the time of shutin. The Horner solution assumes the pressure is decreasing at the time of shutin.

The 1970s may be known in the history of transient pressure behavior as the curve-fitting era. Reportedly, McKinley proposed a very practical, useful curve-fitting method for determining the time when afterflow in a shutin pumping well becomes negligible so the pressure-buildup data can be used for analysis.[7] Overnight the idea of curve fitting was expanded into the analysis of every conceivable form of transient pressure data. McKinley, Ramey et al., and Earlougher may be considered the leaders, but many others were involved.[8,9,10,11]

Curve fitting involves matching log-log curves. Anyone who has tried to do this will immediately recognize the mechanical difficulty involved. Older reservoir engineers may recall the frustration of trying to obtain a unique solution when log-log extrapolations and curve fitting were attempted with hyperbolic decline curves. Modern curve-fitting methods present similar problems. However, most reservoir engineers now seem to recognize the limitations of the curve-fitting method and see its practical use as one of qualitative, data-checking, last-resort-type of analysis.[12]

Unchanging pressure at shutin. Pressure is seldom exactly unchanging, but we treat it that way if no appreciable or measurable change would have occurred during the shutin time if the well had not been shut in. This then serves as our definition of *unchanging*. The concept of superposition presented in chapter 3 clearly shows the advantage of having a well whose pressure would not have changed during the time of shutin if the well were not shut in. Using this concept, we must superimpose the effect of the negative rate of shutting in the well on the pressure that would have existed in the reservoir if the well had not been shut in.

A negative rate change causes a pressure increase. The effect is the same as the effect of injecting into the reservoir since an injection rate is the same as a negative production rate. Consequently, the pressure increase caused by a negative rate, $-q$, acting for a time, Δt, is the same absolute value as the pressure drop indicated for a constant rate drawdown ($p_{w,t} - p_i$) in Eq. 4.9:

$$\Delta p_q = \frac{0.141 q\mu}{kh}\left[\tfrac{1}{2}(\ln \Delta t_D + 0.809) + S\right] \tag{4.24}$$

Where the reduced time is:

$$\Delta t_D = \frac{6.33 k \Delta t}{\phi \mu c r_w^2} \tag{4.25}$$

Eq. 4.24 applies only after Δt_D becomes greater than 100, and it can be applied only until the effect of the shutin reaches the outer drainage boundary. On the basis of the well radius, the reduced time generally becomes greater than 100 in a matter of seconds or minutes. Since the stabilization time is seldom less than a few hours, Eq. 4.24 generally describes the pressure buildup long enough to permit an analysis. Computer studies have in effect investigated the upper time limit of the applicability of Eq. 4.24.[13,14] However, simply using the distance to the nearest drainage boundary to calculate the stabilization time using the t_s equation, Eq. 3.50, should give a sufficiently accurate time limit.

Since the negative rate effect is superimposed on the pressure that would have existed in the reservoir if the $-q$ rate change had not occurred, we can state the well pressure during the time of shutin as:

$$p_w = p_{wf} + \frac{0.141 q\mu}{kh}\left[\tfrac{1}{2}(\ln \Delta t_D + 0.809) + S\right] \tag{4.26}$$

Eq. 4.26 may be clearer when it is presented graphically as in Fig. 4–7.

Eq. 4.26 can be put in a more meaningful form if we substitute the expression of Eq. 4.25 for Δt_D, convert from natural logs to logs with a base 10, substitute Δp_{skin} for ($0.141 q\mu S/kh$), and rearrange the equation to give:

$$p_w = p_{wf} + \frac{0.1625 q\mu}{kh}\left[\log \frac{6.33k}{\phi \mu c r_w^2} + \frac{0.809}{2.3}\right] + \Delta p_{skin}$$

$$+ \frac{0.1625 q\mu}{kh} \log \Delta t \tag{4.27}$$

Recognizing that the first three terms of this equation do not change with time, it can be written as:

$$p_w = \text{constant} + \frac{0.1625 q\mu}{kh} \log \Delta t \tag{4.28}$$

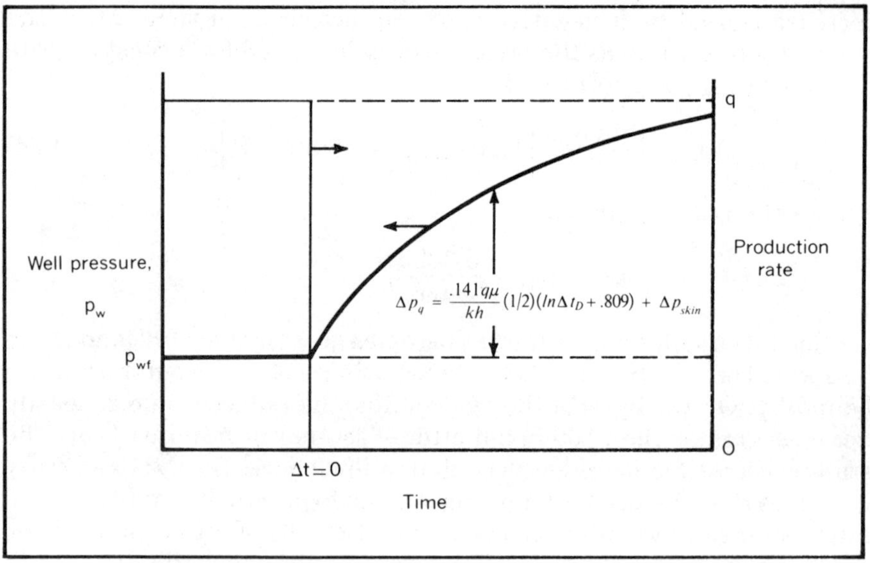

Fig. 4–7 Linear pressure buildup in a steady-state well or with an unchanging well pressure at shutin

Eq. 4.28 shows that a plot of the well pressure, p_w, versus the log of the shutin time, log Δt, results in a straight line whose slope equals:

$$m = \frac{0.1625 q \mu}{kh} \tag{4.13}$$

This slope is the same as the slope obtained for the semilog plot of the well pressure versus producing time for the constant-rate drawdown. A pressure-buildup plot of the shutin well pressure versus the log of the shutin time is illustrated in Fig. 4–8. This figure is generally known as the Miller, Dyes, and Hutchinson pressure-buildup plot, or the MDH method.[15] The shape of the Fig. 4–8 plot is characteristic of pressure-buildup plots in general. The early portion of the curve does not fit the predicted straight line for several reasons. The nonideal early data are based on the assumption that we are producing at a constant rate and that this rate is instantaneously reduced to zero. Since the well is shut in at the surface, except on a drillstem test, flow continues across the well radius after flow has been stopped at the surface because the free gas in the tubing continues to be compressed with the increase in well pressure. As the gas is compressed, oil continues to flow across the well radius to affect the reduction in gas volume. This effect is called afterflow. When the gas is sufficiently compressed, the flow rate approaches zero and the afterflow effect is complete.

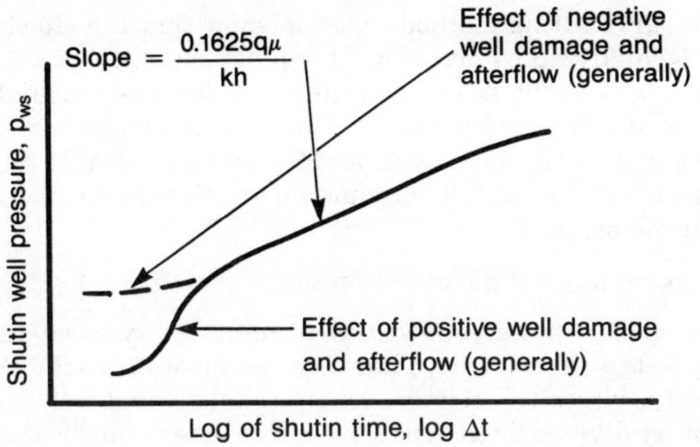

Fig. 4–8 Semilog pressure buildup for a well with an unchanging flowing pressure at shutin

Since the flow rate is essentially zero in the vicinity of the wellbore by the time the straight-line portion of the plot is reached, the slope of the straight-line portion is not affected by the damage around the wellbore. During this period the damaged zone simply serves as a means of transmitting the pressure to the wellbore where the pressure is measured. Consequently, the permeability, k, in the slope equation, Eq. 4.13, is the undamaged permeability. The same conclusion can be reached mathematically by examining Eq. 4.27. The additional pressure drop in the damaged zone, Δp_{skin}, is a function of the rate, $-q$. Since this rate is constant during the shutin, the Δp_{skin} term remains constant and the slope, $0.1625q\mu/kh$, is unaffected by the damaged zone.

The early nonlinear data are also contributed to by the damaged zone. In addition to the requirement that the rate at the well approach zero before the straight-line relationship can be obtained, it is clear that the flow through the damaged zone must be essentially zero before the slope of the data can be a straight line reflecting the undamaged permeability of the formation. The correct slope representing the undamaged permeability may be approached from above rather than from below as in Fig. 4–8. Generally, such a slope indicates the damage is negative rather than positive. Also, the correct slope may increase because of the effect of nearby reservoir barriers and may ultimately decrease because the entire drainage area is affected. These nonideal effects often make it difficult to determine the true straight line that indicates the undamaged permeability.

Using curve-fitting methods, we can show that the effects of afterflow can be analyzed by preparing a log-log plot of the pressure difference, $p_w - p_{wf}$, versus the shutin time, which gives a straight line as long as the afterflow is significant. Once the undamaged permeability has been determined from the slope of the plot, the skin factor, S, can be determined from Eq. 4.26. This equation can then be written as a function of the slope, m:

$$p_w = p_{wf} + 0.867m \left[\tfrac{1}{2}(\ell n\ \Delta t_D + 0.809) + S \right] \tag{4.29}$$

In Eq. 4.29 it is necessary to determine the reduced time corresponding to the shutin time. Therefore, we must know the permeability, porosity, viscosity, effective compressibility, and well radius. The permeability divided by the viscosity can be determined from the slope, m, in Eq. 4.13, and the well radius is known. The other constants, ϕ and c, are generally more difficult to evaluate independently, particularly if a free gas saturation exists in the reservoir. The direct conventional calculation of these terms from core or log analysis and published compressibilities is recommended only as a last resort. If the reservoir is in pseudosteady-state flow at the time of shutin, it is recommended that the terms be evaluated from the equation for the change of pressure with time during pseudosteady state. For radial flow:

$$(\Delta p/\Delta t)_{pseudo} = \frac{1.79q}{\phi h c r_e^2} \tag{3.33}$$

For other flow geometries:

$$(\Delta p/\Delta t)_{pseudo} = \frac{5.615q}{\phi c V_b} \tag{3.34}$$

Where:

V_b = the bulk volume in cubic feet

In some cases the terms ϕc may be evaluated through a knowledge of a nearby reservoir barrier. The distance to a barrier can be calculated by noting the time when the straight-line portion of a buildup-pressure plot deviates from the straight line. This time can be used with the distance to the barrier, r_e, to calculate ϕc from Eq. 3.50.

Eqs. 4.13, 4.24, 4.28, and 4.29 can all be used to analyze a pressure buildup from an unchanging well pressure, regardless of whether the well is infinite acting or is in pseudosteady state or in steady state at the time of shutin. These equations can be used to evaluate pressure falloff tests, such as in injection wells as well as buildups. Also, by substituting the change in rate, Δq, for the rate, q, the equations can be used to analyze pressure tests where the rate is simply reduced rather than shut in. These analyses are known as *two-rate tests*. Thus, these simple equations can probably be used to analyze 90% of the buildup and falloff data. Also, they can be used to analyze drawdown data by substituting p_i for p_{wf}.

Two restrictions to the application of these equations should be noted. The flowing well pressure at the time of shutin must be sufficiently constant so a change in pressure during the ensuing shutin time would be insignificant if the well were not shut in. Also, the shutin time, Δt, must be less than the stabilization time based on the distance to the nearest boundary.

These methods of analysis can only be used when the well pressure at the time of shutin is unchanging. If the well is in steady state at that time, there is no problem in determining that the well pressure is unchanging. However, if the wells are in pseudosteady state or are infinite acting at the time of shutin, the problem of determining if the well pressure can be treated as unchanging is more difficult.

The pseudosteady-state case is probably the easiest to analyze. If the constant change in pressure with time has been established from past pressure surveys or surface-pressure observations, then it simply becomes a matter of multiplying this change of pressure with time under pseudosteady state by the maximum shutin time. This calculation shows how large of a change in the well pressure would have occurred during the shutin time if the well had not been shut in. If such pressure observations have not been made, it is then necessary to estimate the change in pressure with time under pseudosteady-state conditions using Eq. 3.33 or Eq. 3.34. Thus, determining the error introduced by the unchanging assumption for a well in pseudosteady state at shutin is reasonably simple.

Determining the error introduced by the unchanging assumption for a well that is infinite acting at shutin time is somewhat more difficult. We can get an excellent evaluation of the change in pressure that would have occurred if the well had not been shut in by comparing the theoretical value of $(p_w')_{min}$, the pressure that would have existed at the end of the shutin period if the well had not been shut in, with the theoretical flowing pressure at shutin, p_{wf}. Applying Eq. 3.18 using a pressure function for an infinite-acting reservoir and including the additional pressure drop caused by the skin, we can write an expression for the flowing pressure at the producing time, t, at the time of shutin:

$$p_{wf} = p_i - \Delta p_{skin} - \frac{0.141 q\mu}{kh} \tfrac{1}{2}\left(\ell n \frac{\eta t}{r_w^2} + 0.809\right) \tag{4.30}$$

The theoretical well pressure at the end of the shutin time that would have existed if the well had not been shut in for the time, t_{max}, can then be written as:

$$(p_w')_{min} = p_i - \Delta p_{skin} - \frac{0.141q\mu}{kh} \tfrac{1}{2}\left(\ell n \frac{\eta(t + t_{max})}{r_w^2} + 0.809\right) \quad (4.31)$$

The difference between Eqs. 4.30 and 4.31 provides an estimate of the maximum change in the flowing pressure that would have existed if the well had not been shut in:

$$(\Delta p_w')_{max} = \frac{0.1625q\mu}{kh} \log \frac{t + \Delta t_{max}}{t} \quad (4.32)$$

Eq. 4.32 can then be used to determine the error that may be involved in assuming an unchanging pressure for a well that is infinite acting at shutin.

If the change in the well pressure during the shutin period would be significant if the well were not shut in, it is necessary to use more complex analyses.

PROBLEM 4.5: Pressure Buildup from an Unchanging Well Pressure

Fig. 4–9 is a plot of shutin pressure versus the log of shutin time for a well producing from a reservoir with a very active water drive. After 9 hr of shutin time, the pressure reaches a constant value. The well is producing at a rate of 199 stb/d with an unchanging pressure at the time of shutin. Find the undamaged permeability. What is the ϕc product if the effect of the nearest drainage boundary 300 ft away is felt at the well after 7½ hr (log = 0.875). What is the Δp_{skin} and the skin factor, S, if this well has a flowing pressure of 702 psig? What is the flow-rate increase if Δp_{skin} is removed? Assume the same p_w is maintained and the constant shutin pressure is the pressure at the external drainage boundary. The following reservoir data have been determined:

μ = 2.0 cp
r_w = ⅓ ft
B_o = 1.15
h = 22 ft

Finite acting at shutin time. If a well is finite acting at the time of shutin and the pressure is changing too rapidly to permit the unchanging analysis, a simple procedure is still possible to obtain the undamaged permeability and an evaluation of the damage.[16] The equation for the well pressure under these circumstances can be derived in much the same way as the well-pressure equation for the unchanging case is derived. We simply add the pressure increase caused by shutting in the well to the pressure that would have existed if the well had not been shut in. The pressure, p_w', that would have existed if the well had not been shut in is:

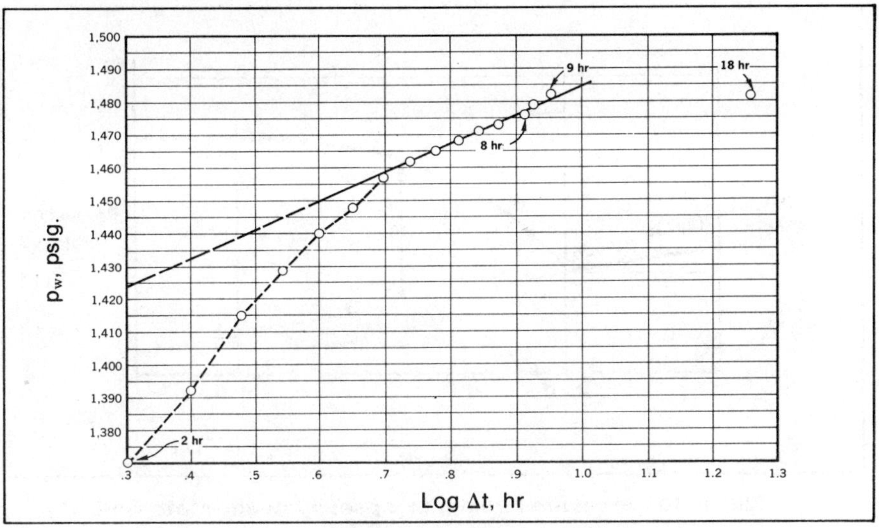

Fig. 4–9 Pressure-buildup data for problem 4.5

$$p_w' = p_{wf} - (\Delta p/\Delta t)_{pseudo}\Delta t \tag{4.33}$$

The pressure increase caused by the negative rate introduced into the reservoir by shutting in the well, Δp_q, is governed by Eq. 4.24. Adding this to the pressure indicated in Eq. 4.33, we obtain:

$$p_w = p_{wf} - \left(\frac{\Delta p}{\Delta t}\right)_{pseudo} \Delta t + \frac{0.141q\mu}{kh}\left[\tfrac{1}{2}(\ell n\ \Delta t_D + 0.809) + S\right] \tag{4.34}$$

The basis for Eq. 4.34 is illustrated graphically in Fig. 4–10.

In Eq. 4.34 there is no simple relationship between the well pressure and time. Both the second and third terms change with time. The change in the second term is negative and varies linearly with the shutin time, Δt. The change in the third term is positive and is proportional to the change in the log of the shutin time. Consequently, there can be no theoretically correct straight line between a plot of the well pressure and a log function of time, nor can there be a straight-line plot between the well pressure and a linear function of time. Therefore, we choose to use a plot of Δp_q versus the log of the shutin time, log Δt, to determine the undamaged permeability and the extent of the damage in the well.

If we expand the Δt_D expression in Eq. 4.24 as in Eq. 4.25 by converting from base e logs to base 10 logs and rearrange the terms, we obtain:

$$\Delta p_q = \frac{0.1625q\mu}{kh}\left[\log \frac{6.33k}{\phi\mu c r_w^2} + \frac{0.809}{2.3} + S\right] + \frac{0.1625q\mu}{kh} \log \Delta t \tag{4.35}$$

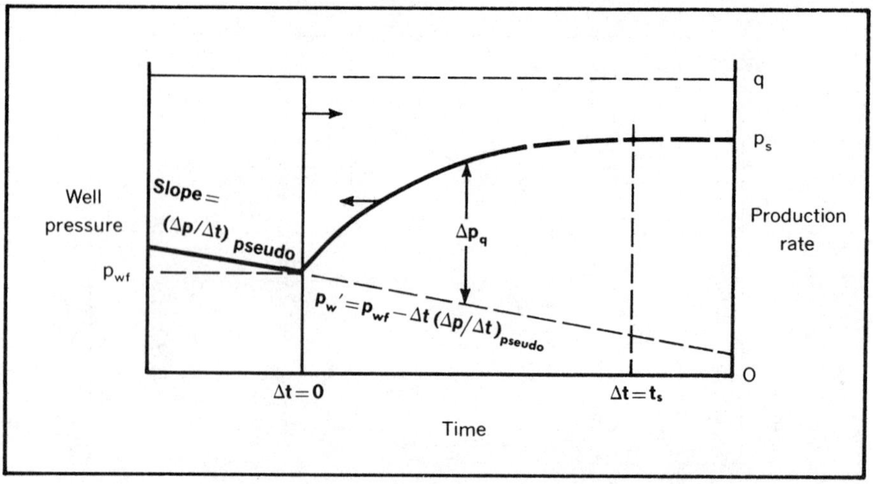

Fig. 4–10 Pressure buildup in a pseudosteady-state well

Note that only the last term changes with time so we can write the equation as:

$$\Delta p_q = \text{constant} + \frac{0.1625 q\mu}{kh} \log \Delta t \tag{4.36}$$

Eq. 4.36 clearly shows that a plot of Δp_q versus log Δt gives a straight line whose slope is again defined as in Eq. 4.13. Furthermore, this slope can be used to calculate the undamaged permeability since the additional pressure drop caused by the damage is included in the skin factor, S:

$$\Delta p_q = 0.867m\ \tfrac{1}{2}\,(\ell n\ \Delta t_D + 0.809) + \Delta p_{skin} \tag{4.37}$$

The data for the Δp_q versus log Δt plot can be obtained by extrapolating the plot of pressure decline versus time and reading the different Δp_q values for each Δt from the plot. Alternatively, the Δp_q values can be calculated for a variety of shutin times by substituting Δp_q for the last term of Eq. 4.34 and rearranging the terms to obtain:

$$\Delta p_q = p_w - p_{wf} + \Delta t\,(\Delta p/\Delta t)_{pseudo} \tag{4.38}$$

Problem 4.6 should help clarify the Δp_q-plot solution. The solution is shown in appendix C.

PROBLEM 4.6: Pressure-Buildup Analysis of an Old Well

A well drilled in a field with uniform 40-acre spacing has produced at 280 stb/d for 10 days when the well is shut in for a pressure-buildup survey. Prior to the shutin, the flowing tubing-head pressure declined about 24 psi/d. The gas-oil ratio has been constant during production. The Δp_q plot is shown in Fig. 4–11.

Assume the well is in pseudosteady state at the time of shutin. Find the product, ϕc, the undamaged permeability, the approximate time when pseudosteady-state flow began, Δp_{skin}, and the skin factor, S. The reservoir data are as follows:

Oil formation volume factor, $B_o = 1.31$
Oil viscosity, $\mu_o = 2.0$ cp
Well radius, $r_w = 0.333$ ft
Net pay thickness, $h = 40$ ft

Pressure-Survey Data

Shutin Time, hr	*Pressure, psia*
0	1,123
2	2,291
4	2,519
8	2,592
12	2,624
16	2,647
20	2,663
24	2,675
30	2,687

It is often convenient to evaluate the damage as a damage ratio rather than as Δp_{skin} or the skin factor, S. To do this, derive an expression for the damage ratio defined as the undamaged permeability divided by the actual permeability. The undamaged permeability can be obtained from the slope of Δp_q versus log Δt. The average or actual permeability can be calculated from the pseudosteady-state radial flow equation using the following pressure at the time of shutin:

$$DR = \frac{k_{undamaged}}{k_{actual}} = \frac{0.1625q\mu/mh}{0.141q\mu[\ell n\,(r_e/r_w) - 0.5]/h(p_e - p_{wf})} \tag{4.39}$$

Eq. 4.39 reduces to:

$$DR = \frac{p_e - p_{wf}}{0.867m[\ell n\,(r_e/r_w) - 0.5]} \tag{4.40}$$

To use Eq. 4.40, it is necessary to determine the pressure at the outer boundary, p_e. We will see that we can determine the average pressure for the drainage area of a well in pseudosteady-state flow at the time of shutin using a rather simple equation. The pressure at the external boundary can then be calculated by recognizing that there is a fixed relationship between the average or static pressure and the pressure at the outer boundary:

$$p_e - p_s = 0.217m \tag{4.41}$$

The basis for Eq. 4.41 is derived from the Craft and Hawkins calculation for average pressure:[17]

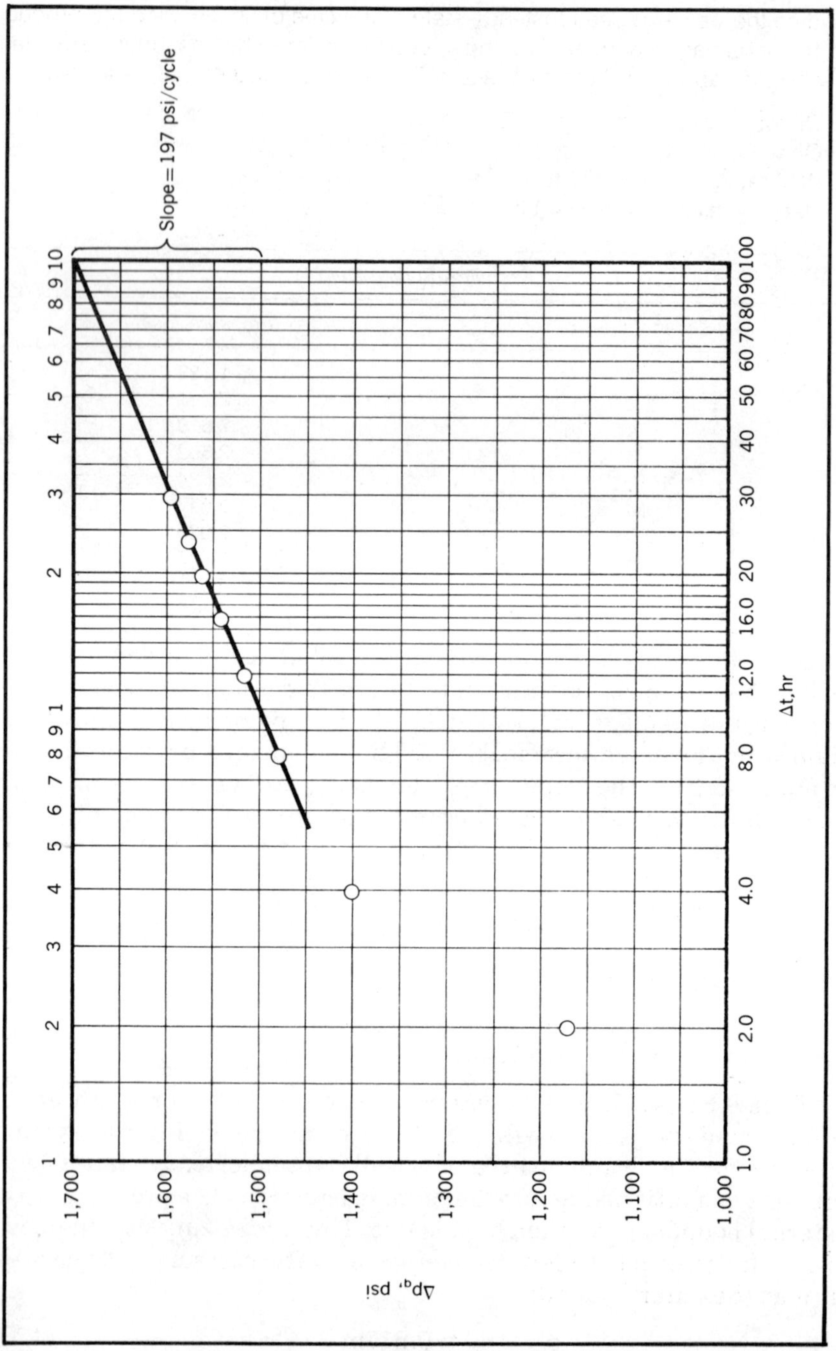

Fig. 4–11 Plot of Δp_q for problem 4.6

$$p_{avg} = \frac{\Sigma(p_r \Delta V)}{V} \tag{4.42}$$

By substituting for ΔV in the radial system $2\pi r \Delta r h \phi$ and for p_r according to the pseudosteady-state radial flow equation (Eq. 3.42), we obtain:

$$p_r = p_w + \frac{\mu q\left(\ell n \dfrac{r}{r_w} - \dfrac{r^2}{2r_e{}^2}\right)}{7.08kh} + \Delta p_{skin} \tag{4.43}$$

By integrating between the external limit and the well limit, we obtain the following equation from Craft and Hawkins:[18]

$$p_{avg} = p_w + \frac{\mu q}{7.08kh}\left(\ell n \frac{r_e}{r_w} - 3/4\right) + \Delta p_{skin} \tag{4.44}$$

Since pseudosteady state assumes a constant compressibility, the volumetric weighted-average pressure must equal the static pressure. Therefore, the expression from Eq. 4.44 is the static pressure, p_s. The pressure at the external radius, p_e, during pseudosteady state can be obtained from Eq. 3.43:

$$p_e = p_w + \frac{\mu q}{7.08kh}\left(\ell n \frac{r_e}{r_w} - 1/2\right) + \Delta p_{skin} \tag{4.45}$$

When Eq. 4.45 is subtracted from Eq. 4.44, we obtain:

$$p_e - p_s = \frac{0.141 q\mu}{kh} \, 1/4 \tag{4.46}$$

When 0.867m is substituted for $0.141q\mu/kh$ and the expression is simplified, the result is Eq. 4.41.

The general technique for analyzing pressure buildup using a Δp_q plot has been discussed for a well that is in pseudosteady state at shutin. However, the same technique can be used for any well if the pressure behavior prior to the shutin time can be extrapolated and negative superposition is used to determine Δp_q at any time. A semilog plot of pressure versus time permits the extrapolation of an infinite-acting well pressure. The pressure extrapolation for a steady-state well is simply the flowing pressure at shutin.

Infinite acting at shutin time. If a well is infinite acting at the time of shutin and the well pressure would have changed significantly during the shutin period if the well had not been shut in, the following methods may be used for the pressure-buildup analysis. If the pressure in the well would not have changed significantly during shutin if the well had not been shut in, then the engineer may choose the analysis described for unchanging pressure at shutin.

The classical method of analyzing a pressure buildup for a well that is in an infinite-acting state at the time of shutin is the Horner method.[19] The derivation used here is similar to that employed to determine the expression for the shutin pressure for a well with an unchanging pressure and for a well whose pressure is changing according to pseudosteady state at the time of shutin. The basic idea is to add the pressure increase that is caused by the negative rate, $-q$, introduced when the well is shut in, to the pressure that would have existed if the well had not been shut in. The pressure increase caused by the negative rate is the same expression used for the unchanging and pseudosteady-state derivations:

$$\Delta p_q = \frac{0.141 q\mu}{kh}\left[\tfrac{1}{2}\left(\ell n\, \Delta t_D + 0.809\right) + S\right] \tag{4.24}$$

As illustrated in Fig. 4–12, the well pressure during shutin is the sum of the Δp_q described by Eq. 4.24 and the pressure that would have existed if the well had not been shut in, p_w'. Since the well is infinite acting, the pressure that would exist without shutin simply equals the well pressure resulting from the pressure drop according to the constant-rate solution to the radial diffusivity equation for an infinite-acting system. If we restrict this calculation to the time when the

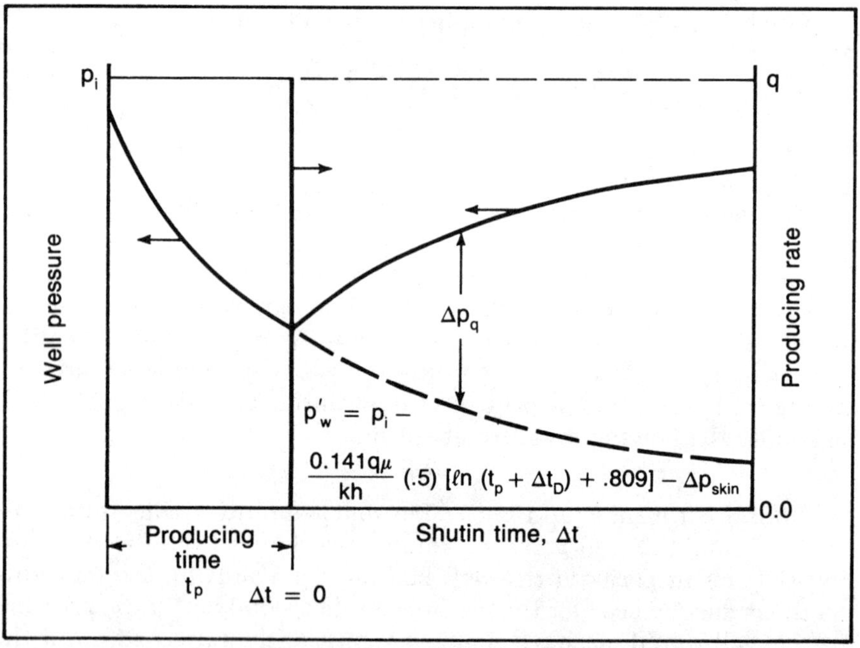

Fig. 4–12 Pressure buildup in an infinite-acting well

reduced time is greater than 100, we can use the log solution. Then the well pressure without shutin is:

$$p_w' = p_i - \frac{0.141 q\mu}{kh} \tfrac{1}{2} \left[\ell n(t_{Dp} + \Delta t_D) + 0.809 \right] - \Delta p_{skin} \quad (4.47)$$

The reduced time $(t_{Dp} + \Delta t_D)$ can be used because the positive rate, q, is effective for the total time of production and shutin. When the increase in pressure caused by the negative rate, Δp_q according to Eq. 4.24, is added to the pressure that would have existed without the negative rate, p_w according to Eq. 4.47, we obtain the shutin well pressure:

$$p_w = p_i - \frac{0.141 q\mu}{kh} \tfrac{1}{2} \left[\ell n\, (t_{Dp} + \Delta t_D) - \ell n\, \Delta t_D \right] \quad (4.48)$$

Note that the Δp_{skin} term and the 0.809 constant drop from the equation because the signs in Eqs. 4.24 and 4.47 differ. Then the difference in the two logs in the brackets is equal to the log of the ratio of the arguments, $(t_{Dp} + \Delta t_D)/\Delta t_D$. Furthermore, the ratio of the reduced times is exactly the same as the ratio of the real times since the reduced time is simply the real time multiplied by the diffusivity constant and divided by the well radius squared. These constants cancel out of the ratio of the reduced times, and we are left with the ratio of the real times. Eq. 4.48 can then be written as:

$$p_w = p_i - \frac{0.141 q\mu}{kh} \tfrac{1}{2}\, \ell n \left(\frac{t_p + \Delta t}{\Delta t} \right) \quad (4.49)$$

When Eq. 4.49 is written in terms of the log to the base 10 and all of the numerical constants are combined, we obtain the Horner equation:[20]

$$p_w = p_i - \frac{0.1625 q\mu}{kh} \log \left(\frac{t_p + \Delta t}{\Delta t} \right) \quad (4.50)$$

Eq. 4.50 indicates that a plot of the shutin well pressure versus the log $[(t_p + \Delta t)/\Delta t]$ gives a straight line whose slope is $0.1625 q\mu/kh$. This expression is the absolute slope of the pressure-buildup and drawdown plots previously examined, and it satisfies Eq. 4–13. This plot is often referred to as a Horner plot.

An example of such a plot is illustrated in Fig. 4–13. Note that it exhibits the same early afterflow and skin effect that have characterized the other pressure-buildup analysis plots. Since Fig. 4–13 represents the pressure-buildup data for a drillstem test, the early-time anomaly includes another nonideal characteristic that will be discussed under the section on drillstem test interpretation.

Engineers are sometimes confused by the fact that the log of

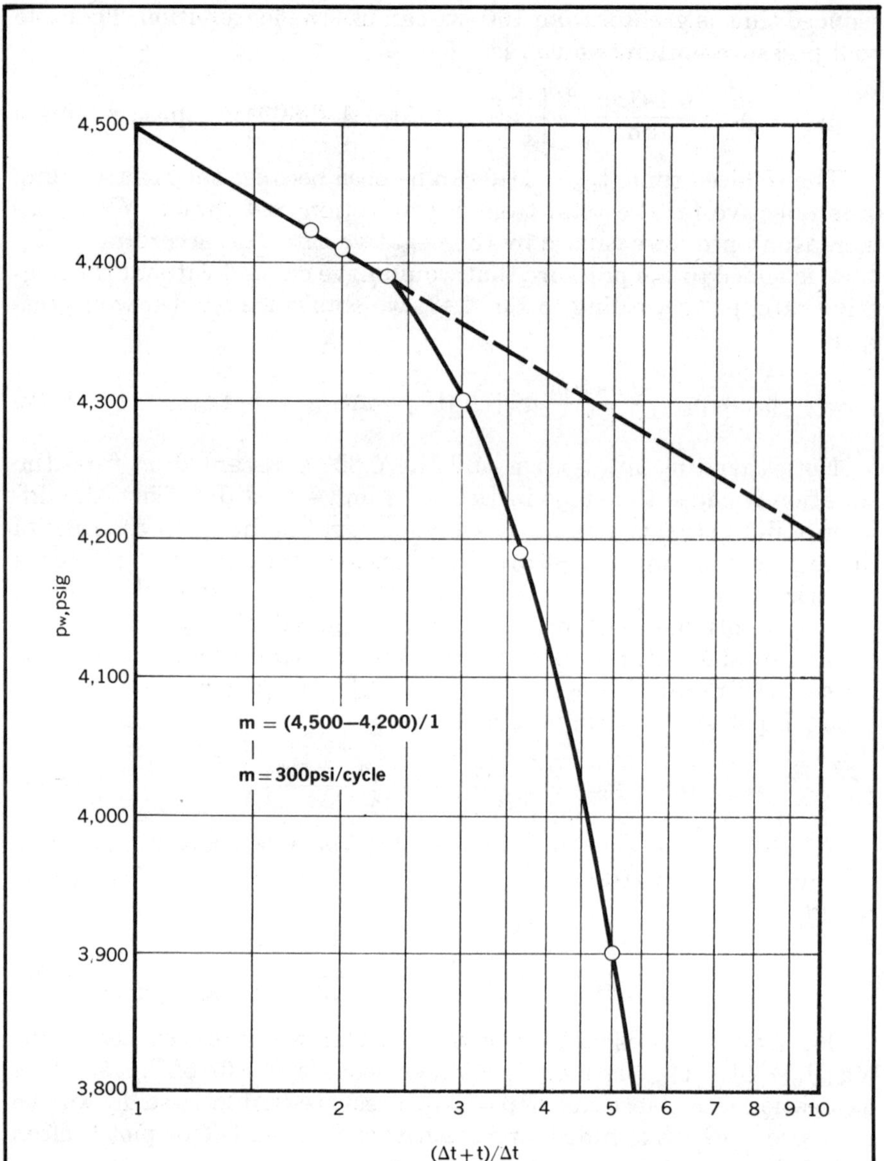

Fig. 4—13 A Horner plot

$\Delta t/(t_p + \Delta t)$ is often plotted rather than the log of $(t_p + \Delta t)/\Delta t$, which changes the sign of the slope. The validity of this procedure can be verified by examining Eq. 4.48. Note that we can change the sign of the second term on the right-hand side of this equation along with the sign of the two natural log expressions in the brackets without affecting the equality of the equation. Such changes make the sign of the second term in Eq. 4.50 positive, and the time ratio becomes $\Delta t/(t_p + \Delta t)$.

The engineer may also watch for reversed scales in a Horner plot. Many engineers prefer to increase the shutin time to the right on a Horner plot and may actually reverse the time-ratio scale to attain this quality.

An analysis of the Horner plot to obtain the well damage must use the well pressure at the time of shutin—the final flowing well pressure. This pressure is the pressure equivalent of the constant-rate infinite-acting solution to the radial diffusivity equation. It may be simpler to obtain an expression for this flowing well pressure by examining Eq. 4.47, which describes the well pressure that would be obtained if the well had not been shut in. This equation can be used to write an expression for the flowing well pressure at shutin by recognizing that this pressure, p_{wf}, is the same as p_w' for a reduced time of t_{Dp} and that 0.867m is 0.141qμ/kh:

$$p_{wf} = p_i - 0.867m \; \tfrac{1}{2} \, [\ell n \, t_{Dp} + 0.809] - \Delta p_{skin} \qquad (4.51)$$

Assuming that the initial reservoir pressure, p_i, is known—and it will be shown that it can always be simply obtained—this equation can be used to determine Δp_{skin}. Then the skin factor, S, can be calculated as shown from Eq. 2.22. In this technique, the value of the mobility, k/μ, can be obtained from the slope, −m, to evaluate the reduced time. The reduced time should be based on the total producing time. This approach to evaluating the skin damage is simple, theoretically correct, and gives the same results as any other method that is theoretically sound. At first it appears that when Eq. 4.51 is used to determine well damage, it relies too much on the accuracy of the producing time, which often must be determined by dividing the cumulative production at the time of shutin by the last producing rate. Another approach, which at first appears to avoid reliance on the accuracy of the producing time, uses an equation that includes the producing time in positive and negative terms, so the effect of the producing time appears to be offset. Such a mathematical expression can be obtained when we recognize that in Equation 4.51 the sum of the second and third terms on the right-hand side is the difference between the initial pressure and the flowing well pressure at the time of shutin, $p_i - p_{wf}$. Also, the second right-hand term in Eq. 4.50 is the difference between the initial pressure and the well pressure at some shutin time Δt, $p_i - p_w$. When these two expressions are subtracted, the initial pressure disappears and we obtain the equation for the difference between the flowing pressure at the time of shutin and the well pressure at some shutin time, Δt:

$$p_w - p_{wf} = 0.867m \ ½ \ (\ell n t_D + 0.809) + \Delta p_{skin} - m \log \frac{t + \Delta t}{\Delta t} \quad (4.52)$$

Eq. 4.52 combines logs to the base e and logs to the base 10, but it provides a forthright, easily understood expression from which to calculate the pressure drop caused by the well damage, Δp_{skin}. As mentioned, the skin factor, S, can be calculated from the Δp_{skin} using Eq. 2.22. Any pressure on the straight-line portion of the Horner plot can be used to calculate the additional pressure drop caused by well damage from Eq. 4.52. The corresponding shutin time must of course be used. This equation appears to be less sensitive to the producing time, t_p, since the first and last terms contain the log of the producing time and have opposite signs. However, Eqs. 4.51 and 4.52 give identical values for Δp_{skin} because according to the Horner equation, Eq. 4.50, if we substitute m for 0.1625q/kh, we obtain:

$$p_w = p_i - m \log \left(\frac{t_p + \Delta t}{\Delta t} \right) \quad (4.53)$$

Note that p_i equals p_w plus the last term of Eq. 4.53. If we substitute p_i for this expression in Eq. 4.52, we of course obtain Eq. 4.51. Thus, Eqs. 4.51 and 4.52 give identical expressions for Δp_{skin} if the Horner method is applied only to infinite-acting reservoirs where it is theoretically correct.

Researchers apparently decided that the practicing engineer could not understand and apply an equation such as Eq. 4.52 and that he needed a more specific calculation to determine the skin factor, S. Consequently, an expression such as Eq. 4.52 was expanded and "simplified" to provide an expression that can be used to calculate the skin factor directly.[21]

$$S = \frac{1.151(p_{1\,hr} - p_{wf})}{m} - 1.151 \log \frac{q_{stb} B_o}{10.4 m c \phi h r_w^2} \quad (4.54)$$

Eq. 4.54 can be obtained from Eq. 4.52 in the following manner:

1. Write the reduced time, t_D, as a function of the reservoir parameters
2. Write Δp_{skin} as a function of the skin factor, S
3. Substitute a function of m for the mobility, k/μ, that appears in the reduced time term

4. Assume that the log of t is equal to the log of t + Δt
5. Use the pressure at 1 hr as the well pressure and $\Delta t = 1/24$ day
6. Solve the resulting expression for the skin factor, S.

If the log of t equals the log of t + Δt, the pressure at the time of shutin is unchanging. Thus, Eq. 4.54 can also be obtained from the equation for Δp_q, Eq. 4.24. Eq. 4.54 is not recommended for determining the skin factor; it is presented to show its shortcomings. If a well has an unchanging pressure at shutin, the unchanging methods previously described are easier to use and understand. Wells that are infinite acting at shutin generally have not produced for long periods of time, and the log of the producing time does not approximate the log of ($t_p + \Delta t$). Therefore, Eq. 4.54 may not give an accurate S. If the engineer is misapplying the Horner method to a finite-acting well, Eq. 4.51 can still be used by making p*, the extrapolated pressure at $\frac{t_p + \Delta t}{\Delta t} = 1.0$, equal to p_i.

When considering the Horner method, the engineer should remember that it theoretically applies only to a well that is infinite acting because the natural log solution used in the derivation is limited to infinite-acting reservoirs (Eq. 4.47). Also, remember that a well pressure plot versus any log function of time does not theoretically give a straight line for a finite-acting reservoir. In addition, bear in mind that there is no reason to apply the Horner method to finite-acting reservoirs since the pressure-buildup methods described for unchanging pressure and pseudosteady state are mechanically simpler.

For those who insist on the use of the Horner method under all circumstances, Ramey and Cobb and Cobb and Smith present data for different-shaped drainage areas that show when a Horner plot starts deviating from the correct slope by as much as 5%.[22,23]

To emphasize the applicability of the Horner method for those engineers who rely heavily on SPE publications, note Earlougher's comments in SPE Monograph No. 5 under "Choice of Analysis Techniques."[24]

- Use the MDH (unchanging pressure) method as a first-pass method unless $t_p < t_{pss}$, t_s, or unless the system can be approximated by a well in the center of a square with constant-pressure boundaries, as in a five-spot, filled-up waterflood. Earlougher's t_{pss} is the time to reach pseudosteady state, t_s.
- Use the Horner method for a second pass if circumstances dictate or as a first pass if t_p is small. An example application of the Horner method is given in the DST interpretation section.

If the engineer still chooses to use Eq. 4.54 to calculate the skin factor, S, he should be careful to read the pressure at 1 hr of shutin from an extrapolation of the straight-line portion of the Horner plot if the straight-line portion does not extend through the shutin time of 1 hr. The straight-line portion of many buildups does not encompass the 1-hr shutin time.

Determining the average drainage-area pressure. As noted, the undamaged permeability and a measure of the well damage may be determined using the same equations whether the well is in pseudosteady state or steady state or is infinite acting at the time of shutin, provided the well pressure qualifies as unchanging. However, this is not true for determining the average pressure that exists in the well drainage area at the time of shutin. To obtain this average pressure, we must know what regime is governing flow at the time of shutin. The methods used are different, depending on whether the well at the time of shutin is infinite acting, in pseudosteady state, or in steady state. The present methods for determining the average pressure in the drainage area of a well appear ridiculous unless we consider their development historically. Consequently, we should first consider the Horner method.

If the Horner semilog plot of well pressure versus $(t + \Delta t)/\Delta t$ is extrapolated to a value of $(t + \Delta t)/\Delta t = 1.0$, this represents a shutin time of infinity. Infinity is the only value of shutin time that mathematically results in a time ratio of 1.0. At a time of infinity, it is then logical to assume that the pressure recorded is the average reservoir pressure. On the surface it appears that this concept may be at odds with the mathematics of the situation. If we examine Eq. 4.53, the Horner equation, we see that a time ratio of 1.0 results in the log of the time ratio being zero, which makes the term containing the log equal to zero. Therefore, the well pressure at this time is equal to the initial pressure, not the average or static pressure.

This calculation seems to represent an inconsistency until we recognize that the average reservoir pressure for an infinite-acting reservoir is mathematically equal to the initial pressure. To determine the weighted-average pressure in a reservoir that is truly infinite, the average pressure in the portion of the reservoir where the pressure has declined is weighted by some finite value. In the portion of the reservoir where the pressure has not yet declined, the pressure, p_i, is weighted by infinity. The weighted-average pressure would then be p_i.

The practical significance of this discussion is that the well pressure on a Horner plot extrapolation to a time ratio of 1.0 is not the average reservoir pressure but is always the initial well pressure if the well is infinite acting. Theoretically, the Horner plot does not apply to a well that is finite acting at the time of shutin.

Mathematically, it appears irrefutable that the Horner equation does not apply to a finite-acting reservoir. However, engineers have always attempted to apply the method to finite-acting reservoirs. Some of the leading experts in this field still talk about the time limits that exist when applying the Horner method to a finite-acting reservoir. They justify this approach by showing that computer-manufactured and actual well data give a straight-line Horner plot over specified time ranges. The range of the apparent straight-line portion must largely be a function of the scale used for the plot.

This approach to the pressure-buildup analysis in a finite-acting reservoir would be admirable if there were no theoretically accurate or easy method of analyzing finite data, but the methods previously described are both theoretically sound and simpler to apply than a Horner analysis. The principal point to be made at this time is that engineers will probably continue to use a Horner plot for their finite-well buildups. However, they should be warned that the well pressure at a time ratio of 1.0 is not the average well pressure but approximates the initial pressure that would be necessary for the well drainage area—which must be larger—to be infinite acting. Matthews, Brons, and Hazebroek recognized that a Horner plot applied to a finite-acting well does not give either the initial pressure or the average pressure when extrapolated to a time ratio of 1.0.[25] Consequently, they coined the term p* to identify this particular pressure. They recognized that p* is equal to the initial pressure if the reservoir is still infinite acting. Then they determined the relationship between p* and the average reservoir pressure, p_s, that exists if the well is finite acting at the time of shutin.

The relationship between p* and the initial pressure, p_i, has been shown to be a function of the reduced time and the shape of the drainage area. These relationships can be found in extensive plots in SPE Monograph No. 1.[26] Such plots are not included in this text because it does not appear that they are needed for pressure-buildup analysis.

To use the Matthews, Brons, and Hazebroek data, we must know the average compressibility for the reservoir to the point in time represented by the producing time at the time of shutin. We must also know the average porosity, the shape of the reservoir, and the thickness. Using these reservoir properties, we can calculate the average pressure by material balance:

$$p_i - p_s = \frac{5.615qt}{\phi c V_b} \tag{4.55}$$

In Eq. 4.55 the symbol V_b is the well drainage volume in cubic feet, and c is the average compressibility. Therefore, the Matthews, Brons, and Hazebroek data has little if any advantage over a straight volumetric calculation of the average reservoir pressure, p_s. For an infinite-acting reservoir there is certainly no advantage.

Consequently, for an infinite-acting reservoir it is recommended that we simply use Eq. 4.55 to calculate the average pressure. It is recognized that this method really represents no practical solution in most cases because the reservoir average pressure is generally needed for material-balance calculations. Nevertheless, this is the only reasonable approach to the problem. If production has not affected the outer boundary of the drainage area, it is illogical to expect a pressure buildup that does not affect the outer boundary to indicate the average pressure in the drainage area, which must be a function of the reservoir size.

The Matthews, Brons, and Hazebroek method misapplies the Horner method to a well that is no longer infinite acting in order to extrapolate the Horner plot to a $(t + \Delta t)/\Delta t$ of 1.0 and, thus, determine p^*.[27] Deitz devised a similar method of determining the average pressure of a drainage area without this misapplication of the Horner method.[28] The Deitz method is also simple to apply, but it uses some undesirable assumptions.

Deitz showed that the average pressure in a drainage area and the pressure behavior during shutin can be compared so the average pressure can be read directly from an extrapolated plot of the shutin pressure versus the log of the shutin time.[29] We can use a circular drainage area with the well in the center as an example of the analysis that can be accomplished if all of the necessary data are obtained and the questionable assumptions are acceptable.

An average pressure equation, Eq. 4.44, for radial pseudosteady-state flow has been derived and can be rearranged as:

$$p_w = p_s - \frac{q\mu}{7.08kh}\left(\ell n \frac{r_e}{r_w} - 3/4\right) - \Delta p_{skin} \qquad (4.44a)$$

If the flowing pressure in pseudosteady state did not change, we can obtain the shutin pressure at any time by superimposing the effect of the negative rate change, Δp_q, on this flowing well pressure. We have previously used Eq. 4.24a as the expression for Δp_q:

$$\Delta p_q = \frac{0.141q\mu}{kh}\left[1/2\ (\ell n\ \Delta t_p + 0.809)\right] + \Delta p_{skin} \qquad (4.24a)$$

Then the superposition results in:

$$p_w = p_s - \frac{0.141q\mu}{kh}\left(\ell n \frac{r_e}{r_w} - 3/4\right) - \frac{0.141q\mu}{kh}(1/2)(\ell n\ \Delta t_D + 0.809) \qquad (4.56)$$

Eq. 4.56 reduces to:

$$p_s - p_w = \frac{0.141q\mu}{kh}\left[\left(\ell n\frac{r_e}{r_w} - 3/4\right) - \left(1/2\ \ell n\ \Delta t_D + 0.809\right)\right] \quad (4.57)$$

Note that when the expression in the brackets is equal to zero, p_w is equal to p_s. It can then be shown that this condition occurs when the shutin time is:

$$t_{ps} = \frac{0.0157\phi\mu c{r_e}^2}{k} \quad (4.58)$$

Now we can extrapolate a plot of p_w versus the log of shutin time to t_{ps} and read the average pressure, p_s.

We must note, however, that Eq. 4.58 is not rigorous. Eq. 4.56 assumes that the well pressure under pseudosteady state would not change if the well were not shut in. We know that this situation is not the case. Therefore, the magnitude of the error involved can be determined by recognizing that during the time t_{ps} the flowing well pressure declined an amount equal to $(\Delta p/\Delta t)_{pseudo}$ times t_{ps}. Using the expression for t_{ps} from Eq. 4.57 and the expression for the change in pressure with time under pseudosteady state according to Eq. 3.32, we can obtain an expression for the magnitude of the error:

$$\text{Error magnitude} = \left(\frac{0.0157\phi\mu c{r_e}^2}{k}\right)\frac{1.79q}{\phi hc{r_e}^2} \quad (4.59)$$

Stated as a function of the pressure-buildup slope, m, we obtain:

$$\text{Error magnitude} = \frac{0.0281q\mu}{kh}\ \frac{m}{(0.1625q\mu/kh)} = 0.73m \quad (4.60)$$

The calculation in Eq. 4.60 is not the exact error but is simply an indication of the magnitude of the error that can be expected. Since m is commonly in the range of 100–200, the error can be sizable. Consequently, the shape factors necessary for applying the Deitz method to other geometries have not been included.

Note that the error in the Deitz method can be avoided by directly using Eq. 4.44 to calculate the average pressure. The only additional data required is the Δp_{skin}, which can be obtained from the pressure buildup. For determining the average well pressure, Eq. 4.44 can be stated in a more convenient form:

$$p_s = p_{wf} + 0.867m\left(\ell n\ \frac{r_e}{r_w} - 3/4\right) + \Delta p_{skin} \quad (4.61)$$

The average pressure in the drainage area of wells that are in steady state at the time of shutin can be found using an equation similar to Eq. 4.61 that applies to steady state rather than to pseudosteady state:

$$p_s = p_{wf} + 0.867m\left(\ell n\,\frac{r_e}{r_w} - \tfrac{1}{2}\right) + \Delta p_{skin} \tag{4.62}$$

Eq. 4.62 is derived in an identical manner to the procedure followed to obtain Eq. 4.61 for pseudosteady state. An equation identical to Eq. 4.44 is derived by Craft and Hawkins for the average pressure in steady-state flow, except the constant ¾ is changed to ½.[30] This difference is carried to Eq. 4.62, which is similar to pseudosteady-state Eq. 4.61 but uses the different numerical constant.

Thus, to obtain the average pressure for the drainage area of a well that is in steady state, we first must determine the buildup slope, m, and Δp_{skin} as outlined under the unchanging method and use these values with the effective external well radius, r_e, to calculate the static pressure from Eq. 4.62. Since steady-state flow generally exists only in a reservoir with a strong water drive and the average reservoir pressure is generally not as important in such a reservoir as it is in a depletion-type reservoir, this problem is not given much consideration.

Odeh apparently recognized the shortcomings of the methods developed by Matthews, Brons, and Hazebroek and by Deitz. He solved the problem by relating the difference between the initial pressure and p* to the difference between the initial pressure and the static pressure, p_s. Then the magnitude of both are normalized by dividing by the slope of the buildup curve, m.[31] These Odeh data are shown in Figs. 4–14 and 4–15. By handling the relationship in this way, Odeh was able to avoid the need to know the system volume and compressibility because they affect the two pressure differences equally.

To use the Odeh data, it is necessary to know the initial pressure, whereas this was not needed in the previous methods. However, it is much easier to obtain an accurate initial pressure than it is to determine the volumetric and compressibility data. The Odeh data might be considered the ideal method of obtaining the average pressure in a drainage area that is in pseudosteady state at the time of shutin except that it is based on p*, which requires the misapplication of the Horner method.

To minimize the error involved in this misapplication, it is recommended that p* be calculated from one of the earliest points on the straight line of the p_w versus log Δt or of the Δp_q plot, which ever is used to analyze the pseudosteady-state well. The value of p* represents the well pressure for a $(t_p + \Delta t)/\Delta t$ of 1.0 or for a log $[(t_p + \Delta t)/\Delta t]$ of zero. Therefore, to extrapolate from some particular shutin well pressure p_w, we must extrapolate to a log $[(t_p + \Delta t)/\Delta t]$ of zero. The pressure

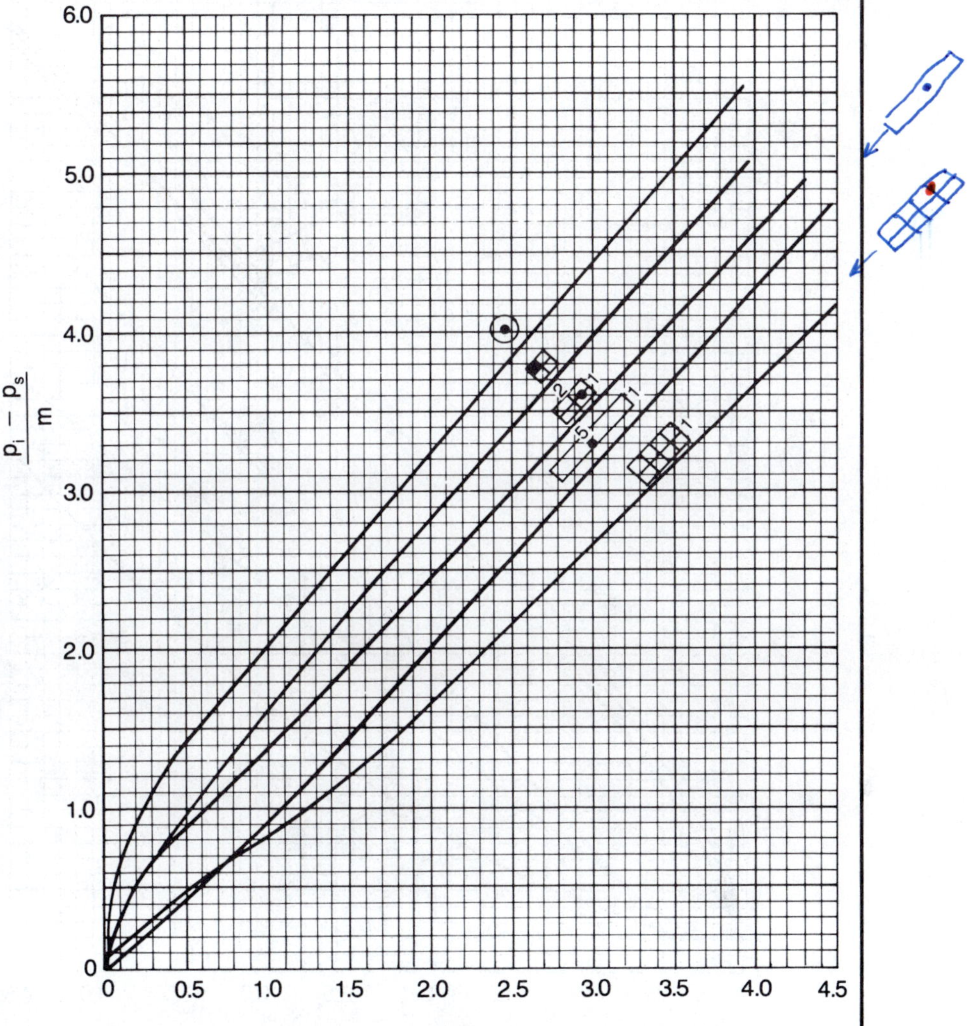

Fig. 4–14 Evaluation of static pressure from p* (after Odeh and Al-Hussainy, "A Method for Determining the Static Pressure of a Well from Buildup Data," courtesy *JPT*, May 1971, © SPE-AIME)

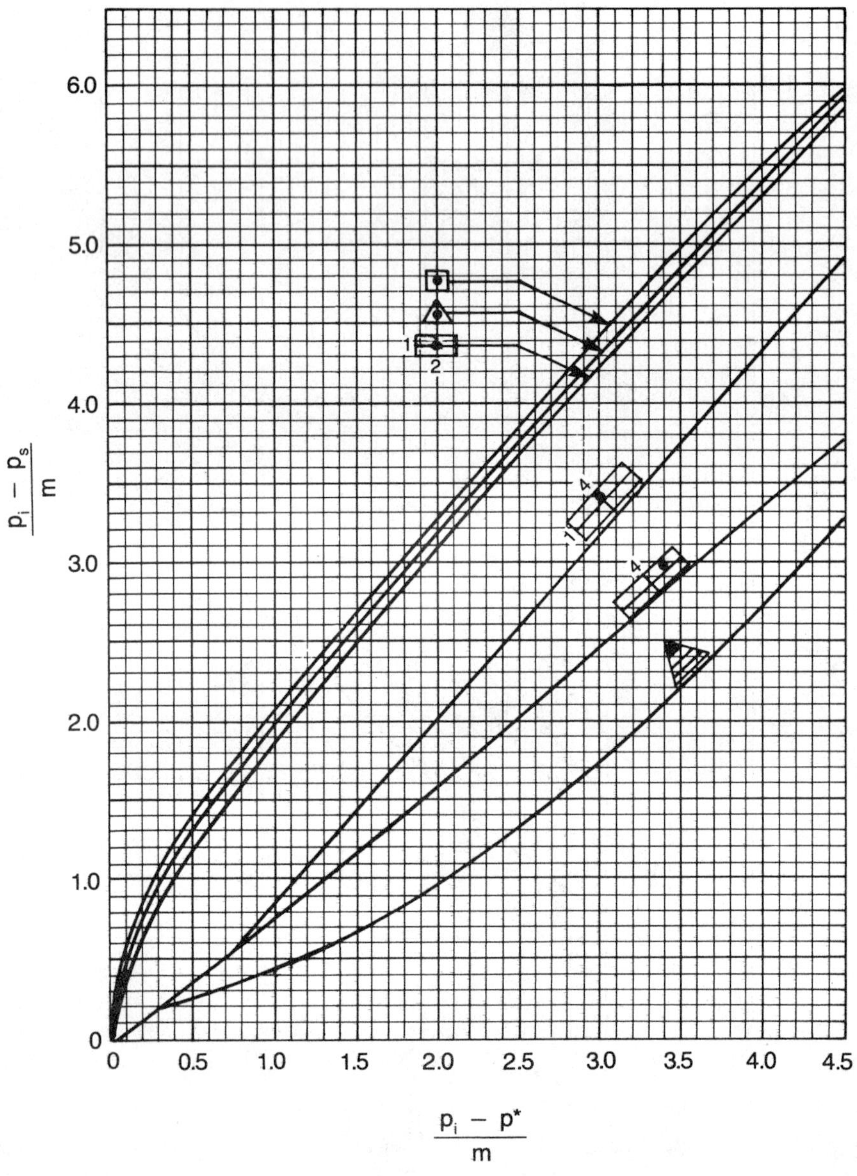

Fig. 4–15 Evaluation of static pressure from p* (after Odeh and Al-Hussainy, "A Method for Determining Static Pressure of a Well from Buildup Data," courtesy *JPT*, May 1971, © SPE-AIME)

increases a value, m, for every cycle or change in log of 1.0, and the total change in the log to a value of zero equals the log $[(t_p + \Delta t)/\Delta t]$. Thus, the value of p^* is:

$$p^* = p_w + m \log [(t_p + \Delta t)/\Delta t] \tag{4.63}$$

In order for a Horner plot of data from a shutin well in pseudosteady state or steady state at the time of shutin to appear to give a straight line, there must be no significant change in the log $(t_p + \Delta t)$ value and, of course, Δt must be less than the time to feel the effects of the nearest drainage boundary at the well. The first point on the straight-line plot of p_w versus log Δt or of Δp_q provides the best possibility for these conditions to be met.

Thus, it is recommended that p^* be calculated by determining a p_w and a corresponding Δt from the earliest portion of the straight-line plot of p_w versus log Δt or of Δp_q. These values can be used in Eq. 4.63 to calculate p^*. Then $(p_i - p^*)/m$ can be used to determine $(p_i - p_s)/m$ from Fig. 4–14 or Fig. 4–15 in order to calculate the average pressure, p_s.

As noted, the Deitz method assumes that there would be no significant change in the flowing well pressure during the shutin time if the well had not been shut in. When a well is near the center of its drainage area, a similar but more accurate approach can be used. The equations are actually derived for radial flow, but experience has shown that little error is involved if an equivalent radius is calculated for a well draining a square, a hexagon, or any similar regular area. In this method the well pressure that would be attained at a shutin time equal to the stabilization time ($\Delta t = t_s$) is used as the basis for the static-pressure calculation. At this time the well pressure reaches the static pressure. It may be easier to understand this phenomenon by writing an expression for the shutin well pressure as a function of the initial pressure rather than as a function of the flowing pressure at the time of shutin:

$$p_w = p_i - \frac{0.141 q\mu}{kh} (\Delta p_D)_{t_p + \Delta t} + \frac{0.141 q\mu}{kh} (\Delta p_D)_{\Delta t} \tag{4.64}$$

No Δp_{skin} term appears in Eq. 4.64 because the additional pressure drop caused by the positive producing rate, q, is equal and opposite in sign to the additional pressure increase caused by the negative rate, $-q$. The equation indicates that the pressure drop caused by the positive rate acts in the reservoir for a time, $t_p + \Delta t$, and that the pressure increase caused by the negative rate resulting from shutting in the well acts for the time of shutin, Δt. Note that once the producing time becomes greater than the stabilization time, the change in the second term with time is equal to the change in pressure with time under the pseudosteady-state flow regime, $(\Delta p/\Delta t)_{pseudo}$. Also, when the shutin time, Δt, exceeds the stabilization time, t_s, the change in this term with time is a rate equal to the change in pressure with time under pseudo-steady state, $(\Delta p/\Delta t)_{pseudo}$. If the second term of Eq. 4.64 is decreasing at the same rate that the third term is increasing, the well pressure is no longer changing. When the well pressure reaches a state where it is no longer changing, the pressure must surely be static.

Interference from other wells generally prevents a direct measurement of the static pressure. Even if this subject well were the only well in the center of the reservoir, the stabilization time would probably be too large in most cases for it to be practical to leave the well shut in until the static pressure is reached. Consequently, it is generally necessary to obtain the static pressure by extrapolating the Δp_q plot to a time equal to the stabilization time, t_s. The resulting pressure term can then be used to calculate the corresponding well pressure by using the previously derived equation:

$$\Delta p_q = p_w - p_{wf} + t(\Delta p/\Delta t)_{pseudo} \tag{4.65}$$

In applying this expression, Δp_q is evaluated at the stabilization time and can be appropriately subscripted. The well pressure becomes the static pressure, p_s, and the shutin time, Δt, is the stabilization time, t_s. We can also write expressions for the stabilization time and the change in pressure with time under pseudosteady-state flow according to Eqs. 3.50 and 3.33. With these substitutions, Eq. 4.65 becomes:

$$p_s = p_{wf} - \frac{0.04\phi\mu c r_e^2}{k} \frac{1.79q}{\phi h c r_e^2} + (\Delta p_q)_{ts} \tag{4.66}$$

When like factors are canceled and numerical constants are combined, the second term reduces to a function of $q\mu/kh$. A function of the pressure buildup slope, m, can then be substituted for this group of terms to obtain:

$$p_s = p_{wf} - 0.439m + (\Delta p_q)_{ts} \tag{4.67}$$

Note that all three methods—the Matthew, Brons, and Hazebroek, the Deitz, and the Δp_q extrapolation—require a knowledge of the drainage area, the porosity, and the average compressibility for the well drainage area before the average pressure can be determined. When the drainage area is below the saturation pressure, the value of the average pressure calculated by these methods is questionable. This question arises because it is necessary to perform material-balance calculations to determine saturations before the average compressibility, c, can be determined and it is necessary to know the average reservoir pressure before the material balance can be carried out. An alternative method to determining the average compressibility is using the change in pressure with time under pseudosteady-state conditions.

If the engineer dealing with the field personnel insists on accurate recordings of data, the decline in well pressure with time will be available from successive bottom-hole pressure tests or from the decline in tubing-head pressures. These data provide a good estimate of the change in pressure with time under pseudosteady-state conditions, $(\Delta p/\Delta t)_{pseudo}$. From this value we can use the equation for the pseudosteady-state change of pressure with time to evaluate the current $\phi c {r_e}^2$ group of terms:

$$(\Delta p/\Delta t)_{pseudo} = \frac{1.79q}{\phi hc{r_e}^2} \tag{3.33}$$

As noted, wells that are in steady state at the time of shutin have different average pressures in the drainage areas than wells that are in pseudosteady state or that are infinite acting at the time of shutin. Many engineers seem to be using the pseudosteady-state methods for determining average pressures for steady-state reservoirs. Methods for determining the average pressure for steady-state wells are not as well developed as those for calculating the average pressure of a pseudosteady-state well.

One of the biggest difficulties in determining the average pressure in a steady-state well is determining the drainage area of the well. Fig. 4–16 indicates the drainage areas of the individual wells of a hypothetical reservoir.[32] Note that the drainage area of each well is connected with the constant-pressure line, the aquifer. Unfortunately, methods for estimating well drainage areas under steady state are not commonly available. It appears that estimating drainage areas in the same manner used for pseudosteady state results in acceptable average pressures; however, this conclusion cannot be defended on a theoretical basis.

Assuming that we can determine the drainage area of a well, we still are faced with a sizable problem in determining the average pressure of that drainage area. When the well is near the center of its drainage area, the radial flow relationship given between the flowing well pressure and the average pressure can be used as in Eq. 4.62. Otherwise, there seems to be little we can do in a meaningful manner to determine the average pressure in a steady-state system at this time. It appears that it would be a simple matter to relate the effective well pressures, i.e., well pressure plus Δp_{skin}, to the average pressure in a particular part of a particular steady-state reservoir using computer-modeling methods. Since we are dealing with steady state, the modeling seems to be relatively simple. Once such a relationship is established for a particular reservoir, it is not necessary to shut in the wells to obtain the average pressure except when a buildup or drawdown is needed periodically to ensure that the skin factor has not changed.

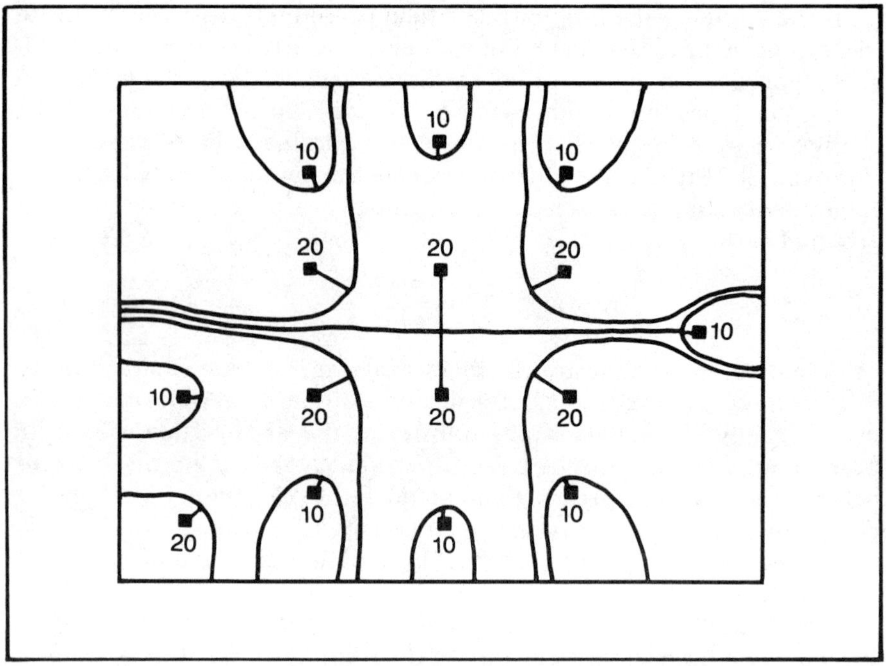

Fig. 4–16 Steady-state drainage regions for a 15-well, nonideal water-drive reservoir; wells producing at rates shown; dotted boundary at constant pressure (after Ramey, Kumar, and Gulati, "Gas Well Test Analysis under Water-Drive Conditions," courtesy *AGA*, 1973)

Steady-state reservoirs do present one common problem of pressure behavior misinterpretation. Many reservoirs with strong water drives are very permeable and porous. These reservoirs tend to have short stabilization times and, thus, often reach constant well pressure soon after being shut in. Most engineers erroneously interpret this constant pressure as the average pressure in the drainage area at the time of shut in. However, this constant pressure is the pressure near the outer drainage area of one of the surrounding wells. When a well is shut in, the area that has been drained by the shutin well soon comes under drainage by the offset wells. As soon as this new drainage system has stabilized, the pressure distribution throughout the reservoir becomes constant. Then the observed well pressure becomes constant and represents one of the pressure points in the drainage system of an offset well.

In summary, the following methods are recommended for determining the average pressure in the drainage area of a well at the time of shutin:

- If a well is not in steady state or pseudosteady state at the time of shutin, the average pressure should be calculated by the material-balance equation, Eq. 4.55. This calculation includes wells that are infinite acting at shutin and wells that are in transition between infinite acting and steady state or pseudosteady state at shutin.

- If a well is situated near the center of its drainage area and is in pseudosteady state or steady state at shutin time, the average pressure in the drainage area at shutin can be calculated by Eq. 4.61 or Eq. 4.62, respectively. These calculations do not require a knowledge of the effective compressibility and porosity.
- If a well is situated near the center of its drainage area, if it is in pseudosteady state at shutin time, and if the change in pressure with time under pseudosteady state is available, the average pressure of its drainage area can be calculated from Eq. 4.67.
- If a well is not situated near the center of its drainage area and is in pseudosteady state at shutin time but the change in pressure with time under pseudosteady state is not available, the Odeh method should be used to determine the average pressure at the time of shutin.
- When a well is not near the center of its drainage area and the well is in steady state at shutin time, it may be necessary to use a steady-state computer model to determine the relationship between the flowing well pressure, Δp_{skin}, and the average pressure of the drainage area.

Problem 4.7 examines two calculations for determining the average pressure in the drainage area. Solutions are shown in appendix C.

PROBLEM 4.7: Determining the Average Pressure in the Drainage Area of a Pseudosteady-State Well

In problem 4.6 the pressure buildup for a well in pseudosteady-state flow at shutin time is described. The tubing-head pressure decline prior to shutin is 24 psi/d. The solution to the problem established the slope, m, as 197; the Δp_{skin} as 415 psi; the stabilization time, t_s, as 3.65 days; and the undamaged permeability as 15.1 md. The Δp_q plot is shown in Fig. 4–10. The initial pressure in this drainage area is 2,960 psia. Find the average pressure for the 40 acres drained by this well using Eq. 4.44 and the Odeh data (2 answers).

Two-rate buildup tests. A two-rate buildup test uses the effect of the negative rate change $(q_1 - q_2)$ to cause an increase in well pressure. This situation is illustrated in Fig. 4–17. An analysis of this pressure increase can provide a means of determining the undamaged permeability, the additional pressure drop caused by damage, and the average drainage-area pressure.

Many problems associated with pressure-buildup tests can be avoided by reducing the producing rate and observing the increase in pressure without completely shutting in the well. The improved char-

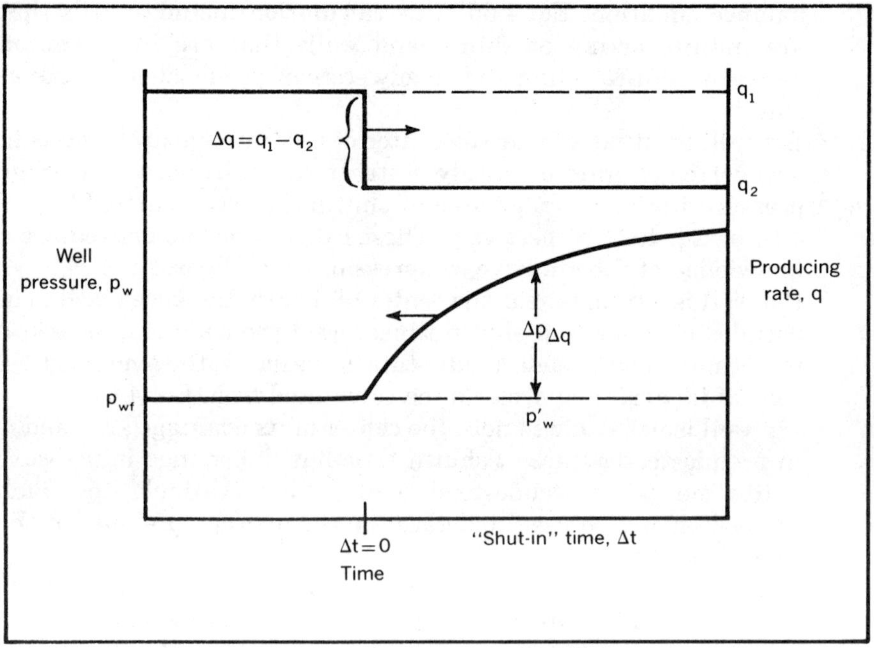

Fig. 4–17 Two-rate buildup data

acteristics of the buildup data result from a more rapid stabilization of the rate change. The engineer should be aware of the difficulty normally experienced when trying to obtain an instantaneous reduction in flow rate from a rate q to zero.

Afterflow effects result mainly from gas in the tubing or casing at the time of shutin, depending on whether a packer is run in the well. Clearly, the worst situation that can exist for afterflow would be to have the producing system filled with low-pressure gas. This situation would exist in a high-pressure gas reservoir producing at a high drawdown rate at the time of shutin, causing a low pressure in the producing system at shutin. Pressure buildups conducted under such conditions often give afterflow effects that last for days and virtually prohibit the type of analyses previously discussed.

Afterflow difficulties grade from this extreme to the ideal situation in which there is no gas in the producing system at shutin and there is virtually no afterflow. If the flow rate from a well is simply reduced, the reduced rate reaches a steady value quickly as compared with a complete shutin. Consequently, the resulting pressure increase is generally much easier to analyze, especially if the well pressure at the time the flow rate is reduced is unchanging.

Remember that an unchanging well pressure is defined as a pressure that would have changed an insignificant amount during the shutin period if the well were not shutin. With the two-rate test we can say that the pressure is unchanging if the pressure would change an insignificant amount during the time the buildup data is analyzed if the rate were not changed. Under these conditions we can see that the equations developed for the pressure-buildup analysis from an unchanging well pressure can be used to analyze the two-rate test using the rate change as the rate in the subject pressure-buildup equations.

The two-rate pressure buildup test also may be used to overcome some of the other difficulties commonly encountered in pressure-buildup data. These situations are called *pressure-buildup anomalies*. One such anomaly is commonly termed *inversion*. In this situation the redistribution resulting from the gravity of the gas and oil trapped in the tubing at shutin often causes an abnormally high pressure recording early in the shutin. This high pressure has in some cases been known to exceed the average shutin reservoir pressure.

The cause of this phenomenon is more completely discussed in the section on pressure-buildup anomalies. However, for our purposes we need to recognize that it is caused by the free gas that is more or less uniformly distributed throughout the oil in the tubing at the time of shutin. This gas completely segregates following shutin so there exists in the well a column of oil containing no free gas and a column of gas over the oil column. By reducing the flow rate rather than completely shutting in the well, this separation of liquid and gas does not occur, and thus, the inversion anomaly is avoided.

The two-rate test may also be used to avoid or minimize the difficulties caused by the existence of two or more beds of vastly different permeabilities or of a reservoir with a matrix permeability much different from the permeability of the fracture or vugular system. In these cases shutting in the well may result in cross flow from the highest reservoir pressure to the lowest reservoir pressure. If the rate is reduced in such a way that the well pressure is still less than the lower reservoir pressure, both the high and low pressure portions of the reservoir continue to flow toward the well and a total kh value can be determined.

It is impossible to provide the engineer with specific guidelines as to when a two-rate test should be employed rather than a conventional buildup test. Most competent reservoir engineers use the buildup test in all cases where good interpretable results are obtained and go to the two-rate test only when difficulties are encountered in running or interpreting the standard buildup. The reason for preferring the buildup is probably more a function of the history of this technology than it is a result of the advantages afforded by the buildup-type test.

That is, the buildup test has been in use much longer than the two-rate test; thus, engineers are more familiar with the test and prefer to use it instead of the two-rate test, which is not as widely understood.

One last practical advantage of the two-rate test should be noted before considering the specific analysis equations employed for a two-rate test. If there is concern about the lost production that results from a buildup test, we may want to use the two-rate test. This is particularly true if the subject well has an excess producing capacity. It is obvious that reducing the production rate does not result in as large of a loss in production as shutting in the well. However, note that with an excess producing capacity we may be able to produce the well at a higher-than-normal rate until it reaches steady state or pseudosteady state. Then when the rate is reduced, no net loss in production results for that month. We can do the same sort of thing in preparing for a buildup, but the quantitative possibilities are much more restrictive. This is no substitute for the engineer's ability to justify the cost of obtaining reservoir data from a profitability standpoint, but the approach has been found helpful in obtaining permission for well tests.

In this section we will only consider methods that are applicable to wells that have unchanging well pressures. More complex methods are available that are applicable to the more general case. However, the need for a two-rate test arises from the fact that considerable amounts of gas are being produced, and as the amount of free gas in the reservoir increases, the tendency toward an unchanging well pressure increases. Consequently, the engineer finds that in most cases where two-rate tests are necessary, an unchanging well pressure will exist in the reservoir and the simple methods described herein will suffice as a means of analysis.

Consider again the illustration depicting the conditions of a two-rate pressure buildup test, Fig. 4–17. This figure represents a well in which the pressure would not have changed significantly if the flow rate had not been changed. The well pressure that would have existed without a change in rate is shown as p_w'. The pressure increase that is superimposed upon this pressure, p_w', is indicated as $\Delta p_{\Delta q}$, which is simply the increase in pressure caused by a negative producing rate change, $q_1 - q_2$, for a time Δt, the time the second rate has been effective. This time is governed by the constant-rate infinite-acting solution to the radial diffusivity equation until Δt is greater than the stabilization time, t_s (Eq. 3.50). This same expression is used to define the Δp_q expression, Eq. 4.24a, except that the change in rate, Δq, is substituted for the rate, q:

$$\Delta p_{\Delta q} = \frac{0.141 \Delta q \mu}{kh} \left(\tfrac{1}{2}\right)(\ln \Delta t_D + 0.809) + \Delta p_{skin} \qquad (4.24b)$$

In Eq. 4.24b the reduced time, Δt_D, has the same value as that described in Eq. 4.25. Eq. 4.24b must be added to p_w' to obtain an expression for the well pressure after the rate reduction. The well pressure at time of shutin, p_{wf}, is p_w'. Eq. 4.24b is added to p_{wf}, and the expression is expanded with all of the terms that do not change with time grouped into a constant term (in the same manner that Eq. 4.28 is derived from Eq. 4.24). We then find that the well pressure at some time after the rate change, Δt, is:

$$p_w = \text{constant} + \frac{0.1625\Delta q\mu}{kh} \log \Delta t \tag{4.28a}$$

Eq. 4.28a is the same as that of Eq. 4.28, which describes the well pressure for a well with an unchanging well pressure at shutin, except that the change in rate, Δq, is substituted for the rate, q. Eq. 4.28a shows that a plot of the well pressure, p_w, versus the log of the time since the rate change gives a straight line whose slope is the familiar $0.1625\Delta q\mu/kh$. The rate, q, has been used previously in the buildup slope expressions; however, we now must use the rate change, Δq:

$$m = \frac{0.1625\Delta q\mu}{kh} \tag{4.68}$$

The permeability in Eq. 4.68 is the undamaged permeability since the additional pressure drop caused by the damage around the wellbore is included in the Δp_{skin} term.

To find the additional pressure drop caused by the damage, Δp_{skin}, we can simply use Eq. 4.24b with a function of m substituted for the group of terms $0.141\Delta q\mu/kh$ to give:

$$\Delta p_{\Delta q} = 0.867m\,(\tfrac{1}{2})(\ell n\,\Delta t_D + 0.809) + \Delta p_{skin} \tag{4.69}$$

Eq. 4.69 is used by reading a value of $\Delta p_{\Delta q}$ from the data. Then the corresponding time since the rate change, Δt, is used as the basis for calculating the reduced time, Δt_D, which is calculated according to Eq. 4.25. As in other applications, mobility, k/μ, can be evaluated from the slope, m, and it is advisable to determine the ϕc product from reservoir behavior if at all possibile. The most popular method of determining ϕc by performance is to calculate it from the equation for the pseudosteady-state change in pressure with time, Eq. 3.33 or Eq. 3.34. The skin factor, S, can be calculated from Eq. 2.22 by substituting $0.867m$ for $\dfrac{0.141q\mu}{kh}$.

In applying Eq. 4.69, we should remember that the additional pressure drop, Δp_{skin}, represents the pressure drop caused only by the rate change. Consequently, we may want to calculate the additional pressure drop resulting from damage at a normal producing rate. This can be done by first calculating the skin factor, S, and using this factor in Eq. 2.22 to calculate the additional pressure drop for any producing rate, q.

Problem 4.8 should help clarify the calculating procedure. The solution is shown in appendix C.

PROBLEM 4.8: Analysis of a Two-Rate Pressure Buildup

A well in a Vicksburg 8,000-ft sand stabilized at a rate of 78 stb/d for a week, at which time the bottom-hole pressure and surface pressure appear to be unchanging. The well rate is then reduced to 64 stb/d with the pressure history indicated in Fig. 4–18. Find k_o, Δp_{skin}, S, and p_s if the following data apply to this well:

c_e = 1.379×10^{-4}/psi (must be calculated considering the gas, oil, and water saturations)
B_o = 1.322
μ_o = 0.39 cp
h = 20 ft
ϕ = 0.2
r_w = 0.265 ft
r_e = 1,490 ft (based on the estimated drainage area)

Pressure falloff tests. Typical data for a pressure falloff test is shown in Fig. 4–19. A negative rate applies to injection; thus, falloff-test analysis applies to injection wells or to disposal wells. Injection

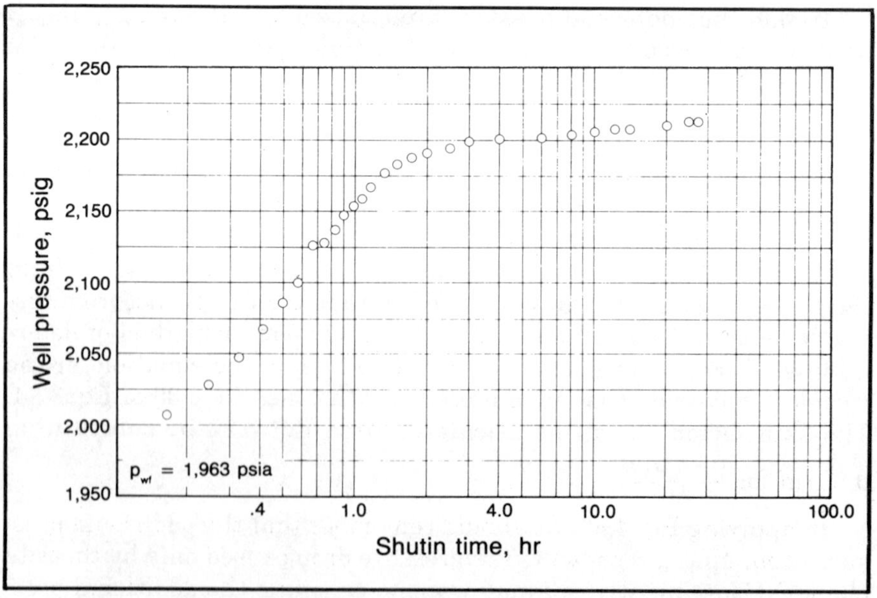

Fig. 4–18 Pressure data for two-rate buildup, problem 4.8

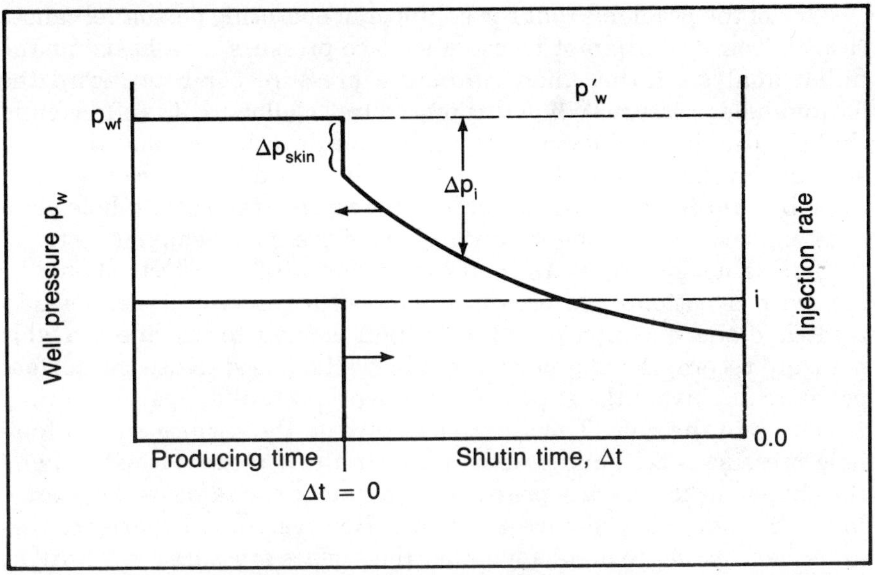

Fig. 4–19 Pressure falloff data

wells generally reach the unchanging well pressure state quickly. Therefore, the equations dealing with pressure buildup for wells with an unchanging well pressure can be applied in most cases if care is taken to use the proper sign on the terms that are a function of the well rate. Some additional problems and some simplifications are typical of pressure-falloff analysis.

Since we normally deal with liquid injection, we find that afterflow effects are almost nonexistent in pressure falloffs unless there is trapped gas someplace in the injection system or the surface pressure goes on vacuum. Otherwise, the very small compressibility of liquid lets the rate go to zero almost immediately after the injection well is shut in at the surface. In fact, there is often an instantaneous drop in the injection pressure that approximates the additional pressure drop caused by well damage (Δp_{skin}) since the flow in the damaged zone also is abruptly terminated. This effect is analogous to having flow through a small-diameter, long pipe that contains an orifice or other extreme flow restriction. If a valve on the upstream side of the orifice is closed, there is an immediate drop in the pressure on the upstream side of the valve that is equal to the flowing pressure drop across the orifice. Then the pressure begins declining more slowly. In the case of the injection well, the damaged zone is analogous to the orifice, and the same pressure behavior occurs when the well is shut in at the surface unless the damage radius of the system is substantial.

One of the problems that the engineer or operating personnel causes results from the attempt to use a surface pressure as a basis for the falloff analysis rather than running a pressure bomb to record the bottom-hole pressure. When this procedure is followed, it is often found that the usable pressure record is of limited duration because the pressure at the surface falls to zero. Obviously, this difficulty can be overcome by running a pressure bomb and recording the bottom-hole pressure as is normally done in other types of pressure-behavior tests.

This is not meant as an implication that all falloff tests should be run with a bottom-hole pressure bomb. If a reservoir is sufficiently tight and the injection pressure is high enough to provide a usable pressure record, the engineer would be well advised to use the surface pressure and avoid the expense and always-present danger of running a bomb into the hole. The difference between the surface and bottom-hole pressure after shutin without gas in the system is constant, and the change in the surface pressure with time is the same as the change in the bottom-hole pressure with time. However, only experience permits the engineer to predict whether the surface pressure record will be of sufficient duration to permit an accurate analysis. When in doubt, run a bottom-hole pressure bomb and repeat the test.

However, remember that afterflow in a pressure falloff test can be very severe when the surface pressure goes to a vacuum. Consider the extreme case where we are injecting into a highly permeable zone at 5,000 ft with only a surface pressure of 50 psig. If salt water is being injected, the hydrostatic pressure of 5,000 ft of water can be 2,500 psi, making the total effective injection pressure 2,550 psig. If the reservoir pressure is 1,500 psig, the effective pressure drop is about 1,050 psi. Then when the well is shut in, the well pressure is reduced to about 2,500 psig since only the surface pressure is affected. The effective pressure drop is reduced to 1,000 psi, and the injection rate is only reduced by about 5%. The injection rate continues after shutin, and the surface pressure becomes a vacuum immediately.

As noted, we can obtain the equations for the analysis of a falloff test by following the same procedures developed in the section on pressure-buildup analysis captioned, "Unchanging Pressure at Shutin." Referring again to Fig. 4–19, note that the well pressure during falloff is the sum of the pressure, p_w', that would have existed without shutin and the pressure decline caused by the positive rate change resulting from a discontinuation of the injection, Δp_i. The Δp_i equivalent of Eq. 4.24 then is:

$$\Delta p_i = \frac{0.141 i \mu}{kh} \left(1/2 \right) (\ell n\ \Delta t_D + 0.809) + \Delta p_{skin} \tag{4.70}$$

Where:

i = Injection rate, b/d

In Eq. 4.70 the Δt basis for the reduced time Δt_D is the time the injection well has been shut in. When this expression is added to p_w'—which is p_{wf} for an unchanging injection pressure—we obtain the equivalent of Eq. 4.26:

$$p_w = p_{wf} - \frac{0.141 i\mu}{kh}\left(\tfrac{1}{2}\right)(\ell n\ \Delta t_D + 0.809) - \Delta p_{skin} \quad (4.71)$$

When Eq. 4.71 is expanded, the equivalent of Eq. 4.28 in the unchanging section is:

$$p_w = \text{constant} - \frac{0.1625 i\mu}{kh}\log \Delta t \quad (4.72)$$

Eq. 4.72 shows that the slope of the plot of well pressure, p_w, versus the log Δt is:

$$m = \frac{0.1625 i\mu}{kh} \quad (4.73)$$

Once m has been evaluated, a form of Eq. 4.71 with 0.867m substituted for $0.141 i\mu/kh$ can be used to evaluate Δp_{skin}:

$$p_w = p_{wf} - 0.867m\ (\tfrac{1}{2})\ (\ell n\ \Delta t_D + 0.809) - \Delta p_{skin} \quad (4.74)$$

In using Eq. 4.74, the reduced time must be calculated by Eq. 4.25, which means that k/μ must be determined from the slope, m, using Eq. 4.73 and the porosity and compressibility must be determined independently. Fortunately, this is not too difficult for an injection well because, with only liquid flowing, the effective compressibility can be determined from empirical data shown in appendix B relatively accurately.

Generally, the average pressure in the drainage area of an injection well is relatively unimportant, since the engineer is much more concerned with displacement efficiencies under such conditions than with material balance and resulting saturations at various stages of injection. If the engineer does desire an average pressure for the drainage area, it can be obtained with reasonable accuracy using Eq. 4.62, which is the average pressure for a radial steady-state flow system. When the equation is applied to a pattern flood, remember that the drainage area of the injection well does not include the drainage area of the producing wells. For example, when applied to an injection well in a 20-acre five-spot pattern, the drainage area of the injection well would only be 10 acres. The effective external well radius would then be calculated on the basis of 10 acres.

Problem 4.9 should help to explain the details of a falloff analysis. The solution is shown in appendix C.

PROBLEM 4.9: Pressure Falloff Analysis

Given the falloff pressure data of Fig. 4–20 and the reservoir parameters listed for an injection well in a five-spot pattern flood, find the undamaged permeability, the pressure drop resulting from the well skin, and the skin factor, S. What is the maximum radius of investigation at a shutin time of 15 min? What is the injection rate if the skin factor is reduced to zero by acidizing? Assume no pressure loss in the tubing is caused by friction, water specific gravity equals 1.0, the producing wells are kept pumped down, pressure is atmospheric, and Δp_{skin} at the producing wells is zero. The reservoir data area as follows:

$h = 50$ ft
$c = 5 \times 10^{-6}/\text{psi}^{-1}$

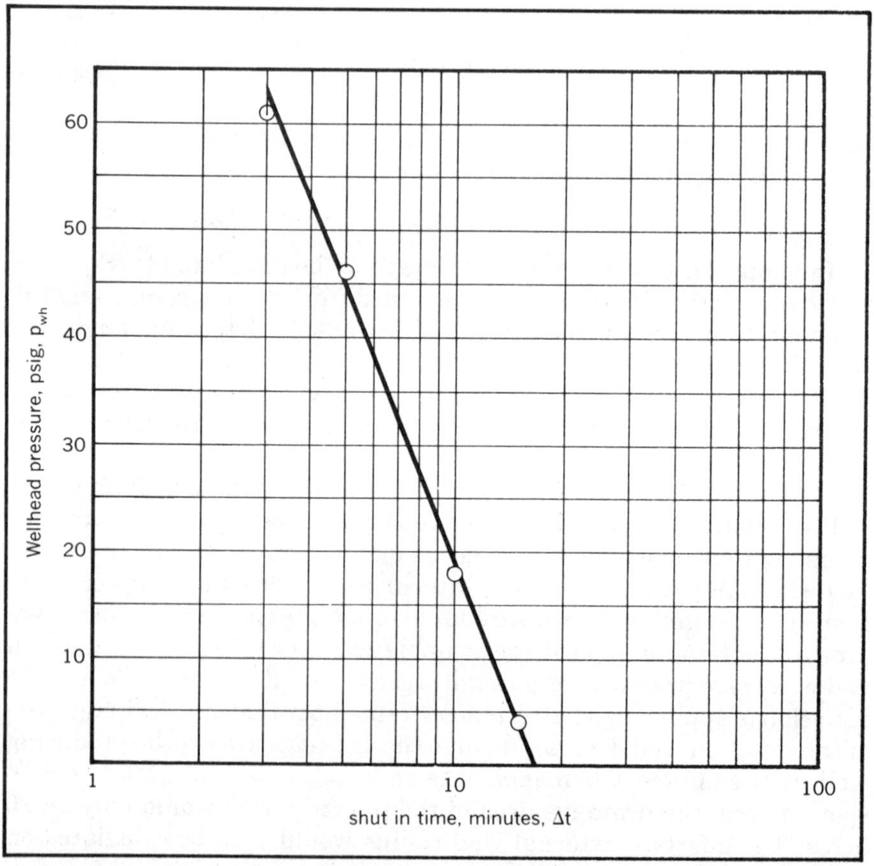

Fig. 4–20 Well pressure versus shutin time for a falloff test, problem 4.9

μ = 1.0 cp
r_w = 1/4 ft
Average sand depth = 3,000 ft
ϕ = 20%
Surface injection pressure at shutin = 500 psig
Injection rate before shutin = 250 b/d

The danger of using too short of a shutin time as a basis for the pressure data analyzed is emphasized in problem 4.9 where the calculated radius of investigation is very small. This problem is present in most types of well tests. However, it is particularly dangerous in the case of pressure falloff determined from surface-recorded data because the surface pressure falls to zero and becomes unusable quickly in many cases.

If the engineer experiences difficulties in obtaining a usable pressure falloff record, he should consider using a constant-rate pressure buildup similar to the constant-rate pressure falloff test for producing wells. The author has found this calculation to be necessary on a number of occasions.

Pressure buildup anomalies. Unexpected pressure behavior is often encountered when pressure testing wells. We often refer to these unexpected data as *anomalies.* Since we believe we can explain and understand some of these data, it may not be grammatically correct to refer to this behavior as anomalies. However, it does provide a convenient classification under which to discuss this type of phenomena.*

In this section we wish to make note of (1) the humping effect caused by the redistribution of the gas and oil following shutin, (2) the effect of a fault or nearby boundary, (3) the effect of stratification with widely varying permeabilities, and (4) the effect of two widely different parallel permeabilities in a reservoir, such as vugular and matrix permeability in limestone or lateral increase or decrease in mobility as would be encountered with a gas cap or water drive.

A typical case of humping caused by phase separation following shutin is illustrated in Fig. 4–21. It has been shown theoretically and

*Adoption of this definition of anomalies may mean that we should include afterflow and the effect of the skin damage in this section. However, since we have already discussed this matter, the material will not be repeated here. Furthermore, the objective of this section is to identify some of the anomalies without going into detailed explanations. The subject is treated in this way because numerous papers have been written on these problems—possibly many more than the subject warrants—and excellent, more detailed, discussions are available in SPE Monograph No. 1, *Pressure Buildup and Flow Tests in Wells,* by C.S. Matthews and D.G. Russell, and Monograph No. 5, *Advances in Well Test Analysis,* by R.C. Earlougher Jr.

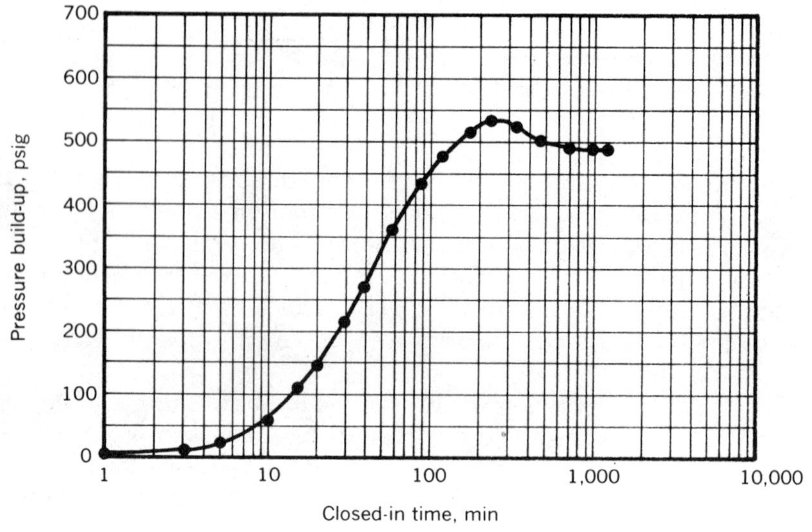

Fig. 4–21 Humping in a pressure buildup (after Matthews and Russell, *Pressure Buildup and Flow Tests in Wells*, Monograph No. 1, courtesy SPE 1967, © SPE-AIME)

in laboratory experiments that such an anomaly results from the phase redistribution in the tubing caused by the subsequent rise of the gas bubbles in the oil column following shutin. Our intuition again serves us poorly because most engineers erroneously conclude that, if the contents of a vessel remain the same, pressure on the bottom of the vessel must be constant regardless of the distribution of the phases in the vessel. Fig. 4–22 shows the erroneous nature of this conclusion. One portion of the diagram indicates the bottom-hole pressure that would exist in this well if it contained 1.0 cu ft of air at atmospheric pressure above a 2,910-ft column of fluid with a pressure gradient of 0.5 psi/ft. This column of fluid has trapped beneath it 1.0 cu ft of air at a pressure of 1,470 psia. To simplify the calculations, assume that the temperature is standard throughout the column and the cross-sectional area of the well is 1.0 sq ft.

Under these conditions the air in the bottom of the hole amounts to 100 scf so the total mass of air in the well is 101 scf. If all of this air occupies the 2.0 cu ft at the top of the well, the resulting pressure is 742 psia. When this amount is added to the pressure caused by the 2,910 ft of fluid, the bottom-hole pressure is 2,197 psia, an increase of 727 psi. The lower bottom-hole pressure that follows the abnormally high pres-

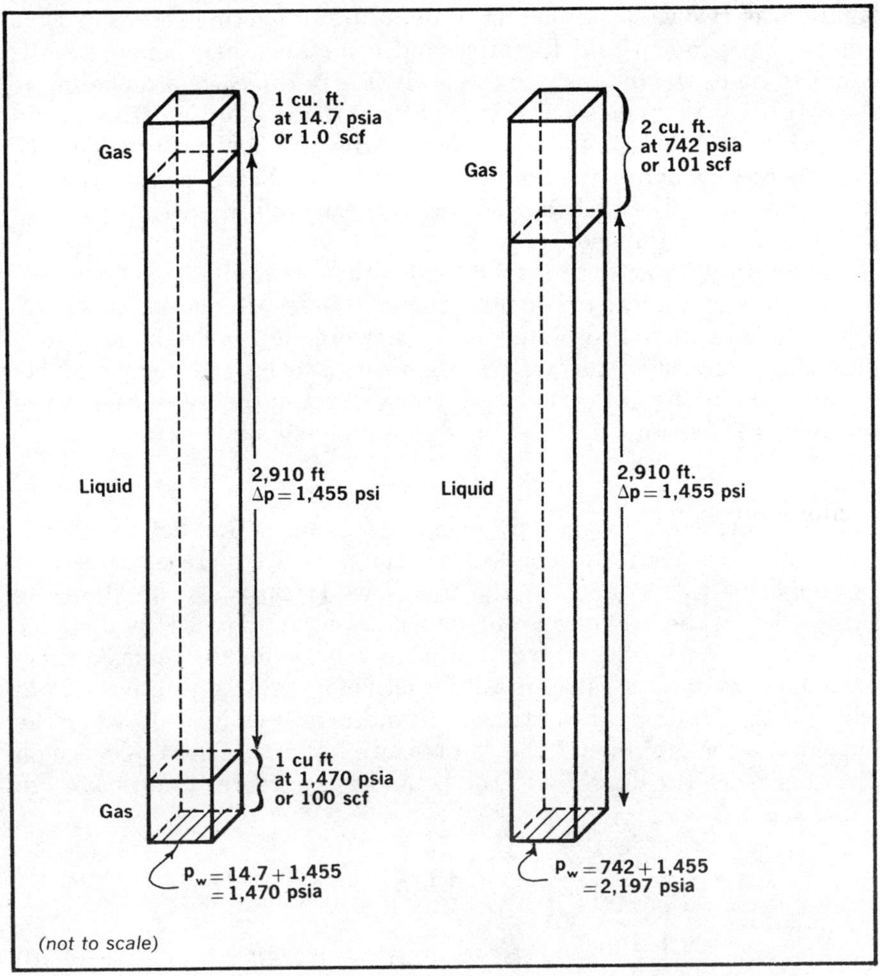

Fig. 4–22 Effect of phase distribution on the bottom-hole pressure

sure is caused by the flow of fluid back into the shutin well as a result of this high pressure. Consequently, it seems likely that abnormally high pressure occurs much more often than we realize because a reservoir must have a reasonably high permeability with a minimum of well damage to permit flow back into the reservoir. Unless flow back into the reservoir occurs, we do not observe the abnormally high pressure caused by the phase redistribution. The difficulties caused by phase redistribution can often be avoided by using a two-rate test as previously defined.

When a plane barrier or boundary exists in one direction from a well, the pressure behavior of the well is the same as the pressure behavior of two wells producing from an infinite-acting reservoir with identical producing-rate histories and a distance between the wells equal to twice the distance to the fault (Fig. 4–5). Such a modeling of the actual case provides a drainage pattern in the infinite-acting model for each well that is identical to the actual well's flow pattern. That is, there is no flow across the boundary in the actual case, and there is no flow across the line midway between the two infinite-acting wells in the infinite-acting model.

In the infinite-acting model consider that the wells have been producing long enough so an unchanging well pressure results at shutin. Then the shutin well pressure is the flowing well pressure at shutin plus the increase in the well pressure caused by the negative rate change of shutting in the real well, plus the well pressure caused by the negative rate change of shutting in the image well:

$$p_w = p_{wf} + \Delta p_q + (\Delta p_q)_{image} \tag{4.75}$$

In expanding Eq. 4.75, the Δp_q term is the same as that defined by Eq. 4.24. However, the term representing the pressure increase caused by the negative rate change in the image well, $(\Delta p_q)_{image}$, involves the calculation of the pressure drop at a radius equal to twice the distance to the fault. With such a large radius as a basis for the reduced time, Δt_D, there may be a substantial period before Δt_D is greater than 2, which is the arbitrary lower time limit indicated by Fig. 3–11 when the log equation approximates the Ei-pressure function. When t_D based on a radius twice the distance to the boundary is greater than 2, we can write Eq. 4.75 as:

$$p_w = p_{wf} + \frac{0.141 q\mu}{kh}\left(\tfrac{1}{2}\right)(\ell n\ \Delta t_D + 0.809) + \Delta p_{skin} + \frac{0.141 q\mu}{kh}\left(\tfrac{1}{2}\right)[\ell n\ (\Delta t_D)_{image} + 0.809] \tag{4.76}$$

We can expand both of the terms for log of reduced time by breaking them into the log of the diffusivity constant divided by the radius squared, plus the log of the shutin time, Δt. After doing this, we again use the technique of placing all of the resulting terms that do not change with time into one constant term:

$$p_w = \text{constant} + \frac{0.141 q\mu}{kh}\left[\tfrac{1}{2}(\ell n\ \Delta t + \ell n\ \Delta t)\right] \tag{4.77}$$

In Eq. 4.77 the two $\ell n\ \Delta t$ terms are obtained from the two log-of-reduced-time terms in Eq. 4.76. When Eq. 4.77 is written in terms of base 10 logs and a function of the normal buildup slope, m, is substituted for $q\mu/kh$, we obtain:

$$p_w = \text{constant} + 2m \log \Delta t \tag{4.78}$$

Eq. 4.78 then shows that the ultimate effect of a plane boundary on an otherwise infinite-acting pressure shutin effect is to double the normal pressure buildup slope, m. This effect has been verified for an unchanging well pressure. The same procedure can be followed to show that this effect is experienced if the well is in pseudosteady state or is infinite acting at shutin. However, to obtain the double slope, the reduced shutin time based on a radius of twice the distance to the fault must reach 2 before the shutin time in days becomes greater than $r_e^2/4\eta$, where r_e is the distance to the next closest boundary. This represents the approximate time when the effects of the closest boundary affect well pressure. At this time Δp_q is no longer based on the log equation.

The deviation from the buildup slope, m, can be used to calculate the distance to the closest boundary in many cases, regardless of whether the slope doubles. This technique is discussed fully in the section captioned "Reservoir Limit Tests." The point to be made at this time is that a boundary near the shutin well results in an increase in the slope of the buildup data, which may double if there is sufficient difference in the distance to the closest and the second-closest drainage boundaries. An example is shown in Fig. 4–23.

Stratification of a reservoir does not generally result in as many anomalies in pressure buildup as may be expected.*

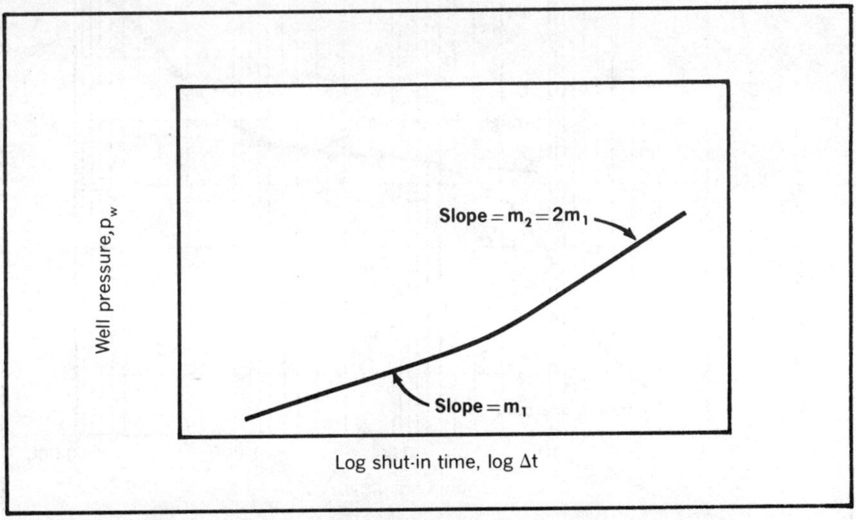

Fig. 4–23 Effects of a plane barrier on pressure buildup

*See chapter 10 of SPE Monograph No. 1, *Pressure Buildup and Flow Tests in Wells,* by C.S. Matthews and D.G. Russell (New York: 1967).

From a practical standpoint we must realize that stratified reservoirs behave as a homogeneous reservoir with the average characteristics of the stratified reservoir if there is unrestricted communication or permeability between the reservoir strata. However, if the strata are in communication only at the wellbore, they act like two separate reservoirs produced through a common system, which is exactly what they are.

In the latter case it should be realized that the two or more noncommunicating strata are depleted during production at constantly changing rates, even though we maintain the total producing rate from a well as a constant value. Therefore, at any particular time of shutin, the different strata are at different stages of depletion. This results in the strata being at different pressures and having different saturations. Under these conditions the resulting buildup may be useless.

Fig. 4–24 illustrates a buildup in a two-layer reservoir. The early part of the buildup is probably affected most by the more permeable strata, which has been more thoroughly depleted than the strata of lesser permeability and, thus, has a lesser reservoir pressure. After flattening the pressure history, which probably reflects the magnitude of the pressure in the more permeable strata, note that the pressure

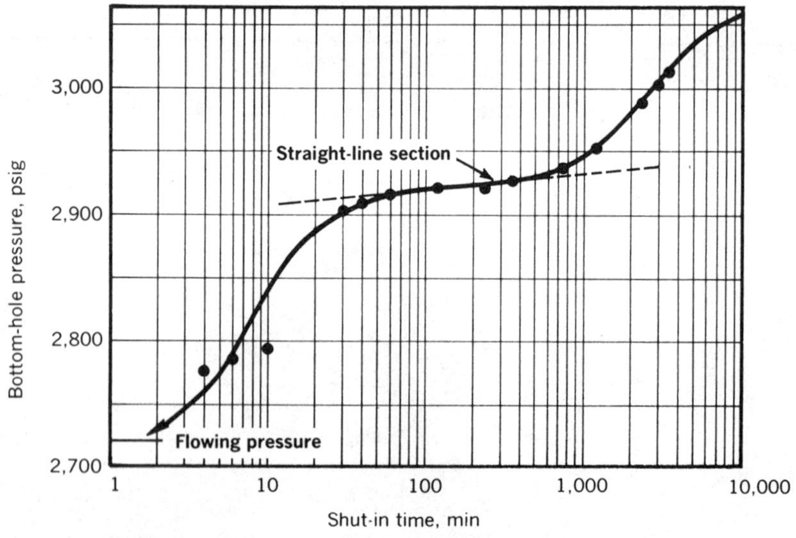

Fig. 4–24 Effects of a two-layer reservoir on pressure buildup (after Matthews, "Analysis of Pressure Buildup and Flow Test Data," courtesy *JPT*, September 1961, © SPE-AIME)

starts to increase more rapidly again. This buildup probably results from the less permeable strata, which has a higher reservoir pressure.

The engineer working in non-U.S. areas should be particularly careful about encountering multilayer effects in buildup data because commingling production from two or more reservoirs is not generally controlled by governmental agencies as it is in the U.S. Many engineers mistakenly believe that it is less expensive to commingle production and, thus, avoid dual completions and multiple wells at the same location. This conclusion does not appear to consider the reduction in the production rate that occurs because the commingled reservoirs cannot be efficiently drawn down if they have greatly different permeabilities. Thus, the company is unable to produce the field at the highest possible rate prior to its operations being nationalized. It also means that the ultimate recovery may be reduced because reliable production data are not available to show the amount of production coming from each reservoir. This problem is of lesser economic importance in non-U.S. areas since nationalization generally occurs long before the secondary or tertiary life of a reservoir is entered.

A reservoir that contains two vastly different pore structures, such as a jointed or fractured reservoir or a vugular limestone, may give pressure buildup characteristics similar to those for a reservoir with two noncommunicating strata. Whether the characteristics exhibited resemble this type or a regular homogeneous reservoir appears to be a function of the degree of porosity and permeability differences, the ratio of pore volumes represented by the two pore systems, and the degree of depletion.

Regardless of whether a buildup anomaly such as that in Fig. 4–24 is caused by noncommunicating strata or two widely different pore structures in the reservoir, it has been found that using a two-rate flow test often permits an evaluation of average reservoir characteristics. By simply reducing the rate, the flow remains in the same direction in the reservoir at all times whereas a rate of zero, as is experienced with a shutin, may result in flow between the strata or different pore systems. In the latter case flow in one strata or system may be taking place in one direction, while in another strata or pore system the flow at the same instant may be in the opposite direction. Regardless of the reason, experience has shown that a two-rate test often gives a pressure behavior that can be interpreted while a buildup test gives uninterpretable pressure data. A constant-rate drawdown test provides similar advantages, as discussed.

Another type of anomaly is the change in the buildup slope that may take place when the buildup begins to be affected by part of the reservoir that has a decidedly different saturation. This is the sort of anomaly that is often observed in the presence of a water-oil or gas-oil contact. The engineer simply needs to be aware of the fact that the presence of a contact can cause such an anomaly. There does not appear to be any way that such an anomaly can be avoided. Therefore, the engineer should limit the analysis to that portion of the buildup obtained before the change in saturation affects the buildup data.

Spherical pressure buildup. In chapter 2 a hemispherical steady-state flow equation is presented. This equation shows that, when production is from a small portion of the top or bottom of a thick formation, there is a tendency to experience hemispherical flow. At very small producing times the behavior approximates cylindrical, or radial, flow with the effective thickness being the perforated thickness. However, as more of the reservoir is affected, the flow is predominantly hemispherical. Then after the entire thickness of the reservoir is affected, the behavior again becomes cylindrical with the effective thickness being the total reservoir thickness.

In very thick reservoirs such as those often encountered in non-U.S. reservoirs, the spherical or hemispherical flow may represent a substantial period of time, the initial cylindrical flow period may be too brief to analyze, and the final cylindrical flow may be at too large of a time to be useful. Moran and Finklea also suggest that a spherical flow analysis is necessary in analyzing pressure data from a wire-line formation tester where flow is from a single perforation.[33]

Culham developed an infinite-acting spherical flow equation that when converted to the units of this book can be written as:[34]

$$p_{r,t} = p_i - \frac{0.141q\mu}{kr}\left(\tfrac{1}{2}\right)\text{erfc}\left(\frac{1}{4t_D}\right)^{0.5} \tag{4.79}$$

Where erfc is the complementary error function, a mathematical expression that occurs often in such applications. When the argument of erfc, $(\tfrac{1}{4}t_D)^{0.5}$ is less than 0.1 or t_D is greater than 25:

$$\text{erfc}\left[(\tfrac{1}{4}t_D)^{0.5}\right] = 1 - \frac{1}{\sqrt{\pi t_D}} \tag{4.80}$$

Then Eq. 4.79 becomes:

$$p_{r,t} = p_i - \frac{0.141q\mu}{kr}\left(\tfrac{1}{2}\right)\left(1 - 1/\sqrt{\pi t_D}\right) \tag{4.81}$$

Eq. 4.81 can be used to analyze a constant-rate drawdown or to derive an expression for analyzing a pressure buildup. If we use the same type of superposition employed in the derivation of the Horner equation, we can show that a well producing at a constant rate for a period t_p has a producing pressure of:

$$p_w' = p_i - \frac{0.141q\mu}{kr}\left(\tfrac{1}{2}\right)\left[1 - 1\Big/\sqrt{\pi\,(t_{pD} + \Delta t_D)}\right] \qquad (4.82)$$

This pressure occurs after a shutin time of Δt if the well had not been shut in. The pressure that would have existed if there had been no shutin is superimposed on the effect of the negative rate caused by the shutin:

$$\Delta p_{q\ \text{spherical}} = \frac{0.141q\mu}{kr}\left(\tfrac{1}{2}\right)\left(1 - 1\Big/\sqrt{\pi\Delta t_D}\right) \qquad (4.83)$$

The sum of Eqs. 4.82 and 4.83 then gives the shutin pressure:

$$p_r = p_i - \frac{0.141q\mu}{kr}\left(\tfrac{1}{2}\right)\left(\frac{1}{\sqrt{\pi}}\right)\left(\frac{r^2}{\eta}\right)^{0.5}\left[\frac{1}{\sqrt{\Delta t}} - \frac{1}{\sqrt{t_p + \Delta t}}\right] \qquad (4.84)$$

When expanded, Eq. 4.84 gives:

$$p_r = p_i - \frac{0.0158q\ \phi^{0.5}\mu^{1.5}c^{0.5}}{k^{1.5}}\left[\frac{1}{\sqrt{t_p}} - \frac{1}{\sqrt{t_p + \Delta t}}\right] \qquad (4.85)$$

Thus, a plot of the well pressure versus $[(1/\sqrt{t_p}) - 1/(\sqrt{t_p + \Delta t})]$, as in Fig. 4–25, gives a straight line whose slope is:

$$m = 0.0158q\phi^{0.5}\mu^{1.5}c^{0.5}/k^{1.5} \qquad (4.86)$$

To obtain a value for the Δp_{skin}, Eq. 4.81 can be modified to include the damage effect:

$$p_{r,t} = p_i - \frac{0.141q\mu}{kr}\left(\tfrac{1}{2}\right)\left(1 - \frac{1}{\sqrt{\pi t_D}}\right) - \Delta p_{skin} \qquad (4.87)$$

Then the spherical skin factor can be defined in a manner similar to that for the cylindrical skin factor as:

$$S_{spherical} = \Delta p_{skin}/(0.141q\mu/kr) \qquad (4.88)$$

Then Eq. 4.87 becomes:

$$p_{r,t} = p_i - \frac{0.141q\mu}{kr}\left[\tfrac{1}{2}\left(1 - \frac{1}{\sqrt{\pi t_D}}\right) + S_{spherical}\right] \qquad (4.89)$$

Therefore, to evaluate the skin factor in a spherical flow system, we simply read any point from the buildup plot of the well pressure versus $(1/\sqrt{t_p} - 1/\sqrt{t_p + \Delta t})$ and use the corresponding parameters in Eq. 4.89 to calculate the skin factor.

Note that the evaluation of the undamaged permeability from the spherical buildup plot does not require the effective spherical well radius, r. However, an evaluation of the spherical skin factor requires the effective spherical well radius. Since the well is physically a perforated cylinder, the value to be used for an equivalent spherical well radius is

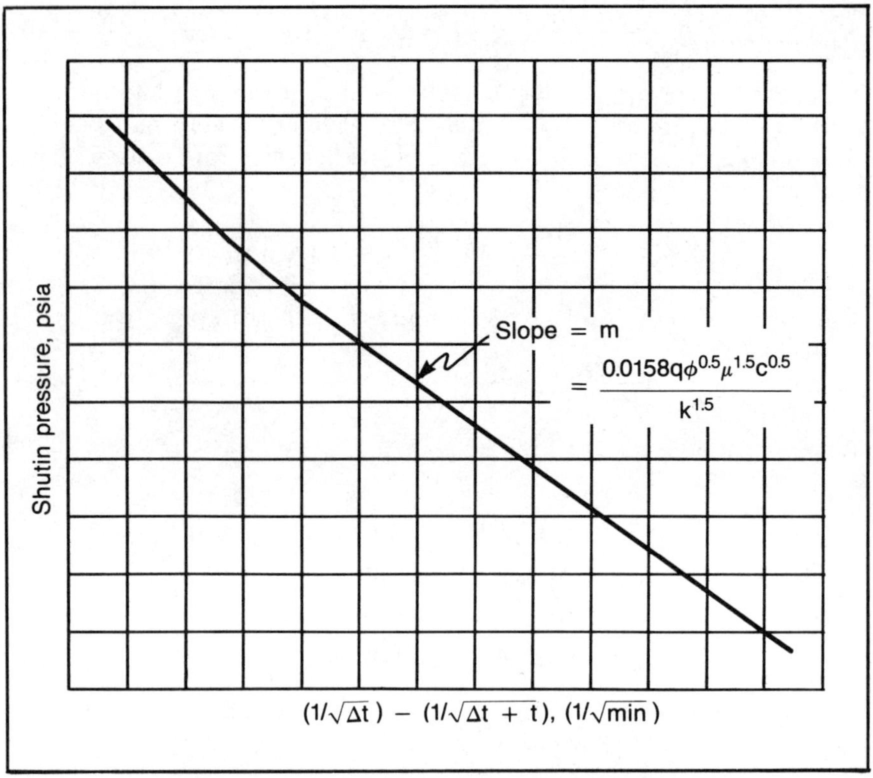

Fig. 4–25 Spherical pressure buildup analysis

not readily apparent. Culham appears to have made a detailed study of this matter since he changed the expression he chose to use for the effective spherical well radius between the time he presented his SPE paper and the time it was published.[35] Culham uses the expression proposed by Moran and Finklea, which is listed here without explanation:[36]

$$r_{ew} = b/2\ \ell n\ (b/r_w) \tag{4.90}$$

Where:

b = the perforated thickness

It is recommended that Eq. 4.90 be used to determine the well radius, r, for the spherical flow equations.

Work problem 4.10 and check the solution against the one in appendix C to understand spherical flow equations better.

PROBLEM 4.10: Pressure Buildup in a North Sea Well Demonstrating Spherical Flow

A DST on a cased and perforated North Sea well has effectively produced for 15.4 hr at a rate of 4,510 stb/d prior to its last buildup, which provided pressures as indicated:

Shutin Time, hr	*Pressure, psia*
0.000	5,264.613
0.059	5,492.078
0.119	5,587.871
0.178	5,604.078
0.238	5,613.918
0.297	5,621.152
0.357	5,626.648
0.476	5,635.332
0.595	5,641.121
0.774	5,647.488
0.952	5,652.406
1.190	5,656.456
1.369	5,659.062
1.846	5,663.691
2.441	5,668.035
3.453	5,672.374
4.347	5,675.269
8.098	5,681.058
10.421	5,683.082
13.636	5,685.109
16.137	5,685.687

The well is perforated from 11,950 to 11,990 ft and is produced from a reservoir with an estimated thickness of 266 ft. Prepare a log-log plot of the pressure change versus shutin time to determine when afterflow becomes negligible. Analyze the spherical behavior to determine the undamaged permeability. The reservoir data are as follows:

Oil viscosity = 0.4 cp
Oil formation volume factor = 1.734
Porosity = 15%, estimated
Effective compressibility = 20×10^{-6}/psi

If the well pressure is unchanging at the time of shutin the pressure-buildup analysis can be greatly simplified. Then p_w' is p_{wf}, the flowing pressure at the time of shutin, and the equivalent of Eq. 4.85 is:

$$p_r = p_{wf} + \frac{0.0158 q \phi^{0.5} \mu^{1.5} c^{0.5}}{k^{1.5}} \left(\frac{1}{\sqrt{\Delta t}} \right) \qquad (4.91)$$

A plot of the shutin pressure versus the reciprocal of the square root of the shutin time gives a straight line whose slope is m, according to Eq. 4.86. The skin factor can then be calculated from a modification of Eq. 4.91:

$$p_r = p_{wf} + m\left[\left(1/\sqrt{\Delta t}\right) + S_{spherical}\right] \tag{4.92}$$

Interpreting Drillstem-Test Data

The use of drillstem tests (DST) to predict the performance and production potential of wells seems to have steadily declined in recent years, although the author cannot cite any specific statistics on the subject. The apparent decline in drillstem tests probably results from many factors. Geophysical logging has been refined to the point that good producers can be predicted with confidence on the basis of logs alone. Furthermore, marginal producers rely on well-treating techniques, such as fracturing and acidizing, that cannot precede the DST. Therefore, the DST is considered useless except in rare cases.

However, it is the author's opinion that many DST companies are reluctant to provide the producer with production-rate estimates, which has also contributed immensely to the decline in the popularity of the DST as an evaluation tool. When a DST company does provide some sort of an analysis of the data, it is limited to determining the undamaged permeability and some abstract measure of the skin damage such as the skin factor or an infinite-acting damage ratio, which is very misleading. The producer wants and needs some estimate of the well's production rate. Although the technology we have available is not well suited to analyzing DST data, estimates of production rates can be made from this information that are very helpful in predicting the ultimate behavior of a well.

Historically, the DST has been a means of determining whether a well will produce, not at what rate it will produce. As noted, logging and well-treating developments have nearly relieved the DST of this function in modern drilling. However, a carefully interpreted DST can provide a useful estimate as to the rate of production expected from a well under various circumstances, which cannot be provided by logging. Thus, the DST can be a primary factor in determining whether a completion is economically advisable.

Mechanically, a DST is simply a means of temporarily relieving the producing formation of the static mud pressure so the reservoir fluid can flow into the wellbore. Originally, a DST consisted of running a dry string of drillpipe into the well, setting a packer to relieve the prospective producing formation of the static mud pressure, permitting the well to flow into the dry drillstem for a period, shutting in the well to obtain the reservoir pressure, and pulling the drillstring with the produced fluid held in the drillpipe by a retaining valve. However, a modern DST can be run with several flow periods and pressure buildups, and the produced fluid can be circulated to the surface before the tool is pulled from the hole.

To predict the settled production rate of a well from a DST, we must use the pressure-buildup and fluid-recovery data from the test. The drillstem test provides a unique opportunity to attempt to measure directly the initial pressure in the well drainage area. Consequently, it is strongly recommended that the engineer make an attempt to obtain this initial pressure. It is recognized that the engineer will not always be successful. However, the short time invested in the attempt seems well worthwhile when it is realized that this is probably the only possibility for obtaining such a direct measurement.

The problem with direct measurement of the initial pressure is that the formation in the vicinity of the wellbore is generally charged with some filtrate lost from the drilling mud and has a pressure higher than the static reservoir pressure. Consequently, it is necessary to permit enough of the reservoir fluid to enter the wellbore to relieve this abnormally high reservoir pressure before shutting in the well to record the initial pressure. Otherwise, the recorded pressure may be too high. However, if too much fluid is permitted to enter the wellbore before shutin, a regular pressure buildup occurs and the data must be extrapolated to the initial pressure using the Horner method. This test then requires some experience and judgment on the part of the engineer, since a tight reservoir should be permitted to produce for a longer time than the more permeable reservoir.

Generally, a flow period of 5–20 min allows a successful direct measurement of the initial pressure. The recorded bottom-hole pressure versus time provides a good clue as to whether the hoped-for initial pressure has been accurately recorded. If the flow period is too short, the pressure moves abruptly to an unchanging pressure without rounding the pressure record. If the flow period is too long, there is a long, rounded pressure record, but the recorded pressure still increases, after 10–15 min of shutin. Ideally, the record is well rounded but reaches a constant value in 5–10 min.

Fig. 4–26 shows a pressure record for a typical DST, although procedures vary so widely we probably cannot identify any as typical. Also, this pressure record is not in the form that the engineer normally receives it from the service company. Fig. 4–26 has increasing time from left to right and increasing pressure from the bottom to the top. An actual record has at least one of these scales reversed, and they overlap with the same x-axis point representing two or even three times on the scale.

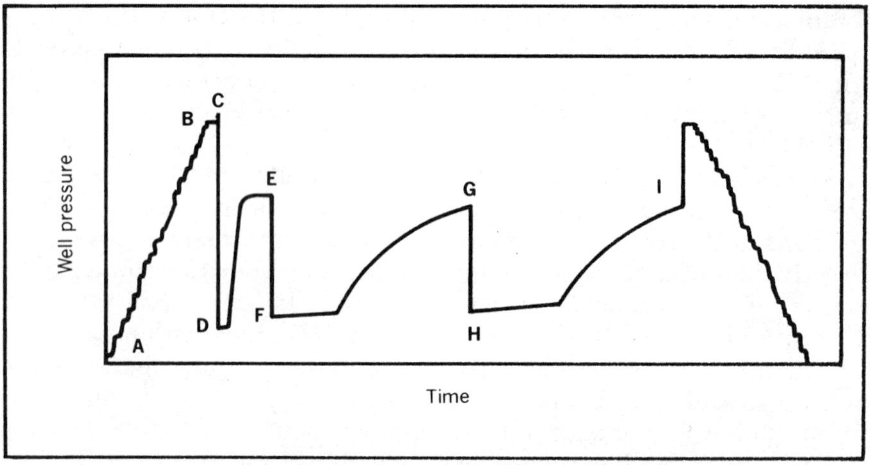

Fig. 4–26 Example of DST pressure data

Point A to point B in Fig. 4–26 indicates the pressure increase that occurs as a result of lowering the element into the column of mud in the hole. The pressure at point B is the total static mud pressure and can be checked against the mud weight to ensure that the service company is using the correct calibration chart for this particular pressure bomb.

Point D in Fig. 4–26 is the reduced pressure that results from setting the packer element and opening the flow valve so the formation is relieved of the static mud pressure. Point C is an abnormally high pressure, sometimes discernible, caused by compressing the mud trapped below the packer before the flow valve opens. The portion of the pressure chart at point E illustrates the ideal smooth, rounded initial buildup to a constant pressure that should represent the initial reservoir pressure. At point F the well has been permitted to flow into the wellbore. This represents the primary flow period and normally covers 30 min to 2 hr although longer flow periods are not uncommon.

Following this initial flow period, the initial pressure-buildup data are recorded at point G. These data are probably the most useful for calculating the undamaged permeability and the additional pressure drop caused by well damage. Following the initial pressure buildup, this particular test employed an additional flow period at point H and another buildup as a result of a shutin at point I. More than one flow period and pressure buildup are normally used to evaluate pressure depletion that may be occurring. The remainder of the pressure record shows the effect of unseating the packer and removing the pressure element from the column of mud along with the drillpipe.

A drillstem test represents one of the few times when the pressure behavior in a well strictly represents infinite-acting behavior. Consequently, it is recommended that at the present time a Horner type of analysis be used to evaluate the pressure-buildup data from the DST. The DST conditions do not exactly fit the Horner method, but the apparent accuracy of the calculated results using the Horner method seems acceptable.

In constructing the Horner plot, it is necessary to determine the producing time, t. Normally, it is recommended that in a Horner application the producing time be calculated by dividing the cumulative production by the rate at time of shutin. However, in the application to DST data, it is recommended that the producing time be used as the actual producing time and the average producing rate be calculated since we only know the total production from recovery in the drillstem or flow to the surface. A useful relationship in this connection is that the drillpipe volume in barrels per thousand feet is about equal to the inside diameter in inches squared.

Once the Horner plot has been constructed, the slope can be used to determine the undamaged permeability. The rate in stock-tank barrels is determined as discussed. The reservoir thickness can be determined from electric and/or radioactivity logs, and the fluid viscosity and oil formation volume factor can be determined from empirical data such as that found in appendix B. In evaluating the slope, consideration should be given to using the initial pressure—determined from the initial shutin pressure—as a point plotted at $(t + \Delta t)/\Delta t = 1.0$. Due to the shortness of many of the DST-buildup periods, the nonideal application of the Horner method, the change in Δp_{skin} as a result of clean up, and the nonconstant flow rate, this additional control of the slope is desirable. It is not suggested that the plot be forced through the initial pressure. However, it should be remembered that the plot should go through this point unless depletion has occurred, or the initial pressure as determined from the initial shutin is in error.

Once the slope and the undamaged permeability have been determined, the additional pressure drop caused by the well damage can be determined from Eq. 4.51:

$$p_{wf} = p_i - 0.867m(\tfrac{1}{2})(\ell n\, t_{Dp} + 0.809) - \Delta p_{skin} \qquad (4.51)$$

In evaluating the reduced time, t_{Dp}, based on the producing time prior to shutin, the porosity can be determined from log analysis. The compressibility can be determined more accurately from empirical data than is normally the case because there generally is no gas saturation to deal with initially in an oil well and no oil in a gas well. The bit size normally suffices for the well radius.

In estimating the settled production rate, the indicated productivity index using the flow rate and pressure drop, $p_i - p_w$, are meaningless. Some of the service companies refer to this value in predicting the settled rate. However, since only a very small portion of the reservoir has been affected during a DST, the so-called productivity index has no meaning. If this value is used to predict the flow rate under settled conditions with no skin damage, an extremely optimistic rate is obtained.

It is recommended that the settled flow rate be based more or less on theoretical considerations and the parameters determined from the pressure-buildup analysis. To use this approach, we must decide whether the well will reach steady state or pseudosteady state and what the drainage area will be. We must also estimate the skin factor that will be obtained by treating the well. When these estimates have been made, we can calculate the flowing or pumping rates that are obtainable.

Unless there is good reason to believe otherwise, it is normally a good practice to assume that pseudosteady-state flow will prevail. If we write the pseudosteady-state equation with the pressure drop corrected for Δp_{skin} and the external well pressure assumed equal to the initial pressure, p_i, we obtain:

$$q_{stb} = \frac{7.08kh(p_i - p_w)}{B_o\mu[\ell n(r_e/r_w) - ½ + S]} \tag{4.93}$$

Eq. 4.93 can be stated in a more convenient form by substituting a function of the slope, m, and the DST flow rate in stock-tank barrels for $7.08kh/B_o\mu$.

$$0.867m = \frac{q_{DST}B_o\mu}{7.08kh} \tag{4.94}$$

$$q_{stb} = \frac{q_{DST}(p_i - p_w)}{0.867m\ [\ell n(r_e/r_w) - ½ + S]} \tag{4.95}$$

Eq. 4.95 can be used to estimate production for a flowing well or a pumping well by simply adjusting the well pressure, p_w.

In estimating the well pressure, minimum errors are introduced by assuming that the pressure drop in the tubing as a result of friction is equal to the decrease in bottom-hole pressure caused by gas being liberated in the tubing as the oil moves to the surface. This is equivalent to calculating the difference between the bottom-hole pressure and the surface pressure as being the static-pressure difference based on liquid alone:

$$p_w = 0.433\gamma_o\ \Delta D + 15 \tag{4.96}$$

The constant, 15, in Eq. 4.96 is atmospheric pressure, and ΔD is the depth in feet between the bottom-hole pressure datum and the top of the unsupported column of oil. If gauge pressure is desired, the constant term, 15, should be omitted. If a well is flowing, ΔD is simply the average depth of the producing formation. However, if the well is being pumped, ΔD is the difference between the pump depth and the average depth of the producing formation.

One of the principal difficulties with a DST is that the flow periods may be so short that a very small amount of the reservoir affects the result of the test. This condition gives very poor statistical weight to the test. To obtain some idea as to how little of the formation is affected by a test, we use the term *radius of investigation,* which seems to have been borrowed from logging. We can simply use the stablization time equation to estimate the radius of investigation, r_i:

$$t = \frac{0.04\phi\mu c r_i^2}{k} \tag{4.97}$$

In using Eq. 4.97, we must arrive at estimations of the porosity, ϕ, and compressibility, c. We should use the mobility, k/μ, that is obtained from the slope of the Horner plot. The porosity should be obtained from logs, and the effective compressibility can be accurately determined from empirical data since only one phase is normally present in the active pore volume during a DST. We will see later that the service companies, in an effort to oversimplify this calculation, provide average values that in some cases even include averages for the length of the flow period. It is the author's opinion that an engineer on the job should be able to obtain more accurate values for the porosity, compressibility, and flow time than is afforded by the use of a statistical average.

To make sure the details of a DST analysis are clear, work problem 4.11 and check the solution against the one in appendix C.

Problem 4.11: Analyzing DST Data

A drillstem test recovers 7,500 ft of clean 40° API (sp gr = 0.825) oil in an 80-min flow test from 20 ft of net pay at 9,985–10,015 ft using 4½-in. drillpipe (ID = 3.83 in.) in 9.8 lb/gal of mud. The following readings are obtained from the pressure recorder:

Buildup Pressure, psig	Shutin Time, min
3,460	10
3,900	20
4,190	30
4,300	40
4,390	60
4,410	80
4,422	100

Hydrostatic mud pressure = 5,080 psig
Initial closed-in pressure, ICIP = 4,500 psig

Initial flowing pressure, IFP = 250 psig
Final flowing pressure, FFP, after 80-min open = 2,700 psig

Assume the following:

1. The lightening of the oil by gas coming out of solution equals the friction loss in the tubing (a pessimistic assumption) during flow
2. $r_w = 1/4$ ft, obtained from bit size
3. $c_{eff} = 5 \times 10^{-5}$/psi
4. $\phi = 20\%$, estimated from logs
5. $B_o = 1.75$, estimated by assuming reservoir is initially at the saturation pressure
6. $\mu_o = 0.4$ cp, determined from empirical data by assuming $p_i = p_s$
7. The well can be treated to obtain $\Delta p_{skin} = 0$

A. Calculate the Horner ratio for shutin times of 60, 80, and 100 min. Determine the initial well pressure, p_i.
B. Check the hydrostatic mud pressure and determine the undamaged permeability.
C. Calculate the skin factor, S.
D. Find the initial stabilized, pseudosteady-state flowing rate if the well drainage radius is 745 ft (40-acre spacing).
E. Estimate the effected radius during the flow period.

Some DST companies may recommend a damage ratio calculation as a means of determining well damage. This parameter, as previously discussed, has many advantages over the use of the skin factor, S, and Δp_{skin} when properly applied. However, in applying the damage ratio concept to a DST, it is necessary to evaluate the average permeability from an infinite-acting test. This factor is somewhat meaningless since the damage ratio determined in this way is a function of the length of the flow period. In other words in the presence of positive damage, the damage ratio continues to decrease as the length of the flow period increases.

This concept can be readily understood if we look at the definition of the damage ratio as being the undamaged permeability divided by the average or damaged permeability. The undamaged permeability is fixed and is independent of the producing time. However, the average permeability, which is an average of the reduced permeability and the undamaged permeability outside the damaged zone, continues to increase as the amount of effected reservoir increases. This increase in the average permeability results from the fact that the amount of damaged formation is fixed, whereas the amount of undamaged formation in the flowing area continues to increase as the effected area increases. The continued increase in the average permeability results in a continued decrease in the damage ratio. In addition, the conventional DST does not provide a stabilized, pseudosteady-state or steady-state flow rate that can be corrected by the damage ratio.

In addition to the fact that the damage ratio is not a constant, the equations proposed for calculating the radius of investigation of a DST also incorporate average values for permeability, porosity, viscosity, compressibility, and length of the flow test. This practice cannot be criticized as long as it is clear that the equations are based on average values. However, the ready acceptance of these equations by practicing engineers who are in a position to determine values for these assumed reservoir parameters with an accuracy much greater than that of an average value is at best confusing.

Use of the Horner method in analyzing the pressure-buildup data of a DST leaves much to be desired. It is generally assumed that, theoretically, the Horner method fits the DST ideally because there is no question that the well behavior is infinite acting during the DST. However, if we review the Horner equation, we see that it is derived for the case where well pressure is declining at the time of shutin and that, if the well had not been shut in, the pressure would have continued to decline. The effect of the negative rate change superimposed on this declining presure base, p_w', results in the Horner equation. Contrast this with the actual situation that exists in most DST's. A DST that does not flow to the surface prior to shutin has a well pressure that is increasing at the time of shutin. Consequently, the effect of the negative rate on the reservoir when the well is shut in is not superimposed on a declining pressure base but is rather superimposed on an increasing pressure base.

This is probably why the buildup data for a DST take so long to reach a straight line on a Horner plot. The effect of the increasing pressure at shutin decreases as the shutin time increases. Eventually, a straight-line Horner plot results. Since the well is shut in at the bottom of the hole on a DST, afterflow is negligible. However, it is only fair to note that most drillstem tests are run in the presence of considerable well damage since the well has not yet been treated. This tends to aggravate the afterflow well-damage anomaly that delays the straight-line portion of the buildup.

Some DST service companies offer a type of curve fitting to overcome the difficulties of afterflow during the buildup portion of the DST. It appears that they are correcting for a phenomenon that does not exist. Since the well is shut in near the producing face, the only afterflow is caused by the compressibility of the fluid between the flow valve and the formation, a distance of 100–200 ft, which is filled with oil (no gas) during the test of an oil reservoir. Thus, the afterflow on a DST appears to be negligible.

Since difficulty in reaching the Horner-plot straight line is normally experienced in analyzing a DST buildup, the initial pressure plotted at $(t + \Delta t)/\Delta t$ of 1.0 used to control the slope m of the plot becomes very important. Consequently, we should try to measure the p_i directly, as suggested previously, and avoid the advice of some of the DST companies that would have us determine p_i from the initial buildup. Obviously, this will not give the correct answer unless the correct slope is used, and the correct slope generally requires a knowledge of p_i. A series of 5-min flow periods and shutins may be used to guarantee the direct measurement of p_i in almost all cases.

It appears that in the future a calculating procedure will be available for analyzing DST-buildup data that will account for the increasing well pressure at the time of shutin and will thus permit utilizing much earlier shutin pressures. In the meantime, the methods illustrated in problem 4.11 are strongly recommended.

A series of flow tests and shutin periods have been used to evaluate pressure depletion of a reservoir as a result of the fluid removed during a drillstem-test flow period. Obviously, any measurable decline in the average reservoir pressure caused by the small amount of fluid removed would mean that the reservoir is too small to justify completion costs of the well. The conclusion that pressure depletion is occurring should be reached only after repeated drillstem tests so the engineer is certain that the removal of some small amount of reservoir fluid results in a decrease in the average reservoir pressure. In this instance the analysis can be based on p^*, the pressure extrapolated to a time ratio of one, since the decline in this value is similar to the decline in the average reservoir pressure. The note of caution—to make certain that pressure depletion is occurring—is probably unnecessary. It is generally impossible to convince an engineering supervisor that a well with a substantial initial rate of production should be plugged because the reservoir is too small to justify the completion costs even when conclusive test data are available. However, much money has been wasted on completion costs that could have been avoided.

Accurate reservoir engineering analysis is impossible without good reservoir data, and the conscientious reservoir engineer must be ready at all times to justify the need for data economically. In a DST, accurate data generally do not add to the cost of the test. In this case it seems important to make it known that we want a quantitative analysis of the data. Generally, this message is all that is required to get the service company to obtain the pressure data as accurately as possible. However, some degree of reliability is added to the data if we ask for the date the pressure bomb is calibrated for the pressure and temperature ranges anticipated. As a rule of thumb, we may request the use of another bomb if the calibration date is more than 6 months.

The engineer will find that a blanked-off pressure recorder, a recorder outside the flow string, in addition to the recorder inside the flow string helps detect mechanical difficulties. Another rule of thumb the engineer may keep in mind in connection with a DST is don't be hasty. Unless you anticipate difficulties in unseating the packer, attempt to record the initial reservoir pressure, use at least three flow periods and a final flow period of sufficient length, and measure the gas, oil, and water production carefully even when the test appears to be noncommercial. The final flow period should be at least 2 hr if the well capacity is anticipated to be less than 10 md-ft and 30 min to 1 hr if the capacity is anticipated to be more than this value.

Sidewall fluid sampling devices are in extensive use and provide an excellent means of determining water-oil and gas-oil contacts. However, the pressure records obtained are not generally of sufficient accuracy to permit a quantitative interpretation.

Type-Curve Matching

Type-curve matching techniques have developed in the last 10 years to the extent that it appears that we can find a type curve supposedly to fit buildups, drawdowns, interference tests, pulse tests, and about every other type of transient pressure behavior for all sorts of geometries including fractured wells, for all conceivable drainage shapes, and for all degrees of well damage or improvement. The general form is a log-log curve plot whose shape is matched by another plot such as a log of pressure, flow rate, or cumulative flow versus a log of time, cumulative flow, or the Horner ratio. Once a shape match has been obtained between the log-log type curve and the log-log data plot, a comparison of the two scales makes it possible to determine a variety of reservoir parameters.

It of course is difficult to match two log-log plots accurately when they do not exactly fit. The older reservoir engineer may recall the difficulty of using a log-log plot of production rate versus time for hyperbolic-decline curve analysis and the insensitivity of the method. Similar difficulties are experienced with type-curve matching. Earlougher states: "For single-well testing, type-curve matching should be used only when conventional analysis techniques. . .cannot be used."[37] This statement accurately describes this author's (Slider's) opinion of the type-curve matching technique. Consequently, since the stated objective of this book is not to cover every method proposed or used in reservoir engineering but simply to provide practical methods of solving most reservoir problems, the subject of type-curve fitting is not covered in detail here. However, type-curve fitting techniques are often the only methods that provide problem solutions. In such cases the engineer should consult the Earlougher work.[38] This section provides the reader with a general understanding of the technique so he can use it with the aid of other references to solve problems that are otherwise unsolvable.

One very useful technique has emerged from type-curve matching. It appears that the first efforts to use type-curve matching in reservoir analysis were by McKinley.[39] He reportedly was concerned initially with analyzing pressure-buildup data from a pumping well where afterflow effects are most severe since the total pressure drop, $p_e - p_w$, is not substantially reduced immediately after the pump is shut down. The bottom-hole-pressure increases after shutin are caused by the increase in the column of fluid in the hole, and the producing rate is not substantially reduced. This gives a linear plot of the well pressure versus time.

Then since the increase in pressure is proportional to the increase in time as long as the flow rate is constant, the fractional increase in the bottom-hole pressure, $\Delta p/p$, is equal to the fractional increase in the shutin time, $\Delta(\Delta t)/t$. Since $\Delta p/p$ is equivalent to the $\ell n\ p$ and since $\Delta(\Delta t)/t$ is equivalent to the $\ell n\ \Delta t$, a plot of $\ell n\ p$ versus $\ell n\ \Delta t$ gives a straight line whose slope is 1.0 as would a plot of log p versus log Δt. Thus, as long as a log-log plot of p_w versus time is a straight line with a slope of 1.0, we can conclude that the change in the pressure is caused only by the increase in the column of fluid in the hole, or the afterflow.

Subsequent work by many different researchers shows that the same technique can be used to determine when the increase in shutin well pressure in gas wells and in oil and gas wells is dominated by afterflow. In other words as long as a log-log plot of the shutin well pressure versus time is a straight line with a slope of 1.0, we can conclude that the pressure change is just a result of afterflow and there is no need to analyze these straight-line data. The same technique can be used to analyze the afterflow effects in a constant-rate drawdown test.

Since there is often some question as to which apparent straight line should be used in pressure-buildup and drawdown analysis, the log-log plot of well pressure versus time is very helpful in eliminating at least a portion of the poor data.

Perhaps the most common type curves are those constructed from pressure function data. The Ei-pressure function plot, Fig. 3–9, is similar to the type curve in the Earlougher monograph, except the plot in the monograph is much more extensive.[40] The idea is that the shape of the pressure-function plot, i.e., that of the exponential pressure function versus dimensionless time, has the same shape as the log-log plot of reservoir pressure at some point versus time.

For example, the reservoir limit test section presented a method of determining the distance to a barrier by determining ($p_w' - p_w$) from an extrapolation of the early straight-line relationship between the well pressure recorded while the well produces at a constant rate and the actual well pressure once the boundary is felt at the producing well. This difference ($p_w' - p_w$) is set equal to the pressure drop stated as a function of the exponential pressure, and the expression is solved for the exponential pressure function. Then the corresponding t_D is determined from Fig. 3–9, and the distance to the barrier is calculated.

It was noted that, in order to distinguish the barrier effect from similar behavior such as the change in saturation near the wellbore during the flow test, it is necessary to repeat the calculations using different values of ($p_w' - p_w$) and the corresponding producing time.

Type-curve matching provides an alternative to this method. A log-log plot can be prepared of $p_w' - p_w$ versus the producing time using the same cycle sizes as those of Fig. 3–9. Then this plot can be matched with the type curve in Fig. 3–9 so it overlays the type curve. Once this is accomplished, the relationship between the two scales can be determined by reading any convenient point from the match. Then Eq. 4.18 can be adapted to the match point interpretation by using it as:

$$(p_w' - p_w)_{match} = \frac{0.141 q\mu}{kh}\left[\left(\tfrac{1}{2}\right)\left(-\mathrm{Ei}\,\frac{-1}{4t_{Di}}\right)\right]_{match} \tag{4.98}$$

The permeability/viscosity ratio can be calculated from Eq. 4.98. Knowing this, the other half of the match can be used with the adaptation of Eq. 4.19 to calculate the distance to the barrier:

$$(t_{D\,image})_{match} = \frac{6.33(k/\mu)\,t_{match}}{\phi c(2d)^2} \tag{4.99}$$

Knowing the match values, the k/μ determined from Eq. 4.98, and the porosity-compressibility product, the distance to the barrier, d, can be calculated. When this method is used to determine the distance to the barrier, there is no need to repeat calculations or to verify the doubling of the semilog-plot slope in order to validate the barrier effect. Verification comes from the curve fit.

In addition to the application of type-curve fitting in interference testing, Earlougher also discusses type-curve matching for testing constant pressure and drawdown, for estimating permeability, porosity-compressibility product, skin factor, and wellbore storage, for analyzing vertically and horizontally fractured wells, and for interpreting vertical interference tests.[41] The evaluation of the permeability, porosity-compressibility product, skin factor, and wellbore storage is discussed for a variety of conditions. The mechanical process of type-curve matching is illustrated in Fig. 4–27.

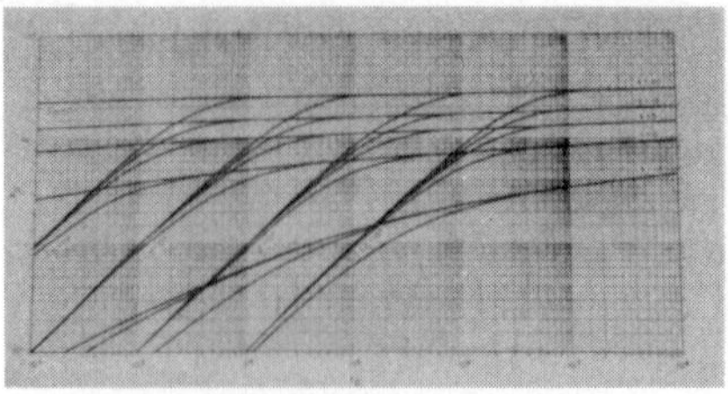

(a) Choose a type curve.

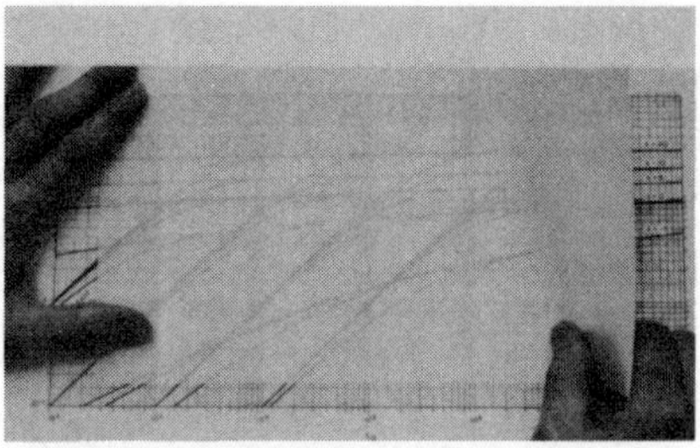

(b) Overlay with tracing paper.

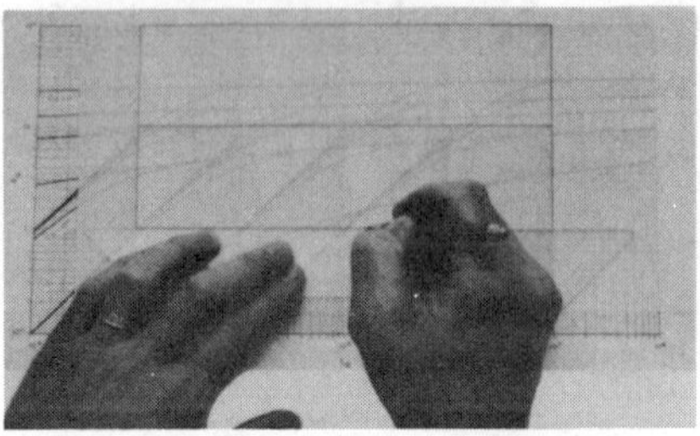

(c) Trace major grid lines.

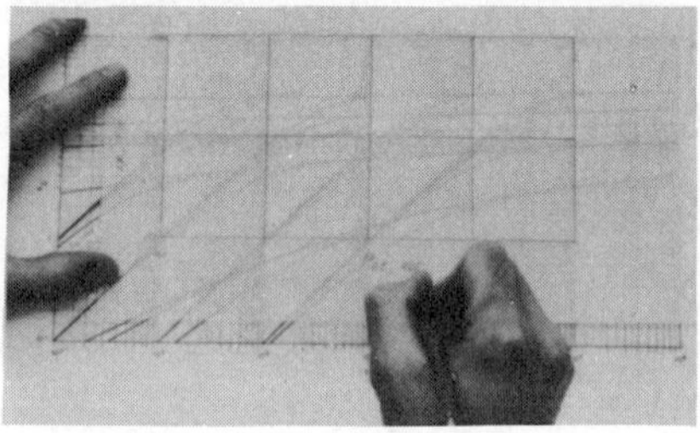

(d) Label axes.

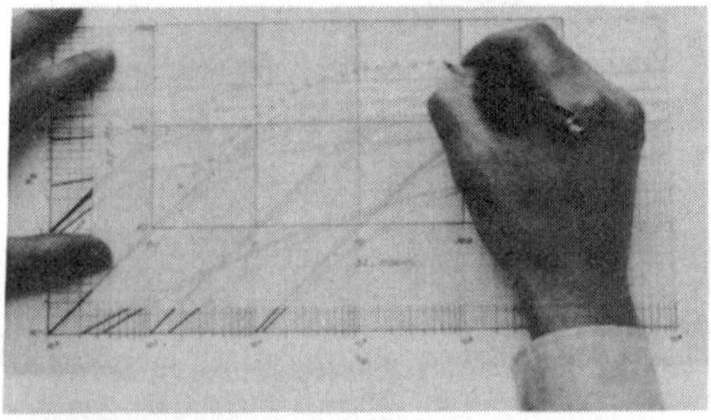

(e) Plot observed data using type-curve grid.

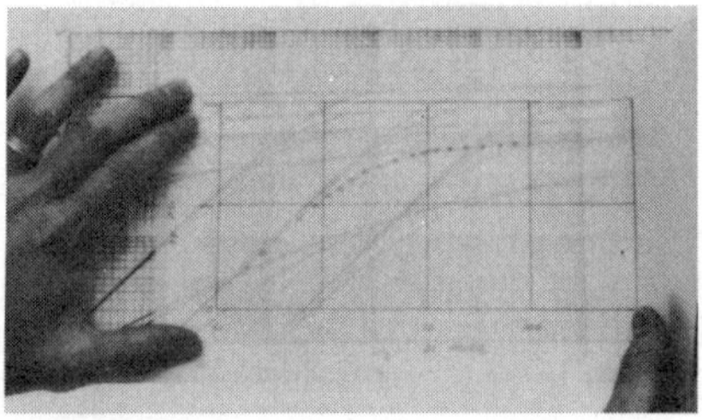

(f) Slide tracing paper to match a type curve.

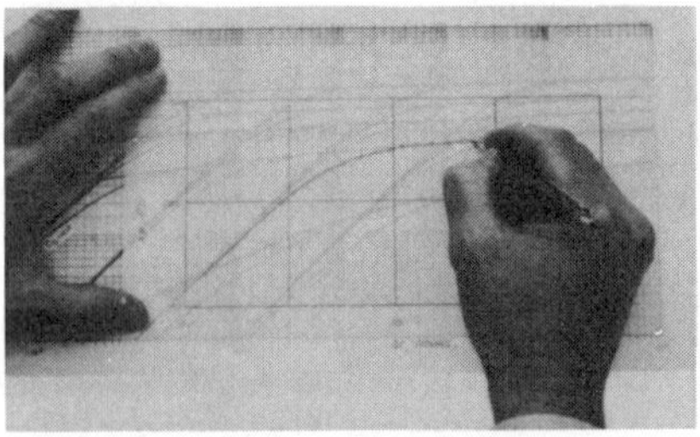

(g) Trace the matched curve.

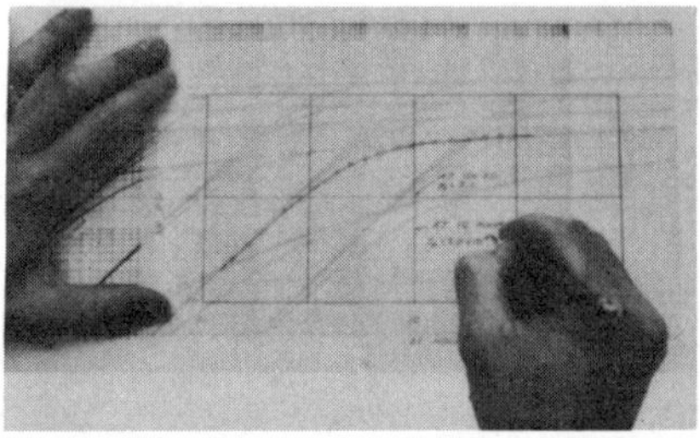

(h) Pick a match point.

Fig. 4–27 Steps in type-curve matching (after Earlougher, *Advances in Well Test Analysis*, Monograph No. 5, © 1977, SPE-AIME)

It should be reemphasized that this procedure should be used only as a last resort when other techniques are not feasible. Earlougher has been quoted as indicating that type-curve matching should be used only when conventional techniques cannot be used for single-well testing. Perhaps the best known transient-pressure expert, Henry J. Ramey Jr., also said: "Type-curve matching should be done in emergency or as a checking device."[42]

Pulse-Test Interpretation

Flow tests and buildup tests indicate the permeability thickness product only in the vicinity of the well. The average permeability and thickness for the reservoir is not indicated by such tests. An alternative to the single-well test is to use a signal or disturbance in one well, called the active well, and observe the effects of this disturbance in another well. The magnitude of this disturbance, the pressure change, at any particular time is a function of the permeability thickness. The time that it takes the disturbance to reach the observation well indicates the porosity-compressibility product of the system as shown in the t_s equation, Eq. 3.50. Since these effects result from the travel of the disturbance through the reservoir, the permeability, thickness, porosity, and compressibility indicated are influenced by the reservoir conditions between the wells.

Using an active well and an observation well in running a test is referred to as an *interference test,* since we actually measure the interference of the active well with the observation well. If the distance between wells is normal, it takes a substantial amount of time to cause measurable pressure changes, even with the quartz-type gauge that measures pressures to less than 0.01 psi. Consequently, it is often helpful to use a series of injection or production rates of the same magnitude that are separated by shutin periods. This series gives a cycle of pressure changes that can sometimes be more accurately interpreted than those using a single, long injection or production rate change.

Using a series of injection or production and shutin periods to determine the pressure effects in an offset well or wells is referred to as a *pulse test*. In this section we consider an analysis method that can be used for pulse-test data, and we explain why it works in qualitative terms. The subject is so involved and so many different approaches have been devised to analyze the pulse test that the exact theoretical background is not covered. The instructions and data presented should be sufficient for the engineer to solve and understand simple pulse results.*

Fig. 4–28 indicates the nature of a pulse test. It assumes that the flowing pressure at the time of shutin is unchanging, i.e., it would not have changed during the pulse test if the rate changes had not been

*Earlougher's *Advances in Well Test Analysis* (SPE, 1967) provides many additional ideas on pulse testing and an extensive list of references on the subject.

instituted. Two characteristics of the pulse-test response are generally used in analyzing the test. The first of these characteristics is the *pressure pulse,* the magnitude of the pressure response associated with a particular pulse, Δp_{pj}, as illustrated in Fig. 4–28 for Δp_{p1} and Δp_{p4}. The

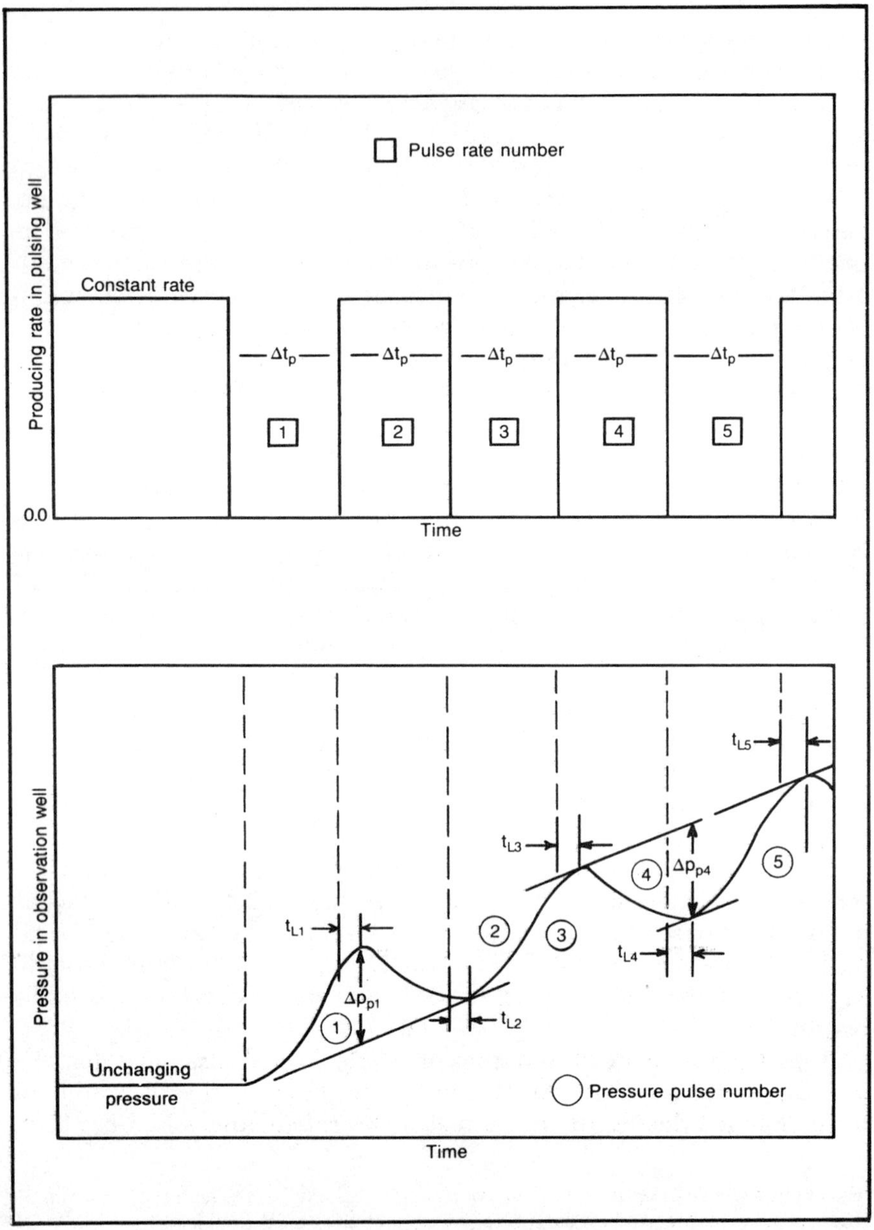

Fig. 4–28 Pulse test rate and pressure response

j subscript refers to the pulse number. This pressure-pulse magnitude varies with the pulse number, but there are groups of Δp_{pj} values that we can treat as being of the same magnitude. The other characteristic that is used in the quantitative analysis is the time lag associated with a particular pulse, t_{Lj}. The time lag for the first pulse, t_{L1}, is illustrated in Fig. 4–28. Note that this represents the time required for the second pulse to reverse the pressure trend at the observation well. The term t_{Lj} is the time measured from the beginning of a pressure pulse, $j + 1$, to the time when the pressure trend is reversed in the observation well.

The Δp_{pj} and t_{Lj} quantities can probably be more exactly defined by the mechanical methods used to determine the quantities. The end of the time lag, the point when the pressure trend is reversed, is determined by drawing a tangent to adjacent peaks or valleys on the pressure record and then drawing a parallel tangent to the valley or peak in between, which is caused by the subject pulse. This point of tangency is the time when the pressure trend is reversed. The pressure pulse, Δp_{pj}, is then simply the vertical, pressure, distance between the two parallel lines.

For example, to measure t_{L4} from the record of Fig. 4–28, we must first draw a line on the pressure record tangent to the peaks associated with pulse 3 and pulse 5. Once this tangent is established, we draw a tangent to the pressure peak associated with pulse 4 that is parallel to the previously constructed tangents. The point of tangency associated with pulse 4 is the time when the pressure trend is reversed, and t_{L4} is this time minus the time when pulse 4 ended at the active well and pulse 5 began. Then Δp_{p4} is the vertical distance between the two parallel tangents as shown. The term $0.5[-\text{Ei}(-1/4t_D)]$ is used for the pressure function to calculate the pressure changes in the observation well caused by the rate pulses in the active well. Then we can use the methods in chapter 3 to write an expression for the pressure in the observation well during the first pulse if the flowing pressure at the time of shutin, p_{wf}, is unchanging:

$$p_t = p_{wf} + \frac{0.141q\mu}{kh}\left\{\tfrac{1}{2}[-\text{Ei}(-\tfrac{1}{4}t_D)]\right\} \tag{4.100}$$

Then by using superposition, we can show that during the second pulse, or production, the pressure in the observation well is:

$$p_t = p_{wf} + \frac{0.141q\mu}{kh}\left[\tfrac{1}{2}\left(-\text{Ei}\frac{-1}{4t_D}\right)\right] - \frac{0.141q\mu}{kh}\left\{\tfrac{1}{2}\left[-\text{Ei}\frac{-1}{4(t_D - \Delta t_{Dp})}\right]\right\} \tag{4.101}$$

Where:

t_D is based on the total time
Δt_{Dp} is based on the pulse period
Δt_p, assuming all pulse periods are equal

Thus, the first term is the pressure buildup caused by the original shutin, and the second term is the pressure drawdown caused by resuming the producing rate after the first shutin period. Similarly, the pressure during the third pulse, or shutin, is:

$$p_t = p_{wf} + \frac{0.141q\mu}{kh}\left[\tfrac{1}{2}\left(-\mathrm{Ei}\frac{-1}{4t_D}\right)\right] - \frac{0.141q\mu}{kh} \times \left\{\tfrac{1}{2}\left[-\mathrm{Ei}\frac{-1}{4(t_D - \Delta t_{Dp})}\right]\right\} + \frac{0.141q\mu}{kh}\left\{\tfrac{1}{2}\left[-\mathrm{Ei}\frac{-1}{4(t_D - 2\Delta t_{Dp})}\right]\right\} \tag{4.102}$$

Again, assume that all of the pulses are of the same length.

It should be apparent then that we can write an expression for the pressure in the observation well during any pulse, j, and solve it for the dimensionless pressure drop:

$$\Delta p_{Dt} = \frac{p_t - p_{wf}}{(0.141q\mu/kh)} = \tfrac{1}{2}\left[-\mathrm{Ei}\frac{-1}{4t_D}\right] - \tfrac{1}{2}\left[-\mathrm{Ei}\frac{-1}{4(t_D - \Delta t_{Dp})}\right] + \tfrac{1}{2}\left[-\mathrm{Ei}\frac{-1}{4(t_D - 2\Delta t_{Dp})}\right] \cdots \pm \tfrac{1}{2}\left\{-\mathrm{Ei}\frac{-1}{4[t_D - (j-1)\Delta t_{Dp}]}\right\} \tag{4.103}$$

Eq. 4.103 then shows that pulse tests employing the same Δt_{Dp}, dimensionless pressure-pulse length, give the same Δp_D versus t_D relationship, as shown in Fig. 4–29, including the Δp_{Dpj} and t_{DLj} characteristics defined as:

$$\Delta p_{Dpj} = \frac{\Delta p_{pj}}{(0.141q\mu/kh)} \tag{4.104}$$

$$t_{DLj} = \eta t_{Lj}/r^2 \tag{4.105}$$

Brigham goes a step further to show that any pulse tests with the same ratio between the observed time lag for a particular pulse and the pulse length, $t_{Lj}/\Delta t_p$, give the same Δp_{Dpj} and t_{DLj} values.[43] The logic of this conclusion is apparent when we recognize that the lag time is a function of the group of terms $k/\phi\mu cr^2$, just as the time travel relationship shows that $t_s = 0.04\phi\mu cr^2/k$. Since the same group of terms is used to relate the real time to the dimensionless time as in Eq. 4.105 (η =

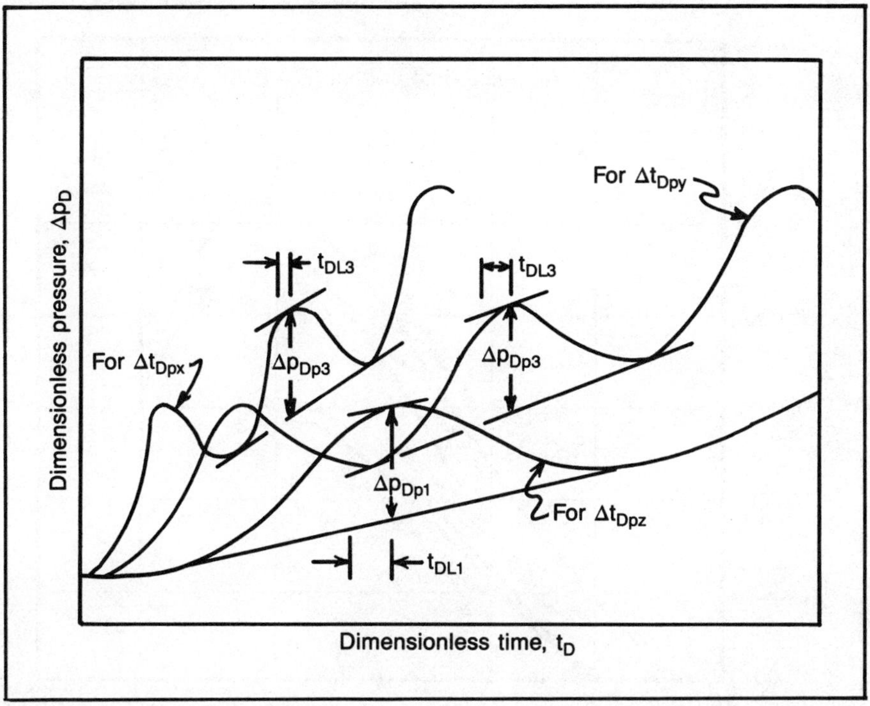

Fig. 4–29 Pulse test of dimensionless pressure versus dimensionless time

6.33k/$\phi\mu$c), it is logical that pulse tests with the same time lag to pulse length ratios have the same dimensionless pressure pulses and dimensionless time lags as shown in Fig. 4–30 for Δp_D and Fig. 4–31 for t_{DL} values. Note that, except for the first odd and first even pulses, all of the odd pulses give the same Δp_D and t_{DL} values and all of the even pulses give the same Δp_D and t_{DL} values. The term, t_c, in these figures is the time for a full pulse cycle, one odd pulse plus one even pulse. Plotting $\Delta p_D(\Delta t_c/t_L)^2$ rather than Δp_D simply makes it possible to present the data on a smaller plot.

To this point we have considered only pulse tests where all of the pulses are of equal length. Kamal and Brigham show that the magnitude of the pressure pulse can be maximized by using unequal shutin and production, or injection, periods characterized by $F' = \Delta t_{odd}/\Delta t_c$, where all of the odd cycles are of the same length and all of the even cycles are of the same length.[44] As noted, Δt_c is the sum of one odd and one even cycle. Kamal and Brigham also found that for a particular F′ after the first even and the first odd pulses, all of the even pulses exhibited about the same dimensionless pressure pulse and dimensionless

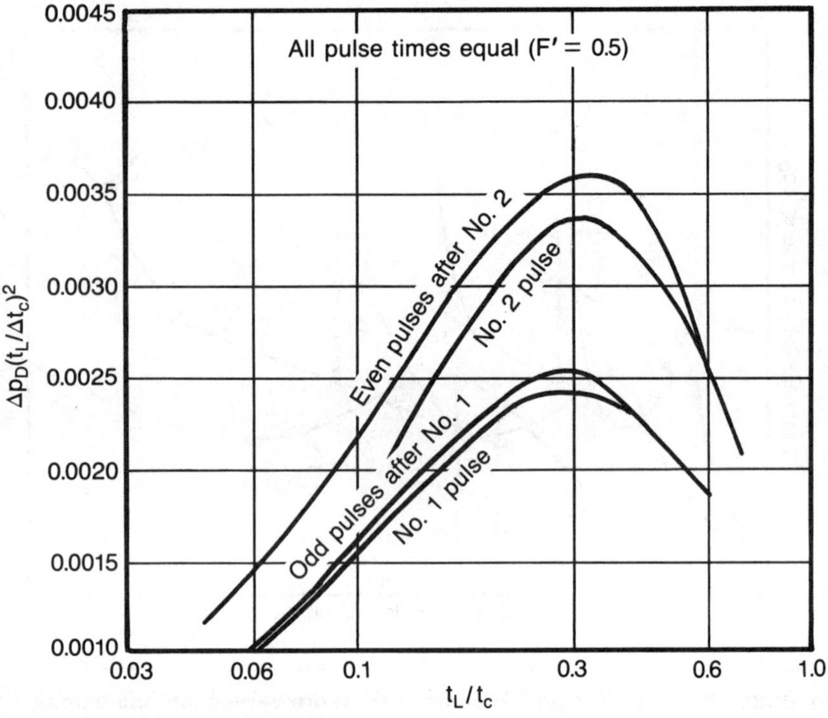

Fig. 4–30 Dimensionless pulse magnitude (Δp_D) versus time lag/cycle length (t_L/t_c) ratio. (Use Figs. 4–36 to 4–39 for quantitative values.)

lag time values and all of the odd pulses had about the same dimensionless pressure pulses and dimensionless time lags.[45] The result of the Kamal and Brigham work is presented in Figs. 4–32 through 4–39.

Analysis of pulse test data is then obvious. First, choose a portion of the pressure data that appears to be realistic and reliable. (One of the difficulties with pulse testing is that the required pressure accuracy is often in the range of unexplained normal pressure anomalies, masking the desired characteristics.) After determining the portion of the pressure data judged best, pursue the tangent construction previously described to determine a time lag, t_{Lj}, and the pressure pulse, Δp_{pj}, for a particular pulse, j. Then calculate $t_{Lj}/\Delta t_c$ and F′ and determine $\Delta p_D(t_L/\Delta t_c)^2$ from the appropriate curve of Figs. 4–36 through 4–39. Calculate Δp_D from the expression $\Delta p_D(t_L/\Delta t_c)^2$. Use Δp_D in Eq. 4.104 as Δp_{Dpj} and the Δp_{pj} obtained from the tangent constructions to calculate k/μ or k. Determine t_{LD} from the appropriate curve in Figs. 4–32 through

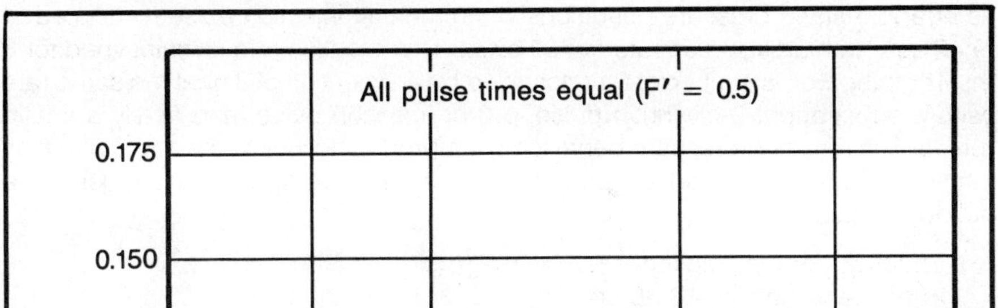

Fig. 4–31 Dimensionless time lag (t_{LD}) versus time lag/cycle length (t_L/t_c) ratio. (Use Figs. 4–32 to 4–35 for quantitative values.)

4–35. Use $(t_L)_D$ as t_{DLj} in Eq. 4.105 along with the time lag, t_{Lj}, determined from the tangent construction and r equal to the distance between the active and observation well to calculate any of the parameters in $\eta = 6.33k/\phi\mu c$ that are unknown, the compressibility or the compressibility-porosity product. These calculations should be repeated for all pulses with reliable pressure records.

To understand pulse testing better, work problem 4.12 and review the solution in appendix C.

PROBLEM 4.12: Pulse-Test Analysis*

Analysis of the performance of an enhanced oil recovery project requires some knowledge of the directional permeabilities in a reservoir. For this purpose a pulse test is designed with the pulsing well being an injection well in a five-spot pattern and the four offsetting production wells being the responding wells. The reservoir

*Data in problem 4.12 are provided courtesy of Texaco Inc.

is at about static pressure conditions when the first injection pulse is initiated at 9:40 a.m. with an injection rate of 700 bw/d. This injection rate is maintained for 3 hr. Then the well is shut in for 3 hr, followed by an injection of 3 hr at the same rate with a subsequent 3-hr shutin pulse, a 3-hr injection pulse, and finally a shutin pulse. The resulting pressure behavior in the producing well to the north, 300 ft, is shown in Table 4–1. Find the permeability and the average thickness using the following data:

Effective compressibility = 9.6×10^{-6}/psi
Effective viscosity = 0.86 cp
Porosity = 16%

TABLE 4–1 Pressure Behavior of Producing Well

Time	*Pressure, psig*	*Time*	*Pressure, psig*	*Time*	*Pressure, psig*
9:40 a.m.	390.1	2:23 p.m.	411.6	11:22 p.m.	425.1
10:10 a.m.	390.6	2:30 p.m.	411.6	12:13 a.m.	429.3
10:30 a.m.	392.0	2:45 p.m.	411.4	12:40 a.m.	431.3
10:40 a.m.	393.0	3:02 p.m.	411.3	1:21 a.m.	433.9
10:48 a.m.	393.8	3:30 p.m.	411.0	1:53 a.m.	433.6
11:05 a.m.	395.8	4:05 p.m.	410.8	2:35 a.m.	432.0
11:15 a.m.	396.8	4:30 p.m.	412.0	3:15 a.m.	430.2
11:30 a.m.	398.6	5:00 p.m.	413.2	3:55 a.m.	428.5
11:45 a.m.	400.7	5:35 p.m.	416.4	4:32 a.m.	428.8
12:15 p.m.	403.8	6:00 p.m.	418.9	5:08 a.m.	430.6
12:30 p.m.	405.8	6:35 p.m.	422.3	5:53 a.m.	434.5
12:47 p.m.	407.8	7:05 p.m.	424.6	6:30 a.m.	437.4
1:00 p.m.	409.1	7:33 p.m.	425.3	6:58 a.m.	440.3
1:20 p.m.	410.7	7:59 p.m.	425.1	7:30 a.m.	440.9
1:32 p.m.	411.3	8:31 p.m.	423.9	7:58 a.m.	440.7
1:45 p.m.	411.7	9:01 p.m.	423.1	8:28 a.m.	439.6
2:00 p.m.	411.9	9:38 p.m.	421.8	8:57 a.m.	438.6
2:15 p.m.	411.9	10:26 p.m.	421.4	9:45 a.m.	437.0

For petroleum engineers this chapter is probably the most useful. Most petroleum engineers are concerned at one time or another with the analysis of pressure and/or rate data from a well. Although much has been written on these analysis methods, it has been very confusing to know what method to use and when the more exotic techniques should be employed. This chapter gives specific instruction for the best methods to use and the common errors that are widely made in analyzing pressure and rate data. No effort is made to cover every conceivable type of analysis, but it is believed that careful use of this material will enable the engineer to analyze 90–95% of the data encountered accurately when a solution is possible. Moreover, the engineer will be able to avoid the misapplication of methods and the resulting misleading answers.

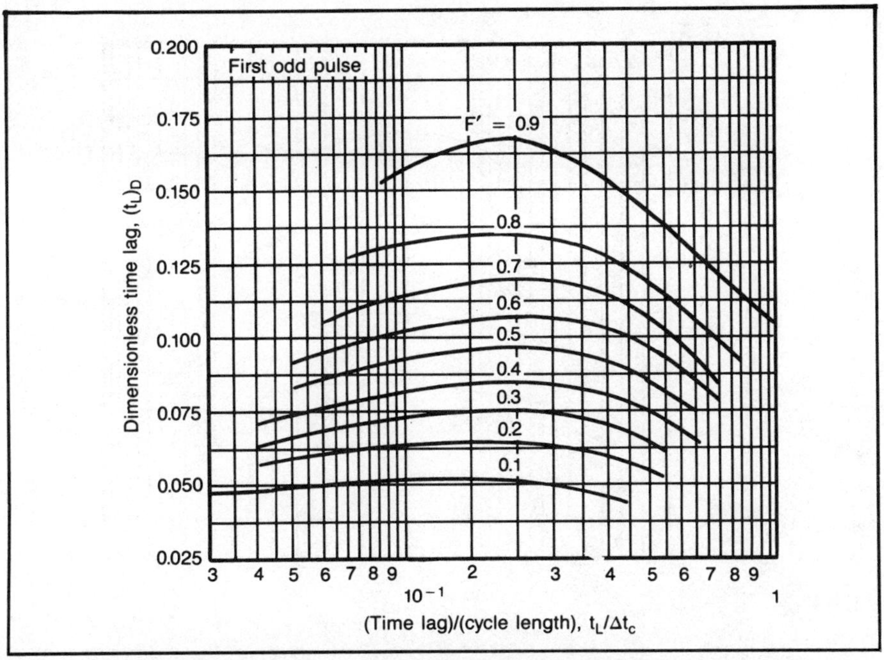

Fig. 4–32 Pulse testing: relation between time lag and cycle length for first odd pulse (after Kamal and Brigham, see ref. 44)

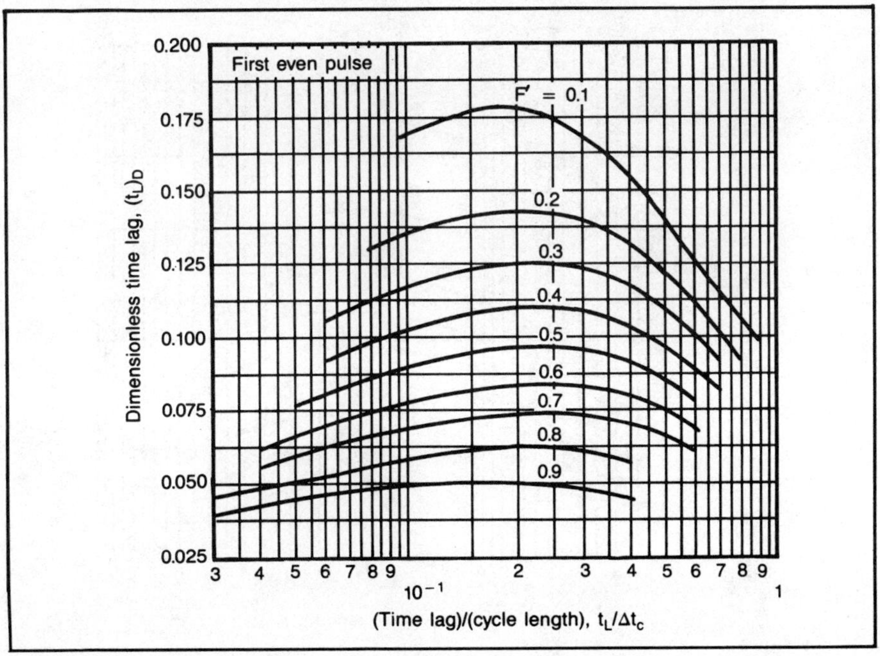

Fig. 4–33 Pulse testing: relation between time lag and cycle length for first even pulse (after Kamal and Brigham, see ref. 44)

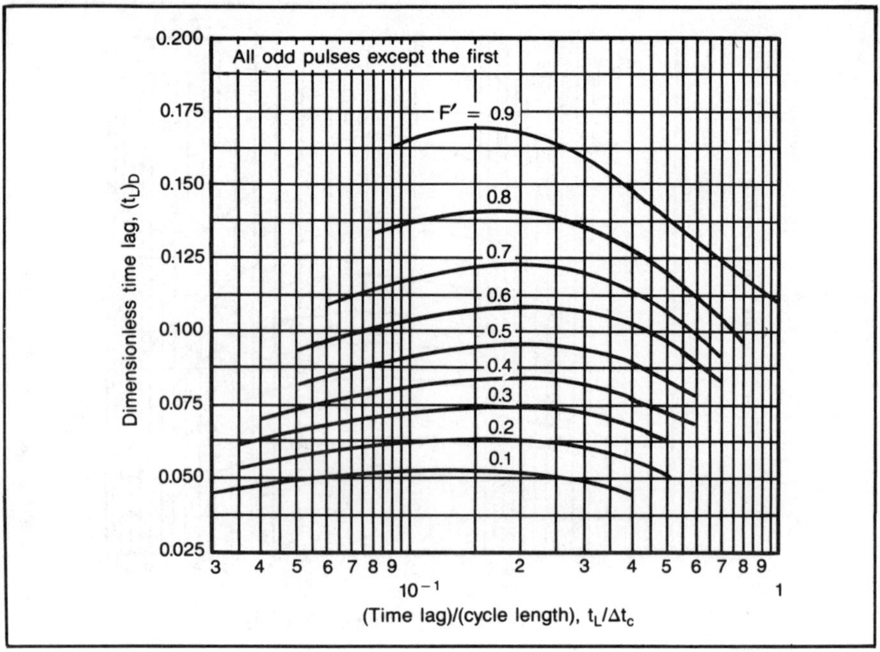

Fig. 4–34 Pulse testing: relation between time lag and cycle length for all odd pulses after the first (after Kamal and Brigham, see ref. 44)

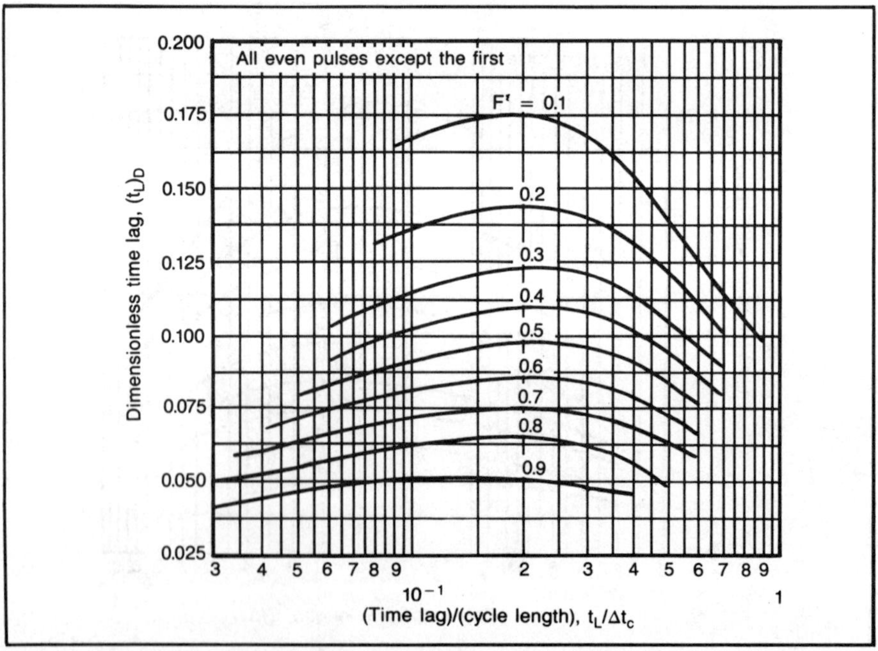

Fig. 4–35 Pulse testing: relation between time lag and cycle length for all even pulses after the first (after Kamal and Brigham, see ref. 44)

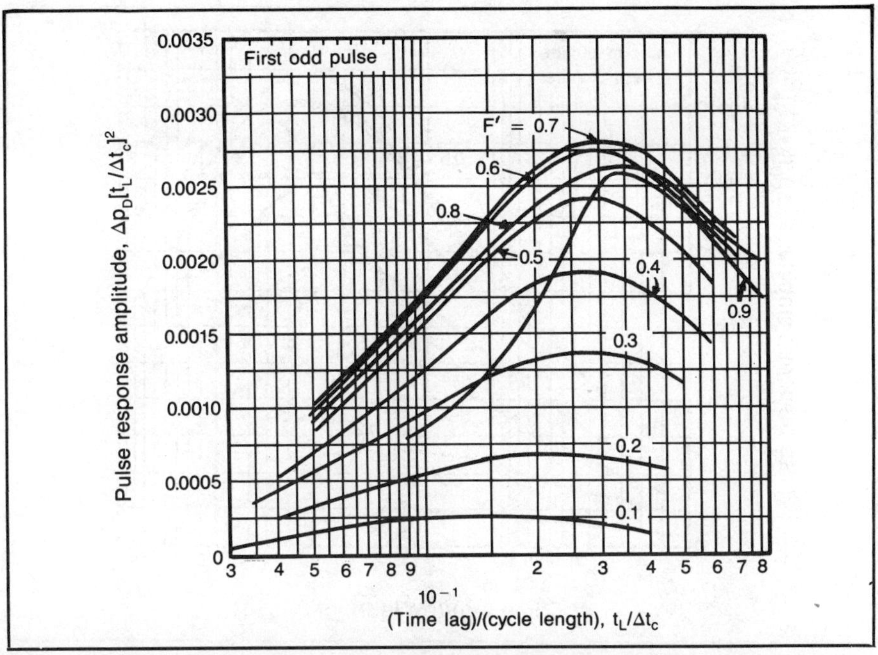

Fig. 4–36 Pulse testing: relation between time lag and response amplitude for first odd pulse (after Kamal and Brigham, see ref. 44)

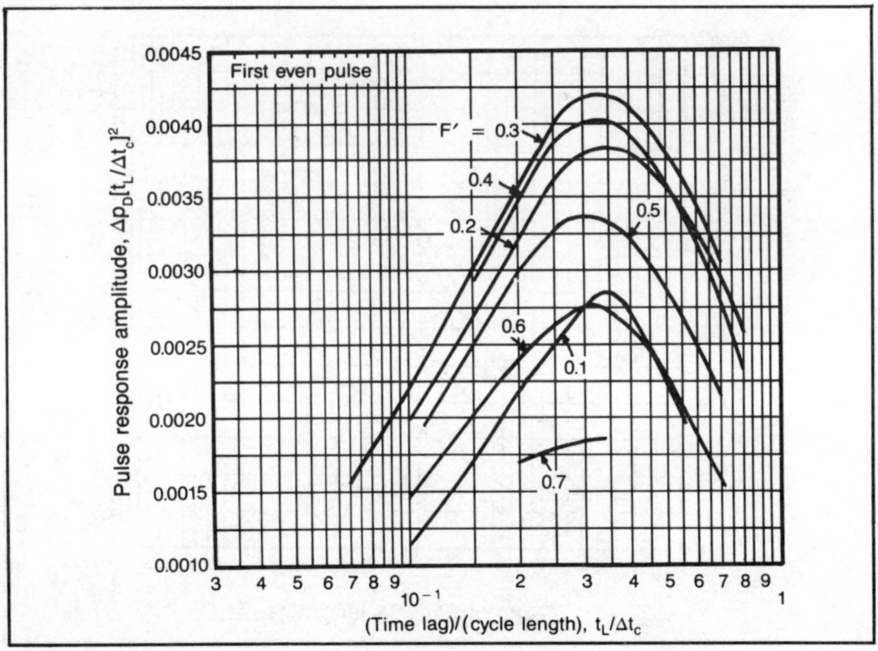

Fig. 4–37 Pulse testing: relation between time lag and response amplitude for first even pulse (after Kamal and Brigham, see ref. 44)

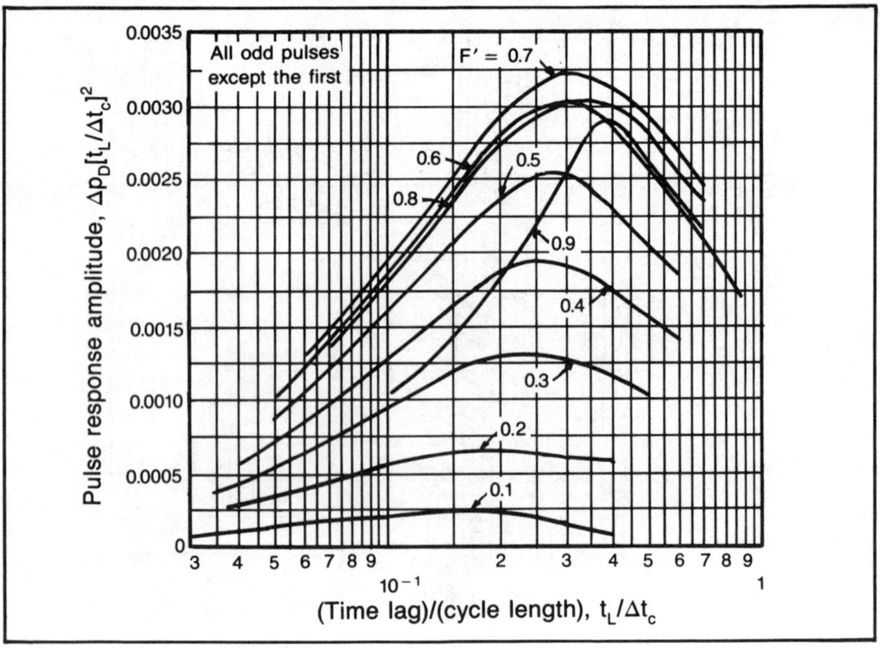

Fig. 4–38 Pulse testing: relation between time lag and response amplitude for all odd pulses after the first (see ref. 44)

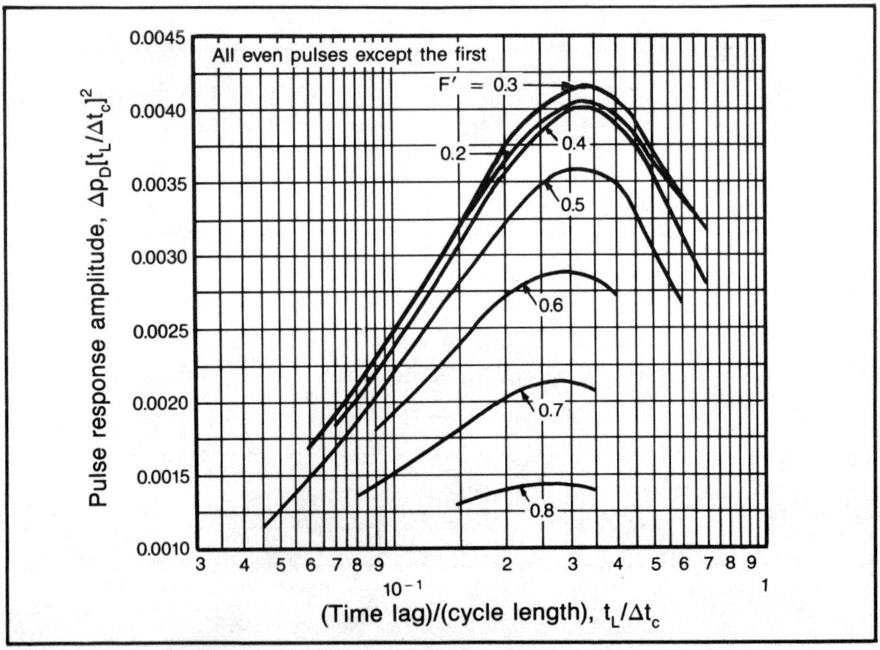

Fig. 4–39 Pulse testing: relation between time lag and response amplitude for all even pulses after the first (see ref. 44)

Additional Problems

4.13 A well has a shutin bottom-hole pressure of 2,300 psia and flows 215 bo/d under a drawdown of 500 psi. The well produces from a formation of 36-ft net productive thickness. Use the following reservoir data:

r_w = 6 in.
r_e = 660 ft
μ_o = 0.88 cp
B_o = 1.32 bbl/stb

A. What are the productivity index and specific productivity index of the well?*
B. What is the average permeability of the formation?*
C. What is the capacity of the formation?*
D. If a 500-bo/d capacity pump is run in the well and is operated without gas-lock difficulties, what is the bottom-hole pressure if the productivity index is constant?
E. Rework part D if the Vogel IPR curve is applicable.

4.14 A well with a water cut of 90% is subjected to a productivity-index test with resulting data as follows:

Bottom-hole pressure, psia	3,200	3,028	2,886	2,689	2,304
Total fluid rate, b/d	0	910	1,730	2,690	3,740

What size capacity submersible pump should be used in this well? Assume the water cut is unaffected by the rate increase.

4.15 The pressure drawdown data in Table 4–2 are obtained from a well producing at a rate of 800 stb/d with the initial reservoir pressure at 1,895 psia.** Find the undamaged permeability and the skin factor, S. Also estimate the drainage area and the furthest distance to the drainage boundary. The following reservoir data apply:

μ_o = 1.0 cp
ϕ = 0.1
h = 8 ft
r_w = 0.33 ft
p_i = 1,895 psia

*Taken from *Applied Petroleum Reservoir Engineering,* by B.C. Craft and M.F. Hawkins (Englewood Cliffs: Prentice-Hall, 1959).
**Taken from *Pressure Buildup and Flow Tests in Wells,* by C.S. Matthews and D.G. Russell, Monograph No. 1 (New York: SPE, 1967), p. 142–145.

TABLE 4–2 Pressure Drawdown Data

Time, min	*Flowing Pressure, psia*
10	1,855
20	1,791
30	1,754
40	1,727
50	1,707
60	1,690
70	1,676
90	1,653
100	1,643
150	1,595
200	1,559
300	1,500
400	1,452
600	1,373
900	1,279
1,200	1,200
1,500	1,121
1,800	1,042
3,000	726

$c_e = 17.7 \times 10^{-6}$/psi
$B_o = 1.25$
$S_w = 0.35$

4.16 The solution to problem 4.4 in appendix C shows that a reservoir barrier exists 229 ft from a specific point. The first answer is based on well pressure read after 10 days of constant rate production. Verify that the anomaly observed results from a plane barrier by repeating the calculations based on pressures at 30 days and 4 days. What is m based on the last three plotted points? What conclusions can be reached concerning the distance to the next closest barrier if we assume we can detect a pressure anomaly of 3 psi on our Amerada gauge? What would the distance be if calculated from the t_s equation, Eq. 3.50?

4.17 At the time of shutin, a well has produced 2,682 stb of oil and is producing at a rate of 149 stb/d with a bottom-hole pressure of 2,300 psia. After the shutin, the well pressures are as follows:

Shutin Time, hr	Bottom-Hole Pressure, psia
2	2,359
4	2,392
8	2,423
12	2,441

Shutin Time, hr	Bottom-Hole Pressure, psia
16	2,453
20	2,461
24	2,468

Find the undamaged permeability, Δp_{skin}, and the skin factor assuming the well pressure is unchanging. The reservoir data are as follows:

Oil formation volume factor = 1.31
Thickness = 40 ft
Oil viscosity = 2 cp
Effective compressibility = 15×10^{-6}/psi
Porosity = 10%
Well radius = 0.333 ft
External drainage radius = 660 ft

4.18 Rework problem 4.17 assuming the well is infinite acting at the time of shutin. Put the Horner plot on the same graph as the one made for 4.17, making the pressure scale common.

4.19 Calculate the production time required to reach pseudosteady state in problem 4.17. Repeat the pressure-buildup analysis without making any assumptions concerning the well being infinite acting or the flowing pressure at shutin being unchanging. Put the Δp_q plot on the pressure plot from problems 4.17 and 4.18, using a common log Δt scale.

4.20 Calculate the static pressure for the drainage area of problems 4.17, 4.18, and 4.19, using the Odeh data and Eq. 4.44 if the initial pressure is 2,740 psia (two answers).

4.21 The DST data are as shown in Tables 4–3 to 4–6. Determine the initial reservoir pressure, Δp_{skin}, and the apparent skin factor. Estimate the initial stabilized flow rate if the well can be acidized to obtain a negative skin factor of about −5. The reservoir data are as follows:

Well pressure = 200 psia
Average gas gravity = 0.013 cp
Average z = 0.895
T_f = 530°R
Net pay thickness = 28 ft
Average drainage radius = 1,490 ft
Well radius = 0.328 ft, 7.825-in. hole
Average gas-filled porosity = 6.8%
Gas compressibility = 0.00122/psi

TABLE 4–3 Fluid Sample Data for Problem 4.21

Fluid Sample Data		Date	Ticket Number
Sampler pressure	psig at surface	Kind of Job: Open hole	Halliburton District
Recovery: cu ft gas		Tester	Witness
cc oil		Drilling Contractor	
cc water			
cc mud			
Total liquid cc			
Gravity	API at °F		
Gas/oil ratio	cu ft/bbl		

Legal Location Sec. – Twp. – Rng.

Equipment and Hole Data

Formation tested	Devonian	
Elevation	480.1	ft
Net productive interval	20 (1,730 – 1,750)	ft
All depths measured from	Kelly bushing	
Total depth	1,750	ft
Main hole/casing size	$7\frac{7}{8}$ in.	
Drillcollar length	477 ft ID $2\frac{1}{2}$ in.	
Drillpipe length	1,202 ft ID 2.764 in.	
Packer depth(s)	1,702	ft
Depth tester valve	1,694	ft

	Resistivity		Chloride Content
Recovery water	at	°F	ppm
Recovery mud	at	°F	ppm
Recovery mud filtrate	at	°F	ppm
Mud pit sample	at	°F	ppm
Mud pit sample filtrate	at	°F	ppm
Mud weight 9.2	Viscosity 45 cp		

Cushion	Type	Amount	Ft	Depth Back Pres. Valve	Surface Choke	Bottom Choke
		None		None	3/4 in.	3/4 in.

Recovered	145	ft of	water gas cut mud
Recovered		ft of	
Recovered		ft of	
Recovered		ft of	
Recovered		ft of	

Mco. from Tester Valve

Field Area

Remarks: See Production Test Data Sheet

County

Temperature	Gauge No. 236 Depth: 1,695 ft		Gauge No. 225 Depth: 1,716 ft		Gauge No. Depth: ft		Time	
Est. °F	12 Hour Clock Blanked Off NO		12 Hour Clock Blanked Off YES		Hour Clock Blanked Off		Tool Opened	am 6:56 pm
Actual 77 °F	Pressures		Pressures		Pressures		Opened Bypass	am 10:49 pm
——	Field	Office	Field	Office	Field	Office	Reported min	Computed min
Initial Hydrostatic	849.1	852	854.4	861				
First Period – Flow – Initial	41.5	52	54.7	70			—	—
First Period – Flow – Final	103.6	110	109.3	115			35	35
First Period – Closed in	780.8	793	786.6	794			45	44
Second Period – Flow – Initial	117.5	136	129.8	141			—	—
Second Period – Flow – Final	142.3	144	143.5	148			62	61
Second Period – Closed in	794.4	799	806.9	799			91	93
Third Period – Flow – Initial							—	—
Third Period – Flow – Final								
Third Period – Closed in								
Final Hydrostatic	828.6	832	840.8	837			—	—

State

TABLE 4–4 Data for Problem 4.21

Casing perfs ________ Bottom choke 3/4 in ________ Surface temp ________ °F Ticket No. 879250

Gas gravity ________ Oil gravity ________ GOR ________

Specific gravity ________ Chlorides ________ ppm Res. ________ at ________ °F

Date Time am pm	Choke Size, in.	Surface Pressure, psi	Gas Rate, Mcf	Liquid Rate, b/d	Remarks Orifice Well Tester 1½-in. Orifice Plate
6:52	¾				Tool on bottom.
6:56	1½				Tool opened to the blow hose with a good
					blow off the bottom of the bucket with
					gas to the surface in 4 min. Gas
					turned to the blow line—orifice well
					tester set up 268,000 cu ft in 24 hr
					through 1½-in. orifice plate.
7:31					Tool was closed for a 44-min first
					closed-in pressure.
8:16	1½				Tool opened to the blow line and orifice
					well tester at 8:18 with 332,000 cu ft
					in 24 hr. Gas kept increasing and
					leveled off after tool was open for 50
					min. Gas leveled off to 385,000 cu
					ft in 24 hr through 1½-in. orifice plate.
9:18					Tool closed for a 93-min second
					closed-in pressure.
10:49					Tool off bottom.

TABLE 4–5 Data for Problem 4.21

Gauge No.	236					Depth 1,695 ft		Clock No. 12189			12 hr		Ticket No.		
	First Flow Period		First Closed-In Pressure			Second Flow Period		Second Closed-In Period			Third Flow Period		Third Closed-In Pressure		
	Time Defl., 0.000 in.	psig Temp. Corr.	Time Defl., 0.000 in.	$\text{Log}\ \frac{t+\theta}{\theta}$	psig Temp. Corr.	Time Defl., 0.000 in.	psig Temp. Corr.	Time Defl. 0.000 in.	$\text{Log}\ \frac{t+\theta}{\theta}$	psig Temp. Corr.	Time Defl., 0.000 in.	psig Temp. Corr.	Time Defl., 0.000 in.	$\text{Log}\ \frac{t+\theta}{\theta}$	psig Temp. Corr.
0	0.000	52	0.000		110	0.000	136	0.000		144					
1	0.0334	61	0.0266		757	0.0738	124*	0.0395		753					
2	0.0668	77	0.0532		774	0.1408	126	0.0790		774					
3	0.1002	91	0.0798		781	0.2078	131	0.1185		783					
4	0.1336	102	0.1064		786	0.2748	135	0.1580		787					
5	0.1670	106	0.1330		788	0.3418	139	0.1975		790					
6	0.2004	105	0.1596		789	0.4090	144	0.2370		790					
7	0.2340	110	0.1862		790			0.2765		792					
8			0.2128		790			0.3160		792					
9			0.2394		791			0.3555		793					
10			0.2660		792			0.3950		794					
11			0.2930		793			0.4545		796					
12								0.4740		796					
13								0.5135		798					
14								0.5530		798					
15								0.6130		799**					
Gauge No.			225			Depth 1,716 ft		Clock No. 12188			12 hr				
0	0.000	70	0.000		115	0.000	141	0.000		148					
1	0.0331	66	0.0269		756	0.0739	127*	0.0402		753					
2	0.0662	81	0.0538		775	0.1411	130	0.0804		774					
3	0.0993	96	0.0807		782	0.2083	135	0.1206		783					
4	0.1324	107	0.1076		786	0.2755	139	0.1608		787					
5	0.1655	111	0.1345		788	0.3427	143	0.2010		789					
6	0.1986	109	0.1614		789	0.4100	148	0.2412		790					
7	0.2320	115	0.1883		790			0.2814		791					
8			0.2152		791			0.3216		792					
9			0.2421		792			0.3618		793					
10			0.2690		793			0.4020		794					
11			0.2960		794			0.4422		795					
12								0.4824		796					
13								0.5226		797					
14								0.5628		798					
15								0.6230		799**					
Reading Interval	5		4			10		6							min

Remarks: *Interval = 11 min **Interval = 9 min

TABLE 4–6 Data for Problem 4.21

	OD, in.	ID, in.	Length	Depth, ft
Drillpipe or tubing				
Reversing sub				
Water cushion valve				
Drillpipe	3.5	2.764	1,202 ft	
Drillcollars	6	2.5	477 ft	
Handling sub & choke assem.	3.5	2.5	54 in.	
Dual CIP valve	5	0.87	49 in.	1,687
Dual CIP sampler				
Hydrospring tester	5	0.75	60 in.	1,694
Multiple CIP sampler				
Extension joint				
AP running case	5	3.75	50 in.	1,695
Hydraulic jar				
VR safety joint				
Pressure equalizing crossover				
Packer assem.–6¾ in. rubber	5	1.53	67 in.	1,702
Distributor				
Packer assembly				
Flush joint anchor	4.5	3.5	11 ft	
Pressure equalizing tube				
Blanked-off BT running case	5	3.75	60 in.	1,716
Drillcollars pipe	3.5		30 ft	
Anchor pipe safety joint				
Packer assembly				
Distributor				
Packer assembly				
Anchor pipe safety joint				
Sidewall anchor				
Drillcollars				
Flush joint anchor				
Blanked-off BT running case				
Total depth				1,750 ft

4.22 In problem 4.12 pulse data from a well and an offset 330 ft to the north are analyzed. Concurrent with the recording of the pressure data in the north offset well, the pressure data from the east offset of the pulsing well are recorded with results in Table 4–7. Find the permeability and average thickness. This offset is also 330 ft from the pulsing well.

TABLE 4–7 Pressure Data

Time	*Pressure, psig*	*Time*	*Pressure, psig*	*Time*	*Pressure, psig*
9:40 a.m.	435.2	2:45 p.m.	450.5	12:10 a.m.	463.2
10:10 a.m.	435.5	3:03 p.m.	451.1	12:55 a.m.	466.1
10:30 a.m.	435.8	3:30 p.m.	451.1	1:30 a.m.	467.9
10:46 a.m.	436.2	4:00 p.m.	451.5	1:50 a.m.	469.1
11:05 a.m.	437.0	4:25 p.m.	451.6	2:30 a.m.	469.7
11:30 a.m.	438.2	4:53 p.m.	452.3	3:08 a.m.	469.8
11:45 a.m.	439.1	5:40 p.m.	453.7	3:55 a.m.	469.6
11:50 a.m.	439.5	6:05 p.m.	455.3	4:30 a.m.	469.0
12:00 p.m.	439.9	6:30 p.m.	457.3	5:07 a.m.	469.5
12:15 p.m.	441.0	6:55 p.m.	458.5	5:55 a.m.	471.7
12:31 p.m.	442.5	7:25 p.m.	460.1	6:34 a.m.	473.5
12:48 p.m.	443.8	7:55 p.m.	461.3	7:03 a.m.	474.9
1:10 p.m.	445.8	8:28 p.m.	461.4	7:33 a.m.	476.4
1:28 p.m.	447.0	9:00 p.m.	461.4	8:01 a.m.	477.2
1:48 p.m.	448.1	9:31 p.m.	461.1	8:32 a.m.	477.4
2:06 p.m.	449.1	10:32 p.m.	461.0	8:58 a.m.	477.4
2:22 p.m.	449.8	11:32 p.m.	462.2	9:45 a.m.	476.9

Notes

1. T.E.W. Nind, *Principles of Oil Well Production* (New York: McGraw-Hill, 1964), p. 58.

2. C. Gatlin, *Petroleum Engineering—Drilling and Well Completions* (Englewood Cliffs: Prentice-Hall, 1960), p. 261.

3. J.V. Vogel, "Inflow Performance Relationships for Solution-Gas Drive Wells," *JPT* (January 1968), pp. 83–92.

4. C.S. Matthews and D.G. Russell, *Pressure Buildup and Flow Tests in Wells,* SPE Monograph No. 1 (New York: SPE, 1967).

5. Matthews and Russell, 1967.

6. R.C. Earlougher Jr., *Advances in Well Test Analysis,* SPE Monograph No. 5 (New York and Dallas: SPE, 1977).

7. R.M. McKinley, "Wellbore Transmissibility from Afterflow Dominated Pressure Buildup Data," *JPT* (July 1971), p. 222a.

8. McKinley, 1971.

9. Ram G. Agarwal, Rafi Al-Hussainy, and H.J. Ramey Jr., "An Investigation of Wellbore Storage and Skin Effect in Unsteady Liquid Flow: I. Analytical Treatment," *SPEJ* (September 1970), pp. 279–290.

10. Robert A. Wattenbarger and H.J. Ramey Jr., "An Investigation of Wellbore Storage and Skin Effect in Unsteady Liquid Flow: II. Finite Difference Treatment," *SPEJ* (September 1970), pp. 291–297.

11. R.C. Earlougher Jr. and Deith M. Kersch, "Analysis of Short-Time Transient Test Data by Type-Curve Matching," *JPT* (July 1974), pp. 793–800.

12. H.J. Ramey Jr., "Practical Use of Modern Well Test Analysis," SPE paper 5878, Annual Fall Technical Conference (October 1976).

13. W.M. Cobb and J.T. Smith, "An Investigation of Pressure Buildup Tests in Bounded Reservoir" (abridged version), *JPT* (August 1975), pp. 991–996.

14. A. Kumar and H.J. Ramey Jr., "Well-Test Analysis for a Well in a Constant-Pressure Square" (abridged version), *SPEJ* (April 1974), pp. 107–116.

15. C.C. Miller, A.B. Dyes, and C.A. Hutchinson Jr., "Estimation of Permeability and Reservoir Pressure from Bottom-Hole Pressure Buildup Characteristics," *Trans.,* AIME (1950), pp. 91–104.

16. H.C. Slider, "A Simplified Method of Pressure Buildup Analysis for a Stabilized Well," *JPT* (September 1971).

17. B.C. Craft and M.F. Hawkins, *Applied Petroleum Reservoir Engineering* (Englewood Cliffs: Prentice-Hall, 1959).

18. Craft and Hawkins, 1959.

19. D.R. Horner, "Pressure Buildup in Wells," Proceedings of the Third World Congress (1951), p. 503.

20. Horner, 1951.

21. API Mid-Continent District Study Committee on Completion Practices, "Selection and Evaluation of Well Completion Methods," *Drilling and Production Practices* (API, 1955), p. 421.

22. H.J. Ramey Jr. and W.M. Cobb, "A General Buildup Theory of a Well in a Closed Drainage Area," *JPT* (December 1971), pp. 1493–1505.

23. Cobb and Smith, 1975.

24. Earlougher, 1977.

25. C.S. Matthews, F. Brons, and P. Hazebroek, "A Method for Determination of Average Pressure in a Bounded Reservoir," *Trans.* AIME (1954), pp. 182–191.

26. Matthews, Brons, and Hazebroek, 1954.

27. Matthews, Brons, and Hazebroek, 1954.

28. D.N. Dietz, "Determining Average Reservoir Pressure from Pressure Buildup Surveys," *JPT* (August 1965), pp. 955–959.

29. Dietz, 1965.

30. Craft and Hawkins, 1959.

31. A.S. Odeh and R. Al-Hussainy, "A Method for Determining the Static Pressure of a Well from Buildup Data," *JPT* (May 1971), pp. 621–624.

32. H.J. Ramey Jr., A. Kumar, and M.S. Gulati, "Gas Well Test Analysis under Water-Drive Conditions," *AGA* (1973).

33. J.H. Moran and E.E. Finklea, "Theoretical Analysis of Pressure Phenomena Associated with the Wireline Formation Tester," *JPT* (August 1964), pp. 899–908.

34. W.E. Culham, "Pressure Buildup Equations for Spherical Flow Regime Problems," *JPT* (December 1974), pp. 545–555.

35. Culham, 1974.

36. Moran and Finklea, 1962.

37. Earlougher, 1977.

38. Earlougher, 1977.

39. McKinley, 1971.

40. Earlougher, 1977.

41. Earlougher, 1977.

42. Ramey, 1976.

43. W.E. Brigham, "Planning and Analysis of Pulse Tests," *JPT* (May 1970), pp. 618–624.

44. M. Kamal and W.E. Brigham, "Pulse-Testing Response for Unequal Pulse and Shut-in Periods," *SPEJ* (October 1975), pp. 399–410.

45. Kamal and Brigham, 1975.

Additional References

Johnson, C.R.; Greenkorn, R.A.; and Woods, E.G. "Pulse Testing: A New Method for Describing Reservoir Flow Properties between Wells." *JPT*, September 1966, pp. 1599–1604.

Kumar, A. "Well Test Analysis for a Well in a Constant Pressure Square." SPE paper 4054, San Antonio: October 1972.

Matthews, C.S. "Analysis of Pressure Buildup and Flow Test Data." *JPT*, September 1961.

Muskat, M. *The Flow of Homogeneous Fluids Through Porous Media*. New York: McGraw-Hill, 1937, p. 341.

Odeh, A.S.; and Selig, F. "Pressure Build-up Analysis, Variable-Rate Case," *JPT*, July 1963, pp. 790–794.

Uren, L.C. *Petroleum Production Engineering—Oil Field Exploitation*. New York: McGraw-Hill, 1939.

5

Gas Reservoir Engineering

Industry engineers have less experience with gas reservoir engineering than they have with reservoir engineering in general because it has only been in recent years that natural gas attained a price that commanded attention. Previously, no one worried about how much gas they had in a reservoir or how long it would take to produce it. With the increase in the value of gas, more attention has been given to gas. However, total gas reserves and total deliverability became a problem only in recent years when gas was found to be in short supply. Gas transmission companies have always had to know the total amount of gas and the rate at which it could be produced to be certain that they showed a profit on a particular pipeline. However, even the transmission companies were not greatly concerned with sophisticated reservoir predictions. They had so many prospects for production that they had to use only the most profitable, which were the low-pressure, high-permeability reservoirs.

These reservoirs led to simplifications in gas technology that introduce gross inaccuracies when applied to the high-pressure, low-permeability reservoirs that typify the commercial gas reservoir produced today in the U.S. The low pressures typical of gas reservoirs produced when gas technology was being developed led to the assumption that the gas compressibility could be adequately approximated by the reciprocal of the pressure—a gross inaccuracy for a high-pressure reservoir. The high permeabilities led to the assumption that a gas well would reach pseudosteady-state flow, or would stabilize, in a few hours and that an average reservoir pressure could be recorded by shutting in a well for a short period—typically 48 hr. The fallacies of these assumptions when applied to very low permeability reservoirs have undoubtedly cost many gas-producing companies millions of dollars.

Reportedly, the unexpected shortfall of gas supplies in the early 1970s was largely caused by the inability of engineers to predict accurately the deliverability of gas wells. Some major oil company divisions were forced to reduce their gas reserves by as much as 10%.

On the surface it would appear that gas reservoir engineering would be much simpler than oil reservoir engineering because there is no need for data such as oil formation volume factors, gas in solution, and saturations. Also, the volumetric behavior of gas seems to be much better defined than the behavior of a crude oil-solution gas system. However, reservoir engineering technology was developed for oil reservoirs because at that time there was little profit to be made from the production of gas. Consequently, by the time there was a demand for applying the previously derived reservoir engineering fundamentals to the production of gas, the reservoir engineering technology was developed to the point that it was not easy to modify it to the flow in gas reservoirs. This unfortunately is the situation in gas reservoir engineering today. We are taking technology developed originally for the production and analysis of oil reservoirs and are trying to apply it to gas reservoirs. We must do this by rather inexact mathematical methods.

The application would not present much of a problem except that the two fluids are so greatly different in their characteristics. Gas densities are very low compared to liquids. Gas viscosities may be only a small fraction of a liquid viscosity under similar conditions. Therefore, volumetric flow rates of gas may be 100 times the flow rate of liquid under similar circumstances. Thus, with gas we often have turbulent flow around the wellbore, but we seldom have this problem with oil. Since the Darcy equation was derived for streamline (viscous) flow, we must in some way modify all of the basic flow equations to account for the effect of turbulence around the wellbore.

In addition, we must continue accounting for the gas deviation factor, which is a function of both the temperature and pressure. Another difficulty is presented by the compressibility of the fluid. We find that the compressibility of oil is both small and practically constant, whereas the compressibility of gas is about 100 times that of liquid and varies inversely with the pressure. In summary, we find that we must generally use equations designed for the flow of oil and apply them in a very inexact manner to the flow of gas by trying to adapt and modify them for the vast differences in the fluids.

The objective of this chapter is to present methods that permit the engineer to predict accurately the rate versus time behavior of gas wells or reservoirs under a variety of conditions. In order to accomplish this objective, it is necessary for the engineer to have an understanding of the fundamentals of unsteady-state and pseudosteady-state flow as presented. In this chapter we first discuss the properties of natural gas, stressing the best methods of determining these properties. Then material balance and graphical material balance—cumulative gas production versus the producing rate—are discussed with the emphasis on the difficulties encountered in such applications and the prediction of reservoir behavior with an active water drive.

Material balance provides the means of predicting the reservoir behavior as a function of the reservoir pressure. However, to predict the rate versus time behavior, we must consider fluid flow fundamentals. We can show that there are very few steady-state fluid flow applications in gas reservoirs and that pseudosteady state is the basis for the normal, stabilized flow tests required by most states to determine the absolute open flow (AOF) potential used to allocate gas production from a gas reservoir. We can see that stabilized flow tests are seldom usable today. Most wells must be tested under infinite-acting conditions (an isochronal test), and the data must be interpreted in terms of an equivalent stabilized test.

The effect of equipment capacity on a well's deliverability also is considered. Three pieces of information—the deliverability test, the material-balance relationships, and the equipment capacity—are used to predict the rate versus time behavior of the well or reservoir. Alternative methods of determining undamaged permeability, well damage, and turbulence factors from buildup and drawdown tests and using these for predictions also are considered, as well as the use of real gas potentials that treat the pressure, viscosity, and gas deviation factor as one parameter.

Natural Gas Properties

The gas equation used to predict the behavior of gas is:

$$pV = znRT \tag{5.1}$$

Where:

p = Pressure
V = Volume
T = Temperature, absolute units
n = Number of mols of gas
R = Numerical constant that makes the equation correct for a particular set of units
z = Gas deviation factor

Eq. 5.1 has been used previously as Eq. 2.31. This equation represents a rather abstract expression for most engineers. It can be derived from basic considerations as is done in thermodynamics courses, but it still seems to remain somewhat ethereal to most engineers.[1] Consequently, it is helpful to relate the expression to another equation that does seem to be readily understood. This more understandable equation relates the pressure, volume, and absolute temperature of a given mass of ideal gas at different conditions:

$$\frac{p_1 V_1}{T_1} = \frac{p_2 V_2}{T_2} \tag{5.2}$$

Eq. 5.2 simply tells us that the volume of a gas is proportional to the absolute temperature and inversely proportional to the pressure. If we combine this equation with the observation that 1 lb-mol of any gas at 60°F and atmospheric pressure occupies 379 cu ft, we can derive Eq. 5.1.* Let the left-hand side of Eq. 5.2 represent standard conditions and let the right-hand side show reservoir or other conditions. Then the pressure and temperature at standard conditions are 14.7 psia and 520°R. The volume is the number of mols, n, multiplied by 379 cu ft/mol. The right-hand side remains as the general pressure, temperature, and volume:

$$\frac{(14.7)(n \times 379)}{520} = \frac{pV}{T} \tag{5.3}$$

$$pV = n \times 10.73T \tag{5.4}$$

$$pV = nRT \tag{5.5}$$

Eq. 5.5 is the gas equation for an ideal gas. The gas constant, R, is 10.73 for the units employed in this derivation. However, actual gases do not follow Eq. 5.5 at high pressures such as we have in a reservoir. Solving for volume, we obtain:

$$V = \frac{nRT}{p} \tag{5.6}$$

However, except at very low and very high pressures, we find that this theoretical volume of gas according to Eqs. 5.5 and 5.6 is more than actually exists in practice. The gas is actually compressed more than we would have anticipated. To make Eqs. 5.5 and 5.6 accurate, we then add a correction factor, which accounts for the fact that the gas deviates from the theoretical prediction. This factor, z, is the gas deviation factor. Its correct use is shown in the right-hand side of Eq. 5.1.

Most practicing reservoir engineers are capable of evaluating a gas deviation factor for a given set of conditions, but most of them do not know whether this value is the most accurate one available. We can put this into proper perspective by noting the relative accuracy of the various methods of obtaining gas deviation factors.

*Engineers trained in other engineering regimes will recognize the volume of 359 cu ft as the volume of 1 lb-mol of gas at atmospheric pressure and 32°F (0°C).

Evaluating z factors. On the surface it would seem that the most accurate values of z factors available for a particular gas would be those measured in the laboratory for that particular gas. However, most laboratory equipment used for this purpose is not sufficiently accurate to provide better data than that obtainable from empirical methods. A study of pure gases consisting of only one component—for example, methane or ethane—showed that there is a well-established relationship between the z factors and the pressure, temperature, critical pressure, and critical temperature of the pure gas. This relationship takes the form of Fig. 5–1. More extensive z-factor data are available in

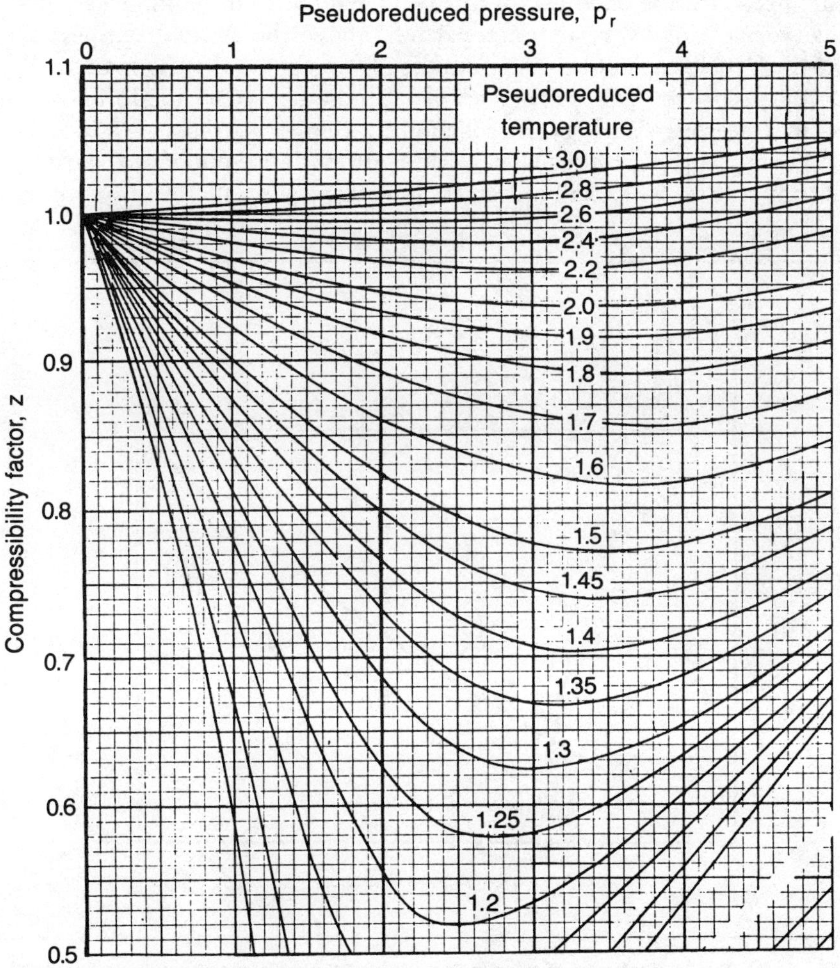

Fig. 5–1 Gas deviation factors (after Standing and Katz, *Trans.*, AIME, 1942, © SPE-AIME)

appendix B, Fig. B8. These figures consist of plots of the gas deviation factor versus the reduced pressure, p_R, for various temperatures, T_R. The reduced pressure and temperature are related to the critical pressure, p_c, and critical temperature, T_c, as:

$$p_R = \frac{p}{p_c} \text{ and } T_R = \frac{T}{T_c} \tag{5.7}$$

The critical pressure and critical temperature are defined as the conditions above which liquid and vapor phases of the compound cannot be distinguished. This is illustrated by the phase diagram in Fig. 5–2. At temperatures below the critical temperature, there is a particular pressure for each temperature at which liquid changes to gas or vice versa. However, at temperatures above the critical temperature, the gas phase cannot be distinguished from the liquid phase regardless of the pressure. The same sort of statement can be made about the critical pressure and the corresponding temperatures.

For a gas that contains only one hydrocarbon component, data such as that in Fig. 5–1 can be used to determine the z factor at any partic-

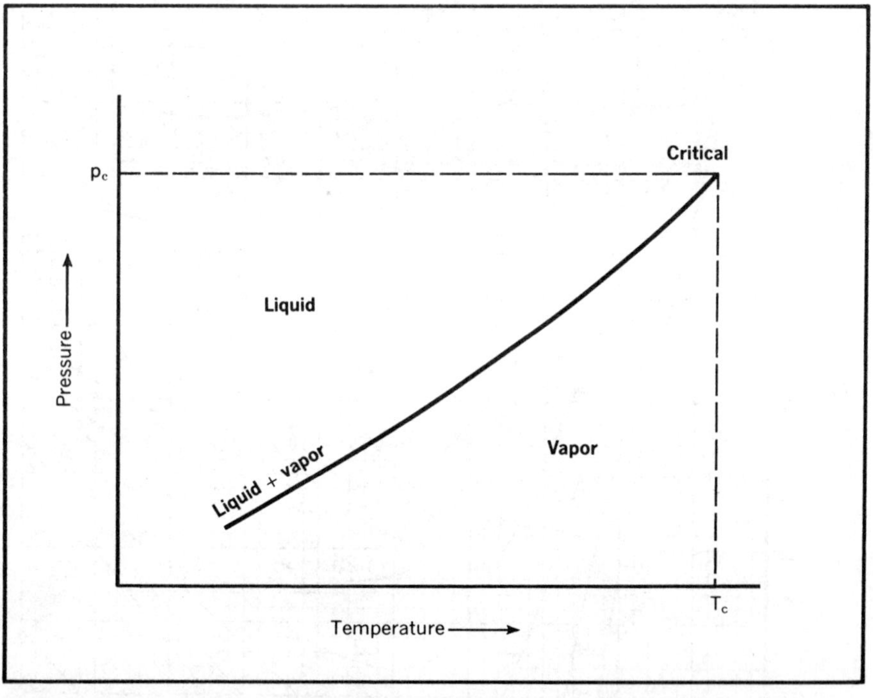

Fig. 5–2 Typical phase diagram for a single component system (after Burcik, *Properties of Petroleum Reservoir Fluids*, courtesy John Wiley and Sons, 1959)

ular pressure and temperature. The critical pressures and temperatures of various hydrocarbon compounds are listed in Table B5 in the appendix.

Unfortunately, natural gases are a mixture of different hydrocarbons, and consequently, they do not have a true critical pressure and temperature as we have defined it. For example, if a natural gas is a mixture of methane, ethane, and propane and the pressure on the mixture in the liquid state is reduced isothermally until the first bubble of gas is formed, this condition would occur somewhere in the vicinity of the bubble point of the methane, the most volatile of the pure gases in the mixture. If the pressure decline continued until the last drop of the mixture changed to gas, this situation would occur somewhere near the bubble-point pressure of the propane, the least volatile of the gas components. Thus, all of the components in the mixture would not change to gas at one pressure as is the case with a gas consisting of only one component. A phase diagram for such a gas is illustrated in Fig. 5–3.

Therefore, the critical pressure and temperature terms lose much of their physical significance when applied to a gas that is a mixture of

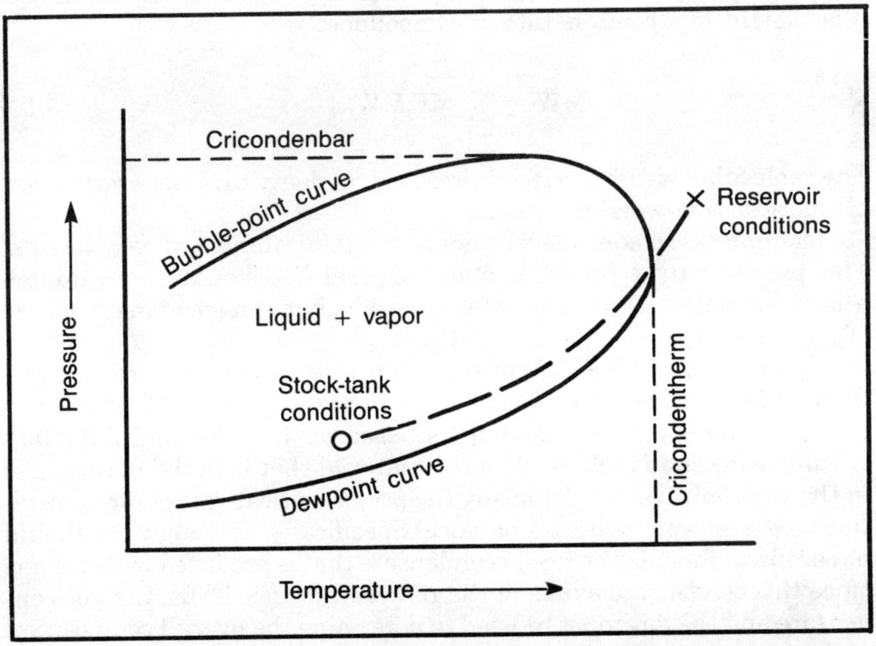

Fig. 5–3 Typical phase diagram for a multiple-component system (after Burcik, *Properties of Petroleum Reservoir Fluids*, courtesy John Wiley and Sons, 1959)

several different components as are all natural gases. This then would mean that the z-factor correlations of Fig. 5–1 cannot be used for a mixture of gases. However, it has been found that use of a pseudocritical pressure and temperature permit the application of the z-factor data to a mixture of gases as though the critical pressure and temperature were equal to the pseudocritical pressure and temperature.[2] The pseudocritical pressure and temperature are mol-fraction-weighted critical pressures and temperatures. Using a subscript to denote the pseudo part of the critical pressure and temperature and recognizing that the mol fraction of a gas is equal to the volume fraction of the gas at atmospheric pressures where gas analyses are run, we may state the pseudocritical pressure and temperature as:

$$_{p}p_{c} = \sum_{j=1}^{j=n} (MF_{j}p_{cj}) = \sum_{j=1}^{j=n} (VF_{j}p_{cj}) \tag{5.8}$$

$$_{p}T_{c} = \sum_{j=1}^{j=n} (MF_{j}T_{cj}) = \sum_{j=1}^{j=n} (VF_{j}T_{cj}) \tag{5.9}$$

The logical nature of Eqs. 5.8 and 5.9 can be recognized when we compare the equations with the equation used to calculate the molecular weight of a combination of compounds:

$$MW = \sum_{j=1}^{j=n} MF_{j}MW_{j} \tag{5.10}$$

The molecular weight, critical pressure, and critical temperature are all mol-fraction-weighted values.

The n in these equations refers to the total number of components. The pseudocritical pressure and temperature should be evaluated according to these equations when possible, but empirical methods are often necessary.

If gas is to be sold, a component analysis is always made of that gas. If the gas produces no condensate, or liquid, it is a simple matter to calculate the pseudocritical values based on Eqs. 5.8 and 5.9. Care should be taken to include all of the material that is in the natural gas in the reservoir when calculating the pseudocritical values for application to a reservoir material balance. Specifically, the engineer should be careful to include the liquid condensate that is produced with the gas since this condensate exists in the reservoir as gas. Thus, the gas condensate and the gas must be used to determine the overall composition of the gas in the reservoir. This component composition must then be used in calculating the critical values of the reservoir gas and the corresponding gas deviation factor. If it is not clear how the calculation of the overall composition in the reservoir is made from the pipeline gas composition, the condensate composition, and the condensate-gas ratio, refer to Katz et al., where a detailed example can be found.[3]

It should be noted that the z data in Figs. 5–1 and B8 apply theoretically only to hydrocarbons. However, nonhydrocarbon nitrogen, carbon dioxide, and hydrogen sulfide, often found in natural gases, can be included with a minimal (perhaps 2%) affect on the accuracy if the total nonhydrocarbon is 2% or less. For larger nonhydrocarbon contents the engineer may wish to obtain lab measurements of the gas deviation factors.

When a component analysis is not available for a gas or the condensate, empirical data are available that provide the pseudocritical pressure and temperature as functions of the specific gravity of the gas. These data are presented in Fig. B9. To use the data, simply enter the plot with the specific gravity of the reservoir gas and read the pseudocritical pressure and temperature. Corrections are made for nonhydrocarbons as indicated in Fig. B9.

To obtain the specific gravity of the reservoir gas, the specific gravity of the pipeline gas must be corrected for the condensate that is produced with the gas as mentioned. However, in this case there is little prospect for combining the compositions of the two materials into one composition of the reservoir material since we probably do not have the component analysis of the pipeline gas or we would not be using the specific gravity—critical value data. However, we can use the Standing correlation shown in Fig. B12 to obtain the reservoir gas gravity from the pipeline gas gravity. When using this correlation, note that the pipeline gas gravity is called the trap gas gravity and the well fluid gravity refers to the reservoir gas gravity. Furthermore, the correlation is based on the empirical relationship between the molecular weight and the specific gravity of the condensate stated as degrees API.

Referring again to the correlation of the gas gravity and the critical values in Fig. B9, note that there are two correlation lines for both the critical pressure and critical temperature labeled "miscellaneous gases" and "condensate well fluids." Actually it is found that the curves for miscellaneous gases define the dry gases rather accurately. However, the gases that contain a considerable amount of condensate provide more of a problem in applying the data in Fig. B9. For example, a gas that has a condensate content of 10 bbl/MMcf of dry gas has a much different pseudocritical pressure and temperature than a similar gas that contains 80 bbl/MMcf. Thus, the correlations leave much to be desired when applying them to gases that yield condensate. However, for gases that yield little condensate, the miscellaneous gases correlation provides reasonably accurate critical values.

To summarize, we have two common methods of determining gas deviation factors when the nonhydrocarbon content is less than about 2%. The preferred method is to calculate pseudocritical pressures and temperatures with Eqs. 5.8 and 5.9 using the component analysis of the reservoir gas. Then use the calculated reduced pressure and temperature in Fig. B8 to determine the z factors. The second method should be used only when time or data are insufficient to permit use of the first method. In the second method the reservoir gas gravity is used to obtain the pseudocritical pressure and temperature from Fig. B9. Then evaluate the gas deviation factor from Fig. B8. When the nonhydrocarbon content is greater than 2%, the engineer should consider obtaining measured lab data for the gas.

The engineer who must work with gas reservoirs and has access to a digital computer may find it very helpful to have the gas deviation data from Fig. B8 stored in the computer. Many gas reservoir engineering methods require iteration on the pressure with a corresponding change or iteration on the z factor. Thus, it is very helpful to have the data from Fig. B8 in storage. Then a simple subroutine can be used to interpolate the data to obtain the desired gas deviation factor. The z-factor data is digitized in Table A–2 of the Katz et al., *Handbook of Natural Gas Engineering*.[4] A computer program for storing and interpolating the Katz et al., digitized z-factor data is shown here in Table 5–1.

For most reservoir engineering studies it is easier to use a curve-fitting program for obtaining z factors. Table 5–2 presents such a program.

Since the z factor is a function of the pressure and the pressure is often the parameter for which we are solving, most engineers generally find it useful to prepare a plot of z versus p/z for a gas reservoir as well as a plot for z versus pressure. Such plots are illustrated in Fig. 5–4. With such data plots it is a simple matter to obtain a gas deviation factor for any reservoir pressure or to obtain a reservoir pressure for any p/z value.

Gas formation volume factor. The engineer should be familiar with the oil formation volume factor, the ratio of the volume of oil in the reservoir to the volume of oil that results when this oil is taken from reservoir conditions to stock-tank conditions. The gas formation volume factor is similar. It relates the volume of gas at reservoir conditions to the volume at standard conditions. The nature of the gas formation volume factor is somewhat different because both volumes in the ratio represent the same mass. In the oil formation volume factor the mass of the volume in the reservoir is different from the mass in the stock tank because the liberation of gas accompanies the change in conditions.

TABLE 5–1 Fortran Program for Storing and Interpolating Gas Deviation Factors

```
   COMMON Z(297,20)
   CALL ZREAD
   READ(5,11) TEMP,PRESS,GRAV
11 FORMAT(3F10.4)
   CALLZCALC(TEMP,PRESS,GRAV,ZOUT)
   WRITE(6,10)ZOUT
10 FORMAT(E20.8)
   CALLEXIT
   END
   SUBROUTINE ZREAD
   COMMON Z(297,20)
   DO 3 I=1,297
 3 READ(5,6) (Z(I,J),J=1,20)                          ZREAD
 6 FORMAT(20F4.3)
   RETURN
   END
   SUBROUTINE ZCALC (TEMP,PRESS,GRAV,ZOUT)
   DIMENSION P(297), T(20)
   COMMON Z(297,20)
   P(1)=0.2
   DO20I=2,297
20 P(I)=P(I-1)+0.05
   T(1)=1.05
   DO21 I=2,10
21 T(I)=T(I-1)+0.1
   DO24 I=16,20
24 T(I)=T(I-1)+0.2
   TC=173.+310.*GRAV
   PC=695.-40.*GRAV
   TR=TEMP/TC
   PR=PRESS/PC                                         ZCALC
   PP=(PR-0.15)/0.05
   M1=PP
   M2=M1+1
   DIF=PR-P(M1)
   DO30I=1,20
   IF(TR-T(I))31,31,30
31 N1=I-1
   N2=I
   DIFI=TR-T(I-1)
   GOTO32
30 CONTINUE
32 CONTINUE
   PINT=DIF/0.05
   TINT=DIFT/(T(N2)-T(N1))
   ZOUT=Z(M1,N1)+PINT*(Z(M2,N1)-Z(M1,N1))+TINT*(Z(M1,N2)
   -Z(M1,N1))
   RETURN
   END
```

TABLE 5–2 Program for Evaluating Gas Deviation Factors

```
      SUBROUTINE ZMIXT (NCOMP, X,PC,TC,P,T,Z,IO)
C
C     STANDING-KATZ Z-FACTOR CHART BY YARBOROUGH-HALL
C        EQUATION OF STATE INCLUDING CO2 AND H2S
C        CORRECTION OF WICHERT AND AZIZ
C     CALLING PARAMETERS
C        NCOMP— NUMBER OF COMPONENTS
C        X      — MOL FRACTIONS OF COMPONENTS—ORDER OF
C                    FIRST TWO IS X-CO2 AND X-H2S
C        PC     — COMPONENT CRITICAL PRESSURE
C        TC     — COMPONENT CRITICAL TEMPERATURES
C        P      — PRESSURE, ANY UNITS
C        T      — TEMPERATURE, K OR R
C        Z      — COMPRESSIBILITY
C        IO     — UNIT OUTPUT DEVICE NUMBER FOR PRINTER
C
      IMPLICIT REAL*8(A–H,O–Z)
      REAL X,PC,TC,P,T,Z
  100 FORMAT (75H0*** WARNING FROM ZMIXT — TR LESS THAN 1.0.
     THE VARIABLE IS T, TPC, TR FOLLOW ,E20.7,2E15.7)
      DIMENSION PC(1),TC(1),X(1)
      TCMA=0.0D0
      PCMA=0.0D0
      DO 1 N=1,NCOMP
      TCMA=TC(N)*X(N) + TCMA
    1 PCMA=PC (N)*X(N) + PCMA
      AWA=X(1) + X(2)
      CWA=120.D0*(AWA**0.9D0 – AWA**1.6D0) + 15.0D0* SQRT (X(1))
     – X(1)
    1 **4
      TPC=TCMA – CWA
      PPC=PCMA*TPC/ (TCMA + X(1)*(1.0D0 – X(1))*CWA)
      RRT=TPC/T
      IF(1.0–RRT) 45,50,50
   45 TR=T/TPC
      WRITE(IO,100) T,TPC,TR
   50 A=0.06125D0*RRT*DEXP(–1.2D0*(1.0D0 – RRT)**2)
      B=RRT*(14.76D0 – 9.76D0*RRT + 4.58D0*RRT*RRT)
      C=RRT*(90.7D0 – 242.2D0*RRT + 42.4D0*RRT*RRT)
      D=2.18D0 + 2.82D0*RRT
      Y=0.001D0
      RP=P/PPC
      DO 2 J=1,30
      IF (Y–1.0D0) 20,20,10
   10 Y=0.6D0
   20 F=–A*RP + (Y + Y*Y + Y*Y*Y – Y**4)/(1.0D0–Y)**3 – B*Y*Y +
     C*Y**D
      IF(DABS(F) – 0.5D–06) 4,4,3
    3 DFDY=(1.D0 + 4.0D0*(Y + Y*Y – Y**3) + Y**4)/(1.0D0–Y)**4 –
     1 2.0D0*B*Y + D*C*Y**(D–1.0D0)
```

TABLE 5–2 continued

```
      2 Y=Y - F/DFDY                                          ZMIXT 49
      4 Z=A*RP/Y                                              ZMIXT 50
        RETURN                                                ZMIXT 51
        END                                                   ZMIXT 52
$
```

Source: H. Hershey, Ohio State University, after Hall and Yarborough, "How to Solve the Equation of State for z Factor," *Oil & Gas Journal,* February 18, 1974.

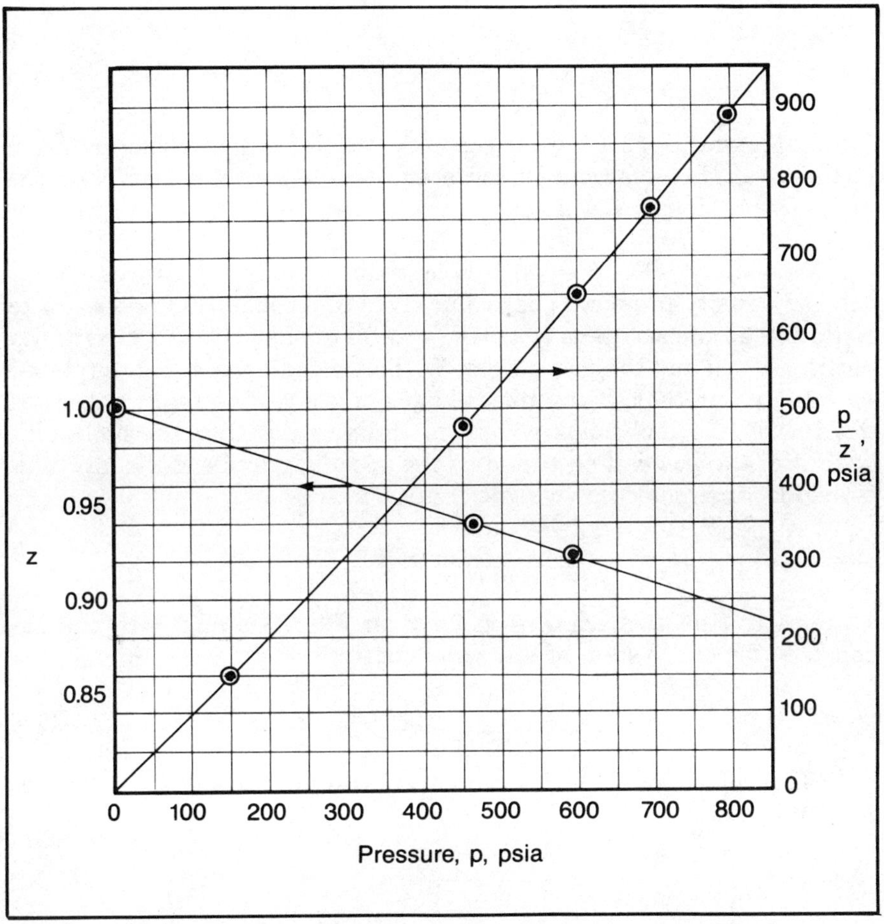

Fig. 5–4 Gas deviation factor data for problem 5.6

Since the gas formation volume factor simply relates the volume of gas at reservoir conditions to the volume of the same mass at standard conditions, we can use Eq. 5.1 to derive an expression for the gas formation volume factor. The units of the gas formation volume factor can vary, but most engineers prefer to state this factor as reservoir barrels per standard cubic foot. Since 1 scf of any gas is equal to 1/379 of a mol, we can substitute 1/379 into Eq. 5.1 and solve for the volume at the reservoir pressure and temperature. Dividing the result by 5.615 (the number of cubic feet in a barrel) changes the volume to barrels and provides the gas formation volume factor, B_g, in reservoir barrels per standard cubic foot:

$$B_g = V/5.615 = \frac{z(1/379)\,(10.73)T}{5.615p} \tag{5.11}$$

$$B_g = \frac{0.00504zT}{p} \tag{5.12}$$

In applying Eq. 5.12, it is necessary to determine the gas deviation factor, z, at the reservoir pressure and temperature, p and T, by the methods previously discussed.

Gas density. Eq. 5.1 can also be used to derive an expression for the gas density in pounds per cubic foot. To do this, it is necessary to write the number of mols of gas, n, as a function of the specific gravity of the gas, γ_g, and the gas weight, W. This calculation is accomplished by recognizing that the number of mols of gas is the weight of the gas divided by the molecular weight of the gas and that the molecular weight of the gas can be stated as the specific gravity of the gas multiplied by the molecular weight of air. Thus:

$$n = \frac{W}{MW} \text{ or } n = \frac{W}{29\gamma_g}$$

Substituting this expression for n in Eq. 5.1 along with the gas constant for our system of units, we obtain:

$$pV = z\frac{W}{29\gamma_g}(10.73)T \tag{5.13}$$

When solved for W/V, Eq. 5.13 gives the gas density, ρ_g, in pounds per cubic foot:

$$\rho_g = \frac{2.7\ \gamma_g p}{zT} \tag{5.14}$$

Eq. 5.14 can be used to calculate the gas density of any gas at any particular pressure and temperature after first evaluating the gas deviation factor at those conditions.

Gas compressibility. Compressibility of gas is defined as the fractional change in the volume per unit of pressure change:

$$c_g = -\frac{\Delta V/V}{\Delta p} \tag{5.15}$$

In practical units the compressibility is the fractional change in the volume per psi, or since the fraction is dimensionless, the unit of compressibility is simply 1/psi or psi^{-1}. When gas reservoir engineering technology was first developed, most gas reservoirs being produced were low-pressure reservoirs in which the change in the gas deviation factor with a change in pressure did not greatly affect the compressibility of the gas. Under these conditions it was sufficiently accurate to approximate the compressibility of the gas as being equal to the reciprocal of the pressure, 1/p. Consequently, many equations used in gas technology have had this expression substituted for the compressibility. However, the relatively high pressures of reservoirs being produced today precludes the approximation of the compressibility by this expression. The compressibility is only equal to the reciprocal of the pressure when an ideal gas is under consideration.

To find an accurate expression for gas compressibility, differentiate Eq. 5.6 with respect to the pressure:

$$V = \frac{znRT}{p} \tag{5.6}$$

$$\frac{\Delta V}{\Delta p} = nRT\left(\frac{\Delta z}{p\Delta p} - \frac{z}{p^2}\right) \tag{5.16}$$

If Eq. 5.16 is substituted into Eq. 5.15 with the right-hand side of Eq. 5.6 for the volume, V, we obtain:

$$c_g = -nRT\left[\frac{\Delta z}{p\Delta p} - \frac{z}{p^2}\right] \Big/ \frac{znRT}{p} \tag{5.17}$$

When simplified, Eq. 5.17 gives an expression for the gas compressibility:

$$c_g = \frac{1}{p} - \frac{\Delta z/z}{\Delta p} \tag{5.18}$$

From Eq. 5.18 we can see that the compressibility is only equal to the reciprocal of the pressure when the second term is zero. Therefore, $(\Delta z/\Delta p)$ must be zero. Thus, the gas must be an ideal gas for the compressibility to be equal to the reciprocal of the pressure.

If we plot the pressure versus the gas deviation factor for each gas we are considering, we can simply use this plot to calculate the compressibility at any particular pressure. The slope of the tangent to the curve at the pressure of interest is $\Delta z/\Delta p$, and the calculation can then be completed using Eq. 5.18.

However, it is often unnecessary to prepare a plot of the pressure versus the gas deviation factor, so it is often inconvenient to evaluate the gas compressibility in this way. We do have a plot of the gas deviation factor versus the reduced pressure in Fig. B8 or Fig. 5–1. Consequently, we can change Eq. 5.18 to use a slope $\Delta z/\Delta p_R$ instead of $\Delta z/\Delta p$. This is accomplished by substituting $p_R p_c$ for p in the equation:

$$c_g = \frac{1}{p_c p_R} - \frac{\Delta z/z}{p_c \Delta p_r} \tag{5.19}$$

In this form the slope of z versus p_R can be used as $\Delta z/\Delta p_R$ to calculate the compressibility. However, Trube performed these operations for the engineer and presented the results as a reduced compressibility plotted versus the reduced pressure where each curve represents a different reduced temperature (Fig. B6).[5] The reduced compressibility is defined as the gas compressibility multiplied by the critical pressure, $c_g p_c$. Thus, Eq. 5.19 can be converted to a definition of the reduced compressibility by rearranging it to obtain:

$$c_R = \frac{1}{p_R} - \frac{\Delta z}{z \Delta p_R} \tag{5.20}$$

Thus, to obtain the gas compressibility at a particular reservoir pressure and temperature, we determine the reduced pressure and temperature, evaluate the reduced compressibility from Fig. B6, and calculate $c_g = c_R/p_c$.

Gas viscosity. By using a rolling-ball pressure viscosimeter in a temperature bath, the viscosity of a specific natural gas can probably be most accurately determined in the laboratory. However, the empirical data available for determining the viscosity of gas is sufficiently accurate that an engineer seldom finds it necessary to request a laboratory measurement of the viscosity. Considering the inaccuracy of some of the other data that are used in the same equations with the gas viscosity, the viscosity may well be the most accurate parameter even when it comes from empirical data. Examples of such data are shown in Figs. B13, B19, and B20. These data can be used to determine the gas viscosity under various conditions within the data range if they are used carefully. Fig. B13 is very simple to use when there are no nonhydrocarbons in the gas and the other parameters fall within the range of the data. For data that fall outside the range of Fig. B13, the more general correlations of Figs. B19 and B20 can be employed.

To demonstrate the application of the methods described in determining the characteristics of gas, work problem 5.1 and compare the solution with the one in appendix C.

PROBLEM 5.1: Determining Gas Characteristics

To avoid repetitious calculations, we are considering a hypothetical gas of 80% methane, 15% ethane and 5% normal butane by volume. The gas is at 3,000 psia and 170°F.

A. Find the reduced pressure and temperature using Eqs. 5.8 and 5.9 to obtain the pseudocritical pressure and temperature.

B. What is the gas deviation factor?

C. Find the gas specific gravity.

D. What is the pseudoreduced pressure and temperature based on Fig. B9? What is the z factor?

E. What is the gas formation volume factor for this reservoir?

F. Find the gas density at reservoir conditions.

G. What is the compressibility of the gas? What is the compressibility according to $c = 1/p$?

H. Estimate the gas viscosity at reservoir conditions.

I. If 55°API gravity condensate is produced with this 0.7 gravity gas at a ratio of 40 bbl/MMcf, what is the reservoir gas deviation factor?

Material Balance

The material balance for a gas reservoir is undoubtedly the simplest material balance the reservoir engineer encounters. Nevertheless, we must proceed cautiously when applying the material balance if we are to avoid sizable errors. We should be aware that the original gas in the reservoir contains water vapor and condensate, which are produced with the gas, and that the normal gas material balance does not include automatic corrections for these factors. Thus, the gas produced must include the condensate and water vapor.

We should also be aware of the fact that the early average reservoir pressures may be in error if the reservoir is of very low permeability and if pressure-buildup data are relied on for determining average reservoir pressures. The calculation of water encroachment into a gas reservoir also presents special problems as a result of the long delay before the pressure drop reaches the original water-gas contact. Considerable error can be incurred by assuming that the decline in the pressure at the original water-gas contact is the same as the decline in the average pressure in the reservoir. Reservoirs that approach overburden pressures often experience compaction during the early life of the reservoir, which causes material-balance anomalies. These and other problems involved in the application of material balance to a gas reservoir must be considered.

The gas material-balance equation. This discussion of material balance of a gas reservoir may be considered as an introduction to the subject since it is concerned with the simplest case of material balance and the simplest material-balance equations. A general discussion of material-balance technology is provided in chapter 6.

Material-balance equations equate the volume of a mass of material at a particular pressure and temperature to the volume of the same mass of material at some different pressure and temperature. In a petroleum reservoir, including a gas reservoir, we define the original free-gas volume in the reservoir as G and state this volume in standard cubic feet. Thus, the original reservoir volume of free gas in barrels is the product of G and the initial gas formation volume factor, B_{gi}, stated in reservoir barrels per standard cubic foot. If the gas reservoir is not subjected to a water drive and no interstitial water is produced, the reservoir volume occupied by the gas normally remains constant for all practical purposes. Thus, the same initial reservoir volume can be stated in terms of the gas remaining in the reservoir at any particular time after some standard cubic feet of gas, G_p, has been produced.

At this time the volume of gas remaining in the reservoir is $(G - G_p)$ in standard cubic feet, and the reservoir volume is $(G - G_p)B_g$ in barrels when the gas volume is converted to barrels at the current reservoir pressure. Therefore, the gas formation volume factor used to make this conversion must be based on the average reservoir pressure that exists after G_p standard cubic feet of gas has been produced. We can then equate the reservoir gas volume stated in these two different ways to obtain the material-balance equation for a gas reservoir that does not have a water drive and where no interstitial water is produced:

$$GB_{gi} = (G - G_p)B_g \tag{5.21}$$

If a gas reservoir is subjected to a water drive, the gas volume in the reservoir is reduced as the water encroaches into the original gas-bearing portion of the reservoir. The barrels of gas in the reservoir initially, GB_{gi}, less the barrels of gas in the reservoir at some time after a volume of gas, G_p, has been produced, $(G - G_p)B_g$, is equal to the barrels of water that have entered the original gas-bearing pores of the reservoir if none of the encroached water has been produced. If W_p barrels of the encroached water has been produced, the difference is equal to the total number of barrels of water that have encroached into the original gas-bearing portion of the reservoir, W_e, less the produced water, W_p:

$$GB_{gi} = (G - G_p)B_g + (W_e - W_p) \tag{5.22}$$

Eq. 5.22 assumes that the formation of liquid condensate in the reservoir and the evaporation of connate water cause insignificant errors in the equality. There are notable exceptions to the assumption that no significant liquid condensate forms in the reservoir, but the assumption appears to be valid that the change in the gas reservoir volume caused by water evaporation is negligible. The original gas in place can be calculated volumetrically, and the water encroachment can be evaluated using the constant-pressure solution to the radial diffusivity equation. A lack of understanding of the time delay associated with the water encroachment into a gas reservoir appears to cause much difficulty in some segments of the gas-producing industry.

The gas production, G_p, to be used in Eq. 5.22 should include the produced condensate and for maximum accuracy the produced water that was in the original reservoir gas.

Material balance gas production. When dry natural gas at high pressure and temperature is removed to a low pressure and temperature, some of the heavier hydrocarbons generally change to liquid or condensate. This condition is illustrated in Fig. 5–3, in which the stock-tank conditions are in a portion of the phase diagram where gas and liquid exist in equilibrium. The dotted line simply shows the possible pressure-temperature path the hydrocarbons may travel in going from far out in a reservoir to surface conditions. Since Eq. 5.22 only provides for the production of dry gas from the reservoir, it is necessary to include the condensate production in the produced gas term, G_p, as equivalent dry gas.

Using Eq. 5.1, we can derive an expression for the amount of gas in standard cubic feet that is the equivalent of 1 stb of condensate. This can be stated as a function of the specific gravity and molecular weight of the condensate. Since we want the volume at standard conditions, the pressure, temperature, and gas deviation factor are fixed at 14.7 psia, 520°R, and 1.0, respectively. The gas deviation factor is 1.0 for all gases at standard conditions. Also, these units fix the gas constant, R, at 10.73. This leaves the number of mols, n, to be stated as a function of the specific gravity of the condensate, γ_L, and the molecular weight, MW. The weight of 1 bbl of fresh water is 350 lb, so the weight of 1 bbl of a liquid whose specific gravity is γ_L is $350\gamma_L$. Substituting these values into Eq. 5.1, we can derive the expression for the gas equivalent of the condensate, GE, in standard cubic feet per stock-tank barrel:

$$(14.7)(V) = (1.0)\frac{(350\gamma_L)}{MW}(10.73)(520) \tag{5.23}$$

$$GE = V = 133{,}000\gamma_L/MW \tag{5.24}$$

For the greatest accuracy in determining the equivalent gas volume of the produced condensate, Eq. 5.24 should be used and the molecular weight should be calculated from a component analysis or a laboratory measurement. However, the component analysis and laboratory molecular weight are seldom available for a condensate. In such cases the data in Table B15 are very helpful. This table provides an empirical relationship between the condensate gravity in °API and the gas equivalent, GE, in standard cubic feet per stock-tank barrel. The table is based on the equations shown in the footnotes. These equations change °API to the specific gravity, γ_L; calculate the molecular weight, MW, from the specific gravity using an empirical equation; and calculate the gas equivalent, GE, from Eq. 5.24.

We know that all petroleum reservoirs contain some water saturation because the pore volume was at some time in geologic history completely occupied by water. Then when the hydrocarbons migrated into the reservoir, they were incapable of displacing all of the water. Hence, all petroleum reservoirs contain connate water, although it may be immobile for the entire producing life of the reservoir. The gas in a gas reservoir has been in the presence of this connate water for billions of years. Consequently, we can safely assume that the gas is initially saturated with water vapor in the reservoir. Since some of this water vapor will probably condense at surface conditions and most of the water vapor must be removed before the gas is metered for sale, the metered dry gas does not include the water vapor that is part of the original gas in place in the reservoir. Consequently, we must correct the produced dry-gas volume for the water vapor before it is used in a material-balance calculation.

Fig. B17 shows the water content of natural gas in the presence of liquid water at various pressures and temperatures. It can be used to determine the pounds of water in 1 MMscf of gas. The main portion of the graph gives the water content of fresh water, but in the lower right-hand corner a graph of correction factors is included to account for the effect of dissolved solids on the water content of the gas. In order to be as accurate as possible, the original water-vapor correction should be applied to all produced gas before the figures are used in material-balance calculations.

However, the exact handling of the water-vapor correction is not quite this simple. For example, the connate or interstitial water may become mobile as the reservoir pressure declines. In many cases the reservoir exists in a transition zone, where the water saturation is greater than the irreducible water but is less than 100%. The interstitial water is mobile from the moment of first production. Regardless of the reason, it remains that some of the interstitial water may be produced. Care must be taken to ensure that this produced liquid water is not assumed to be produced water vapor because it must be handled differently in the material balance. Produced water that was originally liquid in the reservoir should be included in the material-balance calculations as produced water. W_p, whether there is a water drive on the gas reservoir.

Examination of the gas material-balance equation, Eq. 5.22, shows that the equation still holds true although there is no water encroachment. The production of interstitial water increases the reservoir volume of gas. If W_e in Eq. 5.22 is zero, the equation says that the original reservoir volume of gas, the left-hand side of the equation, is equal to the reservoir volume of gas after G_p standard cubic feet of gas have been removed, minus the produced interstitial water. Thus, the reservoir gas volume after production is equal to the original gas volume plus the interstitial water produced.

The interstitial water produced may be distinguished from the water vapor produced on the basis of the water salinity since water vapor is fresh water and interstitial water produced as liquid is brine. However, some of the interstitial water produced may actually have passed through the vapor state and, thus, may be produced without its original salt content. Fig. B17 helps explain this situation.

Note that as the reservoir pressure decreases, the equilibrium water-vapor content increases. Therefore, part of the interstitial water can vaporize as the pressure declines. The evaporation of the interstitial water makes more space available for the reservoir gas; however, it may be produced as fresh water since it is produced in the vapor phase. Consequently, the author recommends that all of the produced water in excess of that originally in the produced gas in the vapor state in the reservoir be treated as W_p in Eq. 5.22. Therefore, the engineer should calculate the amount of water vapor that was originally in the produced gas in the reservoir. Then this amount should be added to the metered dry gas, the equivalent gas of the produced condensate, and the vent gas from the condensate storage to obtain the produced gas value, G_p:

$$\begin{aligned} G_p = {} & \text{(dry gas sold)} + \text{(gas equivalent of condensate produced)} \\ & + \text{(original water vapor in produced gas)} \\ & + \text{(vent gas from condensate storage)} \end{aligned} \tag{5.25}$$

The difference between the total produced water and the original water vapor in the produced gas must be treated as produced liquid water, W_p, in Eq. 5.22. Note that this treatment still is not exact. The difference between the produced water and the original water-vapor content of the produced gas provides some correction for the additional pore volume available to the gas in the reservoir because interstitial-water evaporation accompanies the decline in the reservoir pressure.

However, this procedure represents only part of the correction that is theoretically necessary as a result of this phenomena because all of the evaporated water is not produced but remains in the reservoir. Therefore, more space is available to the gas phase in the reservoir, which is not accounted for in the material-balance equations discussed. This additional pore space is offset to a large degree by the fact that a larger gas-phase mass is present in the reservoir than is indicated in the material-balance equations considered because of the interstitial-water evaporation. Consequently, the net correction to the equations is small. By accounting for the effect of the vaporized water produced, we are accounting for the portion of the reservoir evaporation phenomena that is not offset by an additional gas mass in the reservoir.

The small equation error that remains can be taken into account except for the kinetics effect that accompanies the vaporization. It is uncertain how much time is required for equilibrium to be obtained at each pressure. Gas production is continuous so equilibrium cannot be reached at each pressure. Also, the water is in the smallest pores, which provides a very small interface between the gas and water. Consequently, the evaporation rate would be difficult to establish. Thus, it is recommended that Eq. 5.22 be used in material-balance calculations. The gas production, G_p, should be calculated by Eq. 5.25, and the water production, W_p, should include the produced water in excess of the original water vapor in the produced gas.

To test our knowledge of these corrections, work problem 5.2 and check the solution against the one in appendix C.

PROBLEM 5.2: Determining Gas Production for Use in Material Balance

Calculate the reservoir gas production, G_p, and water production, W_p, for material-balance equations using the following data:

Separator gas production = 10,000 MMscf
Condensate production = 150,000 stb
Condensate gravity = 55°API (0.759 sp gr)
Stock-tank gas production (vent gas) = 30,000 Mcf
Freshwater production = 15,000 bbl
Initial reservoir pressure = 4,000 psia
Reservoir temperature = 220°F
Formation water salinity = 150,000 ppm
GE for water from Eq. 5.24 = 7,390 scf/bbl

Determining the reservoir pressure. The most accurate means of determining the average reservoir pressure in a gas reservoir is to run a pressure bomb in a permanently shutin well. In the past it was almost a certainty that some of the wells in a gas reservoir in good mechanical condition would be idle for long periods each year. During the idle periods these wells would afford an ideal opportunity to measure the static pressure directly in a gas reservoir. However, the recent gas shortage has made it much more unlikely that such shutin wells can be found in gas reservoirs.

When idle gas wells are not available, the reservoir engineer should attempt to obtain static reservoir pressures through the conventional pressure-buildup analysis. Such tests are described in chapter 4, and the difficulties of applying them to gas reservoirs also is discussed further in this chapter. In addition to the general problems of determining average reservoir pressures, pressure buildup in a gas well is so complicated by afterflow effects that it may be impossible to obtain an interpretable pressure-buildup survey. Nevertheless, the engineer should realize that there is nothing magical about a 48-hr shutin pressure to make it represent the static pressure in the gas reservoir. This period seems to have been an acceptable standard in the past; i.e., the shutin pressure after 48 hr is the static reservoir pressure. Until better methods are devised, we must still make our best effort to interpret pressure-buildup data to determine the average reservoir pressure.

Many situations exist today that make it undesirable to run a pressure bomb in a well. The reservoir may be so deep or at such a high pressure that management deems it too high of a risk to justify running a bomb in a well. The reservoir gas may be so dry that management believes accurate, static bottom-hole pressures can be obtained from surface-pressure measurements. These measurements can be done with sufficient accuracy if there are no fluids standing in the bottom of the hole and a calculating method is used that accounts for the variation in temperature and gas deviation factors from the surface to the bottom of the hole. Two methods are presented that provide such accuracy. One of these requires a digital computer to perform the iterative calculations, and the other is in the form of empirical data that has some limitations as to the maximum depths to be considered and the accuracy with which the charts can be read. We also examine the basis of the equation generally recommended by state regulatory bodies, which requires average temperatures and average z factors.

Static pressure differences for a column of gas are more difficult to evaluate than similar static pressure differences for a column of liquid. Gas density is a function of pressure, and the pressure in a gas column varies because of static pressure differences. The pressure gradient at any particular point of pressure and temperature in a gas column can be calculated from the density of the gas at those conditions. If the density is in pounds per cubic foot, this measurement is the same as a pressure gradient in pounds per square foot per foot of difference in depth. If we desire the pressure gradient in the more conventional psi per foot of depth, we can simply divide the density in pounds per cubic foot by the number of square inches in a square foot, 144. Thus, we can obtain the pressure gradient in psi per foot from Eq. 5.14, which defines the density in pounds per cubic foot:

$$\frac{\Delta p}{\Delta D} = \frac{\rho_g}{144} = \frac{2.7\gamma_g p}{zT} \Big/ 144 \tag{5.26}$$

$$\frac{\Delta p}{\Delta D} = \frac{0.01875\gamma_g p}{zT} \tag{5.27}$$

Eq. 5.27 can then determine the change in pressure over some small increment of depth, using the arithmetic average pressure as p in the equation:

$$\Delta p = \frac{0.01875\gamma_g p}{zT}\,\Delta D \tag{5.28}$$

For hand calculations Eq. 5.28 must be assumed to be trial and error in its application. However, when a digital computer is used to calculate a bottom-hole pressure from a surface pressure by applying Eq. 5.28 to very small increments of depth, there is practically no limit on the size of the depth increments that can be used. A maximum increment size below which there is no significant increase in the accuracy of the calculations can be determined. A simple program for calculating the bottom-hole pressure from surface pressure, gas gravity, surface temperature, temperature gradient, critical pressure, critical temperature, and gas deviation factor is shown in Table 5–3. This program is written in conversational language, PL 1. The program is designed to permit the submission of all data through a typewriter terminal. The procedure calculates the pressure at the bottom of an increment using the average temperature in the increment to represent the entire increment and using the average pressure, p, as the pressure at the top of the increment plus half the pressure change, Δp, for the depth increment just above the subject increment. By starting the calculations at the top of the hole where the surface pressure is known, it is then possible to use the pressure calculated for the bottom of each increment as the pressure at the top of the next lower increment and continue calculations until the depth at which the pressure is desired is reached. The program is sufficiently simple that a programmable calculator can be used for the calculation.

As is the case with all digital computer programs presented in this text, the program in Table 5–3 is not a polished program but is simply presented to provide the engineer or programmer with the fundamental ideas so a polished program can be developed. If the program is used in a Fortran language with gas deviation data in storage and a subroutine available to interpolate the data as previously defined, the program input and use would be greatly simplified.

TABLE 5–3 Computer Program for Calculating the Static Bottom-Hole Pressure of a Gas Well*

```
seq from 1 through. . . from 1 through. . .
 1.            PUT IMAGE(1)(head);
 2.  head:     IMAGE;
               CALCULATION OF STATIC BHP IN A GAS WELL
 3.            GET LIST(sg,dx,psurf,tsurf,td,ttd,pc,tc,ntr,npr);
 4.            /*TEMPERATURES MUST BE IN DEGREES RANKINE*/;
 5.            /*sg is gas specific gravity; dx is calculating interval desired;psurf is surface
               pressure
 6.            /*td is total depth; ttd is temperature at total depth; pc is critical pressure;*/;
 7.            /*tc is critical temperature; ntr is the number of tr values to be entered in
               program
 8.            /*npr is the number of pr values to be entered in the program; prntin is the
               desired pressure
interval*/;
 9.            /*pr is the reduced pressure; tr is the reduced temperature*/;
10.            ;
11.            /*prntin/dx and td/dx must be whole numbers*/;
12.            GET LIST(prntin);
13.            DECLARE pr(20), tr(10), z(10,20);
14.            ;
15.  loop1:    DOi=1 TO npr;
16.            GET LIST(pr(i));
17.            END loop1;
18.  loop2:    DO j=1 TO ntr;
19.            GET LIST(tr(j));
20.            END loop2
21.  loop3:    DO i=1 TO npr;
22.  loop4:    DO j=1 TO ntr;
23.            GET LIST(z(j,i));
24.            END loop4;
25.            END loop3;
26.            t=tsurf;
27.            p=psurf;
28.            ddd=0;
29.            dp=0;
30.            d=0;
31.            tgrad=(ttd-tsurf)/td;
32.            PUT IMAGE(1)(slip);
33.     slip:  IMAGE;
     Depth, ft.   Pressure, psia   Temperature, Rankine   Reduced temp.   Reduced
                  press.   z
34.  again:    ta=t+tgrad*(dx/2);
35.            tra=ta/tc;
36.            pra=(p+dp/2)/pc;
37.            i=0;
38.            i=i+1;
39.            h=i+1;
40.            IF pra>=pr(i)&pra<=pr(h) THEN GO TO tra
41.            IF h>npr THEN GO TO wrong;
42.            GO TO iinc;
43.     trac:  j=0;
```

TABLE 5–3 *continued*

```
jay:     j=j+1;
         k=j+1;
         IF tra>=tr(j)&tra<=tr(k) THEN GO TO zprltc;
         GO TO jay;
zprltc:  zprltr=z(j,i)+(tra-tr(j))/(tr(k)-tr(j))*(z(k,i)-z(j,i));
         zprhtr=z(j,h)+(tra-tr(j))/(tr(k)-tr(j))*(z(k,h)-z(j,h));
         zprtr=zprltr+(pra-pr(i))/(pr(h)-pr(i))*(zprhtr-zprltr);
         dp=.01875*sg*(p+dp/2)*(zprtr*ta);
         d=dx+d;
         t=t+tgrad*dx;
         p=p+dp;
         ddd=ddd+1;
         IF ddd=prntin/dx|d>=td THEN GO TO print;
         GO TO again;
slider:  IMAGE;
 -----.--   -----.--   ---.--   --.--   --.--   -.---
         IF d<td THEN GO TO again;
print:   PUT IMAGE(d,p,t,tra,pra,zprtr)(slider);
         ddd=0;
         IF d<td THEN GO TO again;
         STOP;
wrrng:   PUT IMAGE(1)(iml);
iml:     IMAGE;
                                    DATA ARE INSUFFICIENT
         STOP;
```

*In PL-1 conversational language

The Texas Petroleum Research Committee published a booklet of nomographs suitable for determining bottom-hole pressures.[6] These nomographs evaluate static pressures to depths of about 15,000 ft with reasonable accuracy. Some data are presented for depths of as much as 24,000 ft. The nature of these data are illustrated in Figs. 5–5 through 5–8. Random checks of the Bulletin 72 static pressures against the calculated values using the computer program of Table 5–2 showed excellent agreement. The bulletin also provides a means of calculating bottom-hole flowing pressures from surface pressures, but no check of these values has been made by the author.

The most often used method for calculating static bottom-hole pressure from surface pressure is probably the exponential form of Eq. 5.28 based on an average gas deviation factor and average temperature. This equation is obtained by treating the gas deviation factor and temperature as constants and applying Eq. 5.28 to all of the vertical increments, ΔD, from total depth to surface. If this calculation is done and all equations are summed, we obtain:

$$\sum_{p_1}^{p_2} \frac{\Delta p}{p} = \frac{0.01875\gamma_g}{z_{avg}T_{avg}} \sum_{0}^{D} D \tag{5.29}$$

$$\ell n \frac{p_2}{p_1} = \frac{0.01875\gamma_g D}{z_{avg}T_{avg}} \tag{5.30}$$

In exponential form:

$$p_2 = p_1 e^{\left(\frac{0.01875\gamma_g D}{z_{avg}T_{avg}}\right)} \tag{5.31}$$

PROBLEM 5.3: Calculating Static Bottom-Hole Pressure from Surface Pressure Measurements

A reservoir contains a 0.68-sp-gr dry gas at a temperature of 110°F. The average depth of the formation is 3,676 ft, the average wellhead temperature is 90°F, and the static surface pressure in the well is 2,715 psia. Estimate the corresponding average reservoir pressure using Eq. 5.31. The pressure obtained from the computer program in Table 5–3 is given in the appendix C solution for comparison.

When depths are shallow and pressures are relatively low, Eq. 5.31 gives sufficiently accurate results for many purposes. However, the simplicity of the computer program necessary for a more accurate calculation seems to make it unwise and unnecessary to use an equation that must assume an average gas deviation factor and average temperature.

Use of Eq. 5.31 is not as straightforward as it may appear, even if we are willing to accept the inaccuracy of using an average gas deviation factor and temperature. When we attempt to apply Eq. 5.31 to calculate the bottom-hole pressure, we find immediately that the pressure base for the average gas deviation factor is not known. Since we are calculating the bottom-hole pressure, we do not know the arithmetic average pressure until we have determined the bottom-hole pressure. Using Eq. 5.31, we cannot evaluate the bottom-hole pressure until we know the average gas deviation factor, which requires a knowledge of the arithmetic average pressure. Thus, a trial-and-error application of Eq. 5.31 is necessary. An average pressure is assumed, the gas deviation factor is determined, the bottom-hole pressure is calculated, the arithmetic average pressure is calculated, the gas deviation is determined, etc., until there is no further change in the average pressure. In all fairness it should be noted that this trial-and-error solution is not as bad as it sounds. Normally the solution converges to an acceptable accuracy very quickly.

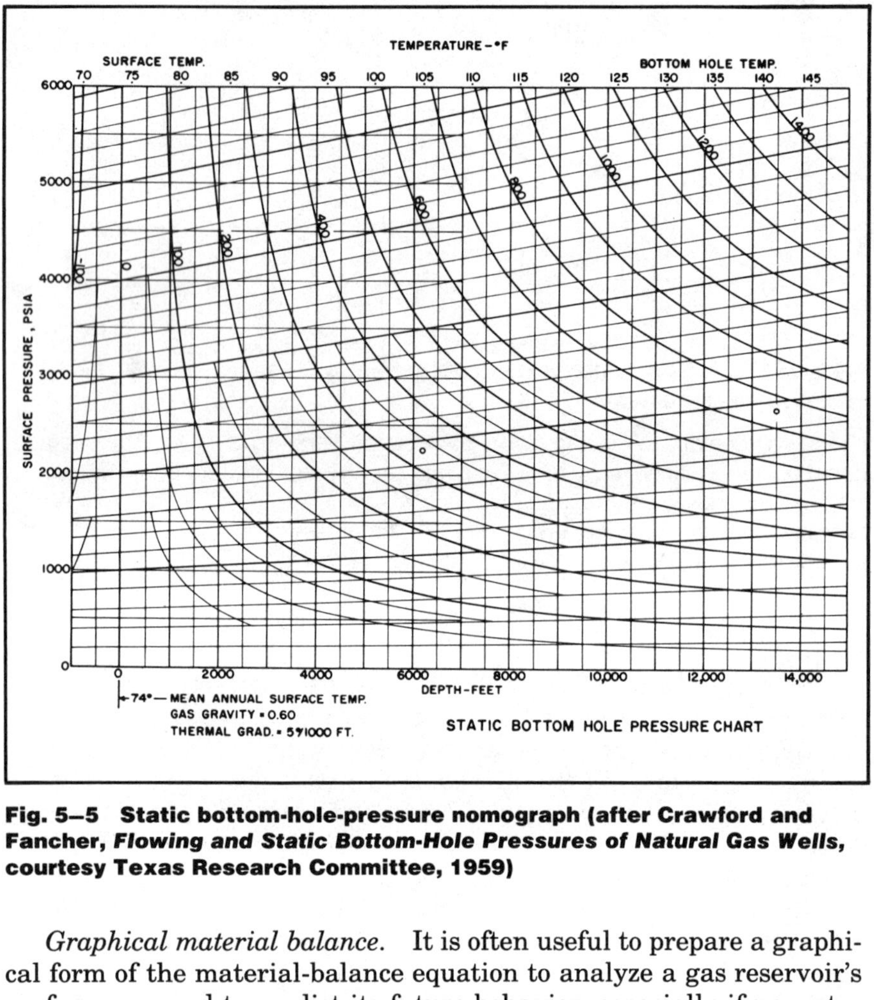

Fig. 5–5 Static bottom-hole-pressure nomograph (after Crawford and Fancher, *Flowing and Static Bottom-Hole Pressures of Natural Gas Wells*, courtesy Texas Research Committee, 1959)

Graphical material balance. It is often useful to prepare a graphical form of the material-balance equation to analyze a gas reservoir's performance and to predict its future behavior, especially if no water drive is present. Eq. 5.21, the material-balance equation applicable in the absence of a water drive, can be placed in a form more useful for graphical interpretation by substituting for the gas formation volume factors according to Eq. 5.12:

$$GB_{gi} = (G - G_p)B_g \tag{5.21}$$

$$G\,\frac{0.00504z_iT}{p_i} = (G - G_p)\,\frac{0.00504zT}{p} \tag{5.32}$$

$$G_p = G - \frac{p}{z}\,\frac{Gz_1}{p_i} \tag{5.33}$$

Note that a plot of the cumulative gas production, G_p, versus p/z should give a straight line that can be extrapolated to predict the

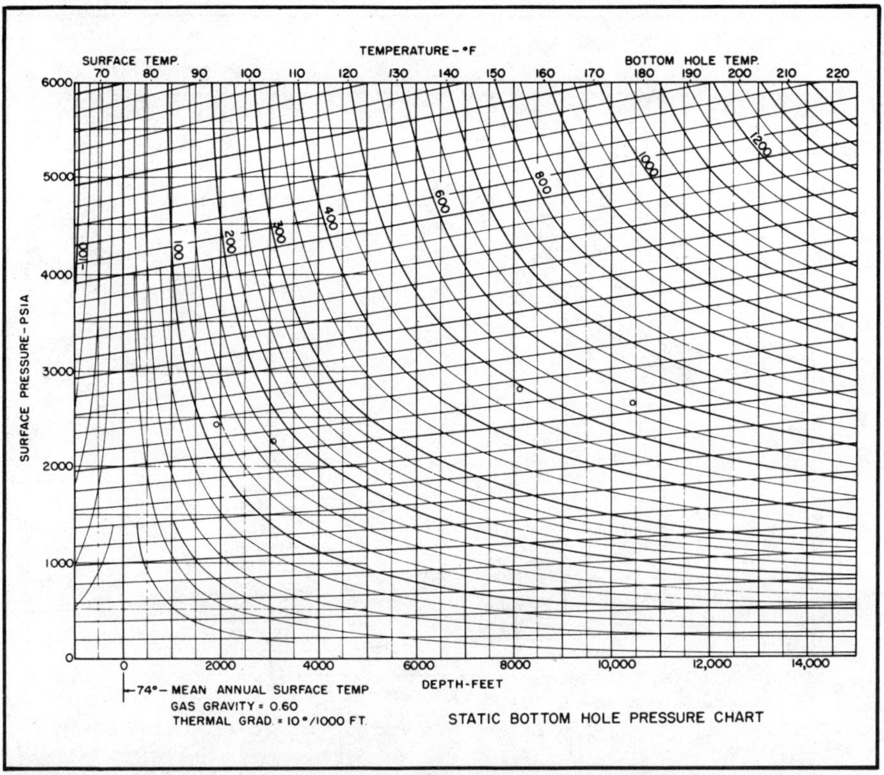

Fig. 5–6 Static bottom-hole-pressure nomograph (after Crawford and Fancher, *Flowing and Static Bottom-Hole Pressures of Natural Gas Wells*, courtesy Texas Research Committee, 1959)

cumulative production at any future average reservoir pressure, p. Further note that when p/z is zero, G_p equals the original gas in place, G. When the gas production, G_p, is zero, p/z equals p_i/z_i. These relationships are demonstrated graphically in Fig. 5–9.

Note in Eq. 5.33 that if the gas in the reservoir were a perfect gas, such that the gas deviation factor were always one, a straight-line plot would be obtained for cumulative production versus the pressure alone. It may seem unnecessary to make such a note, but the practice still persists of simply plotting the pressure versus the cumulative production and expecting it to be a relatively straight line. Since there are no perfect gases in underground reservoirs, any straight lines obtained with such a plot are caused by compensating errors.

Eq. 5.33 and Fig. 5–9 appear to be deceivingly simple. However, even the apparently simple task of solving for the initial pressure, p_i, or the pressure obtained when a particular cumulative production is

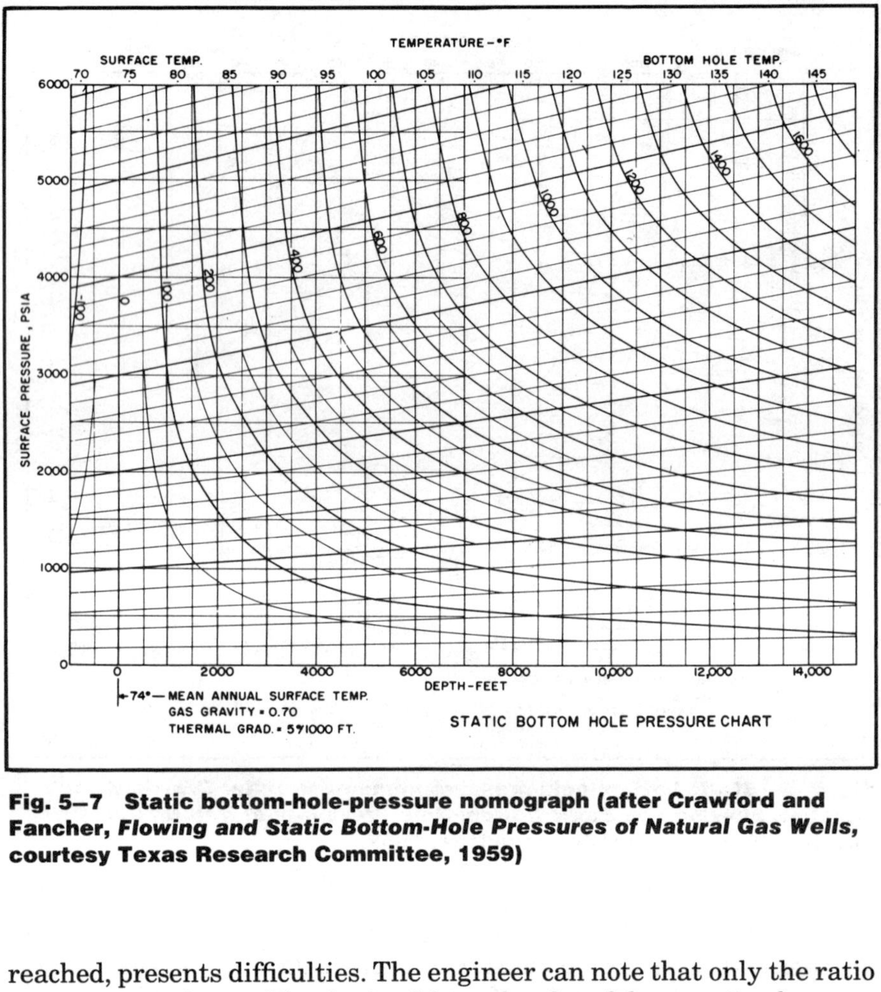

Fig. 5–7 Static bottom-hole-pressure nomograph (after Crawford and Fancher, *Flowing and Static Bottom-Hole Pressures of Natural Gas Wells*, courtesy Texas Research Committee, 1959)

reached, presents difficulties. The engineer can note that only the ratio p/z or p_i/z_i can be readily obtained from the plot of the equation because the gas deviation factor, z, is a function of the pressure, which is the unknown.

If a computer is used that has the gas deviation factors in storage, the calculation of pressure presents no difficulty since the pressure can be determined by trial and error. However, if the engineer is performing hand calculations, the task of solving for the pressure is more difficult. If a plot of p/z versus p such as that illustrated in Fig. 5–4 has been prepared for the subject reservoir, the task of determining the pressure from a p/z value is an obvious and simple one. However, if such a plot has not been prepared, the engineer is encouraged to make use of the plot of (p_r/z) versus z for various reduced temperatures as presented in Fig. B10. These data facilitate the direct evaluation of the

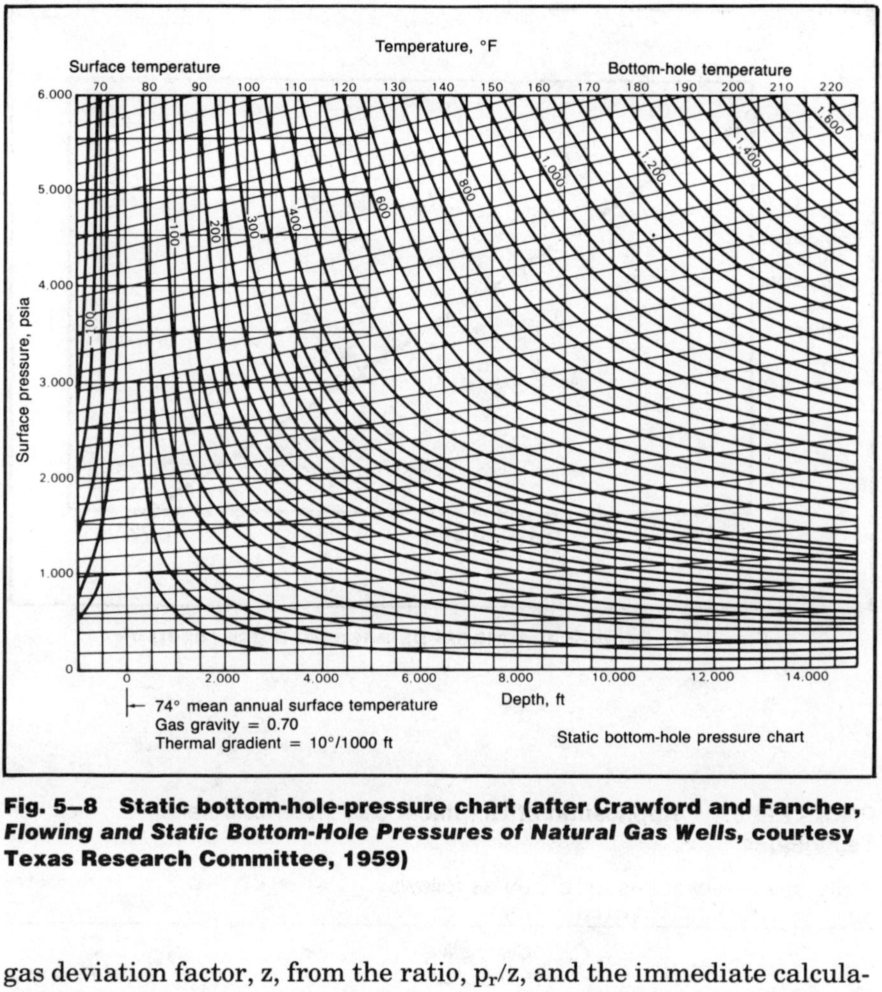

Fig. 5–8 Static bottom-hole-pressure chart (after Crawford and Fancher, *Flowing and Static Bottom-Hole Pressures of Natural Gas Wells*, courtesy Texas Research Committee, 1959)

gas deviation factor, z, from the ratio, p_r/z, and the immediate calculation of the pressure using z and the critical pressure without resorting to trial and error. The subject plot, p_r/z versus z, closely resembles the basic z, p_r, T_r correlation data presented in Fig. B8.

Solving problem 5.4 provides a check of knowledge of the graphical gas material-balance technique. The solution can be compared with the solution in appendix C.

Difficulties with p/z versus cumulative gas production. The theoretical straight line of a plot of p/z versus G_p is often difficult to obtain for many different reasons. An unexpected water drive may exist, average reservoir pressures may be inaccurate, or the pore volume may be changing in an unpredictable fashion as a result of abnormally high reservoir pressures.

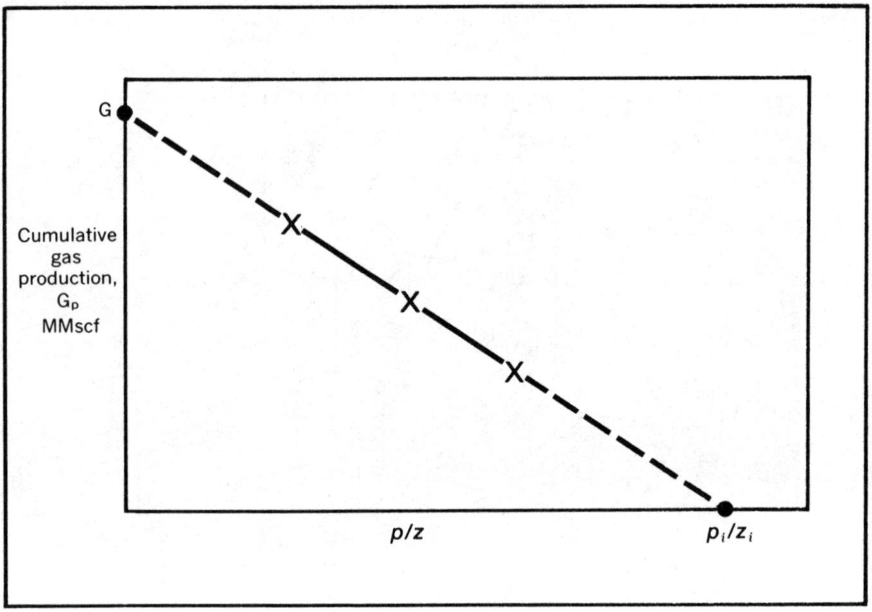

Fig. 5–9 Graphical material balance for a gas reservoir

PROBLEM 5.4: Application of Graphical Gas Material-Balance Techniques

A dry gas reservoir has produced as follows:

Date	*Cumulative Production, MMcf*	*Static Reservoir Pressure, psia*
7-1-78	0	
7-1-79	1,809	3,461
9-1-80	3,901	3,370
10-1-81	5,850	3,209
11-1-82	9,451	3,029

Reservoir temperature = 100°F
Gas gravity = 0.68

Determine the original reservoir pressure and original gas in place. What will be the average reservoir pressure at the completion of a contract calling for delivery of 20 MMcfd for 5 years (in addition to the 9,450 MMscf produced to 11-1-82)?

The most obvious reason for a plot of p/z versus G_p to deviate from a straight line is the presence of an unanticipated water drive. This changes the pore volume occupied by the gas in the reservoir and causes the change in p/z to be too small.

An unexpected water drive can be particularly confusing if it is a delayed type of water encroachment. As a result of the low permeability characteristic of many of the commercial gas reservoirs, it may take considerable time for the production from the reservoir to cause a pressure drop at the original gas-water contact. During this period it may appear that the plot of p/z versus G_p is a straight line. However, once a substantial pressure drop occurs at the original water-gas contact, water encroachment takes place and the slope of p/z versus G_p begins to increase.

A delayed water drive is not as apt to be recognized in an oil-producing reservoir because the much smaller compressibility and much higher permeability necessary for commercial production rates tend to reduce the time factor greatly in the reservoir. When a delayed water drive is apparent in a gas reservoir, many engineers want to use the straight line of p/z versus G_p prior to the effect of the water drive to determine the original gas in place. It is recommended that we resist this temptation because the same time factor that delays the water drive tends to make the measured average pressures inaccurate and the extrapolation meaningless. We cannot expect to measure the average pressure in a well that has not been affected by the entire reservoir.

The same time factor that results in delayed water drives in some gas reservoirs results in inaccurate average pressure measurements in many gas reservoirs, especially during the early producing life. If an engineer is assuming that the average reservoir pressure is being reached after two or three days of shutin, the change in these average pressures may become reasonably accurate once the entire potential drainage area of a well is being drained, i.e., when pseudosteady state is reached. However, during the early producing life when the drainage area is continuing to expand, 2–3-day shutin pressures do not give the correct change in the average reservoir pressure. The change in p/z with G_p is too large. In such a case a first estimate of the ultimate gas recovery may be obtained using the later straight-line slope of the apparent p/z versus G_p plot through the initial p/z value, as illustrated in Fig. 5–10.

To get some idea of the amount of shutin time necessary to measure the average pressure directly in a gas-well drainage area, the engineer should calculate the stabilization time using Eq. 3.50. In most cases it is found that an average pressure cannot be measured directly and that

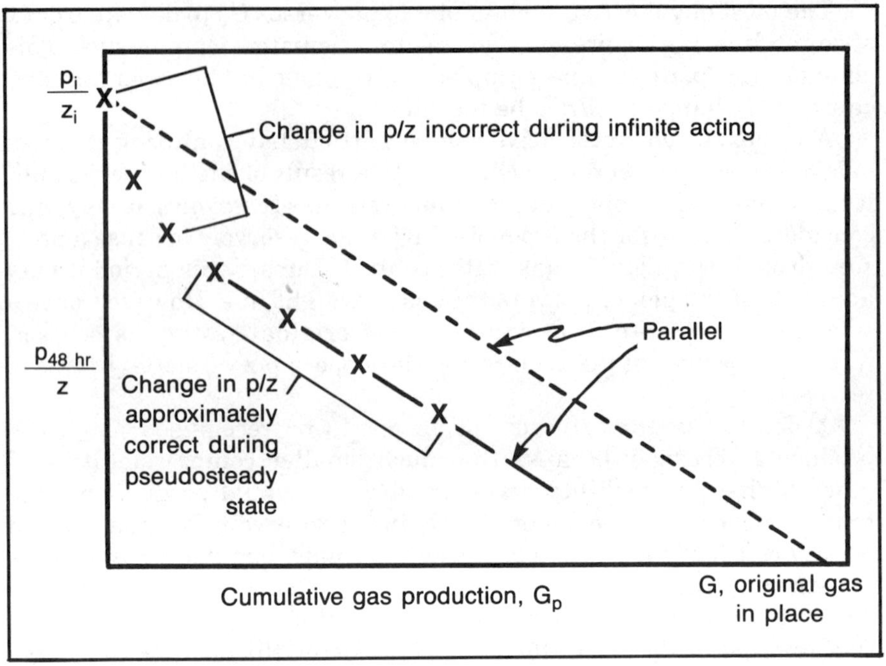

Fig. 5–10 Use of 48-hr shutin pressure (schematic)

it is necessary to calculate an average pressure by interpreting the pressure-buildup behavior as suggested in chapter 4.

Effects of abnormally high reservoir pressures. In chapter 3 we noted that a small change in the pore volume takes place as the reservoir pressure declines and that this change in pore volume is important when the reservoir is above the saturation pressure. It is also noted that this change in pore volume is caused by a very small change in the reservoir thickness and the expansion of the reservoir solids, or grain volume. However, it appears that much larger changes in the pore volume may occur as a result of other phenomena in reservoirs where the initial reservoir pressure is near the overburden pressure. Such reservoirs have abnormally high reservoir pressures, or they are *geopressured reservoirs*.

When the reservoir pressure in the reservoir pores is equivalent to the overburden weight, the overburden weight is not being supported by the mechanical strength of the reservoir rock but may be entirely supported by the fluid in the reservoir. When this occurs, the pore volume may change an amount equal to the amount of withdrawal of fluid from the reservoir without any change in the reservoir pressure. This condition can continue until the solids of the reservoir actually begin supporting the overburden weight and the pore volume becomes constant; i.e., the change in pore volume becomes a function of the change in reservoir pressure and the conventional formation compressibility, c_f.

To illustrate how such a situation occurs, assume that grains of an unconsolidated sand are initially oriented as shown in Fig. 5–11a when production from a reservoir begins and that the sand grains are initially free to move. As fluid is removed from the reservoir, the overburden pressure tends to rearrange the sand grains until they reach an arrangement that cannot be further altered without physical failure of the sand. This latter position is illustrated in Fig. 5–11b. Of course, such a situation can take place only if the initial reservoir pressure is the equivalent of the overburden pressure and the sand grains can move in such a way that the pore volume can be reduced by such a rearrangement. If the reservoir pressure is not originally equivalent to the overburden pressure, the overburden weight must be supported to some degree by the sand grains. They then are unable to move and reduce the pore volume. However, if the sand grains cannot move in such a way as to reduce the pore volume, there is no change in the pore volume even if the initial reservoir pressure is equal to the overburden pressure.

Thus, the reservoir pressure must be about equal to the overburden pressure in order for a pore volume change to take place without a substantial change in the reservoir pressure. However, a reservoir with an initial pressure near the overburden pressure does not necessarily have a change in the pore volume without a corresponding change in the reservoir pressure. As a result of this phenomena, gas reservoirs with abnormally high pressures may have a plot of p/z versus G_p as shown schematically in Fig. 5–12, which exhibits a very small change in pressure with G_p initially. Then when the pore volume becomes constant, a normal pressure decline is exhibited.

Several problems exist in the operation of a reservoir with characteristics such as those in Fig. 5–12. First, it is difficult or perhaps presently impossible to predict at what p/z or G_p value the pore volume quits changing and a volumetric prediction can be made, i.e., where the curve slope changes. Perhaps analogy is the best means presently available for such a prediction. Some papers have been published on this subject, but they appear to take a pseudoanalogy approach to the problem.*

Other problems are associated with the operation of abnormally pressured gas reservoirs. Apparently, when the reservoir pressure has

*One such paper is "Predicting Gas Reserves in Abnormally Pressured Reservoirs," by D.J. Hammerlindl, SPE paper 3479, 1971.

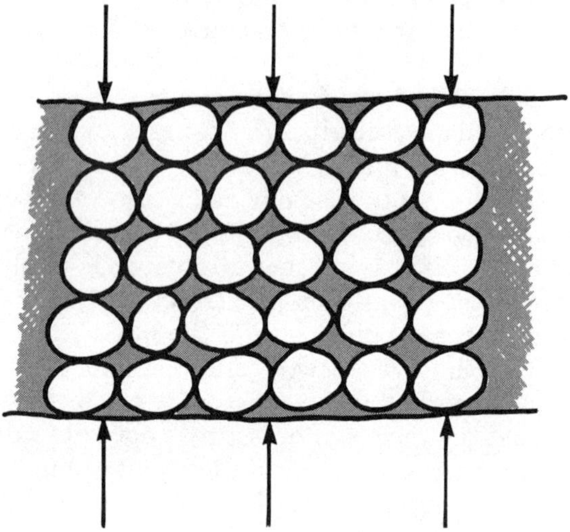

(a) Initial sand-grain arrangement

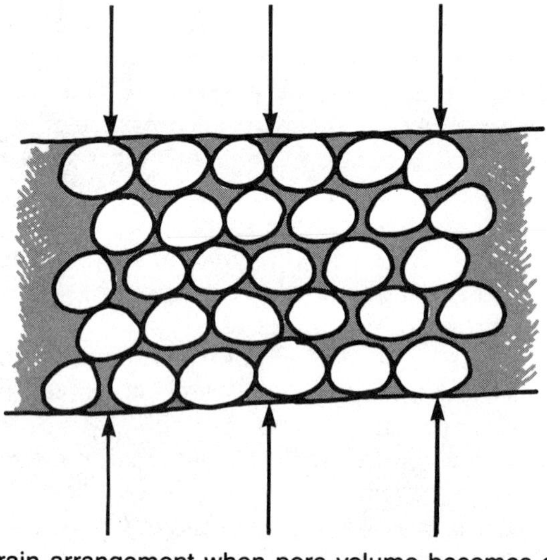

(b) Sand-grain arrangement when pore volume becomes constant

Fig. 5–11 Change in reservoir pore volume without a pressure change

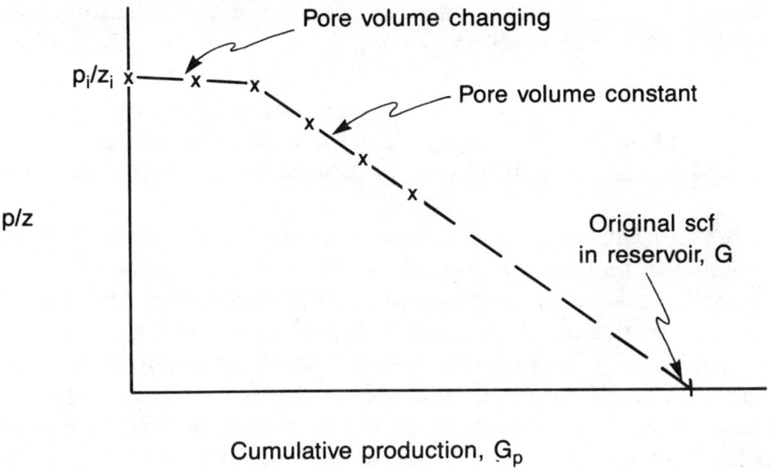

Fig. 5–12 Plot of p/z versus cumulative gas production with initial pore volume change (schematic)

been reduced in some wells, the physical strength of the sand formation and perhaps the sand grains has failed. This failure results in crushing of the sand formation, reducing the formation permeability, and drastically reducing the well producing rate. In other words, in abnormally pressured gas reservoirs the overburden pressure is supported by both the reservoir pressure and the physical strength initially. Then as the pressure in the reservoir is reduced, the sand strength is subjected to a larger crushing stress, which may cause a physical failure of the formation or the sand grains. Some operators have experienced a very sudden, drastic rate reduction in the wells of an abnormally pressured reservoir. It appears that the only present means of avoiding such an occurrence is to reduce the pressure drawdown on wells in reservoirs where this problem has existed. Since minimum reservoir pressures exist at the well, this is the point where formation failure is anticipated.

Determining water encroachment. Water encroachment into a gas or an oil reservoir can be accurately determined in most cases using the constant-pressure solution to the radial diffusivity equation. Since the next section of this chapter deals with fluid flow in a gas reservoir, the reader may wonder why this discussion is placed under the material-balance section. The problem of determining water encroachment would not be appropriate under fluid flow in a gas reservoir because the flow associated with water encroachment is actually taking place in the aquifer or water-bearing portion of the reservoir. Consequently, there is no basic difference between the calculation of water encroachment into a gas reservoir and water encroachment into an oil reservoir. However, the problem of calculating water encroachment into a gas reservoir has some peculiarities that should be recognized by the engineer.

If the engineer is not aware of the nature of unsteady-state flow, the general solutions that are available, and particularly the general constant-pressure solution to the radial diffusivity equation, he should review this material in chapter 3. From a practical standpoint, the constant-rate solution permits the calculation of the cumulative flow of fluid across a particular radius in the reservoir where the pressure has been maintained at a constant value, p_r, which in turn has caused a constant pressure drop, $p_i - p_r$, maintained at this radius for some period of time, t.

Mechanically, the engineer calculates the dimensionless time, t_D, that corresponds to the producing time, t, using Eq. 3.11, which has been modified slightly to apply to our application:

$$t_D = \frac{\eta t}{r^2} \tag{3.11a}$$

When Eq. 3.11a is applied to calculate the cumulative flow of water across the original water-oil contact (WOC) or (water-gas contact) (WGC), the well radius, r, for the aquifer is the internal radius of the aquifer or the external radius of the gas or oil reservoir. Once the dimensionless time, t_D, has been evaluated, it is used to determine a corresponding Q_{tD} from Table 3–1 or Table 3–2 or Figs. 3–15 through 3–18. If the reservoir is determined to be finite acting at the time of interest, the dimensionless reservoir size, $r_{De} = r_e/r$, must be used to determine Q_{tD}. When the function Q_{tD} has been evaluated, the number of barrels of water that has flowed across the original WGC radius, r, can be calculated from another equation modified from chapter 3:

$$Q = 1.12\phi h c r^2\, \theta \Delta p\, Q_{tD} \tag{3.57a}$$

Eq. 3.57 did not contain θ. This symbol represents the fraction of a total circle (360°) through which flow, or water encroachment, is taking place. It is seldom that a water drive occurs throughout the full 360°.

Eq. 3.57a can be used to calculate the water encroachment only if a constant pressure drop, $\Delta p = p_i - p_r$, is maintained for the entire period, t. However, we know that once the pressure at the original WGC begins to decline as a result of gas production, the pressure at this radius probably continues to decline if the gas reservoir is being produced at any rate approaching the capacity of the reservoir. Gas with its very low viscosity can flow much more rapidly than water. That is, water encroachment cannot maintain a steady-state condition in the gas reservoir.

When the gas reservoir is not being produced at a rate approaching the reservoir capacity, it is very likely that the production will be sporadic so that in this case also the pressure at the original WGC varies with time. Consequently, Eq. 3.57a must be applied by superposition because the pressure varies at the original WGC. The technique recommended is to take the original WGC pressure history and approximate it with a sequence of finite pressure drops. The effective time for each pressure drop then is the remaining time to the time of interest.

For example, if a particular pressure drop is assumed to exist at a time of 1 year and we are calculating the cumulative water influx for a time of 18 months, the effective time to be considered for that particular pressure drop is 6 months. An additional pressure drop that occurs at a period of 15 months has an effective time period of 3 months. Once the finite pressure drops and the corresponding times that approximate the actual pressure history have been determined, it is possible to calculate the cumulative water encroachment caused by each individual pressure drop acting for its appropriate time using Eq. 3.57. The actual cumulative water influx at that time then is the sum of all Q's calculated.

Since the group of constants, $1.12\phi hcr^2\theta$, appears in each individual calculation, it is simpler to apply the calculations using Eq. 3.64 modified to include θ:

$$Q = 1.12\phi hcr^2\theta \sum_{j=1}^{j=n}(\Delta p_j Q_{tDj}) \tag{3.64a}$$

Since the reservoir parameters in Eq. 3.64a apply specifically to the water-bearing portion of the reservoir, they are normally evaluated only as a result of dry holes and other errors. Since the accuracy of these data is always questionable, it is best to determine the group of terms, 1.12 $hcr^2\theta$, from the past reservoir performance. If we call the group of terms B, the aquifer constant, Eq. 3.64a becomes:

$$Q = B \sum_{j=1}^{j=n}(\Delta p_j Q_{tDj}) \tag{3.64b}$$

When Eq. 3.64b is applied to the history, material balance can be used to determine Q. Thus, everything is known except the constant, B, which can be calculated from Eq. 3.64b. Once the constant B is known, Eq. 3.64b also can be used to predict future water influx. We should not be unduly concerned with the fact that we do not know, with any great accuracy, the reservoir parameters necessary for the calculation of dimensionless time, t_D. These parameters—the permeability, porosity, water viscosity, and effective compressibility—will not be known with any more accuracy than the parameters that make up the constant B. However, they have a negligible effect on the accuracy of the predicted water encroachment because they are used in both the evaluation of B and the prediction of the water encroachment. Thus, compensating errors are introduced into the calculating procedure. Also, their effect is roughly proportional to the log of the values so the inaccuracy is further reduced.

If sufficient history is available, the engineer may wish to make a plot of Q calculated from material balance versus the term $\sum_{j=1}^{j=n} (\Delta p_j Q_{tDj})$ and evaluate the slope that is the constant B (Eq. 3.64b). This plot gives more statistical weight to the evaluation of B and tends to minimize the importance of the early pressure data, which are often inaccurate for various reasons.

Engineers generally assume that the change in pressure at the original WOC or WGC is the same as the change in the average pressure in the gas- or oil-bearing portion of the reservoir. When oil reservoirs are considered, this assumption seldom introduces a significant error. However, when calculating the water influx into a gas reservoir, the engineer should carefully evaluate the time before the pressure at the original WGC is affected. As noted, disturbances travel very slowly in many of the gas reservoirs; thus, a considerable amount of time may transpire before any substantial change in reservoir pressure occurs at the original WGC. Until a pressure drop occurs at this point in the reservoir, no water encroachment takes place. Consequently, care should be taken to estimate as accurately as possible the time when the first pressure drop occurs at the original WGC, and this should represent the zero time for the calculation of the water influx.

Work problem 5.5 to understand the method of calculating water encroachment better. Compare the solution with that in appendix C.

We should note the difficulty of determining the average predicted reservoir pressure for the basis of the water-influx calculation as in problem 5.5. Under the best conditions these pressures can only be

PROBLEM 5.5: Calculating Water Encroachment in a Gas Reservoir

A gas reservoir with a strong water drive has the following pressure history:

Time, months	*Average Reservoir Pressure, psia*
0	2,500
6	2,450
12	2,356

Reservoir studies have shown the following:

- The pressure at the original water-gas contact is 50 psi greater than the average pressure after initial conditions. Six months of production is necessary to cause a significant pressure drop at the original WGC.
- When t = 6 months, $t_D = 5.0$.
- The aquifer can be treated as being infinite.
- Material balance indicates that 192,000 bbl of water have encroached into the gas reservoir at the end of the first 12 months of production ($p_{avg} = 2{,}356$ psia).

Find the cumulative water encroachment after 18 months and after 24 months if average reservoir pressures of 2,276 and 2,216 psia, respectively, are anticipated at the end of these periods.

determined by trial and error. The simplest situation is the one in which a known rate of production is to be maintained. As a place to start, we may calculate the resulting reservoir pressure at the end of some period by the material-balance equation, Eq. 5.22, assuming water encroachment is zero during that period. Then based on the calculated average reservoir pressure, calculate the water encroachment using Eq. 3.64b.

This water encroachment makes the originally calculated reservoir pressure incorrect, and this procedure is repeated using the last calculated water encroachment. A new calculated pressure would result in a new calculated water encroachment, which would change the pressure, etc. Each successive round of calculation introduces smaller changes until the changes become insignificant. At this point the true value of the average pressure and water encroachment have been reached.

Problem 5.6 is concerned with an unusually small, low-pressure gas reservoir that lends itself to textbook analysis because of its size and the good well capacity. Work this problem and compare the results with the solution in appendix C to understand better the nature of the trial-and-error solution that results from having to solve for the reservoir pressure and water encroachment simultaneously.

PROBLEM 5.6: Simultaneous Solution of Reservoir Pressure and Water Encroachment

A small gas reservoir initially at 807 psia and 583°R is produced for 500 days at a rate of 1,000 Mcfd. Calculate the amount of encroached water after 100 and 300 days of production. The deviation plots in Fig. 5–4 are applicable. Assume the aquifer is infinite acting and the reservoir has the following characteristics:

Average permeability = 14 md
Average porosity = 11%
Effective water compressibility = 3.3×10^{-6}/psi
Viscosity = 1.0 cp
Reservoir thickness = 28 ft
Original oil in place = 1,118 MMscf
Average radius of original WGC = 1,702 ft

If we assume maintenance of a maximum production rate, the solution becomes even more difficult because with each sequence of calculations the gas production during the subject period changes in accordance with the deliverability characteristics as a result of the reduction in the reservoir pressure.

Fluid Flow in Gas Reservoirs

Reservoir fluid flow of gas presents additional problems above and beyond those concerned with the flow of liquid in the reservoir. These difficulties take several forms. What makes it most difficult is that the basic fluid flow equations were derived for the flow of liquids. At that time the production of gas presented no problems since there was a plentiful supply of gas and its price was small compared to the price of oil. Consequently, by the time the problems of gas production reached economic significance, the technology of liquid flow was reasonably well established. Thus, it was natural for the engineer to attempt to adapt the liquid flow technology to gas flow.

The difficulty with this approach is that the liquid flow equations use several assumptions that simply are not valid for gas flow. Turbulence, which seldom causes a problem in liquid flow, presents a continual problem in gas flow into a well. Fluid compressibility, which is constant and very small for liquids, varies widely for gas and is quite large. In addition, we are continually faced with the task of evaluating the gas deviation factor to determine the flow rate.

It would appear that as a result of these great differences between liquid and gas, gas flow equations should be derived from fundamentals to avoid the simplifying assumptions. Unfortunately, this was not done. Instead, the present state of the technology is to continue to use most of the equations derived for liquid flow with the inappropriate assumptions noted. Arbitrary and generally poorly defined limits on the applications tend to avoid the inherent errors caused by these inappropriate assumptions. Consequently, gas reservoir fluid flow is a very complex subject.

Turbulence in reservoir fluid flow. As noted, one of the difficulties we experience in applying reservoir flow equations to the flow of gas is that turbulence becomes important, whereas in liquid flow turbulence is seldom, if ever, significant. Oil wells that flow at rates great enough to experience turbulence are generally such good producers that there is no reason to be involved with their analysis.

In considering the modification of flow equations to account for turbulence, we should first note that turbulent effects in reservoir flow are not the same as turbulent effects in pipe, channel, and other more conventional types of fluid flow where turbulence can be closely related to the Reynold's number. Efforts to relate the onset of turbulence in reservoir fluid flow to the Reynold's number have been largely unsuccessful. Katz and others have shown (or speculated) that this results from the fact that in reservoir flow the pore sizes are continually changing in cross section. In pipe and channel flow any change in cross section is at least well defined. Also there is a continuous change in direction associated with flow through porous media, but in pipe or channel flow such abrupt and continuous changes in direction do not occur. Since the turbulent flow behavior in a reservoir does not fit the conventional predictions associated with the Reynold's number, many engineers, researchers, and educators prefer to refer to the turbulence effect in reservoir fluid flow as *non-Darcy flow*. In this book the author continues to use the term "turbulence" to describe the situation where the pressure drop is not proportional to the flow rate stated at the average pressure.

Unfortunately, the methods used to correct for turbulence in reservoir gas flow are not consistent. We may first modify the Darcy equation by adding a term to account for the additional pressure drop caused by turbulence and then use the modified Darcy equation to derive the working reservoir flow equations. Also, we may empirically modify the nonturbulent flow equation by adding an empirical exponent to the difference in the pressure-squared term—e.g., $(p_e^2 - p_w^2)^n$. In another method we may add a term that represents the additional pressure squared resulting from turbulence that uses a proportionality constant. Unfortunately, there is no theoretical means of determining one of these corrections from the other.

The Darcy equation with turbulence. We originally noted that the Darcy equation was derived empirically for viscous flow and that it would not apply to turbulent conditions. Several attempts have been made to modify the Darcy equation so it can be applied to turbulent flow. The pressure gradient for viscous flow stated in terms of Darcy units would be:

$$\frac{\Delta p'}{\Delta x'} = \frac{\mu v'}{k} \tag{5.34}$$

It is recognized that the pressure drop in turbulent flow is proportional to the velocity squared. Many engineers and researchers have proposed various schemes to modify the Darcy equation to account for turbulence and other nonideal conditions. The form that we choose is probably no more accurate than some of the other schemes, but we can see that it meshes well with the additional pressure-drop idea introduced in discussing the skin effect. Consequently, it appears to be somewhat better suited for our reservoir fluid flow technology than some of the other methods.

To account for the additional pressure drop caused by turbulence, another term is added to Eq. 5.34 to yield:

$$\frac{\Delta p'}{\Delta x'} = \frac{\mu v'}{k} + \beta' \rho (v')^2 \tag{5.35}$$

Where:

ρ = Density, g/cc, or specific gravity relative to water

β' = Porous-media constant that describes the effect of turbulence in a particular porous media

Beta can be determined by appropriate measurements in the laboratory, but it is more often evaluated from empirical data such as that presented in Fig. B11.[7] The engineer should note that these data present β in reciprocal feet rather than in reciprocal centimeters as is necessary for Eq. 5.35.

The engineer is reminded that the velocity in the Darcy equations, Eqs. 5.34 and 5.35, is an apparent, rather than an actual, pore velocity. Thus, in comparing Eq. 5.35 with any equation that uses a velocity in a Reynold's number expression, the difference in the two velocities should be accounted for.

Steady-state gas flow. For steady state to prevail in gas flow, it is necessary for a mass flow rate to remain constant throughout the reservoir. We saw previously that a strong water drive helps liquid flow in a reservoir approach steady state. Often, the water enters the oil-bearing portion of the reservoir at the same rate that the oil is produced. However, steady-state flow of gas in the reservoir is seldom, if ever, achieved in the U.S. today. Since the shortage of gas developed, most reservoirs are produced at nearly peak production rates. Under these circumstances it is virtually impossible for water to enter a water-driven gas reservoir as rapidly as the gas is produced because of the difference in viscosities.

Nevertheless, the steady-state radial gas flow equations are often useful because the conditions around the wellbore closely approximate steady state. Since the bulk of gas production comes from far out in the reservoir, the flow rate near the wellbore is reasonably constant at all radii near the wellbore. Since most pressure drops occur near the wellbore, it is then possible to use the steady-state gas flow equation to estimate the effect of turbulence and well damage.

Eq. 5.35 can be used to derive a radial flow equation, but it is quite complex because of the turbulence term. When the resulting equation is solved for the pressure drop, we get:

$$p_1^2 - p_2^2 = \frac{1.424 \mu_{avg} z_{avg} T_f q_g \ \ell n(r_1/r_2)}{hk} + \frac{3.161\,(10^{-12})\ \beta \gamma q_g^{2} z_{avg} T_f \left(\frac{1}{r_2} - \frac{1}{r_1}\right)}{h^2} \qquad (5.36)$$

An examination of Eq. 5.36 shows that if the second term on the right-hand side is ignored, the equation is exactly the same as the steady-state flow equation resulting when turbulence is ignored, e.g., Eq. 2.39, which gives the $\Delta(p^2)$ term. The constant, 1.424, is simply the reciprocal of the constant in the steady-state gas flow equation without turbulence, 0.703.

Once we recognize that the first term of the right-hand side of Eq. 5.36 is the $\Delta(p^2)$ value caused by stream-line or nonturbulent flow of gas, we can then assume that the additional pressure drop caused by turbulence is represented by the second right-hand term. We can then use Eq. 5.36 to estimate the additional pressure drop caused by turbulence.

This steady-state equation can be used for this purpose although we know that at a particular time a well system is in either unsteady state or pseudosteady state. This equation can be used because the conditions around the wellbore closely approximate steady state even during unsteady-state or pseudosteady-state flow and because all pressure drops caused by turbulence occur near the wellbore. The latter point becomes clear when we recognize that turbulence is directly attributable to velocity and that the velocity near the wellbore is extremely high compared with the flow velocity further removed from the wellbore.

Eq. 5.36 shows this mathematically. As noted, the second term of the equation represents the additional $\Delta(p^2)$ caused by turbulence. Also, note that the magnitude of the second term is proportional to the difference in the reciprocals of the radii at which the pressures are measured. Therefore, the bulk of the pressure drop caused by turbulence is concentrated at the wellbore. For example, consider a well with a radius of 0.5 ft and an external drainage radius of 500 ft. When the reciprocal term is applied to the entire 0.5–500 ft, its value is (1/0.5) – (1/500), or about 1.998. If we now apply the term to the 5–500-ft zone, we obtain (1/5) – (1/500), or 0.198. Thus, according to the steady-state gas flow equation, about 90% of the total $\Delta(p^2)$ caused by turbulence in the subject well occurs in the 5 ft around the wellbore [(1.998 – 0.198)/1.998]. Similarly, it can be shown that about 99% of the $\Delta(p^2)$ resulting from turbulence in this well occurs between the wellbore and a radius of 50 ft.

Using Eq. 5.36 to estimate the additional pressure drop caused by turbulence is extremely valuable because a standard pressure-buildup analysis of a gas well results in a Δp_{skin} that includes turbulence as well as the damage or improvement around a wellbore. Consequently, by using pressure buildup to determine the total additional pressure drop caused by well damage and turbulence and using Eq. 5.36 to determine the additional pressure drop caused by turbulence, it is possible to differentiate between well damage, which we may be able to repair, and the effect of turbulence, which we generally can do nothing about.

The modifier "generally" is used to qualify the conclusion that nothing can be done about the pressure drop resulting from turbulence because massive fracture treatments may change the pattern from essentially radial flow to something approaching linear flow. When such treatments are used, the velocity may be reduced enough to decrease the pressure drop caused by turbulence.

Problem 5.7 helps the reader understand the use of Eq. 5.36 to differentiate between the pressure drop caused by damage and that caused by turbulence. To find the pressure drop resulting from turbulence, we simply calculate the well pressure that would result if there were no turbulence, using the first term of Eq. 5.36, and subtract from that the well pressure calculated by using the entire equation.

PROBLEM 5.7: Turbulent Effects in Gas Wells

A well is producing at a rate of 3.9 MMcfd before shutin for a pressure-buildup survey. The conventional buildup analysis indicates an undamaged permeability of 1.5 md, a Δp_{skin} of 1,400 psia, and a pressure at the external drainage radius, an average of 550 ft away, of 4,583 psia. The following reservoir data are applicable:

r_w = 0.333 ft
ϕ = 5.0%
μ_{avg} = 0.27 cp

h = 30 ft
γ = 0.76
T_f = 712°R
z_{avg} = 0.97

A. What would be the flowing well pressure if there were no well damage or turbulence?

B. What would the well pressure be with turbulence and no well damage?

C. What are the actual additional pressure drop caused by well damage and the actual additional pressure drop caused by turbulence?

In problem 5.7 the gas deviation factor and the gas viscosity are given to simplify the problem. Note that a similar practical problem would not be this simple. The viscosity of gas and the gas deviation factor vary with the pressure of the gas. In Eq. 5.36 it is indicated that the viscosity and the gas deviation factor should be evaluated at the average pressure. Since this equation was not derived herein, it is not obvious what is meant by the term "average." In practice, the average pressure should be taken as the arithmetic average, $(p_1 + p_2)/2$. This same average pressure is used to evaluate gas viscosities and gas deviation factors throughout the book.

It can be shown that the linear pressure drop for the flow of a perfect gas is proportional to the flow rate stated at the arithmetic average pressure. First, note that a given mass of ideal gas under isothermal conditions exhibits a constant-pressure volume product, pq, if the rate is stated in volumes at the corresponding pressure. Thus, we can say that:

$$(q \times p) = (q_m \times p_m)$$

Where:

q_m = Volumetric flow rate, res b/d, stated at the average pressure, p_m

p_m = $(p_2 + p_1)/2$

If we now use this relationship to derive an equation for the steady-state flow of gas using the Darcy equation, we obtain:

$$q = -\frac{1.127kA}{\mu}\frac{\Delta p}{\Delta x} \tag{2.3}$$

For radial flow $A = 2\pi rh$ and $x = r$:

$$\frac{q_m p_m}{p} = \frac{1.127k\,(2\pi rh)}{\mu}\frac{\Delta p}{\Delta r}$$

$$q_m p_m \sum_{r_1}^{r_2}\left(\frac{\Delta r}{r}\right) = \frac{7.08kh}{\mu}\sum_{p_1}^{p_2} p\Delta p$$

$$q_m p_m \ell n \frac{r_2}{r_1} = \frac{7.08kh}{\mu}\left(\frac{p_2{}^2}{2} - \frac{p_1{}^2}{2}\right)$$

$$q_m p_m \ell n \frac{r_2}{r_1} = \frac{7.08kh}{\mu}(p_2 - p_1)\left(\frac{p_2 + p_1}{2}\right)$$

$$q_m = \frac{7.08kh(p_2 - p_1)}{\mu \quad \ell n\left(\frac{r_2}{r_1}\right)} \tag{5.37}$$

Eq. 5.37 shows that the gas flow rate stated at the arithmetic average pressure, p_m, is proportional to the pressure drop. Conversely, the pressure drop is proportional to the rate stated at the average pressure; thus, the rate-related parameters should be evaluated at the arithmetic average pressure. Eq. 5.37 is referred to often since it is frequently necessary to convert from a liquid flow rate in reservoir barrels to a gas flow rate in Mscfd, or 1,000 scfd.

Although there is seldom any use for a steady-state linear gas flow equation, it should be noted for completeness that Eq. 4.35 can also be used to derive such an equation. The result is:

$$p_1{}^2 - p_2{}^2 = \frac{z_{avg}\ T_f \mu q_g L}{0.112Ak} + \frac{1.254\ (10^{-10}) z_{avg} T_f\ q_g{}^2\ \beta L \gamma}{A^2} \tag{5.38}$$

In Eq. 5.38, as in Eq. 5.36, the first term represents the $\Delta(p^2)$ that would occur in the absence of turbulence. The second term represents the additional $\Delta(p^2)$ caused by turbulence.

Pseudosteady-state gas flow. In applying the pseudosteady-state equations to gas, the engineer should remember that the equations represent true, unsteady-state flow and that they are derived for fluids with constant compressibility, whereas the compressibility of gas varies inversely with the pressure. Thus, the application of unsteady-state flow equations to gas, including pseudosteady-state flow equations, lose their accuracy as the variation in pressure increases both radially and timewise. This inaccuracy can be remedied by treating the reservoir as a sequence of different reservoirs in series and breaking the time interval, t, into smaller time segments.

In chapter 3 we considered a well that has produced at a constant rate for a time in excess of the stabilization time defined as:

$$t_s = 0.04\ \phi \mu c\ r_e{}^2/k \tag{3.50}$$

Then the pressure at all points in the reservoir begins declining at the rate:

$$(\Delta p/\Delta t)_{pseudo} = 1.79\ q/\phi h c_{sta} r_e{}^2 \tag{3.33}$$

It is necessary to evaluate the compressibility at the static pressure; hence, the symbol, c_{sta}, is used. Furthermore, at this time the following rate equation stated in terms of the external radius and the pressure at this radius prevails:

$$q_w = \frac{7.08kh(p_e - p_w)}{\mu[\ell n(r_e/r_w) - 0.5 + S]} \tag{3.43a}$$

When we think in terms of applying these equations to the flow of gas, remember that the gas compressibility is about proportional to the reciprocal of the pressure. Consequently, Eq. 3.33 cannot be exact for gas since the pressure varies with the radius and the compressibility varies with the pressure. Thus, the change in pressure with time for gas cannot be the same throughout the reservoir. Nevertheless, we can show by material balance that Eq. 3.33 does indicate the average change in pressure with time if the compressibility, c, is evaluated at the average (static) pressure and we recognize the rate, q, as being the volumetric rate in barrels per day at the same average pressure.

In spite of these limitations, we find that when the pseudosteady-state equations are converted to gas flow rates stated in Mscfd and a modification for turbulence is introduced, the resulting equation provides a reasonable description of the behavior of a gas reservoir after it has produced at a constant rate for a period of time governed by Eq. 3.50. In this equation the compressibility is based on the pressure at the external boundary during time t_s.

To convert Eq. 3.43a to a gas flow rate, we recall that according to Eq. 5.37 the pressure drop in the flow of gas is proportional to the rate stated at the arithmetic average pressure. Then according to Eq. 2.33:

$$q = \frac{5.04\ q_g z_{avg} T_f}{(p_e + p_w)/2} \tag{5.39}$$

Substituting for q in Eq. 3.43a, we obtain:

$$\frac{5.04 q_g z_{avg} T_f}{(p_e + p_w)/2} = \frac{7.08kh(p_e - p_w)}{[\ell n(r_e/r_w) - 0.5 + S]} \tag{5.40}$$

When Eq. 5.40 is solved for the flow rate, q_g, we obtain:

$$q_g = \frac{0.703kh(p_e^2 - p_w^2)}{\mu z_{avg} T_f [\ell n(r_e/r_w) - 0.5 + S]} \tag{5.41}$$

Remember that the Darcy equation used to derive Eq. 3.43a and hence Eq. 5.41 must be modified to account for turbulence if it is to give reasonably accurate answers. This correction can be carried out empirically by the addition of a fractional exponent to the $p_e^2 - p_w^2$ term. In effect, the fractional exponent increases the pressure drop necessary to provide a particular flow rate. Eq. 5.41 then becomes:

$$q_g = \frac{0.703kh(p_e^2 - p_w^2)^n}{\mu z_{avg} T_f [\ell n(r_e/r_w) - 0.5 + S]} \tag{5.42}$$

Eq. 5.42 is often written with the constant 0.5 combined with the ℓn term as:

$$q_g = \frac{0.703kh(p_e^2 - p_w^2)^n}{\mu z_{avg} T_f [\ell n(0.606 r_e/r_w) + S]} \tag{5.43}$$

Eq. 5.43 is often called the gas deliverability equation, and it is used as the basis for conventional deliverability tests. The turbulence constant is evaluated empirically from the flow tests and probably accounts for many nonideal conditions in addition to turbulence. For example, our equation assumes isothermal flow, yet we know that the large pressure drops that accompany the radial gas flow and the attendant expansion of the gas has a cooling effect. In fact, temperature surveys in the wellbore are used to indicate which portions of a formation are producing gas. Where gas is being produced, large temperature anomalies exist in the wellbore.

In very tight formations the turbulence constant may account for the change in the formation permeability with the change in the reservoir pressure. It may also account for the change in the gas viscosity and gas deviation factors that accompany any pressure change. We see that as a result of the empirical manner in which the constant is measured, it may tend to correct the prediction for any or all of these factors. In fact, some experts suggest that these other factors may be the dominating effect rather than the turbulence. It should be specifically noted that the constant n does not account for well damage.

As Eqs. 5.42 and 5.43 are written, the permeability, k, represents the undamaged permeability. If we prefer to write the equations using the average permeability, the skin factor, S, can be dropped because the drainage radius is continually increasing and the average permeability is changing with time:

$$q_g = \frac{0.703\ k_{avg} h(p_e^2 - p_w^2)^n}{\mu z_{avg} T_f\ \ell n(0.606 r_e/r_w)} \tag{5.44}$$

Conventional back-pressure tests. When considering gas-well testing, the engineer should remember that there are two reasons for well testing. The first reason is to satisfy government requirements for such regulatory bodies as the Texas Railroad Commission, the Louisiana Office of Conservation, and the Kansas Corporation Division. The second reason is to obtain data for predicting the behavior of a well or a gas reservoir. There is a natural tendency to try to satisfy both requirements with one test or one set of tests.

It is the author's opinion that trying to satisfy both objectives with one test is a mistake that may result in the producer obtaining a lower allowable. Generally, a company should follow government instructions for testing a gas well explicitly, regardless of whether the required procedure is good or bad. Then the company will get its fair share of the production pie. The engineer can run additional tests in the best way possible as a basis for predicting the rate versus time behavior of a well or reservoir.

If the engineer tries to improve the state's testing recommendations to obtain a more realistic apparent open flow, he invariably finds that the more realistic apparent open flow is lower than the one obtained using the state's methods. Thus, the instructions given in this chapter for testing gas wells are for obtaining data to determine the best possible prediction of the gas well behavior and not for the purpose of conducting well tests for a government agency.

The most common method of testing a gas well for predicting gas producing rates under various conditions is to flow the well at a particular rate until a constant well pressure is obtained. Then flow the well at another rate until a new constant pressure is reached. Continue this procedure until flowing well pressures have been recorded for a variety of rates. If three or four rates are used, the test may be referred to as a three-point or four-point test. If run and interpreted properly, such tests may be used to predict deliverability at various drawdown pressures, at various states of depletion, and for a variety of spacing conditions. However, it is often impossible or impractical to run a conventional back-pressure test properly because of the time involved.

Note that since Eq. 5.44 is basically a pseudosteady-state equation, it is necessary for each flow rate to be maintained long enough to affect the entire well drainage area. Therefore, the producing time must be greater than the stabilization time, t_s, as defined in Eq. 3.50. Until the entire drainage area is affected, the drainage radius at a particular rate is continuing to increase and Eq. 5.44 is not yet applicable.

However, if the test is run properly, a straight-line plot generally is obtained when the log of the producing rate is plotted versus the log of the pressure drawdown term $(p_e^2 - p_w^2)$. The basis for this straight line can be shown by taking the log of both sides of Eq. 5.44 to obtain:

$$\log q_g = \log \frac{0.703 k_{avg} h}{\mu z_{avg} T_f \,\ell n(0.606 r_e / r_w)} + n \log (p_e^2 - p_w^2) \qquad (5.45)$$

The right-hand side of Eq. 5.45 is grouped in this way to show that the first term contains factors that can generally be treated as constants. Then we can write the equation as:

$$\log q_g = \log C + n \log (p_e^2 - p_w^2) \tag{5.46}$$

Where:

$$C = \frac{0.703 k_{avg} h}{\mu z_{avg} T_f \ \ell n(0.606 r_e/r_w)} \tag{5.47}$$

Thus, a plot of the log of q_g versus the log of $(p_e^2 - p_w^2)$ provides a straight line whose slope is n if C is a constant. We must then consider whether the group of terms representing C can be expected to remain unchanged during a flow test. Remember that in most cases a drawdown test on a gas well does not take more than a few days to run, and most flow tests are completed in 2–3 days. The length of time becomes important because we can conclude that, during the flow test, the average (static) well pressure and the pressure at the drainage boundary, p_e, do not change significantly. Thus, we may conclude that the pressure-affected parameters in the constant C do not change during the flow test.

However, we must remember that we are dealing with a flow equation. We showed previously in Eq. 5.37 that the pressure drop is proportional to the flow rate stated at the arithmetic average pressure, $(p_e + p_w)/2$. Thus, μ and z should be evaluated at the arithmetic average pressure, which does vary somewhat as a result of the variation of p_w with the rate. These two factors—the viscosity and the gas deviation factor—are the parameters in C that have the most obvious variation. However, at relatively low pressures—less than the pressure corresponding to the trough or lowest z value on the plot of z versus p_r—we find that the z factor increases while μ decreases with a decrease in pressure. Therefore, the product, μz, tends to remain constant. Of course, this situation does not apply at higher pressures where the variation is much more complex.

Fig. 5–13 illustrates a deliverability curve. Experienced engineers should note that this plot may be reversed from the plot normally used or the one required by some state regulatory bodies. The author prefers to plot the deliverability data in this way so the slope is directly measurable as n.

Note that if the skin factor, S, and the undamaged permeability had been used in Eq. 5.45 instead of k_{avg}, the skin factor would become part of the constant, C, and the straight-line quality of the deliverability plot would still be maintained. To illustrate the application of the deliverability curve to practical problems, work problem 5.8 and check the solution against the one in appendix C.

Throughout the consideration of the practical problems of gas reservoir engineering, the engineer must determine what pressure base to use for evaluating reservoir parameters that are considered constant mathematically but that actually vary with the pressure. In problem

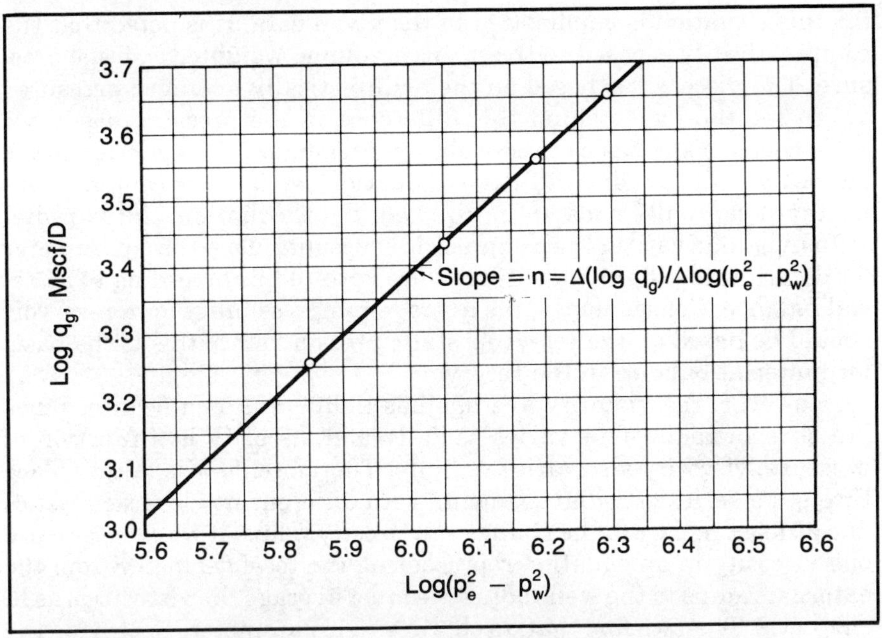

Fig. 5–13 Deliverability curve from stabilized drawdown data

PROBLEM 5.8: Conventional Gas-Well Back-Pressure Test

The reservoir in problem 5.4 totals 4,500 acres and is being produced by two wells. Back-pressure tests are conducted on one of the wells with the following results:

Stabilized BHP, psia	*Producing Rate, MMscfd*
2,798	0
2,669	1.81
2,591	2.71
2,499	3.59
2,426	4.51

Reservoir characteristics are as follows:

Permeability = 74 md
Porosity = 15%
Gas viscosity = 0.021 cp, based on average pressure
$c = 3.57 \times 10^{-4}$, based on static pressure

A. What must the bottom-hole pressure be in this well to produce at a rate of 5 MMscfd when p_e has declined to 2,000 psia, if the constant C of Eq. 5.46 is not significantly affected?

B. How long must each rate be maintained to reach stabilized conditions during the test?

5.8 this situation is emphasized in the given data. It is noted that the compressibility is based on the static or volume-weighted average pressure. The viscosity is based on the arithmetically averaged pressure.

To see the logic behind this difference in the pressure bases, we must consider the basic purpose of each parameter. If we go back to the derivation of the radial diffusivity equation, we find the compressibility that eventually ends up in Eq. 3.50, the t_s equation. In the radial diffusivity derivation, the compressibility is introduced as a parameter for determining the change in pressure with radius according to material balance. Consequently, the average compressibility for a reservoir should be based on the reservoir static pressure since this is the basis for material balance in the reservoir.

However, the viscosity average has a different significance. Since pressure in radial flow varies radially and viscosity is a function of pressure, viscosity also varies radially. Therefore, flow is taking place through a series of radial segments with different flow characteristics due at least in part to the change in viscosity. Thus, if we choose to use one viscosity in an equation representing the total geometry from the external radius to the well radius, we must average the viscosities as in chapter 2. Then we find that, to obtain single parameters to provide the correct total pressure drop in a gas system, we must evaluate the flow parameters at the arithmetic average pressure for either radial or linear flow.

Later, we see that there are mathematical means of letting the viscosity and gas deviation factor vary with the pressure. This calculation is done using the real gas potential, which is a function of pressure, viscosity, and gas deviation factor. However, this mathematical approach does not lend itself to use of the exponent n, and thus, it does not appear to be widely used by the practicing engineer.

Depletion and well-spacing effects on the deliverability curve. We have discussed the effect of a change in the average pressure in the drainage area of a well on the parameters in the constant C of Eqs. 5.46 and 5.47 during a well test. Although the effective viscosity and gas deviation factor do vary with the change in the average pressure, a drawdown test run over a realistic range of rates usually exhibits a straight-line plot of the log of the rate versus the log of $(p_e^2 - p_w^2)$. However, when the average reservoir pressure or p_e changes as a result of pressure depletion, a larger change in the average pressure occurs and a significant change in the viscosity and gas deviation factor results. This may cause a change in the constant C and a shift in the deliverability curve. The empirical constant n is apparently affected almost entirely by the turbulence, which in turn is a function of the flow rate, q_g. Thus, the slope of the deliverability curve remains unchanged with a change in C. Therefore, the change in the deliverability curve caused by a change in C is simply a vertical shift of the curve equal to the change in the log C.

Note that a change in the spacing changes r_e, which also affects a change in the constant C. Consequently, the effect of a change in well spacing on deliverability can be evaluated by calculating the change in C caused by the change in r_e and by noting the shift in the deliverability curve. Eq. 5.47 shows that the constant, C, is inversely proportional to the viscosity, the gas deviation factor, and the log term containing r_e. Thus, a ratio of C's caused by depletion or a change in well spacing is:

$$\frac{C_2}{C_1} = \frac{\mu_1 z_{avg\,1}\,\ell n(0.606 r_{e1}/r_w)}{\mu_2 z_{avg\,2}\,\ell n(0.606 r_{e2}/r_w)} \tag{5.48}$$

Using Eq. 5.48, a test run at one well spacing and state of depletion with an arithmetic average testing pressure can be used to predict the deliverability at other well spacings and states of depletion. To do this, it is not necessary to know all of the reservoir parameters in the expression for C, Eq. 5.47. The constant C can most accurately be evaluated by calculating it from the observed deliverability curve. First, calculate the constant n from the slope. Then read the values for $p_e^2 - p_w^2$ and q_g corresponding to any point on the deliverability curve. When these values are put in Eq. 5.46 with the value of n, the constant C_1 can be calculated. Once C_1 is known, C_2 can be calculated by Eq. 5.48. Using C_2 for C in Eq. 5.46 and any assumed value for q_g, a corresponding value of $p_e^2 - p_w^2$ can be calculated. This point then appears on the new deliverability curve, which can be constructed parallel to the original deliverability curve.

Work problem 5.9 and check the solution against the one in appendix C to understand this correction procedure.

PROBLEM 5.9: Adjustment of a Gas Deliverability Curve for a Change in Well Spacing or Depletion

The deliverability curve resulting from problem 5.8 is Fig. 5–13, which is obtained with a drainage area of 2,250 acres and a pressure at the external boundary, p_e of 2,800 psia. In problem 5.4 we found that the static reservoir pressure would be 1,568 psia at completion of the proposed contract. Assume that additional wells are drilled so a total of 5 equally spaced wells drain the 4,500-acre reservoir when the contract for problem 5.4 is completed, i.e., p_s = 1,568 psia. What must the bottom-hole pressure be for the well to produce its share, (1/5), of the 20 MMscfd required by the contract? Note that problems 5.4 and 5.8 apply to the same reservoir and the well radius is 0.25 ft.

Other deliverability curve difficulties. The engineer should also note that other parameters in the constant C may change during a test or with depletion and may cause difficulty. As the pressure declines either around the wellbore or as a result of depletion, there is a tendency for a liquid hydrocarbon, condensate, to form. Since the lowest pressure in the drainage area is at the well, the condensate tends to form at this point. A change in the liquid saturation around the wellbore of course changes the relative permeability to the gas, which in turn causes a change in C. This does not appear to cause a problem as often as the engineer may expect—perhaps because the liquid saturation around the wellbore remains at an irreducible value most of the time as a result of the high velocity gas flow.

Another parameter in C that may change is the producing thickness, h. We normally assume the thickness cannot change, but under some circumstances the flowing thickness does change. Assume a well is producing from two zones of greatly different average permeabilities. The more permeable zone depletes more rapidly than the less permeable zone, so at a particular time the pressure in the more permeable zone is much less than the pressure in the less permeable zone. If we then proceed to test this well, a well pressure may be maintained that is equal to or even greater than the pressure in the most permeable zone. Then the most permeable zone would not flow at all, thus changing the flowing thickness. We should also add that some tight formations appear to have a threshold pressure drop below which no flow takes place but above which flow seems to follow conventional theory. If such a zone existed in a drainage area, it would then be possible for the flowing thickness to change with the pressure drawdown.

We should also add that the effective formation temperature in the constant C may change with a change in the flow rate. When gas expands as it approaches the wellbore, a cooling effect takes place that results in a substantial temperature reduction. Of course, the cooling effect is a function of the amount of gas that expands and the magnitude of the expansion. Thus, at larger flow rates more gas is expanding, and the well pressure is less so the magnitude of the expansion is greater. Therefore, the cooling effect is greater. It appears that this variation in temperature with the flow rate causes little difficulty during a flow test.

It should be noted that the reason for the apparent insensitivity of the drawdown test to such changes in temperature, viscosity, and gas deviation may be a result of the empirical nature of the constant, n. We assume that n is caused by turbulence, but it may also be affected by the temperature, viscosity, gas deviation factor, and other systematic variations in the reservoir parameters.

The engineer should carefully remember that this analysis of a conventional back-pressure test is based on the assumption that stabilized or pseudosteady-state flow is obtained at each rate. The fact that the data obtained appears to give a straight-line deliverability curve does not mean that the flow rates have stabilized at each rate. If each flow period is about the same length, it can be shown that the deliverability curve does closely approximate a straight line. This situation will be better understood after the discussion of isochronal testing.

The engineer is urged to make a rough calculation of the stabilization time, t_s, in Eq. 3.50, using approximations of the necessary reservoir parameters to ascertain that the producing times are sufficient. The author has seen data from very tight, widely spaced gas wells that indicate stabilization times in excess of 1.5 years. While this time is not considered typical, it does indicate the magnitude of the gas-well testing problem. When stabilization times are excessive, it is necessary to use some other type of gas-well test to predict behavior. For this purpose the technology of infinite-acting unsteady-state flow is necessary.

The radial diffusivity equation for gas. It may seem illogical to discuss the pseudosteady-state flow of gas before we discuss the more general case of unsteady state. However, the equations for pseudosteady-state flow are much simpler than those for unsteady-state flow, and the application of these equations in conventional gas-well testing is much simpler than the concepts of testing under infinite-acting, or isochronal, testing. Thus, it appears easier to understand the material when it is presented in this sequence.

As is the case with fluid flow in general, the equations usually used for the flow of gas are very similar to those governing the flow of liquids. The radial diffusivity equation for the flow of a slightly compressible fluid such as a liquid has been derived as:

$$\frac{\Delta p}{\Delta t} = \frac{6.33k}{\phi\mu c}\left[\frac{1}{r}\frac{\Delta p}{\Delta r} + \frac{\Delta(\Delta p/\Delta r)}{\Delta r}\right] \tag{3.8}$$

In Eq. 3.8 the group of constants, $6.33k/\phi\mu c$, is the diffusivity constant, η.

A similar equation can be derived in much the same manner for gas flow, except mass flow rates must be used rather than volumetric flow rates:*

$$\frac{\Delta(p^2)}{\Delta t} = \frac{6.33k}{\phi\mu c}\frac{1}{r}\frac{\Delta(p^2)}{\Delta r} + \frac{\Delta[\Delta(p^2)/\Delta r]}{\Delta r} \tag{5.49}$$

This gas diffusivity equation for radial flow differs from the equation applicable to gas only to the extent that the pressure, p, in Eq. 3.8 is in every case replaced by the pressure squared, p^2. Consequently, a

*The derivation of Eq. 5.49 can be found in the *Handbook of Natural Gas Engineering*, by D.L. Katz, New York: John Wiley and Sons, 1957.

dimensionless solution of the radial gas diffusivity equation results in the same numerical answers as those obtained for a dimensionless solution of the liquid radial diffusivity equation. Thus, any dimensionless solutions for the liquid radial diffusivity equation can also be applied to gas, providing they are converted to appropriate gas units. Instead of the general solutions being in terms of pressure, they are in terms of the pressure squared. Thus, the constant-rate solutions permit a calculation of the change in p^2, and the constant-pressure solutions are based on a constant change in p^2.

We should strongly emphasize that Eq. 5.49 treats the gas compressibility as a constant. The engineer sometimes gets the impression that, since the equation is derived using mass units rather than volumetric units, (as in Eq. 3.8), the variation in compressibility is accounted for. However, Eq. 5.49 shows that the compressibility is still treated as a constant. In fact, either radial diffusivity equation can be applied to the flow of gas with the same resulting solutions. The conversion equations applied to Eq. 3.8 are then based on the volumetric flow rate at the arithmetic average pressure. Therefore, a separate calculation of this volumetric flow rate at the average pressure must precede the application of the equation that is based on volumetric flow.

Since the radial diffusivity equation for gas assumes the compressibility of gas is a constant, we must be careful that the compressibility variation throughout the reservoir is small. Actually, from a practical point of view, this limitation is not severe because most of our applications are for infinite-acting reservoirs where it is not difficult to determine the mathematical average compressibility.

Mathematically, for any reservoir that is still infinite acting, the reservoir outer boundary is infinite. Thus, the average pressure is the pressure in the unaffected portion of the reservoir, which is the initial pressure, p_i. Consequently, the average compressibility, c, is based on the initial pressure. We find that the infinite-acting solutions to the diffusivity equation give accurate answers when applied to gas reservoirs and compared with more exact numerical computer solutions. When the reservoir is no longer infinite acting, we can treat it as a pseudosteady-state system and can obtain adequate accuracy.

When the unsteady-state gas flow equations of this text are compared with similar gas flow equations, the engineer should be aware that the gas compressibility of this text may be replaced with the reciprocal of the pressure in other sources. This substitution was widely made when gas reservoir engineering technology was developed. At that time all of the commercially productive reservoirs were low-pressure reservoirs, and the approximation that the gas compressibility equalled the reciprocal of the pressure was approximately correct. The gas compressibility does equal the reciprocal of the pressure for an ideal gas. However, as we produce higher-pressured gas reservoirs, we reach the point where this approximation introduces errors that may approach 40%.

Unsteady-state gas flow—constant-pressure solution. We are first considering the constant-pressure solution to the radial diffusivity equation. The engineer should recall that this solution is for a well in the center of a circular reservoir whose pressure is initially uniform throughout and in which a constant pressure is maintained from time zero throughout the producing life. The general solution is in the form of cumulative production, combined with a group of reservoir constants, Q_{tD}, versus time combined with a different group of reservoir constants, t_D. The relationship between Q_{tD} and t_D is shown in Tables 3–1 and 3–2. These reduced terms are defined as:

$$t_{Dw} = \frac{\eta t}{r_w^{\ 2}} \tag{3.11}$$

$$\eta = \frac{6.33k}{\phi \mu c} \tag{3.9}$$

$$Q = 1.12\ \phi h c r_w^{\ 2}\ \Delta p_w Q_{tD} \tag{3.57b}$$

Eq. 3.57b is a slightly modified form of Eq. 3.57. The subscripts w have been added to make it clear that the radius for Δp and t_D in Eq. 3.57b is the same well radius. The Q is stated in reservoir barrels. If we are to apply this equation to gas, it is convenient to state Q in Mscf. This can be done by substituting the equivalent of Q, stated as a function of Q_{Mscf}, which is the cumulative gas flow in thousands of standard cubic feet. An expression for the relationship between reservoir barrels and thousands of standard cubic feet is derived in chapter 2 and is applied to rates. Modified slightly to make it applicable to cumulative volumes, we obtain:

$$Q = \frac{5.04 Q_{Mscf} T_f\ z_{avg}}{p} \tag{2.33a}$$

From Eq. 2.33a we see that some pressure base must be used to convert from reservoir barrels to Mscf. Remember that the Q_{tD} functions can be applied with reasonable accuracy only to infinite-acting gas reservoirs and the well pressure in an infinite-acting reservoir varies from the initial pressure, p_i, at the current drainage radius to the constant well pressure, p_w. Therefore, we use the arithmetic average pressure for the volume conversion. This pressure is in line with the previously noted observation that for steady-state gas flow the total pressure drop is proportional to the gas flow rate stated at the arithmetic average pressure. When $(p_i + p_w)/2$ is substituted for p in Eq.

2.33a and the resulting expression is substituted for Q in Eq. 3.57b, we can show that:

$$Q_{Mscf} = \frac{0.111\ \phi h r_w^2\ c(p_i^2 - p_w^2)Q_{tD}}{z_{avg}T_f} \tag{5.50}$$

Eq. 5.50 has not been used to any great extent by practicing engineers, but it appears to have considerable potential. After we have completed our consideration of the problem of predicting stabilized flow rates, we see that a reasonably accurate prediction of such rates can be made in most circumstances. However, as noted, many gas wells require considerable lengths of time before they reach stabilized flow. Eq. 5.50 can be used to predict these flow rates prior to stabilization. In applying the equation, be sure to note that the equation does not contain a correction for the additional pressure drop caused by well damage or permeability improvement ($-\Delta p_{skin}$), nor does it account for the additional pressure drop caused by turbulence. Consequently, the well radius used in Eq. 5.50 should be $r_{we} = r_w e^{-S}$. This expression is derived in chapter 2 to account for the well damage as indicated by the skin factor, S. The term r_{we} should be used in determining t_D and in Eq. 5.50. In addition, $(p_e^2 - p_w^2)^n$ should be used for the Δp^2 expression in Eq. 5.50 with the constant n having been previously determined from a well test.

Eq. 5.50 then becomes the more explicit equation:

$$Q_{Mscf} = \frac{0.111\phi h r_{we}^2\ c(p_i^2 - p_w^2)^n\ Q_{tD}}{Z_{avg}\ T_f} \tag{5.51}$$

Where:

$$t_D = \eta t / r_{we}^2$$

As noted, the compressibility should be evaluated at p_i, which is the mathematical average reservoir pressure for an infinite-acting reservoir. A similar analysis could be achieved using the q_{tD} values from appendix B.

If we want to predict the producing rate under constant-pressure conditions prior to stabilization, we probably must run a pressure-buildup or constant-rate drawdown analysis to obtain the necessary reservoir parameters, especially the skin factors and turbulence. Eq. 5.51 produces excellent agreement with numerical computer solutions as long as the reservoir is infinite acting. When the reservoir is no longer infinite acting, it can be assumed that it is in pseudosteady-state flow with the further cumulative production being calculated by these equations or material balance. Thus, the unsteady-state, infinite-acting, constant-pressure solution together with pseudosteady-state flow should give an adequate prediction of cumulative gas flow versus time for constant-pressure conditions.

Unsteady-state gas flow—constant-rate solution. The constant-rate solution has found more practical application in gas flow applications than the constant-pressure solution. As noted, the same numerical solutions apply to the constant-rate solution to the radial diffusivity equation as apply to the same case for liquid flow. Instead of using the solution to predict the change in pressure, we can use the same pressure functions to predict the change in the square of the pressure, p^2. Consider the general pressure function definition rearranged to include the skin factor, S, and a specific pressure drop:

$$(p_i - p_{r,t}) = \frac{0.141q\mu}{kh}\left[(\Delta p_D) + S\right] \tag{3.12a}$$

In Eq. 3.12a, as with most equations used previously, the rate, q, is stated in volumetric units of reservoir barrels per day. When we write this expression with the rate stated as a function of the mass flow rate in Mscfd, q_g, as indicated in Eq. 2.33a, we obtain:

$$(p_i - p_{r,t}) = \frac{0.141\mu}{kh}\frac{5.04q_g T_f\, z_{avg}}{p}(\Delta p_D + S) \tag{5.52}$$

As when deriving the expression to convert from Q_{tD} to Mscf, it is necessary to determine some pressure base to use for the pressure, p, in Eq. 5.52. Thus, we again use the idea that the linear pressure drop is proportional to the rate stated at the arithmetic average pressure, $(p_i + p_{r,t})/2$. When this substitution is made in Eq. 5.52 and the expression is solved for the square of the pressure at the radius, r, and time, t, we obtain:

$$p_{r,t}^{\,2} = p_i^{\,2} - \frac{1.424q_g\mu z_{avg}T_f}{kh}\left[(\Delta p_D) + S\right] \tag{5.53}$$

Eq. 5.53 still does not account for the additional pressure drop caused by turbulence. In infinite-acting behavior, which represents most of the applications for Eq. 5.53, the affected radius continues to increase. Therefore, the empirical exponent n for the Δp^2 term, which we used for pseudosteady-state and steady-state flow where r_e is constant, does not appear to be sound. Consequently, we include the additional pressure drop caused by turbulence as a separate term, $Bq_g^{\,2}$. In this expression B is simply a proportionality constant that is determined by testing the well. This B should not be confused with the parameter used in the steady-state flow equations, Eqs. 5.36 and 5.38. In turbulent gas flow through pipes and channels, it is well known that the drop in the pressure squared is proportional to the velocity squared. Consequently, in reservoir flow we expect the additional pressure drop caused by turbulence to be proportional to the flow rate squared. Then B is simply the proportionality constant. Thus, Eq. 5.53 can be written to include the additional pressure drop caused by turbulence as:

$$p_{r,t}^2 = p_i^2 - \frac{1.424 q_g \mu z_{avg} T_f}{kh}(\Delta p_D + S) + Bq_g^2 \tag{5.54}$$

Eq. 5.54 can also be written with the specific pressure functions, p_{tD}, or the Ei function expression substituted for the pressure function, Δp_D, to obtain conversion equations similar to those used for liquid:

$$p_{r,t}^2 = p_i^2 - \frac{1.424 q_g \mu z_{avg} T_f}{kh}\left[(p_{tD}) + S\right] + Bq_g^2 \tag{5.55}$$

$$p_{r,t}^2 = p_i^2 - \frac{1.424 q_g \mu z_{avg} T_f}{kh}\left[\left(\tfrac{1}{2}\right)\left(-\mathrm{Ei}\frac{-1}{4t_D}\right) + S\right] + Bq_g^2 \tag{5.56}$$

Eqs. 5.55 and 5.56 have the same limitations as their counterparts for liquid flow. Eq. 5.55 can only be used to calculate the pressure at a radius where the flow rate is known—for all practical purposes the well radius. Eq. 5.56 can only be used when the reservoir is infinite acting and the point-source solution is applicable, $\eta t/r_w^2 > 100$. We must also add a practical limit to Eq. 5.55. As a principle result of the invalid assumption that the fluid compressibility is constant, it has been found that the p_{tD} solution can be used with confidence only when the reservoir is infinite acting. This factor does not present a serious limitation because a finite-acting reservoir producing at a constant rate is in pseudosteady-state flow and can be analyzed using these equations.

Treating the additional $\Delta(p^2)$ caused by turbulence as a steady-state effect in an unsteady-state equation introduces no significant errors because most of the turbulence occurs near the wellbore. At this point gas velocities are extremely high because the cross-section area is very small. Consequently, the turbulence pressure-drop term can be treated as a steady-state effect in much the same way that the pressure drop caused by the well damage is treated as a steady-state effect.

Eq. 5.36, which describes the steady-state flow of gas with turbulence, also indicates the validity of assuming all of the additional pressure drop caused by turbulence occurs at the well. As we noted, the second term of the radial steady-state flow, Eq. 5.36, represents the additional pressure drop caused by turbulence. This term contains the difference between the reciprocals of the well radius and external radius. Since the well radius is a fraction of a foot, the limitation or effect of the reciprocal of the external radius on the accuracy of the equation is negligible. For example, if the well radius is 1/3 ft, reciprocal 3.0, the reservoir outside a radius of 10 ft, reciprocal 0.1, affects the additional pressure drop caused by turbulence only about 3%.

Isochronal testing. As noted, it is often difficult or impossible to test a gas well in the conventional back-pressure test manner to obtain usable data for predicting gas-well deliverability. When the stabilization time is excessive, it is still possible to test the well under infinite-acting conditions and interpret the data to predict stabilized deliverability. This procedure is generally carried out in a test known as an *isochronal test.*[8]

Ideally, the procedure is started with the drainage area at static pressure throughout. The well is then produced for some relatively short period of time, t*, and the bottom-hole flowing pressure is noted. The well is shut in until the static pressure is reached, after which it is produced at a different rate for the same time period, t*. The resulting well pressure at this rate is noted. The well is shut in until the static pressure is reached and is then produced at a third rate for a time t*, with the resulting well pressure noted. This procedure can be repeated to obtain well pressures resulting from flow at a variety of rates for a period of time, t*.

When a log-log plot is prepared of q_g versus $p_e^2 - p_w^2$, a straight line is obtained. This plot is shown to be the same as the stabilized deliverability curve of a particular drainage radius less than the actual ultimate drainage radius. The effective (particular) drainage radius is a function of the producing period, t*, and the reservoir parameters. As shown for the conventional back-pressure test, once a stabilized deliverability curve has been obtained for one drainage radius, it can be corrected to a stabilized deliverability curve for any other drainage radius desired. This testing procedure is called isochronal, or equal-time, testing.

In effect, in isochronal testing we evaluate the well under infinite-acting conditions and interpret the data for an equivalent stabilized system. To do this, we determine the reservoir size that would give the same pressure drop term $(p_e^2 - p_w^2)^n$ as is experienced by an infinite-acting reservoir that has produced for a time t*, $(p_i^2 - p_w^2)^n$. To obtain this relationship, we first solve Eq. 5.43 for the pseudosteady-state pressure-drop term:

$$(p_e^2 - p_w^2)^n = \frac{1.424 q_g \mu z_{avg} T_f \left[\ell n(0.606 r_e/r_w) + S\right]}{kh} \tag{5.57}$$

Eq. 5.57 can also be written with the additional pressure drop accounted for in a Bq_g^2, term rather than in the empirical exponent n, as:

$$(p_e^2 - p_w^2) = \frac{1.424 q_g \mu z_{avg} T_f \left[\ell n(0.606 r_e/r_w) + S\right]}{kh} + Bq_g^2 \tag{5.58}$$

Next, consider Eq. 5.55 written for the well pressure in an infinite-acting reservoir and solved for the pressure-drop term. Since the reservoir is infinite acting, we can use the log equation for the pressure function, p_{tD}:

$$(p_i^2 - p_w^2) = \frac{1.424 q_g \mu z_{avg} T_f \, \frac{1}{2}(\ell n \, t_D^* + 0.809 + 2S)}{kh} + Bq_g^2 \quad (5.59)$$

The dimensionless time, t_D^* is based on the isochronal producing time, t*. When we equate the right-hand sides of Eqs. 5.58 and 5.59, we obtain a relationship between the infinite-acting producing time and the external radius, r_e, of a stabilized or pseudosteady-state system that gives the same pressure drop term:

$$\frac{1}{2} (\ell n \, t_D^* + 0.809) = \ell n \, (0.606 r_e^*/r_w) \quad (5.60)$$

When t_D^* is expanded to $(6.33kt^*/\phi\mu c r_w^2)$, we can obtain:

$$r_e^* = (38.5kt^*/\phi\mu c)^{0.5} \quad (5.61)$$

Eq. 5.61 simply says that an infinite-acting well producing for a time t* would have the same pressure-drop term as a similar well producing under pseudosteady-state conditions with an external drainage radius, r_e^*.

Note carefully that this is not a stabilization-time equation. Many engineers confuse Eq. 5.61 with the stabilization-time equation since they are exactly the same except for the numerical constant. When we solve Eq. 5.61 for the time, we obtain:

$$t^* = \frac{0.026\phi\mu c r_e^2}{k} \quad (5.62)$$

Compare the numerical constant of Eq. 5.62, 0.026, with the numerical constant in the stabilization-time equation, 0.04. This comparison tells us that the pressure-drop term for an infinite-acting well is equal to the pressure-drop term for a pseudosteady-state system whose external boundary, r_e^*, satisfies Eq. 5.61 before a system with the same radius, r_e^*, reaches pseudosteady-state conditions. Therefore, the pressure at the external boundary, r_e, of a drainage system has dropped prior to the time the system reaches pseudosteady state.

This situation is reasonable from a physical standpoint if we recognize that, after the pressure at the outer boundary of a reservoir has been affected by production, the pressure at the well continues to be infinite acting until the effect of the boundary travels back to the well through the reservoir. During this interim when the pressure at the outer boundary is first affected and when the pressure at the well feels the effect of the outer boundary, the pressure at the outer boundary continues to decline. Consequently, it is logical for the pressure drop in a reservoir to be equal to the pressure drop that is ultimately reached under pseudosteady state before the reservoir system actually enters pseudosteady state.

As indicated, once r_e^* is calculated by Eq. 5.61 to represent the isochronal data, the data can be corrected to any desired reservoir drainage using Eq. 5.48 to shift the performance curve. Note that changing the drainage radius, r_e, simply shifts the performance curve without changing the slope. Thus, if we establish the slope by an isochronal test, we need only one stabilized point to establish a new performance curve. Then the new curve can be drawn parallel to the previously established performance curve. Since most commercial gas wells ultimately reach pseudosteady state, this stabilized point should be used as the basis for positioning the stabilized performance curve for an existing well spacing when possible.

Eq. 5.48 is still useful for predicting performance under various well spacings or at various states of depletion, but its use in predicting the performance for a presently existing spacing should be limited to the period needed to get one stabilized point. This time may be substantial when low permeability, low porosity, and wide spacing exist. It also may be impossible to reach pseudosteady state in a well if the stabilization time is so large that a constant producing rate cannot be approximated, e.g., a well that is produced during the heating season and is shut in during the summer months with t_s greater than 9 months.

Work problem 5.10, which demonstrates the use of isochronal data in predicting the stabilized deliverability of a gas reservoir, and compare the solution with the one in appendix C.

PROBLEM 5.10: Using Isochronal Data

Isochronal back-pressure test data are obtained in the reservoir of problem 5.4 with two wells producing. The tests are run by producing at 1,810 Mscfd for 60 min and shutting in the well until the pressure is again 2,798 psia, producing at 2,710 Mscfd for 60 min and shutting in the well until the pressure is 2,798 psia, producing at 3,590 Mscfd for 60 min and shutting in the well until the pressure is 2,798 psia, and finally producing at a rate of 4,510 Mscfd for 60 min. The results are as follows:

Mscfd	*BHP at q_g after 60 min of production, psia*
0	2,798
1,810	2,710
2,710	2,659
3,590	2,605
4,510	2,545

k = 74 md
c = 0.000357/psi
μ_g = 0.021 cp
ϕ = 0.15

A. What is the equivalent stabilized drainage radius for the isochronal data?
B. Calculate and plot the 1-hr isochronal curve and the stabilized back-pressure curve for a drainage area of 2,250 acres.
C. The well continued to produce at 4,510 Mscfd and reached a constant bottom-hole pressure of 2,425 psia. Plot the stabilized back-pressure curve based on this point.

Designing an isochronal test. One difficulty that arises in conducting an isochronal test is the problem of shutting in the well until it reaches static pressure before continuing the flow test at a new rate. Theoretically, it may appear that this presents an insurmountable problem, since we know that it would take a very long shutin time before the well would approach the static pressure. Furthermore, this problem is worse when we have the greatest need for isochronal tests. To explain, remember that we need isochronal tests when the stabilization time is such that it becomes impractical to test under stabilized conditions. The identical conditions that make the stabilization time too high have a similar effect on the shutin time necessary to have the well pressure approach the static pressure. Consequently, we need some means of circumventing the necessity of shutting in the well until the pressure approaches the static pressure.

By using the negative superposition principle to obtain the $\Delta(p^2)$ value caused by a particular producing rate, we can greatly shorten the necessary shutin time. Fig. 5–14 illustrates the technique that can be used. It shows the pressure change that has occurred from flowing a well at a rate of 4 MMscfd for 1 hr and the resulting pressure buildup from the shutin. Ideally, we would shut in the well until the pressure again reaches the initial pressure, 2,000 psia. The solid extrapolation indicates the shutin pressure that would have been obtained if the new producing rate, 3 MMscfd, had not been initiated after 1 hr of shutin. The dotted line shows the pressure drawdown that is obtained as a result of the new producing rate of 3 MMscfd. Consequently, the difference between the pressure that would have been obtained at the end of 3 hr if the 3-MMscfd rate had not been initiated, 1,980 psia, and the actual pressure obtained at the 3-MMscfd rate, 1,474 psia, is the effect of the new rate alone.

Once the engineer understands this situation, he can simply leave the well shutin until he can predict with confidence what the pressure would be during the next flow period if the new flow rate were not initiated. Then the difference between the extrapolated pressure and the actual pressure at the end of the next flow period is the desired isochronal pressure drop.

It must be noted that superposition can only be applied to the portion of the $\Delta(p^2)$ that is caused by viscous flow and the well damage. In

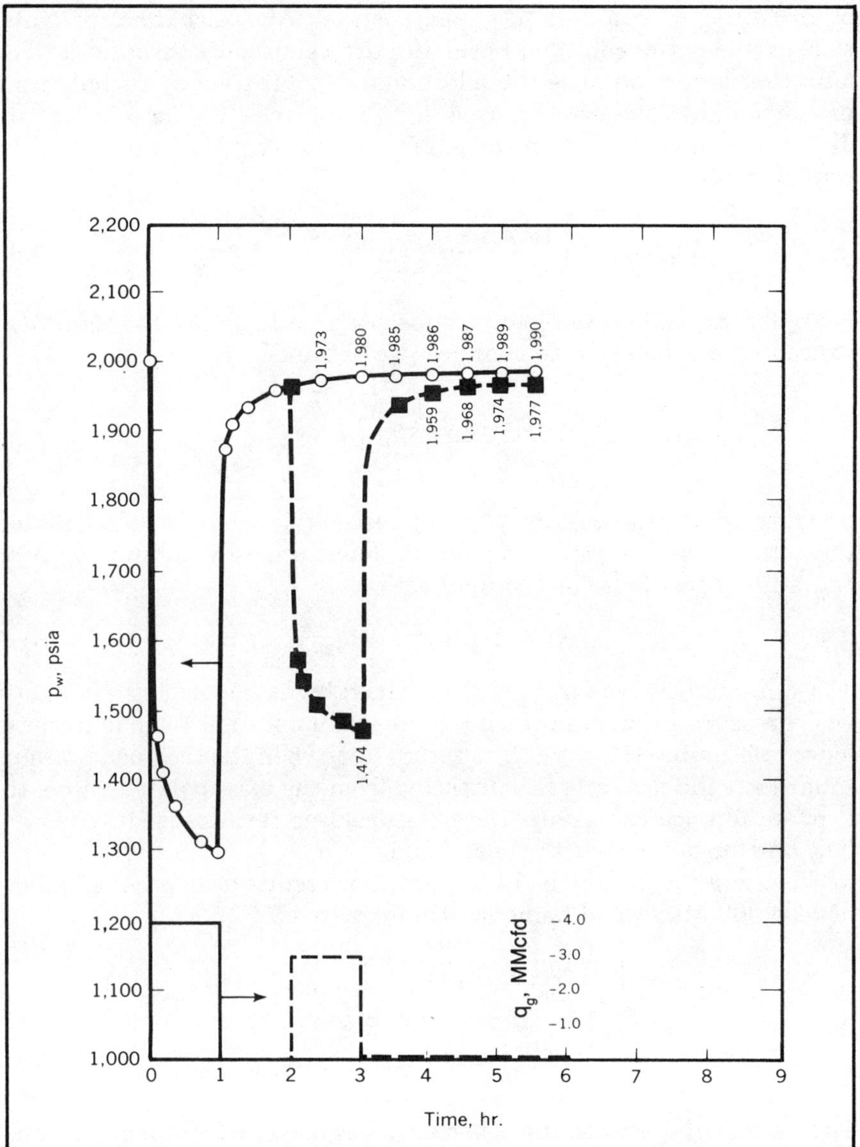

Fig. 5–14 Slow buildup to average pressure (after Slider, "Gas Well Testing Using Negative Superposition," paper 3875, © 1972, SPE-AIME)

other words we can only use superposition to predict the $\Delta(p^2)$ value that excludes the additional pressure drop caused by turbulence. This situation occurs because the additional $\Delta(p^2)$ caused by turbulence is proportional to the rate squared, q_g^2. If we write Eq. 5.53 in a finite-difference form, we obtain the expression for the $\Delta(p^2)$ caused by viscous flow alone.

$$\Delta(p^2)_{visc} = \left[\frac{1.424\mu T_f z_{avg}\,(\Delta p_D + S)}{kh}\right] \Delta q_g \qquad (5.63)$$

When Eq. 5.63 is applied by superposition to obtain the total $\Delta(p^2)$ caused by a series of rate changes, we obtain:

$$\Delta(p^2)_{visc} = \frac{1.424\mu T_f z_{avg}}{kh} \sum_{j=1}^{j=n} (\Delta q_{gj})\,(\Delta p_{Dj} + S) \qquad (5.64)$$

Once this pressure-drop term by viscous flow has been calculated, the total pressure-drop term can be determined by adding the $\Delta(p^2)$ caused by turbulence at the final rate:

$$\Delta(p^2) = \Delta(p^2)_{visc} + Bq_{gn}^2 \qquad (5.65)$$

Consequently, the p_w^2 value obtained by extrapolating the shutin well pressure prior to initiating a new rate does not include a turbulence affect since the rate, q_g, is zero. Thus, when the recorded pressure squared at the new rate is subtracted from the extrapolated value, the $D(p^2)$ difference represents the pressure-drop term caused by viscous flow and turbulence at that particular rate.

For example, in Fig. 5–14 the Δp^2 that results from producing for 1 hr at 4,000 Mscfd and being shutin for 2 hr is:

$$p_e^2 - (p_w')^2 = \frac{1.424\mu T_f z_{avg}}{kh}(4{,}000)\,(\Delta p_{D3\,hr} + S) - \frac{1.424\mu T_f z_{avg}}{kh}(4{,}000)\,(\Delta p_{D2\,hr} + S) + 0.0 \qquad (5.66)$$

The 0.0 results because there is no $\Delta(p^2)$ caused by turbulence because the last rate is 0.0. Eqs. 5.64 and 5.65 can be applied to determine $\Delta(p^2)$ caused by producing at 4,000 Mscfd for 1 hr, being shut in for 1 hr, and producing at 3,000 Mscfd for 1 hr:

$$p_e^2 - p_w^2 = \frac{1.424\mu T_f z_{avg}}{kh}(4{,}000)\,(\Delta p_{D3\,hr} + S) - \frac{1.424\mu T_f z_{avg}}{kh}(4{,}000)\,(\Delta p_{D2\,hr} + S)$$

$$+ \frac{1.424\mu T_f z_{avg}}{kh}(3{,}000)(\Delta p_{D1\,hr} + S) + B(3{,}000)^2 \qquad (5.67)$$

When these two expressions, Eqs. 5.66 and 5.67, are subtracted, we obtain:

$$(p_w')^2 - p_w^2 = \frac{1.424\mu T_f z_{avg}}{kh}(3{,}000)(\Delta p_{D1\,hr} + S) + B(3{,}000)^2 \qquad (5.68)$$

This expression describes the $\Delta(p^2)$ caused by producing at 3,000 Mscfd for 1.0 hr.

A modified isochronal test has been proposed that simply uses the pressure at the end of the shutin period as p_i for the purposes of calculating the $\Delta(p^2)$.[9] However, the proposed negative-superposition analysis makes the inaccuracy of the modified isochronal test unnecessary.

In addition to the design of the shutin period during a drawdown test that facilitates extrapolation, there are some other ideas that may be useful to the engineer. There is a big temptation on the part of the engineer to make the isochronal flow period ridiculously small. In most gas wells a 15–30-min isochronal flow period appears to give a straight-line log-log plot. However, the engineer should remember that such a short flow period seldom affects much of the reservoir and, thus, does not give as reliable of a test as a longer flow period. The engineer can estimate the amount of the reservoir affected during the flow period using the t_s equation, Eq. 3.50, to calculate the effected radius, r_e, for t_s equal to the flow period. Thus, the engineer should be encouraged to use as long of a flow period as possible for an isochronal test. Of course, if it is possible to use an isochronal flow period equal to or greater than the stabilization time, there is no need for an isochronal test because a stabilized flow test can and should be employed.

Quite often an engineer designs an isochronal flow test for four different rates with successively higher flow rates. Then when he gets to the greatest flow rate, he cannot maintain the constant flow rate for the isochronal period desired. This situation can be avoided by using the greatest flow rate first. If we can maintain the maximum flow rate for the desired isochronal period, we are of course assured of being able to maintain the lesser flow rates for the isochronal period. Do not be misled by government's desire for all flow tests to be run at increasing rates. There is no theoretical basis for this restriction. It appears that the restriction of using only increasing flow rates during gas well tests is a result of common misapplication of the stabilized flow test. When all of the misapplied stabilized flow tests are run with increasing rates, the errors involved in the tests in one reservoir are of similar magnitudes. Then the allocation of production to different wells or operators on the basis of apparent open flow is more realistic.

As noted, we should resist the temptation to provide a government agency with a better test than they have requested, unless we are absolutely certain that so doing results in a greater indicated apparent open flow than would be obtained by following the government forms. Generally, providing the government agency with an accurate measurement is detrimental to the best interests of the operator because a stabilized flow test that does not stabilize at each rate results in too large of an AOF. Then the subsequent allocation of production would be greater than it would be if we reported the more accurate but lesser AOF. Consequently, well tests should be designed to fulfill one of two purposes: (1) to satisfy government requirements or (2) to provide the best basis for predicting the rate versus time behavior of the well or reservoir. It is a rare situation where we can obtain both objectives with the same test.

We should perhaps anticipate that we may receive some negative response from field personnel when a constant-rate flow test of any kind is requested. When field personnel say they cannot run a constant-rate test, we should visit the wellsite with the foreman or roustabout pusher. After inspecting the equipment on the well, show this person what valve must be turned to maintain what pressure in order to approximate a constant rate. If we can find a person who is willing to turn a valve to maintain a constant pressure at some appropriate point in the flow system, we can approximate a constant flow rate. Constant-rate valves or constant back-pressure regulators can greatly simplify a constant-rate flow test.

Determining isochronal data from continuous flow data. State regulatory agencies recommend or require flow-after-flow tests without intervening shutin periods. Consequently, most gas producers have many such flow-test records in the files that do not represent stabilized data. Thus, it seems wise to cover methods of interpreting such data.

Flow-after-flow data that do not stabilize can be interpreted in terms of stabilized performance data in most cases. First, recognize the theoretical relationship between the actual pressure drops measured in flow-after-flow tests and the desired pressure drops. The pressure-drop term, Δp^2, caused by a sequence of rates in an infinite-acting reservoir can be obtained by combining Eqs. 5.64 and 5.65 with the log equation substituted for the pressure function, Δp_D:

$$(p_i^2 - p_w^2)_{actual} - Bq_{gn}^2 = \frac{1.424\mu T_f z_{avg}}{kh} \sum_{j=1}^{j=n} (\Delta q_g)_j \times [0.5(\ell n\, t_{Dj} + 0.809) + S] \qquad (5.69)$$

However, the term that we desire is the pressure drop caused by the last rate acting for the isochronal time, t*:

$$(p_i^2 - p_w^2)_{qgn} - Bq_{gn}^2 = \frac{1.424\mu T_f z_{avg}}{kh} q_{gn}[0.5(\ell n\ t_D{}^* + 0.809) + S] \tag{5.70}$$

Then divide Eq. 5.70 by Eq. 5.69 to obtain:

$$\frac{(p_i^2 - p_w^2)_{qgn} - Bq_{gn}^2}{(p_i^2 - p_w^2)_{actual} - Bq_{gn}^2} = \frac{q_{gn}\ [0.5(\ell n\ t_D{}^* + 0.809) + S]}{\sum_{j=1}^{j=n} (\Delta q_{gh})\ [0.5(\ell n\ t_{Dj} + 0.809) + S]} \tag{5.71}$$

To put the effect of the well damage as measured by the skin factor, S, in its proper perspective, we can rewrite Eq. 4.63 with the pressure drop caused by damage written as a separate term:

$$\frac{(p_i^2 - p_w^2)_{qgn} - Bq_{gn}^2}{(p_i^2 - p_w^2)_{actual} - Bq_{gn}^2} = \frac{q_{gn}\ [0.5(\ell n\ t_D{}^* + 0.809)] + \Delta(p^2)_{skin}}{\sum_{j=1}^{j=n} (\Delta q_{gj})\ [0.5(\ell n\ t_{Dj} + 0.809)] + \Delta(p^2)_{skin}} \tag{5.72}$$

Note from Eq. 5.72 that if the two pressure-drop terms in the left-hand side of the equation are nearly the same, the turbulence term, Bq_{gn}^2, which is the same in both the numerator and denominator, has little effect on the ratio. Similarly, the pressure drop caused by the skin damage is the same in the numerator and denominator of the right-hand side, and their effect on that ratio is negligible. This conclusion is also based on the assumption that the pressure drop caused by turbulence and the skin is small relative to the total pressure drop. Consequently, if these assumptions are met, the equation simplifies to:

$$\frac{(p_i^2 - p_w^2)_{qgn}}{(p_i^2 - p_w^2)_{actual}} \approx \frac{q_{gn}\ [(\ell n\ t_D{}^* + 0.809)]}{\sum_{j=1}^{j=n} (\Delta q_{gj})\ [(\ell n\ t_{Dj} + 0.809)]} \tag{5.73}$$

Eq. 5.73 provides a convenient, useful technique for correcting conventional flow-after-flow data that do not stabilize into isochronal data. By using the observed pressure, rates, and times and assuming reservoir parameters necessary for the evaluation of the dimensionless times, it is possible to arrive at more meaningful data than that represented by the original test. Since the assumed reservoir parameters necessary for the evaluation of the dimensionless times appear in the numerator and denominator as logs, the accuracy of these parameters does not greatly affect the results of the calculations. Furthermore, no difficulty is generally encountered in meeting the assumptions that the ratio of the observed and desired pressure-drop terms is near 1.0 and that the additional pressure drops caused by turbulence and the skin effect are small compared to the total pressure-drop terms.

If enough pressure data are available for the first pressure drawdown—the pressure decline caused by the initial producing rate—the assumptions inherent in Eq. 5.73 can be avoided. This is accomplished by evaluating the group of constants in the right-hand side of Eqs. 5.69 and 5.70, $\mu T_f z_{avg}/kh$, from the slope of the drawdown plot of p_w^2 versus the log of producing time. The slope of such a plot is:*

$$m' = (1.638\mu T_f z_{avg} q_g'/kh) \quad (5.78)$$

Once this group of terms has been evaluated, Eqs. 5.69 and 5.70 can be subtracted and rearranged to obtain:

$$(p_i^2 - p_w^2)_{qgn} = (p_i^2 - p_w^2)_{actual} - 0.867\,(m'/q_g')\left[\sum_{j=1}^{j=n} (\Delta q_{gj} \log t_j) - q_{gn} \log t^*\right] \quad (5.74)$$

Thus, by using the initial drawdown data to evaluate m′, Eq. 5.74 can be used to find the isochronal data without relying on the assumptions of Eq. 5.73. However, this technique is seldom practical because the initial constant-rate drawdown is usually not recorded during a conventional stabilized flow test. If we are designing a flow test for predicting the behavior of a gas well, we probably prefer not to run the test in this manner.

Work problem 5.11 and check the solution against the appendix C solution to test understanding of the methods presented for correcting flow-after-flow data to isochronal data.

A major drawback to these methods of correcting flow-after-flow data to isochronal data is that the testing procedure need not be related to the isochronal data being calculated. Eq. 5.73 indicates this clearly. From the basic flow-test data any isochronal rate and time data can be used in the equation, and the pressure changes can be calculated. The final total flow rate for each measured pressure must be the same as the isochronal flow rate calculated for the equations to be valid. Otherwise, the additional pressure drop caused by turbulence would not cancel out. However, turbulence terms can be added to the equation to account for this difference. Therefore, there may be little relationship between the pressures measured during the flow-after-flow test and the calculated isochronal performance.

*This expression is derived later.

PROBLEM 5.11: Determining Approximate Isochronal Data from Conventional Drawdown Data

A well in the reservoir in problem 5.10 is tested by flowing it at a rate of 1,810 Mscfd for 60 min, flowing at 2,710 Mscfd for 60 min with no intervening shutin, flowing at 3,590 Mscfd for 60 min, and flowing at 4,510 Mscfd for 60 min. This is an old well test with no additional pressure data available. Determine the approximate 1-hr isochronal test data and prepare a log-log plot of these data, using the following information:

Total Testing Time, min	*Rate during Preceding 60 min, Mscfd*	*BHP, psia*
0	0	2,800
60	1,810	2,710
120	2,710	2,653
180	3,590	2,596
240	4,510	2,532

Determining the skin and turbulence (B) constants. To this point the methods used to predict the deliverability of a well have relied on the empirical constant n. The relationships used have not involved the skin constant, S, nor the turbulence constant, B. Although these parameters are used to derive some of the equations in the analysis of flow data, the working equations did not contain either S or B. This was possible because we were always concerned with data where the drainage radius was fixed as in pseudosteady-state or isochronal methods. Thus, the average permeability can be used and the fraction of the total reservoir drainage where turbulence is significant remained constant. Furthermore, it is unclear just what the constant n represents because of the variation of the gas viscosity, deviation factor, temperature, and compressibility with the pressure or flow rate. For these reasons many engineers prefer to use the regular unsteady-state and pseudosteady-state equations that incorporate the skin factor and turbulence proportionality constant as the basis for their performance predictions.

In such a case some technique must be used to evaluate S and B. The conventional stabilized drawdown test or isochronal test does not provide a means of determining these parameters. They can, however, be determined from multiple constant-rate drawdowns, multiple pressure buildups preceded by different flow rates, or specially designed flow-after-flow tests.

Constant-rate drawdown. As shown, the constant-rate solution to the gas radial diffusivity equation, Eq. 5.49, is governed by Eq. 5.59 as long as the behavior is infinite acting. When solved for the well pres-kure and written for a general time, it becomes:

$$p_w^2 = p_i^2 - \frac{1.42 q_g \mu z_{avg} T_f \left[(1/2)(\ell n\, t_D + 0.809) + S\right]}{kh} - Bq_g^2 \quad (5.75)$$

The natural log of t_D is equal to the natural log of η/r_w^2 plus the natural log of the time, t. Therefore, when we group the terms on the right-hand side of Eq. 5.75 that do not change with time into one constant, we can write the equation as:

$$p_w^2 = \text{constant} - \frac{1.424 q_g \mu z_{avg} T_f\, (1/2)\, \ell n\, t}{kh} \quad (5.76)$$

Then substituting 2.303 log t for ℓn t and combining numerical constants, we obtain:

$$p_w^2 = \text{constant} - \frac{1.638 q_g \mu z_{avg} T_f}{kh} \log t \quad (5.77)$$

Eq. 5.77 shows that a plot of p_w^2 versus log t is a straight line whose slope is:

$$m' = \frac{1.638 q_g \mu z_{avg} T_f}{kh} \quad (5.78)$$

Then Eq. 5.75 can be written as a function of m′:

$$p_w^2 = p_i^2 - 0.867\, m' \left[1/2(\ell n\, t_D + 0.809) + S\right] - Bq_g^2 \quad (5.79)$$

Recognizing that the additional $\Delta(p^2)$ caused by the well damage is 0.867 m′S, we can write Eq. 5.79 as:

$$p_w^2 = p_i^2 - 0.867\, m' \left[1/2(\ell n\, t_D + 0.809)\right] - \Delta(p^2)_{skin} - Bq_g^2 \quad (5.80)$$

Note that once m′ has been determined from a plot of p_w^2 versus log t, we can use Eq. 5.80 to determine the additional pressure drop caused by turbulence and well damage, $\Delta(p^2)_{skin} + Bq_g^2$. This is generally regarded as a conventional drawdown test analysis, and of course, we are unable to differentiate between the pressure drop caused by well damage and the pressure drop caused by turbulence. Previously, we referred to this total pressure drop as "$\Delta(p^2)_{skin}$." Note that we can write this expression as:

$$\text{"}\Delta(p^2)_{skin}\text{"} = 0.867\, m'S + Bq_g^2 \quad (5.81)$$

By running two drawdowns at different rates and evaluating m′ and the apparent delta pressure squared caused by the skin for each, we have two equations with two unknowns, S and B, which can be solved for S and B. These parameters can then be used together with the basic flow equations to predict the behavior of a gas well under a variety of reservoir conditions.

Unfortunately, a plot of the well pressure squared versus the log of the producing time during an infinite-acting constant-rate drawdown does not always result in a straight line. When the gas reservoir pressure is high enough for a particular gas and reservoir temperature, the gas acts more like a liquid. The viscosity increases with a decrease in temperature, and the compressibility is relatively constant compared to that at low pressures. Thus, for high reservoir pressures the engineer may find that liquid flow equations best fit the gas-well behavior. In such cases a plot of the well pressure (not squared) versus the log of the producing time for a constant-rate infinite-acting drawdown gives a straight line.

The analysis then is the same as that for a constant-rate drawdown for an oil well, except the flow rate in reservoir barrels per day should be evaluated for the gas using the arithmetic average pressure and a turbulence term, Bq_g^2, should be added to the equations. Note that the arithmetic average pressure, the pressure at the external drainage boundary plus the well pressure divided by two, varies during the drawdown because the well pressure is declining. Consequently, determining the average pressure to be used can be a problem, and we may wish to use different average pressures with corresponding flow rates and viscosities for different portions of the drawdown test.

Use of the pseudo gas potential. As illustrated by the pressure drawdown analysis, using an average viscosity, gas deviation factor, and gas compressibility can be difficult. Consequently, in many cases it is necessary to combine the pressure with the viscosity and gas deviation factor and combine the three as a single parameter, a function of $(p/\mu z)$. This approach has been developed by a number of different researchers and educators.[10,11] The parameter is:

$$m(p) = 2 \sum_{o}^{p} \left(\frac{p}{\mu z}\right) \Delta p \tag{5.82}$$

Eq. 5.82 has been called the gas potential, the real gas potential, the real gas pseudopressure, the modified pressure squared, and the pseudopressure.[12] We choose to use the title *pseudo gas potential* to identify m(p).

To demonstrate the nature and use of the pseudo gas potential, again consider the problems involved in deriving the radial steady-state gas flow equation without turbulence, Eq. 2.39. This equation is derived by substituting Eq. 2.33 for q in reservoir barrels per day:

$$q = \frac{5.04\ q_g T z}{p} \tag{2.33}$$

In the Darcy equation $2\pi rh$ is substituted for A and r for x:

$$q = \frac{1.127kA}{\mu}\frac{\Delta p}{\Delta x} \tag{2.3}$$

From this substitution we obtain Eqs. 2.34 and 2.35:

$$\frac{5.04\ q_g T z}{p} = \frac{1.127k\ (2\pi rh)}{\mu}\frac{\Delta p}{\Delta r} \tag{2.34}$$

$$p\Delta p = \frac{5.04\ q_g T z \mu}{7.08\ kh}\frac{\Delta r}{r} \tag{2.35}$$

To obtain Eq. 2.39, we assume that an average value can be used for μ and z, although we recognize that both vary in a nonlinear fashion with the pressure. Then the $\Delta r/r$ values and the $p\Delta p$ values can be summed between r_w and r_e and p_w and p_e, respectively.

$$\sum_{p_w}^{p_e} p\Delta p = \frac{5.04\ q_g T z \mu}{7.08\ kh} \sum_{r_w}^{r_e} \frac{\Delta r}{r} \tag{2.36}$$

$$\frac{p_e^2}{2} - \frac{p_w^2}{2} = \frac{5.04\ q_g T z \mu}{7.08\ kh} (\ell n\ r_e - \ell n\ r_w) \tag{2.38}$$

$$q_g = \frac{0.703\ kh\ (p_e^2 - p_w^2)}{\mu z T\ \ell n\ (r_e/r_w)} \tag{2.39}$$

To account for the variation of μ and z with the pressure p, we can rewrite Eq. 2.35 as:

$$\left(\frac{p}{\mu z}\right)\Delta p = \frac{5.04\ q_g T}{7.08\ kh}\frac{\Delta r}{r} \tag{2.35a}$$

Then writing Eq. 2.35a for each Δr and summing all of the equations, we obtain:

$$\sum_{p_w}^{p_e} \left[\left(\frac{p}{\mu z}\right)\Delta p\right] = \frac{5.04\ q_g T}{7.08\ kh} \sum_{r_w}^{r_e} \frac{\Delta r}{r} \tag{2.36a}$$

Recognize that the left side of Eq. 2.36a is the area under a plot of $(p/\mu z)$ versus p between $p = p_w$ and $p = p_e$, as illustrated in Fig. 5–15. Note that if we use a general reference pressure other than p_w, we can write:

$$\sum_{p_w}^{p_e} \left(\frac{p}{\mu z}\right)\Delta p = \sum_{p_{ref}}^{p_w} \left(\frac{p}{\mu z}\right)\Delta p - \sum_{p_{ref}}^{p_w} \left(\frac{p}{\mu z}\right)\Delta p \tag{5.83}$$

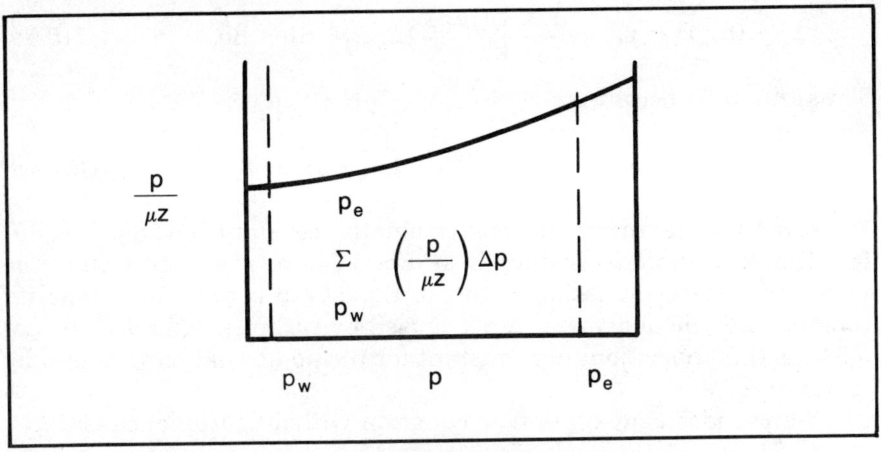

Fig. 5–15 Plot of p/μz versus pressure

Eq. 5.82 can be substituted for the summation terms:

$$m(p) = 2\sum_{p_{ref}}^{p}\left(\frac{p}{\mu z}\right)\Delta p \tag{5.82}$$

Then we obtain:

$$\sum_{p_w}^{p_e}\left(\frac{p}{\mu z}\right)\Delta p = \frac{m(p_e)}{2} - \frac{m(p_w)}{2} \tag{5.84}$$

Substituting the right-hand side of Eq. 5.84 for the left-hand side of Eq. 2.36a and $\ell n\ (r_e/r_w)$ for the summation of $\Delta r/r$, we obtain the equivalent of Eq. 2.39:

$$\frac{m(p_e)}{2} - \frac{m(p_w)}{2} = \frac{5.04q_g}{7.08kh}\ T\ \ell n\ \frac{r_e}{r_w} \tag{5.85}$$

$$q_g = \frac{0.703kh}{T}\frac{[m(p_e) - m(p_w)]}{\ell n\left(\frac{r_e}{r_w}\right)} \tag{2.39a}$$

Note that the constant 2 in the m(p) expression makes it possible to use the same numerical constant in Eqs. 2.39 and 2.39a.

Generally, we can substitute m(p) for any corresponding $(p^2/\mu_{avg}z_{avg})$ term:

$$m(p) = \frac{p^2}{\mu_{avg}z_{avg}} \tag{5.86}$$

The constant-rate solution to the radial diffusivity equation is:

$$(p_{r,t})^2 = p_i^2 - \frac{1.424 q_g \mu z T}{kh}[\Delta p_D + S] - B q_g^2 \tag{5.54}$$

Thus, Eq. 5.54 becomes:

$$m(p_{r,t}) = m(p_i) - \frac{1.424 q_g T}{kh}[\Delta p_D + S] - B' q_g^2 \tag{5.54a}$$

Note that the turbulence proportionality constant B in Eq. 5.54 differs from the similar constant B′ in Eq. 5.54a by some function of the viscosity and gas deviation factor. In Eq. 5.54 B is the proportionality constant for the additional $\Delta(p^2)$ caused by turbulence, and B′ in Eq. 5.54a is the proportionality constant for the additional m(p) caused by turbulence.

The pseudosteady-state flow equation without turbulence is:

$$q_g = \frac{0.703\ k_{avg} h\ (p_e^2 - p_w^2)}{\mu z T\left[\ell n\ \frac{r_e}{r_w} - \frac{1}{2} + S\right]} \tag{5.41}$$

When converted to the pseudo gas-pressure form, Eq. 5.41 becomes:

$$q_g = \frac{0.703 k_{avg} h\ [m(p_e) - m(p_w)]}{T\left[\ell n\ \frac{r_e}{r_w} - \frac{1}{2} + S\right]} \tag{5.41a}$$

However, note that we cannot use the empirical turbulence constant n as an exponent of m(p) because the pseudo gas potential includes μ and z and because n loses its significance. Thus, the pseudo gas potential does not help us interpret a conventional stabilized drawdown or isochronal test. However, we should note that rather complex methods are available for taking conventional drawdown data and interpreting them in terms of a plot of $[m(p) - B'q_g^2]$ versus q_g. This plot does not change with a change in μ and z.*

There appears to be little justification for the analysis of conventional drawdown or isochronal data using the real gas potential. If it is known before the well tests are run that it will be necessary to use real gas potentials for the data analysis, a more direct approach is to run a constant-rate drawdown at different rates and evaluate the S and B parameters directly. If more statistical strength is desired, additional rates can be used.

*These methods may be found in the Canadian Energy Resources Conservation Board manual, *Theory and Practice of the Testing of Gas Wells* (Alberta: 1975). A similar technique that uses p^2 values and results in S and B values is in "Analysis of Modified Isochronal Tests to Predict the Stabilized Deliverability Potential of Gas Wells without Using Stabilized Flow Data," by Brar and Aziz, *JPT* (February 1978).

Turbulence and skin evaluations from continuous flow data. Previously, we discussed the approximation of isochronal data from conventional flow-after-flow (without shutin) type of gas-well test data. This analysis shows that the pressure change following a rate change is principally caused by the rate change itself and is normally affected very little by what happened previously in the well or the absolute value of the rate following the rate change. Using negative superposition, we can directly measure the effect on the well pressure of a particular rate change. That is, by extrapolating the pressure decline established prior to a particular rate change and subtracting the pressure resulting from the rate change, the effect of the rate change on the well pressure at that particular rate can be determined by difference.

Consequently, if turbulence is negligible, we can run a test at a variety of rates chosen in such a sequence that they provide various rate changes such as those in Fig. 5–16. The resulting pressure data can be interpreted with pressures extrapolated as in Fig. 5–17. The resulting differences between the extrapolated and measured pressure squared, when plotted against the change in the rate, provides an isochronal data plot as indicated in Fig. 5–18. However, carefully note that this is true only if the turbulence effect is negligible. When the turbulence effect is not negligible, the resulting data can be used to

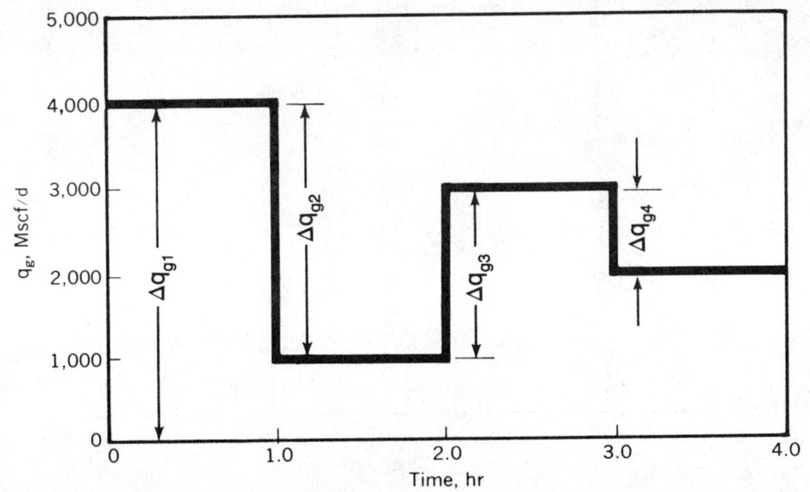

Fig. 5–16 Rate history for pressure history of Fig. 5–17 (after Slider, "Gas Well Testing Using Negative Superposition," paper 3875, © 1972, SPE-AIME)

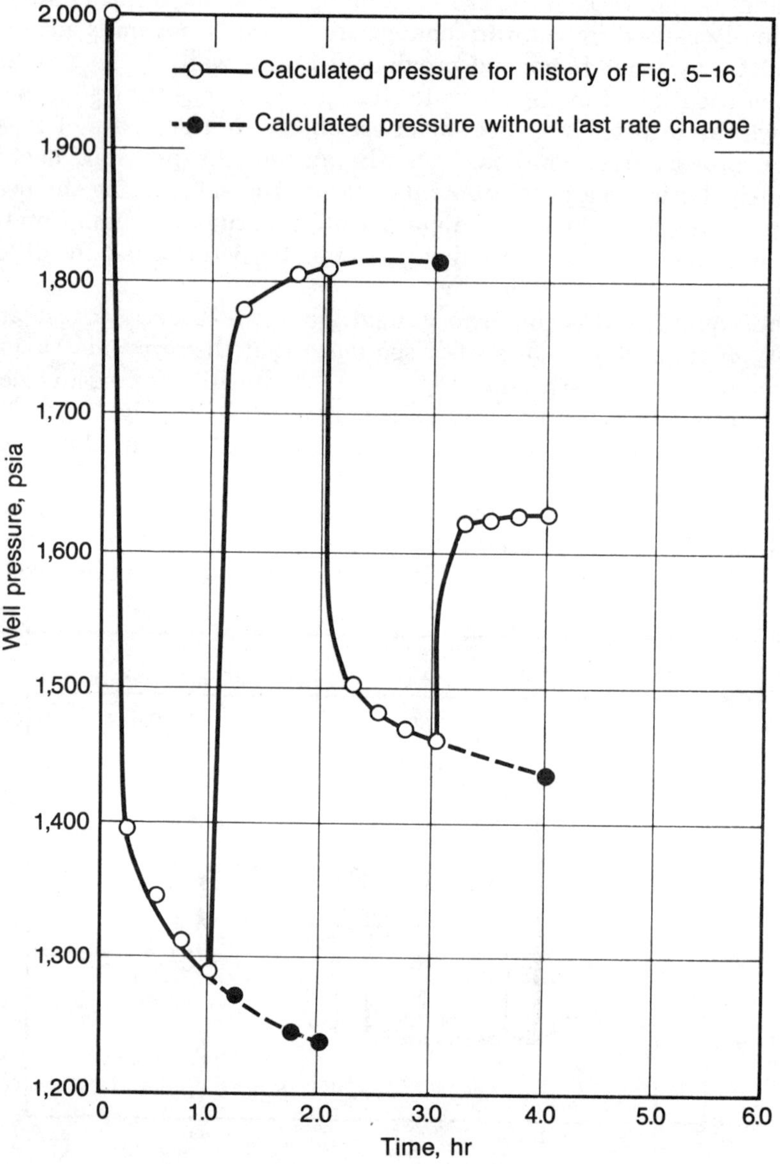

Fig. 5–17 Pressure history for rate history of Fig. 5–16 (after Slider, "Gas Well Testing Using Negative Superposition," paper 3875, © 1972, SPE-AIME)

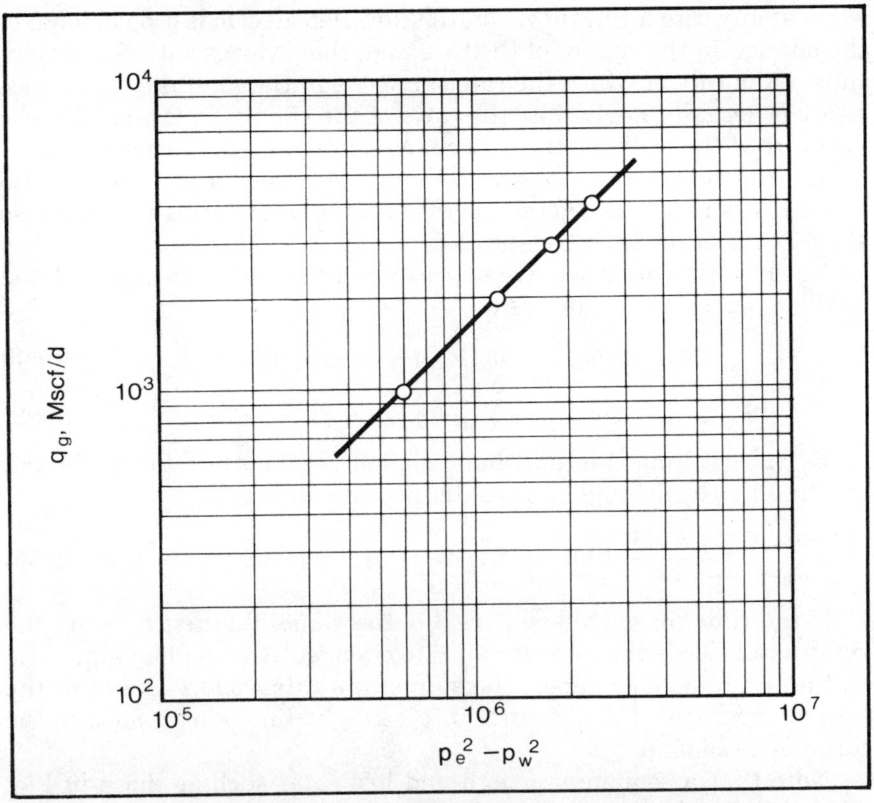

Fig. 5–18 One-hour isochronal data from negative superposition interpretation of data in Fig. 5–17 (after Slider, "Gas Well Testing Using Negative Superposition," paper 3875, © 1972, SPE-AIME)

evaluate the additional pressure drop caused by turbulence and the data can be corrected to a $\Delta(p^2)_{visc}$ plot. This plot can then be used to predict the well pressure at any production rate.

When turbulence is significant, a rate change at one rate level does not result in the same $\Delta(p_w{}^2)$ that would be caused by the same rate change at some other rate level. Eq. 5.69 indicates that the change in the square of the well pressure as a result of a single rate change is:

$$\Delta(p_w{}^2) = \frac{1.424\mu T_f z_{avg}}{kh} \Delta q_g[0.5(\ell n\ t_D + 0.809) + S] + B\Delta(q_g{}^2) \quad (5.87)$$

Eq. 5.87 can be converted to an equivalent m(p) equation according to Eq. 5.86:

$$\Delta m(p_w) = \frac{1.424T_f}{kh} \Delta q_g[0.5(\ell n\ t_D + 0.809) + S] + B'\Delta(q_g{}^2) \quad (5.87a)$$

The first term is proportional to the rate change and would be the same at any rate level. However, the turbulence term is proportional to the change in the square of the rate and, thus, varies with the magnitude of the rate at which the rate change is initiated. If the turbulence effect is actually negligible, the ratio of the change in the pseudo gas pressure at the well and the rate change, $m(p_w)/\Delta q_g$, are constant. Furthermore, we can see that when this ratio varies, a plot of the ratio versus the sum of the before and after rates gives a straight line whose slope is the turbulence constant, B.

To show this, first note that a rate change from q_{g1} to q_{g2} is related to the change in the squares as:

$$q_{g1}{}^2 - q_{g2}{}^2 = (q_{g1} - q_{g2})\,(q_{g1} + q_{g2}) \tag{5.88}$$

$$\Delta(q_g{}^2) = (\Delta q_g)\,(q_{g1} + q_{g2}) \tag{5.89}$$

By substituting the right-hand side of Eq. 5.89 into Eq. 5.87a and dividing by the change in the rate, Δq_g, we obtain:

$$\frac{\Delta m(p_w)}{\Delta q_g} = \frac{1.424T_f}{kh}\,[0.5\,(\ell n\ t_D + 0.809) + S] + B'(q_{g1} + q_{g2}) \tag{5.90}$$

If the time for each flow period is the same, the first term on the right-hand side is the same for all rate changes. Also, a plot of the ratio on the left-hand side versus the sum of the rates before and after the rate change, $q_{g1} + q_{g2}$, results in a straight line whose slope is the turbulence constant, B′.

Note that a sequence of high and low rates such as those in Fig. 5–16 does not lend itself to an accurate evaluation of the turbulence constant because the sum of the before and after rates varies very little when high and low rates are used in sequence. For example, $q_{g1} + q_{g2}$ for the four flow periods in Fig. 5–16 is 4, 5, 4, 5. Consequently, it is recommended that the sequence of rates be designed so a small rate change is attempted at low and high rates. To accomplish this and still obtain a variety of rate changes, we may use a sequence of rates such as that exhibited in Fig. 5–19. With approximately the same rate change of 1.0 used at a high (5–4) and low (0–1) level, the high and low test values can be substituted into Eq. 5.90. Then the simultaneous equations can be solved to obtain a direct calculation of the turbulence constant, B′:

$$B' = \frac{[\Delta m(p_w)/\Delta q_g]_{high} - [\Delta m(p_w)/\Delta q_g]_{low}}{(q_{g1} + q_{g2})_{high} - (q_{g1} + q_{g2})_{low}} \tag{5.91}$$

If the effect of the turbulence is significant, it is then necessary to correct each observed $\Delta m(p)$ value to a $\Delta(p^2)_{visc}$ value using Eq. 5.65, the evaluated constant B′, and Eq. 5.86. The engineer often finds that the turbulence effect is negligible. Thus, we can use the simplified

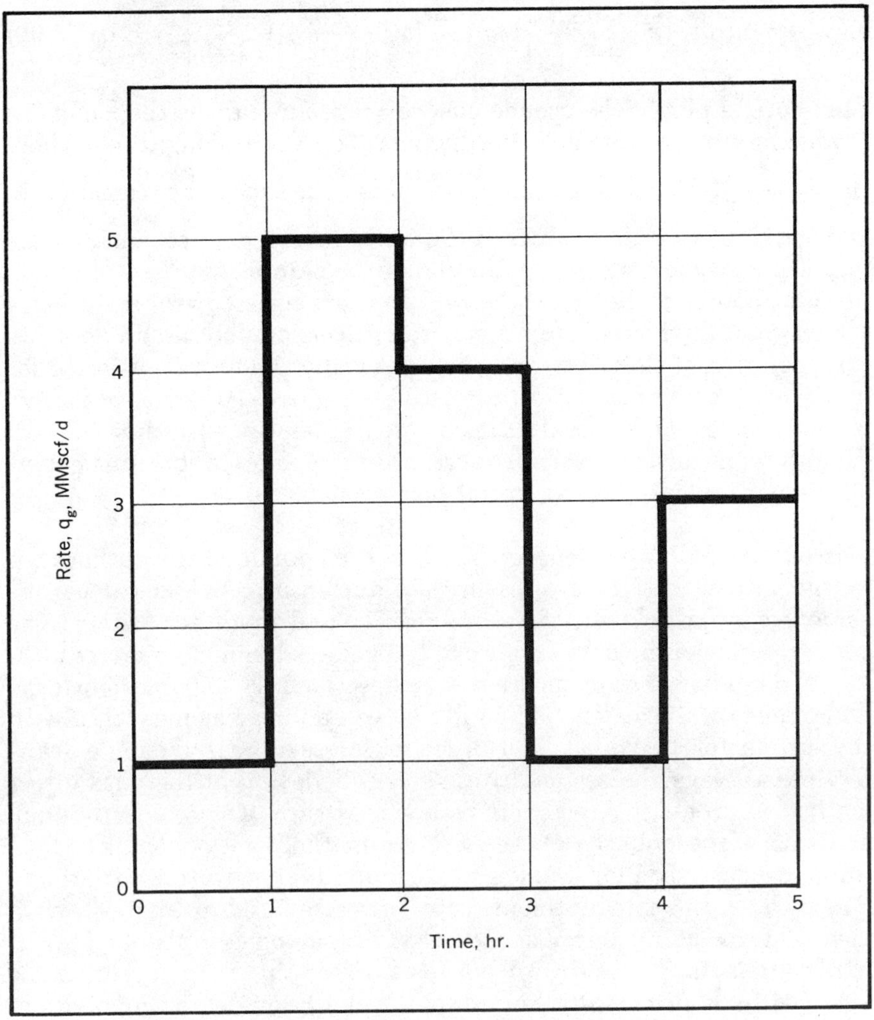

Fig. 5–19 Rate sequence for turbulence evaluation

analysis of flow-after-flow type of deliverability tests.[13] However, when turbulence effects are significant, the approach outlined can be used without much additional difficulty to predict the deliverability at any flow rate.

Once the turbulence constant, B′, has been evaluated, the skin factor, S, and the corresponding pressure drop caused by the well damage can be evaluated. Probably the simplest way to obtain the skin factor is to apply Eq. 5.53a to the initial drawdown data.

As long as the drawdown data is infinite acting, $\Delta p_D = 0.5\ (\ell n\ t_D + 0.809)$ and we can show that:

$$m(p_w) = m(p_i) - \frac{1.639q_gT}{kh}\left[\log t + \log\frac{6.33k}{\phi\mu c r_w^2} + \frac{S}{1.15}\right] - B'q_g^2 \quad (5.92)$$

Therefore, a plot of the pseudo gas-well pressure versus the log of the producing time at the initial producing rate gives a straight line whose slope is $\frac{1.639_gT}{kh}$. Substituting these values, the turbulence constant, B, and the corresponding values of the well pressure and producing time into Eq. 5.92 permits the evaluation of the skin factor, S.

Also, note that the plot of the well pressure squared versus the log of the original producing time is generally needed to obtain an accurate extrapolation of the pressure behavior during the first flow period for use in evaluating flow-after-flow data by negative superposition. During subsequent flow periods, the change in pressure with time is small enough to permit a reasonably accurate extrapolation, but this generally is not the case for the initial flow period.

Pressure buildup in gas wells. It is the opinion of the author that the general objectives of pressure buildup in a gas reservoir can be achieved by other methods more readily and more accurately. The deliverability of the a reservoir can be predicted in a more direct fashion through flow testing as previously described. The evaluation of the undamaged well capacity, kh, and the well damage as measured with the skin factor, S, can be determined with less difficulty using a drawdown test. Nevertheless, many engineers believe that the most direct method of predicting reservoir behavior is through pressure-buildup analysis. Consequently, those analysis methods most useful for gas wells are described for readers who are familiar with this situation.

Much of the enchantment with pressure-buildup analysis as a means of predicting performance in a gas reservoir is probably a result of dissatisfaction with drawdown tests. Since this disenchantment is believed to be principally caused by a lack of understanding of drawdown data interpretation, it is believed that the reader who has carefully used the preceding material will be convinced that pressure-buildup testing in a gas reservoir is an unnecessary evil. Nevertheless, the author believes that the methods that follow best serve the purposes of those engineers who prefer pressure-buildup analysis in gas reservoirs.

Essentially, the same equations discussed under pressure buildup in chapter 4 can be used for pressure buildup in gas reservoirs. In addition to the problems with turbulence, gas deviation factors, and variations in compressibility, pressure buildup in gas wells presents some additional problems that are more or less unique to this particular part of reservoir engineering technology. A constant reservoir rate prior to shutin is nearly nonexistent for a gas well, and the same applies to constant-rate drawdown. Also, the afterflow effects often make pressure-buildup data unusuable for extremely long periods following shutin. One plus factor for buildup in gas reservoirs also is noted. Due to the same high compressibility that causes many problems with gas flow analysis, reservoir disturbances are infinite acting for much longer periods than are similar disturbances in liquid-filled reservoirs. Therefore, the infinite-acting constant-rate solutions to the radial diffusivity equation that are so useful in transient-pressure analysis are effective for much longer periods when we are working with a gas reservoir. The time during which a disturbance in a gas reservoir is infinite acting can be best investigated quantitatively by examining the stabilization time equation:

$$t_s = 0.04\phi\mu c r_e^2/k \tag{3.50}$$

The compressibility may be 100 times greater than that of an oil reservoir. In addition, gas wells are characterized by wide spacing, which tends to make the external drainage radius very large and permeabilities extremely small. All of these factors tend to make the stabilization time for a gas reservoir very large. For completeness, we should note that the porosity and viscosity of gas reservoirs are characteristically low, which tends to reduce the stabilization time.

Since conditions change slowly in a gas reservoir, most pressure-buildup data for gas reservoirs—except that obtained during a drill-stem test—can be analyzed by the unchanging method described for oil and water reservoirs in chapter 4. The basic idea is that the flowing pressure at the time of shutin would not change significantly during the time of shutin if the well had not been shutin. Consequently, the pressure increase caused by the negative rate change of shutting in the well is superimposed on the flowing pressure at the time of shutin, p_{wf}. Thus, for the flow of liquid, the well pressure is:

$$p_w = p_{wf} + \frac{0.141q\mu}{kh}\left[(0.5)\,(\ell n\ \Delta t_D + 0.809) + S\right] \tag{4.26}$$

When Eq. 4.26 is applied to gas, we recognize that it is necessary to convert the gas flow rate normally stated in Mscfd to reservoir barrels per day at the well by using the following equation from chapter 2:

$$q = \frac{5.04 q_g z_w T_w}{(p_{wf} + p_e)/2} \tag{2.33a}$$

Note that a constant flow rate in Mscfd does not result in a constant reservoir flow rate at the wellbore because the pressure at the wellbore is changing constantly. Although this is interesting and important when the well pressure is changing rapidly at the time of shutin (such as during a drillstem test), reservoir conditions normally change so slowly that the well pressure and producing rate at shutin are nearly constant. Consequently, it is recommended that Eq. 2.33a be used to calculate the rate, q, which can then be used to make an unchanging pressure-buildup analysis as outlined in chapter 4.

This analysis can be obtained using Eq. 5.87 to describe the change in pressure from the flowing well pressure, p_{wf}, at shutin to some shutin pressure at a shutin time, Δt. Then an expression similar to Eq. 4.26 is obtained:

$$p_w^2 = p_{wf}^2 + \frac{1.639 q_g \mu z_{avg} T_f}{kh} \times \left[\log\Delta t + \log(\eta/r_w^2) + 0.809 + (S/1.15)\right] + Bq_g^2 \quad (5.93)$$

Mathematically, Eq. 5.93 indicates that a plot of the well pressure squared versus the log of the shutin gives a straight line whose slope is described by Eq. 5.78, for the slope of a drawdown curve. Note that all of the other terms, including the pressure drop caused by turbulence, remain constant.

It appears to be mathematically impossible for the well pressure squared versus the log of the shutin time and the well pressure versus the log of the shutin time to give a straight line. This is true since different assumptions have been made in deriving Eq. 4.26 and 5.93 and applying them to a gas well. Note that when we use Eq. 4.26 and 2.27a, we are assuming that the well pressure is constant. When we use Eq. 5.93, we are assuming that the average well pressure is $(p_w + p_{wf})/2$. Obviously, neither is correct.

Generally, the engineer finds that the well pressure plot gives the best straight line. However, for reservoirs with lower reservoir pressures, the engineer may find that the well pressure-squared plot is easier to interpret.

If neither pressure-buildup analysis approach is adequate, the engineer may wish to try a plot of $m(p_w)$ versus the log of the shutin time. By using Eq. 5.53a, $\Delta p_D = 0.5(t_D + 0.809)$ for infinite-acting behavior, and superposition with an unchanging pressure at shutin, we can show that:

$$m(p_w) = m(p_{wf}) + \frac{1.639 q_g T}{kh} \times \left[\log \Delta t + \log(n/r_w^2) + 0.809 + \frac{S}{1.15}\right] + B' q_g^2 \quad (5.94)$$

According to Eq. 5.94, a plot of $m(p_w)$ versus log Δt gives a straight line whose slope is $1.693 q_g T/kh$. By running two buildups preceded by different rates, B′ and S can be determined. Some companies reportedly prepare a plot of the viscosity and gas deviation factor product versus pressure and apply pressure versus the log of time applications for the lesser pressures when the product is constant. Pressure-squared applications are used when the plot becomes a straight line at larger pressures, and m(p) is applied at the inbetween times when the product is not constant and the plot is not a straight line. All three of the suggested pressure-buildup approaches are difficult to apply to most gas-well pressure buildups because of afterflow. Afterflow is extremely severe in gas wells because of the high compressibility of the gas. As the pressure in the wellbore increases following shutin, the gas is compressed and more gas flows into the wellbore to satisfy the reduced volume. Serious afterflow in gas wells may last for days and may effectively prohibit the analysis of the pressure buildup by conventional methods. The greater the gas reservoir pressure is and the lower the permeability is, the more severe the afterflow problem becomes.

If an engineer is determined to evaluate a gas reservoir using pressure buildup, it is strongly recommended that the data be obtained by the two-rate method rather than by conventional shutin. By simply reducing the flow rate and observing the increase in the well pressure caused by the negative rate change, an engineer can avoid much of the afterflow difficulty. Generally, a new rate of production can be established in a relatively short period of time.

The analysis of a two-rate test is exactly the same as the analysis of a shutin buildup, except the rate used in all of the equations is the rate change rather than the rate at the time of shutin. The engineer is faced with the same choice of methods as described, but he can usually avoid the difficulty of extremely long periods of afterflow.

Several papers have been presented on the effect of the reservoir pressure on the permeability. We know that a change in reservoir pressure causes a small change in the pore volume of a reservoir. Consequently, it should not be surprising that the small change in the pore volume or porosity can have an effect on the permeability of a formation.[14] The effect is only important for gas reservoirs since economic gas producing rates can be obtained from tight reservoirs because of the low viscosity of gas. Consequently, it may be necessary to use different pressure-buildup methods when tests are run in tight reservoirs.

However, it appears that the turbulence factor, n, of the pressure deliverability equations may include this variation in permeability with pressure. Vairog and Rhoades show that there tends to be a straight-line relationship between the log of $\Delta(p^2)$ and the log of the permeability to gas.[15] It appears that such a relationship may be automatically included in the n factor. Then the outlined methods of predicting deliverability based on flow tests using the deliverability equation automatically include the prediction of the variation of permeability with the reservoir pressure.

Equipment Capacity Limitations on Deliverability

Many reservoir engineers think only in terms of the capacity of the reservoir to produce when they are predicting the deliverability of a well. We should never forget that production must also pass through the tubing, separators, dehydrators, meter run, and flow line to the pipeline. Some pressure drop is associated with each one of these pieces of equipment, and the pressure drop is a function of the flow rate. Consequently, we find that in many cases the production rate is limited by the capacity of the equipment rather than the capacity of the reservoir to produce. When such a situation arises, it may be possible to install larger-diameter equipment. However, the point is that the ability of a well to produce is a function of both the reservoir capacity and the equipment capacity.

This situation can be shown graphically by plotting the bottom-hole pressure versus the rate of flow for the reservoir, assuming that no pressure drop occurs in the equipment. Then assume that we can remove all of the flow equipment from the well and run flow tests so that for some given pipeline, or downstream, pressure we can obtain a plot of the flow rate versus the entrance pressure into the system. This plot is the bottom-hole pressure curves on the same axes as illustrated in Fig. 5–20.

The reservoir capacity curve represents a particular state of depletion or external reservoir pressure, p_e. The equipment capacity curve represents a particular equipment setup and pipeline pressure. Under these conditions note that, as the bottom-hole pressure increases, the rate of flow on the basis of the reservoir capacity decreases. However, an increase in the bottom-hole pressure actually results in an increase in the flow rate through the equipment. Consequently, at relatively low rates the flow rate of a well may be limited by the capacity of the reservoir, and at relatively high rates the flow rate of a well may be limited by the capacity of the flow equipment. In the latter case we say that the reservoir can produce at a rate exceeding the capacity of the equipment.

Note that for a particular set of equipment, pipeline pressure, and state of reservoir depletion, there is some maximum rate that can be produced represented by the intersection of the two capacity curves. At this point the reservoir flow results in a bottom-hole pressure that matches the pressure drop needed for flow through the production equipment at this rate. At any other rate the capacity of the well to produce is limited by either the reservoir or the equipment capacity. If a reservoir engineer is to provide accurate predictions of deliverability under various conditions, he must consider the capacity of the equipment.

Since equipment capacity generally falls into the responsibilities of a mechanical or production engineer, we must often rely on them for the actual equipment capacity curve used to base our predictions. How-

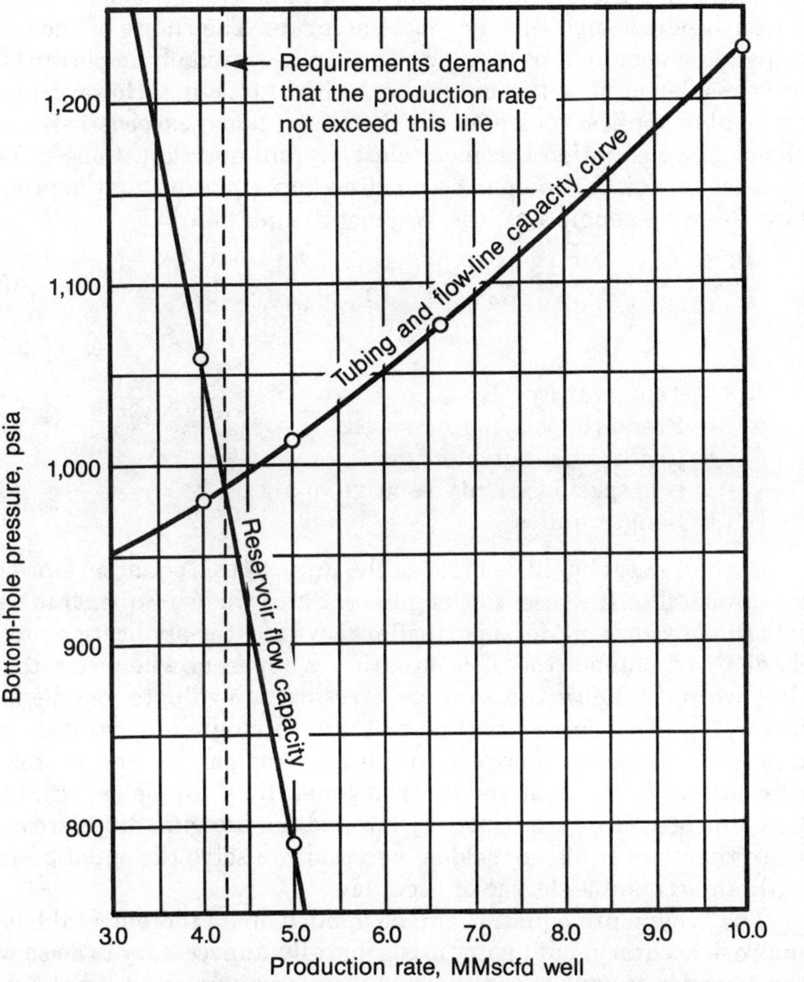

Fig. 5–20 Example relationship between reservoir and equipment capacity

ever, we must frequently generate our own curves, so it is necessary to discuss some of the techniques of obtaining them. Equipment suppliers can generally provide the capacity of separators, dehydrators, and other pieces of equipment; therefore, we have limited our discussion to methods of predicting the capacity of the tubular goods involved in the producing equipment.

Flow-line capacity. Many different equations have been proposed for predicting the pressure drop in horizontal pipes at various flow rates. Each of these equations has a particular range of operating conditions under which they are most accurate. The choice of such equations to be used in a particular situation is extremely important if we are considering flow through many miles of pipeline. However, in our particular application where the flow line seldom exceeds 1–2 mi, the choice of an equation becomes relatively unimportant. Consequently, we base our discussion on the pipeline flow equation that is probably best known to engineers, the Weymouth equation.[16]

$$q_{scf/hr} = \frac{18.062T_{sc}}{p_{sc}}\left[\frac{(p_1^2 - p_2^2)d_{in.}^{16/3}}{\gamma T_{avg}L_{mi}z_{avg}}\right]^{(0.5)} \tag{5.95}$$

Where:

T = Temperature, °R
p = Pressure, psia
$d_{in.}$ = Inside pipe diameter, in.
γ = Gas specific gravity relative to air
L_{mi} = Length, miles

Careful note should be made of the units in this equation since they are unusual to the reservoir engineer. Since we use Eq. 5.95 to calculate the pressure, p_1, for a particular flow rate, the application must be made by trial and error. This situation is apparent when we recognize that we must know the average pressure to evaluate gas deviation factor, z_{avg}. This same situation exists in solving many gas flow situations for a pressure. However, in this case and particularly in this specific situation, the trial and error is generally a simple one. Since gas flow-line pressures are relatively low and the pressure drops are small, more than two trials are seldom necessary to solve the equation for p_1 with an acceptable degree of accuracy.

The Weymouth equation can be applied on an incremental basis to improve accuracy, but again, it is generally unnecessary to do so when the equation is applied to the typical low-pressure, short flow line. The general form of the Weymouth equation can be derived analytically, but the friction factor included in the equation must be included in an empirical form.[17]

Many nomographic solutions of the Weymouth equation using a variety of units and assumptions may be found. Fig. 5–21 is an example of one such nomograph. Empirical gas deviation factors have been used with fixed standard conditions and gauge pressures and an average flowing temperature of 60°F. When nomographic solutions are used, the assumptions should be carefully noted to determine the effect on the accuracy in the application.

Tubing or casing capacity. The problem of predicting the pressure drop through a vertical flow line, such as tubing or casing, is a difficult one. This can be seen qualitatively when the engineer realizes that Eq. 5.95 and a static pressure-difference equation, such as Eq. 5.31, must be satisfied simultaneously under these conditions. Many different efforts have been made to obtain such equations. As an example, consider the equation developed by R.V. Smith:[18]

$$q_g = \frac{200d^5(p_2{}^2 - e^s p_1{}^2)s}{T_{avg}z_{avg}f\Delta X(e^s - 1)} \tag{5.96}$$

Where:

$$s = \frac{0.0375\ \Delta X\gamma}{T_{avg}z_{avg}} \tag{5.97}$$

In addition to the complexity of these equations, note that it is necessary to evaluate the friction factor, f, independently which is a function of the flow rate, the depth of the well, the gas gravity, and the gas viscosity. Even with such a complex set of equations, the solution is still made by trial and error since z_{avg} is a function of the pressure, which is unknown in our application. If an equation such as this is used through a computer program, the engineer should make certain that the results obtained are acceptable for the particular range of pressures, temperatures, depths, and gas gravities where it is applied. From a practical point of view, such reliability can be determined only on a reservoir-to-reservoir basis. In other words the calculations should be compared with measured pressures for a particular reservoir before they are relied on in that reservoir.

If an engineer must make predictions of flowing bottom-hole pressures without the benefit of a computer program, the data in *Bulletin 72* of the Texas Petroleum Research Committee are recommended. However, the same restriction is suggested for these data as for the computer programs.

These data are recommended previously in this chapter as a means of determining the static bottom-hole pressure. In the case of the static pressure data, each basic figure from *Bulletin 72* represents a particular gas gravity and temperature gradient. The figures representing the

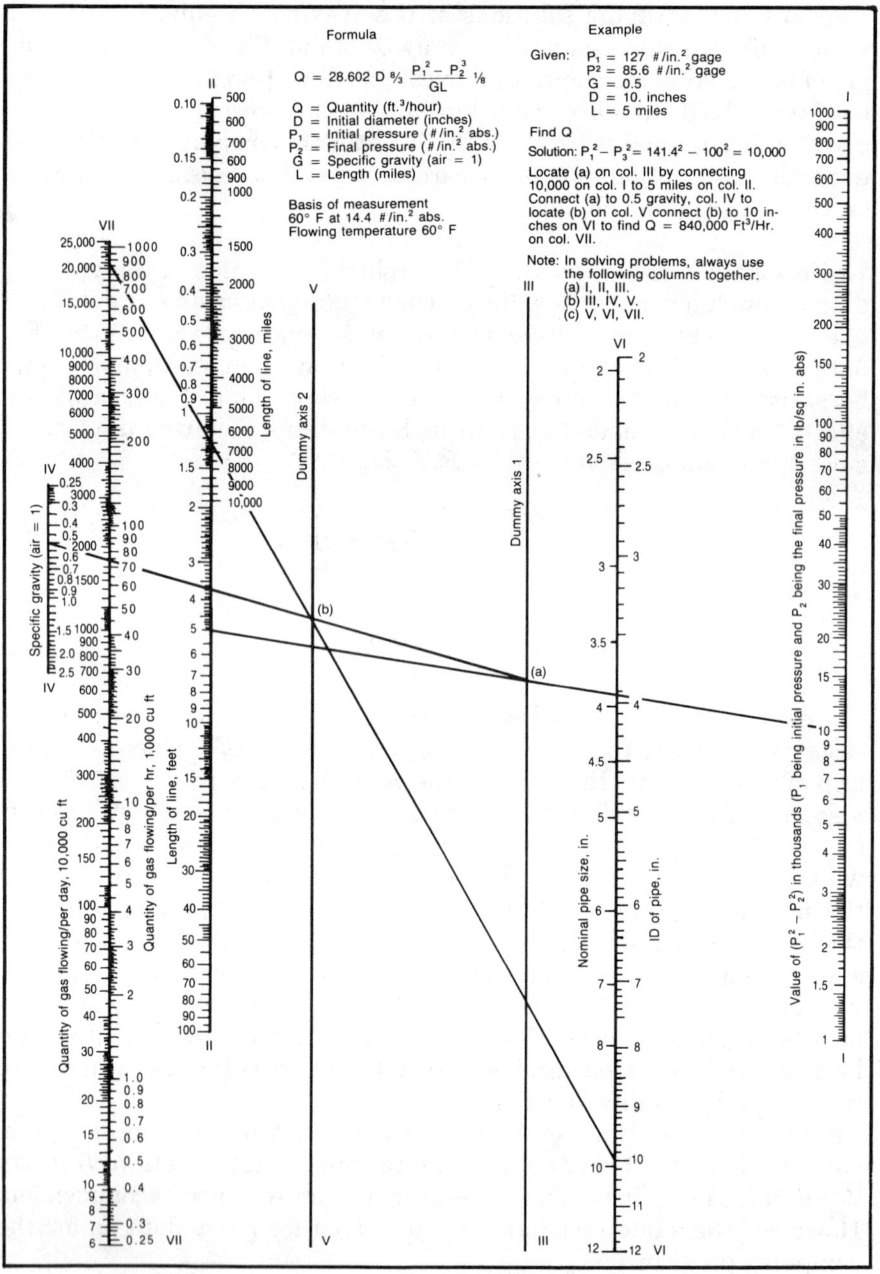

Fig. 5–21 Alignment chart for solution of Weymouth's formula of gas flow in small high-pressure lines (*Engineering Data Book*, courtesy Natural Gasoline Supply Men's Association, April 1951)

flowing bottom-hole pressures have the additional parameter of flow rate fixed for each figure. Also, all of the flowing pressure drops represent the flowing pressure drop through 2½-in. tubing. *Bulletin 72* monographs for flowing wells are illustrated in Figs. 5–22 through 5–25. These figures simply represent examples of the *Bulletin 72* data. Any practical application would require the entire report.

The application technique to determine the pressure drop in 2½-in. tubing is exactly the same as the technique used to determine the pressure difference in a static column of gas, except an interpolation must be made for three parameters instead of two. Interpolations are necessary for the gas gravity, the temperature gradient, and the flow rate. In addition, if the flowing pressure drop is required in some other tubing or casing, it is necessary to determine the equivalent flow rate for the 2½-in. tubing before making the calculations. In other words it is nec-

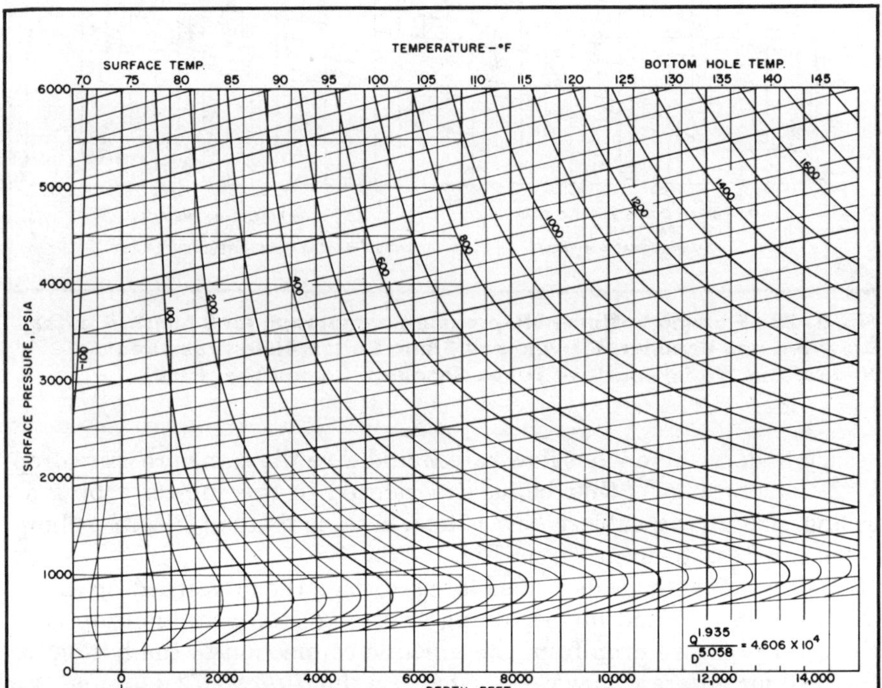

Fig. 5–22 Flowing bottom-hole-pressure nomograph for 4 MMscfd (after Crawford and Fancher, *Flowing and Static Bottom-Hole Pressures of Natural Gas Wells*, courtesy, Texas Research Committee, 1959)

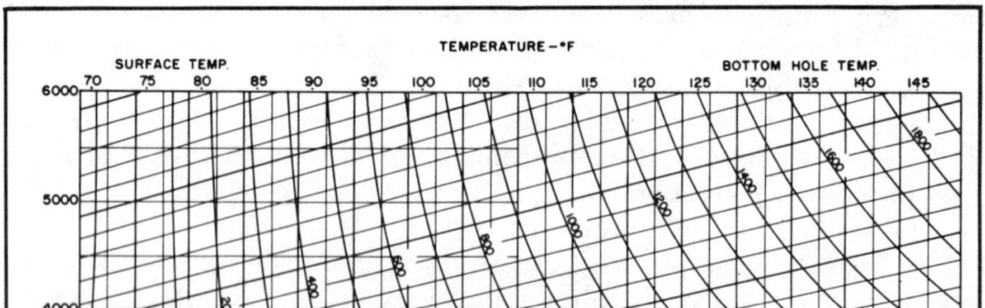

Fig. 5–23 Flowing bottom-hole-pressure nomograph for 8 MMscfd (after Crawford and Fancher, *Flowing and Static Bottom-Hole Pressures of Natural Gas Wells*, courtesy Texas Research Committee, 1959)

essary to determine what flow rate would give the same pressure drop in 2½-in. tubing as that being experienced in the subject tubing or casing size at the subject flow rate. These equivalent 2½-in. tubing rates can be determined from Fig. 5–26.

Using the methods described for predicting the pressure drop in the surface flow line and in the tubing or casing, it is possible to predict the total pressure drop from the pipeline connection to the bottom of the well for different flow rates. Now, use the *Bulletin 72* data and the Weymouth equation to solve problem 5.12 and check the solution against the one in appendix C.

PROBLEM 5.12: Pressure Drops in the Producing System

The average well in the gas reservoir of problems 5.4 and 5.8 is equipped with 3,700 ft of 8-in. tubing (ID = 2.992 in.) and a 1.0-mile, 3.058-in. ID flow line to the pipeline. The gas temperature at the pipeline is 70°F.; at the wellhead, 90°F; and at

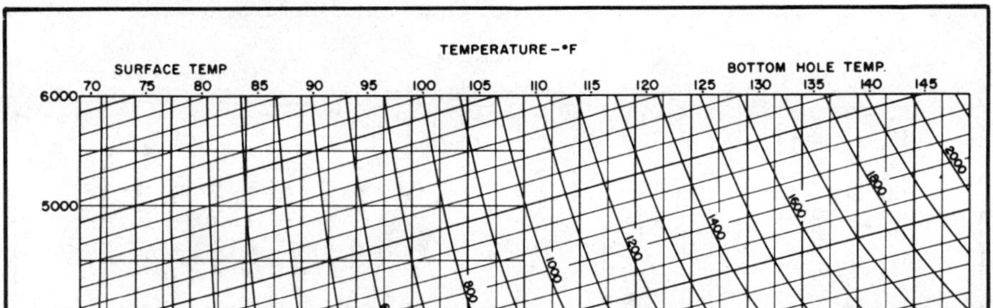

Fig. 5–24 Flowing bottom-hole-pressure nomograph for 12 MMscfd (after Crawford and Fancher, *Flowing and Static Bottom-Hole Pressures of Natural Gas Wells*, courtesy Texas Research Committee, 1959)

the reservoir, 110°F. Find the minimum bottom-hole pressure necessary to provide a pipeline delivery of 4 MMcfd, 5 MMcfd, 6.67 MMcfd, and 10 MMcfd at 800 psia.

To simplify interpolations in calculating the pressure drop in the tubing, use the temperature gradient as 5°/1,000 ft and the gas gravity as 0.7. Also, note that the static pressure data represents the bottom-hole pressure for a rate of 0.0 (Figs. 5–5 through 5–8).

Predicting Reservoir Performance

To predict the production history of a reservoir, it is necessary to consider the capacity of the reservoir to produce, the capacity of the equipment, and the state of depletion of the reservoir as predicted by material balance. Since the average reservoir pressure is a function of

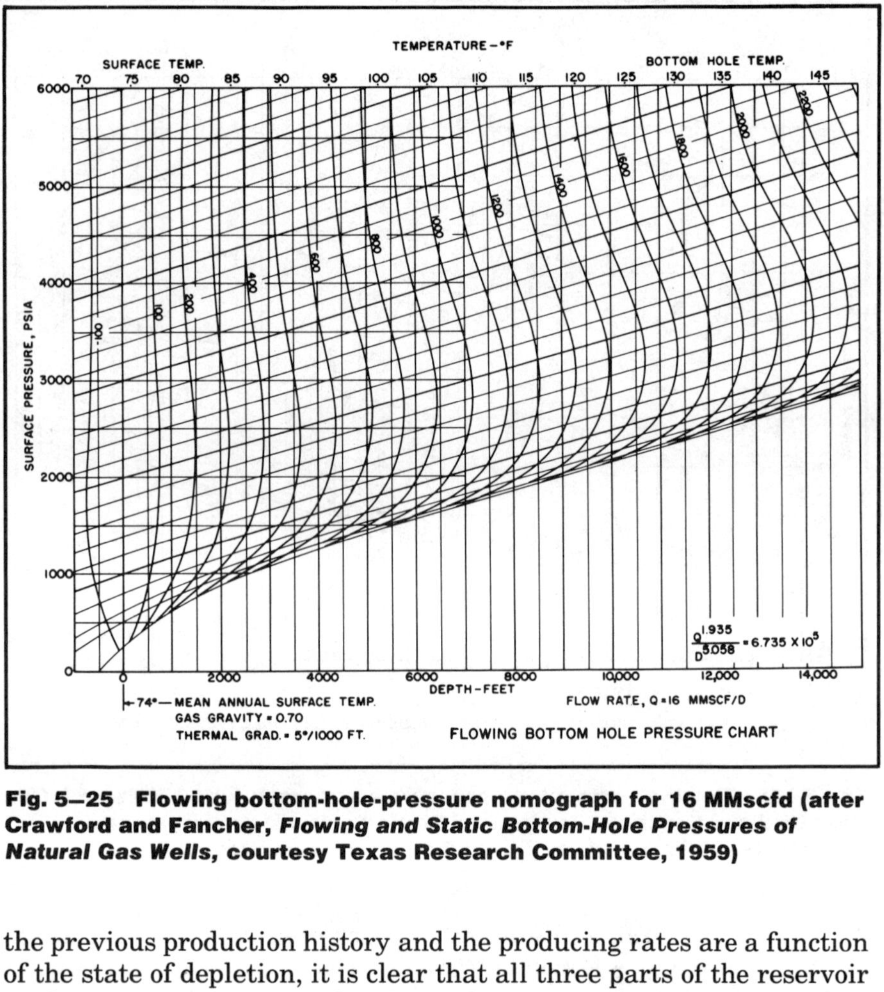

Fig. 5–25 Flowing bottom-hole-pressure nomograph for 16 MMscfd (after Crawford and Fancher, *Flowing and Static Bottom-Hole Pressures of Natural Gas Wells*, courtesy Texas Research Committee, 1959)

the previous production history and the producing rates are a function of the state of depletion, it is clear that all three parts of the reservoir prediction must be interrelated to achieve a prediction for the reservoir.

Using various testing procedures, we learned that we can predict the capacity of a reservoir to produce under various conditions. This reservoir capacity can be represented as a plot of the bottom-hole pressure versus the producing rate for a particular state of depletion, p_e. Such a curve is considered in Fig. 5–20. This figure also shows a similar capacity curve for the producing system. As the bottom-hole pressure declines, the flow rate through the reservoir increases since a drop in bottom-hole pressure means an increase in the pressure drop in the reservoir. However, as the bottom-hole pressure declines, the flow-rate capacity through the tubing/flow-line producing system declines because, with a fixed pressure at the pipeline, a decline in the bottom-hole

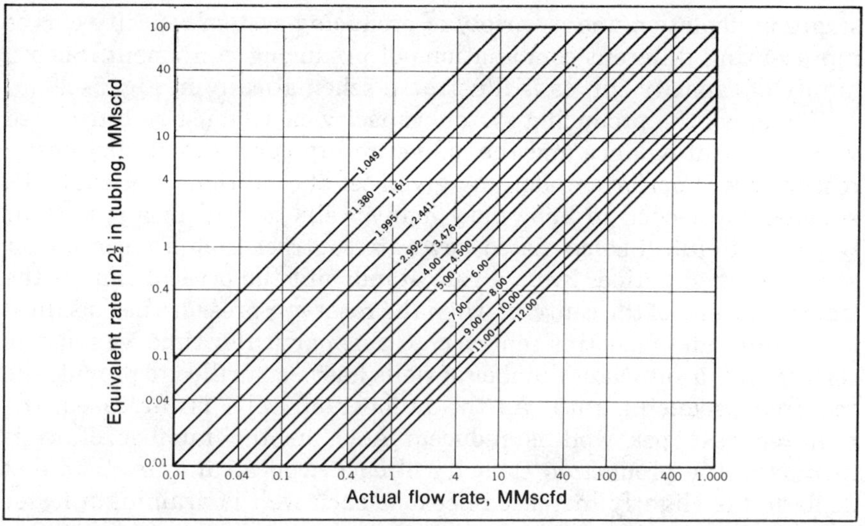

Fig. 5–26 Equivalent rate chart (after Crawford and Fancher, *Flowing and Static Bottom-Hole Pressures of Natural Gas Wells*, courtesy Texas Research Committee, 1959)

pressure means that the pressure drop in the producing system declines.

When the capacity of the reservoir and the producing system are considered together, it is easily seen that the producing capacity of a well may be limited by the capacity of the producing system rather than by the capacity of the reservoir. For example, using the data in Fig. 5–20, consider what would happen if we attempted to flow the well at a bottom-hole pressure of 787 psia. Note that the reservoir alone without any back pressure or resistance from the producing system would produce at a rate of 5.0 MMscfd. However, this flow rate through the producing system requires a bottom-hole pressure of 1,015 psia. Consequently, with the particular tubing/flow-line producing system represented by the capacity curve of Fig. 5–20, the well could not produce at this rate. In fact, the well could not produce at any rate greater than 4.25 MMscfd. This is the rate represented by the intersection point of the two capacity curves. At this point the bottom-hole pressure of 988 psia provides a flow rate through the reservoir of 4.25 MMscfd and is sufficiently high to provide the same flow rate through the producing system.

The engineer should bear in mind that Fig. 5–20 represents only one stage of depletion and one combination of tubing, separator, dehydrator, flow line, and pipeline pressure. Such a plot can be made much more general by adding other reservoir capacity curves for different stages of depletion and a variety of producing-system capacity curves representing different combinations of producing equipment. Such a family of capacity curves is illustrated schematically in Fig. 5–27.

As noted, the producing-system capacity curve must be considered with material balance and the deliverability curves to predict how a reservoir will perform under any given set of conditions. For example, suppose we needed to know how many wells to drill in a particular reservoir to fulfill some stated flow-rate contract from a reservoir for some specified period. From a rate standpoint the critical time is the time at the end of the contract when the reservoir pressure has declined to a minimum under this contract. At this particular time we must be certain that a sufficient number of wells has been drilled to provide the required producing rate. As the number of wells is increased, the required rate per well is reduced. Also, as the number of wells increases, the production capacity of each well at any fixed state of depletion is slightly increased because each well is draining a lesser volume of the reservoir.

This situation where the individual well capacity is increased with each additional well drilled seems to violate the intuition of many engineers. These engineers feel that the individual well capacity should be reduced because the drainage area of each well is reduced. It is true

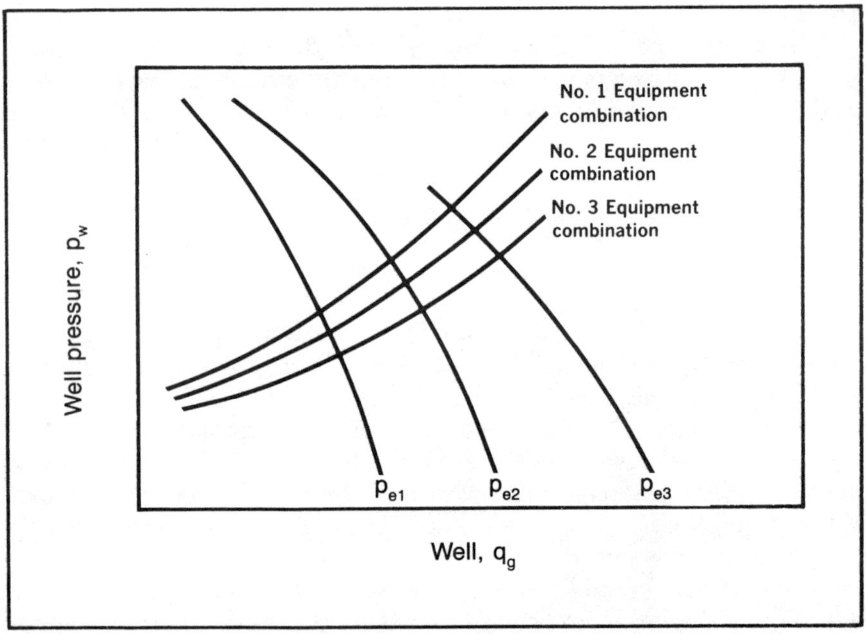

Fig. 5–27 Family of reservoir equipment capacity curves

that the ultimate recovery per well is reduced, but the lesser drainage area reduces the effective r_e/r_w and, thus, increases the producing rate of the well for any particular $p_e^2 - p_w^2$.

The hypothetical problem must be solved by trial and error. With the state of depletion fixed by the contract length and total reservoir rate, we can determine the average pressure or the pressure at the external drainage boundary of each well. If we assume a number of wells, we can use the basic deliverability curve and can determine the deliverability curve for the resulting spacing. With the number of wells assumed, we in effect have assumed the rate per well since the contract fixes the total reservoir rate. Based on the per-well rate, we can determine the bottom-hole pressure necessary to supply that rate from the deliverability curve. Then this bottom-hole pressure can be used together with the equipment capacity curve to determine if the equipment capacity is sufficient to supply the required per-well rate at the subject bottom-hole pressure. If the equipment cannot supply the rate, more wells are considered. The number of wells must be bracketed before the engineer can be certain that he has determined the most economical solution to this problem.

It is suggested that the reader work problem 5.13 and check the solution against the one in appendix C.

PROBLEM 5.13: Determining Gas-Well Spacing for Completing a Pipeline Purchasing Contract

The gas reservoir in problems 5.4 and 5.8 must produce 20 MMcfd for the next 5 years (from November 1, 1981) to meet contract requirements. If the physical equipment on each well is as described in problem 5.12 and has the capacity indicated in Fig. 5–28, how many equally spaced wells are needed to meet this contract? Assume the p_e at the contract completion is 1,568 psia as determined in problem 5.4. Also use the stabilized performance curves in Fig. 5–29.

Since the calculations of this problem involve a trial-and-error procedure with a considerable amount of calculation in each trial, it is advantageous to use a digital computer. A program is included in Tables 5–5 and 5–6 with symbols defined in Table 5–4. To simplify the calculations, this program uses the Smith equations for calculating the pressure drop in the tubing.[17] This avoids the need to store data and interpolate it in the program. The program simply calculates the bottom-hole pressure resulting from the flow of fluid at a particular rate and drainage volume and compares that with the bottom-hole pressure necessary to provide the same flow rate through the equipment.

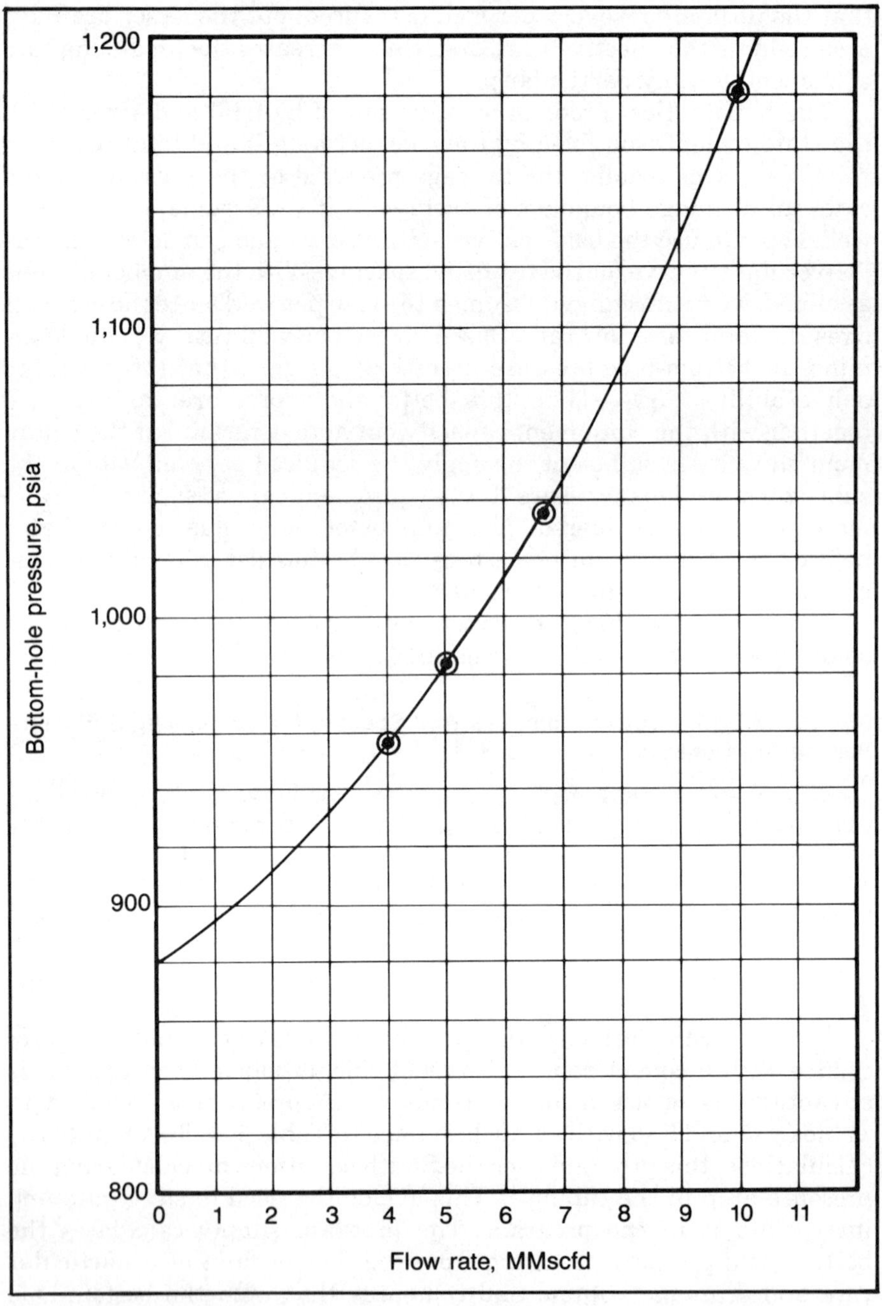

Fig. 5–28 Tubing and flow-line capacity data for problem 5.13 and 5.14.

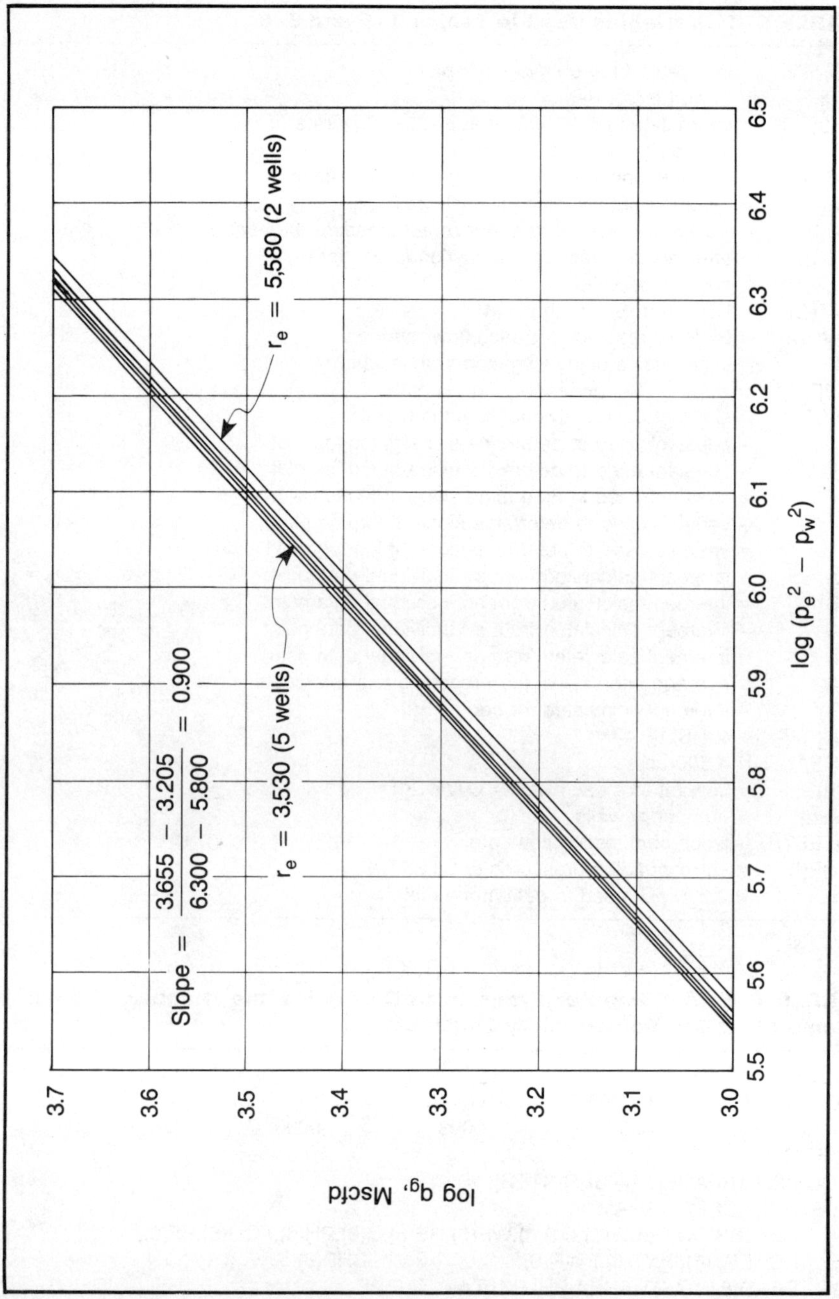

Fig. 5–29 Stabilized deliverability curves for different well spacings (results of problem 5.9, data for problems 5.13 and 5.14)

TABLE 5–4 Variables Used in Tables 5–5 and 5–6

B(I)	= intercept of log q vs log ΔP^2 plot
F(I)	= friction factor data
FF(I)	= interpolated friction factor at specific flow rate
QT	= total daily flow rate = MMcfd
QG(I)	= effective flow rate for calculating friction factor
RE(I)	= reservoir outer radius for each well
PH(I)	= wellhead pressure from horizontal pressure drop equation
FQG(I)	= effective flow rate data in friction factor data
I	= number of wells
PWT(I)	= BHP from Smith equation
PWR(I)	= BHP for reservoir for each flow rate
REI	= initial radius of investigation from isochronal testing
SLIP	= slope of line (average); log-log plot
SLOP	= slope of each individual line; log-log plot
K	= constant used to determine slope of log-log plot
L	= constant used to determine slope of log-log plot
N	= constant used to determine slope of log-log plot
X	= variable used to determine slope of log-log plot
Y	= variable used to determine slope of log-log plot
BA	= intercept calculated from an individual data point
BB	= intercept calculated from an individual data point
BC	= intercept calculated from an individual data point
BD	= intercept calculated from an individual data point
BO	= average intercept of the preceding four intercepts
B(I)	= intercept calculated for each RE(I)
EX	= 1.1 SLIP
AREA	= 4,500 acres
Q(I)	= flow rate for each well = QT/WE(I)
WE(I)	= number of wells
QTEST(I)	= isochronal testing flow rate
BHP(I)	= corresponding pressures to QTEST(I)
PE	= 2,800 psi, used to get different intercepts

TABLE 5–5 A Computer Program to Determine the Number of Wells to Complete a Gas Deliverability Contract*

```
// JOB T
LOG DRIVE    CART SPEC    CART AVAIL    PHY DRIVE
  0000         0003         0003          0000
// FOR
*IOCS(2501READER,1403PRINTER)
*LISTSOURCEPROGRAM
      DIMENSION B(6),F(6),QG(6),WE(6),RE(6),Q(6),PH(6),FQG(6),FF(6),
      CQTEST(6),PWT(6),PWR(6)                  ,BHP(6)
      READ(8,2) Z,QT,AREA,PHI,PEI,TAW,TAP,PE
      READ(8,2) D,POUT,DIST,U,G,WD,RW,PERM
      WRITE(5,8)
    8 FORMAT(20X,17THE INPUT DATA IS,)
      DO3 J=1,5
```

TABLE 5–5 *continued*

```
      READ(8,4) QG(J),F(J),QTEST(J),BHP(J),WE(J)
      WRITE(5,9)QG(J),F(J),QTEST(J),BHP(J),WE(J)
    9 FORMAT(5F20.5)
    4 FORMAT(5F10.5)
    3 CONTINUE
    2 FORMAT(8F10.5)
      REI=SQRT(38.5*PEI*PERM/(PHI*U*24.))
      RE(1)=SQRT(AREA*43560./(WE(1)* 3.14))
      SLIP=0.
      DO10 L=1,3
      K=L-1
      DO10 I=2,4
      IF(K-1)16,17,18
   17 IF(I-3)16,16,10
   18 IF(I-2)16,16,10
   16 N=I+1+K
      X=ALOG(PEI**2-BHP(          N)**2)-ALOG(PEI**2-BHP(I)**2)
      Y=ALOG(QTEST(               N))-ALOG(QTEST(I))
      SLOP=Y/X
      SLIP=SLIP+SLOP
   10 CONTINUE
      SLIP=SLIP/6.
      WRITE(5,21)SLIP
   21 FORMAT(//11H THE SLOPE=,F7.5)
      BA=QTEST(2)*1,000./((PEI**2-BHP(2)**2)**SLIP)
      BB=QTEST(3)*1,000./((PEI**2-BHP(3)**2)**SLIP)
      BC=QTEST(4)*1,000./((PEI**2-BHP(4)**2)**SLIP)
      BD=QTEST(5)*1,000./((PEI**2-BHP(5)**2)**SLIP)
      BO=(BA+BB+BC+BD)/4.
      B(1)=BO*ALOG(.606*REI/RW)/ALOG(.606*RE(1)/RW)
      EX=1./SLIP
      WRITE(5,6)
    6 FORMAT(//9X,12HFRICT. FCTR.,12X,10HRES. PRESS,7X,18HWELL
      HEAD PRESCSURE,5X, 9HEQUIP.BHP,3X,9HNO. WELLS,//)
      DO100 I=1,10
    C CALCULATING PW FOR RESERVOIR
      RE(I)=SQRT(AREA*43560./(WE(I)* 3.14))
      B(I)=B(1)*ALOG(.606*RE(1)/RW/ALOG(.606*RE(I)/RW)
      Q(I)=QT/WE(I)
      PM=(Q(I)/B(I))**EX
      PEE=PE**2
      IF(PEE-PM)100,100,61
   61 PWR(I)=SQRT(PEE-PM)
      CALCULATING P DROP FOR HORIZONTAL PIPE
      QH=QT/24.
      QH=(QT*14.7/(WE(I)*24.*18.062*520.))**2

      H=G*TAP*DIST*Z/D**5.3333
      PH(I)=SQRT(QH*H+POUT**2)
      CALCULATING P DROP FOR VERTICAL PIPE
      FQG(I)=QT*G/(WE(I)*U*1000.)
```

TABLE 5–5 continued

```
      IF(FQG(I)-QG(1))31,31,32
   32 IF(FQG(I)-QG(2))33,34,35
   35 IF(FQG(I)-QG(3))36,37,38
   38 IF(FQG(I)-QG(4))39,40,41
   41 IF(FQG(I)-QG(5))42,43,43
   31 FF(I)=F(1)
      GOTO44
   34 FF(I)=F(2)
      GOTO44
   37 FF(I)=F(3)
      GOTO44
   40 FF(I)=F(4)
      GOTO44
   43 FF(I)=F(5)
      GOTO44
   33 FF(I)=F(2)+(F(1)-F(2))*(QG(2)-FQG(I))/(QG(2)-QG(1))
      GOTO44
   36 FF(I)=F(3)+(F(2)-F(3))*(QG(3)-FQG(I))/(QG(3)-QG(2))
      GOTO44
   39 FF(I)=F(4)+(F(3)-F(4))*(QG(4)-FQG(I))/(QG(4)-QG(3))
      GOTO44
   42 FF(I)=F(5)+(F(4)-F(5))*(QG(5)-FQG(I))/(QG(5)-QG(4))
   44 S=.0375*G*WD/(TAW*Z)
      PWT(I)=SQRT((QT/(WE(I)*200000.))**2*G*TAW*Z*FF(I)*WD*(2.72**S-1)
      C/(D**5*S)+2.72**S *PH(I)**2)
      WRITE(5,12) FF(I),PWR(I),PH(I),PWT(I),I
   12 FORMAT(4F20.5,19)
      IF(PWT(I)-PWR(I))51,51,100
  100 CONTINUE
   51 WRITE(5,7) I
    7 FORMAT(//30HTHE OPTIMUM NUMBER OF WELLS IS,13)
      STOP
      END

FEATURES SUPPORTED
ICCS

ORE REQUIREMENTS FOR
COMMON     0 VARIABLES     248 PROGRAM 1334

END OF COMPILATION

/ XEQ
```

*This table is presented as a basis for our discussion. It is not recommended for use in this form.

Tables 5–5 and 5–6 represent only one type of capacity problem. The more general problem today is probably concerned with the situation where a pipeline company prepares a contract for purchase of gas from a reservoir for some particular rate schedule that is not necessarily a constant rate. Then the engineer must determine the rate sched-

TABLE 5–6 Example Data

THE INPUT IS				
50000.00790	0.01630	0.00000	2800.00049	1.00000
100000.01580	0.01510	1800.00024	2710.00049	2.00000
200000.03161	0.01470	2700.00049	2659.00049	3.00000
300000.06323	0.01430	3600.00049	2605.00049	4.00000
400000.06323	0.01420	4500.00098	2545.00049	5.00000
THE SLOPE=0.90418				

FRICT. FCTR.	RES. PRESS	WELL HEAD PRESSURE	EQUIP. BHP	NO. WELLS
0.01463	105.29957	888.46582	1020.70739	3
0.01485	935.72119	850.89465	963.39013	4
0.01498	1171.67578	832.93090	935.51806	5

ule and agreement terms at which the company can show the greatest profit. The more wells drilled, the greater is the development cost. However, this results in the quickest return of the revenue from the project. The fewer wells drilled, the lower is the development cost but the longer it takes to realize the profit from the project. Consequently, it is possible to optimize the number of wells drilled to maximize the deferred profit, depending on a company's financial situation and the profitability evaluation parameters the company believes are most realistic for them.

The rate versus time behavior of a reservoir is best predicted by estimating the rate versus cumulative relationship and converting it to a rate versus time relationship. The rate versus time relationship is determined by calculating the time increments necessary for the production of increments of cumulative production at appropriate average producing rates. The rate versus cumulative relationship can be determined from the curve of p/z versus cumulative production obtained by material balance, as discussed, and the reservoir deliverability relationships of q_g versus $(p_e^2 - p_w^2)$.

We can either assume a rate and determine the corresponding cumulative production or assume a cumulative production and find a corresponding rate. Normally, the capacity to produce exceeds the initial pipeline capacity, and it is necessary to find the cumulative production and time when the reservoir will no longer be able to produce at or above the pipeline capacity. Knowing the pipeline capacity, we can determine the $(p_e^2 - p_w^2)$ necessary for this producing rate. An equipment capacity curve gives the p_w at this rate, permitting the calculation of p_e. Then assuming p_e is equal to the average pressure for the reservoir, a p/z ratio can be determined and a corresponding cumulative production, G_p, can be found. Using this G_p and the maximum producing rate, the time at which the reservoir rate will start to decline can be calculated, $G_p/q_g\max$.

The same procedure can be followed to determine the G_p corresponding to any lesser reservoir rate. Then assuming capacity production, the increment of time required to produce some increment $G_{p2} - G_{p1}$ of production can be determined by dividing the increment by the average producing rate $(q_{g1} + q_{g2})/2$.

To test understanding of this procedure, work problem 5.14 and compare the solution with the one in appendix C.

PROBLEM 5.14: Predicting Rate versus Time Behavior for a Gas Reservoir

A pipeline will take gas from the reservoir in problem 5.13 at a maximum rate of 30 MMscfd. If the reservoir is developed with 6 wells, when will the producing rate start to decline? When will the rate be 4 MMscfd/well? 3 MMscfd/well? Plot the rate versus time. Measure times from the start of the contract, i.e., when G_p = 9,450 MMscf. The z-factor data are given in Fig. 5–30. Use the equipment capacity and the reservoir deliverability relationships from Figs. 5–28 and 5–29. Also, the problem 5.4 solution shows that:

$$G_p\text{, MMscf} = [4{,}458 - (p/z)_{avg}]/0.053$$

Predicting the Behavior of Microdarcy Reservoirs

The increase in the price of natural gas in the late 1970s made it economically possible to produce gas reservoirs that were previously classified as impermeable. These reservoirs are in the microdarcy (10^{-6} darcy) range and are produced by completing the well using a massive hydraulic fracture treatment, which we hope results in a fracture in the formation that is several hundred feet long. Some of these commercial wells may produce at no more than 30 Mcfd. However, they may produce at such a rate for years with no noticeable decline because of the large amount of gas in place relative to the producing capacity. These conditions place serious restrictions on the methods previously discussed.

It is impossible to run a stabilized flow test on such a well, and considerable care must be exercised in interpreting the data from an isochronal flow test. If we consider the induced fracture as being of infinite permeability, we can approximate the amount of the reservoir affected by a constant-rate flow test. Calculate the distance from the fracture that has been affected using the t_s equation, Eq. 3.50, to determine the distance, r_e, using the producing time as t_s. Then at succeeding times the drainage boundary can be illustrated as in Fig. 5–31.

After relatively short periods of flow, the well has affected a short distance when compared to the length of the fracture so flow is essentially linear. However, once the well has produced long enough to affect

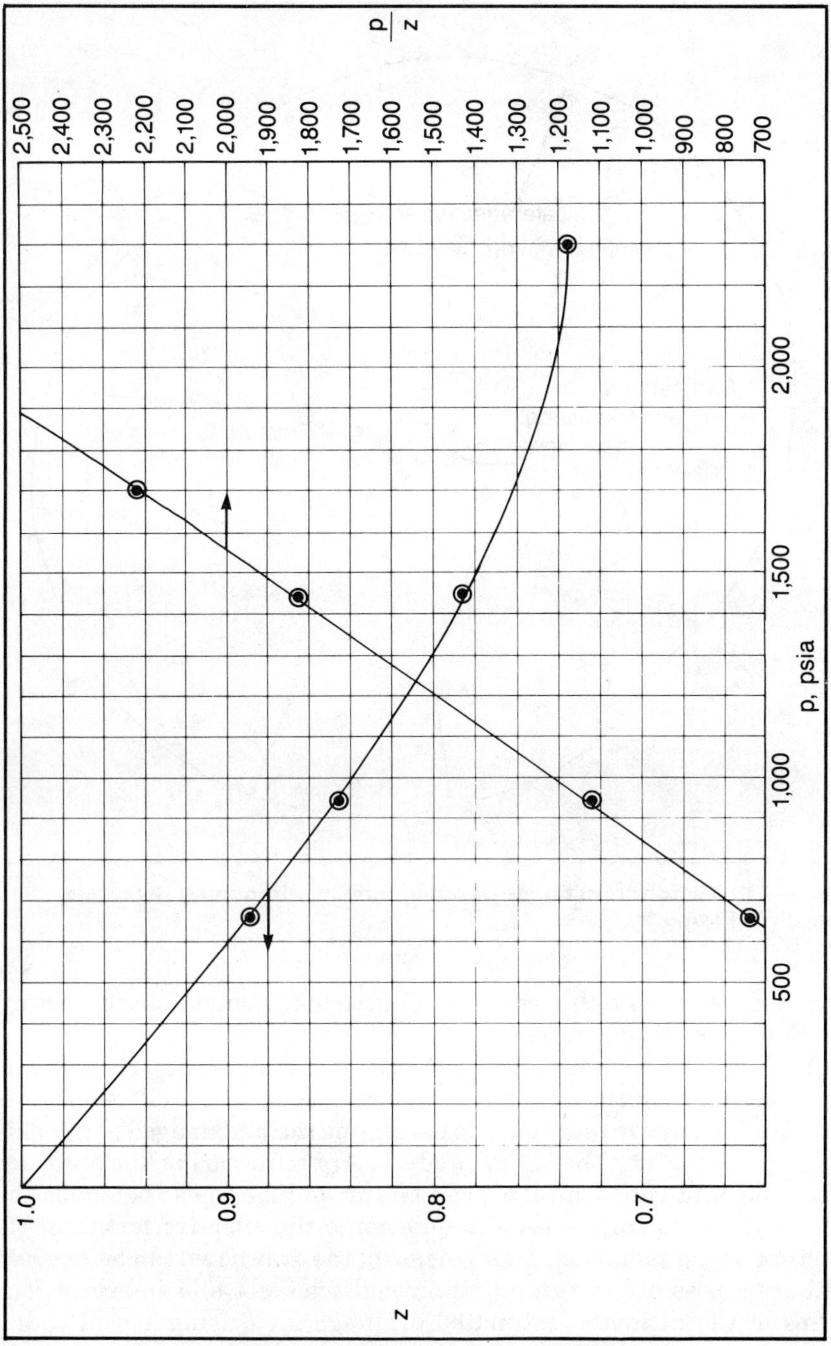

Fig. 5–30 The z-factor data for problem 5.14

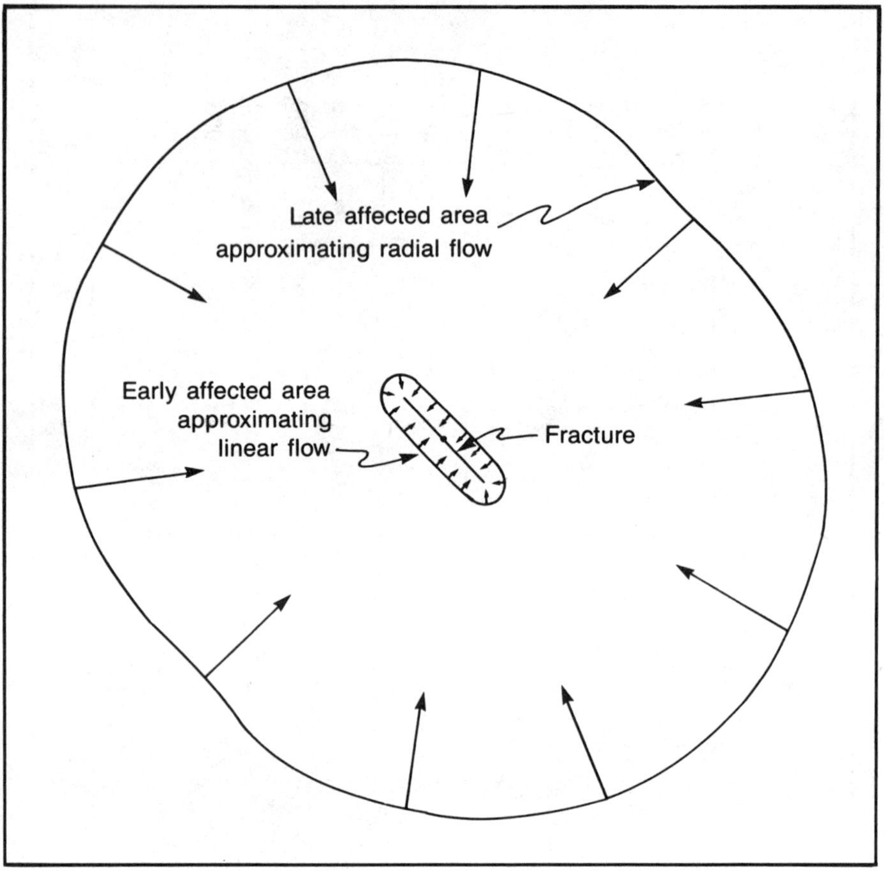

Fig. 5–31 Switch from linear to radial flow in microdarcy reservoir containing large fracture

a large distance into the reservoir relative to the length of the fracture, the flow approximates radial flow.

Perhaps the easiest way to predict the behavior of such a well or reservoir is to run a constant-rate flow test on the well before it is fractured. Then use these data to determine the undamaged permeability or permeability thickness product. After fracturing the well, another constant-rate drawdown can be run and the linear behavior portion of the data can be used to determine the effective length of the fracture. If the radial behavior portion of the drawdown can be reached, it then is possible to run an isochronal-type test and interpret it in terms of the ultimate, estimated drainage area using a well radius equivalent to the fracture. If one stabilized point can be obtained, the well radius and drainage radius approximations can be avoided.

Many of the steps in the suggested analysis may be impossible to obtain in many microdarcy reservoirs. Many wells do not have sufficient capacity to permit a flow test before treatment. With a very small well capacity the engineer must be particularly careful of afterflow effects. It is easy to think we are running a constant-rate drawdown at a very small rate, when in fact we are simply producing gas from a pressured tubing or casing and an equal amount of gas is not entering the well. A material balance on the gas in the tubing or casing based on observed surface and bottom-hole pressures can clarify the afterflow situation. The engineer may wish to consider the use of very small diameter tubing, such as the continuous rolled tubing of some workover rigs, to minimize afterflow effects.

If a flow test cannot be performed before the well is treated, the engineer should consider using a small fracture to get enough productive capacity to run the flow test. However, keep the fracture size small enough to permit obtaining radial flow behavior in a short period. After running this initial constant-rate drawdown to obtain the undamaged permeability, the well can be refractured in the manner desired.

If it is impossible or undesirable to run an isochronal test because of the time required to reach radial flow, the pseudosteady-state rate can still be predicted using the pseudosteady-state equation and the well radius equivalent of the fracture. Prats showed that a fracture of infinite permeability acts like a well with a radius of one-fourth of the total length of the fracture once the flow approximates radial behavior.[18]

One of the positive factors we obtain from a massive hydraulic fracture in a microdarcy reservoir is that the additional pressure drop caused by turbulence is often negligible. The cross-sectional area at the well is so large as a result of the very large fracture that the flow velocity is too small to cause turbulence. Also, after fracturing there appears to be little damage around the well, or fracture. This should not be confused with the fact that the effect of a fracture created by a small fracture treatment can be considered as a negative skin that is determined by using the well radius as the internal radius of the casing.

The analysis of the radial-behavior constant-rate drawdown either before the fracture treatment or after the treatment is the same as any conventional gas-well drawdown analysis. However, the analysis of the linear behavior of the constant-rate drawdown requires the unsteady-state, linear, infinite-acting equation:

$$p_w = p_i - \frac{1.259q}{hL}\left(\frac{\mu}{k\phi c}\right)^{0.5}(t_{days})^{0.5} \tag{3.68}$$

Remember that the q in Eq. 3.68 has the units of reservoir barrels per day, and when applied to gas, it would be at the arithmetic average pressure. Eq. 3.68 shows that a plot of the well pressure versus the square root of time gives a straight line whose slope is as follows if the unit of the producing time is days:

$$m_L = -\frac{1.259q}{hL}\left(\frac{\mu}{k\phi c}\right)^{0.5} \tag{5.98}$$

To make sure we understand the analysis of microdarcy gas reservoirs containing massive hydraulic fractures, work problem 5.15 and compare the solution with the one in appendix C.

PROBLEM 5.15: Predicting the Pseudosteady-State Flow Rate for a Microdarcy Gas Reservoir Containing a Massive Hydraulic Fracture

A well with an estimated net-pay thickness of 10 ft is fractured with 80,000 gal of fluid. (This does not quite qualify as a massive hydraulic fracture, which is normally defined as a fracture larger than 100,000 gal.) No prefrac drawdown test has been run on this well. The initial drawdown test at a constant rate of 143 Mcfd results in the pressure data plotted in Fig. 5–32. The early pressure data are also plotted in Fig. 5–33 as a function of the square root of the producing time in minutes. The gas flow rate is estimated as 342 res b/d at the arithmetic average pressure. Find the undamaged permeability, the fracture length, and the initial pseudosteady-state flow rate using the data in Figs. 5–32 and 5–33. The following reservoir characteristics are applicable:

Viscosity = 0.014 cp
Gas compressibility = 0.0008787/psi
Ultimate drainage radius = 915 ft
Reservoir temperature = 100°F
Porosity = 11%
z_{avg} = 0.86

The literature contains many type curves, or log-log plots, for a variety of fracture configurations and fracture conductivities, including fractures with conductivities varying from the well to the end of the fracture. These type-curve applications suffer from the same difficulties as type curves in general, as discussed in chapter 4. However, as with other type curves if no other method of analysis can be found, we may wish to try this application.

Determining the Size of a Gas Reservoir

The problem of determining the size of a gas reservoir seems to be grossly misunderstood by managers and engineers in general. Given some pressure measurements and a few hours of production, the average manager or engineering supervisor seems to think that a reservoir engineer should be able to determine the drainage volume of the sub-

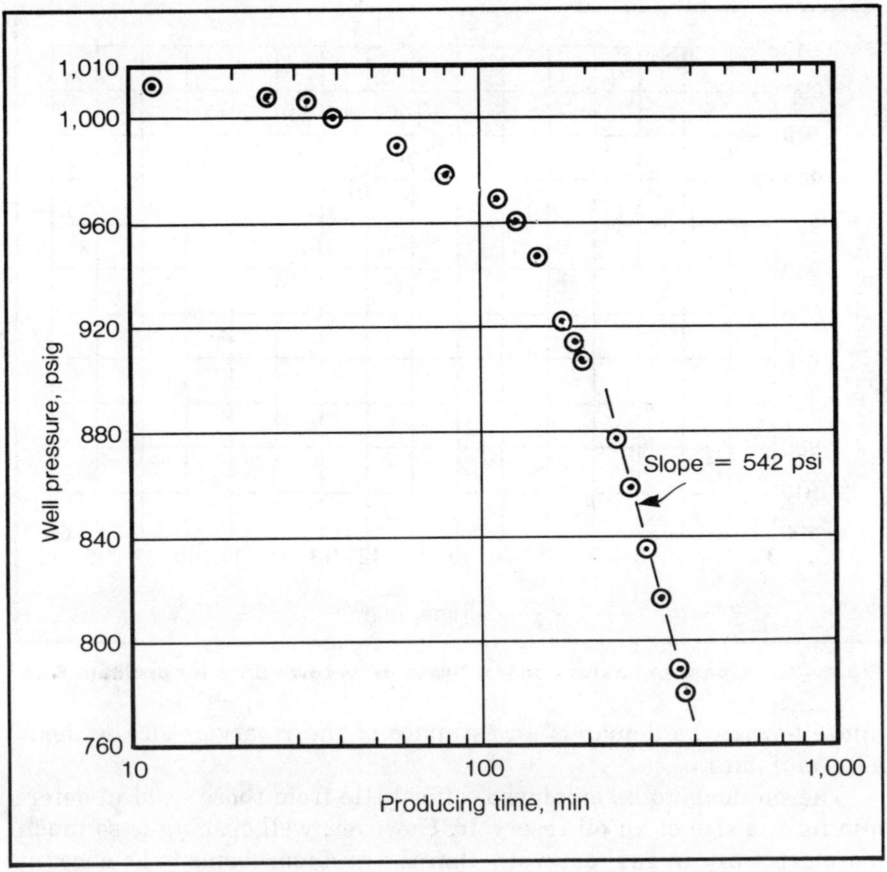

Fig. 5–32 Constant-rate drawdown data for a microdarcy gas reservoir containing a large fracture—for problem 5.15

ject well. In fact, if the engineer can tell the reservoir size from such meager tests, the reservoir is probably so small that it will be uncommercial. In determining the size of a gas reservoir from test and pressure data, there is no substitute for time.

One practical approach to this problem is for the reservoir engineer to determine the minimum production time necessary to calculate a reservoir size. This can be determined using the stabilization-time equation and the estimated reservoir parameters. Fig. 5–34 may be useful in this regard. Remember that this technique provides only a minimum reservoir size estimate. If the test is run improperly or the production rate is too small to result in interpretable pressure drops, the data may still be useless. However, the main point to be made is that there is no substitute for time. A manager is simply misleading

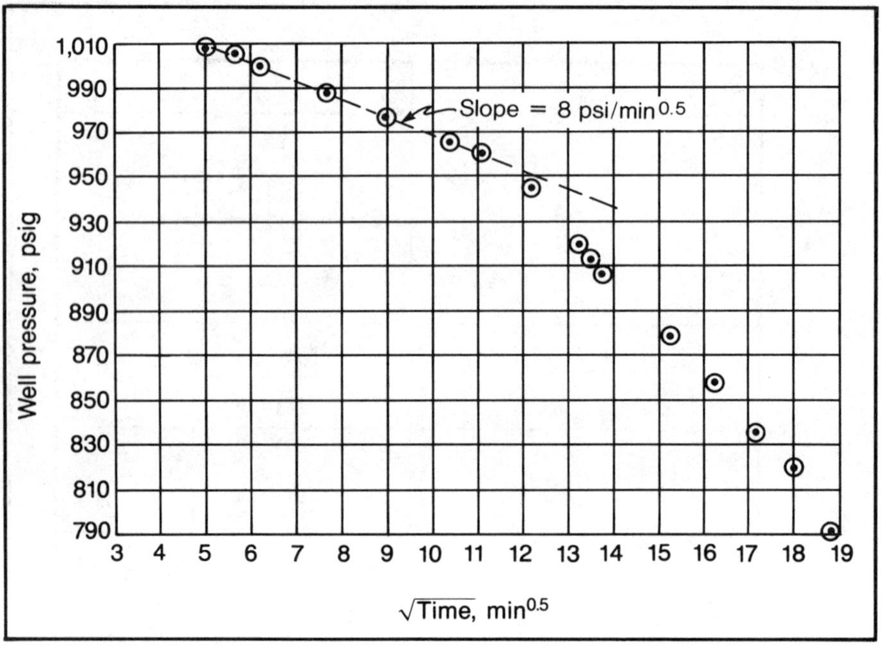

Fig. 5–33 Linear behavior constant-rate drawdown data for problem 5.15

himself when he demands an estimate of the reservoir size without adequate data.

The methods to be employed differ little from those used in determining the size of an oil reservoir. However, well spacing is so much more arbitrary in gas reservoirs that the problem seems to be encountered more often.

The use of drawdown data to determine the distance to the nearest boundary in a reservoir has been discussed extensively in chapter 4 for oil reservoirs. The techniques apply in a similar way for gas reservoirs. Eq. 5.75 gives the well pressure resulting from production at a constant rate q_g for a time, t. A plot of p_w^2 versus the log t gives a straight line with a slope, m′, that satisfies Eq. 5.78 as long as the well pressure is infinite acting. However, once the nearest boundary is felt at the well, the plot is no longer a straight line. The difference between the extrapolation of the straight line and the actual pressure recorded is:

$$(p_w')^2 - p_w^2 = \frac{1.424 q_g \mu z_{avg} T_f}{kh}(0.5)[-Ei(-\tfrac{1}{4}t_{Di})] \qquad (5.99)$$

The dimensionless time, t_{Di}, is a function of the distance to the closest reservoir boundary, d:

$$t_{Di} = (6.33k/\phi\mu c)t/(2d)^2 \qquad (4.19)$$

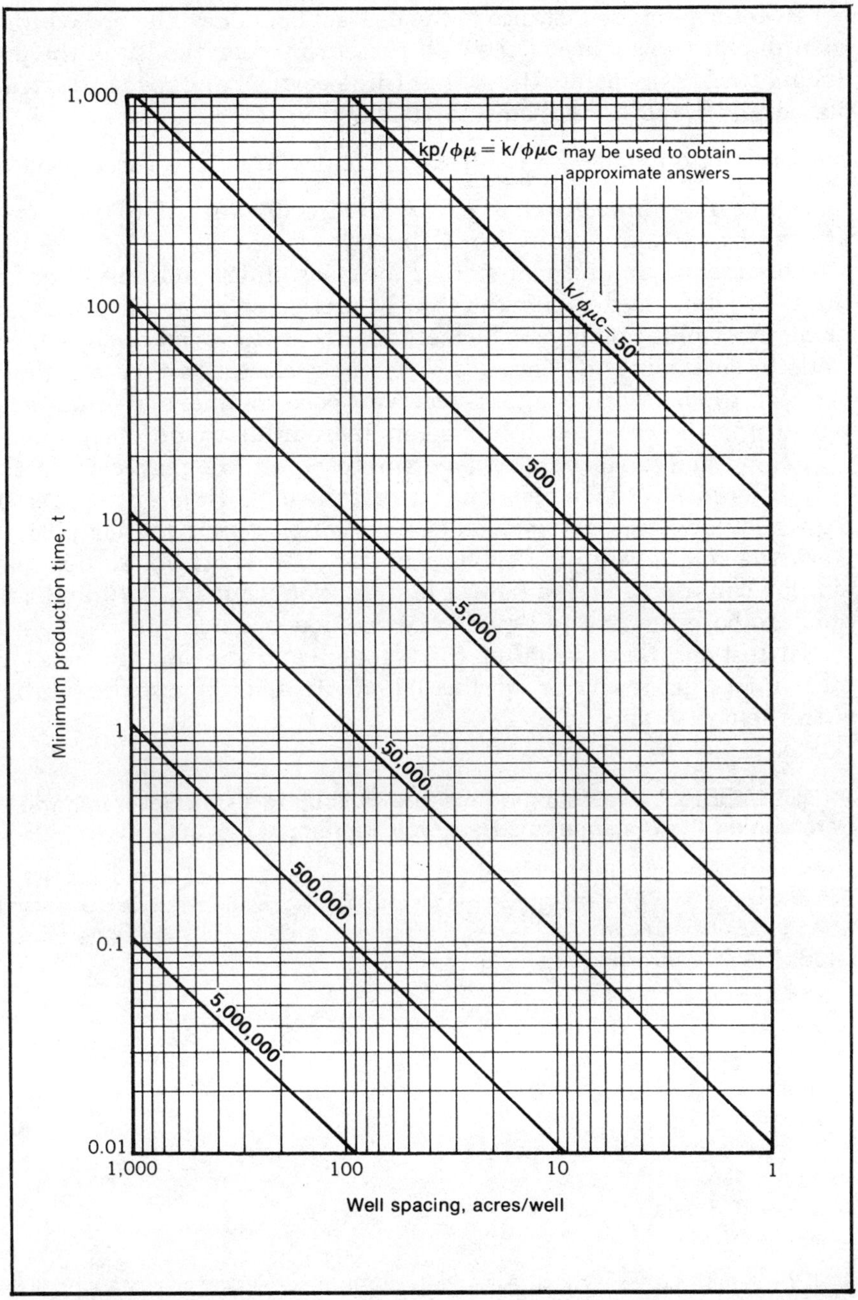

Fig. 5–34 Well spacing provable by a production time (based on $t = 0.04\ \phi\mu c r_e^2/k$)

To determine the distance to the nearest boundary, the procedure is then to plot the square of the well pressure versus the log of the producing time, extrapolate the straight-line portion, and read the extrapolated $(p_w')^2$ and p_w^2 at some particular time, t. Then using Eq. 5.99, the Ei function, $0.5\left(-Ei\,\frac{-1}{4t_{Di}}\right)$, can be calculated; the corresponding t_{Di} can be determined from Fig. 3–9; and the distance, d, to the boundary can be calculated from Eq. 4.19.

This technique and many other methods of determining reservoir limits are discussed in chapter 4 in the section on reservoir limit tests along with some problems. The methods in chapter 4 can be used in gas wells to determine the distance to the next closest barrier, the angle between the barriers, the total reservoir pore volume, and other ideas concerning the reservoir limits when appropriate conditions prevail.

In applying the techniques for an oil reservoir to a gas reservoir, the only difference is that a plot of p_w^2 versus the log of time, or m(p) versus log t, may be necessary instead of a plot of p_w versus the log of time. Also, the relationship is governed by Eq. 5.75 and 5.78, rather than similar equations for the flow of liquids. Note that Eq. 4.19 and 3.31 apply to both the flow of liquids and the flow of gas.

To test our knowledge of this section and the one on pressure buildup in a gas reservoir, work problem 5.16 and compare the solution with the one in appendix C.

PROBLEM 5.16: Evaluating a Pressure Buildup in a Gas Reservoir and Determining the Distance to a Reservoir Barrier

A discovery well in a gas reservoir is produced for a period of time during which 248 MMscf of 0.76-specific-gravity gas is produced. The final rate is 3.9 MMscfd when the well is shut in with the resulting pressure behavior as indicated in Fig. 5–35. The reservoir data are as follows:

γ = 0.76
ϕ = 5.0%
T_f = 253°F
p_o = 4,990 psia
μ_g = 0.027 cp
h = 30 ft
r_w = 4 in.
p_{wf} = 998 psia
S_{wc} = 0.35

A. If the initial reservoir pressure is 4,990 psia, is this reservoir infinite acting at the time of shutin? (Find indicated p_i and compare with actual p_i.)

B. What is the distance to the nearest boundary?

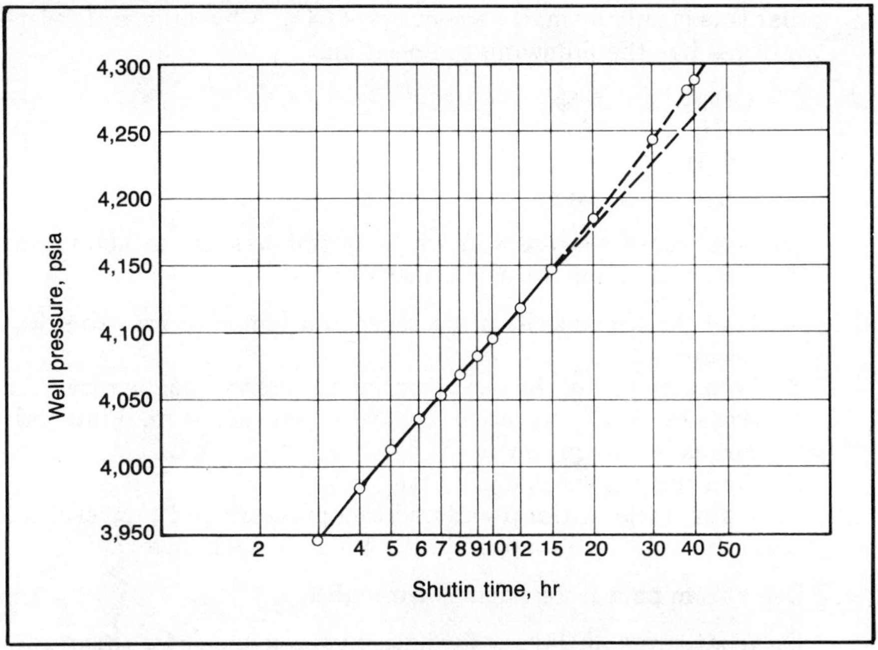

Fig. 5–35 Gas-well pressure-buildup data for problem 5.16

C. What is the gas pore volume if at the time of shutin the pressure is declining about 6 psi/day? Assume no water drive and $p_s = 4{,}600$.

D. Calculate the static pressure by material balance.

In the 1970s the importance of gas reservoir engineering increased greatly. Prior to that time, there appeared to be an abundance of gas available so exact gas reservoir engineering was unnecessary. However, with gas in short supply, it became necessary to predict accurately the rate versus time behavior of gas wells and reservoirs. Also, gas production was initiated in microdarcy reservoirs and from very high-pressure reservoirs that had not been produced previously. Many engineers found that they did not have the technical knowledge to handle these problems. Therefore, we have discussed the proven methods of predicting the behavior of gas wells under a variety of conditions.

Additional Problems

5.17 This problem demonstrates the effect of CO_2 on gas characteristics. Compare the answers with those in problem 5.1. Remember that this is only a small percentage of CO_2. A hypothetical reservoir gas has the following composition:

CH_4	78%
C_2H_6	15%
nC_4H_{10}	5%
CO_2	2%

The gas occupies a reservoir whose initial pressure is 3,000 psia. The reservoir temperature is 170°F.

A. Find the pseudocritical pressure and temperature using Eq. 5.8 and 5.9.
B. Prepare a plot of the gas deviation factors versus the reservoir pressure and p/z versus the reservoir pressure. Plot both curves on one graph.
C. Find the gas specific gravity.
D. What is the initial pseudoreduced pressure and temperature based on Fig. B9, appendix B? What is the z factor?

Use z from part B for parts E through I.

E. What is the initial gas formation volume factor for this reservoir?
F. Find the gas density at initial reservoir conditions.
G. What is the initial compressibility of the gas? What is the initial compressibility according to c = 1/p?
H. Estimate the gas viscosity at the initial reservoir conditions.
I. If 55°API gravity condensate is produced with this 0.72-gravity gas at a ratio of 40 bbl/MMcf, what is the reservoir gas deviation factor?

5.18 A. A shutin dry-gas well has a surface pressure of 2,000 psia. The 0.9-specific-gravity gas has a critical temperature of 452°R and a critical pressure of 659 psia. If the temperature gradient is 10°F/1,000 ft, find the pressure at 8,500 ft using Eq. 5.28 and depth increments of 1,000, 1,000, 1,000, 2,000, 2,000, and 1,500 ft. Surface temperature is 74°F. (In practice, smaller depth increments, perhaps 50 ft, should be used to obtain sufficiently accurate bottom-hole pressures.)
B. Calculate the 8,500-ft pressure using Eq. 5.31 and a single depth increment of 8,500 ft.

5.19 The following production and pressure data are available for a reservoir:

Date	Cumulative Pipeline Gas Sales, MMcf	Cumulative Condensate Production, stb	Static Reservoir Pressure, psia
7–1–79	0	0	?
7–1–80	1,809	75,000	3,461
9–1–81	3,901	160,000	3,370
10–1–82	5,850	237,000	3,209
11–1–83	9,450	378,000	3,029

Condensate gravity = 50°API
Reservoir temperature = 110°F
Gas gravity = 0.7

A. Determine the reservoir gas withdrawal, G_p, for each date to use in material-balance calculations. Assume the water vapor and vent gas corrections are negligible.
B. Determine the original reservoir pressure and the original gas in place. What is the average reservoir pressure at the completion of a contract calling for delivery of 20 MMcfd for 5 years, in addition to the 9,450 MMscf produced to 11–1–83?

5.20 Extend problem 5.6 by calculating the water encroachment after 500 days of production.

5.21 To demonstrate the effect of permeability and porosity on turbulence, work problem 5.7 using an undamaged permeability of 40 md and a porosity of 10%. The additional pressure drop caused by turbulence in this case is very small. This does not mean that turbulence effects in such reservoirs normally are negligible. Such wells would be produced at much greater drawdowns, resulting in greater rates and turbulence.

5.22 In problem 5.8 two additional wells are drilled in a reservoir so each well drained 1,125 acres and the stabilized flow test is repeated on one of the wells with the results as indicated:

Stabilized BHP, psia	Producing Rate MMscfd
2,800	0
2,676	1.81
2,604	2.71
2,525	3.59
2,445	4.51

The reservoir characteristics at the time of the test are as follows:

Permeability = 74 md
Porosity = 15%
Gas viscosity = 0.021 cp at the average pressure
c_e = 3.57×10^{-4}/psi
z = 0.742 at the average pressure

A. What must the bottom-hole pressure be in this well to produce at a rate of 5 MMscfd when p_e has declined to 1,568 psia? The contract is completed as in problem 5.4. The effective viscosity and gas deviation factor have changed to 0.018 cp and 0.75, respectively.
B. How long must each rate be maintained to reach stabilized conditions during the test?

5.23 In problem 5.22 the stabilized deliverability curve for one of the four wells in a 4,500-acre reservoir is determined for the time when p_s = 1,568 psia. From this curve calculate the stabilized deliverability curve if six wells existed in this reservoir. What must the bottom-hole pressure be for a well to produce its share of the 20 MMscfd if six wells exist in the reservoir at that time? Note that problems 5.4, 5.8, 5.9, 5.22 and 5.23 all apply to the same reservoir. The well radius is 0.25 ft.

5.24 To demonstrate the sensitivity of isochronal data to errors in reservoir parameters, work parts A and B of problem 5.10 using permeability of 18.5 md, or one-fourth of the permeability used in problem 5.10. The well radius is 0.25 ft. Find the well pressure if the flow rate is 5,000 Mcfd and the permeability is (a) 18.5 md and (b) 74 md.

5.25 The following data are obtained during a normal four-point flow test:

Cumulative Time, min	*Rate during Preceding Time, Mcfd*	*Bottom-Hole Pressure, psia*
0	0	807
26	635	800
52	1,124	793
79	2,351	759
101	3,867	694

Calculate and plot an isochronal curve from the four-point drawdown data. Note that the turbulence can be interpreted as varying from zero ($n = 1.0$) at the lowest rate to a maximum of $n = 0.6$. The reservoir characteristics are as follows:

$k = 0.014$ darcies
$c = 0.0012195$/psi
$\mu = 0.013$ cp
$r_w = 0.328$ ft
$\phi = 10.9\%$

5.26 Rework problem 5.13; assume the contract is changed to deliver a constant rate of 12.5 MMcfd for 8 years. The total gas produced for the length of the contract remains the same.

5.27 Compare the results of problem 5.14 with the results obtained when the number of wells drilled is 3 and the maximum pipeline rate is 15 MMcfd.

5.28 A well in the very tight Vicksburg formation in the McAllen Ranch field is subjected to a constant-rate drawdown where the rate varied from 1,703 Mcfd to 575 Mcfd during a 2-hr flow test. The initial shutin bottom-hole pressure is 10,602 psia, and the resulting flowing bottom-hole pressures are as indicated in Fig. 5–36. After a massive hydraulic fracture (exact treatment unknown), the well is flowed for about 2 weeks at rates varying from 700–180 Mcfd. It is then shut in for about 7 days with results as indicated on the plots of bottom-hole pressure versus the square root of the shutin time and bottom-hole pressure versus the log of the shutin time. Use the data in Figs. 5–37 to 5–39 and the following reservoir characteristics:

Reservoir temperature = 292°F
Gas gravity = 0.628
Gas-occupied porosity = 10%

A. What is the permeability thickness product indicated by the prefrac drawdown? The post-frac buildup? Also, determine S.
B. The total perforated interval is 61 ft. Using the permeability thickness product from the prefrac drawdown, what is the fracture length if the producing interval is the total perforated interval of 61 ft? What is the fracture length if the formation permeability is 0.1 md?
C. If this well drains 160 acres, what is the initial pseudosteady-state producing rate. (Find S from the radial flow portion of

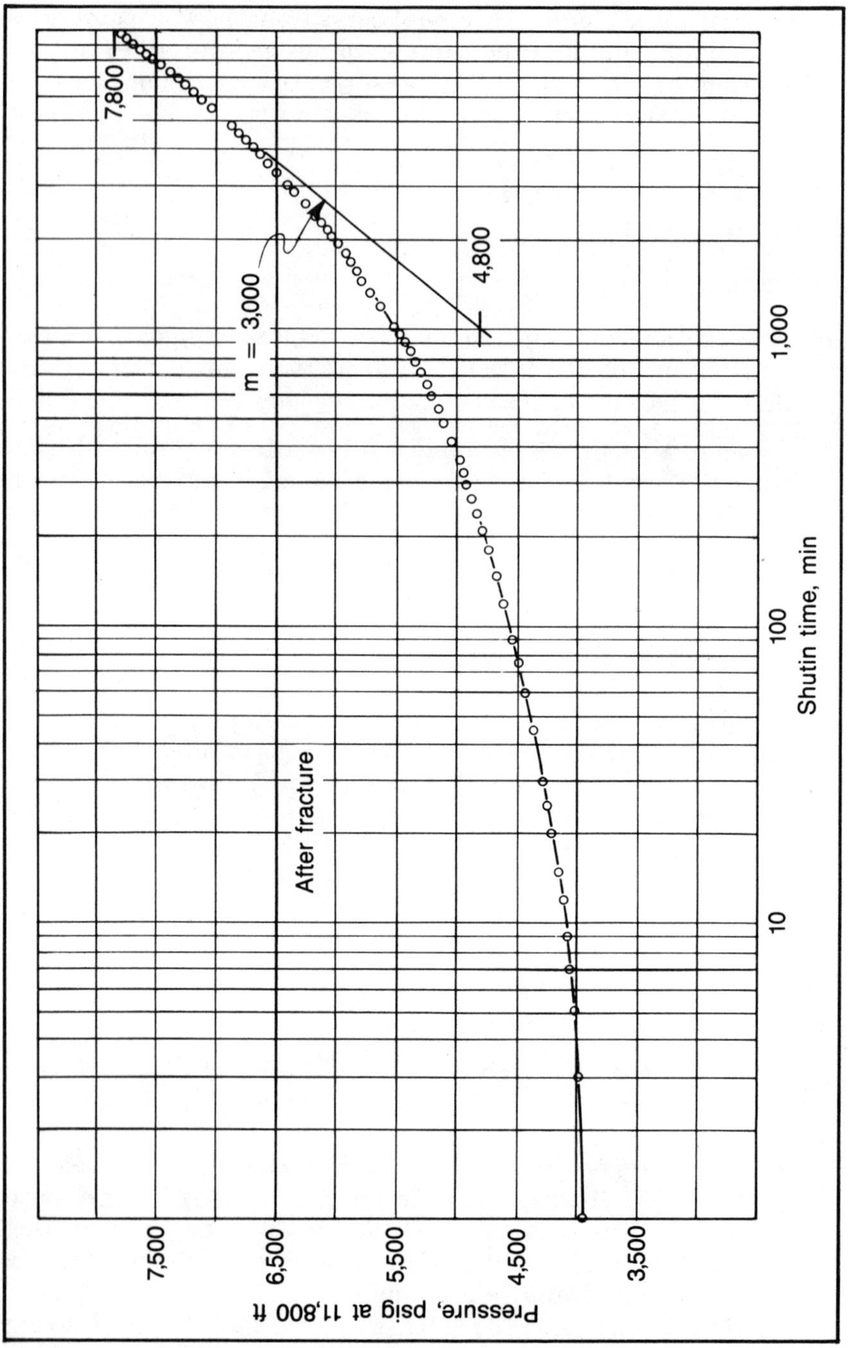

Fig. 5–36 Data for problem 5.28

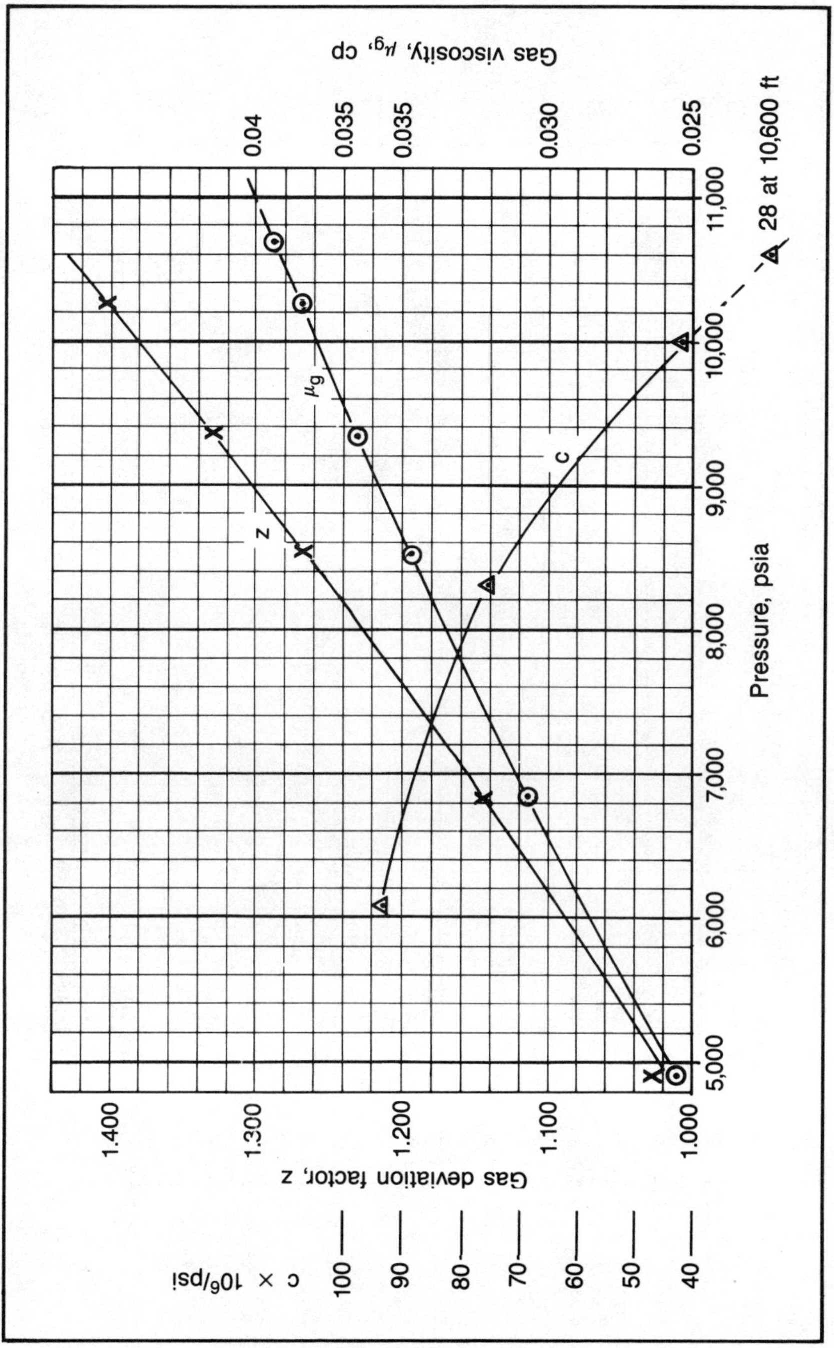

Fig. 5–37 Vicksburg well data for problem 5.28

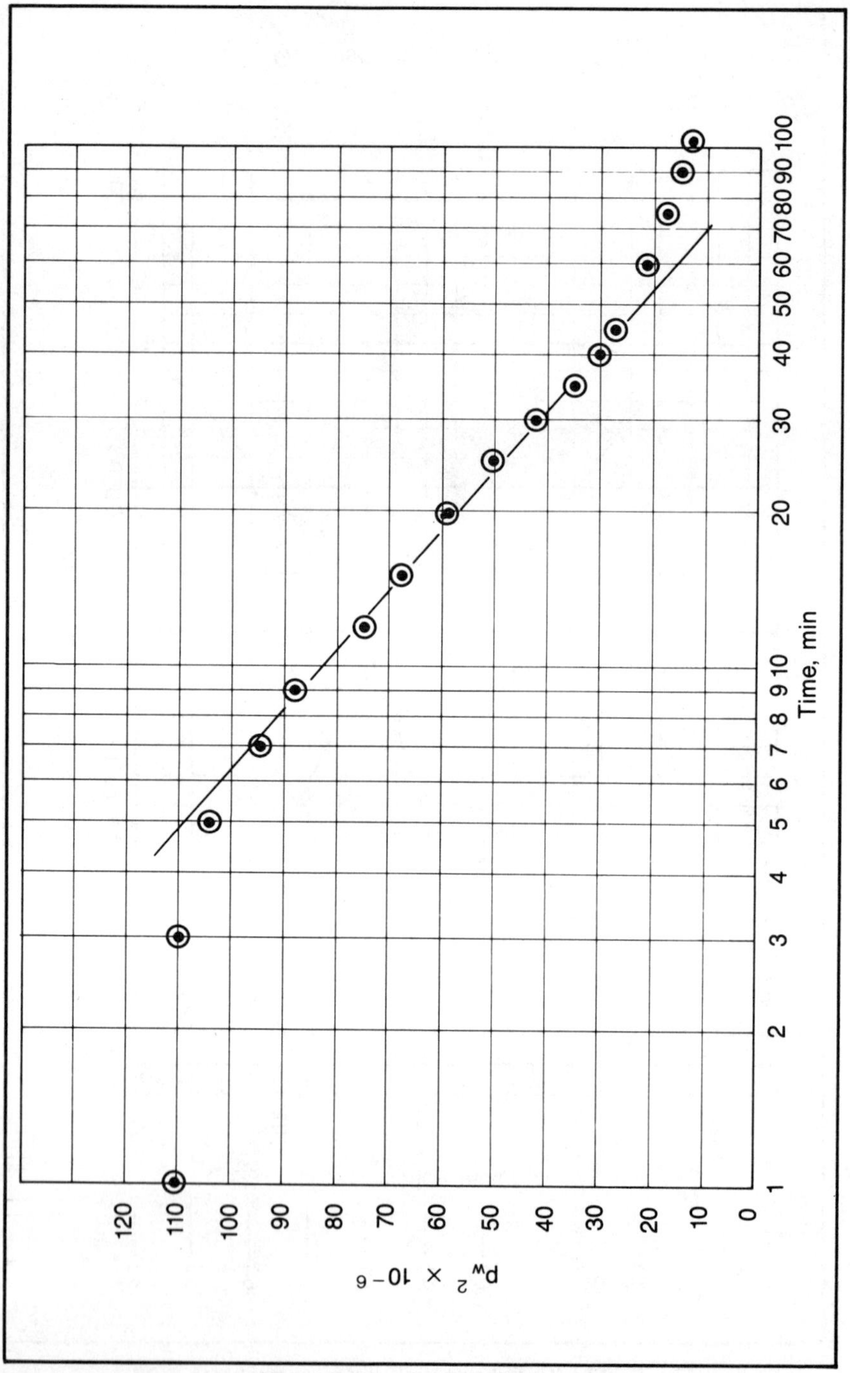

Fig. 5–38 Vicksburg well constant-rate drawdown before fracture for problem 5.28

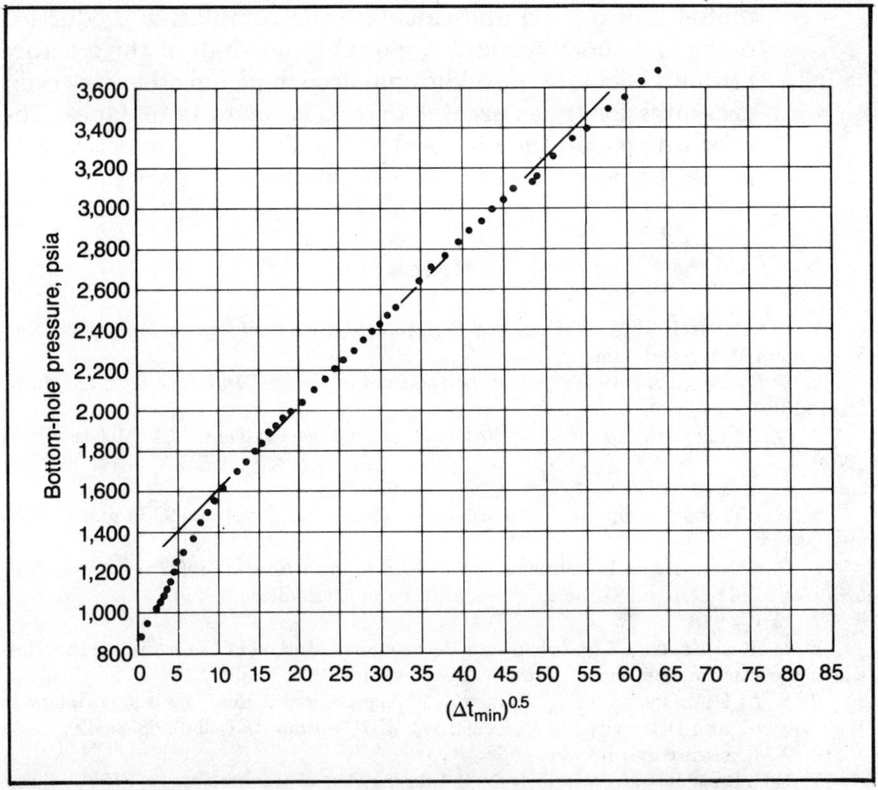

Fig. 5–39 Vicksburg well pressure-buildup after fracture linear flow portion for problem 5.28

the pressure-buildup data. Then use S in the pseudosteady-state equation to calculate the rate.) Use the following reservoir characteristics:

k_g = 0.1 md
p = 2,000 psia
h = 61 ft
μ_g = 0.0318 cp
Gas-occupied porosity = 10%
r_w = 0.333 ft
c_g = 0.000086/psi

D. What is the producing rate when the cumulative production is 1 MMcf if the original gas in place is 14.8 MMcf and z-factor data are as indicated in Fig. 5–37? To complete this calculation, determine the cumulative production when the well first approximates pseudosteady-state. Calculate t_s using the estimated k of 0.1 md and calculate the cumulative production using Q_{tD} functions and r_w equal to one-half of the fracture radius. Calculate the additional drop in p_e and other reservoir pressures for times greater than t_s by material balance. The resulting p_e can then be used to calculate the production rate using the pseudosteady-state equations as in part C.

Notes

1. O.A. Househ, K.M. Watson, and R.A. Ragatz, *Chemical Process Principles* (New York: John Wiley and Sons, 1943).
2. W.B. Kay, "Density of Hydrocarbon Gases and Vapors," *Ind. Eng. Chem.*, volume 28 (1936), 1014–1019.
3. D.L. Katz et al., *Handbook of Natural Gas Engineering* (New York: McGraw-Hill, 1959).
4. Katz et al., 1959.
5. A.S. Trube, "Compressibility of Natural Gases," *JPT,* volume 9 (January 1957), 69.
6. P.B. Crawford and G. Fancher, *Flowing and Static Bottom-Hole Pressures of Natural Gas Wells* (Austin: Texas Petroleum Research Committee, 1959).
7. Kay, 1936.
8. M.H. Cullender, "The Isochronal Performance Method of Determining the Flow Characteristics of Gas Wells," *Trans.,* AIME, volume 204 (1955), 137.
9. R. Al-Hussainy and H.J. Ramey Jr., "Application of Real Gas Flow Theory to Well Testing and Deliverability Forecasting," *JPT,* volume 18 (1966), 637–642.
10. Al-Hussainy and Ramey, 1966.
11. D.G. Russell et al., "Methods for Predicting Gas Well Performance," *JPT,* volume 15 (1966), 1365–1369.
12. Energy Resources Conservation Board, *Theory and Practice of the Testing of Gas Wells,* 3rd edition (Calgary: 1975).
13. C.S. Matthews and D.G. Russell, *Pressure Buildup and Flow Tests in Wells,* Monograph No. 1 (Dallas: SPE, 1967).
14. J. Vairogs and V.M. Rhoades, "Pressure Transient Tests in Formations Having Stress-Sensitive Permeability," SPE paper 4050, presented at the annual meeting (1972).
15. Vairogs and Rhoades, 1972.
16. L.C. Uren, *Petroleum Production Engineering—Oil Field Exploitation* (New York: McGraw-Hill, 1934).
17. R.V. Smith, "Determining Friction Factors for Measuring Productivity of Gas Wells," *Trans.,* AIME, 189:73 (1950).
18. Cullender, 1955.

Additional References

Ballantyne, Wayne. A computer program design assignment from an undergraduate course, Ch.E. 743.01, Ohio State University, February 1969.

Burcik, E.J. *Properties of Petroleum Reservoir Fluids.* New York: John Wiley and Sons, 1957.

Clark, C.H. "Application of the van Everdingen and Hurst Calculation Procedure to the Back Pressure Test." Research Conference on Flow of Gas from Reservoirs, Ann Arbor: University of Michigan, June 30, 1955.

Crawford, P.B. "Calculating Bottom-Hole Pressures from Surface Measurements." *The Petroleum Engineer,* November 1959.

Engineering Data Book. Natural Gasoline Supply Men's Association, April 1951.

Hall, K.R.; and Yarborough, L. "How to Solve the Equation of State for z-Factor." *OGJ*, February 18, 1974.

Hammerlindl, D.J. "Predicting Gas Reserves in Abnormally Pressured Reservoirs." paper 3479, SPE Annual Fall Meeting, 1971.

Slider, H.C. "Gas Well Testing using Negative Superposition." paper 3875, Omaha: SPE Northern Plains Section Regional Meeting, 1972.

Standing, M.B., and Katz, D.L. "Density of Natural Gases." *Trans.*, AIME (1942).

6

Fluid Distribution and Frontal Displacement

We have discussed the reservoir without recognizing the vertical variation in the distribution of the reservoir fluids. We have at times referred to a water drive or gas drive, but we have treated this phenomena as though the saturation changed abruptly in a vertical direction from the saturations of the oil-bearing portion of the reservoir to the saturations of a gas cap or a water-drive zone. We noted occasionally the variation in a gas saturation laterally or radially as a result of the variation in pressure. However, we did not note the variation caused by the encroachment of gas-cap gas or water into the original oil-bearing portion of the reservoir. This does not mean that the methods previously discussed are impractical. However, we must apply these methods to reservoir situations where they are not unduely affected by saturation variations either vertically or horizontally. Therefore, a thorough understanding of fluid distribution and frontal displacement enable the engineer to apply these methods intelligently and accurately.

In addition, this material specifically permits the engineer to determine the saturation distribution in the reservoir, either vertically or laterally, initially or after producing the reservoir for some period. It provides the fundamental understanding of the relationships between fluid-saturation distribution, gravity forces causing segregation, and the different viscous forces associated with the movement of the different fluids in the reservoir. These fundamentals can then be applied to the practical problems of evaluating the saturation distribution in the reservoir initially and during the course of reservoir displacement. The fundamentals are directly applicable to the problems of the displacement of oil by a natural water drive or gas-cap drive, including coning, fingering, and hydrodynamic tilt of the water-oil or gas-oil contact. In addition, the frontal displacement fundamentals are directly applicable to most secondary recovery methods that employ a frontal displacement.

We begin with a discussion of the initial vertical distribution of the reservoir fluids and the means of evaluating this vertical saturation. We also calculate the saturation distribution from laboratory-measured capillary pressure data. This technology is presented for its practical application and to give an excellent physical understanding of the relationship between the microscopic saturation distribution, the pore-size distribution, wettability, and gravity segregation. Sometimes, the initial saturation can be determined more readily and realistically from well logs, but the complete physical understanding of the phenomena comes from the capillary pressure relationships.

Following the discussion of the initial saturation distribution, methods of determining the saturation distribution during a frontal displacement are presented. These methods include the Buckley-Leverett and Welge procedures. This material is presented in a general manner so the equations can be applied to any immiscible fluid displacement in porous media, including the displacement of gas by oil as encountered in waterflooding and other secondary processes where an oil bank is formed. The treatment also includes those special cases where unbalanced viscous forces exceed the gravity forces and result in fingering, coning, and hydrodynamic tilt.

The fluid displacement discussion does not apply to the internal gas drive that results from the formation of free gas from solution gas as the pressure declines in a solution-gas-drive reservoir. However, the fundamentals may be useful in evaluating the vertical movement of gas and oil in such a reservoir. This movement often results in the formation of a secondary gas cap. As indicated, the methods apply theoretically only to immiscible fluids. However, it has been observed that miscible displacements have the same qualitative characteristics as would be predicted for an immiscible system having the same viscosity ratio and a relative permeability ratio of 1.0. Thus, the displacement analysis methods presented have some limited application to miscible systems, although they are theoretically rigorous only for the immiscible systems.

Initial Saturation Distribution in a Reservoir

Engineers often assume that there is an abrupt change in the saturation vertically when we consider movement from a water zone to an oil zone or from an oil zone to a gas cap. However, this is seldom the case. A more normal initial-saturation distribution would be that depicted in Fig. 6–1. In this illustration the saturations grade smoothly from 100% water saturation in the water zone to an irreducible water

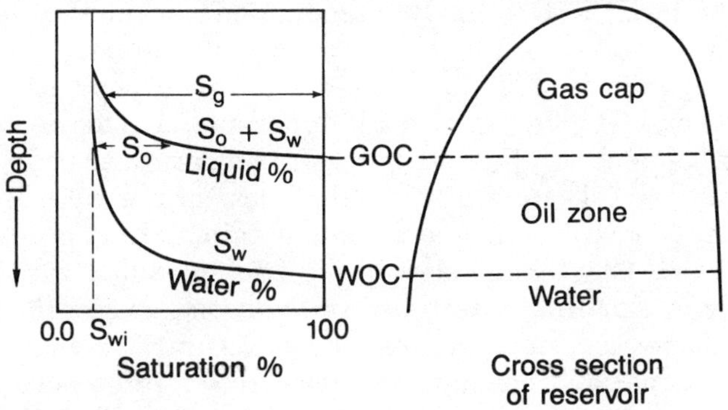

Fig. 6–1 Normal initial saturation distribution in a combination-drive reservoir

saturation some vertical distance above the water zone. The total liquid saturation grades smoothly from 100% in the oil zone to 0.0 in the gas cap. This would of course result in a similar gradual transition in the oil saturation and in the gas saturation if these values were plotted separately. Consequently, the simplified diagram on the right-hand portion of Fig. 6–1 does not properly convey the actual saturation distribution that exists initially in the reservoir.

Fig. 6–1 can serve as a definition of what is generally meant by a *gas-oil contact* or a *water-oil contact*. These terms mean different things to different engineers. For the purposes of clarity, we define the water-oil contact (WOC) as the uppermost depth in the reservoir where a 100% water saturation exists. Note that this also serves as a definition of a gas-water contact. Some engineers prefer to define the WOC and water-gas contact (WGC) as the uppermost depth at which only water is produced. Due to relative permeability considerations, these two depths are not the same.

Similarly, the gas-oil contact (GOC) is defined in this book as the minimum depth at which a 100% liquid—oil plus water—saturation exists in the reservoir. Again, some engineers prefer to define this as the greatest depth at which free gas is produced.

Determining the original WOC and GOC in a reservoir. Many different methods are used to determine the initial WOC and GOC in a reservoir. Most of these methods are more qualitative than quantitative to the extent that the engineer seldom attempts to distinguish between the 100% water level and the level at which 100% of the production is water. A similar statement can be made about the usual evaluation of the GOC. Obviously, a production test can then be used to determine the water-oil contact and the gas-oil contact.

A more sophisticated production test for the purposes of determining the WOC and GOC is the sidewall fluid sampler. This device uses a packer arrangement together with one or two perforations to obtain a reservoir fluid sample from a small thickness of the well. By taking a series of these samples, it is possible to obtain a direct measurement of the WOC and GOC. The sidewall fluid sampler is a wire-line tool that is run while drilling a well before the casing is set. Interpreting data from these tests is complicated by the mud filtrate that has entered the formation and that may constitute a large portion of the small (about 6 gal) sample that is taken. However, using the data to determine the WOC and GOC is generally conclusive.

Geophysical logs, such as electrical and radioactivity logs are one of the most reliable methods of determining the WOC and GOC. These logs give the added attraction of providing the actual saturation distribution. However, as with most log analysis, there is generally some element of doubt associated with the interpretation. Consequently, it is best to keep an open mind about the WOC and GOC levels and consider the evaluations determined from logs as tentative conclusions subject to verification by flow testing.

Core analysis represents one of the most reliable methods of determining the WOC and GOC, but the methods of interpreting these data to determine the WOC and GOC deserve some consideration. The engineer should remember that the saturation data obtained from a conventional core and conventional core analysis seldom represent the true saturations of the reservoir. If they do represent the actual saturations in the reservoir, that reservoir is probably so tight that it will be uncommercial to produce unless the formations are very thick.

When a conventional core is cut using a standard, diamond, or wireline core barrel, the core is subjected to some degree of flushing by the drilling fluid or the drilling fluid filtrate. Consequently, when the core is cut and it still remains at the reservoir pressure and temperature in the bottom of the hole, the core generally contains mud filtrate, connate water, and the residual oil or gas left after flushing it with the filtrate. While the core is being removed to the surface, the pressure and temperature are continually reduced until it reaches atmospheric pressure and temperature. The gas in solution in the oil that remains in the core after flushing with the mud filtrate is liberated during the pressure reduction. The gas expands and forces mud filtrate, oil, and possibly connate water from the core. Consequently, a core from the oil-bearing portion of the reservoir contains gas, some small amount of oil, and a large amount of water, which is mud filtrate and connate water. Similarly, a core from a gas zone contains some gas, no oil, and a large amount of water. A core from the water zone contains almost 100% water and no oil. Thus, although the saturation values are of little use from a quantitative standpoint, the change in the general characteristics of the saturation data from one zone to another provides a generally reliable means of evaluating the WOC and GOC.

The use of conventional core-analysis data to determine the WOC and GOC can be visualized by studying Fig. 6–2. Although this core has unusually low water saturations, it is easy to pick the GOC at about 4,829 ft. It is also possible that a WOC exists at 4,849 ft, although the change in the general saturations at this point are masked somewhat by the change in the permeability and porosity. Unusually low water saturations, as indicated in this core, are sometimes caused by excessive weathering of the core; i.e., the core samples have not been canned or in some other way sealed from the atmosphere quickly enough and much of the water has evaporated.

Devices now permit cutting a core and maintaining it at the bottom-hole pressure at which it was cut until it is analyzed in the laboratory. Pressure coring, or the use of a pressure-core barrel as it is known in the industry, obviously eliminates the flushing of the core liquids by liberated gas. It is claimed that the device also minimizes the flushing of the reservoir fluids by the mud filtrate when cutting the core.

Capillary pressure data. When geophysical log interpretations can be relied on, these should be the basis for establishing the initial vertical saturation distribution in a reservoir. In many cases the accuracy of log interpretations is insufficient to permit the evaluation of the change in saturation with depth. In the latter situation it is often useful or necessary to evaluate the initial vertical saturation distribution from capillary pressure data. A knowledge of this technique also provides the engineer with an excellent physical understanding of the vertical variation in saturation in the reservoir.

We should first examine the nature of capillary pressure data. Naturally occurring porous media contains a variety of pore sizes. Most engineers are aware of the fact that a wetting fluid rises in a small (capillary) tube because capillary attraction exists between the fluid and the tube. The magnitude of the capillary rise is a function of the size of the tube, the angle formed between the wetting phase and the tube, and the interfacial tension that exists at the surface of the wetting phase. Although most engineers have probably been exposed to

Tabular data and interpretation

No fracture — NF

Vertical fracture — VF

Sample number	Depth ft	Perm. md	Porosity %	Residual saturation % pore space: Oil	Residual saturation % pore space: Total water	
29	4805.5	0.0	7.5	0.0	68.0	NF
30	06.5	0.0	12.3	0.0	78.0	NF
31	07.5	2.5	17.0	0.0	43.0	NF
32	08.5	59	20.7	0.0	29.0	NF
33	09.5	221	19.1	0.0	31.4	NF
34	10.5	211	20.4	0.0	38.7	VF
35	11.5	275	23.3	0.0	34.7	NF
36	12.5	384	24.0	0.0	26.2	NF
37	13.5	108	23.3	0.0	30.9	NF
38	14.5	147	16.1	0.0	29.2	NF
39	15.5	290	17.2	0.0	34.3	VF
40	16.5	170	15.3	0.0	24.2	NF
41	17.5	278	15.9	0.0	26.4	NF
42	18.5	238	18.6	0.0	39.8	NF
43	19.5	167	16.2	0.0	39.5	VF
44	20.5	304	20.0	0.0	38.0	VF
45	21.5	98	16.9	0.0	34.3	NF
46	22.5	191	18.1	0.0	34.8	VF
47	23.5	266	20.3	0.0	31.1	VF
48	24.5	40	15.3	0.0	22.9	VF
49	25.5	260	15.1	0.0	13.9	VF
50	26.5	179	14.0	0.0	21.4	VF
51	27.5	312	15.6	0.0	28.8	NF
52	28.5	272	15.5	0.0	34.8	VF
53	29.5	395	19.4	6.2	25.3	NF
54	30.5	405	17.5	13.1	25.7	VF
55	31.5	275	16.4	17.7	22.5	NF
56	32.5	852	17.2	19.8	19.2	VF
57	33.5	610	15.5	21.9	21.3	VF
58	34.5	406	20.2	16.3	22.3	VF
59	35.5	535	18.3	19.7	24.6	VF
60	36.5	663	19.6	19.4	16.3	VF
61	37.5	597	17.7	17.5	19.8	VF
62	38.5	434	20.0	14.0	27.5	VF
63	39.5	339	16.8	20.8	19.7	VF
64	40.5	216	13.3	18.1	23.3	VF
65	41.5	332	18.0	15.6	15.6	VF
66	42.5	295	16.1	19.3	15.5	VF
67	43.5	882	15.1	19.2	21.2	NF
68	44.5	600	18.0	20.6	22.2	VF
69	45.5	407	15.7	15.3	13.4	VF
70	4847.5	479	**17.8**	20.8	14.6	VF
71	48.5	0.0	**9.2**	14.1	8.7	NF
72	49.5	139	**20.5**	0.0	77.1	NF
73	50.5	135	**8.4**	0.0	57.2	NF
74	51.5	0.0	**1.1**	0.0	63.6	VF

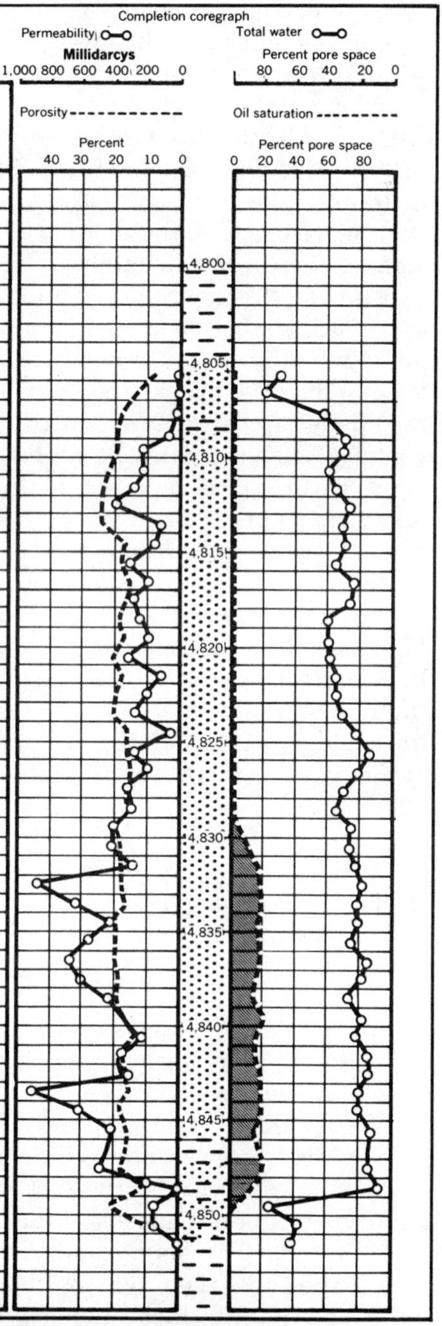

Fig. 6–2 Sample core-analysis report (after Gatlin, *Petroleum Engineering*, courtesy Baroid Petroleum Division, NL Industries Inc.)

the quantitative aspects of the relationship between the capillary rise and these parameters, it seems wise to refresh our memories concerning this phenomena.

In Fig. 6–3 a schematic core represented by three pore sizes is illustrated as being in a container of water, with the water attracted to a different level in each pore size. The capillary forces are indicated in the largest pore with the force, F, acting along the surface in the water, which forms the angle, θ, with the wall of the tube. The force is proportional to the energy required to maintain the interface between the gas and the liquid. This is called the *surface tension* for a liquid-gas interface, or for a liquid-liquid interface we use the term *interfacial tension*. Surface or interfacial tension is the units of force per unit of length. The length over which this force is applied in a capillary tube is the circumference of a circle of radius r or $2\pi r$. Thus, the total capillary force is $2\pi r\sigma$ and the vertical force can be shown to be $2\pi r\sigma \cos\theta$. When this is expressed as a pressure by dividing the total force by the cross-sectional area, πr^2, we obtain the expression for the capillary pressure:

$$p_c = 2\,\sigma \cos\theta / r \tag{6.1}$$

We can also state the capillary pressure, p_c, as a function of the unbalanced column of fluid of height, h, that it causes. We know that the pressure gradient for a column of water is 0.433 psi/ft and that the pressure gradient for a column of fluid of specific gravity, γ, is 0.433γ.

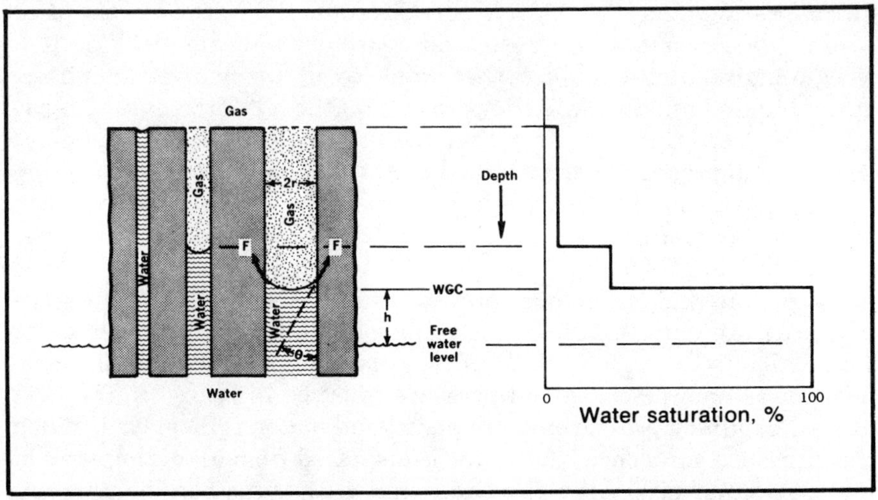

Fig. 6–3 Relationship between saturation distribution and pore-size distribution

Thus, the pressure equivalent of the unbalanced column of fluid is equal to the capillary pressure:

$$p_c = 0.433\,(\Delta\gamma)h \tag{6.2}$$

The $\Delta\gamma$ in Eq. 6.2 is the difference in the specific gravities of the two fluids. If we equate Eqs. 6.1 and 6.2, we can show that the height of the capillary rise is proportional to the interfacial tension and the cosine of the wetting angle. It also is inversely proportional to the size of the capillary and the difference in the specific gravities of the wetting and nonwetting phases:

$$h \sim \frac{\sigma \cos \theta}{r\,(\Delta\gamma)} \tag{6.3}$$

This expression is helpful in providing a qualitative knowledge of the effect of these parameters on the magnitude of the transition zone in the reservoir. The transition zone may be defined as the vertical thickness over which the saturations range from 100% water to irreducible water in the case of a WOC and from 100% liquid to an irreducible water saturation in the case of a GOC.

The general laboratory procedure used to measure the capillary pressure characteristics of a formation is to saturate the core sample with a wetting phase. Then measure how much of the wetting phase is displaced from the sample when it is subjected to some given pressure of a nonwetting phase. To define this process more accurately, we assume that the schematic core of Fig. 6–3 is taken into the laboratory and saturated with water, which we assume is a wetting phase. We further assume that the relative pore sizes and the fraction of the total pore volume represented by each pore size are indicated in Fig. 6–4. Now, imagine that we subject the sample to oil, the nonwetting phase, at the left end and increase the pressure on the oil until some water is displaced from the core. Note that the first displacement takes place when the oil pressure just exceeds the capillary pressure corresponding to the largest pore. In other words the capillary force holds the water in the largest pore until the oil pressure is larger than the capillary pressure of the largest pore.

Also, note that the amount of water that is displaced at that particular pressure represents the pore volume of all pores of that particular size, which in this case is 16/21 of the pore volume. Once this displacement has taken place, the oil pressure must be increased to the next largest capillary pressure before additional water is displaced. When this pressure is reached, the water is displaced from all of the pores of that particular size. After the additional 4/21 of the pore volume has been filled with oil, the oil pressure again must be increased if additional water is to be displaced. However, in our schematic core the

Fractional pore volume	Pore size
1/21	1/7
4/21	2/7
16/21	4/7

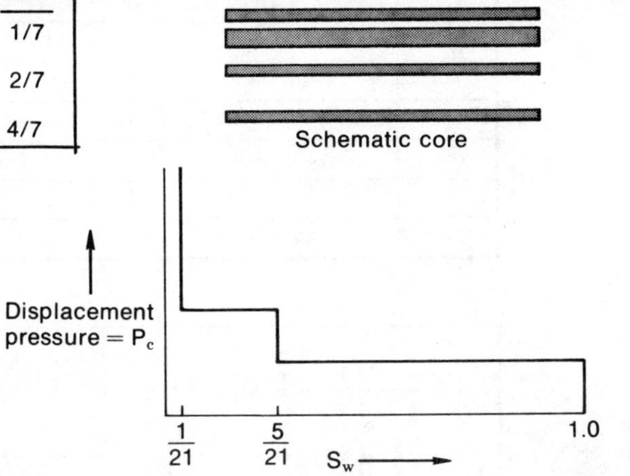

Fig. 6–4 Relationship between pore-size distribution and a capillary pressure curve

smallest pores are so small that the capillary pressure is infinite, and the water from these pores cannot be displaced. A plot of the displacement pressure versus the water displaced then represents a plot of capillary pressure versus the percent of the pores with a capillary pressure greater than the subject capillary pressure.

Remember that a reservoir rock contains a continuous variety of pore sizes from the largest pore in the formation to the smallest pores. Consequently, the capillary pressure curve obtained for an actual reservoir rock does not have discontinuities such as those in the capillary pressure curve in Fig. 6–4, which represents a hypothetical reservoir rock with only three pore sizes. Instead, the capillary pressure curve of an actual reservoir rock is a continuous smooth curve as shown in Fig. 6–5. Note that the capillary pressure curve also can be calibrated to represent a plot of pore size versus the percent of pores whose size is less than the subject pore size. This plot can be made because the interfacial tension and wetting angle of Eq. 6.1 do not change during the test.

In the laboratory many different combinations of fluids have been used to measure capillary pressure. The most common combination is probably water and air, but consider the element of time involved in the laboratory measurements. The capillary forces that are responsible for the rearrangement of the saturations in the core sample are very

Zone	Depth, ft.	Avg. Perm., md	Avg. Porosity
I	3,998 - 4,007	564	0.272
II	4,007 - 4,019	166	0.208
III	4,019 - 4,031	591	0.273
	4,031 - 4,037	Shale	
IV	4,037 - 4,046	10.2	0.142
V	4,046 - 4,055	72	0.191

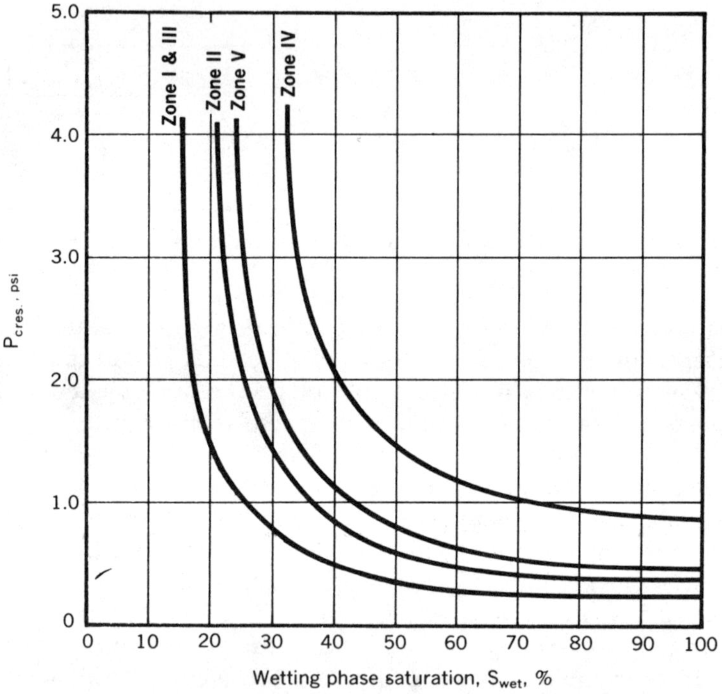

Fig. 6–5 Example capillary pressure curves for a stratified reservoir

small. As the saturation rearrangement approaches equilibrium, the driving forces become smaller and smaller and, of course, become zero at equilibrium. However, the time to approach equilibrium can be very large. It is not unusual to run a core capillary test for three months on a single core sample when water and air are used as the fluids. If fluids with a large interfacial tension such as mercury and air are used, the time to reach equilibrium is vastly improved. In fact, mercury-air capillary pressure curve can probably be run in a half hour. However, some reservoir engineers contend that an air-mercury system is too different from the actual water-oil reservoir system to be representative of reservoir capillary pressure characteristics. The author believes that the additional accuracy gained by a water-air or water-oil system in the laboratory is not justified when the expense of running these tests is considered. When both systems of tests have been run in a reservoir, the data examined by the author have seemed to correlate in an acceptable fashion.

Eq. 6.1 shows that the capillary pressure is proportional to the product of the interfacial tension, σ, and the cosine of the wetting angle, θ. Consequently, capillary pressures measured using different fluids are related as:

$$p_{c1}/p_{c2} = (\sigma \cos \theta)_1/(\sigma \cos \theta)_2 \tag{6.4}$$

Of course, the capillary data must be converted to equivalent reservoir capillary pressure data before it can be used, unless a J-function analysis is made.

Calculating the initial saturation distribution from capillary pressure data. The most direct method of determining the saturation distribution in a reservoir from capillary pressure data is to interpret the family of laboratory capillary pressure curves run for a particular reservoir in view of the general nature of the reservoir. Then one or more laboratory capillary pressure curves can be obtained to represent the reservoir. This may require J functions, permeability distribution data, and a geologic understanding of the general nature of the reservoir. Next, it is necessary to convert the lab data to reservoir data by recognizing the difference between the characteristics of the fluids used in the laboratory and the actual reservoir fluids. Then the reservoir capillary pressure curves can be interpreted in terms of a saturation distribution versus depth curve. In considering this general problem of predicting the saturation distribution from capillary pressure data, we consider the technology in reverse order in the hope that it is more readily understood when presented in this manner.

Once a capillary pressure curve has been determined that is representative of the reservoir, it can be directly interpreted to provide the saturation distribution versus depth. The first problem is to determine the depth of the free-water level since capillary pressure data is measured from this point. Fig. 6–3 shows that there is a difference between the free-water level and the minimum depth at which the 100% water saturation exists. This difference represents the capillary rise that is characteristic of the largest pore size in the reservoir. If this pore size is so large that there is no capillary rise in the pore (for example, in vugular limestone), the free-water level and the 100% water-saturation level are the same. However, in most reservoirs these two levels are different because even the largest pore size is small enough to cause some capillary rise.

In very tight reservoirs the difference in these two levels may be many feet. The feet of difference can be obtained by reading the minimum capillary pressure that corresponds to the 100% water saturation on the capillary pressure curve. Then convert this pressure to a height above the free-water level using Eq. 6.2. This height can be subtracted from the WOC—determined from well and test data as described—to calculate the free-water level in the reservoir.

The engineer should carefully note that any discussion of the saturation above the free-water level and the determination of the distance between the free-water level and the 100% water level applies equally to the calculation of the saturation distribution in the gas cap above the 100% liquid saturation. The difference in this case represents the height between the actual 100% liquid saturation level that exists in the reservoir and the 100% liquid saturation level that would exist if the reservoir contained a pore size so large that there were no capillary rise in the largest pores. Capillary pressure data would then be measured from this datum rather than from the free-water level as with the water-oil reservoir capillary pressure data.

Once the free-water level—or free 100% liquid level in the case of the gas cap—has been determined, it is then a simple matter to determine the saturation at any point above this level. The distance above the free-water level, h, can be entered in Eq. 6.2, and the corresponding reservoir capillary pressure can be calculated. The saturation corresponding to this capillary pressure can then be read from the reservoir capillary pressure curve.

To test our understanding of these concepts, work problem 6.1 and compare the solution with the one in appendix C.

PROBLEM 6.1: Determining the Static Saturation Distribution from Reservoir Capillary Pressure Data

The capillary pressure curves of Fig. 6–5 are given for the various zones of the OSU Sand. Verify the plot of water saturation versus depth for this reservoir as depicted in Fig. 6–6. The 100% water saturation depth determined from core analysis is at 4052.5 ft, the water density is 65.3 lb/cu ft and the oil density is 56.2 lb/cu ft.

A well-stratified reservoir with much different permeabilities in the various strata can result in some unexpected saturation distribu-

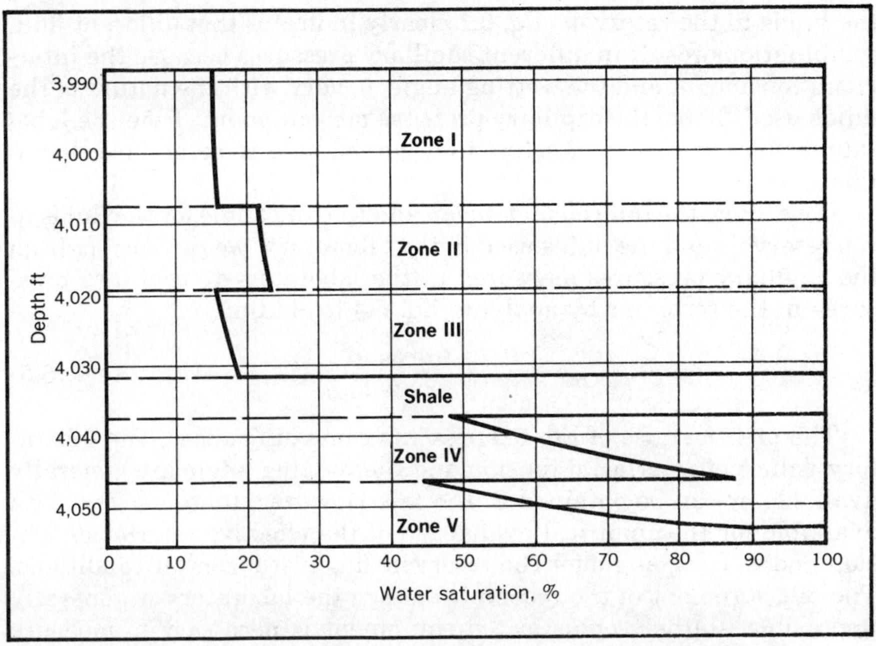

Fig. 6–6 Saturation distribution for reservoir with characteristics of Fig. 6–5

tions. The capillary pressure characteristics and zonation depicted in Fig. 6–5 and fluid characteristics enumerated in problem 6.1 result in a saturation distribution in the stratified reservoir as shown in Fig. 6–6. Note that it is entirely possible to have substantially higher water saturations located vertically above the lower water saturations. This phenomena may explain some of the reports from cable-tool drillers of water production above oil productive zones and gas production below oil zones in continuous reservoirs.

We should also note that differing reservoir characteristics at the free-water level in a reservoir can cause a large difference in the vertical position of the WOC observed in various portions of a reservoir. The spread of the reservoir capillary pressures representing the 100% water saturation points in a suite of capillary pressure curves such as Fig. 6–5 give some insight into the variation in the observed WOC that may be expected in a particular reservoir.

Calculating reservoir capillary pressure data from lab data. As discussed, many different combinations of fluids are used in the laboratory for measuring the capillary pressure characteristics of a reservoir sample. Whether oil-water, water-air, or mercury-air fluid systems are used for the measurements, the laboratory fluids are not the same as the fluids in the reservoir. Eq. 6.1 clearly indicates that different fluid combinations result in different capillary pressures because the interfacial tension, σ, and the wetting angle, θ, vary with the nature of the fluids used. Thus, the capillary pressure measurements from the laboratory must be corrected before they can be used in reservoir calculations.

If we know the interfacial tension and wetting angle for the fluids in the reservoir and the fluids used in the laboratory, we can correct from the capillary pressures measured in the laboratory to capillary pressures in the reservoir by applying Eq. 6.4 to obtain:

$$p_{c\ res} = p_{c\ lab} \frac{(\sigma \cos \theta)_{res}}{(\sigma \cos \theta)_{lab}} \tag{6.5}$$

The practical use of Eq. 6.5 presents some difficulties. The laboratory values of interfacial tension and the wetting angle are generally available or can be obtained in the lab. However, there are few data available for the empirical evaluation of the effective interfacial tension and wetting angle for the reservoir fluids at reservoir conditions. The measurement of the reservoir data in the laboratory is generally impossible. Rather sophisticated equipment is necessary to measure the interfacial tension with gas in solution in the oil at a relatively high reservoir pressure and temperature. Some data that can be used on an analogy basis can be found in the *Handbook of Natural Gas Engineering,* by Katz et al.[1]

Note that the units of the interfacial tension, σ, used in Eq. 6.5 are immaterial since a ratio is employed. One of the difficult problems that confronts the reservoir engineer in choosing capillary pressure data to represent the reservoir is what we may call the *zonation* of the reservoir. The problem is one of characterizing the reservoir as far as recognizing vertical changes in the reservoir permeability and porosity. We talk about a layered reservoir as one in which, for example, the 100-md zone is at the top, the 50-md zone is in the middle, and the 500-md zone is in the bottom of the reservoir thickness. However, from well to well in this reservoir, we may find one of the zones missing, or worse, the sequence of permeabilities from top to bottom may be changed from low-middle-high to high-low-middle or middle-high-low. In many cases no effective zonation exists, but in others we may find an average zonation with wide swings around this average from well to well.

There are some statistical methods of mathematically determining the best zonation to consider in a well, but the author believes that, at this stage in the development of reservoir description methods, the reservoir engineer can arrive at the best zonation treatment by simply examining the data from well to well. Then use good judgment as to how to group the reservoir permeability and porosity variations into zones with average permeabilities and porosities representing each zone.

Problem 6.2 is provided to test our knowledge of the conversion of laboratory capillary pressures to reservoir data and to familiarize ourselves with the problem of zonation. The solution is given in appendix C.

PROBLEM 6.2: Conversion of Laboratory Capillary Pressure Data to Reservoir Capillary Pressure Data

The laboratory water-air capillary pressure curves are given in Fig. 6–7, and the core analysis for OSU well 1 are listed in the following table. Verify the capillary pressure data and zonation in Fig. 6–5 by calculating the reservoir capillary pressure for each zone. The water-oil interfacial tension from this reservoir is estimated to be 28 dynes/cm, and the wetting angle is 0°. Note that the core-analysis data are given as an average for every 3 ft to simplify the problem.

Depth	*Permeability, md*	*Porosity, fraction*
3,998–4,001	578	0.272
4,001–4,004	542	0.264
4,004–4,007	559	0.277
4,007–4,010	184	0.212
4,010–4,013	196	0.204
4,013–4,016	111	0.207
4,016–4,019	172	0.210
4,019–4,022	641	0.279
4,022–4,025	648	0.266
4,025–4,028	567	0.271
4,028–4,031	508	0.276
4,031–4,034	Shale	—
4,034–4,037	Shale	—
4,037–4,040	9	0.140
4,040–4,043	14	0.164
4,043–4,046	8	0.121
4,046–4,049	63	0.175
4,049–4,052	81	0.196
4,052–4,055	72	0.202

Averaging capillary pressure data. The permeability and porosity of reservoir samples whose capillary pressures are measured in the laboratory seldom match the characteristics of the average permeability and porosity whose capillary pressure is desired. In most cases a suite of capillary pressures is run on a variety of reservoir samples with a wide range of permeabilities, but none of these samples corresponds as closely to those desired as the ones in problems 6.1 and 6.2. Consequently, the engineer is generally faced with a mass of capillary pressure data, none of which is exactly what is wanted. Furthermore, the method to be used in averaging the available data is not always appar-

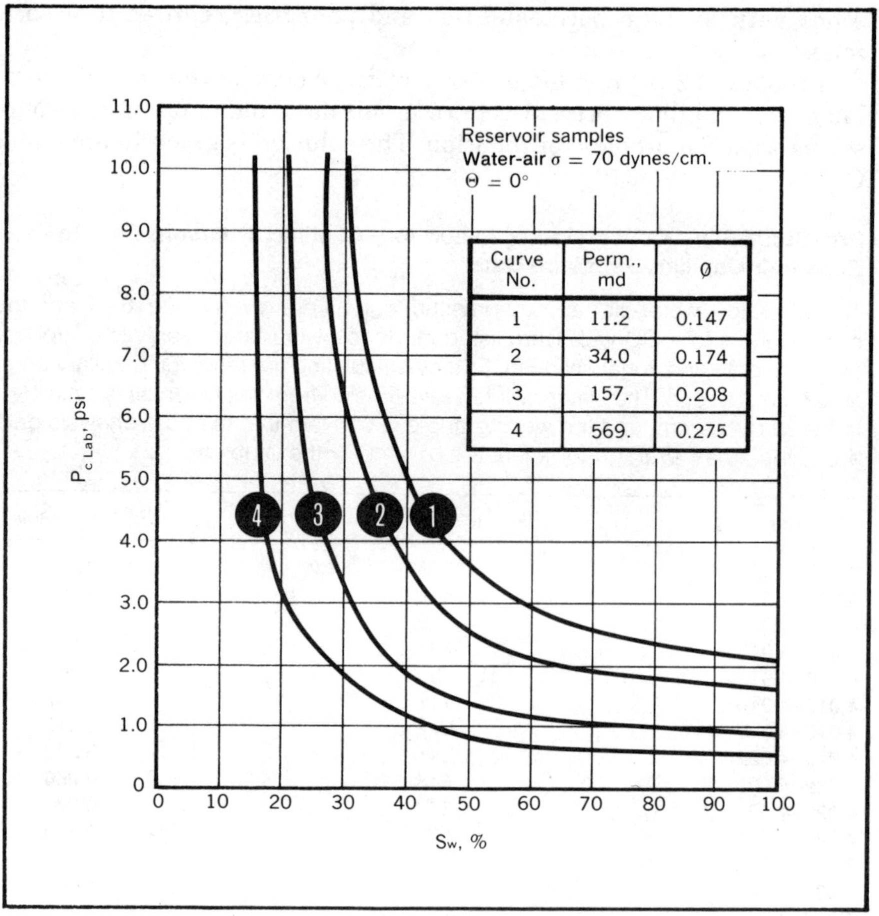

Fig. 6–7 Capillary pressures of reservoir samples

ent. The suite of capillary curves may cover the entire graph when they are plotted on a single set of coordinates as in Fig. 6–7.

One of the classical methods of averaging capillary pressure data is using a J function. Sometimes, the criteria necessary to make this technique accurate are apparently met, and the procedure is very useful. However, other sets of data do not appear to conform at all to the assumptions of the method, and the procedure is useless. Nevertheless, the procedure provides a good understanding of the relationship between the capillary pressure characteristics and the permeability and porosity of the porous media. Since it also is often a useful averaging technique, the J function warrants our consideration.

The J-function concept is based on the theoretical relationship between the permeability and porosity that can be derived by assuming that a porous media can be characterized as a bundle of capillary tubes that are not interconnected. The flow rate through one capillary tube according to the Poisuelles equation is:[2]

$$q' = \frac{\pi (r')^4 \, \Delta p'}{8\mu \, L'_{cap}} \tag{6.6}$$

Where:
q' = Volumetric flow rate in Darcy units

Consequently, the flow rate through a number of capillary tubes, n, is:

$$q' = \frac{n\pi (r')^4 \, \Delta p'}{8\mu \, L'_{cap}} \tag{6.7}$$

Note that the porosity of a formation made of a bundle of capillaries is:

$$\phi = n\pi r^2/A \tag{6.8}$$

The permeability defined by the Darcy equation is:

$$k = q' \, \mu \, L'_{core}/A'\Delta p' \tag{6.9}$$

If ϕA is substituted for $n\pi r^2$ and a function of the permeability, k, is substituted for q' in Eq. 6.6, we can show that the permeability and porosity are related:

$$(r')^2 = (8k/\phi)\,(L'_{cap}/L'_{core}) \tag{6.10}$$

Note that the length of the capillary is recognized as being different from the core length. This becomes obvious when we recognize that a particular capillary path through a core is not a direct route from one face to another. It actually tends to wind around in the porous media and follow a tortuous path. Consequently, this hypothetical ratio between the core length and the capillary length is often defined as the *tortuosity*.[3] If we assume that the tortuosity is constant for a particular reservoir and we combine this with the numerical constant of Eq. 6.10, we can show that the radius is related to the permeability and porosity as:

$$r = (\text{constant})(k/\phi)^{0.5} \tag{6.11}$$

Substituting this definition of the radius into the definition of the capillary pressure, Eq. 6.1, we obtain:

$$p_c = 2\sigma \cos\theta/(\text{constant})\,(k/\phi)^{0.5} \tag{6.12}$$

When we solve this expression for the constant and the numerical constant of the equation, we obtain the J function:

$$J = \frac{p_c (k/\phi)^{0.5}}{\sigma \cos\theta} \tag{6.13}$$

Physically, by normalizing a capillary pressure measurement for the differences in the permeabilities and porosities of the samples and the differences in the fluids used to measure the capillary pressure, a function of the capillary pressure is obtained that is independent of the permeability, porosity, interfacial tension, and the wetting angle. Furthermore, this function can be applied to a reservoir with any permeability and porosity that contain a combination of wetting and nonwetting fluids characterized by a particular interfacial tension and wetting angle.

Consequently, this function of the capillary pressure can be calculated from capillary pressure data and the characteristics of the samples and fluids used. One single curve is obtained representing all of the capillary pressure data for all of the samples measured. Therefore, a single J-function curve represents many different permeabilities and porosities, and the water-air, water-oil, and mercury-air laboratory systems can be represented in one curve. Several sets of reservoir data have been observed by the author that fit the J-function theory. One such example is illustrated by the J-function curve in Fig. 6–8.

However, always keep in mind that the J function is based on the assumptions that the porous media acts like a bundle of capillaries and that the ratio between the length of the capillaries and the core length is a constant. Apparently, there are many reservoirs that do not fit these assumptions since much data that are almost useless for interpretation purposes are often obtained when a J-function plot is made for a particular reservoir. Another method of averaging capillary pressure data is presented, but first it is suggested that the engineer work problem 6.3. The solution is given in appendix C.

PROBLEM 6.3: Using the J Function to Average Capillary Pressure Data

Given the capillary pressure data in Fig. 6–7, verify the J-function plot of Fig. 6–8 by calculating the J function at the 50% saturation point for each curve in Fig. 6–7. Using the J-function curve in Fig. 6–8, calculate and plot a reservoir capillary pressure curve for zone II of problem 6.1.

If the J-function curve looks like a buckshot spread or fails to give a well-defined relationship, it is obvious that the reservoir with which we are working does not satisfy the assumptions inherent in the J-function method. The engineer should then look to other averaging techniques. One approach that often gives meaningful data is the method of weighing the capillary pressure data on the basis of the statistical permeability distribution. In chapter 2 the fact is recognized that even in the most uniform-looking, naturally porous media, the measurement of permeability of small samples results in a wide range of measured per-

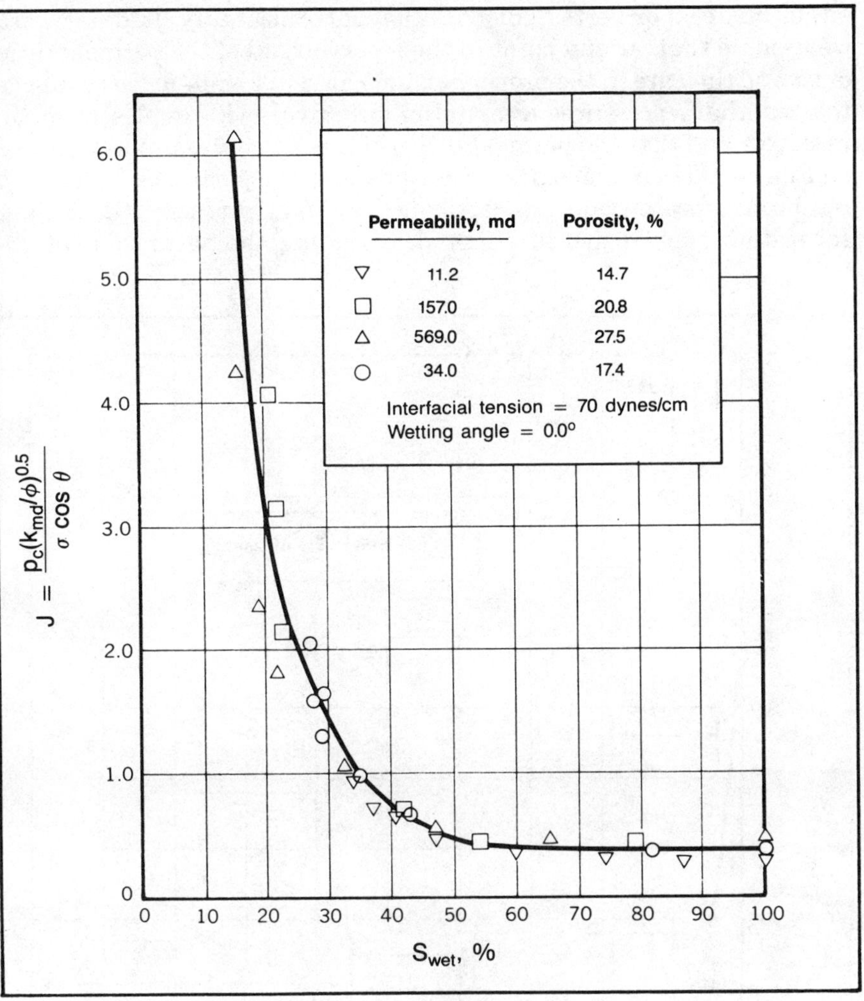

Fig. 6–8 A J-function curve

meabilities. Example data was presented in Fig. 2–1. Such data can be converted to statistical form by changing the x-axis scale from "thickness" to "percent of reservoir." Then it is possible to determine the percentage of the reservoir that should be represented by each capillary pressure sample on the basis of the permeability of that sample and the permeability distribution of the entire reservoir.

This simple technique can best be described by referring to some example data. Suppose that the statistical study of the permeability measurements for a reservoir indicates a permeability distribution such as that in Fig. 6–9. Also, assume that there is no indication of either an areal or vertical distribution of permeability. In other words we assume that, at any point in the reservoir, all of the permeabilities exist and they are in the proportion indicated in Fig. 6–9. Now, suppose that capillary pressures are run on only five core samples from this reservoir and that the permeabilities of these samples are as indicated on Fig. 6–9. Then it is logical to assume that at a particular level in the reservoir represented by a particular capillary pressure, the average saturation can be measured by determining the saturation at this

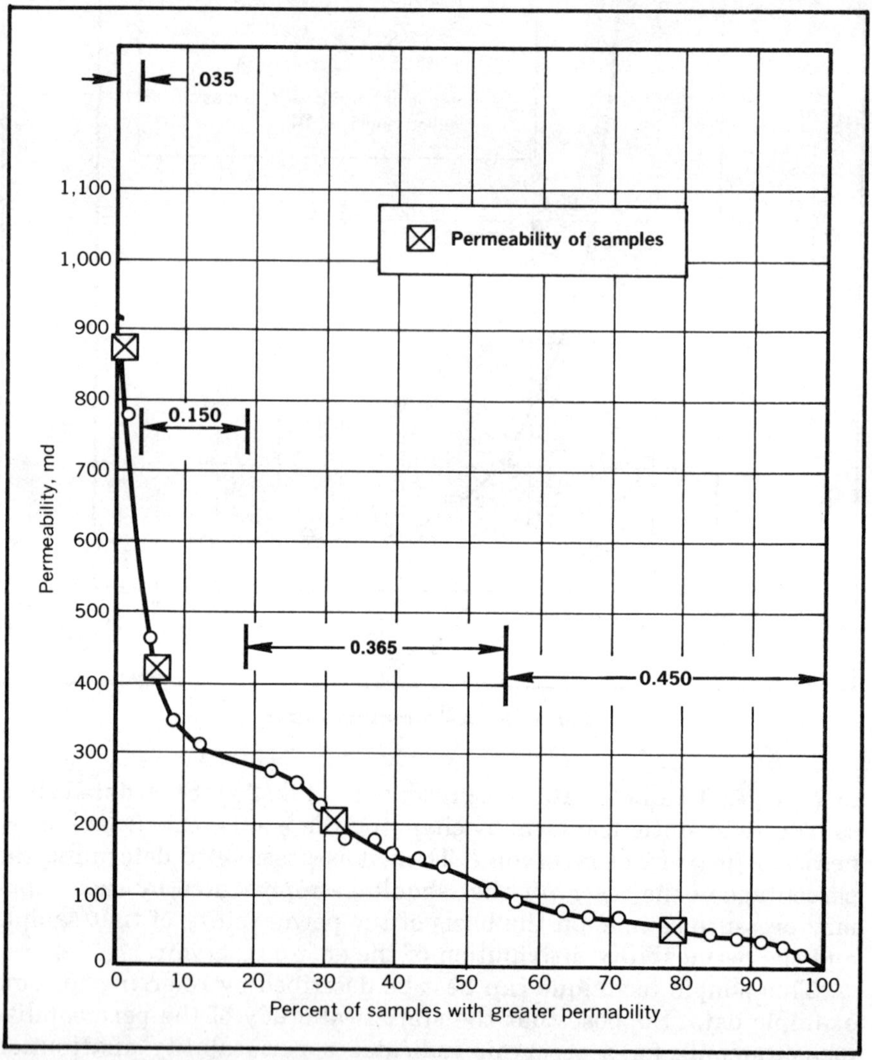

Fig. 6–9 Determining weighting factors from permeability distribution

level, or capillary pressure, for each of the samples. Then a weighted average can be calculated on the basis of the percent of the reservoir that can be considered to be represented by each sample.

For example, the relationship between the permeability distribution of Fig. 6–9 and the permeabilities of the capillary samples indicates that the sample permeabilities of 875, 420, 210, and 55.7 md represent 0.035, 0.150, 0.275, and 0.450 fractions of the reservoir, respectively. Consequently, the saturations for a particular capillary pressure of reservoir level should be weighted by these fractions.

Other capillary pressure uses. Capillary pressure data may have more practical qualitative uses than they have quantitative uses. As noted, these data can be used quantitatively to determine the initial saturation distribution in the reservoir and to provide an understanding of the relationship between the permeability and porosity of the porous media through the J function. Capillary pressure characteristics also provide a qualitative means of determining the pore-size distribution and an understanding of the phenomena that account for the nonwetting-phase irreducible saturation in porous media.

From the equation for capillary pressure, $p_c = 2\sigma\cos\theta/r$, we see that a knowledge of the interfactial tension and the wetting angle, θ, permits the calculation of a pore radius that is characteristic of any particular capillary pressure, p_c. Thus, a capillary pressure curve can be easily converted to a pore-size distribution curve. In fact, this method is routinely used to determine the pore-size distribution of porous materials.

Of even more interest to the reservoir engineer is the indication of the pore-size distribution rather than the absolute values of pore size. For example, in Fig. 6–7 we note that curve 4 contains nearly 43% of its pores with almost the same pore size indicated by a range of capillary pressures from about 0.6–0.7 psi. Curve 1 has only 8% of its pores with a comparable uniform pore size from 2.1–2.2 psi. Such a knowledge of a reservoir is very useful. For example, if these two core samples are flooded, we would expect to have about 43% of the oil recovered from sample 4 before water breakthrough. A similar flood of sample 1 would only recover 8% of the oil before water breakthrough.

The engineer is sometimes confused by the fact that all of the nonwetting-phase oil cannot be displaced by the wetting-phase water, which should appear simply to imbibe into the pores of the ordinary sandstone displacing all of the oil. However, we find that when the nonwetting-phase oil becomes discontinuous in the porous media, the capillary attraction can work against the displacement of the oil. For example, consider Fig. 6–10, which represents a pore with a change in radius. Suppose we are trying to displace the oil to the right with water.

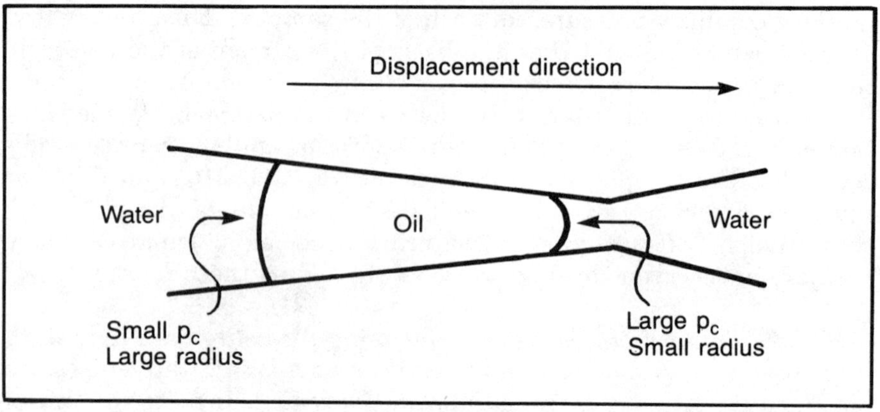

Fig. 6–10 Residual oil trap

Since the interconnected pores in the reservoir are different sizes, the water and oil flow through these pores at different velocities, which causes the oil and water to mix at the intersections. Some of the oil also is broken into discontinuous globules. Then when a drop of oil tries to move through a pore with a changing radius as in Fig. 6–10, it is subjected to two capillary pressure forces. The capillary pressure on the right acting in a direction to the left is based on the smaller pore size at this point. Thus, it is a larger capillary pressure than the capillary pressure on the left acting in a direction to the right because the pore radius on the left is larger. Consequently, the difference in capillary pressure forces tends to move the drop of oil to the left. Thus, in order to move the oil to the right through the restriction, a pressure gradient must exist that is greater than the difference in the two subject capillary pressures. Otherwise, the oil does not move through the restriction, and it becomes a part of the residual oil.

Reversing the direction of the displacement causes the subject drop of oil to be freed. However, we must remember that most of the drops of oil freed by reversing the direction of the displacement probably encounter a similar restriction in the opposite direction. Thus, they still remain as residual oil. It has been reported that changing the direction of flow in a waterflood, i.e., making the injection wells producers and vice versa, has resulted in some small amount of additional production.

This discussion should make it clear that reducing the interfacial tension between the water and oil reduces the residual oil. Thus, this reduction results in an increase in the ultimate recovery as a result of flood. Furthermore, making the interfacial tension zero displaces all of the oil contacted by the water. Such a displacement with a zero interfacial tension is termed a *miscible displacement*.

Reservoir Saturation Distribution during Displacement

In our consideration of the initial saturation distribution, we noted that an engineer's first assumption without any theoretical knowledge of the situation is that initial saturations are uniform throughout the water zone, oil zone, or gas cap of a reservoir. The same engineer may assume that, as production takes place in a reservoir, the gas cap expands or the water encroaches and the saturation in the invaded portions of the reservoir becomes uniform. Such an assumption is often referred to as *piston-like displacement.* Just as uniform saturation distribution seldom exists in a reservoir, piston-like displacement seldom takes place in the reservoir.

Actually, in the frontal displacement of oil by advancing gas or water, a more or less uniform saturation distribution is formed with a discontinuity only at the furthermost advance of the gas or water. Our objective is to consider methods of determining the saturation distribution along the displacement stream lines. The concepts developed can be directly useful as a means of determining the saturation distribution in water-drive and gas-cap-drive reservoirs, and they also provide general concepts concerning the displacement of one immiscible fluid by another in a porous media. These ideas should be particularly helpful in the consideration of secondary recovery.

We first consider the general characteristics of the displacement and the general procedure to be followed in calculating the saturation distribution at any particular time. Once this procedure has been established, the detailed steps in the analysis and some useful simplifying short cuts are developed.

General characteristics of fluid displacement. To illustrate what happens when one fluid displaces another in the reservoir, consider a water-drive reservoir such as that shown in the cross section in Fig. 6–11. In this figure water is encroaching updip at a relatively slow rate as the oil is produced near the top of the structure. Now, consider the saturation in the horizontal slices of the reservoir as the displacement proceeds. We do this by identifying each horizontal slice as being some distance, X, from the initial minimum depth of the 100% water saturation with the distance, X, measured along the bed dip of the formation. When the initial saturation in these horizontal slices is plotted versus the X position of each slice, we obtain a relationship such as that shown in Fig. 6–11.

Consider the nature of the saturation distribution after one year when a considerable amount of oil has been produced from this reservoir and a like amount of water has encroached into the initial oil-bearing portion of the reservoir. The resulting saturation distribution

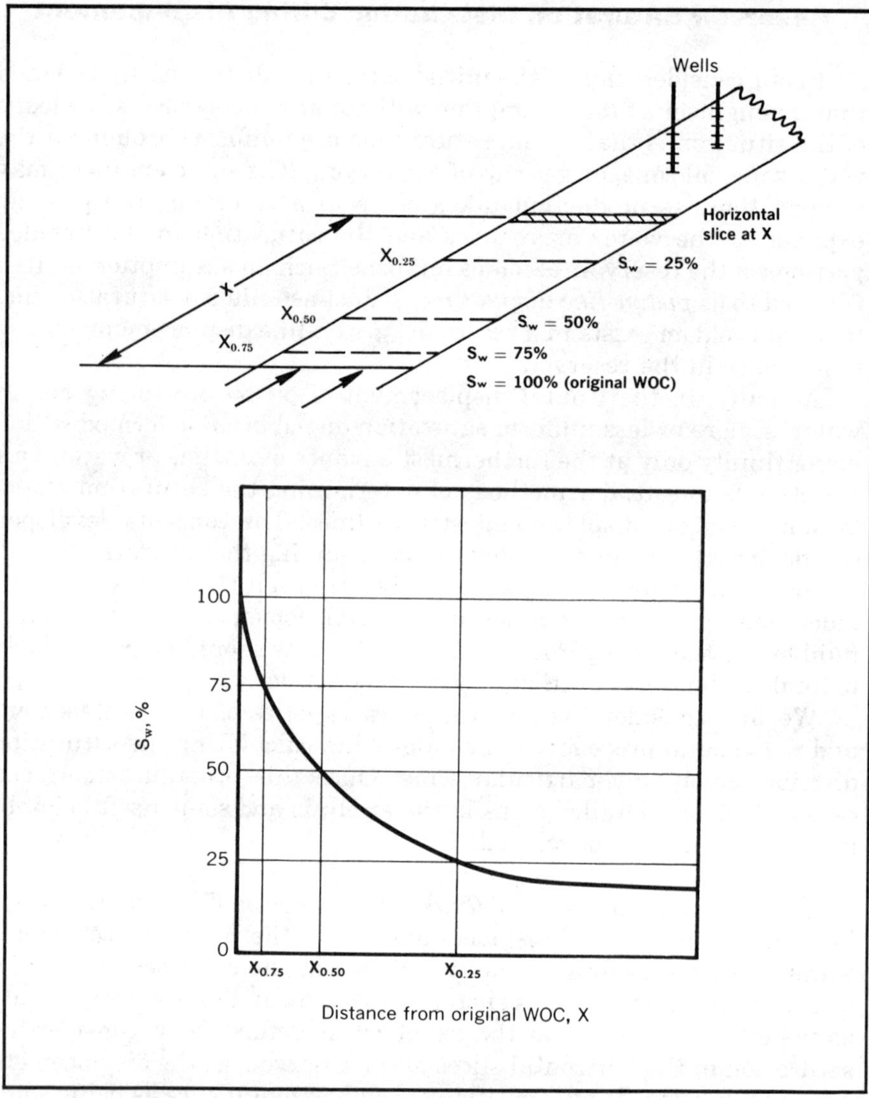

Fig. 6–11 Saturation profile in a water-drive reservoir

is shown in Fig. 6–12 with the initial saturation distribution and the saturation distribution at later times of 2 and 3 years. This figure has been generalized by using the displacing-fluid saturation symbol, S_d. Also, the position of the original WOC is labeled as the "influx face," and the position of the nearest producer is termed the "producing face."

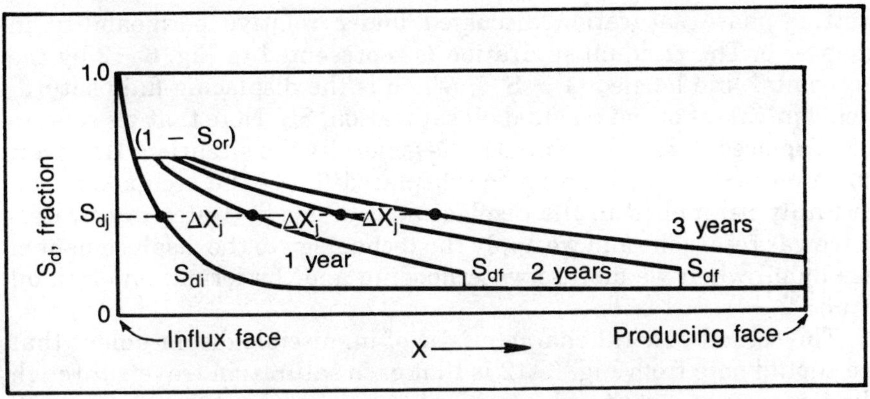

Fig. 6–12 Fluid-displacement characteristics with initial saturation distribution

Notice that the only saturation discontinuities that exist are at the front or furthermost advance of the displacing fluid. Also, note that the displacing-phase saturation is increasing at all times at most points in the reservoir behind the advancing front. In other words the advancing front does not displace all of the mobile oil as it moves through the reservoir. Instead, it acts more like a very inefficient piston. The front of the displacing fluid corresponds to the first stroke of the inefficient piston, which displaces some fraction of the mobile oil. As water continues to flow through the same pore volume, it acts like successive piston strokes with some percentage of the mobile oil that is left being displaced. Finally, after many pore volumes have flowed through the same pore space—many piston strokes have taken place—all of the mobile oil has been displaced. The zone behind the displacing front is sometimes referred to as the *drag zone,* which seems to be fairly descriptive of what takes place physically in this part of the displacement.

The engineer should be careful not to become confused as to the meaning of Fig. 6–12. A common error is to interpret this figure as indicating some vertical distribution of fluid. The y axis is then mistaken to be the vertical distance so there is water or displacing fluid underriding the oil. This of course is an erroneous impression. Remember that the diagram shows the average saturation in horizontal slices of the reservoir plotted against the distance from the original 100% displacing-phase saturation measured along the bedding plane.

We should note two general characteristics of fluid displacements in porous media that are clearly indicated in Fig. 6–12. First, there is a saturation below which the saturation of the displaced (oil) phase cannot be driven regardless of the amount of displacing fluid that passes through the porous media. This is consistent with the irreducible nonwetting-phase saturation discussed under relative permeability in chapter 5. The residual saturation is represented in Fig. 6–12 by the horizontal line labeled $(1 - S_{or})$, which is the displacing-fluid saturation equivalent of the residual oil saturation, S_{or}. Note that we refer to the displaced fluid as oil since this is generally the situation. However, any immiscible fluid can be the displaced fluid. The technology can certainly be applied to the displacement of gas by water in a water-drive gas reservoir, and we apply the technology to the displacement of gas by oil when we discuss waterflooding and the formation of an oil bank.

The second general characteristic of immiscible displacement that we should note from Fig. 6–12 is that each saturation travels through the reservoir at some fixed velocity as long as the displacement rate, q_t, is constant. In other words if oil is displaced by a natural water drive, as originally suggested, and the production and encroaching water rates are constant, the 50% water saturation would move through the reservoir at some constant rate; the 60% water saturation would move through the reservoir at a constant but greater rate; and the 65% water saturation would travel at a constant but still greater rate. This characteristic is illustrated in Fig. 6–12 by indicating that a particular saturation, S_{dj}, moves a distance ΔX_j through the reservoir during each year. Carefully note that during the first year this distance of movement is measured from the initial position of the S_{dj} saturation in the reservoir and is not measured from the 100% S_d saturation. This is pointed out specifically because such an error in application is very common.

We should also note that the frontal saturation does not remain constant when an initial saturation profile such as the one in Fig. 6–12 is encountered. This point is made to contrast Fig. 6–12 with the situation that exists in Fig. 6–13 where the frontal saturation does remain constant. The difference in the conditions of Figs. 6–12 and 6–13 is that in Fig. 6–13 the saturation initially is uniform throughout. This situation can represent a waterflood being initiated in a reservoir that does not have a natural water drive or gas-cap drive. It also can represent the situation that is approximated when the pores are so large in a reservoir that the transition zone can be treated as being of negligible thickness. In either case and in many others, the initial saturation conditions can be treated as being uniform throughout the reservoir.

Note that having an initial uniform saturation throughout the reservoir actually means that a very large, abrupt discontinuity exists in the reservoir at the influx face. To the left of the influx face shown in Fig. 6–13, we know that the displacing-phase saturation is 100%. To the right the saturation is S_{di}, the initial displacing-fluid saturation that is uniform throughout the oil-bearing portion of the reservoir.

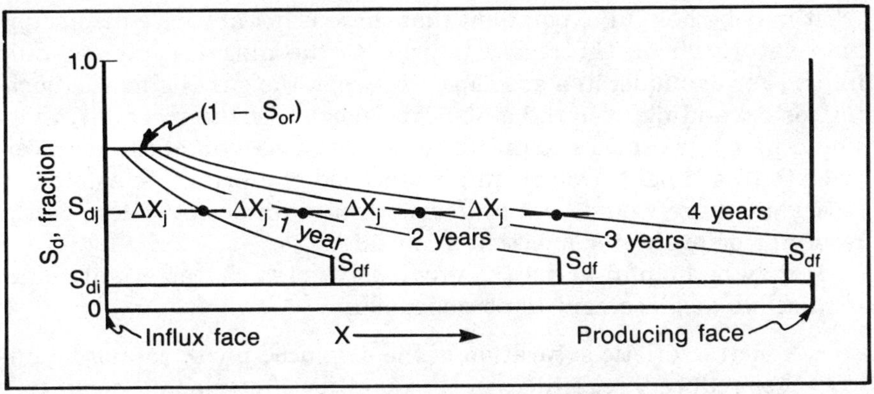

Fig. 6–13 Fluid-displacement characteristics with initial uniform saturation

Therefore, all of the S_d saturations from S_{di} to $(1 - S_{or})$ exist at the same point in the reservoir at the influx face.

Fig. 6–13 shows that the results of the displacing-fluid encroachment into a reservoir with a uniform initial saturation are very similar to the results of a displacing-phase encroachment into a reservoir that has a nonuniform saturation distribution initially. Note that there is still a lower limit on the oil saturation that cannot be further reduced regardless of how much of the displacing fluid travels through the reservoir. Also, the velocity of a particular saturation is constant and different from the velocity of other saturations. However, it should be remembered that the initial position of all of the saturations is the influx face in contrast to the situation in Fig. 6–12, in which each saturation has a different initial position.

Some general observations should be made about Fig. 6–13 that are peculiar only to a displacement occurring in a reservoir where the displacing-fluid saturation is uniform throughout the reservoir initially. In this case the frontal saturation remains constant as noted, until it reaches the producing face. When this constant frontal saturation is coupled with the constant velocity of each saturation starting from the same point in the reservoir (the influx face), we can show that the average saturation behind the displacing front remains constant. This portion of the displacing-phase system is generally referred to as the *displacing-fluid bank*. For example, in a waterflood it is called the *water bank*. Although the displacing-fluid bank continues to increase in size as the displacement continues, remember that the average saturation in this bank remains constant until the front reaches the producing face if the initial displacing-fluid saturation is uniform.

It may be helpful to point out that there is not always a displacing-fluid saturation in the reservoir prior to the initial displacing-fluid influx. For example, in a gas-cap drive when the gas begins encroaching, or expanding, into the oil-bearing portion of the reservoir, there would be no gas saturation in the oil-bearing portion of the reservoir prior to this time. However, in a waterflood of a primary-depleted oil reservoir, there would be a uniform connate-water saturation in the reservoir before water injection is initiated.

It may be helpful to list the previously noted characteristics of an immiscible displacement in porous media:

- A characteristic saturation of the displaced phase cannot be further reduced, regardless of the quantity of displacing phase that flows through the porous media.
- As long as the influx rate remains constant, each displacing fluid saturation moves through the reservoir at a particular velocity that is characteristic of that particular saturation and different from the velocity of other saturations.

If the displacement takes place in a reservoir whose displacing-fluid saturation is initially uniform throughout the reservoir, we also note that:

- The saturation at the displacing-fluid front remains constant until it reaches the producing face.
- The average saturation in the displacing-fluid bank remains constant until the front reaches the producing face.

Calculating the saturation distribution during a displacement. The fraction of the total fluid flowing that is displacing fluid varies with the distance behind the displacing-fluid front because the saturation in the displacing-fluid bank varies with the distance behind the front. This can be better understood by examining the nature of two-phase relative permeability data representing the displaced and displacing fluids. For example, examine Fig. 6–14, which is the water-oil relative permeability data for a water-displacement problem. Note that when the saturation in the reservoir is 55% water and 45% oil, the relative permeabilities to water and oil are the same. Consequently, if the viscosities of the two fluids are equal, the flow rates of the fluids are equal. Then, at a point in the reservoir with this saturation, half of the flowing fluid is water and half is oil.

Consider the situation at a point in the reservoir where the saturations are about 45% water and 55% oil. From the relative permeability data note that the ratio of the permeabilities is about 3:1 with oil as the larger. Consequently, if we again assume the viscosities to be equal,

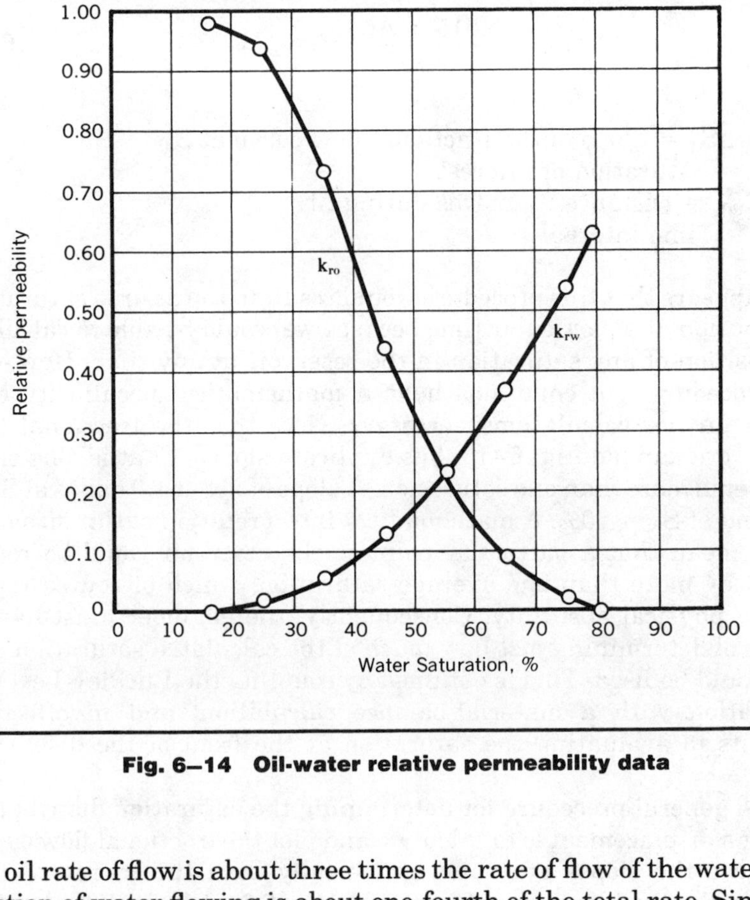

Fig. 6–14 Oil-water relative permeability data

the oil rate of flow is about three times the rate of flow of the water. The fraction of water flowing is about one-fourth of the total rate. Since the fraction of the displacing phase that is flowing varies with the saturation, which in turn varies with the distance behind the displacing-phase front, the saturations behind the front must continually change with time until the irreducible saturation is reached.

Consequently, the first step in calculating the saturation distribution during a displacement is to determine the fractional flow curve based on the relative permeabilities, viscosities, and other reservoir parameters. Such a relationship is a plot of the fraction of the total flow rate that is displacing fluid versus the saturation as illustrated in Fig. 6–15. This fractional flow curve can then be used to evaluate the change in the fractional flow with the change in saturation at any particular saturation ($\Delta f_d/\Delta S_d$). This value in turn is used in the

Buckley-Leverett equation to calculate the position of a saturation in the reservoir at a particular time. The Buckley-Leverett equation for linear flow is:[4]

$$\Delta X_{Sdj} = \frac{5.615\ q_t\ \Delta t}{\phi A} (\Delta f_d/S_d)_{Sdj} \tag{6.14}$$

Where:

$\Delta f_d/\Delta S_d$ = Slope of the fractional flow curve at S_{dj}
S_{dj} = Saturation of interest
ΔX_{Sdj} = Distance S_{dj} moves during Δt
Δt = Time interval

It appears that this procedure would result in the desired saturation distribution at a particular time because we would be able to calculate the position of any saturation in the reservoir at any time. However, the procedure and equations have a mathematical peculiarity that makes further calculations necessary. Note that the fractional flow curve, for example Fig. 6–15, has duplicate slopes. That is, one slope represents more than one saturation. A slope of about 1.0 exists at S_w = 32% and at S_w = 70%. A mathematical interpretation of this situation would mean that a particular point in the reservoir could be represented by more than one average saturation, which of course represents a physical absurdity. Consequently, the engineer must have a means of determining just how much of the calculated saturation profile should be used. This is obtained by coupling the Buckley-Leverett calculation with a material-balance calculation, and in effect, it amounts to evaluating the saturation at the front of the displacing bank.

The general procedure for determining the saturation distribution during a displacement is to calculate and plot the fractional flow curve. Then use the slopes evaluated graphically from the fractional flow curve in the Buckley-Leverett equation, Eq. 6.14, to calculate the position of the various saturations at the time of interest. Determine the frontal saturation at the time of interest.

Determining the fractional flow curve. The fractional flow, f_d, is simply defined as the fraction of the total fluid flow that is caused by the flow of the displacing phase. If only two phases are flowing at a particular spot in the reservoir and the displaced phase is oil, we define the fractional flow as:

$$f_d = \frac{q_d}{q_d + q_o} \tag{6.15}$$

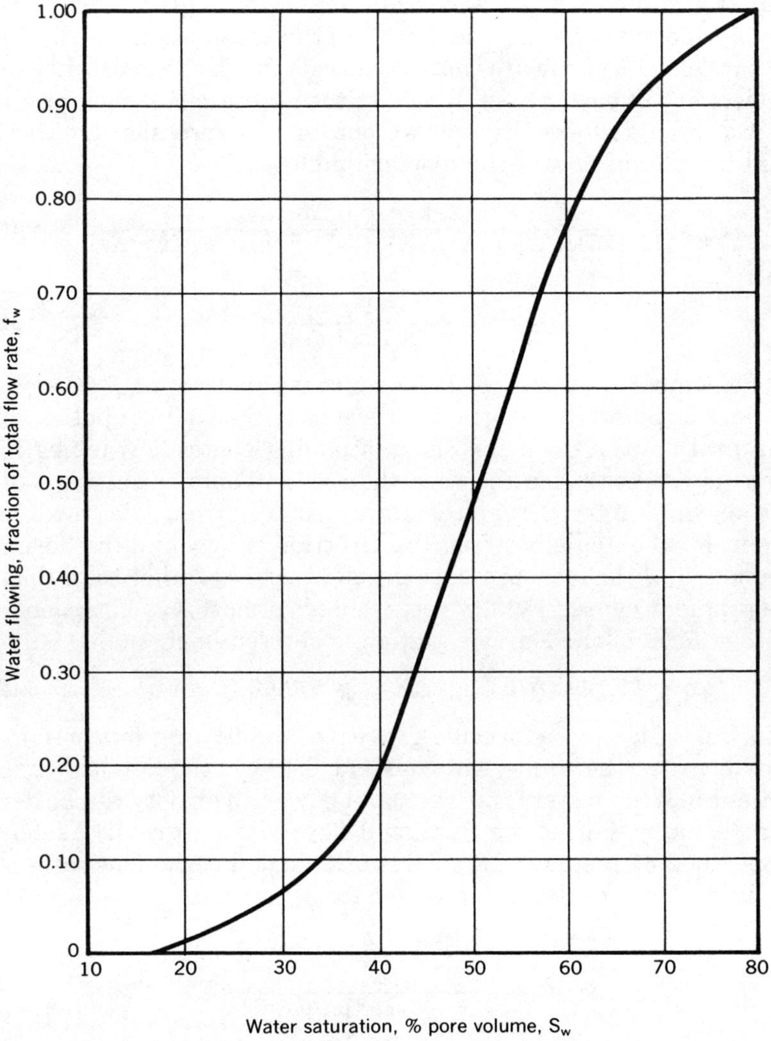

Fig. 6–15 Fractional flow curve

Now, suppose we substitute for the flow rates according to the Darcy equation originally introduced in chapter 2:

$$q = -(1.127kA/\mu)\,(\Delta p/\Delta X) \tag{2.3}$$

If we assume that flow is horizontal and the capillary pressure can be neglected, we can conclude that all of the parameters controlling the flow rate of each phase are the same except for the permeability to and the viscosity of each phase. By substituting the resulting expressions into Eq. 6.15 and simplifying, we obtain an expression for the fractional horizontal flow of the displacing phase:

$$f_{dNO} = \frac{1.127k_dA/\mu_d)(\Delta p/\Delta X)}{(1.127k_dA/\mu_d)(\Delta p/\Delta X) + (1.127k_oA/\mu_o((\Delta p/\Delta X)} \tag{6.16}$$

$$f_{dNO} = \frac{1}{1 + (k_o/k_d)(\mu_d/\mu_o)} \tag{6.17}$$

The symbol f_{dNO} is used to denote that this fractional flow expression does not contain any gravity effects, or as stated, it applies only to horizontal flow or when the effect of nonhorizontal flow is negligible. When flow takes place in other than a horizontal direction, we can show as in chapter 2 that the Darcy equation must be modified to account for the angle between the direction of flow and the horizontal direction, α, if the total pressure gradient is used rather than the pressure gradient caused by flow represented in most flow equations. The modified form of the Darcy equation is derived in chapter 2 is:

$$q = (1.127kA/\mu)[(\Delta p/\Delta X)_{total} \pm 0.433\gamma \sin \alpha] \tag{2.61}$$

In Eq. 2.61 γ is the specific gravity of the flowing fluid relative to water because the numerical constant 0.433 is the vertical pressure gradient in fresh water. The fractional flow with gravity can be derived in the same way that the fractional flow without gravity is derived, except Eq. 2.61 instead of Eq. 2.3 is used to define the flow rates of the two fluids. The resulting expression is:

$$f_d = \frac{1 - \dfrac{0.488k_{ro}A\,|\Delta\gamma|\sin\alpha}{\mu_o\, q_t}}{1 + (k_o/k_d)(\mu_d/\mu_o)} \tag{6.18}$$

Note that Eq. 6.18 is simply a modification of the f_{dNO} equation, Eq. 6.17. The first term, 1.0, divided by the denominator is f_{dNO} according to Eq. 6.17, so the second term is the modification necessary to have the equation apply to nonhorizontal flow. Several items should be noted in connection with the application of Eq. 6.18. The gravity term contains the effective permeability to the oil stated as a function of the absolute permeability, k. This is done so the application of the equation to any particular reservoir permits the simplification of the equation to a form in which all of the variables are a function of the saturation. In this regard it may be helpful to note that the ratio of the effective permeabilities, k_o/k_d, is the same as the ratio of the relative permeabilities, k_{ro}/k_{rd}. Also note that $\Delta\gamma$, the difference between the specific gravities of the displaced and displacing fluids relative to water, has an absolute symbol around it. This symbol comes about mathematically because of the way the author prefers to define α, the acute angle between the direction of flow and the horizontal direction. This angle is the same as the conventional bed-dip angle definition and is always taken as positive. It is simpler and less confusing to most engineers to treat the angle α as always being positive, regardless of the direction of flow. Since this is a mathematical inconsistency, it is counteracted in Eq. 6.18 by considering the difference between the displaced and displacing phases as positive. This artificial method controls the sign of the gravity term, so it is always negative. This situation is normally encountered in practice. Water displaces oil or gas updip and has a greater density than either oil or gas. Therefore, the gravity tends to keep the fluids separated. Similarly, gas displaces oil downdip but it has a density that is less than the oil, so it tends to keep the fluids separated by gravity.

The engineer should note that in deriving Eqs. 6.17 and 6.18, it is assumed that the pressure gradient in the displaced fluid is equal to the pressure gradient in the displacing fluid. However, we noted that the pressure in the wetting phase differs from the pressure in the nonwetting phase by the amount of the capillary pressure. This difference or capillary pressure is a function of the saturation. Consequently, Eq. 6.17 and 6.18 are not rigorous. However, the errors introduced by neglecting the capillary pressure in these fractional flow equations appears to give a negligible error. The effect of the capillary pressure on the saturation distribution is to round out the discontinuities that result from the prediction based on Eqs. 6.17 and 6.18.

Capillary effects are important in laboratory experiments where the tests must be scaled in such a way that the capillary pressure effects are negligible. Generally, laboratory flow rates or pressure gradients must be sufficiently large so the capillary pressure effects are negligible.

To test knowledge of the methods of calculating the fractional flow curve, work problem 6.4 and check the solution against the one in appendix C.

PROBLEM 6.4: Calculating a Fractional Flow Curve

A water-drive reservoir is of such size and shape that water encroachment to the first line of producers can be treated as linear flow. The water drive is sufficiently active that fluid flow is steady state. The withdrawal rate from the reservoir averages 2,830 res b/d. Calculate the fractional flow values for this reservoir corresponding to the saturations listed in Table 6–1. Reservoir data are as follows:

Average formation dip, degrees = 15.5
Average width of reservoir, ft = 8,000
Reservoir thickness, ft = 30
Average cross-sectional area (A), sq ft = 240,000
Permeability, md = 108
Connate water (irreducible water), % = 16
Reservoir oil specific gravity = 1.01
Oil viscosity, cp = 1.51
Reservoir water specific gravity = 1.05
Water viscosity, cp = 0.83

TABLE 6–1 Relative Permeability Data*

S_w	k_{rw}	k_{ro}
79	0.63	0.00 (critical)
75	0.54	0.02
65	0.37	0.09
55	0.23	0.23
45	0.13	0.44
35	0.06	0.73
25	0.02	0.94
16	0.00 (critical)	0.98

*Plotted in Fig. 6–14

Eq. 6.18 has some qualitative utility that may exceed its value for calculating the fractional flow curve. This equation provides an excellent means for determining the effect of various reservoir parameters on the displacement efficiency. The lower the fractional flow of the displacing fluid at a particular saturation, the higher is the displacement efficiency because the fraction of oil flowing must be higher. Note that if gravity does not have an effect on the fractional flow, the displacement efficiency is a function of only the mobility ratio, $(k_d\mu_o/k_o\mu_d)$. The lower this mobility ratio is, the higher is the displacement efficiency. This we know intuitively, or we can reach the same conclusion by studying Eq. 6.17.

But if gravity affects fractional flow, all parameters in the second term of the denominator (Eq. 6.18) affect the efficiency. Since minimizing the fraction of displacing fluid flowing maximizes the fraction of oil flowing, maximizing the gravity term maximizes the displacement efficiency. Thus, increasing the permeability, the specific gravity or density difference, the formation dip, and the mobility of the oil, (k_{ro}/μ_o) improves the displacement efficiency.

The flow rate and the cross-sectional area are probably best thought of as a velocity. Eq. 6.18 then shows that, when gravity effects are significant, the displacement efficiency is maximized by minimizing the velocity, q_t/A. It is also worthy of note that, according to this equation, there is no gravity effect as long as flow is horizontal since the sine of zero is zero.

The Buckley-Leverett equations. As noted, once the fractional flow curve has been evaluated, the slopes from the curve can be used to calculate the position of a particular saturation at a particular time. The equation normally used for this purpose is the linear Buckley-Leverett equation, Eq. 6.14.[5]

This equation can be derived by evaluating the change in the displacing-phase saturation in a Δx increment of the reservoir during some time period, Δt, and relating this expression to the change in the average fraction of displacing fluid flowing into and out of Δx during the time interval, Δt. These relationships may be best visualized by referring to Fig. 6–16. It can be seen that the time interval, Δt, represents the length of time for a particular saturation in the reservoir to move a distance, Δx. Now, if we consider the change in the average saturation during the time period, Δt, to be ΔS_d, it is clear that the change in the number of barrels of the displacing fluid in the Δx segment during time interval Δt is the pore volume of the segment multiplied by the change in the saturation:

$$\Delta \text{ barrels of d in } \Delta x \text{ during the time } \Delta t = (\Delta x A \phi / 5.615)\,\Delta S_d \qquad (6.19)$$

Also notice in Fig. 6–16 that the fraction of fluid flowing at a particular point in the reservoir changes as the saturation at that point changes. Therefore, a plot of fractional fluid flow at the beginning and ending of the time interval would look as shown in Fig. 6–16. Now, note that the average $f_{d\,in}$ multiplied by the cumulative total flow in during the period Δt represents the total displacing fluid entering the Δx increment during the period. Similarly, the average $f_{d\,out}$ during the period multiplied by the cumulative total flow out is the displacing phase leaving during the period. Consequently, the difference in the displacing fluid in and the displacing fluid out is the increase in the displacing fluid in Δx during the period:

$$\Delta \text{ Barrels of d in } \Delta x \text{ during time } \Delta t = q_t\,\Delta t \Delta f_d \qquad (6.20)$$

When Eqs. 6.19 and 6.20 are equated and the expression is solved for Δx, we obtain Eq. 6.14:

$$\Delta x_{Sdj} = (5.615\, q_t \Delta t / \phi A)(\Delta f_d / \Delta S_d)_{Sdj} \qquad (6.14)$$

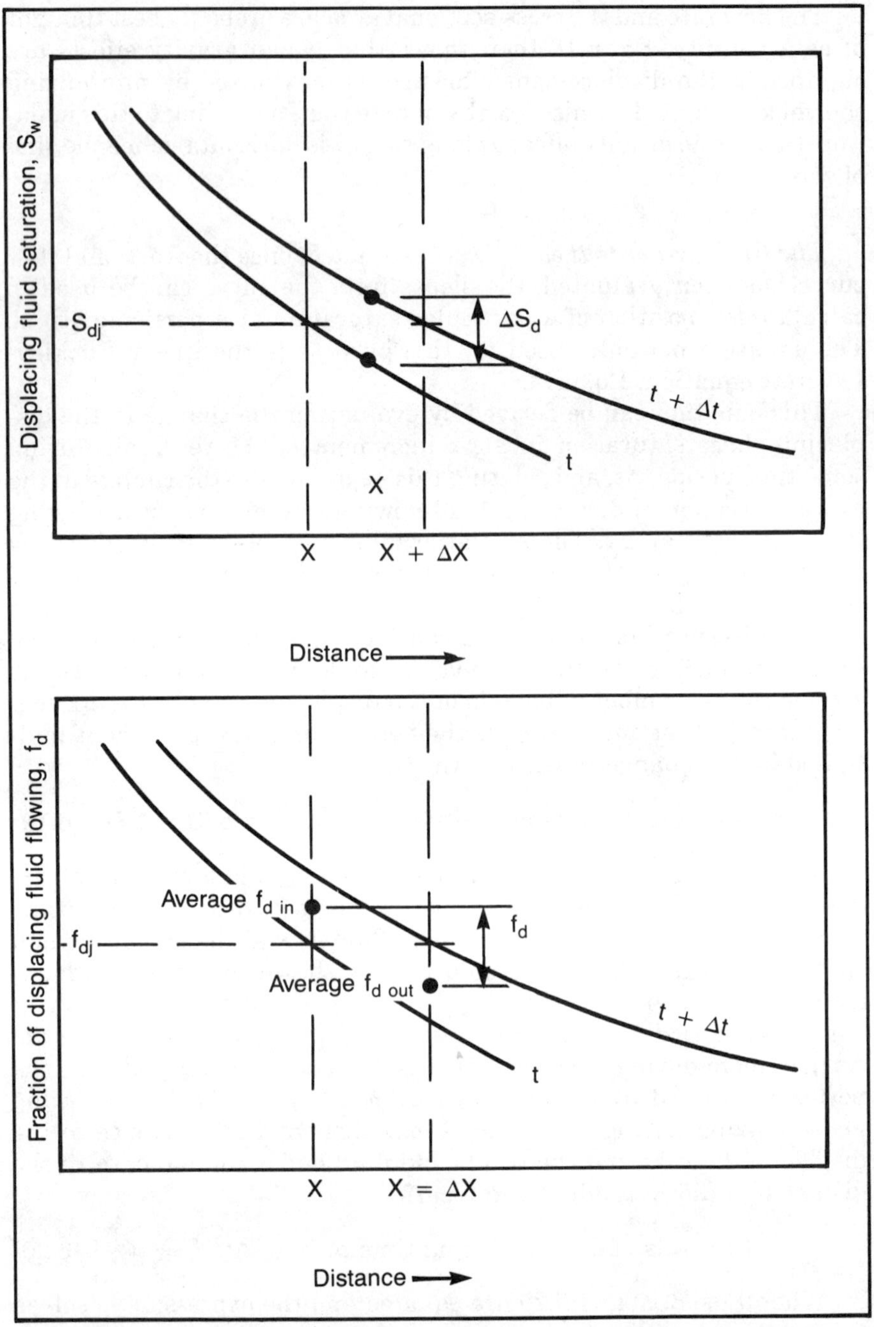

Fig. 6–16 Diagrams for Buckley-Leverett equation derivation

As noted, Eq. 6.14 is used by evaluating the expression, $\Delta f_d/\Delta S_d$, graphically at the saturation of interest. By assuming different values for the saturation and calculating its position at a particular time, it is possible to evaluate the saturation distribution at any particular time.

To test understanding of the application of the Buckley-Leverett equation, work problem 6.5 and compare the solution with the one in appendix C.

PROBLEM 6.5: Application of the Buckley-Leverett Equation

Fig. 6–17 gives the initial saturation distribution for the reservoir in problem 6.4 and the calculated saturation distribution in this reservoir at the end of 0.5, 1.0, and 2.0 years. Verify these saturation distributions by calculating the position of the 0.79, 0.75, and 0.70 water saturations at the times indicated. Use the previously verified fractional flow curve of Fig. 6–15. Do not be concerned with the evaluation of the frontal saturations. The reservoir porosity is 21.5%, and the average distance from the original WOC to the first line of producers is 350 ft.

Determining the frontal saturation by material balance. Examination of Fig. 6–15, which is a typical fractional flow curve, shows that it is meaningless to evaluate the position of every saturation represented on this curve. Since there is a duplication of slopes, the calculated reservoir position of the two different saturations with the same slope would be identical. Consequently, some method is required to deter-

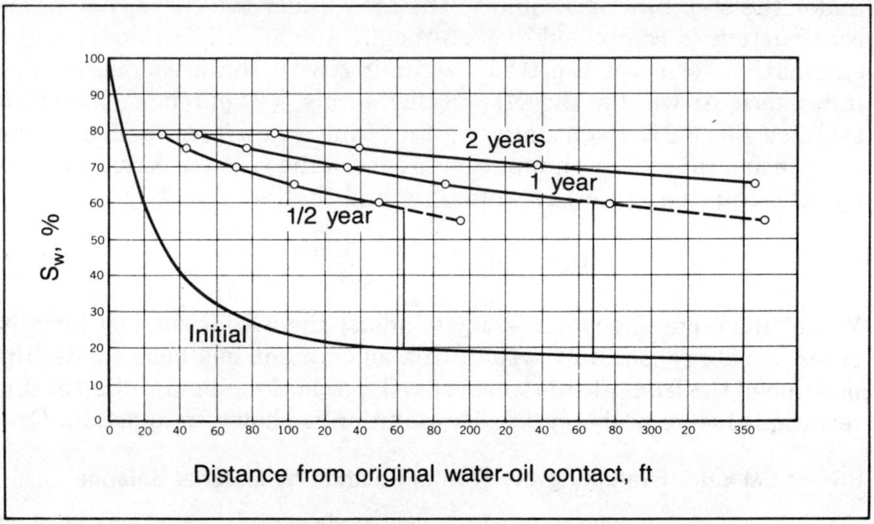

Fig. 6–17 Saturation distribution in a water-drive reservoir

mine how much of the calculated saturation profile is needed. This can be accomplished by material balance. We know that, at any particular time of interest, the total influx into the reservoir is the influx rate multiplied by the time, q_t t. Prior to breakthrough of the displacing front into the producing wells, the total influx is equal to the increased number of barrels of the displacing fluid in the reservoir.

Initially, we know that the number of cubic feet of displacing fluid in any particular Δx increment of the reservoir is $S_{di}\Delta x A \phi$. Thus, the increase in the number of cubic feet of the displacing phase in ΔX is $A\phi\Delta X(S_d - S_{di})$. If we summed the increase in all of the segments from the influx face to the displacing-fluid front, we obtain the total number of cubic feet of displacing-fluid influx during the period of time, t:

$$5.615\, q_t t = \phi A \sum_{0}^{X_f} (S_d - S_{di})\, \Delta X \tag{6.21}$$

This then fixes the frontal position as the X_f necessary to satisfy Eq. 6.21. This expression is somewhat unusual because the unknown normally is not the limit of a summation term. The equation is generally solved graphically. The method is most easily explained by considering an example. Fig. 6–18 shows the initial saturation distribution for the reservoir in problems 6.4 and 6.5. The apparent saturation distribution calculated in problem 6.5 is plotted in Fig. 6–17. The dotted lines in Fig. 6–18 are solution construction lines. The two curves are approximated with step functions. The objective, of course, is to have the area under the step function equal to the area under the curve. After this construction is completed, we can begin numerically evaluating the summation terms of Eq. 6.21, beginning with the area nearest the influx face. As we sum the $(S_d - S_{di})\Delta X$ terms, we approach a sum that satisfies Eq. 6.21. Eventually we find that a new term added to the summation causes the summation to exceed the value necessary to satisfy the equation. In other words, we find that:

$$\Sigma(S_d - S_{di})\Delta X > \frac{5.615\, q_t t}{\phi A} \tag{6.22}$$

When this occurs, it is necessary to adjust the ΔX for the last term in order to satisfy Eq. 6.21. When this adjustment has been made, the position of the front at this time is fixed. To check understanding of this technique, solve problem 6.6. The solution is shown in appendix C.

PROBLEM 6.6: Evaluating the Frontal Position by Material Balance

Use the apparent saturation profile for ½ year as shown in Fig. 6–17 and replotted in Fig. 6–18 to determine the frontal position at this time.

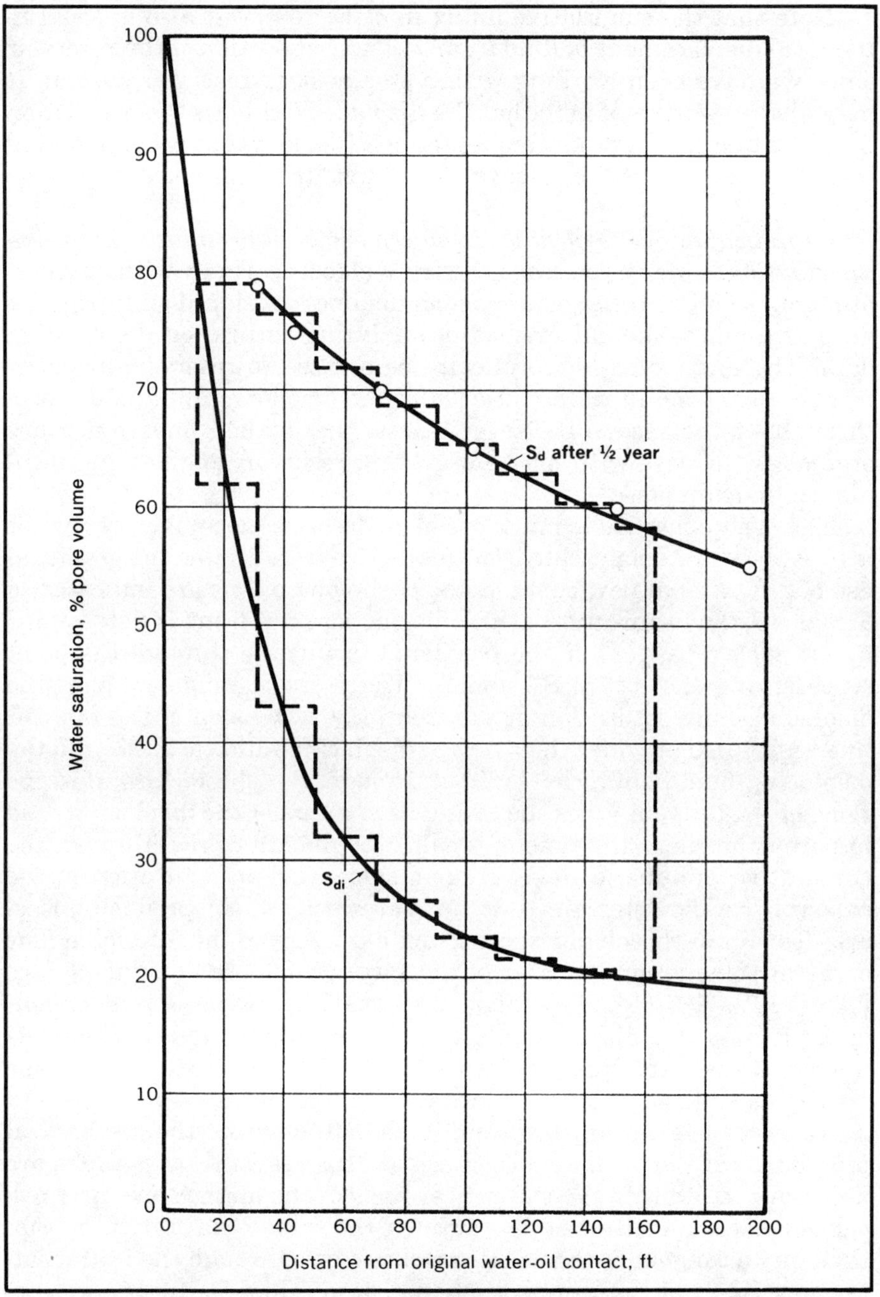

Fig. 6–18 Example graphical integration

Note that the cumulative influx into the reservoir also is equal to the total displacement of fluid from the subject portion of the reservoir since we have been working with a steady-state reservoir system. It may also be worthy of note that the displaced fluid less the production from the first line of producers represents the fluid that is being moved into the reservoir above or beyond the first line of producers.

Graphical analysis of fluid displacement with no initial saturation distribution. Many years ago, Henry Welge (reportedly with an assist at lunch from Dr. Teller of atomic-bomb fame) developed an intriguing graphical-mathematical method of analyzing saturation distribution data. The method can theoretically be applied to reservoir displacements only if the initial saturation of the displacing fluid is uniform throughout the reservoir. Nevertheless, the method finds much use and, as is the case with most interesting reservoir engineering techniques, much misuse.

The Welge method is much easier to describe mechanically than it is to explain mathematically. The method centers around the graphical use of a fractional flow curve. Some of the operations are indicated in Fig. 6–19. The saturation at the displacing-fluid front is determined by drawing a tangent to the fractional flow curve through the point representing the initial S_{di} and f_{di}. The point of tangency fixes the displacing-fluid saturation at the front, S_{df}. Extension of the tangent line to $f_d = 1.0$ provides the average displacing-fluid saturation in the displacing-fluid bank prior to breakthrough, $\overline{S}_d$. The reciprocal of the slope of the tangent gives the cumulative influx of the displacing fluid in pore volumes at the time of breakthrough. After breakthrough the cumulative influx and the average displacing-fluid saturation in the reservoir can be determined for any saturation at the producing face, S_{dPf}. To obtain these values, construct a tangent to the fractional flow curve at some assumed value of the saturation at the producing face identified as S_{dj} in Fig. 6–19. The reciprocal of the slope of this tangent is the cumulative displacing-fluid influx in pore volumes. Where the tangent crosses the line represented by $f_d = 1.0$, we read the average displacing-fluid saturation in the reservoir when $S_{dPf} = S_{dj}$.

Whereas one large paragraph is used to describe the mechanical procedures of the Welge method, we see that several paragraphs are necessary to indicate the theoretical basis of the method. We first recognize that the method of determining the frontal saturation by constructing a tangent to the fractional flow curve through the initial saturation, S_{di}, and the initial fractional flow value, f_{di}, is the same as saying mathematically that:

$$\left(\frac{\Delta f_d}{\Delta S_d}\right)_{fr} = \frac{(f_{dfr} - f_{di})}{(S_{dfr} - S_{di})} \tag{6.23}$$

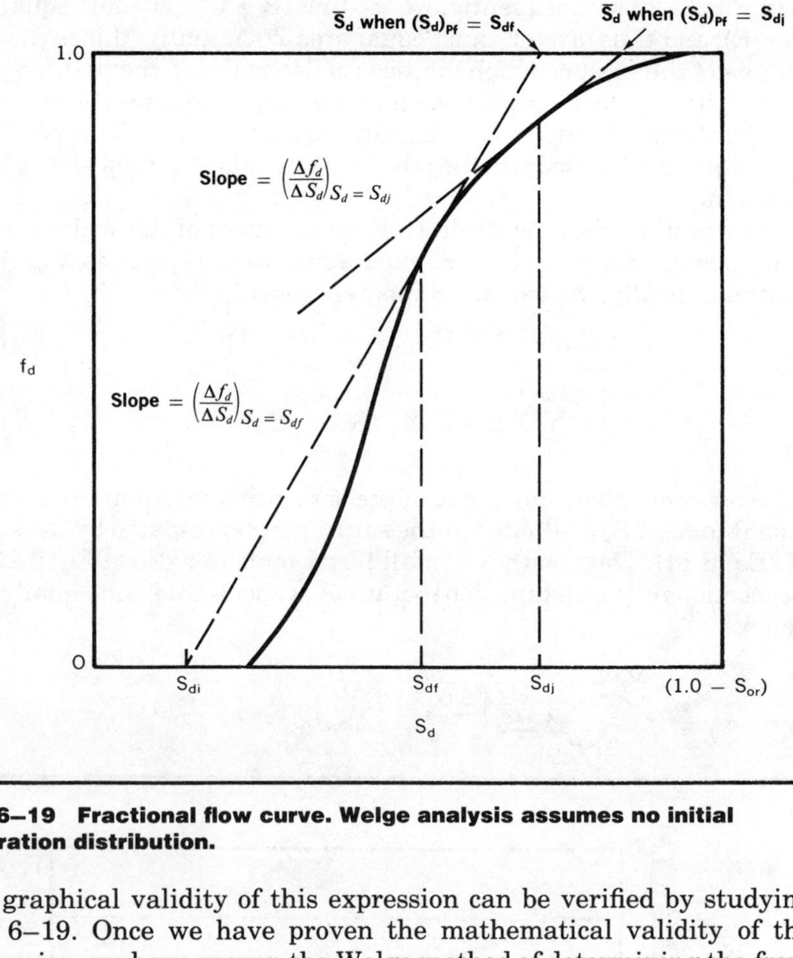

Fig. 6–19 Fractional flow curve. Welge analysis assumes no initial saturation distribution.

The graphical validity of this expression can be verified by studying Fig. 6–19. Once we have proven the mathematical validity of the expression, we have proven the Welge method of determining the frontal saturation.

It has been noted that if the entire fractional flow curve is used to calculate a saturation distribution, a double-valued plot of saturations versus distance from the influx face results over a portion of the plot. This plot is illustrated schematically in Fig. 6–20. In Eq. 6.21 we show that the cumulative displacing-fluid influx is proportional to the area in Fig. 6–20 representing the summation term, $\Sigma(S_d - S_{di})\Delta X$, in Eq. 6.21 and areas B, C, and D in Fig. 6–20. Mathematically, based on the two-valued saturation curve, the cumulative influx can also be written as:

$$5.615\ q_t t = \phi A \sum_{(1 - S_{or})}^{S_{di}} X \Delta S_d \tag{6.24}$$

Note that the summation in Eq. 6.24 is then equal to the area B, C, and E in Fig. 6–20. Consequently, we see that (B + C + D) must equal (B + C + E), and thus, area E must equal area D. Actually, this is the basis for one of the most common methods of determining the position of the front. The engineer simply determines by trial and error the position of the front that results in area E being equal to area D. This particular technique can be used whether the initial displacing-fluid saturation is uniform.

Continuing with the theoretical development of the Welge method, this means that area C + E must equal area C + D in Fig. 6–20. Mathematically, we can state this relationship as:

$$\text{Area (C + E)} = \text{Area (C + D)} \tag{6.25}$$

$$\sum_{S_{di}}^{S_{dfr}} X \Delta S_d = X_{fr} (S_{dfr} - S_{di}) \tag{6.26}$$

Now remember that the X of the saturation distribution is proportional to $\Delta f_d/\Delta S_d$ evaluated at the saturation represented by the subject X (Eq. 6.14). Thus, with X's in all terms on both sides of Eq. 6.26, the proportionality constant, $5.615 q_t \Delta t/\phi A$, cancels and the equation becomes:

$$\sum_{(\Delta f_d/\Delta S_d)_{Sdi}}^{(\Delta f_d/\Delta S_d)_{fr}} \left(\frac{\Delta f_d}{\Delta S_d}\right) \Delta S_d = \left(\frac{\Delta f_d}{\Delta S_d}\right)_{fr} (S_{dfr} - S_{di}) \tag{6.27}$$

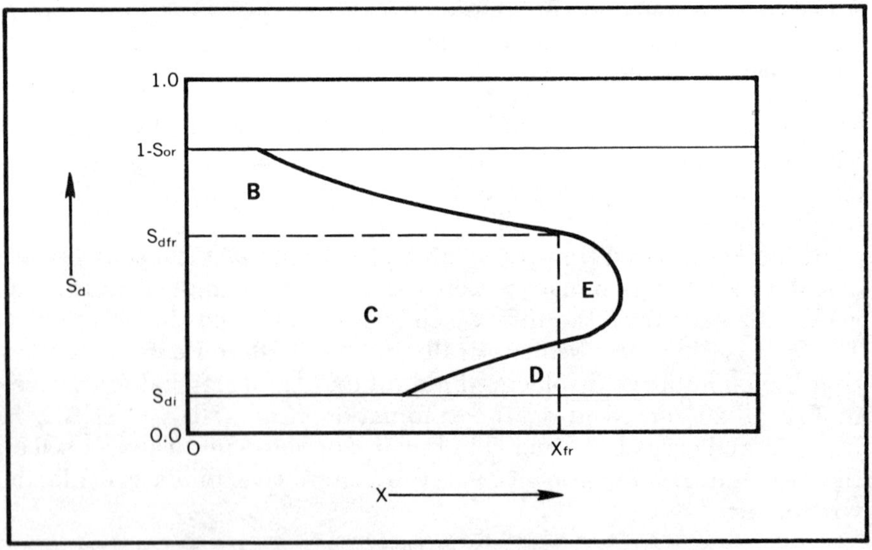

Fig. 6–20 Schematic of Buckley-Leverett S_d versus distance

Eq. 6.27 simplifies to:

$$f_{dfr} - f_{di} = \left(\frac{\Delta f_d}{\Delta S_d}\right)_{fr} (S_{dfr} - S_{di}) \tag{6.28}$$

When Eq. 6.28 is solved for the fractional flow curve slope at the front, we obtain Eq. 6.23, which shows that the frontal saturation can be determined by drawing a tangent to the fractional flow curve through the initial conditions of S_{di} and f_{di}. The frontal saturation is then represented by the point of tangency.

By rearranging the Buckley-Leverett equation, we show that the reciprocal of the slope of the fractional flow curve, when evaluated at the producing-face saturation after breakthrough, is equal to the cumulative displacing-fluid influx in pore volumes. Since we are evaluating the slope at the producing face, the X value in the Buckley-Leverett equation, Eq. 6.14, is equal to the length of the linear system. The pore volume in barrels, V_p, is $\Delta X_{Sdj} A\phi/5.615$. Cumulative influx, Q_i, is $q_t t$. Now, substituting into and rearranging Eq. 6.14, we obtain an expression for the cumulative influx in pore volumes:

$$(Q_i/V_p) = \frac{1}{(\Delta f_d/\Delta S_d)_{Pf}} \tag{6.29}$$

The last part of the graphical method is concerned with evaluating the average displacing-fluid saturation in the reservoir. This evaluation is made by constructing a tangent to the fractional flow curve at the displacing-fluid saturation that exists at the producing face at a particular time, (S_{dPf}), and extending the tangent to the line representing the $f_d = 1.0$ values. The saturation representing this intersection is the average displacing-fluid saturation in the reservoir at that time, $\overline{S}_d$. Fig. 6–19 shows that this graphical technique is equivalent to the equation:

$$\left(\frac{\Delta f_d}{\Delta S_d}\right)_{Pf} = \frac{(1 - f_{dPf})}{(\overline{S}_d - S_{dPf})} \tag{6.30}$$

Thus, if we prove the theoretical validity of Eq. 6.30, we have proven that the graphical method is theoretically correct.

The average saturation, $\overline{S}_d$, for a saturation distribution after breakthrough, as shown schematically in Fig. 6–21, is the area under the saturation curve divided by the length of the reservoir, L:

$$\overline{S}_d = \frac{\sum_{O}^{L} S_d \Delta X}{L} = \frac{(1 - S_{or}) X_{or} + \sum_{X_{or}}^{L} S_d \Delta X}{L} \tag{6.31}$$

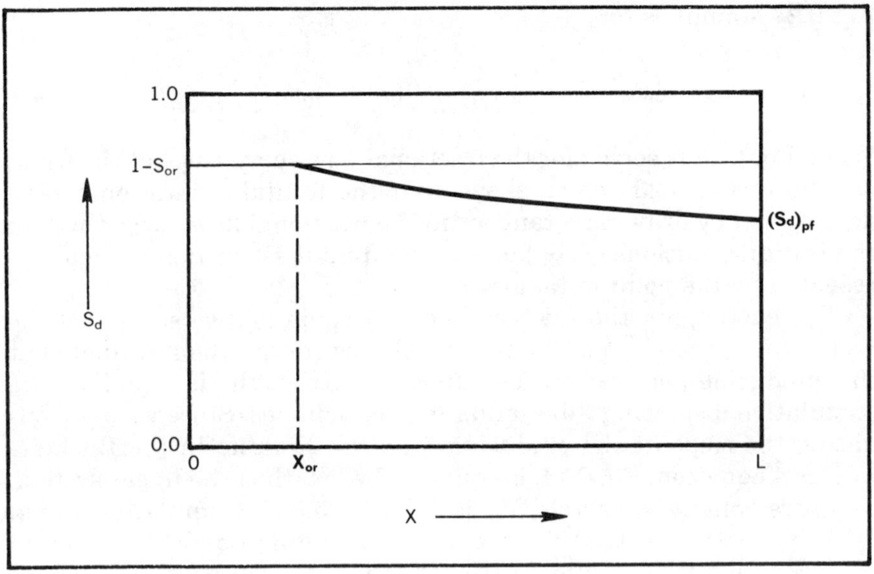

Fig. 6–21 Schematic of S_d versus distance, X, after breakthrough

In this expression the area is broken into two sections to facilitate handling. Note that each term in the numerator and the denominator (L) contains an X value from the saturation distribution curve. According to Eq. 6.14, the X position of a saturation is proportional to the slope of the fractional flow curve at that saturation. Therefore, the proportionality constant, $5.615\, q_t \Delta t/\phi A$ cancels out, and we can simply replace all of the X values in Eq. 6.31 with the slope corresponding to that saturation:

$$\overline{S}_d = \frac{(1 - S_{or})(\Delta f_d/\Delta S_d)_{ro} + \sum_{ro}^{Pf} S_d \Delta(\Delta f_d/\Delta S_d)}{(\Delta f_d/\Delta S_d)_{Pf}} \tag{6.32}$$

To simplify Eq. 6.32 to the desired form, Eq. 6.30, it is necessary to transpose the summation term to one that is stated as a function of ΔS_d rather than as a function of $\Delta(\Delta f_d/\Delta S_d)$. Eq. 6.32 then becomes:

$$\overline{S}_d = \frac{(1 - S_{or})(\Delta f_d/\Delta S_d)_{ro} + S_{dPf}\,(\Delta f_d/\Delta S_d)_{Pf} - S_{dro}(\Delta f_d/\Delta S_d)_{ro}}{(\Delta f_d/\Delta S_d)_{Pf}} - \frac{\sum_{ro}^{Pf} (\Delta f_d/\Delta S_d)\Delta S_d}{(\Delta f_d/\Delta S_d)_{Pf}} \tag{6.33}$$

If the engineer is confused by the mathematical step from Eq. 6.32 to Eq. 6.33, he may better recognize the pure math form of the summation-term transformation as:

$$\int u\,dz = u\,z - \int z\,du \qquad (6.34)$$

Where:
$u = S_d$
$z = \Delta f_d/\Delta S_d$

Note that $S_{dro} = (1 - S_{or})$ so the terms in the numerator of Eq. 6.33 that contain these expressions cancel out. Also, the summation term is simply $f_{dPf} - f_{dro}$, and the fraction of displacing fluid flowing at the residual oil saturation, f_{dro}, is 1.0 since no residual oil can flow. Eq. 6.33 can then be written as:

$$\overline{S}_d = \frac{S_{dPf}\,(\Delta f_d/\Delta S_d)_{Pf} - f_{dPf} + 1.0}{(\Delta f_d/\Delta S_d)_{Pf}} \qquad (6.35)$$

When Eq. 6.35 is solved for $(\Delta f_d/\Delta S_d)_{Pf}$, it becomes Eq. 6.30. Eq. 6.30 is the mathematical equivalent of the mechanical procedure for drawing a tangent to the fractional flow curve at the displacing-fluid saturation at the producing face, extending the tangent to the horizontal line representing $f_d = 1.0$, and reading the average displacing-fluid saturation in the reservoir as the saturation at the intersection.

Fig. 6–19 illustrates the application of this technique to determine the average displacing-fluid saturation at breakthrough. In this case the average saturation at the producing face is the frontal saturation, S_{df}, with the average saturation, S_d, determined as illustrated. It can be shown that this average saturation also represents the average saturation in the displacing-fluid bank at any time prior to breakthrough.

The Welge graphical method of evaluating the saturation distribution during a displacement is a fast, simple technique when it can be used. Even when a substantial transition zone exists in the reservoir, i.e., the initial displacing-fluid saturation is not uniform, the engineer may wish to assume some average uniform initial saturation distribution to obtain a quick analysis before performing the more time-consuming but the more nearly accurate analysis previously outlined.

The engineer should note that the fluid displaced, oil or gas, from the reservoir plus the displacing fluid produced is equal to the displacing fluid entering the reservoir. In other words the increase in the average displacing-fluid saturation is equal to the decrease in the oil and gas saturation.

It is recommended that the reader work problem 6.7 and check the solution against the one in appendix C to understand the Welge graphical method better.

PROBLEM 6.7: Using the Welge Graphical Method to Calculate Displacement in a Waterflood.

A line-drive waterflood is to be performed in a thin, homogeneous, horizontal reservoir with an initial gas saturation of 5%. The equilibrium gas saturation is 0.0. Assume all gas is displaced before the oil is displaced (an oil bank forms). The fractional flow curve for this reservoir is shown in Fig. 6–22. The connate-water saturation is 35%. What is the water saturation at the front prior to water breakthrough into the producing wells? Estimate the oil displaced in pore volumes versus the cumulative water injection in pore volumes. Assume a sweep efficiency of 1.0.

It should be carefully noted that in using the Welge method the tangent to the fractional flow curve must be drawn through the initial conditions in the reservoir and not necessarily through a point on the $f_d = 0.0$ line. This appears to be a common error made in applying the Welge method, but examination of the derivation shows that it is necessary to draw the tangent through the initial S_{di} and f_{di} points that fall on the f_d curve. This is particularly important in determining the effect of an initial saturation that is greater than the irreducible saturation. This problem is discussed more fully in chapter 8.

The radial Buckley-Leverett equation. Our discussion of fluid displacement has been concerned entirely with the linear flow system. Generally, the linear system seems to satisfy the fluid-displacement needs because natural encroachment takes place at a very large radii. Consequently, the behavior is closely approximated by a linear system. For example, if water encroachment takes place across a 10,000-ft radius and the advance of the water eventually reached a radius of 8,000 ft, we can treat this problem as a linear encroachment with a width that varied from $2\pi10{,}000$ to $2\pi8{,}000$, or with an average width of $2\pi9{,}000$. The width is, of course, the circumference of the circle. This gives a variation in the cross-sectional area, A, that is no more than that caused by normal thickness variations.

A radial Buckley-Leverett equation can be derived without undue difficulty. The Fig. 6–16 plots of the displacing-fluid saturation and the displacing-fluid fractional flow curve plotted against distance are used to derive the linear Buckley-Leverett equation. Actually, the same diagrams can be used to derive the radial equation by simply substituting the square of the radius instead of the distance as the basis for

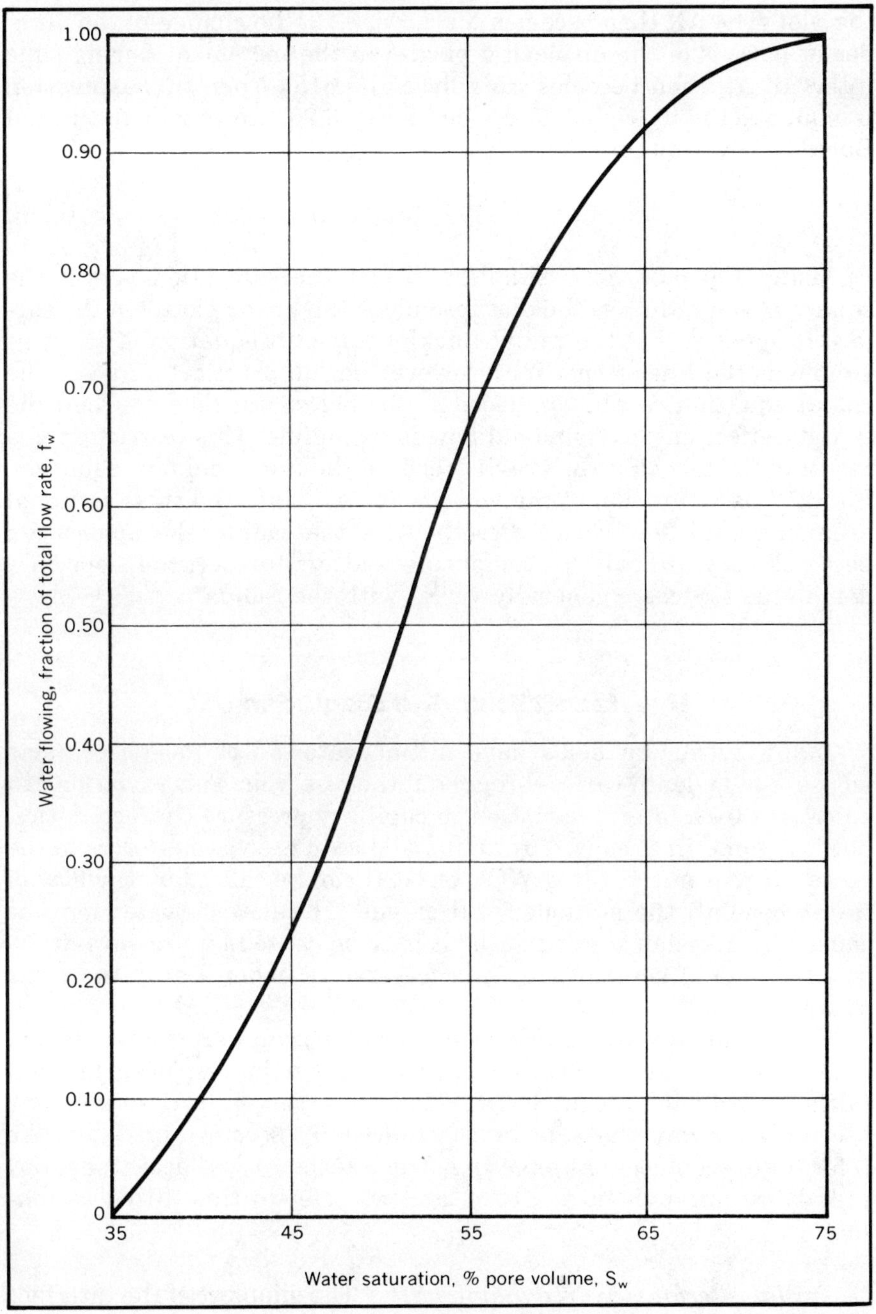

Fig. 6–22 Fractional flow curve

the plot. The ΔX then becomes $\Delta(r^2)$. Eq. 6.19, the change in the number of barrels of the displacing phrase in the increment during time interval, Δt, then becomes $\pi\Delta(r^2)h\phi\Delta S_d/5.615$. When this expression is equated to the right-hand side of Eq. 6.20, we obtain the radial Buckley-Leverett equation:

$$\Delta(r^2)_{Sdj} = \frac{1.79q_t\Delta t}{\phi h}(\Delta f_d/\Delta S_d)_{Sdj} \tag{6.36}$$

Using Eq. 6.36, it is simplest to plot the saturation versus the square of the radius so the plot resembles the linear plots. On the surface it appears that the radial Buckley-Leverett equation is about as simple as the linear equation. However, for all practical purposes the radial equation can be applied only for horizontal flow or when the gravity effect on the fractional flow is negligible. This restriction is a result of the fact that the gravity term in the fractional flow equation, Eq. 6.18, is a function of the velocity (q_t/A). Since the cross-sectional area in radial flow varies directly with the radius, the application becomes very difficult. To complicate matters further, the slope in a dome-type structure generally varies with the radius.

Interface Tilt during Displacement

Many petroleum fields have initial water-oil or gas-oil contacts whose subsea depth varies throughout the reservoir. This variation can be caused by a lateral variation in capillary pressure characteristics, but it is sometimes caused by an imbalance of the viscous forces in the reservoir. An initial tilted WOC or GOC can be caused by the flow of water beneath the stationary oil or gas. The flow of water may be caused by artesian water flow or it may be caused by pressure drawdown in one of several petroleum reservoirs underlain by the same aquifer.

Even when a WOC or GOC is initially horizontal, it is often discovered that the interface becomes tilted during the displacement that takes place during the producing life. In fact, this tilted condition may become so severe that the interface actually becomes unstable. We refer to this condition as *fingering*. The interface tilt is accentuated by radial flow around the wellbore, and we refer to this tilting as *coning*.

Initial interface (hydrodynamic) tilt. The simplest of the interface tilt problems is probably the initial contact tilt in the reservoir caused by the pressure gradient resulting from flow of water beneath stationary oil or gas whose density is different from the density of the water.

Under these conditions the pressure remains constant along any horizontal line in the oil or gas zone because the oil or gas is not flowing and no pressure drop results from this flow. However, along a horizontal line in the water zone, there is a pressure drop caused by flow. Such a situation can exist only when the interface between the two zones tilts. Then the total pressure in the water zone along a horizontal line in the direction of flow decreases because it has less water and more of the less-dense oil above it.

This phenomena was originally analyzed by Shell geologist Hubbert, who proposed the prediction of the location of petroleum reservoirs through an understanding of this phenomena. Fig. 6–23 gives an acceptable explanation of the quantitative relationships accompanying this interface tilt. The figure shows the unlikely situation where two cable-tool wells some distance ΔH apart just miss the oil trap shown. When an effort is made to flow the wells, they fill with water to the level indicated. The difference between these two levels is the pressure drop caused by flow stated in terms of equivalent feet of water, Δpiez. In hydraulics this is known as the *piezometric pressure drop*. This factor can be converted to the pressure drop caused by flow in psi using the

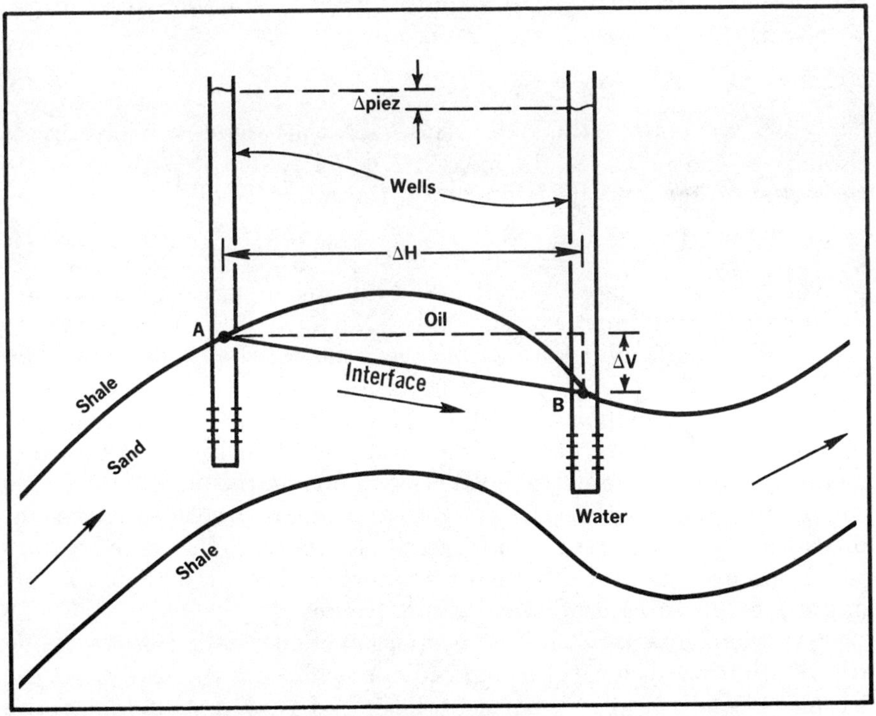

Fig. 6–23 Schematic of hydrodynamic tilt of an interface

pressure gradient of 0.433 psi/ft for fresh water and the water specific gravity relative to fresh water, γ_w:

$$\Delta p_{flow} = 0.433\gamma_w \, \Delta(\text{piez}) \tag{6.37}$$

Now, consider the difference in pressures at points A and B. These points represent pressures at the interface between the oil and the water. Consequently, the pressures are the same whether we measure them in the oil or in the water. We then write an expression for the pressure difference when measured in the water and when measured in the oil and equate these expressions to obtain a quantitative evaluation of the magnitude of the interface tilt.

If the pressure drop is measured in the water, it is the difference between the pressure drop caused by flow and the static pressure difference, since there is an increase in pressure from point A to point B when no flow is taking place:

$$p_A - p_B = \Delta p_{flow} - \Delta p_{static} \tag{6.38}$$

$$p_A - p_B = 0.433\gamma_w \, \Delta(\text{piez}) - 0.433\gamma_w \, \Delta V \tag{6.39}$$

If we measure the same pressure drop on the oil side of the interface, the pressure difference is caused entirely by the static pressure difference in the oil because the oil is not flowing:

$$p_A - p_B = 0.433\gamma_o \, \Delta V \tag{6.40}$$

When we equate the expressions for the pressure drop, divide through by the horizontal distance, ΔH, and solve for $\Delta V/\Delta H$, we obtain the equation for the tilt or tangent of the interface dip:

$$(\Delta V/\Delta H) = [\gamma_w/(\gamma_w - \gamma_o)][\Delta(\text{piez})/\Delta H] \tag{6.41}$$

The initial interface tilt controls oil and gas accumulations because the formation dip must exceed the interface tilt if the oil or gas is to stay in that particular trap. Otherwise, the petroleum accumulation occurs at some higher structural position.

Eq. 6.41 is a convenient mathematical form of the interface tilt, but it is uncommon for the change in pressure with horizontal distance to be recorded as a piezometric head. The change in the piezometric head can of course be calculated from the total pressure drop of two reservoir pressures corrected to the same datum. Correction to the same horizontal datum assures that the pressure difference is caused by flow alone, and Eq. 6.37 can be used to calculate Δ(piez).

The term "interface tilt" can be more explicitly called *isosaturation tilt.* In other words a saturation gradient still exists vertically, but the isosaturation lines are not horizontal. Instead, they are tilted according to Eq. 6.41. The saturation distribution calculated by the Buckley-

Leverett equation remains valid, but it now applies to slices of the reservoir that are parallel to the interface tilt rather than to horizontal slices.

It should be noted that on a worldwide basis artesian flow seldom occurs in an aquifer of such magnitude that the interface tilt is more than a few feet per mile. Such occurrences have been observed in the Rocky Mountain area. It is also suspected that Sumatra (Indonesia) can have such occurrences. A large change in the salinity of the produced water with time has been observed in Sumatra's water-drive reservoirs. This change is undoubtedly caused by water entering these aquifers at the formation outcrop. Produced water is so fresh that it is dumped in the jungle without adversely affecting the vegetation or water supply.

However, the engineer should remember that most reservoir flow of water follows the unsteady-state theory as opposed to the steady-state theory that characterizes artesian flow of water.

Interface tilt during linear displacement. When one fluid is displacing another in a tilted reservoir as in Fig. 6–24, where gas is displacing oil downdip, the interface tends to be tilted because the displacing fluid is less viscous than the oil. Therefore, it tends to move more readily than the oil. Consequently, the gas overrides and bypasses the

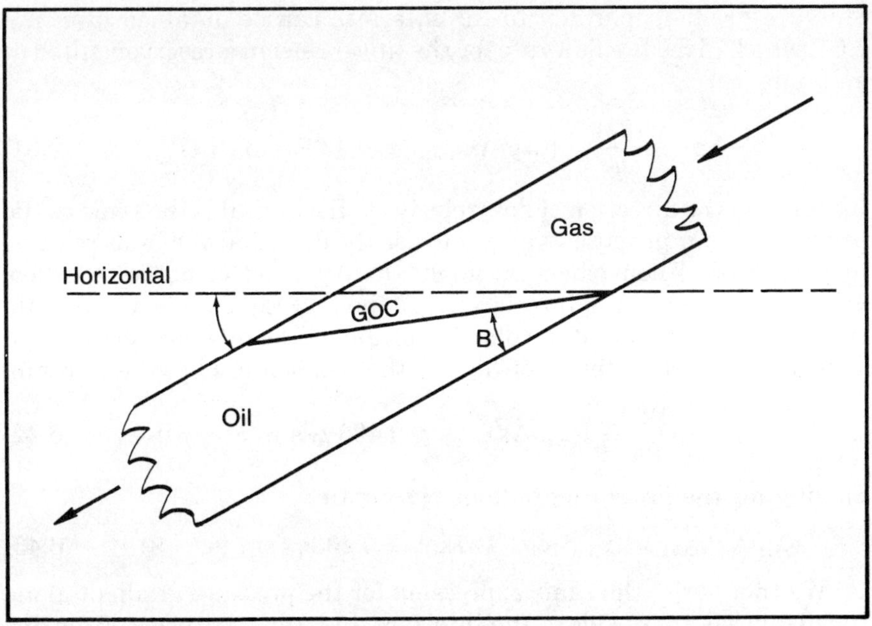

Fig. 6–24 Interface tilt during fluid displacement

oil if the two fluids have the same density. However, the densities are different, and this difference tends to keep the interface horizontal. In fact, the interface remains horizontal as a result of the difference in the densities if the mobilities of the two fluids are equal.

The practical situation of course is one in which both forces are effective. The viscous forces tend to cause the interface tilt, and these tilts are balanced by gravity forces so a stabilized interface tilt may occur during a displacement. The same phenomena occurs when water is displacing oil updip. In this case the less viscous water trys to flow under the oil, but it is opposed by the difference in gravities of the water and the oil.

Fig. 6–24 is referred to in deriving the expression that relates the viscous forces, gravity forces, and stabilized interface tilt. Note the manner in which the angles α and β are measured. The α is simply the formation dip, but β is a measure of the interface tilt relative to the bed dip. Relative to a horizontal position, the interface tilt is the difference, $\alpha - \beta$. Note that when the two angles, α and β, are equal, the interface is horizontal.

We derive the interface tilt equation in much the same way Pirson derives this relationship.[6] This is accomplished by writing an expression for the pressure gradient at the interface in terms of the flow of the displacing fluid and by writing a second expression for the pressure gradient in terms of the displaced fluid, which we consider to be oil. The pressure gradient parallel to the interface can be obtained from Eq. 2.61, which gives the flow rate for the linear flow in a reservoir tilted at an angle, α.

$$q = \frac{1.127kA}{\mu}[(\Delta p/\Delta x)_{total} \pm 0.433\gamma \sin \alpha] \tag{2.61}$$

In this case the direction of flow relative to horizontal is the same as the bed dip. For our purposes we need to use the direction of flow as relative to horizontal. Then when we apply the equation to obtain the flow velocity parallel to the interface, v_{par}, the α of Eq. 2.61 is $\alpha - \beta$. Substituting the apparent flow velocity parallel to the interface for q/A and writing Eq. 2.61 in terms of flow of the displacing phase, we obtain:

$$v_{par} = \frac{1.127k_d}{\mu_d}[(\Delta p/\Delta x)_{total} \pm 0.433\gamma_d \sin(\alpha - \beta)] \tag{6.42}$$

Solving for the pressure gradient, we obtain:

$$(\Delta p/\Delta x)_{total} = (v_{par}\,\mu_d/1.127k_d) \pm 0.433\gamma_d \sin(\alpha - \beta) \tag{6.43}$$

We then write the same expression for the pressure gradient along the displaced-phase side of the interface and, thus, in terms of the displaced phase, oil:

$$(\Delta p/\Delta x)_{total} = (v_{par}\mu_o/1.127k_o) \pm 0.433\gamma_o \sin(\alpha - \beta) \tag{6.44}$$

Note that the velocity parallel to the interface is vectorially related to the velocity parallel to the bed dip, v, so $v_{par} = v \cos\beta$. If we equate the expressions for the pressure gradient parallel to the interface as stated in Eqs. 6.43 and 6.44, substitute ($v \cos\beta$) for v_{par}, and rearrange the expression, we obtain:

$$v \cos\beta[(\mu_o/k_o) - (\mu_d/k_d)] = \pm 0.488(\gamma_o - \gamma_d)\sin(\alpha - \beta) \tag{6.45}$$

To put Eq. 6.45 in a more convenient form, it is necessary to substitute an equivalent trigonometric expression for $\sin(\alpha - \beta)$:

$$\sin(\alpha - \beta) = \sin\alpha\cos\beta - \cos\alpha\sin\beta \tag{6.46}$$

Now substitute $\tan\alpha\cos\alpha$ for $\sin\alpha$. After making these substitutions in Eq. 6.45, divide by $\cos\beta$, substitute $\tan\beta$ for ($\sin\beta/\cos\beta$), and substitute q/A for v. When the resulting expression is solved for $\tan\beta$, we obtain:

$$\tan\beta = \tan\alpha \pm \frac{q[(\mu_o/k_o) - (\mu_d/k_d)]}{0.488A(\gamma_o - \gamma_d)\cos\alpha} \tag{6.47}$$

In Eq. 6.47 the minus sign (−) is used for gas displacing oil downdip, and the plus sign (+) is used for water displacing oil or gas updip. The engineer can keep this straight by simply recognizing that $\tan\alpha$ must always be less than $\tan\beta$ so the overall sign of the last term of Eq. 6.41 must be negative. For water this is accomplished by the specific gravity difference, which is negative.

The engineer should also carefully note that the specific gravities are relative to water in Eq. 6.47, even when it is applied to gas. This of course is contrary to the customary air base normally used for all gases.

The effect of the producing rate, q, on this interface tilt can be evaluated using Eq. 6.47. Remember that the total sign of the second term, which contains the rate, q, normally is negative. Also, note that the interface is horizontal when $\tan\beta$ equals $\tan\alpha$. Consequently, the larger the second term containing the rate, q, is, the larger is the interface tilt. Therefore, the larger the rate is, the greater is the interface tilt.

Obviously, it is possible for the rate term to become so large that $\tan\beta$ is 0.0. Thus, β is zero. When this occurs, the interface tilt is at a maximum, and the interface is parallel to the dip of the formation. Under such conditions we cannot define exactly where the interface is located. Consequently, we define such a condition by saying that the interface is unstable or that fingering takes place. The producing rate at which an interface becomes unstable, or at which fingers form, can be determined by setting tan β in Eq. 6.47 equal to zero and solving for the critical rate, q_{cF}, at which this situation occurs:

$$q_{cF} = \pm\ 0.488(\gamma_o - \gamma_d)A \sin \alpha/[(\mu_o/k_o) - (\mu_d/k_d)] \qquad (6.48)$$

Remember that Eqs. 6.47 and 6.48 represent stabilized conditions when a balance has been reached between the gravity and viscous forces. These equations give no indication as to how long it takes to reach such a stabilized condition. This time factor becomes very important when we recognize that in many cases Eqs. 6.47 and 6.48 indicate a critical rate too small to be economical. There is no simple method known by the author to evaluate or estimate the time required for the reservoir to reach a stabilized condition. This tendency toward adverse fingering is offset to a large degree by the fact that the permeability in a direction perpendicular to the bedding plane is generally much less than the permeability parallel to the bedding plane. This of course tends to retard the interface tilt and the tendency for fingering.

If fingering appears to be a critical problem in a particular reservoir, it is suggested that the phenomena be studied with a digital computer reservoir model. In such a study the difference between the perpendicular and parallel permeabilities can be included, as well as permeability stratification. Eqs. 6.47 and 6.48 should be used only in a qualitative manner or to define limits of reservoir behavior.

Interface tilt in a radial flow system—coning. We have considered interface tilt with only one fluid moving and interface tilt with both fluids moving at the same velocity. Both of these problems are based on linear flow systems with constant velocities and tilt angles. We can now consider the phenomena of interface tilt associated with a radial flow system with one or both fluids flowing. This problem is associated with the radial flow of oil into a well with gas above or water below. In radial flow the velocity varies with the radius. Therefore, we find that the interface tilt varies with the radius and, thus, is not constant. In this section we talk about the concepts of coning by investigating the phenomena for some simplified systems. However, consideration of practical solutions to the coning problem is beyond the scope of this book, assuming that practical solutions do in fact exist. This assumption may be questioned by many experienced engineers.

Fig. 6–25 illustrates the physical aspects of coning of water. This diagram shows the coning situation just before the water breaks into the bottom of the wellbore. We consider this as a stabilized situation with the cone about to reach the well. However, it is stabilized so the water is not flowing. Under these conditions we have the same situation that existed in the previous discussion of hydrodynamic tilt, except in this case the flow is radial rather than linear and oil is flowing

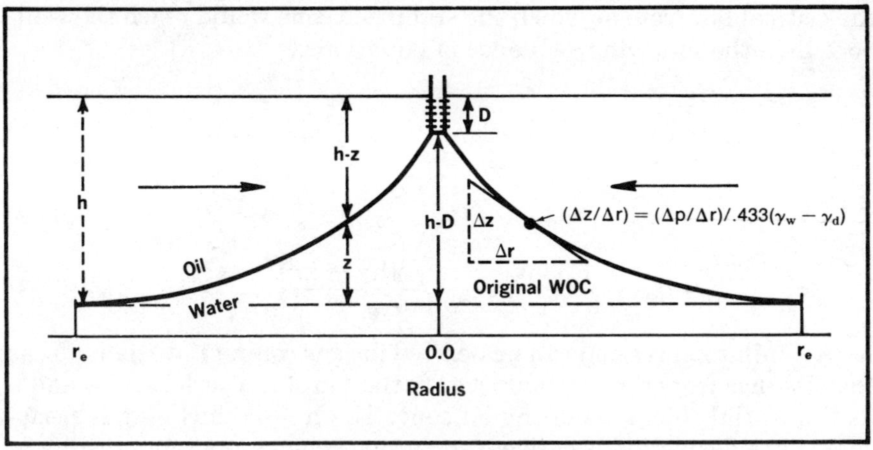

Fig. 6–25 Water coning

instead of water. Thus, we note that the pressure at any particular horizontal position or level in the water is constant because the water is not flowing. Also, in the oil a pressure drop is caused by flow. This situation can exist only if the interface is tilted so the decrease in the pressure in the oil caused by flow is offset by a change in the interface position. Thus, the change in pressure over some distance, Δr, is equal to the change in the position of the interface, Δz, over the same distance Δr, converted to an equivalent pressure change using the specific gravities of the fluids:

$$(\Delta p/\Delta r) = (\Delta z/\Delta r)0.433\ (\gamma_d - \gamma_o) \qquad (6.49)$$

We can use this expression for the pressure gradient for horizontal flow in the Darcy equation to derive the expression for the critical rate at which coning into the well will occur. To use the Darcy equation, we must also evaluate the cross-sectional area, A, at a particular radius. This factor is the surface of a cylinder whose radius is r and whose height is $h - z$. Thus, $A = 2\pi r(h - z)$. When these substitutions are made into the Darcy equation, Eq. 2.3, we obtain:

$$q = (1.127kA/\mu)(\Delta p/\Delta x) \qquad (2.3)$$

$$q = (1.127k/\mu)[2\pi r(h - z)][(\Delta z/\Delta r)0.433(\gamma_w - \gamma_o)] \qquad (6.50)$$

When Eq. 6.50 is applied to all of the Δr increments from the well radius, r_w, to the outer radius, r_e, we find that the sum of all of the $\Delta r/r$ values is $\ell n(r_e/r_w)$. The sum of all of the $(h - z)\Delta z$ values from a value of $z = (h - D)$ at the well (Fig. 6–25) to a value of zero at the external boundary where the interface is unaffected is $(h^2 - D^2)/2$. Performing these operations with Eq. 6.50 and solving the resulting expression for the critical flow rate at which the stabilized cone would reach the wellbore give the following sequence of equations:

$$q\sum_{r_w}^{r_e}(r/\Delta r) = (3.07k/\mu)(\gamma_w - \gamma_o) \sum_{h-D}^{0.0} (h - z)\Delta z \tag{6.51}$$

$$q\ \ell n(r_e/r_w) = (3.07k/\mu)(\gamma_w - \gamma_o)[(h^2 - D^2)/2] \tag{6.52}$$

$$q_{cC} = \frac{1.535k(\gamma_w - \gamma_o)(h^2 - D^2)}{\mu\ \ell n\ (r_e/r_w)} \tag{6.53}$$

A similar expression can be derived for gas coning if we define D as the distance from the original GOC to the top of the perforations and h as the initial thickness of the oil zone. The h − D thickness is perforated. The derivation is carried out in the same manner and results in the expression:

$$q_{cC} = \frac{1.535k(\gamma_o - \gamma_g)(2hD - D^2)}{\mu\ \ell n\ (r_e/r_w)} \tag{6.54}$$

These coning equations have been derived in a manner similar to the techniques used in the Pirson text.[7] Pirson also derives an expression for the optimum placement of h_c feet of perforations in an oil zone with a gas cap above and a water zone below. In this case, D is the distance in feet from the initial GOC to the bottom of the perforations:

$$D = h - (h - h_c)[(\gamma_o - \gamma_g)/(\gamma_w - \gamma_g)] \tag{6.55}$$

Where:
h = Initial thickness of the oil zone

Eqs. 6.53, 6.54, and 6.55 are not based on realistic assumptions. One of the biggest difficulties involves the assumption that the permeability is the same in all directions. As noted, this assumption is seldom realistic. Since sedimentary formations were initially laid down in thin, horizontal sheets, it is natural for the formation permeability to vary from one sheet to another vertically. Therefore, there is generally quite a difference between the permeability measured in a vertical direction and the permeability measured in a horizontal direction. Furthermore, the permeability in the horizontal direction is normally considerably greater than the permeability in the vertical direction. This also seems logical when we recognize that very thin, even microscopic sheets of impermeable material such as shale may have been periodically deposited. These permeability barriers have a great effect on the vertical flow and have very little effect on the horizontal flow, which would be parallel to the plane of the sheets.

The lower vertical permeability tends to minimize the coning. Thus, Eqs. 6.53 and 6.54 tend to give overly pessimistic results. In other words, the predicted critical coning rates are much too high. However, Eq. 6.55 used to predict the optimum position of perforations to produce oil between an oil and water zone does not appear to be affected significantly by the difference between the vertical and horizontal permeability. The error appears to have about the same effect on the gas coning as it has on the water coning, and the position of the oil zone perforations should not be affected.

Also, note that this series of coning equations is based on the assumption of steady state. On the surface this seems to be a logical assumption for a water-drive or gas-cap-drive reservoir, but carefully note the nature of the steady-state system assumed. In a water-drive or gas-cap-drive steady-state system, we normally assume that the oil is being displaced in a more or less vertical direction by gas or water at a rate equal to the production rate. However, in deriving these coning equations we have assumed flow radially with the gas above or water below the radial flow so there is nothing displacing the oil radially. The inadequacy of such a model is obvious. We are assuming that all of the fluid entering the original oil-bearing portion of the reservoir flows horizontally across the cylinder with the r_e radius, and none of the fluid enters vertically with a stabilized interface. In reality, the exact opposite is true. None of the entering fluid flows across the r_e radius cylinder, and all of the fluid enters vertically. Therefore, a stabilized cone is impossible for a bottom-water or vertical gas-cap drive. Nevertheless, in the vicinity of the wellbore, the flow system does approximate horizontal radial flow and the previously derived equations give some qualitative guides for coning.

One further shortcoming of our coning equations should be noted. They give no indication of the amount of time required for a cone to form. Also, as with fingering, the indicated permissible flow rates below the critical rates are generally uneconomical, and the time required for the cone to form is very important.

The practical solution of coning problems is probably limited to computer modeling of the particular problem with which we are concerned. As stated, it is beyond the scope of this book to provide a practical solution of this nature. However, most engineers use a digital computer whose capabilities can solve a simple coning problem. The problem should be modeled to give a realistic flow pattern based on the knowledge of the reservoir. The recommended technique is simply assuming a sequence of steady states with new saturations calculated in each segment at the end of each time interval. The relative permeabilities can then be adjusted on the basis of the new saturations, and they can be used to calculate the next steady-state flow-period behavior. The general technique of generating a matrix of equations describing flow between the various segments and solving the matrix with standard computer techniques is described in chapter 11.

Several simplified and relatively well-known coning analysis techniques can be found that predict critical rates for stabilized conditions.* At least one method attempts to evaluate the time required for a cone to form.[8] However, all of these methods appear to have very serious limitations in their basic assumptions. Consequently, an effort should be made to obtain a computer flow model analysis for any problem that presents serious economic consequences.

Additional Problems

6.8 Given the following capillary pressure curves in Fig. 6–26 and zonation as shown in the following table, construct a water saturation versus depth curve similar to Fig. 6–6. The 100% water saturation determined from core analysis is 4,054 ft, the water specific gravity is 1.08, and the oil specific gravity is 0.825.

Zone	*Depth, ft*	*Permeability, md*	*Porosity, %*
II	3,988–4,007	166	20.8
V	4,007–4,019	72	19.1
Shale	4,019–4,031	Shale	—
III	4,031–4,037	591	27.3
I	4,037–4,046	564	27.2
IV	4,046–4,055	10.2	14.2

6.9 The laboratory water-air capillary pressure curve is given in Fig. 6–27. If the reservoir water-oil interfacial tension is 28 dynes/cm and the wetting angle is 0.0, determine the reservoir capillary pressure curve. Compare this curve with the zone V curve given in Problem 6.8.

6.10 From the capillary pressure data in Table 6–2, calculate and plot a J-function curve and use it to generate the capillary pressure curve for zone V of Fig. 6–5. The reservoir data are as listed:

Permeability = 72 md
Porosity = 19.1%
Interfacial tension = 28 dynes/cm
Wetting angle = 0.0°

*See "Mechanics of Two Immiscible Fluids in Porous Media," by Meyer and Gardner, *Journal of Applied Physics* (1954). Also, see *Petroleum Engineering,* by Gatlin, (Englewood Cliffs: Prentice-Hall, 1960).

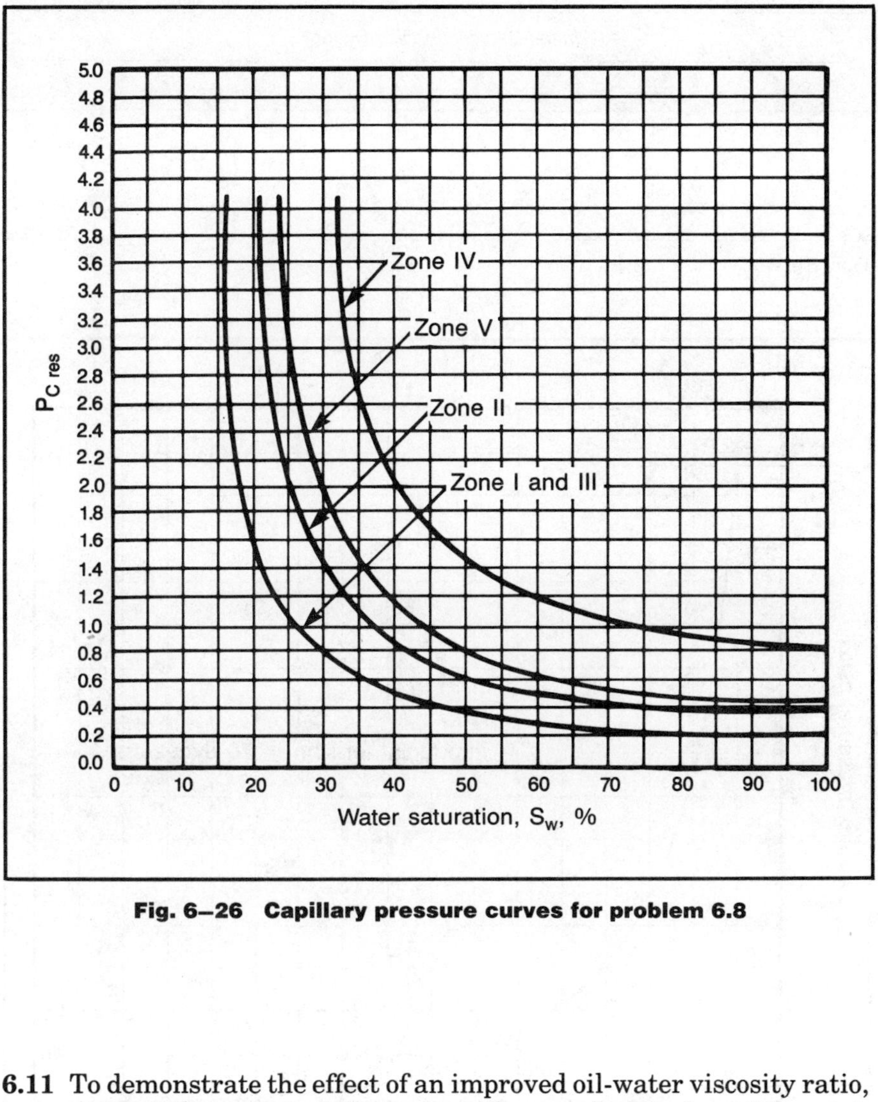

Fig. 6–26 Capillary pressure curves for problem 6.8

6.11 To demonstrate the effect of an improved oil-water viscosity ratio, assume the water viscosity is 1.51 cp and rework problem 6.4. Calculate and plot the fractional flow curve with and without the effect of gravity (2 curves).

6.12 Use the fractional flow curve calculated in problem 6.11. Calculate and plot the saturation versus distance profile at 0.5, 1.0, and 2.0 years using the 0.79, 0.70, and 0.65 water saturations. Porosity is 21.5%, and the average distance from the original WOC to the first line of producers is 350 ft. The resulting plot should be similar to Fig. 6–17. The following data are applicable:

*Initial Saturation Distribution**

S_w	$X_{initial,\ ft}$
0.79	10
0.75	12
0.70	15
0.65	18

*From Fig. 6–17.

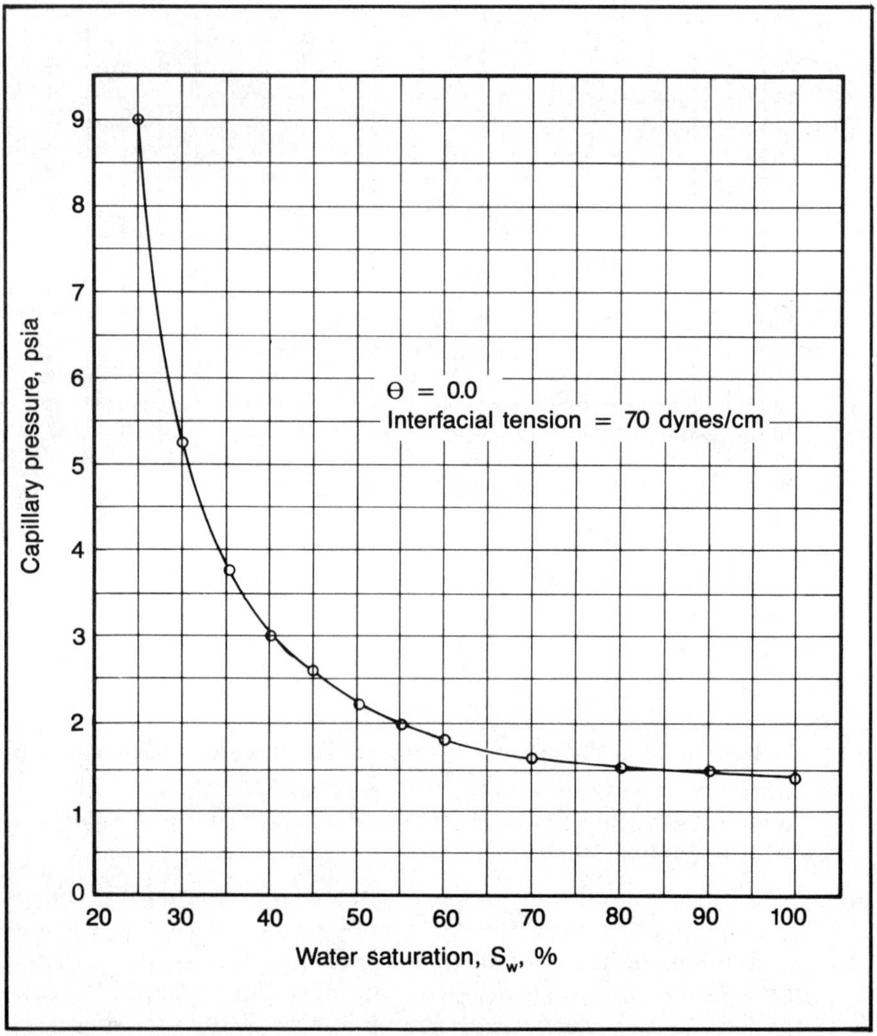

Fig. 6–27 Laboratory capillary pressure curve for problem 6.9

TABLE 6–2 Water-Air Capillary Pressure Data
Hughey Sand
Water-air interfacial tension = 70 dynes/cm
Wetting angle = 0°

Core 1		Core 2		Core 3		Core 4	
k = 11.2 md		*k = 34 md*		*k = 157 md*		*k = 569 md*	
ϕ = 0.147		*ϕ = 0.174*		*ϕ = 0.208*		*ϕ = 0.275*	
S_w, %	p_c, psi	S_w, %	p_c, psi	S_w, %	p_c, psi	S_w, %	p_c, psi
100	2.15*	100	1.60*	100	0.93*	100	0.60*
87	2.27	82	1.79	79	1.00	65	0.67
74	2.59	54	2.28	60	1.19	47	0.87
60	2.95	43	3.25	42	1.77	40	1.15
54	3.50	35	4.91	31	3.28	33	1.63
47	3.85	29	6.55	23	5.46	29	2.31
41	4.82	28	7.92	22	7.93	22	2.80
37	5.69	27	10.25	21	10.25	19	3.59
34	7.45					16	6.52
32	10.25					16	10.25

*Threshold capillary pressure

6.13 A line-drive waterflood is to be performed in a thin, homogeneous, horizontal reservoir with an initial gas saturation of 5% and an average water saturation at the start of injection of 35%. Assume the fractional flow curve of problem 6.11 is applicable and determine the oil displaced in pore volumes versus the cumulative water injection in pore volumes if an oil bank forms such that all gas is displaced ahead of the oil bank. Use the Welge method.

Notes

1. D.L. Katz et al., *Handbook of Natural Gas Engineering* (New York: McGraw-Hill, 1959).
2. B.C. Craft and M.F. Hawkins, *Applied Petroleum Reservoir Engineering* (Englewood Cliffs: Prentice-Hall, 1959).
3. J.M. Amyx, D.M. Bass, and R.L. Whiting, *Petroleum Reservoir Engineering* (New York: McGraw-Hill, 1960).
4. S.E. Buckley and M.C. Leverett, "Mechanism of Fluid Displacement in Sands," *Trans.*, AIME (1942), volume 146, 107–117.
5. Buckley and Leverett.
6. S.J. Pirson, *Oil Reservoir Engineering*, 2nd edition (New York: McGraw-Hill, 1958).
7. Pirson, 1958.
8. D.P. Sobocinski and A.J. Cornelius, "A Correlation for Predicting Water Coning Time," *JPT* (May 1965).

Additional References

Chaney, Noble, Henson, and Rice. "How to Perforate Your Well to Prevent Oil and Gas Coning." *OGJ,* 1956, p. 54.

Gatlin, Carl. *Petroleum Engineering*. Englewood Cliffs: Prentice-Hall, 1960.

Hubbert, M.K. "Entrapment of Oil under Hydrodynamic Conditions." *AAPG Bulletin,* August 1958, p. 1954.

Kaplan, J.A. "A Gaussian Digital Simulation of Bottom Water Coning in a Petroleum Reservoir." M.S. thesis, Ohio State University, 1971.

Kosakowski, M.W. "An Iterative Digital Study of Water Coning in a Bottom Drive Oil Reservoir," M.S. thesis, Ohio State University, 1971.

Meyer and Gardner. "Mechanics of Two Immiscible Fluids in Porous Media." *Journal of Applied Physics,* 1954.

7

Material Balance

Material balance can be more precisely termed *mass balance*. It refers to a group of useful equations that are derived by recognizing the conservation of mass. These equations can be derived by equating masses of reservoir fluids that exist in and out of the reservoir at different times. However, it is generally easier to derive the equations by equating volumes of reservoir fluids at different times.

We are concerned almost entirely with the *tank type* of material balance. Therefore, we treat the reservoir for theoretical purposes as having the same pressure throughout at any particular time or stage of depletion. This assumption is of course unrealistic, since we know that there is a considerable variation in pressures throughout the reservoir when the reservoir is being produced as a result of the flow in the reservoir. However, it is shown that the tank-type analysis accurately predicts the behavior of the reservoir in most cases if accurate average pressures and production figures can be obtained.

There is a tendency today to bypass the fundamental ideas of reservoir material balance because reservoir computer models are widely available within most petroleum companies and from computer software firms. These computer models treat the reservoir as a series of interconnected tanks with running material balances on each tank, and the concurrent calculation of the movement of fluid or flow is made between tanks. When properly used, these computer models can provide a more detailed prediction of the behavior of a reservoir than can a simple single-tank model. However, the problem that appears to have arisen is that the models are so simple to use that the engineer fails to recognize the misapplication of the computer model. Then ridiculous results are often accepted as correct. Some of these results are almost legendary. For example, there have been water-drive reservoir models that assume a constant rate of water encroachment or solution-gas-drive models that produce at a constant gas-oil ratio. In the author's opinion computer reservoir models can only be used intelligently by engineers who have a basic understanding of reservoir engineering fundamentals, especially the fundamentals of material balance.

Consequently, our objectives are twofold. First, we present methods that can be used to predict the behavior of most reservoirs by considering the oil-bearing portion of the reservoir as a single tank. Second, we provide an understanding of the fundamentals of reservoir engineering material balance that permits the engineer to use available reservoir computer models intelligently to obtain a more detailed analysis of a reservoir. It is the author's opinion that the engineer does not have a knowledge of a particular reservoir sufficient for a computer reservoir model application until he has determined, through attempted conventional material-balance analysis, that the reservoir is too complex to analyze with a single tank-type model. Only then will he understand such characteristics as water encroachment, reservoir capacity variations, and past performance sufficiently to permit an intelligent and realistic application of a computer reservoir model to that particular reservoir.

Most reservoir engineering techniques involve some application of material balance or some implied application. The most useful applications of material-balance equations require the concurrent use of flow equations. When material-balance concepts are combined with flow concepts, it is possible to predict the production behavior of an oil or gas reservoir as a function of time. Without the fluid flow concepts, which tell us how rapidly fluids can be withdrawn from the reservoir at various stages of depletion, the material balance simply provides performance as a function of the average pressure in the reservoir. Thus, we may say that material balance provides us with a prediction of cumulative production versus average reservoir pressure for a reservoir. To put this on a cumulative production versus time basis, it is necessary to introduce fluid flow concepts.

This material is presented in the sequence that has proven to be the most effective in obtaining an understanding of the material-balance methods and not in the manner that fits the best classical outline. A general material-balance equation is first developed by considering in sequence a gas reservoir, which provides the basis for understanding the gas cap on an oil reservoir; oil production as a result of the expansion of oil above the bubble point, in which behavior is similar to that of a gas reservoir; gas liberation in an oil reservoir; PVT data, which is necessary to understand gas liberation in an oil reservoir; solution-gas-drive behavior with gas liberation in the reservoir; the effects of encroaching gas-cap gas or water; and the effects of the small change in pore volume that accompanies a decline in the reservoir pressure. Different forms of the material-balance equation that are encountered in the industry, but are not used in this book, are noted to avoid later confusion. Then the general difficulties encountered in applying reservoir material balances are noted. Next, methods of predicting the behavior of solution-gas- and gas-cap-drive reservoirs are presented, including those that are recommended and those that are common in the industry.

Additional difficulties caused by reservoirs being partially saturated and partially undersaturated with gas or having an initial gas cap are covered. The prediction of the behavior of reservoirs that have a water drive is considered with the calculation of water encroachment independent of material balance. Finally, a section is presented that considers the methods of increasing the ultimate primary production through reservoir control.

A General Material-Balance Equation

A general material-balance equation is derived that can be applied to any hydrocarbon reservoir. All of the terms of this equation are never significant at any one time. Nevertheless, by using this general equation, the engineer does not need to be familiar with a multitude of special equations, such as equations for material balance above the saturation pressure, for solution-gas-drive reservoirs, for gas-cap-drive reservoirs, and for water-drive reservoirs. All of these problems can be handled with the one general material-balance expression. The various reservoir types are considered individually simply as a means of developing the general material-balance equation.

Material balance in gas reservoirs. The simplest type of material balance that can be written for a reservoir is that written for a dry-gas reservoir. It is possible to derive this equation by equating the mass of gas in the reservoir initially with the mass of gas in the reservoir at some later time, plus the mass of gas produced. However, it is easier to obtain the desired expression by equating volumes. If there is no water drive and the change in pore volume with decline in pressure is negligible—which it is for a gas reservoir—we can write an expression for the volume of gas in the reservoir that remains constant, stated as a function of the reservoir pressure, p; the standard cubic feet of gas produced, G_p; the original standard cubic feet of gas in the reservoir, G; and the gas formation volume factor, B_g, all evaluated at different times.

The gas formation volume factor, B_g, represents the ratio between the volume of gas at one particular pressure and temperature and the volume of gas at another particular pressure and temperature. The units used are barrels at reservoir conditions per standard cubic foot of gas. The standard cubic foot is, of course, the volume of gas at 60°F and

14.7 psi. Thus, we use the well-known gas equation introduced in chapter 5 to write an expression for the gas formation volume factor:

$$pV = znRT \tag{5.1}$$

Where:
- V = Volume, scf
- z = Gas deviation factor
- n = Number of mols of gas
- R = Gas constant, 10.73
- T = Temperature, °R

We desire an expression for the volume of 1 scf of gas at a reservoir pressure, p, and temperature, T. One scf of gas is equivalent to 1/379 mols of gas since 1 mol of gas at standard conditions occupies 379 cu ft. When the volume of gas at reservoir conditions in standard cubic feet is converted to barrels by dividing by 5.615, we obtain an expression for B_g as follows:

$$V = znRT_f/p \tag{7.1}$$

$$B_g = V/5.615 = z(1/379)(10.73)T_f/(p\ 5.615) \tag{7.2}$$

$$= 0.00504\ zT_f/p$$

Eq. 7.2 is discussed more fully in chapter 5, as is the evaluation of the gas deviation factor, z. Considerable care should be taken in evaluating the gas deviation factor when the gas formation volume factor, B_g, is calculated. Thus, engineers who are not familiar with the evaluation of the gas deviation factor by various methods and the limitations of these methods should study chapter 5 before proceeding with such calculations. Others with a previous knowledge of gas deviation factor analysis can simply refer to appendix B, Figs. B8, B9, and B12 and Table B5 to determine appropriate z factors.

With the gas formation volume factor, B_g, defined, we can proceed with the development of the material balance for a dry-gas reservoir. The original barrels of gas in the reservoir can be equated to the gas in the reservoir at some future time after G_p standard cubic feet of gas have been produced, using Eq. 7.3:

$$GB_{gi} = (G - G_p)B_g \tag{7.3}$$

Where:
- B_{gi} = Gas formation volume factor at initial reservoir pressure and temperature

The gas formation volume factor, B_g, in the right-hand side of Eq. 7.3 is based on the pressure and temperature of the reservoir after G_p standard cubic feet of gas have been produced. Note that the amount of gas remaining in the reservoir in standard cubic feet is the difference between the gas originally in the reservoir and the gas produced, $G - G_p$. Eq. 7.3 is illustrated graphically in Fig. 7–1.

If this gas reservoir is subjected to a water drive, water enters the pore space originally occupied by gas as the pressure in the gas-occupied pore volume is decreased. Then the original pore space occupied by gas is occupied by free gas $(G - G_p)$ and encroached water, W_e after G_p production. Thus, Eq. 7.3 becomes:

$$GB_{gi} = (G - G_p)B_g + (W_e - W_p) \tag{7.4}$$

Where:

W_p = Water produced

Of course, W_p reduces the amount of the original gas pore volume occupied by water. Eqs. 7.3 and 7.4 are used in many different ways to predict the behavior of a gas reservoir in chapter 5. Work problem 7.1 to check understanding of material balance for a gas reservoir. The solution is shown in appendix C.

PROBLEM 7.1: Gas Material Balance

To illustrate the meaning of these material-balance equations, consider a reservoir that contains 400 MMscf of gas at an initial pressure of 3,150 psia with a temperature such that 1 scf of gas occupies 0.0010 bbl in the reservoir, i.e., B_{gi} = 0.0010 bbl/scf. Find the amount of production accumulated at a time when the pressure in the reservoir has declined to 2,900 psia. At this pressure 1 scf of gas occupies 0.0011 bbl of pore space in the reservoir, i.e., B_g = 0.0011 bbl/scf.

Note that if a gas reservoir is attached to an oil reservoir to form a gas-cap reservoir, an expression similar to Eqs. 7.3 and 7.4 can be written to describe the change in the gas-cap volume. The gas-cap volume after some production of gas-cap gas, G_{pc}, that may occur with the oil production—by accident or poor planning, not by design—is still expressed by the right-hand side of Eq. 7.3, $(G - G_{pc})B_g$. Thus, the gas cap may expand into the original oil zone, or the oil may encroach into the original gas cap (Fig. 7–1):

$$\text{Change in gas-cap volume} = GB_{gi} - (G - G_{pc})B_g \tag{7.5}$$

Work problem 7.2 and check the solution against the one in appendix C.

PROBLEM 7.2: Gas-Cap Expansion

Assume the gas reservoir in problem 7.1 with an initial pressure of 3,150 psia and B_{gi} = 0.0010 res bbl/scf is attached to an oil reservoir to form a gas-cap reservoir. If production from the oil-bearing portion of the reservoir causes the gas-cap

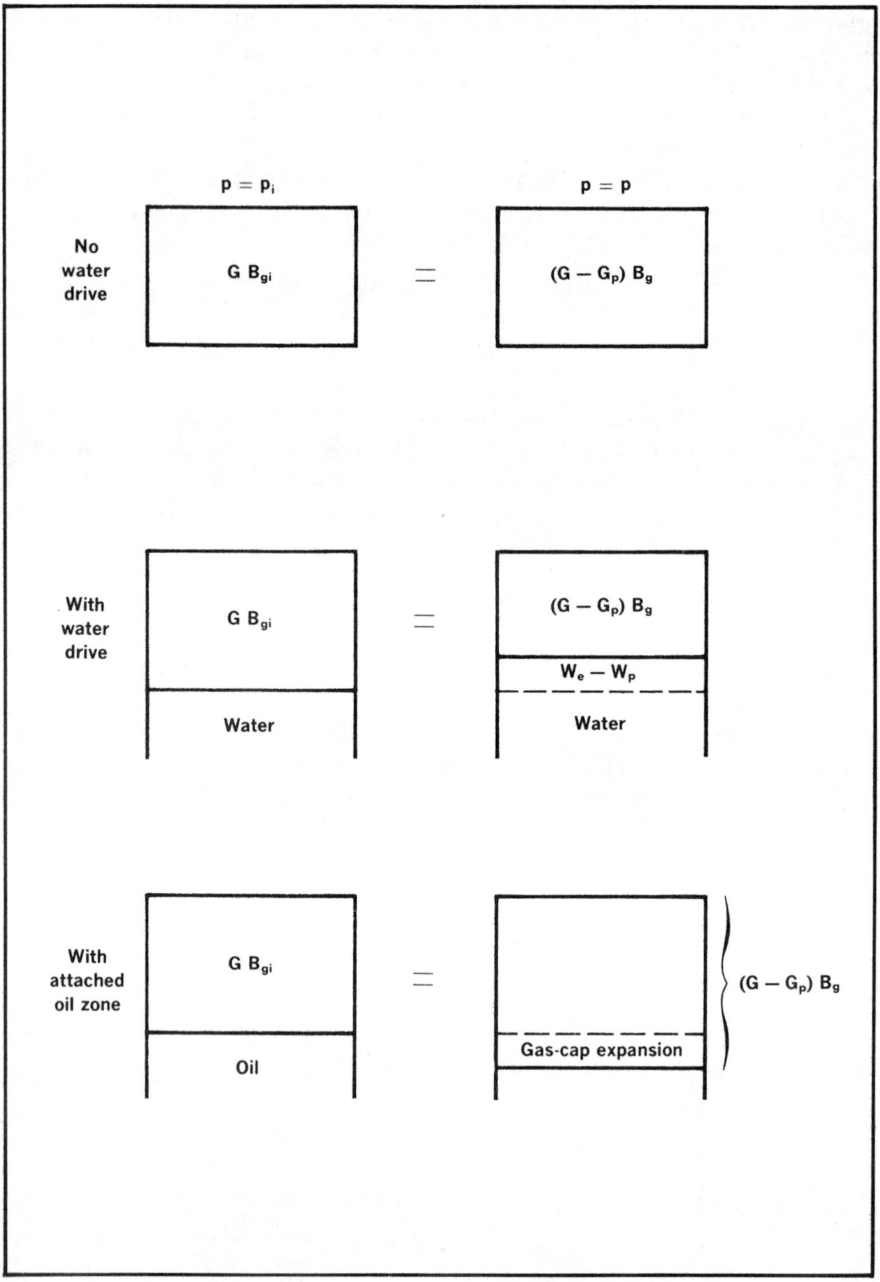

Fig. 7–1 Schematic representation of the dry-gas-reservoir material-balance equations

pressure to decline to 2,900 psia (B_g = 0.0011 res bbl/scf) with no gas-cap production, how many reservoir barrels of gas encroach into the original oil zone? How many standard cubic feet of gas does this represent?

Material balance for liquid expansion. It is surprising the number of oil reservoirs that produce for significant periods by the simple expansion of liquid in the reservoir. These reservoirs are generally very large, but they have limited permeability. Therefore, at the relatively small rates of production that can be obtained, the average pressure in the reservoir remains above the bubble-point or saturation pressure for a considerable time. During this period of course, the production results entirely from the expansion of the oil in the reservoir and the reduction in the pore volume caused by the decrease in the reservoir pressure. For example, West Texas appears to have many such reservoirs. For the purposes of developing the general material-balance equation for oil reservoirs, let us first assume that the change in pore volume is negligible. Carefully note that this is a grossly erroneous assumption. Much of the production for reservoirs producing above the bubble point is a result of the change in pore volume accompanying the decline in reservoir pressure. However, it is convenient to add this production as a separate term to the general material-balance equation.

Assuming that production is caused only by liquid expansion, a material-balance equation for an oil reservoir similar to that of Eq. 7.3 for a gas reservoir can be written:

$$NB_{oi} = (N - N_p)\,B_o \tag{7.6}$$

Where:

N = Volume of oil originally in the reservoir, stb
N_p = Oil produced at some particular stage of depletion, stb
B_o and B_{oi} = Oil formation volume factors

The factors B_o and B_{oi} relate the volume of oil in the reservoir at a particular pressure to the volume of this oil in the stock tank. The reservoir volume of oil shrinks in traveling to the stock tank because a portion of the oil changes to gas as a result of the change in phase that accompanies the change in pressure and temperature. Note that this ratio is not the ratio of the volume of the same mass of oil or hydrocarbon at stock-tank conditions compared with the volume of the same mass at reservoir conditions as with the gas formation volume factor, B_g. With the oil formation volume factor the volume at stock-tank conditions contains less mass than does the related volume at reservoir conditions, which includes the volume of gas in solution.

This phrase, *gas in solution,* is used to define or describe the phenomena of phase change that accompanies the hydrocarbon liquids transition to the lower stock-tank pressure and temperature. The concept of gas in solution is a handy one, particularly since most of the gas liberated from the reservoir oil has a very similar chemical composition. Thus, it can be treated as a constant composition without introducing a significant error.

Fig. 7–2 illustrates the schematic relationship between the oil formation volume factor and reservoir pressure. In this diagram p_i is the initial reservoir pressure, p_s is the saturation pressure or the pressure at which the first bubble of gas forms in the reservoir, and p_{st} is the stock-tank pressure.

In Eq. 7.6 we are considering B_o values at or above the bubble-point pressure. Consequently, the change in B_o in this pressure range is simply a result of the expansion of the oil in the reservoir. Thus, the same mass of oil at pressure p_i represents a lesser volume than it does at the saturation pressure, p_s. The increase in B_o is directly related to the compressibility of the liquid oil. Fig. 7–3 illustrates Eq. 7.6 schematically.

Gas liberation in the reservoir. When the fluid in a reservoir reaches the saturation or bubble-point pressure, gas is liberated in the reservoir. In such a case if the reservoir exists initially at or above the bubble-point pressure, the pore volume originally occupied by oil is occupied at some pressure less than the saturation pressure by liquid and liberated gas. This type of reservoir is shown in Fig. 7–3. The oil

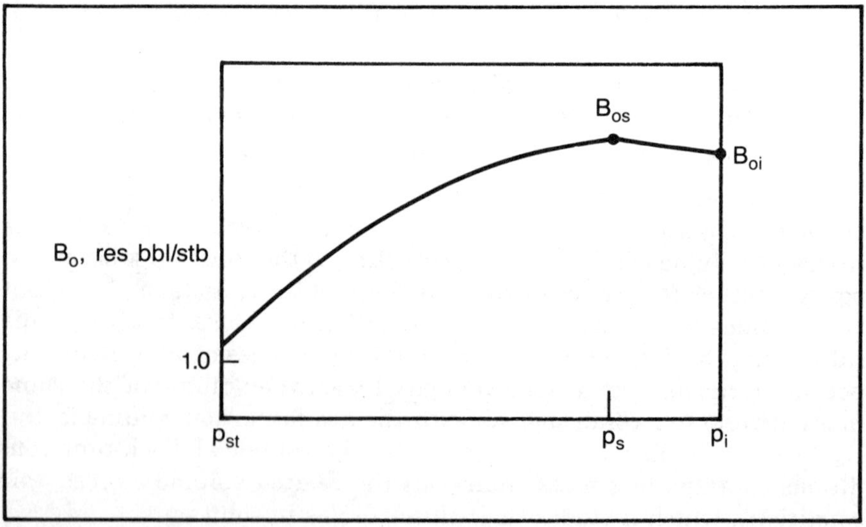

Fig. 7–2 Typical oil formation volume factor versus reservoir pressure relationship

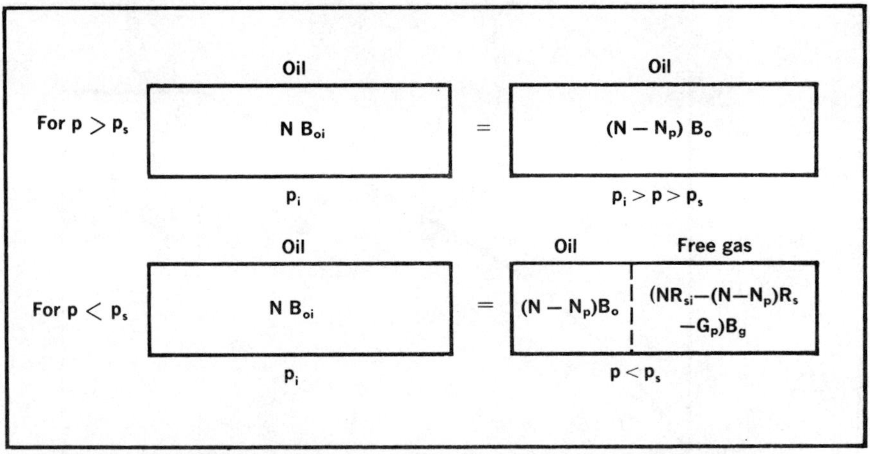

Fig. 7–3 Schematic representation of the solution-gas-reservoir material-balance equations

volume is represented by the right-hand side of Eq. 7.6, but the liberated gas remaining in the reservoir requires some additional analysis.

As the pressure in the reservoir declines because of oil and gas production and falls below the bubble-point pressure, gas is liberated. As the reservoir pressure continues to decline as a result of oil and gas production, additional gas is liberated. As gas is liberated, less gas remains in solution in the oil. If the stock-tank pressure is reached, there is no longer any gas in solution relative to stock-tank conditions and all of the original solution gas has been liberated. This is illustrated in Fig. 7–4 where the gas in solution in the reservoir oil is plotted versus the pressure.

Since the oil formation volume factor is related to stock-tank barrels of oil, it is also convenient to relate the amount of gas in solution to stock-tank barrels of oil. Thus, R_s is stated in standard cubic feet per stock-tank barrel. Note that the liberated gas is the initial gas in solution, R_{si}, less the gas remaining in solution at a particular pressure, R_s. Also, note that gas in solution above the bubble-point pressure does not change with the pressure since none of this gas is liberated, until the reservoir pressure declines to the saturation pressure, p_s.

Pressure-Volume-Temperature (PVT) Relationships

One of the best means of obtaining an understanding of the phase behavior and volume changes that take place in the reservoir during its depletion is to study the similar behavior that takes place in a PVT

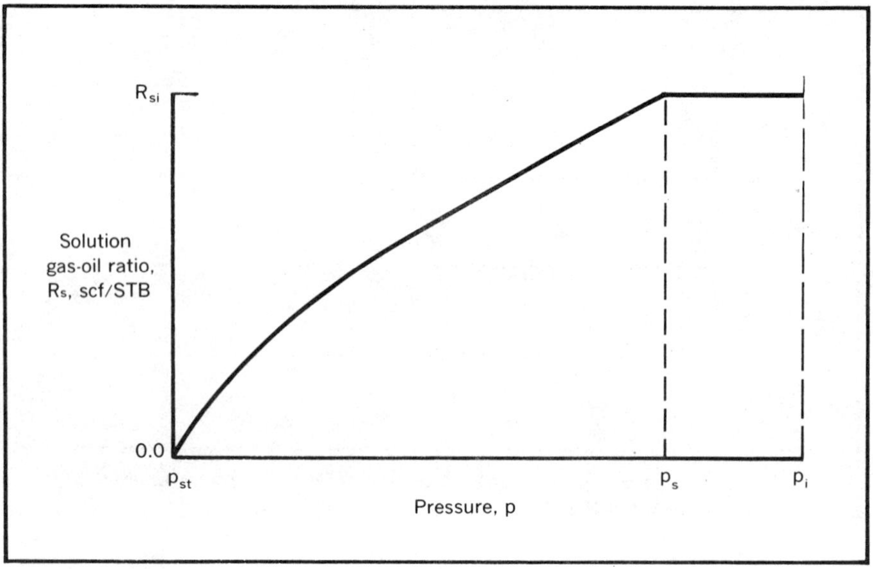

Fig. 7–4 Typical solution-gas-oil ratio versus pressure relationship

cell in the laboratory during the measurement of the PVT characteristics of a reservoir hydrocarbon system. Ideally, a reservoir is sampled immediately following the discovery well in a reservoir. The bottom-hole sampling of the reservoir fluid then occurs at or above the saturation pressure. This sample is taken into the laboratory and transferred to a high-pressure PVT cell.

The pressure, temperature, and volume of this cell can be controlled in such a way as to permit measurement of the oil formation volume factor and the gas in solution at the various pressures to be encountered during the producing life of the reservoir. Ordinarily, the pressure and temperature of the cell are first adjusted to the initial reservoir pressure and the reservoir temperature. The volume of the fluid in the cell at this time is of some arbitrary value, depending on the size of the sample taken, the size of the cell, and the anticipated PVT analysis.

To obtain a reduction in pressure in the cell, the cell volume is increased slightly. This increase is accomplished physically in the laboratory by either removing mercury from the cell or simply moving a piston. If the reservoir is initially above the saturation pressure, the initial change in volume, V, with pressure is very small, as indicated in Fig. 7–5. However, once the pressure has been lowered below the saturation pressure, p_s, gas is liberated and the total volume of the cell, V_t, begins increasing more rapidly.

Use of a visual cell with a window permits observation of the gas-liquid interface and evaluation of the amount of oil and gas in the cell at any particular pressure. It can then be observed that the oil volume is a maximum at the saturation pressure. It declines as more of the oil changes to gas as the pressure in the cell is decreased. Once the cell has reached the stock-tank (atmospheric) pressure, it is permitted to cool to a stock-tank temperature, which results in further thermal shrinkage of the oil to a volume, $(V_o)_{st}$.

Concurrently, the cell volume of gas, indicated in Fig. 7–5 as the total volume, V_t, less the oil volume, V_o, continues to increase with the decrease in cell pressure until it is at stock-tank conditions. Then all of the gas has been liberated relative to stock-tank conditions. This gas is initially in solution when the cell pressure is at or above the saturation pressure, p_s. Consequently, the initial solution-gas-oil ratio is $[(V_t)_{st} - (V_o)_{st}]5.615/(V_o)_{st}$. The numerical constant is used to correct the ratio from volumes per volume or barrels per barrel to standard cubic feet

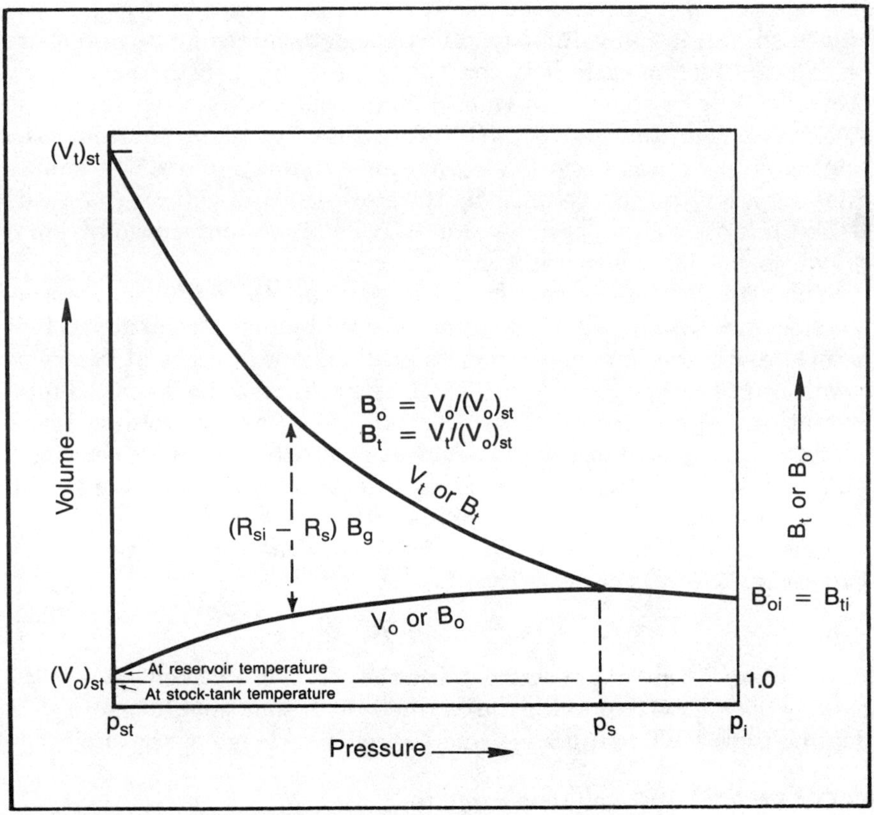

Fig. 7–5 Interrelationship of PVT data

per stock-tank barrel. Note that no gas formation volume factor is necessary to correct to standard conditions since stock-tank conditions are generally taken as equivalent to standard conditions.

At other reservoir pressures the amount of free gas in the cell is the difference between the total gas liberated to stock-tank conditions and the gas still in solution in the oil at the subject pressure. Conversely, the gas in solution at any particular cell pressure is the difference between the total gas liberated to stock-tank conditions and the free gas in the cell at any pressure $(V_t - V_o)$ after this free-gas volume has been corrected to a corresponding volume at standard conditions. When $(V_t - V_o)$ is divided by the stock-tank oil volume $(V_o)_{st}$, which results from the V_o volume of oil, we obtain volumes of gas at reservoir conditions per volume of oil at stock-tank conditions or reservoir barrels of gas per stock-tank barrel of oil. Then the reservoir barrels of gas can be converted to standard cubic feet by dividing by B_g reservoir barrels per standard cubic feet. The resulting scf/stb represents the standard cubic feet of gas liberated per stock-tank barrel of oil. The difference between this figure and R_{si} equals the gas remaining in solution, R_s.

The oil formation volume factor can be determined for any pressure by calculating the ratio between the volume of oil in the cell at any pressure and the stock-tank volume that results when this oil reaches stock-tank conditions, $(V_o)_{st}$. Note that Fig. 7–5 also represents the relationship between various formation volume factors. By simply dividing all of the cell volumes by the stock-tank oil volume $(V_o)_{st}$, the stock-tank oil volume becomes 1.0 and the oil-volume curve becomes the oil formation volume factor, B_o, curve.

The total-volume curve also becomes the total formation volume factor curve. This is a PVT parameter used by many companies. It is defined as the volume of oil and its original solution gas at reservoir conditions per stock-tank barrel of oil. The difference between the total formation volume factor curve and the oil formation volume factor curve is the illustrated volume of gas at reservoir conditions per stock-tank barrel of oil, $(R_{si} - R_s)B_g$. The difference in the solution-gas-oil ratios is the standard cubic feet of liberated gas, and the gas formation volume factor corrects the volume to barrels at reservoir conditions. From Fig. 7–5 it is obvious that:

$$B_t = B_o + (R_{si} - R_s)\,B_g \tag{7.7}$$

Problem 7.3 should be worked and the solution checked against the one in appendix C to provide the engineer with some mental exercise in dealing with PVT parameters and to test knowledge of the subject.

PROBLEM 7.3: PVT Lab Data Exercise

Given the following lab data from PVT analysis, evaluate R_s in scf/stb and B_o and B_T at the stated pressures. Also, find the compressibility factor for oil above the saturation pressure as volume/volume/psi. The gas deviation factors at 1,000 psia and 500 psia have been evaluated as 0.91 and 0.95, respectively.

Cell pressure, psia	2,000	1,500 = p_s	1,000	500	14.7
Oil volume in cell, cc	650	669	650	615	500
Gas volume in cell, cc	0	0	150	700	45,000
Cell temperatures, °F	195	195	195	195	60

Behavior of Produced GOR

The produced gas-oil ratio (GOR) is constant above the saturation pressure. However, once the gas saturation has reached a point that the free gas in the reservoir begins to flow, the behavior of the gas-oil ratio becomes more complicated. The produced gas-oil ratio, R, at any particular time is the ratio of the standard cubic feet of gas being produced at any time to the stock-tank barrels of oil being produced at that same instant. Hence, the name *instantaneous gas-oil ratio* is often applied to the symbol R. Remember that the gas produced includes solution gas and free gas. To say this differently, the gas produced enters the wellbore as gas in solution in the oil entering the wellbore, or as free gas flowing into the wellbore concurrently with the oil. Thus, we can say that:

$$R = R_s + R_{flow} \tag{7.8}$$

Where:
R_{flow} = Flowing gas-oil ratio in the reservoir, scf/stb or (q_{scf}/q_{stb})

This expression can be expanded using the radial flow equation derived in chapter 2:

$$q_r = (1.127kA_r/\mu)(\Delta p/\Delta r)_r \tag{2.7}$$

Writing q_{scf} and q_{stb} in terms of Eq. 2.7 applied at the wellbore with the rates corrected from reservoir volumes to scf and stb, respectively, we obtain an expression for the reservoir flowing GOR and the produced GOR:

$$R_{flow} = \frac{q_{scf}}{q_{stb}} = \frac{(1.127k_gA_w/\mu_g)(\Delta p/\Delta r)_w/B_g}{(1.127k_oA_w/\mu_o)(\Delta p/\Delta r)_w/B_o} \tag{7.9}$$

$$R_{flow} = \frac{k_g}{k_o}\frac{\mu_o}{\mu_g}\frac{B_o}{B_g} \tag{7.10}$$

$$R = R_s + \frac{k_g}{k_o}\frac{\mu_o}{\mu_g}\frac{B_o}{B_g} \tag{7.11}$$

Note that the ratio of the effective permeabilities in Eq. 7.11 is the same as the ratio of the relative permeabilities, k_{rg}/k_{ro}, which is a function of the liquid saturation. Fig. 7–6 shows example data. The water saturation in a gas-drive reservoir without a water drive is nearly constant, but the oil saturation is changing continually. Consequently, in order to determine the relative permeability ratio and the produced gas-oil ratio, it is necessary to evaluate the oil saturation corresponding to any cumulative oil production. This calculation can be done by material balance.

The oil saturation is the remaining reservoir barrels of oil in the reservoir, $(N - N_p)B_o$, divided by the reservoir pore volume in barrels. The pore volume can be determined from the initial oil saturation, $1 - S_{wc}$, and the original reservoir barrels of oil in the reservoir, NB_{oi}. Thus, the material-balance expression for the oil saturation can be written:

$$S_o = \frac{\text{res bbl of oil}}{\text{bbl of pore volume}} = \frac{(N - N_p)B_o}{(NB_{oi})/(1 - S_{wc})} \tag{7.12}$$

$$S_o = \frac{B_o(N - N_p)(1 - S_{wc})}{NB_{oi}} \tag{7.13}$$

A typical gas-oil ratio history for a gas-drive reservoir is shown in Fig. 7–7. As noted, the GOR remains constant as long as the reservoir pressure is above the saturation pressure because the k_g in Eq. 7.11 is zero since there is no free gas in the reservoir. Also, R remains constant above the bubble-point pressure, as shown in Fig. 7–4. Once the saturation pressure is reached, free gas begins to form in the reservoir, but no free gas flows until the equilibrium gas saturation is reached. This relative permeability characteristic is illustrated in Fig. 2–7. The increase in the gas saturation from zero at the saturation pressure to the equilibrium gas saturation is accompanied by a reduction in the reservoir pressure. This decrease in the reservoir pressure below the saturation pressure is accompanied by a decrease in the gas in solution (Fig. 7–4). Thus, with k_g remaining at zero and R_s declining, there is a decline in the producing gas-oil ratio as shown in Eq. 7.11.

Since it is difficult to measure small equilibrium gas saturations in the laboratory, it is the author's opinion that this ratio is generally considered to be much greater than it actually is. In fact, this declining-GOR portion of the history can seldom be distinguished in practice. However, part of this difficulty is undoubtedly a result of the fact that the entire reservoir does not reach the saturation pressure at the same time, tending to mask the GOR decline.

Once gas begins to flow, the GOR increases rapidly. This increase is caused by the rapid increase in the relative permeability ratio with a change in saturation. It is also caused by the extreme difference in the viscosities of the gas and oil. As noted, at low pressures the ratio of the

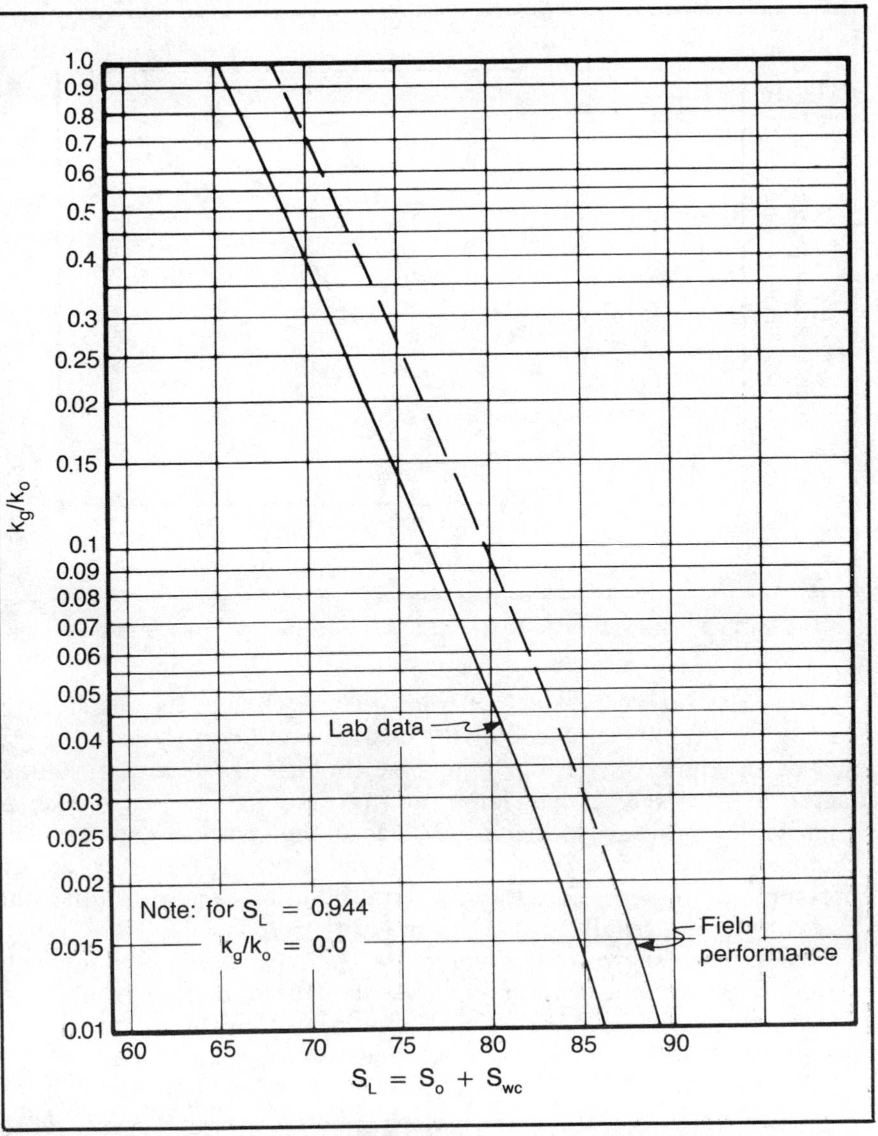

Fig. 7–6 Comparison of laboratory-measured and field-calculated relative permeability data

gas viscosity to the oil viscosity may be 100, and even at high pressures the viscosity ratio is high.

During the late life of the gas-drive reservoir, the gas-oil ratio may decline as indicated in Fig. 7–7. This decline is sometimes difficult for the engineer to understand since the relative permeability ratio and

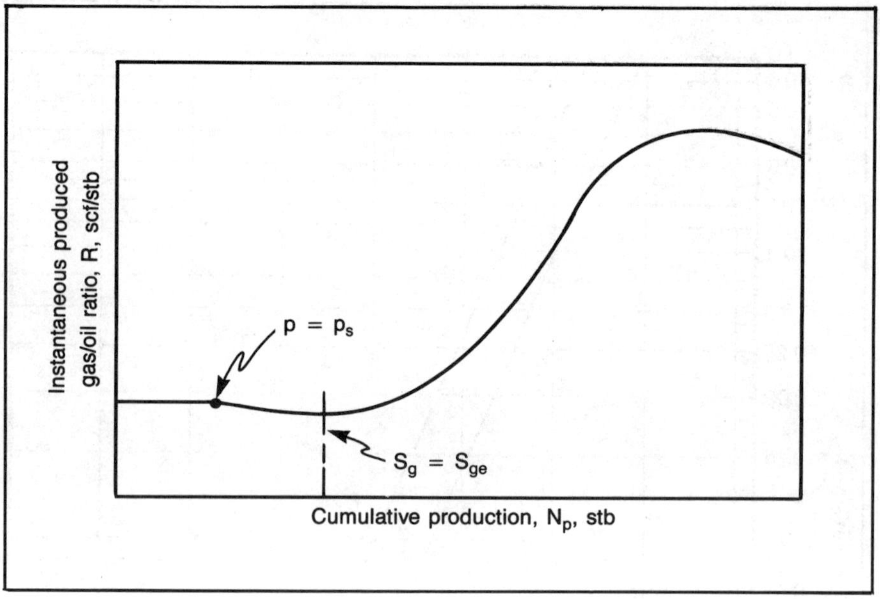

Fig. 7–7 Gas-oil ratio history for a solution-gas-drive reservoir

the viscosity ratio continue to increase. The decline in the gas-oil ratio during the late life is caused by the change that takes place in the gas formation volume factor, B_g. Remember that the gas formation volume factor is inversely proportional to the reservoir pressure. Consequently, B_g continues to increase in size as the reservoir depletes. The change in B_g at high reservoir pressures is small. However, at low pressures it changes more rapidly than the relative permeability and viscosity ratios, resulting in a decline in the produced GOR.

It has been shown that the cumulative or total gas production to some particular time or reservoir pressure is used in material balance. This cumulative gas production, G_p, is related to the instantaneous gas-oil ratio, R, and cumulative oil production by the expression:

$$G_p = \sum_{j=1}^{j=n} (R_j \Delta N_{pj}) \tag{7.14}$$

The validity of this expression can be seen if we recognize that, during some increment of oil production, ΔN_p, the corresponding increment of gas production, ΔG_p, is equal to the average instantaneous gas-oil ratio during the period, multiplied by the oil production increment, $R\Delta N_p$. Therefore, the total gas production at any time is the sum of these increments as indicated in Eq. 7.14.

The cumulative produced gas, G_p, is defined as the net cumulative produced gas for the purposes of material balance. This is necessary because gas is often injected into an oil or gas reservoir to maintain the reservoir pressure and, thus, increase the ultimate recovery. Consequently, we often refer to G_p as the *net produced gas* to distinguish it from the gross or total produced gas. The net produced gas is then determined by subtracting the injected gas from the total produced gas. Consequently, we can also identify gross and net produced gas-oil ratios.

Another parameter that is often used in material balance is the *cumulative produced gas-oil ratio,* which refers to the total (cumulative) gas produced at some particular time, divided by the total oil produced at that same time. The phrase generally infers that gross gas production figures are to be used, but a net cumulative produced gas-oil ratio may also be used.

Problem 7.4 tests our understanding of the GOR concepts. Check the solution against the one in appendix C.

PROBLEM 7.4: Determining Gas-Oil Ratios*

The following production and gas-injection data pertain to a reservoir:

Cumulative Oil Production, N_p, MMstb	*Producing Gas-Oil Ratio, R, scf/stb*	*Cumulative Gas Injected, MMscf*
0	300	—
1	280	—
2	280	—
3	340	—
4	560	—
5	850	0
6	1,120	520
7	1,420	930

To help interpret the meaning of the data, note that the producing gas-oil ratio initially—when the first barrel of oil is produced—is 300 scf/stb. When the one-millionth barrel of oil is produced, the producing GOR is 280. Also, note that gas injection is commenced after 5 MMbbl of oil have been produced. Plot on the same graph the producing GOR, the cumulative produced gas, the net cumulative produced gas, and the cumulative injected gas versus cumulative oil production (4 curves).

*Problem 7.4 is derived from *Applied Petroleum Reservoir Engineering,* by Craft and Hawkins (Englewood Cliffs, Prentice-Hall, 1959), p. 143.

Material Balance with Gas Liberation

With a knowledge of the gas in solution at various pressures, it is possible to write an expression for the free gas in the reservoir at any particular pressure, as shown in Fig. 7–3. The original gas in an oil reservoir initially found above the bubble-point pressure is the gas in solution, NR_{si}. At some later time when the reservoir pressure has declined below the bubble-point pressure as a result of the withdrawal of oil and gas from the reservoir, the original gas in solution is found in three places. Either the gas has been produced as G_p standard cubic feet; it is still in solution in the reservoir oil remaining, $(N - N_p)R_s$ standard cubic feet; or it exists in the reservoir as free gas as opposed to gas in solution in the reservoir. Thus, the free gas in the reservoir can be written as:

$$\text{scf of free gas in reservoir} = \text{original gas in solution} - \text{remaining gas in solution} - \text{produced gas}$$

$$= NR_{si} - (N - N_p)R_s - G_{ps} \quad (7.15)$$

If this volume of free gas is stated in reservoir barrels, it can be added to the remaining reservoir barrels of oil in the reservoir and equated to the original reservoir barrels of oil as illustrated in Fig. 7–3:

$$NB_{oi} = (N - N_p)B_o + [NR_{si} - (N - N_p)R_s - G_{ps}]B_g \quad (7.16)$$

One principle use of material balance is to determine the original stock-tank barrels of oil in the reservoir. Consequently, material-balance equations are generally written in the form where the equation has been solved for the initial stock-tank barrels of oil in the reservoir. When Eq. 7.16 is written this way, we obtain:

$$N = \frac{N_pB_o + B_g(G_{ps} - N_pR_s)}{B_o - B_{oi} + (R_{si} - R_s)B_g} \quad (7.17)$$

Note that the left-hand side of Eq. 7.6 is the original reservoir volume of the oil, and the right-hand side is the same reservoir volume at some later time after a drop in pressure results from the production of N_p stock-tank barrels of oil and G_p standard cubic feet of gas. Consequently, if fluid encroached into the original oil-bearing portion of the reservoir from either an expanding gas cap or water encroachment, the right-hand side of Eq. 7.6 would be smaller than the left-hand side by the amount of encroachment. Therefore, we can use Eq. 7.16 to write an expression for the change in the gas-cap volume and water encroachment in a reservoir where such conditions exist:

$$\text{bbl of change in gas-cap volume} = (G - G_{pc})B_g - GB_{gi} \quad (7.5)$$

$$(W_e - W_p) + \text{bbl of change in gas-cap volume} =$$
$$NB_{oi} - \{(N - N_p)B_o + [NR_{si} - (N - N_p)R_s - G_p]B_g\} \quad (7.18)$$

Work problems 7.5 and 7.6. Check the solutions against the ones in appendix C.

PROBLEM 7.5: Material Balance in a Solution-Gas-Drive Reservoir

An oil reservoir initially contains 4 MMstb of oil at a pressure of 3,150 psia with 600 scf/stb of gas in solution. When the average reservoir pressure has dropped to 2,900 psia, the gas in solution is 550 scf/stb. The initial oil formation volume factor is 1.34, and at 2,900 psia the oil formation volume factor is 1.32. If no initial gas-cap or water drive is associated with this oil reservoir, how many stock-tank barrels of oil can be produced when the pressure has declined to 2,900 psia? The cumulative produced gas-oil ratio at this time is estimated as 600 scf/stb. The cumulative gas-oil ratio is defined as the total gas produced divided by the total oil produced at a particular time. Thus, G_p is 600 N_p. The gas formation volume factor at 2,900 psia is 0.0011 res bbl/scf.

PROBLEM 7.6: Calculating Oil-Zone Shrinkage

If our calculations are correct, we have predicted that an oil reservoir would produce 101,800 stb of oil with an average gas-oil ratio of 600 scf/bbl when the average reservoir pressure declined to 2,900 psia if there is no gas-cap or water drive. However, assume this reservoir actually produced 130,800 stb of oil and 78.480 MMscf of gas when the average reservoir pressure reached 2,900 psia. What unexpected gas-cap gas or water invasion of the original oil zone could cause this additional production?

If the expression for the change in the gas-cap volume in Eq. 7.5 is substituted for barrels of gas-cap expansion in Eq. 7.18, the resulting expression can be solved for the original oil in place to obtain the general material-balance expression that includes all of the possible sources of natural reservoir energy except the change in the pore volume:

$$N = \frac{N_pB_o + B_g(G_{ps} - N_pR_s) - [(G - G_{pc})B_g - GB_{gi}] - (W_e - W_p)}{B_o - B_{oi} + (R_{si} - R_s)B_g} \quad (7.19)$$

In Eq. 7.19 the gas production is broken into gas-cap-gas production, G_{pc}, and solution-gas production, G_{ps}. However, it is normally impossible to distinguish between gas-cap gas and solution gas for any given volume of produced gas. Fortunately, production of either has the same effect on the material balance, and the two terms can then be added together and simply listed as G_p:

$$N = \frac{N_pB_o + B_g(G_p - N_pR_s) - G(B_g - B_{gi}) - (W_e - W_p)}{B_o - B_{oi} + (R_{si} - R_s)B_g} \qquad (7.20)$$

In problem 7.2 we calculated a gas-cap reservoir expansion of 40,000 res bbl for a pressure change from 3,150 to 2,900 psia. In problem 7.6 we calculated an oil-zone shrinkage of 40,000 res bbl for the same pressure change. Now, we will consider this as one reservoir with a pressure change of 3,150 to 2,900 psia in problem 7.7. The solution is shown in appendix C.

PROBLEM 7.7: Determining the Original Stock-Tank Barrels of Oil in the Reservoir

A gas-cap-drive reservoir with no water drive has an original gas cap of 400 MMscf at an initial pressure of 3,150 psia. The initial gas formation volume factor, oil formation volume factor, and gas in solution are 0.001 res bbl/scf, 1.34 res bbl/stb, and 600 scf/stb, respectively. After 78.480 MMscf of gas and 130,800 stb of oil are produced from this reservoir, the pressure is 2,900 psia and the gas formation volume factor, oil formation volume factor, and gas in solution are 0.0011 res bbl/scf, 1.32 res bbl/stb, and 550 scf/stb, respectively. Assume that geologic control is such that the extent of this reservoir is unknown. Find the original stock-tank barrels of oil in place.

Pore-Volume Changes in Material Balance

The compressibility of oil may be in the range of 10^{-5} vol of oil expansion/vol of oil/psia of pressure change. If the pressure is dropped 1 psi on a volume of 100,000 bbl of oil with a compressibility of 10^{-5}/psi, that volume of oil would expand 1 bbl. Although this is a very small expansion, it is interesting to note that many reservoirs produce primarily by the expansion of oil for several years. The compressibility of water is even smaller, about 10^{-6}/psi of pressure change; however, this expansion causes tremendous quantities of water to encroach into an oil reservoir. Thus, the change in pore volume with a change in reservoir pressure is very small, in the range of 10^{-6} fractional volume change/psi. Nevertheless, it can play a primary role in the energy supplied for the production of oil, just as the compressibility of oil and water play large roles in oil and gas production. The reduction in the pore volume that accompanies a decline in the reservoir pressure is caused primarily by two factors: the reduction in the gross volume in the reservoir and the increase in the reservoir solids volume.

A microscopic look at the pore size of a sand reservoir is schematically represented in Fig. 7–8. In this diagram the pore space between the sand grains is indicated by a darkened area. A reservoir thousands of feet underground is subjected to an overburden pressure caused sim-

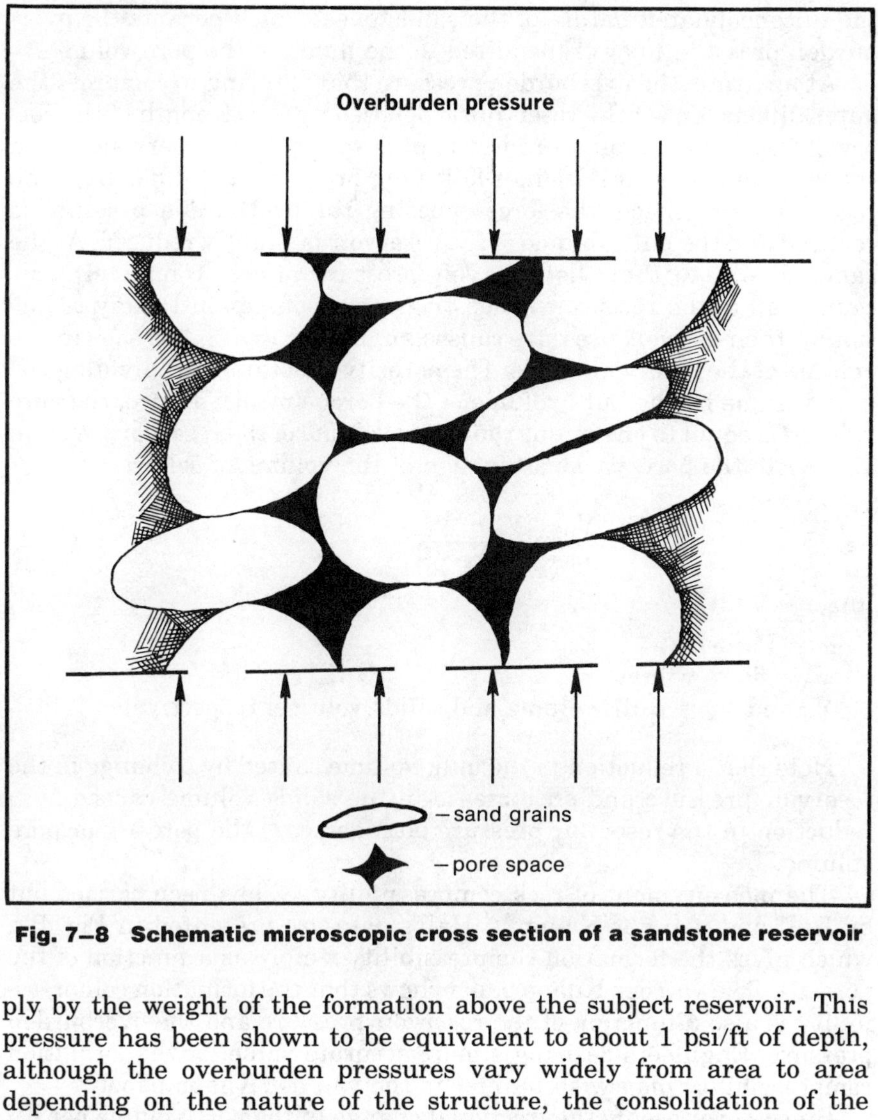

Fig. 7–8 Schematic microscopic cross section of a sandstone reservoir

ply by the weight of the formation above the subject reservoir. This pressure has been shown to be equivalent to about 1 psi/ft of depth, although the overburden pressures vary widely from area to area depending on the nature of the structure, the consolidation of the materials, and the geologic age of the materials. These overburden pressures simply apply a compressive force to the reservoir rock.

The pressure in the pore space does not normally approach that of the overburden pressure. A normal pressure in the pore space, which is known as the reservoir pressure, is about 0.5 psi/ft of depth, provided the reservoir rock is sufficiently consolidated so the overburden pressure is not transmitted to the pressure of the fluids in the pore space. In some areas such as the Gulf Coast of Texas and Louisiana, reservoir pressures may be very close to the overburden pressure as a result of the unconsolidated nature of the sandstones, which permits the overburden pressure to be transmitted to the fluids in the pore volume.

At any rate, the overburden pressure that is trying to compress the outer dimensions of the reservoir is opposed by the strength of the reservoir rock and the pressure in the pore space of the reservoir. When the reservoir pressure declines following production of fluids from the reservoir pore space, the force opposing the overburden pressure is reduced and the bulk volume of the reservoir is slightly reduced. At the same time note that the reservoir pressure almost completely surrounds all of the reservoir solids or grains. Consequently, any reduction in the reservoir pressure causes an accompanying increase in the volume of the reservoir solids. The porosity is obtained by dividing the pore volume by the bulk volume of the porous media. Hence, the pore volume is equal to one minus the solids volume of the reservoir. We can also write the porosity as a function of the volume of solids:

$$\phi = \frac{V_p}{V_b} = \frac{V_b - V_s}{V_b} = 1 - \frac{V_s}{V_b} \tag{7.21}$$

Where:
ϕ = Porosity
V_p = Pore volume
V_b and V_s = Bulk volume and solids volume, respectively

Note that a reduction in the bulk volume caused by a change in the reservoir pressure and an increase in the solids volume caused by a reduction in the reservoir pressure both decrease the porosity or pore volume.

The measurement of rock compressibility, c_f, has been carried out by Hall and van der Knapp.[1,2] Hall's data are presented in Fig. B7, which gives the formation compressibility simply as a function of the porosity. The van der Knapp paper shows that the formation compressibility is also a function of the reservoir pressure and the overburden pressure. Engineers desiring a more accurate value for the formation compressibility may wish to refer to the van der Knapp paper.[3]

Since c_f represents the fractional change in the pore volume per psi pressure change, we can write the pore volume change in barrels as:

$$\Delta V_p = c_f \, \Delta p \, V_p \tag{7.22}$$

As noted, water exhibits a compressibility that is as large or larger than the formation compressibility. Since all known producing formations contain some water, the expansion of this water also contributes to a reduction in pore volume available for the hydrocarbons.

All formations contain some water because sedimentary formations were normally laid down in ancient seas and, thus, were initially filled with water. At some later time oil or gas migrated into these pore spaces and displaced the water. However, since the pore volume is made of a variety of very small pores, the hydrocarbons cannot displace all of the water from the pore space. Consequently, the small pores remain filled with water.

This irreducible water is often called the *connate water,* although other experts feel that any initial water saturation in an oil reservoir should be called connate water. Nevertheless, the water that is immobile to the displacing hydrocarbons generally remains immobile while the hydrocarbons are produced. This connate water can then expand as a result of the reduction in pressure in the reservoir and can cause the corresponding reduction in pore volume. Water compressibilities can be determined from several sources and a wide range of water compressibilities are presented in Fig. B3. The pore volume change resulting from the expansion of the water can then be stated as:

$$\Delta V_p = c_w \, \Delta p \, S_{wc} \, V_p \tag{7.23}$$

Where:

S_{wc} = Connate water saturation

In Eq. 7.23 S_{wc} is multiplied by the pore volume to obtain the volume of connate water. The change in pore volume resulting from the rock compressibility and the water compressibility can be conveniently combined into one term that is the sum of Eqs. 7.22 and 7.23:

$$\Delta V_p = (c_f + c_w S_{wc}) \, \Delta p \, V_p \tag{7.24}$$

Note that this expression for change in pore volume has the same material-balance effect as the water encroachment or gas-cap-gas encroachment. Consequently, the term can be added to Eq. 7.20 with the same sign as the gas-cap expansion and water-encroachment terms. However, the original pore volume is most conveniently stated as a function of the original oil in place:

$$V_p = N \, B_{oi}/(1 - S_{wc}) \tag{7.25}$$

Where:

NB_{oi} = Original res bbl of oil in the reservoir

$(1 - S_{wc})$ = Original oil saturation

When we substitute the V_p expression of Eq. 7.25 in Eq. 7.24 and add the resulting expression to Eq. 7.20, we obtain:

$$N = \frac{N_pB_o + B_g\,(G_p - N_pR_s) - G(B_g - B_{gi}) - (W_e - W_p)}{B_o - B_{oi} + (R_{si} - R_s)B_g} - \frac{(c_f + c_wS_{wc})\Delta p \dfrac{NB_{oi}}{(1 - S_{wc})}}{B_o - B_{oi} + (R_{si} - R_s)B_g} \tag{7.26}$$

When Eq. 7.26 is rearranged to factor out the N from the ΔV_p term, we obtain the general material-balance equation:

$$N = \frac{N_pB_o + B_g(G_p - N_pR_s) - G(B_g - B_{gi}) - (W_e - W_p)}{B_o - B_{oi} + (R_{si} - R_s)B_g + (c_f + c_wS_{wc})\,\Delta p\,B_{oi}/(1 - S_{wc})} \tag{7.27}$$

Modifications of the General Material-Balance Equation

It is strongly recommended that all material-balance applications begin with Eq. 7.27 so none of the significant reservoir energies are omitted in a particular application. We find that all of the terms in the general material-balance equation are seldom of significant size at the same time. Nevertheless, the use of this one equation seems to greatly simplify material balance since it eliminates the hodgepodge of equations necessary to describe each individual reservoir drive and its peculiar circumstances. By using the general material-balance equation, it is unnecessary to have special equations such as expansion oil drive above the bubble point, solution-gas drive below the bubble point, or water drive above the bubble point. Nevertheless, it should be helpful to investigate the effects of these special circumstances because most material-balance equations are limited rather than general.

For reservoirs above the bubble point, note that many of the terms of the equation become zero. For example, the gas production term containing $G_p - N_p R_s$ is equal to the total gas produced G_p. Also, the gas-expansion term containing G equals zero since there can be no free gas in the reservoir above the bubble point. Consequently, G becomes zero. In the denominator the liberated-gas term, $R_{si} - R_s$ is zero because R_s, the gas in solution at any particular pressure, is equal to the gas originally in solution, R_{si}. Note that the symbol G does not include the original gas in solution. This superfluous statement is made because of the frequency of this error among inexperienced engineers.

When Eq. 7.27 is applied to a reservoir that does not contain an initial free-gas saturation or an initial gas-cap, the term containing G is zero as noted. Furthermore, the gross water encroachment, W_e, becomes zero if a reservoir is not affected by a water drive. However, the engineer should continue to carry the water production term, W_p, even when it is apparent that there is no water drive in the reservoir. This term is needed because connate water is often produced as the reservoir pressure declines, even though no new water is entering the reservoir as W_e.

Once a reservoir contains free gas, we find that the change in the pore volume (expansion) term containing c_f and c_w is insignificant as compared with the free-gas terms. This becomes apparent when the compressibilities of water and the formation are compared in magnitude with the compressibility of gas. The water and formation compressibilities, c_w and c_f, are in the range of 10^{-5}–10^{-6}/psi. The gas compressibility—not to be confused with the compressibility factor for gas—is in the range of 10^{-3}–10^{-4} or about 100 times as great as the compressibility of the water and formation. Consequently, a gas saturation of 1% may provide as much energy as the water and formation compressibility terms. Therefore, once the gas saturation is substantial, the pore volume change becomes insignificant.*

It should also be emphasized that the general material-balance equation does not contain restrictions on applications above and below the bubble-point pressure. There is no reason to break a material-balance calculation into a calculation to the bubble point and from the bubble point to values of pressure less than the bubble point as appears to be common practice. The derivation procedures indicate that there is absolutely no restriction on the initial condition applied in the derivation of these equations.

Two reservoir properties are used frequently in connection with material balance that receive only sparing use in this book. These are the total formation volume factor, B_t, previously defined, and m, the ratio of the initial reservoir free-gas volume to the initial reservoir oil volume, GB_{gi}/NB_{oi}. By using the total formation volume factor, B_t, all of the oil formation volume factors, B_o, can be eliminated from the equations so only the simpler B_t parameter is needed. Since all PVT lab procedures include the evaluation of the oil formation volume factor, the advantage of a material-balance equation that avoids the use of B_o seems to be more imaginary than real.

It is more difficult to rationalize a preference for the use of the ratio, m, in material-balance equations since it is necessary to know the original standard cubic feet of free gas in the reservoir in order to determine m. Nevertheless, consider the conversion of the general material-balance equation, Eq. 7.27, to the B_t and m parameters. First, substitute for the gas production, G_p, a function of the cumulative produced gas-oil ratio, N_pR_p. Then add and subtract the term $N_pR_{si}B_g$ to the numerator; substitute mNB_{oi}/B_{gi} for G; and rearrange the equation:

$$N = \frac{N_pB_o + B_gN_pR_{si} - B_gN_pR_s + B_gN_pR_p - B_gN_pR_{si} - (W_e - W_p)}{B_o + (R_{si} - R_s)B_g + (c_f + c_wS_{wc})\,\Delta pB_{oi}/(1 - S_{wc}) + mB_{oi}\,(B_g - B_{gi})/B_{gi}} \tag{7.28}$$

*Effective compressibility is discussed in chapter 3.

Now, substitute B_t for $B_o + B_g(R_{si} - R_s)$ and B_{ti} for B_{oi} where possible and rearrange the equation to obtain:

$$N = \frac{N_p[B_t + (R_p - R_{si})B_g] - (W_e - W_p)}{B_t - B_{ti} + (c_f + c_w S_{wc})\,\Delta p B_{ti}/(1 - S_{wc}) + mB_{ti}(B_g - B_{gi})/B_{gi}} \tag{7.29}$$

General Difficulties in Applying Material-Balance Equations

Many engineers are discouraged with tank-type material-balance predictions because they find the analysis gives unreasonable results. They quickly turn to a computer model when the real problem is not with the basic material-balance equations. Instead, it is the PVT, production, or pressure data or with the lack of understanding of fluid flow fundamentals that leads to inaccurate predictions of water encroachment and producing rates. This section is concerned with these general material-balance difficulties that are common to all reservoir engineering problems.

Determining PVT data. Many problems are involved in applying material-balance calculations to a reservoir. One of the foremost of these is the general lack of pressure-volume-temperature data available for specific reservoirs. Ideally, such PVT data should be determined for the specific reservoir in the laboratory or should be calculated from the specific reservoir hydrocarbon compositions using appropriate equilibrium constants. Under ideal conditions this requires a bottom-hole sample of the reservoir fluid before the flowing pressure at the well falls below the bubble-point pressure. Several sampling devices are available for this problem. However, once the pressure in the well is below the bubble-point pressure, it becomes extremely difficult to obtain a representative sample of oil and free gas.

In some cases it is necessary to take surface samples and recombine the oil and gas in the laboratory. Again, it is difficult to know exactly what gas-oil ratio should be recombined to represent the initial reservoir fluid. One of the primary functions of a reservoir engineer in a producing company should be to make certain that proper data are obtained on reservoirs.*

Since most reservoir engineers of the author's vintage could not convince managements that reservoir engineering data are an economic expenditure, many studies may lack adequate PVT data. If samples have not been taken, much useful information and guidance can be obtained using empirical data. Figs. B16 and B18 provide empirical

*Refer to chapter 1 for problems associated with obtaining data.

data that permit the determination of formation volume factors and gas in solution as a function of the reservoir temperature and pressure and the oil and gas gravities. Similar data for the evaluation of gas deviation factors is in Figs. B8 and B9. Since the trend of PVT data—the change in the data with pressure—is more important than the absolute values, these empirical data often give excellent results. Empirical PVT data do not relieve the reservoir engineer of the responsibilities of gathering the most accurate data possible by obtaining initial bottom-hole fluid samples from reservoirs, but it does tend to minimize the results of such failures.

Even when adequate sampling procedures have been followed, problems exist concerning the evaluation of PVT data. In the laboratory there is a choice between *flash* and *differential* types of liberation procedures to be followed. In the flash procedure all of the liberated gas remains in contact with the oil from which it is liberated as the process continues. However, in the differential process, ideally, all of the liberated gas is immediately separated from the oil from which it has been liberated. The actual reservoir situation is somewhere between these two extremes because all of the gas liberated in the reservoir is not immediately removed from the presence of the remaining oil. However, as the gas is liberated in the reservoir, it has a much greater mobility than the oil. Consequently, it is produced at a much greater rate than the oil itself. Therefore, the oil in the reservoir is in contact with all of its initial solution gas for only a short period. The reservoir liberation process is then somewhere between flash and differential.

However, consider the liberation process that takes place in the tubing as the oil and gas move up the tubing with the pressure being reduced. This does appear to be a flash process, but there is excess gas present. The solution gas is liberated from the oil present, and the free gas that flowed into the well with the oil is present. No such liberation takes place in either the flash or differential process.

Fortunately, the results of these two processes are not greatly different for most reservoir fluids, and the reservoir fluids with great differences do not lend themselves generally to the simplified type of material balance described herein. Conventional material balance assumes that the composition of the gas liberated throughout the life of the reservoir is constant. However, when the hydrocarbon system contains considerable amounts of propane, butane, and pentane (the intermediate hydrocarbons), the assumption of a constant liberated-gas composition is no longer applicable. Then conventional material balance becomes too inaccurate for use. Such reservoirs are called *volatile oil reservoirs,* but they are not common. One international group of companies searched their files for examples of such volatile oil reservoirs but could find only three. There seems to be little question that there are few volatile oil reservoirs such that conventional material balance does not adequately describe their behavior.

Engineers are often confused by the evaluation of PVT data empirically. Since an engineering analysis is no more accurate than the data used, it is important that PVT data be determined as accurately as possible. Consequently, work problem 7.8 and check the answers against the solution in appendix C.

PROBLEM 7.8: Determining PVT Data Empirically

In problem 7.3 we evaluated the oil formation volume factors and gas in solution at reservoir pressures of 2,000, 1,500, 1,000, 500, and 14.7 psia from lab data. Now make the same evaluations empirically if the gas gravity is 0.7 and the oil stock-tank gravity is 40°API. The reservoir temperature and saturation pressure are 195°F and 1,500 psia, respectively. Note that the evaluation of the oil formation volume factor above the saturation pressure requires the use of the oil compressibility data as in Table B4.

Other empirical PVT data are available. Standing's correlations can be found in the Amyx, Bass, and Whiting book *Petroleum Reservoir Engineering*.[4] Methods of calculating PVT data from equilibrium constants and a component analysis of the system can be found in the Amyx, Bass, and Whiting text and in the Katz et al., *Handbook of Natural Gas Engineering*.[5,6]

Accuracy of production data. Another type of data problem that exists in connection with material-balance applications is the general inaccuracy of the gas production and water production. These data greatly affect the behavior of most oil reservoirs. The economic worth of these two commodities when produced from an oil reservoir is often nil since water is never sold and the gas produced with oil may not be sold. Consequently, records of the amounts of gas and water produced may be very poor. It is not at all uncommon for field personnel to simply "boiler house" water and gas production data with little regard for correctness.

When the gas associated with oil production is sold, there can be little question as to the amount of gas produced. However, the reservoir engineer should be fully aware of the possibility that water and gas production figures may be grossly in error. The engineer should be prepared to check with field personnel in any way possible to ascertain the accuracy of these data as well as the accuracy of the allocation of oil production to the individual wells and zones.

Since oil is sold, there is little question as to the accuracy of the total oil volumes produced. However, allocation of this production to different zones and different wells may also be subject to considerable question. The engineer should continually be on the lookout for suspicious data. The author has seen the same water production data for wells reported month after month and year after year, although the oil production rate and gas production rate were varying widely. Also, gas-oil ratios have been reported for solution-gas-drive reservoirs that were the same month after month and year after year. Even when meters are available, we must be careful.

Fluid flow fundamentals. It is impossible for the reservoir engineer to make intelligent applications of material-balance calculations without adequate knowledge of fluid flow fundamentals. The ratio between gas production and oil production depends on the flow characteristics of the reservoir. Furthermore, water encroachment, W_e, also must be evaluated in terms of fluid flow fundamentals. Therefore, it is virtually impossible to use material-balance concepts without a good knowledge of fluid flow fundamentals even though only a prediction based on the average reservoir pressure is desired. Of course, a more extensive knowledge of fluid flow fundamentals is necessary if the prediction is to be based on time rather than on the average reservoir pressure.

The accuracy of reservoir pressure data. The author believes that the biggest difficulty in material-balance applications is the use of inaccurate average reservoir pressures. This subject is treated thoroughly in chapter 4 and to a lesser degree in chapter 5 as it applies to gas reservoirs.

It is common to leave a well shut in for 1–2 days and assume that the pressure has reached the average reservoir pressure for that well's drainage area. With a gas reservoir the engineer may be very cautious and use 3–4-day shutin pressures. In most cases he will not even bother to examine the pressure record to see if the recorded pressure is still changing after the shutin period. Even when the pressure appears to be constant, it is generally continuing to change at a rate that is too small to see over the course of a few hours. However, this change is extremely important over the period of several days. As noted in chapter 5, some gas reservoirs require years to stabilize, which makes it virtually impossible to record the average reservoir pressure directly. As a guideline it is wise to remember that we cannot directly record the average pressure in a drainage area until the well has been shut in for a period equal to or greater than the stabilization time, t_s. This time should be calculated using the greatest distance from the well to the outer drainage boundary as r_e and Eq. 3.50:

$$t_s = 0.04\phi\mu c r_e^2/k \tag{3.50}$$

Remember that the compressibility in Eq. 3.50 is based on the effective compressibility of the system. Therefore, it is greatly affected by the gas saturation. Consequently, the engineer should be especially careful that average pressures taken directly are based on realistic shutin times when substantial amounts of gas saturation exist in the reservoir.

When shutin times exceed the stabilization time, the recorded pressure will not be exactly equal to the average pressure unless this well is the only one in the reservoir or all of the wells in the reservoir are shut in. However, the recorded pressure may closely approximate the average pressure. In this case the recorded pressure after a shutin time greater than the stabilization time is so greatly affected by the offset producers that the pressure in the subject well simply represents a pressure in the new—since shutin—drainage area of an offset well.

Although the technology leaves much to be desired, the most accurate average reservoir pressures are obtained from pressure-buildup analysis. If the engineer does not completely understand the difficulties and recommended methods of determining the average pressure in a well's drainage area, he is strongly encouraged to review this subject in chapter 4.

Predicting Gas-Drive Behavior

When material balance is discussed in detail, it is customary to consider the use of material balance in predicting the behavior of a solution-gas-drive reservoir. The same techniques can be applied to the analysis of a gas-cap-drive reservoir, but the prediction of the produced gas-oil ratio becomes more difficult because of the trouble encountered in controlling the produced gas-oil ratio in actual production operations.

Aside from this latter difficulty, it still is questionable whether a theoretical prediction of reservoir behavior for a gas-drive reservoir based on material balance is more accurate than a prediction based on decline-curve analysis. This question can be better understood after the engineer has considered the problem of determining the relative permeability data that greatly influences the prediction. It is noted that about the only method of accurately determining the actual relative permeability characteristics of a gas-drive reservoir is by calculation from past reservoir behavior and an extrapolation of this behavior into the future. Since past performance is necessary to obtain any accuracy from the theoretical prediction, it can be argued that empirical decline-curve analysis gives as accurate of a prediction of the reservoir behavior.*

*See chapter 8.

This situation is particularly true under full capacity operations. It does seem clear that any prediction of gas-drive reservoir behavior should be reconciled as soon as possible with the decline-curve experience of the reservoir. One point is indisputable: A complete understanding of gas-drive reservoir prediction methods provides the engineer with an excellent understanding of the various factors that control the behavior and ultimate recovery.

There appears to be an infinite variety of methods for solving this problem. Only those that appear to be most useful are presented.

The Schilthuis method. Since the material-balance concepts were originated by Schilthuis, it seems that a method relying basically on the form of his original equation should be so named.[7] Starting with the general material-balance equation, Eq. 7.27, we see that applying this equation to a reservoir below the saturation pressure with no water drive eliminates the water encroachment term, W_e. The pore volume expansion term is also insignificant if the initial reservoir pressure, p_i, is at or near the saturation pressure, p_s. Consequently, for a solution-gas-drive reservoir where the reservoir pressure is about equal to the saturation pressure and for gas-cap-drive reservoirs, Eq. 7.27 becomes:

$$N = \frac{N_pB_o + B_g(G_p - N_pR_s) - G(B_g - B_{gi})}{B_o - B_{oi} + (R_{si} - R_s)B_g} \tag{7.30}$$

Solving Eq. 7.30 for N_p, we obtain:

$$N_p = \frac{N[B_o - B_{oi} + (R_{si} - R_s)B_g] + G(B_g - B_{gi}) - B_gG_p}{B_o - B_g\,R_s} \tag{7.31}$$

Thus, in order to predict the cumulative oil production at a particular stage of depletion, it is necessary to know the original oil and gas in the reservoir and the original reservoir pressure, which fixes N, G, B_{oi}, R_{si}, and B_{gi}. The stage of depletion fixes the reservoir pressure at that time, which in turn sets the other PVT data in the equation, B_o, R_s, and B_g. However, we still must evaluate the gas production, G_p, for this stage of depletion before we can proceed with the calculation of N_p.

The cumulative gas production, G_p, is determined from the produced gas-oil ratio, R, and the cumulative oil production, as discussed:

$$G_p = \sum_{j=1}^{j=n} R_j\,\Delta N_{pj} \tag{7.14}$$

However, we know from Eq. 7.11 that the gas-oil ratio is a function of the relative permeability ratio, which in turn is a function of the oil saturation in the reservoir. Furthermore, Eq. 7.13 shows that the saturation is a function of the cumulative oil production, N_p, which is the value we want to calculate using Eq. 7.31. Consequently, the application of Eq. 7.31 involves a trial-and-error solution.

The most direct procedure for obtaining this solution is as follows:

1. Set the reservoir pressure
2. Guess G_p
3. Calculate N_p by Eq. 7.31
4. Calculate S_o using Eq. 7.13
5. Determine the relative permeability ratio based on the liquid saturation from data such as that in Fig. 7–6
6. Calculate G_p using Eqs. 7.14 and 7.11
7. Compare the calculated G_p with the assumed G_p from step 2
8. Repeat the calculations from step 2 if the assumed and calculated values do not agree to a satisfactory degree or
9. Go back to step 1 and set a new pressure if the assumed and calculated G_p values agree.

In step 2 it is helpful to use a plot of G_p versus pressure based on experienced or previously calculated G_p values. This plot can then be extrapolated to the pressure set in step 1 to obtain the G_p to be assumed. In step 6 first calculate R_n, the instantaneous GOR at the pressure of step 1, using the k_g/k_o value obtained in step 5. Then Eq. 7.14 can be used in the expanded form to obtain G_p at the pressure of step 1:

$$G_{pn} = G_{p(n-1)} + [(R_n + R_{n-1})/2](N_{pn} - N_{pn-1}) \tag{7.32}$$

Where:

Subscript $(n - 1)$ = Previously known or calculated G_p, N_p, or R

The last term of Eq. 7.32 represents the last of the summed terms of Eq. 7.14, with the N_p expression being the ΔN_p and the R expression being the average R.

Although this procedure appears to be the most direct approach to predicting the cumulative oil and gas production versus pressure for a solution-gas-drive reservoir, it is not necessarily the simplest one to use. If a computer is used to solve prediction problems of this type, it makes little difference to the engineer which procedure is followed except for some small difference in computer time that may result. However, if it is necessary to make calculations using a desk calculator, easier methods are available. The problem is which item should be used for the trial-and-error iteration in satisfying Eqs. 7.11, 7.13, 7.14, and 7.31.

The Tarner Method. Tarner appears to be the first person to suggest iteration on the produced gas-oil ratio at the state of depletion to be calculated or at the time when N_p barrels of oil have been produced (which we are calculating).[8] The iteration can be carried out by extrapolating a plot of the instantaneous gas-oil ratio, R, versus the reservoir pressure to the next average reservoir pressure at which the cumulative production of oil and gas is desired. The data for the plot can be previously calculated data or a plot of actual data. In either case the gas-oil ratio determined by extrapolation is used as the assumed gas-oil ratio, R_n, that exists after N_{pn} barrels of oil have been produced. With the gas-oil ratio plot completed, the cumulative gas production, G_{pn}, can be calculated from Eq. 7.32 as if N_{pn}, which we are calculating, were known. Consequently, we can substitute the expression for G_{pn} from Eq. 7.32 into Eq. 7.31 without introducing new unknowns:

$$N_{pn} = \frac{N[B_o - B_{oi} + (R_{si} - R_s)B_g] + G(B_g - B_{gi})}{B_o - B_g R_s} - \frac{B_g\left[G_{p(n-1)} + \left(\frac{R_n + R_{n-1}}{2}\right)\left(N_{pn} - N_{p(n-1)}\right)\right]}{B_o - B_g R_s} \tag{7.33}$$

This expression can be solved for N_{pn}:

$$N_{pn} = \frac{N[B_o - B_{oi} + (R_{si} - R_s)B_g] + G(B_g - B_{gi})}{B_o - B_g R_s + (R_n + R_{n-1})B_g/2} - \frac{B_g[G_{p(n-1)} - (R_n + R_{n-1})\,N_{p(n-1)}/2]}{B_o - B_g R_s + (R_n + R_{n-1})B_g/2} \tag{7.34}$$

The N_{pn} is calculated based on an assumed R_n estimated from an extrapolation of a plot of the produced gas-oil ratio, R, versus the reservoir pressure. Then it is possible to determine the oil saturation in the reservoir at this time, S_{on}, using Eq. 7.13 and the value of N_{pn} calculated from Eq. 7.34. Based on this saturation, the permeability ratio can be determined from given data and R_n can be calculated from Eq. 7.11.

If the assumed and calculated R_n are in satisfactory agreement, the engineer can proceed with the calculation for the next lowest pressure of interest. If the R_n values do not agree sufficiently, it is necessary to adjust the GOR-plot extrapolation accordingly and repeat the calculations until the R_n by extrapolation and the R_n calculated agree.

The procedure for performing a Tarner prediction for a gas-drive reservoir can be summarized as follows:

1. Extrapolate a plot of previously observed or calculated gas-oil ratios versus reservoir pressure to the next reservoir pressure at

which the cumulative production, N_{pn}, is desired. This extrapolation provides a gas-oil ratio, R_n, corresponding to the cumulative production desired, N_{pn}.

2. Use Eq. 7.34 to calculate N_{pn}, based on the previously known cumulative oil and gas production and the produced gas-oil ratio, where indicated by the subscript (n − 1).
3. Use Eq. 7.13 to calculate the oil saturation, S_{on}, corresponding to the cumulative production prediction, N_{pn}.
4. Calculate the produced gas-oil ratio corresponding to N_{pn} and S_{on} using Eq. 7.11. This calculation gives a calculated R_n and should be based on a k_g/k_o ratio corresponding to S_{on}.
5. If the produced gas-oil ratios determined by extrapolation in step 1 and calculated in step 4 agree to an acceptable accuracy, assume a new lower reservoir pressure and repeat steps 1–4. Note that the calculated R_n now is a previously calculated gas-oil ratio and is R_{n-1} for the next sequence of calculations.
6. If the produced gas-oil ratios determined by extrapolation and calculation do not agree to an acceptable accuracy, it is necessary to return to step 1 and adjust the produced gas-oil ratio extrapolation until the calculated R_n and R_n determined by extrapolation agree satisfactorily.

More than two iterations on R_n is seldom necessary to obtain acceptable accuracy. Also, note that subsequent calculations automatically adjust for errors tolerated in the produced gas-oil ratio in previous calculations. With some other popular gas-drive prediction techniques the errors tend to accumulate.

A recommended procedure for predicting performance. One of the biggest difficulties in performing an accurate prediction of the performance of a gas-drive reservoir is obtaining relative permeability data that are characteristic of the producing formation. Several problems exist in this connection. As noted in chapter 2, the relative permeability data obtained differ widely from one sample to another and the problem of determining an average or effective single relationship to represent a particular reservoir has not been satisfactorily solved. Also, the actual procedures followed in the lab leave much to be desired since the size of the sample used is limited by the core diameter. Also, the wettability of the core has been subjected to change by its contact with the drilling fluids and the fluids necessary to shape the core in the laboratory.

Therefore, laboratory-measured relative permeabilities are seldom characteristic of the producing reservoir. Consequently, it is recommended that a prediction of the behavior of a gas-drive reservoir based solely on lab-measured relative permeability data be used only as a last resort. Such a prediction also should be corrected as soon as meaningful reservoir performance data can be obtained. Past reservoir behavior can be used to calculate the effective relative permeability ratios corresponding to various average reservoir saturations. Then these calculated data are extrapolated to represent the data for future reservoir saturations.

To be more specific, the past oil production at various average reservoir pressures is used to calculate the average oil saturations, S_o, using Eq. 7.13. The corresponding relative permeability ratios are calculated from Eq. 7.11, based on the average produced gas-oil ratio at that time. When the oil saturations are corrected to equivalent liquid saturations by adding the connate water saturation and a plot of the log of the relative permeability versus liquid saturation is prepared, the slope of this plot generally is the same as the slope of the similar lab data. However, the field calculated data are generally considerably displaced from the lab data. By drawing a line parallel to the lab data and positioning the plot on the basis of the field-calculated points, a reasonable estimate can be obtained for the relative permeability characteristics of the reservoir.

The recommended procedure for predicting the behavior of a gas-drive reservoir is as follows:

1. Use as much past reservoir production and pressure history as possible to calculate k_g/k_o versus S_o data. Use Eq. 7.11 and the observed average producing gas-oil ratios to determine k_g/k_o. Reservoir oil saturations can be calculated from Eq. 7.13.
2. Correct the reservoir oil saturations to reservoir liquid saturations by adding the connate water saturation. Then plot the liquid saturations versus the permeability ratio data on semilog paper.
3. Plot the lab relative permeability data on the same graph prepared in step 2. Extend the field-calculated permeability data parallel to the lab data.
4. Assume the extrapolated field data from step 3 are the relative permeability characteristics representative of the reservoir. Then proceed with the Tarner method as described.

To check understanding of this procedure, work problem 7.9 and check the solution against the one in appendix C.

PROBLEM 7.9: Predicting the Behavior of a Gas-Drive Reservoir

The following data apply to a solution-gas-drive reservoir:

Original oil in place = 10.025 MMstb
Connate water saturation = 22%

Initial reservoir pressure = 3,013 psia
Saturation pressure = 2,496 psia

p	B_o	R_s	B_g	μ_o/μ_g	R
3,013	1.315	650	0.000726	53.9	650
2,496	1.325	650	0.000796	56.6	650
1,498	1.250	486	0.001335	91.56	1,360
1,302	1.233	450	0.001616	102.61	2,080
1,200	1.224	431	0.001807	108.96	?
1,100	1.215	412	0.001998	115.20	?

A. Calculate S_o and k_g/k_o when the reservoir pressure is 1,302 psia, cumulative production is 1.179 MMstb of oil and 1,123 MMscf of gas. Note that this point falls on the field plot of k_g/k_o versus S_L in Fig. 7–6.

B. Calculate the cumulative oil and gas production when the reservoir pressure reaches 1,200 psia. Base the R_{1200} estimate on extrapolation of the observed GOR in Fig. 7–9. Base the k_g/k_o values on the trend established by the field data in Fig. 7–6.

C. Repeat the calculations of part B for a reservoir pressure of 1,100 psia.

Of course, this analysis lends itself to a simple but very valuable computer program because of the trial-and-error iterations on R_n involved in the Tarner portion of the calculations. The Schilthuis method is also greatly enhanced by programming for computer solution because it includes trial-and-error iterations on the cumulative gas-oil ratio, R_p.

Note that the Schilthuis method is based on a finite equation, but the basic equation for the Tarner method may be considered to be a finite difference form of material balance.

The Muskat Method. The Muskat method is a prediction technique based on a differential form of the material-balance equation:[9]

$$\frac{\Delta S_o}{\Delta p} = \frac{\dfrac{B_g S_o}{B_o}\dfrac{\Delta R_s}{\Delta p} + \dfrac{S_o k_g \mu_o}{B_o k_o \mu_g}\dfrac{\Delta B_o}{\Delta p} + (1 - S_o - S_w) B_g \dfrac{\Delta(1/B_g)}{\Delta p}}{1 + \dfrac{k_g}{k_o}\dfrac{\mu_o}{\mu_g}} \tag{7.35}$$

The theoretical basis for Eq. 7.35 can be seen by recognizing that the equation for the reservoir oil saturation, Eq. 7.13, is stated as a function of the produced oil, N_p. Note that N_p is defined in Eq. 7.31 as a function of the produced gas, G_p. The produced gas is a function of the mobility ratio of the gas and oil, $k_g\mu_o/k_o\mu_g$, through Eqs. 7.11 and 7.14. Consequently, these relationships can be combined into one equation

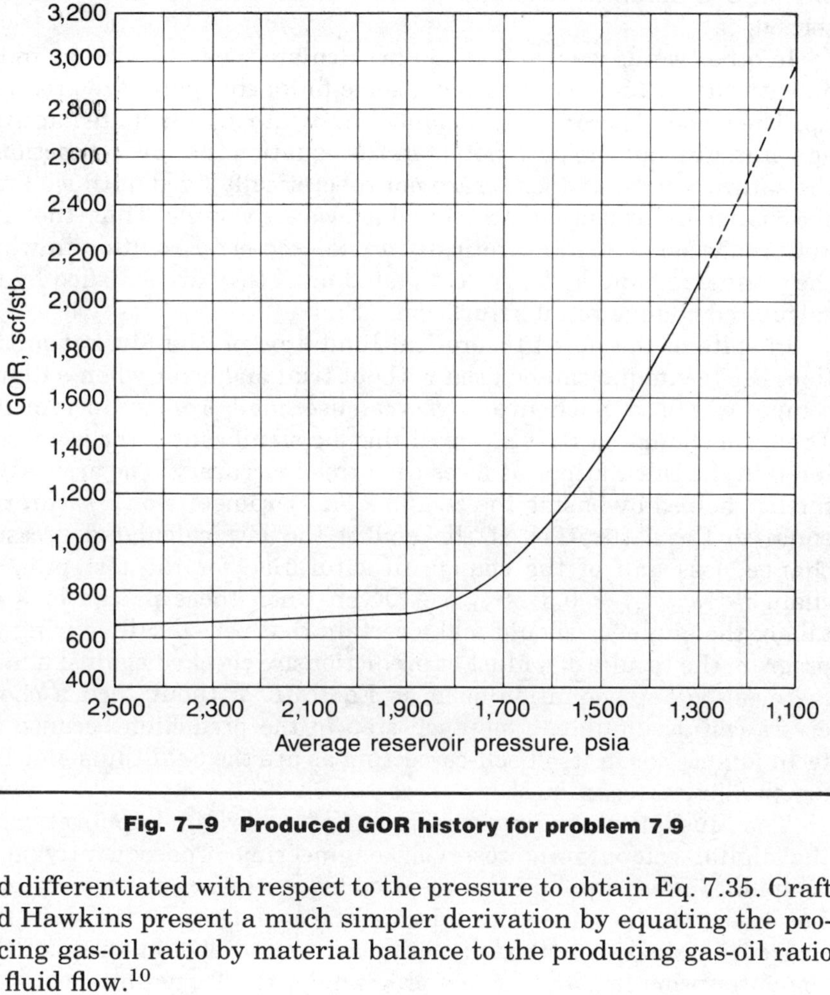

Fig. 7–9 Produced GOR history for problem 7.9

and differentiated with respect to the pressure to obtain Eq. 7.35. Craft and Hawkins present a much simpler derivation by equating the producing gas-oil ratio by material balance to the producing gas-oil ratio by fluid flow.[10]

On the surface it appears that the Muskat equation, Eq. 7.35, avoids the trial-and-error type of solution that is necessary in the Schilthuis and Tarner methods because we normally know everything in the right-hand side of Eq. 7.35. Note that $\Delta R_s/\Delta p$, $\Delta B_o/\Delta p$, and $\Delta(1/B_g)/\Delta p$ represent the slope of a curve of these PVT data versus pressure. Consequently, if we are only interested in the slope of a plot of the reservoir oil saturation, S_o, versus pressure, p, at a particular pressure, we can apply Eq. 7.35 without resorting to trial and error. However, we need a plot of oil and gas production versus the reservoir pressure. Therefore, we must determine the finite change in the reservoir oil saturation for some finite change in the pressure, which means that we must apply the equation on a finite difference basis in so far as $\Delta S_o/\Delta p$ is concerned but on a differential basis in so far as the other Δ terms are concerned.

In other words we use Eq. 7.35 to calculate the finite change in the oil saturation, ΔS_o, accompanying some finite change in the pressure, Δp. The slopes $\Delta R_s/\Delta p$, $\Delta B_o/\Delta p$, and $\Delta(1/B_g)/\Delta p$ are evaluated at average pressure, but the parameters in the equation that are a function of the saturation, S_o and k_g/k_o, are not theoretically fixed until we know the change in saturation, ΔS_o, for which we are solving. Thus, the Muskat prediction is also theoretically a trial-and-error solution in which the S_o average and k_g/k_o are estimated until they are satisfied by the calculated change in oil saturation, ΔS_o.

In spite of the noted theoretical limitation on the Muskat prediction, the technique can be used without trial and error when a digital computer is used. Such an analysis can use small pressure increments. Then the change in S_o is so small that no significant error is encountered in the calculations as a result of this inaccuracy. The accuracy is further helped by basing the S_o and k_g/k_o parameters on a saturation equal to the saturation at the end of the last calculated pressure change, less half of the change in saturation for the last pressure change, $S_{o(n-1)} - 0.5\ \Delta S_{o(n-1)}$. Even when these precautions are taken, the engineer should make certain that periodically during the program the results of a Muskat prediction are checked against a finite material-balance equation such as Eq. 7.31. Without such a check, errors can accumulate from each step in the prediction because the technique is not in itself self-correcting as are the Schilthuis and Tarner prediction techniques.

The Muskat method represents one of the original applications of the digital computer to reservoir engineering. Consequently, most companies have a program that provides a solution-gas-drive prediction based on the Muskat technique. The comments on the shortcomings of the Muskat method do not mean that we should scrap Muskat computer programs and write a program for the Tarner method. However, the author recommends that these programs be checked. Furthermore, if a company does not presently have a computer program for a tank-type prediction of a solution-gas-drive reservoir, the Tarner method is recommended for use because it has a more direct approach to the problem as well as self-correcting characteristics.

Material balance in partially saturated reservoirs. When a reservoir is of considerable thickness or areal extent, there is a period of the productive life when a portion of the reservoir is above the bubble-point pressure and another portion is below the bubble-point pressure. When possible, a material-balance application should avoid these periods of time. In other words analysis or prediction should be for times above or below the period when the reservoir is only partially saturated. However, in reservoirs of considerable thickness or aereal extent, it may be impossible to avoid analysis during these periods. It then becomes necessary to use a special material balance because a discontinuity exists in the PVT data at the bubble-point pressure. Then the average PVT data for the reservoir do not coincide with the average pressure.

This situation can be best illustrated by considering a very thick reservoir. The pressure history for such a reservoir is depicted in Fig. 7–10 and PVT data are shown in Fig. 7–11. Reservoirs of such thickness are not common in the continental U.S., but they are not uncommon in the Middle East. This particular example is concerned with a solution-gas-drive reservoir in which the initial pressure varies from 2,700 psia at 5,000 ft to 3,100 psia at 6,000 ft. The reservoir is 200 psi undersaturated at 5,000 ft and 600 psi undersaturated at 6,000 ft. Such

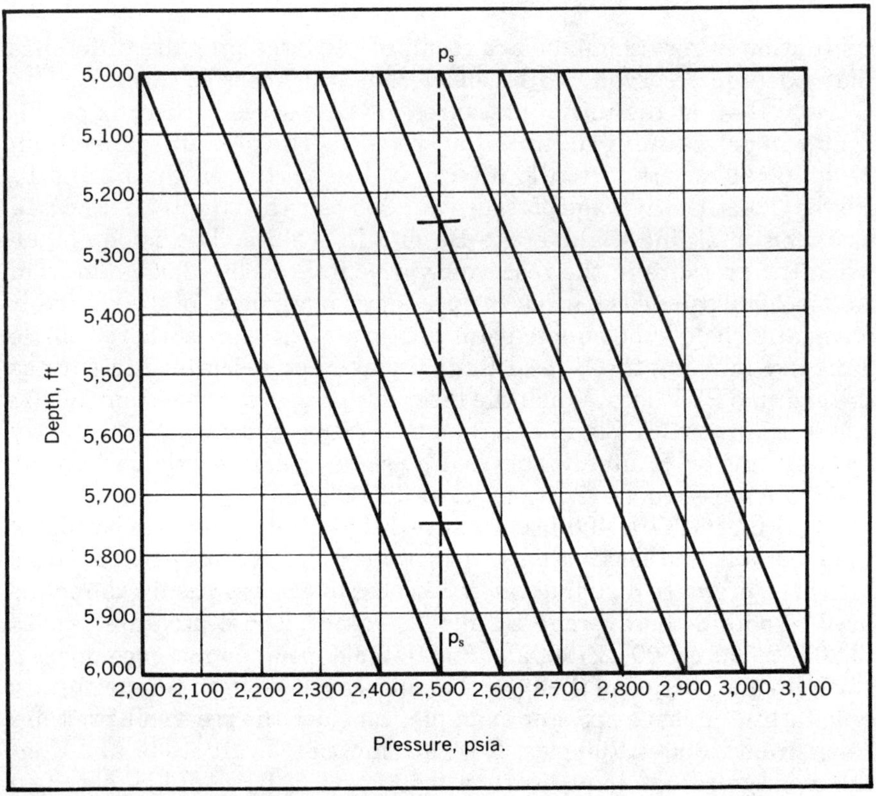

Fig. 7–10 Pressure versus depth for thick reservoir with PVT data as shown in Fig. 7–11

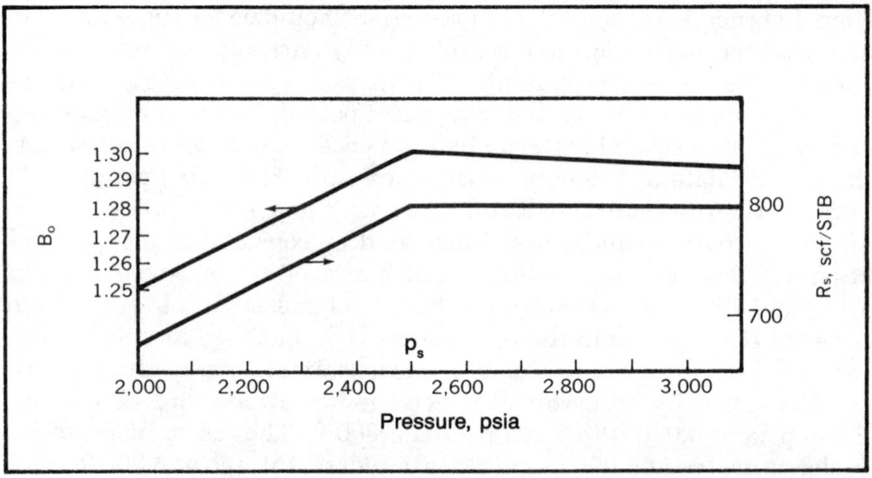

Fig. 7–11 PVT data for reservoir shown in Fig. 7–10

a situation is very common as a result of the large pressure difference that exists in reservoirs with considerable thickness or closure.

Note that, at the initial pressures in the reservoir, there is no difficulty in determining an initial average B_o. The pressure range from 2,700–3,100 covers virtually a straight-line portion of the B_o and R_s curves. Thus, the average pressure of 2,900 psia can simply be used as a basis for obtaining the average B_{oi} and R_{si} values. The same is true when the pressure in the reservoir varies from 2,600–3,000 and from 2,500–2,900 psia. Also, when the pressures in all parts of the reservoir have fallen below the bubble-point pressure of 2,500 psia, there is little difficulty in using the B_o and R_s at the average reservoir pressure as the average PVT data. When the reservoir pressure varies from 2,000–2,400 psia, note in Fig. 7–11 that this range encompasses a nearly straight line on B_o and R_s curves. The same situation exists when the reservoir pressure varies from 2,100–2,500 psia.

Working with the 400-psi pressure differential from top to bottom of the reservoir and looking only at increments of 100-psi decline, there are three pressure distributions where the average pressure cannot be used to obtain the average B_o and R_s values. These pressure ranges (2,200–2,600, 2,300–2,700, and 2,400–2,800 psia) do not encompass a straight-line portion of the B_o and R_s curves but lie in the discontinuity of data that exists at p_s. For example, consider the reservoir pressure range from 2,300–2,700 psia. If we attempted to evaluate B_o and R_s at the average reservoir pressure of 2,500 psia, a B_o would be obtained that would be the maximum for any portion of the reservoir and R_s would still be R_{si}, although half of the reservoir is below the bubble point.

Some companies use rather complicated material balance based on the effective compressibility of all of the fluids in the reservoir and the rock compressibility to analyze a reservoir while it is partly saturated. However, if production is uniformly distributed throughout the reservoir, it is only necessary to determine weighted averages for the PVT data used in the material-balance equations in order to use the same equations we have employed throughout this chapter. The validity of this statement can be understood when the derivation of the material-balance equations are reviewed.

It is seen that the expressions are simply based on the difference in the volumes of the reservoir fluids at two different pressures, with the PVT data used to express the gas and oil volumes at reservoir pressures and temperatures. Originally, we assumed the same average reservoir pressure existed throughout all of the pore volumes, but the same equations can be used if we use appropriately weighted-average PVT data. The gas in solution, R_s, and the oil formation volume factor, B_o, should be weighted by the reservoir oil volumes. The gas formation volume factor should similarly be weighted on the basis of the reservoir gas volume.

For example, assume that the data of Figs. 7–10 and 7–11 represent a depletion-type reservoir—no water drive. Further, assume that the volume of the reservoir in each depth increment from 5,000–6,000 ft is the same and that we want to predict the cumulative oil production when the oil in the reservoir is 2,400 psia at 5,000 ft and 2,800 psia at 6,000 ft.

Since the initial reservoir pressures vary from 2,700–3,100 psia, well above the saturation pressure of 2,500 psia, the initial PVT data should be based on the average pressure of 2,900 psia as noted previously. Thus, B_{oi} and R_{si} would be 1.296 res bbl/stb and 800 scf/stb, respectively. The initial gas volume factor, B_{gi}, would be calculated on the basis of the average pressure of 2,900 psia.

When the reservoir pressure varies from 2,400–2,800 psia, note that the top 250 ft or 25% of the reservoir is below the saturation pressure of 2,500 psia and the bottom 750 ft or 75% of the reservoir is still above the saturation pressure. Since the free-gas volumes in the reservoir are all in the undersaturated portion of the reservoir, the gas formation volume factor, B_g, should be calculated on the basis of the average pressure in the undersaturated zone, or (2,400 + 2,500)/2 = 2,450 psia.

However, the oil formation volume factor and gas in solution representing this state of depletion should be an average weighted on the basis of reservoir oil. The B_o and R_s average for the undersaturated portion of the reservoir can be based on the average pressure in this zone, 2,650 psia, or 1.2985 and 800, respectively. The average B_o and R_s in the portion of the reservoir below the saturation pressure can be based on the same average pressure of 2,450 psia cited as the basis for calculating the B_g. Using this 2,450-psia pressure, B_o and R_s averages for the top 250 ft at this time would be 1.295 and 787, respectively. The appropriately weighted-average oil formation volume factor, B_o, would then be 0.75(1.2985) + 0.25(1.2950), or 1.2976. Similarly, the weighted average gas in solution, R_s, would be 0.75(800) + 0.25(787), or 797 scf/stb. Use of these PVT data and the other appropriate reservoir parameters in a regular material-balance equation would permit the calculation of the cumulative oil and gas production, N_p and G_p.

Calculations for a partially saturated reservoir are normally simplified somewhat because the gas saturation is less than the equilibrium gas saturation. Thus, no free-gas flow takes place in the reservoir. If free gas does flow, the calculations become complicated because the gas saturation varies with depth. It may appear that the corrections are unimportant. However, many examples can be found where the use of PVT data weighted as shown are extremely important to the accuracy of the prediction.

Gas-cap complications. The discussion of gas-drive reservoir predictions has been based for the most part on the assumption that none of the encroaching gas from the gas cap will be produced. This is certainly the most desirable operating condition, but it is generally very difficult to obtain because of the demand for oil. When an engineer can logically foresee that a reservoir will be producing gas-cap gas, the prediction of the reservoir behavior becomes considerably more difficult.

In states where government agencies (e.g., the Texas Railroad Commission) control the maximum gas-oil ratio under which production can take place without a penalty, the reservoir prediction can be based on the assumption that this maximum gas-oil ratio prevails from some particular stage of depletion. Without such an artificial control the assumed gas-oil ratio for various stages of depletion is difficult to determine.

Predicting Water-Drive Reservoir Behavior

When oil or gas is produced by natural water drive, the prediction of reservoir behavior generally becomes more difficult than for a gas-drive reservoir. We have added to the effects of the solution and gas-cap drives the productive force of water encroachment. When there is only one hydrocarbon phase flowing in the reservoir—a gas reservoir or an oil reservoir above the saturation pressure—the analysis is fairly simple. Many discussions of reservoir predictions are based on the assumption that this is the situation.

In fact, much or even most of the oil produced outside the U.S. may be produced from such reservoirs. However, with most wells in the U.S. producing at peak capacity, a domestic water-drive reservoir is seldom found that is producing above the saturation pressure. Consequently, the prediction of the behavior of a water-drive reservoir may be difficult.

To predict performance of a water-drive reservoir, it is necessary to predict water encroachment independent of material balance. Water encroachment is a function of the aquifer flow characteristics and a function of the pressure history and time. To this point our predictions have been independent of time, but a water drive introduces time as an important parameter. Since water-drive reservoirs producing below the bubble point also are producing by solution-gas or gas-cap drive, we must also include the ideas covered under the gas-drive section in a water-drive prediction.

The logical extension of the techniques for predicting the behavior of a gas-drive reservoir to the prediction of a water-drive reservoir is to choose a pressure at which the production is desired. Then iterate in some way on the produced gas-oil ratio as before and iterate a second time (a double iteration) on the time at which the chosen pressure will be reached. However, we find that, generally, it is simpler to set the time and iterate on the pressure. In other words we choose a time at which we desire to know the cumulative oil and gas production, iterate on the pressure to be obtained in the reservoir at that time, and iterate on the produced gas-oil ratio.

Many methods are found for predicting water encroachment.[11] However, in most cases it is the author's opinion that the water encroachment has taken place as predicted by the unsteady-state compressible fluid flow theory. Consequently, only this approach is presented.

The engineer is warned to be especially careful in water-drive-reservoir computer modeling. The tendency here seems to be to use any convenient method of modeling the water-bearing portion of the reservoir as long as it satisfactorily matches the past performance of the reservoir. Normally, this past performance represents only a few years, and the performance prediction obtained with the model represents many years. Therefore, it is possible to obtain an acceptable fit of past data using oversimplified modeling of the aquifer while the long-range prediction falls far short of acceptability. With large non-U.S. water-drive reservoirs, the practice is to model the aquifer with one large outside cell that is treated as having the same pressure throughout at any particular time. Juggling the aquifer size and transmissibility generally permits a short past-history match, but it cannot continue to be accurate for long periods.

When reservoir behavior is dominated by a water drive, the reservoir model should also be dominated by detailed modeling of the aquifer, using several outside cells or segments to represent the aquifer. The trend seems to be to concentrate the details on the producing portions of the reservoir when, given a particular model complexity, detailing the aquifer would result in a more accurate long-range prediction.

This discussion is not meant to imply that the engineer should totally disregard anomalous aquifer behavior. For example, in Sumatra some aquifers appear to exhibit strong evidence of a surface replenishing aquifer. The salinity of the produced water continues to grow fresher timewise, and there is a regional variation in the water salinity that does not correlate with depth. Also, these reservoirs are unusual because they produce oil with a pour point in the 90–100°F range with almost no solution gas (50–60 scf/stb). These reservoirs are very hot with temperatures of 250–300°F from depths of 3,000–4,000 ft at rates of 5,000–10,000 b/d.

Calculating water encroachment. Water encroachment generally follows the constant-pressure solution to the radial diffusivity equation described in chapter 3. This solution appears to have been originally developed by Hurst for the purpose of predicting the behavior of an aquifer.[12] This reservoir application is the only type that has practical use for the dimensionless size, r_e/r_w, reservoirs for which the original data were developed. These data cover dimensionless reservoir sizes of 10 or less, which have no physical meaning unless the well radius, r_w, is the internal radius of an aquifer that is furnishing water encroachment for a hydrocarbon reservoir.

We have learned that a general solution can be obtained for a reservoir producing by unsteady state with a constant pressure maintained in the reservoir at some particular radius, r. This general solution takes the form of a relationship between a dimensionless group of terms that includes the cumulative flow across the radius, r, and another dimensionless group of terms that includes the time corresponding to the cumulative flow. Furthermore, this general solution can be applied to a variety of reservoir problems using the appropriate reservoir constants for the subject reservoir.

To be more specific, the constant-pressure solution relates the dimensionless cumulative flow function, Q_{tD}, to the dimensionless-time function, t_D, as presented in Figs. 3–15 to 3–18 and Tables 3–1 and 3–2 with the functions defined as:

$$Q_{tD} = Q/1.12\phi hcr^2\theta\Delta p \quad (3.51a)$$

$$t_D = 6.33kt/\phi\mu cr^2 \quad (3.11a)$$

Eq. 3.11a is Eq. 3.11 with the diffusivity constant, η, expanded according to Eq. 3.9. Eq. 3.51a is Eq. 3.51 with the parameter, θ, introduced to account for the fraction of a circle represented by the flow system.

Physically, these functions represent the cumulative flow, Q, that takes place across a radius, r, if the pressure at this radius is suddenly dropped an amount, Δp, and this pressure is maintained constant at this radius for a time, t. However, constant pressures are not generally maintained at the original oil-water or gas-water contacts because the pressure drop at this radius is caused by the production of oil or gas from the hydrocarbon-bearing portion of the reservoir. If the wells in the reservoir are being produced at peak capacity, it is not likely that the water encroachment can match the hydrocarbon withdrawal rates. Therefore, the pressure at the original hydrocarbon-water contact continues to decline. In such a case it is necessary to calculate the water encroachment by superposition.*

When Eq. 3.51a is solved for the cumulative flow, Q, we obtain:

$$Q = 1.12\phi hcr^2\theta \, \Delta p \, Q_{tD} \tag{3.51b}$$

When this equation is applied to a sequence of pressure drops (Δp_j) using the dimensionless cumulative flow function (Q_{tDj}) that is appropriate for the effective dimensionless time (t_{Dj}) of each Δp_j, we obtain Eq. 3.64 modified to include the parameter, θ:

$$Q = 1.12\phi hcr^2\theta \sum_{j=1}^{j=n} (\Delta p_j \, Q_{tDj}) \tag{3.64a}$$

Care should be exercised in determining the pressure drops that occur at the original hydrocarbon-water contact, especially if the reservoir has a low permeability. In such a case it is unwise to assume that the drop in the average pressure in the hydrocarbon-bearing portion of the reservoir is the same as the drop in the pressure at the original hydrocarbon-water contact.

The engineer should also evaluate the group of constants in equation 3.64a, $1.12\phi hcr^2\theta$, from past reservoir performance when possible because of the difficulty in determining these reservoir parameters for the aquifer. When the subject group of aquifer constants is treated as one constant, B, Eq. 3.64a becomes:

$$Q = B\sum_{j=1}^{j=n} (\Delta p_j \, Q_{tDj}) \tag{3.64b}$$

*The application of Eq. 3.51a to the calculation of water encroachment using superposition is discussed and illustrated in chapter 5. The more important ideas in this type application are repeated for the sake of convenience.

Using past reservoir performance, the cumulative water influx can be calculated by material balance. Then by plotting the cumulative water influx at various times versus the summation term for those times, a straight-line plot should be obtained whose slope is the constant, B.

Once the aquifer constant has been evaluated, it is then possible to predict the cumulative water influx for any pressure history using Eq. 3.64b.

Undersaturated water-drive reservoirs. Engineers with only U.S. experience may feel that a discussion of undersaturated water-drive reservoirs is a useless exercise since there are few if any such reservoirs in the U.S. However, worldwide, a large percentage of the oil production is produced from such reservoirs.

A water-drive gas reservoir or a solution-gas-water-drive oil reservoir requires a trial-and-error solution to determine pressures that satisfy both the fluid flow and material-balance equations. However, an undersaturated water-drive reservoir can theoretically be analyzed, and a performance prediction made without resorting to trial and error.

If we solve the general material-balance equation, Eq. 7.27, for the water encroachment, W_e, we obtain:

$$W_e = N_pB_o + B_g(G_p - N_pR_s) + W_p + G(B_g - B_{gi}) - N(B_o - B_{oi}) - NB_g(R_{si} - R_s) + (c_f + c_wS_{wc})\,\Delta pB_{oi}/(1 - S_{wc}) \quad (7.36)$$

When the reservoir pressure is above the saturation pressure the second, fourth, and sixth terms on the right-hand side of Eq. 7.36 are 0.0, so the equation becomes:

$$W_e = N_pB_o + W_p - N(B_o - B_{oi}) + (c_f + c_wS_{wc})\,\Delta pB_{oi}N/(1 - S_{wc}) \quad (7.37)$$

Note that the equation for determining the water encroachment, W_e, by fluid flow, Eq. 3.64b, can be written with the summation term stated from $j = 1$ to $j = n - 1$, and a separate term can be added to account for the last pressure drop, Δp_n:

$$(W_e)_{n+1} = B\sum_{j=1}^{j=n-1}(\Delta p_jQ_{tDj}) + \left[B\,\frac{p_{n-1} - p_{n+1}}{2}\right]Q_{tDn} \quad (7.38)$$

It should be noted that to determine the water encroachment after n + 1 periods, $(W_e)_{n+1}$, only n terms are required in the right-hand side of Eq. 3.64b. Also, the last term in Eq. 7.38 assumes that the pressure versus time relationship from p_{n-1} to p_n and from p_n to p_{n+1} is a straight line so half of the sum of these two pressure differences results in Δp_n equaling $(p_{n-1} - p_{n+1})/2$.

If the expression for W_e according to Eq. 7.38 is substituted in Eq. 7.37, we obtain:

$$B \sum_{j=1}^{j=n-1} (\Delta p_j Q_{tDj}) + B\left[\frac{p_{n-1} - p_{n+1}}{2}\right] Q_{tn} = N_p B_o + W_p$$
$$- N(B_o - B_{oi}) + (c_f + c_w S_{wc})\, \Delta p B_{oi}/(1 - S_{wc}) \qquad (7.39)$$

If the aquifer constant, B, has been evaluated note that the cumulative oil production, N_p, can be determined for any final pressure at the original water-oil contact, p_{n+1}. The previously observed or calculated pressure history provides the data for the $\Delta p Q_{tD}$ summation term and p_{n-1}. Then p_{n+1} gives the average pressure in the hydrocarbon-bearing portion of the reservoir—generally they are assumed equal—which in turn fixes B_o and Δp. All of the parameters in Eq. 7.39 should be known except for N_p, which can be calculated.

Eq. 7.39 can also be used to determine the average reservoir pressure when a particular cumulative production value, N_p, will be reached. In this case it is necessary to state B_o as a function of the average reservoir pressure in order to permit the direct calculation of the pressure. However, this should pose no undue difficulty since above the saturation pressure, $(p > p_s)$, there is a straight-line relationship between B_o and the pressure, B_{op}, equals $B_{oi} + c_o(p_i - p)B_{oi}$, where c_o is the oil compressibility.

Work problem 7.10 and check the solution against the one in appendix C to understand the methods of analyzing a water-drive reservoir when the reservoir pressure is above the saturation pressure.

PROBLEM 7.10: Analyzing an Undersaturated Oil Reservoir with a Water Drive

An oil reservoir with an initial pressure of 3,500 psia and saturation pressure of 3,000 psia is being produced at a rate of 1,500 stb/d. The reservoir has an average radius of 2,500 ft and is surrounded by an aquifer with an average radius of about 20,000 ft. Given the following reservoir data, determine the water encroachment after 100 and 200 days of production, or until the saturation pressure is reached. Assume the average pressure in the original hydrocarbon-bearing portion of the reservoir is equal to the pressure at the original WOC.

Porosity = 22%
Reservoir thickness = 60 ft
Formation compressibility = 4×10^{-6}/psi
Water compressibility = 3×10^{-6}/psi
Oil compressibility = 7.7×10^{-6}/psi

Permeability = 100 md
Water viscosity = 0.3 cp
Connate water saturation = 26%
Initial oil formation volume factor = 1.34

Note that in problem 7.10 we simply assumed that we could produce the reservoir at a rate of 1,500 stb/d even though the reservoir pressure had declined. Obviously, this is not necessarily true. Unless it is clear that a well is producing far below its capacity, i.e., the drawdown is very small, it is necessary to use the productivity ratio of the wells to ensure that the proposed rate (1,500 stb/d) can be maintained with the reduced reservoir pressure at some particular time. If it is found that, based on the reservoir pressure resulting from the assumed rate, the assumed rate cannot be maintained at the lesser pressure, it is then necessary to reduce the assumed rate. Then repeat the calculation of the average pressure at the end of some production period. Thus, it is clear that the prediction of the behavior of a water-drive reservoir producing at capacity involves a trial-and-error type of solution.

Water-drive reservoir below the saturation pressure. When we predict the behavior of a water-drive reservoir with the reservoir pressure above the saturation pressure, many of the terms in the general material-balance equation are 0.0 and the oil formation volume factor can be stated as a function of the pressure. When the reservoir pressure is less than the saturation pressure, neither of these simplifying assumptions can be made and the prediction becomes more difficult. In order to make a prediction under these conditions, it is necessary to find a reservoir pressure and corresponding pressure at the original WOC that satisfy the material-balance equation, the fluid flow equation for calculating the water influx, and the production capacity of the reservoir (i.e., the productivity indices). Below the saturation pressure none of the PVT data can be treated as a straight-line relationship with pressure.

The nature of the problem of predicting the behavior of a water-drive reservoir with free gas present in the reservoir is clearly illustrated in chapter 5, specifically in problem 5.6. Before proceeding further, the engineer should review the subject problem and its solution. This problem is less complex than the general solution-gas-water-drive problem because the production of gas and the ratio between the produced gas and the produced oil do not have to be considered in the gas-reservoir water-drive problem.

Recommended solution-gas-water-drive reservoir prediction. The prediction technique preferred by the author is to estimate the pressure that will exist in the reservoir at a particular time and calculate the water influx, Q, using the Q_{tD} functions (Eq. 3.64b) and W_e using material balance. When a pressure is found that makes $Q = W_e$, we can assume this to be the correct pressure and cumulative water encroachment. Then we can predict the pressure for the next desired time.

When stated this way, the procedure sounds simple. However, consider the details of the calculations. As soon as a pressure is estimated, we can calculate Q by fluid flow using Eq. 3.64b. As noted, the aquifer constant should be evaluated from past reservoir performance when possible. The engineer should also be careful to determine the pressure at the original hydrocarbon-water contact and not simply assume that the change in the average pressure in the hydrocarbon-bearing portion is the same as the change in the pressure at the original hydrocarbon-water contact.

Now, consider the calculations necessary for the material-balance calculation of the water encroachment, W_e. We noted that Eq. 7.27, the general material-balance equation, can be solved for the cumulative water encroachment, W_e, to provide a means for making this calculation:

$$W_e = N_pB_o + B_g(G_p - N_pR_s) + W_p - G(B_g - B_{gi}) - N[B_o - B_{oi} + (R_{si} - R_s)B_g + (c_f + c_wS_{wc})\Delta pB_{oi}/(1 - S_{wc})] \quad (7.36)$$

We must first estimate how many barrels of oil will be produced during the next time period in order to determine N_{pn}. Therefore, we must estimate how many wells will be invaded by water during the next period, based on our tentative calculation of Q by fluid flow. Then using productivity indices (chapter 4), a calculated average oil saturation at the producing wells, and relative permeability data, we can determine the number of barrels of oil that will be produced during the subject period. However, to determine the oil saturation, it is necessary to know the production. Therefore, this solution represents a trial-and-error application.

To use Eq. 7.36, we must also know the cumulative gas production at the end of the prediction period, G_{pn}. This entails calculating the instantaneous gas-oil ratio at the end of the prediction period, R_n, and using ΔN_p for the period with the average gas-oil ratio to calculate G_{pn}.

Calculation of the oil saturation in the producing portion of the reservoir presents a real problem. A technique that works in all cases is virtually impossible to derive. Consequently, we simply look at one method and hope that the introduction to the problems involved gives the engineer sufficient background to make similar calculations under other conditions.

First, we assume that wells invaded by water will not be produced. In other words, we are assuming that, as soon as a well starts cutting water, it is shut in. When demand is very high, this may not be a realistic assumption. However, it should be clear after considering the simplified assumptions that the technique can be modified to include the effect of wells producing from the invaded zone.

In addition to assuming that all production takes place from the uninvaded zone, we assume that all bypassed hydrocarbons in the water zone can be represented by oil and that the fractional pore volume of oil bypassed is a constant, S_{oBY}. Note that this is inconsistent with our knowledge of fluid displacement as discussed in chapter 6. Nevertheless, the relatively high pressure normally associated with the water encroachment and the relatively constant average water saturation in the advancing water bank tend to make the inaccuracy of this latter assumption insignificant. Based on this assumption, note that the pore volume in the invaded zone can be stated as:

$$(W_e - W_p)/(1 - S_{oBY} - S_{wc})$$

With this relationship and recognizing that the expression for the total original oil-zone pore volume is $NB_{oi}/(1 - S_{wc})$, the equations for the reservoir oil, pore volume, and oil saturation in the uninvaded zone should be clear:

Res oil in uninvaded zone

$$= \text{total res bbl of oil left in res} - \text{bbl of res oil in invaded zone}$$

$$= B_o(N - N_p) - S_{oBY}(W_e - W_p)/(1 - S_{oBY} - S_{wc}) \quad (7.40)$$

Pore volume in uninvaded zone

$$= \text{total oil-zone pore volume} - \text{invaded oil-zone pore volume}$$

$$= [NB_{oi}/(1 - S_{wc})] - (W_e - W_p)/(1 - S_{oBY} - S_{wc}) \quad (7.41)$$

$$S_{oUN} = \frac{\text{Eq. 7.40}}{\text{Eq. 7.41}}$$

$$= \frac{(N - N_p)B_o - S_{oBY}(W_e - W_p)/(1 - S_{oBY} - S_{wc})}{[NB_{oi}/(1 - S_{wc})] - (W_e - W_p)/(1 - S_{oBY} - S_{wc})} \quad (7.42)$$

In Eq. 7.40 the reservoir oil in the uninvaded zone is equated to the difference between the total oil remaining in the reservoir and that remaining in the water-invaded zone. In Eq. 7.41 the pore volume in the uninvaded zone is equated to the difference between the original pore volume and the pore volume of the invaded zone. The average oil saturation in the uninvaded zone is then the ratio of Eqs. 7.40 and 7.41 as indicated in Eq. 7.42.

Once the oil saturation in the uninvaded zone has been determined, the engineer can estimate the average productivity index, J, for the prediction period. This is done using Eq. 4.6:

$$J_1/J_2 = (k_o/B_o\mu_o)_1/(k_o/B_o\mu_o)_2 \qquad (4.6)$$

With an average productivity index, J_1, known for a particular set of reservoir conditions, the productivity index, J_2, can be calculated for the conditions at the end of the prediction period. Then an average productivity index for the period can be applied to the estimated number of producing wells using the average reservoir pressure during the period. This calculation is used to check the originally estimated value of N_{pn}. If the assumed N_{pn} and the calculated N_{pn} are not nearly equal, this portion of the calculation must be repeated until an acceptable number is obtained.

After an acceptable number has been obtained between the calculated and assumed N_{pn}, based on an assumed average reservoir pressure at the end of the prediction period, calculate the water influx by material balance. Eq. 7.36 can be used for this calculation only after the cumulative gas production at the end of the prediction period has been determined. If the water encroachment by fluid flow, Q, matches the water encroachment by material balance, W_e, we can then conclude that the assumed pressure at the end of the prediction period is correct.

If Q does not match W_e, we must adjust the assumed pressure and repeat all of the calculations until the two values are the same. If the water encroachment by material balance is greater than the water encroachment by fluid flow, a smaller average hydrocarbon reservoir pressure must be assumed. If Q is greater than W_e, it is necessary to assume a larger pressure.

Since this method of calculating the behavior of a water-drive reservoir involves the simultaneous satisfaction of several iterations, it is probably helpful to list the procedure that can be followed in the calculations.

Assume: No free gas exists in the water bank. Piston-like displacement is used. A uniform saturation exists throughout the unaffected zone. Producing wells are shut in when the water front reaches them.

1. Estimate the average reservoir pressure in the uninvaded portion of the reservoir at the end of the next prediction period. Base this pressure on a plot of the average reservoir pressure versus time extrapolated to the end of the prediction period. Determine the corresponding pressure at the original water-oil contact at this time based on some established relationship between the average reservoir pressure and the pressure at the original WOC.

2. Calculate Q using Eqs. 3.64b and 3.11a. The aquifer constant B should have been evaluated from past performance.
3. Estimate which wells will be producing during the next prediction period. Base this on an inspection of the water-encroached area at the start of the period and its advance as a result of the water encroachment Q calculated in step 2 and the bypassed oil in the invaded zone.
4. Estimate N_{pn}. Initially, base this on the past (or predicted) history of the field production rate versus time and an extrapolation of this relationship.
5. Evaluate S_{oUN} at the beginning and the end of the period. If the limiting assumptions are acceptable, use Eq. 7.42 for this purpose. If S_{oUN} increases, N_p is incorrect. Then return to step 4.
6. Determine the average productivity index during the prediction period for the wells that will be producing during this period. Use Eq. 4.6.
7. Based on the results of step 6, the average $(p_e - p_w)$, and the average number of wells producing, calculate N_{pn}. If N_{pn} calculated does not agree sufficiently with N_{pn} estimated in step 4, repeat steps 4–7 until an acceptable agreement is obtained.
8. Calculate G_{pn} based on S_{oUN} at the end of the period. Use Eqs. 7.11 and 7.32.
9. Calculate W_e by material balance. Base the calculation on Eq. 7.36 if the inherent assumptions of this equation are acceptable.

10a. If $Q = W_e$, proceed to the prediction for the next time period.

10b. If $Q > W_e$, repeat steps 1–10 assuming a larger pressure in step 1.

10c. If $Q < W_e$, repeat steps 1–10 assuming a smaller pressure in step 1.

This procedure can be modified to account for most assumptions that are necessary. If gas saturations exist in the water bank, the calculation of the water-bank size must be modified. If the saturation distribution is to be considered along a stream line, other than piston-like distribution, construct a saturation distribution plot in step 3 and check it after the amount of water encroachment has been determined. With this assumption the estimation of the producing rates during the time period also must be based on a different saturation for each well. If wells will be produced to some water cut greater than zero, this must be considered when determining the prediction of production during the prediction period.

One effect that usually must be considered is coning or fingering. These effects are discussed in chapter 6. One or both usually affect the behavior of a water-drive reservoir. In the prediction procedure and in the example problem, we assume that the isosaturation lines during a displacement remain horizontal. When water encroachment is, relatively speaking, from the side or edge of a reservoir rather than from below, the isosaturation lines tend to tilt in the direction of the displacement. In addition, when the front of the water bank approaches the bottom of a well that is producing, a cone tends to form on top of any isosaturation tilt that has already been established. As noted, there is no reliable way of quantitatively predicting such behavior in advance, with the possible exception of computer modeling when considerable reservoir detail is available. Once one or more wells have been invaded by water, it may be possible to use the production and structural data to arrive at parameters to predict the future behavior of the reservoir accurately.

If we can determine the water encroachment volume accurately at the time the water first reaches a producing well, the height of the cone can be determined. Also, by noting the production rate and other reservoir data associated with the cone, a relationship can be developed using the coning equation, which permits the prediction of coning into the wells at higher structural positions. The engineer should remember that the water encroachment also is affected by the permeability stratification in the reservoir. Consequently, the engineer predicting future coning in connection with the prediction of the behavior of a reservoir should understand the stratification effect before proceeding.

Observed breakthrough times can also be used to verify the oil saturation in the water bank. By observing the cumulative water encroachment required to go from some low water cut (e.g., 5%) in one structure well to the same water cut in a well at a higher structural position. The intervening water encroachment and corresponding reservoir volume between the wells permits the determination of the bypassed oil saturation, S_{oBY}.

In many cases where the bypassed oil saturation is extremely important, wells are drilled for the sole purpose of observing the advance in the water bank. Such wells are never produced but are completed with plastic casing opposite the reservoir so logging methods can be used to determine the oil saturation profile at any particular time. Such wells are of course unaffected by coning because they are never produced.

To check our knowledge of this section, work problem 7.11 and compare the solution with the one in appendix C.

PROBLEM 7.11: Analyzing a Combination Solution-Gas-Water-Drive Reservoir

A reservoir with the configuration shown in Fig. 7–12 has an initial reservoir pressure that is at the saturation pressure of 2,500 psia. Fig. 7–13 also shows the bulk reservoir volume above various reservoir depths. The PVT data for the reservoir are as follows:

p, psia	B_o	R_s	B_g	μ_o/μ_g	μ_o
2,500	1.325	650	0.000796	56.60	0.38
2,300	1.311	618	0.000843	61.46	0.39
2,100	1.296	586	0.000907	67.35	0.40
1,900	1.281	553	0.001001	74.33	0.41
1,700	1.266	520	0.001136	81.96	0.42
1,500	1.250	486	0.001335	91.56	0.43

The aquifer constant, B, has been shown to be 225 bbl/psi, the original oil in place in the reservoir is 16.89 MMstb. The six wells produced initially at an average rate of 750 stb/d/well with an average flowing bottom-hole pressure of 2,150 psia that is assumed constant. The average oil saturation in the encroaching water bank is estimated to be 40%. Other reservoir data are as follows:

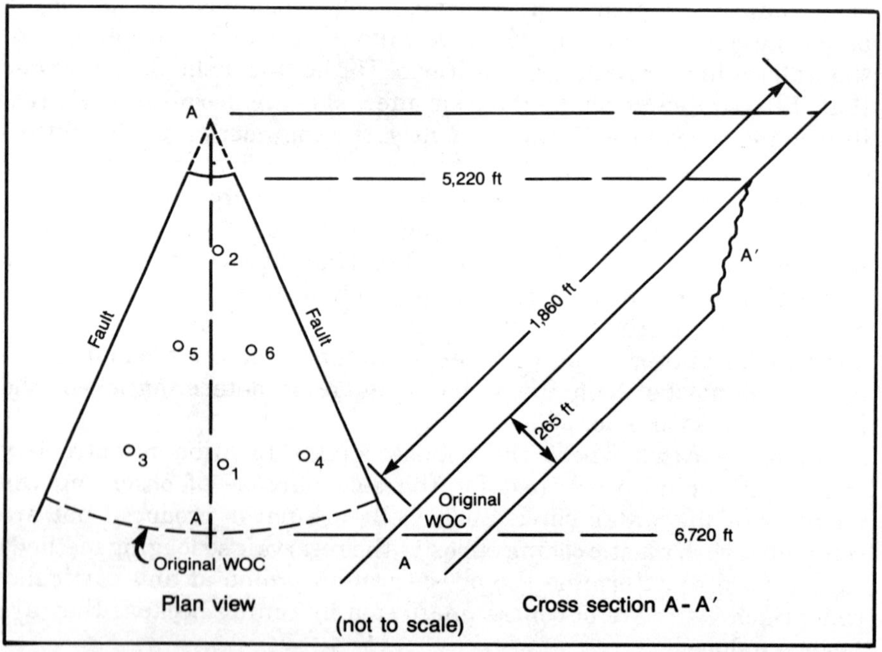

Fig. 7–12 Schematic diagram of reservoir for problem 7.11

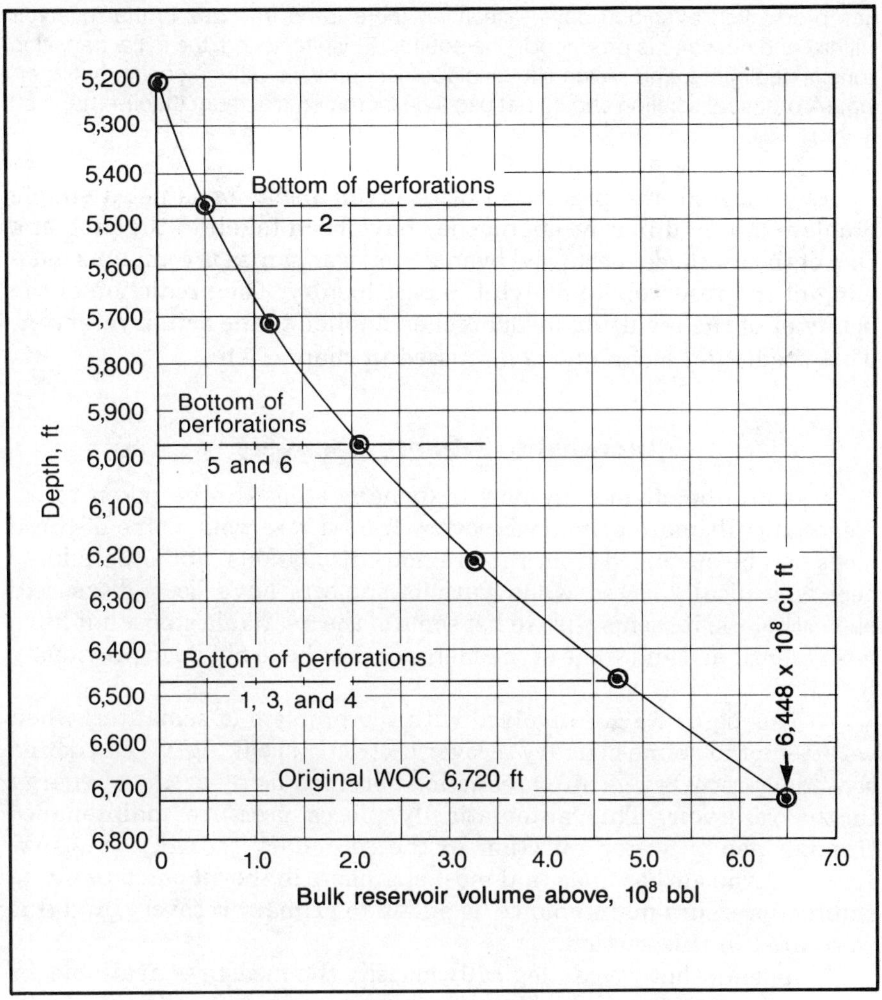

Fig. 7–13 Reservoir volume versus depth for problem 7.11

Permeability = 100 md
Average water saturation in oil zone (initial) = 22%
Oil compressibility = 8×10^{-6}/psi
Formation compressibility = 4×10^{-6}/psi
Water viscosity = 0.3 cp
Porosity = 25%
Water compressibility = 3×10^{-6}/psi

Assume gas-oil relative permeabilities are closely approximated by the Wiley equation for unconsolidated well-sorted sand. Calculate the cumulative oil and gas production after 500 days, assuming there is no free gas in the reservoir initially and no water is produced. The aquifer is infinite acting, the initial transition zone is negligible, and piston-like displacement prevails with no coning or fingering. (A practical solution should not use time increments of more than perhaps 50 days.)

As indicated, the prediction of reservoir behavior is not a simple problem. Many different approaches have been taken to the problem. One of the methods most used by major oil companies is computer modeling of the reservoir to match the past history. The prediction of the behavior of the resulting model is then applied to the actual reservoir. This prediction technique is discussed in chapter 11.

Increasing Primary Recovery

It should be obvious by now that many steps can be taken to increase the ultimate primary recovery from a reservoir. Some of these steps can be surmised from the previous discussions, and others have been specifically noted when various subjects have been discussed. Nevertheless, it seems wise to list some of the available steps that have been mentioned and some of the methods that have not been previously noted.

At this point we get involved with the problem of semantics when we attempt to define primary recovery. Strictly speaking, we can define secondary recovery as any production obtained using artificial energy in the reservoir. This automatically places pressure maintenance through gas or water injection in the secondary recovery category. However, the author finds that most engineers in the oil patch prefer to think of pressure maintenance as an aid to primary recovery, and it is so treated in this section.

It appears that we can logically classify the measures available for improving oil recovery during primary production as *well control procedures* and *reservoir control procedures*. Under reservoir control we discuss mainly pressure maintenance. Many different ideas are discussed under well control.

Well control. It should be stated that any steps taken to increase the oil or gas producing rate from an oil or gas reservoir generally increase the ultimate recovery from that reservoir by placing the economic limit further along the cumulative production scale. We recognize that there is a particular rate of production at which the producing costs equal the operating expenses. Producing a well below this rate results in a net loss. If the productive capacity of a well can be increased, it is clear that additional oil will be produced before the economic rate is reached. Consequently, acidizing, paraffin control, sand control, clean out, and other means actually increase ultimate production from that well.

It is clear that production of gas and water robs an oil reservoir of energy. If the production of gas and water from an oil reservoir is minimized, a larger ultimate production results. The same can be said for minimizing the production of water from a gas reservoir.

Proper control of the individual well rate is a big factor in the control of gas and water coning or fingering. The engineer is referred to chapter 6 for a quantitative and qualitative description of the problem. This general problem is not restricted to water-drive and gas-cap-drive reservoirs. In a solution-gas-drive reservoir it may be possible to produce a well at too high of a rate from an ultimate recovery standpoint because excessive drawdown of the producing well pressure results in an excessive gas-oil ratio and corresponding waste of the solution gas. The engineer should be aware of this possibility and test wells in a solution-gas-drive reservoir to see if the gas-oil ratio is sensitive.

The engineer should also know that excessive drawdown in a solution-gas-drive reservoir through excessive producing rates often causes excessive deposition of paraffin in the tubing and occasionally in the reservoir itself. Keeping gas in solution in the oil by keeping the well pressure as high as possible minimizes the paraffin deposition. Of course, deposition of paraffin in the tubing is not serious when compared with the deposition of paraffin in the reservoir. Given enough time and money, the paraffin can be cleaned from the tubing and flow lines. However, it is problematic whether paraffin deposited in the pores of the formation around the wellbore can be cleaned from these pores. Consequently, the operator should be very careful to avoid such deposition in the formation.

Another adverse effect that may be caused by an excess producing rate is the production of sand. Many unconsolidated formations tend to flow sand through perforations and into the producing system when flow rates are excessive. It may be possible to improve this situation with screens, gravel packing, or consolidating materials.

The proper positioning of wells in a reservoir also plays a big part in the control of gas and water production. It is obvious that wells should be positioned as far as possible from the original gas-oil, water-oil, and gas-water contacts in order to minimize the production of unwanted gas and water. The positioning of the producing wells must, of course, be consistent with the needs for reservoir drainage, the total reservoir producing capacity, and the cost of development.

In determining the proper well spacing to use in a particular reservoir, the engineer should make certain that full recognition is given to the pressure distribution that will prevail in the drainage area of a well when the economic limit is reached. In a continuous reservoir there is no limit on the amount of reservoir that can be affected by one well. However, the engineer should be concerned with the additional oil that can be recovered prior to reaching the economic limit rate by increasing the drainage volume, or radius, of a well. In very tight reservoirs we may be able to accomplish only a small reduction in the reservoir pressure in the additional reservoir volume. This effect may be nearly offset by the reduction of the well rate caused by the increase in the drainage radius. Thus, care should be exercised to ensure that the greatest well spacing possible is also the most economical.

Total reservoir control. The effect of water and gas production on the recovery in an oil reservoir can be shown by solving Eq. 7.27 for the produced oil:

$$N_p = \frac{N[B_o - B_{oi} + (R_{si} - R_s)B_g + (c_f + c_w S_{wc})\,\Delta_p B_{oi}/(1 - S_{wc})]}{B_o - R_s\,B_g}$$

$$- \frac{B_g G_p + G(B_g - B_{gi}) + W_e - W_p}{B_o - R_s\,B_g} \qquad (7.27a)$$

We see that the oil production obtainable at a particular reservoir pressure is almost directly reduced by the reservoir volume of gas and water produced. Furthermore, the derivation of the material-balance equation shows that the cumulative gas production, G_p, is the net produced gas defined as the produced gas less the injected gas. Similarly, if the water encroachment, W_e, is defined as the natural water encroachment, the produced water, W_p, must represent the net water produced, defined as the water produced less the water injected. Therefore, if water or gas can be injected without adversely affecting the amount of water or gas produced, the amount of oil produced at a particular reservoir pressure can be increased.

It is well known that the most efficient natural reservoir drive is water encroachment. The next most efficient is a gas-cap expansion, and the least efficient is solution-gas drive. Consequently, it is important for the reservoir engineer to control production from a reservoir so as little oil as possible is produced by solution-gas drive and as much oil as possible is produced by water drive. However, when two or more drives operate in a reservoir, it is not always clear how much production results from each drive. One convenient method of estimating the amount of production resulting from each drive is to use material-balance drive indices.

Drive indices are most conveniently derived from the material-balance equation written as a function of the ratio of the original reservoir free-gas volume to the original reservoir oil volume, m, and the total formation volume factor, B_t:

$$N = \frac{N_p[B_t + (R_p - R_{si})B_g] - (W_e - W_p)}{B_t - B_{ti} + (c_f + c_w S_{wc}) \dfrac{\Delta p B_{ti}}{(1 - S_{wc})} + \dfrac{m B_{ti}}{B_{gi}}(B_g - B_{gi})} \qquad (7.29)$$

When Eq. 7.29 is cross multiplied and the $(W_e - W_p)$ term is taken to the other side, we obtain:

$$N_p[B_t + (R_p - R_{si})B_g] = N(B_t - B_{ti}) + \frac{N m B_{ti}}{B_{gi}}(B_g - B_{gi}) + (W_e - W_p) + \frac{N \Delta p B_{ti}}{(1 - S_{wc})}(c_f + c_w S_{wc}) \qquad (7.43)$$

In Eq. 7.43 the left-hand side represents the volume of the oil and gas production at the lesser reservoir pressure that exists after this production. Then N_pB_t is the reservoir volume of the oil and the original gas in solution, while $N_p(R_p - R_{si})B_g$ is the reservoir volume of the free gas produced in excess of the original gas in solution, which is included in B_t. This reservoir volume of the hydrocarbon production must be equal to the volume increase of all of the original fluids in the reservoir that occurs if the pressure on these fluids is reduced to the pressure that exists after N_p, G_p, and W_p production plus any accompanying change in the pore volume. The right-hand side of Eq. 7.43 represents such a volume. The first term on the right-hand side is the expansion of the original gas-cap gas; the third term is the increase in the water volume, or water encroachment; and the last term represents the reduction in the pore volume. If the volume changes are stated as a fraction of the reservoir volume of the oil and gas production (the left-hand side of the equation), we obtain:

$$1.0 = \frac{N(B_t - B_{ti})}{N_p[B_t + (R_p - R_s)B_g]} + \frac{N m B_{ti}(B_g - B_{gi})/B_{gi}}{N_p[B_t + (R_p - R_s)B_g]} + \frac{(W_e - W_p)}{N_p[B_t + (R_p - R_s)B_g]} + \frac{N \Delta p B_{ti}(c_f + c_w S_{wc})/(1 - S_{wc})}{N_p[B_t + (R_p - R_s)B_g]} \qquad (7.44)$$

The first term of Eq. 7.44 is then the fraction of the production that results from the expansion of the original oil and the gas in solution—the solution-gas drive—which we refer to as the solution-gas-drive index (SGDI):

$$\text{SGDI} = \frac{N(B_t - B_{ti})}{N_p[B_t + (R_p - R_s)B_g]} \qquad (7.45)$$

The second term in Eq. 7.44 is the fraction of the total hydrocarbon production resulting from the original free-gas expansion, or the gas-cap expansion, which we refer to as the gas-cap-drive index (GCDI):

$$GCDI = \frac{NmB_{ti}(B_g - B_{gi})/B_{gi}}{N_p[B_t + (R_p - R_s)\,B_g]} \tag{7.46}$$

The third term in Eq. 7.44 is the fraction of the total hydrocarbon production resulting from the water encroachment, or the water-drive index (WDI):

$$WDI = \frac{(W_e - W_p)}{N_p[B_t + (R_p - R_s)\,B_g]} \tag{7.47}$$

The last term in Eq. 7.44 is the fraction of production resulting from the change in the pore volume. This term is normally insignificant when the reservoir pressure is less than the bubble-point pressure. When the reservoir pressure is greater than the saturation pressure, this term is very significant. However, it is uncommon to calculate drive indices for reservoirs above the saturation pressure. Thus a pore-volume contraction index is not normally used.

When an oil reservoir has a water drive or a gas cap, the drive indices may be useful in determining how effective the water or gas-cap drive is. Then if it is found that most of the production results from solution-gas drive, the water or gas-cap drive can be supplemented by water or gas injection.

Check understanding of the reservoir drive indices by working problem 7.12. The solution is shown in appendix C.

PROBLEM 7.12: Calculating Drive Indices

Determine the drive indices after 500 days of production for problem 7.11.

It should be noted that drive indices are somewhat less than ideal in providing a quantitative evaluation of the prospects of pressure maintenance. One problem is that the drive indices calculated as indicated are not constant, and thus, the importance of the various drives is not clear.

Injection of gas or water into the reservoir without adversely affecting the amount of produced gas or water is generally most readily accomplished by injecting the produced gas into the original reservoir gas cap and injecting the produced water into the original water-bearing portion of the reservoir. Such procedures are referred to as pressure maintenance since the decline in reservoir pressure with time is reduced or eliminated.

The cost of injecting produced water into the original water-bearing portion of the reservoir is generally relatively small, especially when we consider that the operator must dispose of the produced water in some way. Due to the cost of gas compression and the price of natural gas, a more careful look must be taken at the economic effects of injecting produced gas back into the gas cap. Some operators are using nitrogen to maintain reservoir pressures so the natural gas can be sold. When considering this alternative, remember that we are faced with the cost of separating nitrogen from air for injection and we normally must separate the nitrogen from the produced gas before it can be sold. Most pipelines do not accept natural gas that contains more than a small percentage of nitrogen.

When a gas-cap or water drive does not exist in a reservoir, pressure maintenance as such is generally difficult or impossible to accomplish unless a considerable amount of *closure* exists in the reservoir. The closure of a reservoir can be defined as the difference between the minimum subsea depth in the reservoir body (the top of the formation) and the maximum subsea depth in the reservoir. When the closure is sufficient, it may be possible to use the difference in the densities of the water and oil or gas and oil to keep the phases separated so prohibitively high produced gas-oil or water-oil ratios do not accompany the injection of gas or water into the reservoir. When closure is small, the engineer may find that attempts at pressure maintenance result in conventional secondary recovery by frontal displacement with the accompanying insufficiency of pattern sweep and more adverse effective mobility ratios. Mobility ratio effects in conventional pressure maintenance are minimized by the advantageous gravity, or density difference, effects.

The engineer should give careful consideration to using miscible material in any reservoir where a strong water drive or gas-cap drive exists. In many instances it has been possible and profitable to inject a material that is miscible with both the oil and the formation water between the immiscible oil and water and, thus, greatly improve the displacement efficiency of the system. The same thing has been done in gas-drive reservoirs where a fluid miscible to both the oil and gas has been injected to separate the immiscible oil and gas phases and, thus, increase the displacement efficiency.

In both cases it is necessary to choose a miscible fluid that has a gravity intermediate to the two reservoir fluids being separated—water and oil or oil and gas. Thus, in very permeable reservoirs where gravity forces are large compared to viscous flow forces, the injected miscible material flows by gravity to a position between the two immiscible reservoir fluids. Oil displacement is caused by the injected miscible fluid, and the injected fluid is displaced by the water or gas. The efficiency of a miscible displacement is very high. However, if the formation has a low permeability, the injected miscible material does not form a layer between the two immiscible phases quickly enough. The system is then much less effective. The engineer should also take care in determining the amount of miscible fluid necessary to maintain the miscibility of the system, a process that is still largely controversial. The problem is discussed in chapter 10.

The engineer may encounter the term *maximum efficient rate*. Theoretically, this refers to the maximum rate at which a reservoir can be produced without adversely affecting the ultimate recovery of the reservoir. Many reasons may be cited for the existence of a maximum efficient rate for a particular reservoir. There may be critical rates at which coning or fingering takes place, at which the produced gas-oil ratio in a solution-gas-drive reservoir is adversely affected, or at which paraffin deposition occurs in the reservoir. However, the last two examples are actually maximum efficient well rates rather than reservoir rates.

The term maximum efficient rate, or MER, seems to have originated from a different reservoir phenomena. It can be shown in the laboratory that the residual oil saturation that exists after displacement of the hydrocarbons in a reservoir by water is less if the reservoir contains a free-gas saturation prior to the displacement. The theory is that the free gas takes the place of some of the residual oil. Thus, the displacement results in a larger overall recovery.

For this phenomena to work in the reservoir as it does in the laboratory, the gas saturation at the moving water-oil contact must be maintained at or just below the equilibrium gas saturation. It would be very difficult to maintain a gas saturation at the advancing water-oil contact that is significant.

When the gas saturation at the advancing water-oil contact exceeds the equilibrium gas saturation, an oil bank forms and reduces the gas saturation in front of the advancing oil bank to the equilibrium gas saturation. However, as the oil bank moves past this immobile gas, the reservoir pressure at this point increases because the water bank comes closer and the mobile phase is now entirely liquid instead of part mobile gas. As the pressure increases, the equilibrium gas tends to go back into solution in the oil. By the time the water bank reaches this point, the equilibrium gas probably has disappeared.

This conjecture is made by the author. Nevertheless, it does appear unlikely that it is possible from a practical viewpoint to reduce the residual oil in a water bank by increasing the gas saturation in the water bank.

If we are to predict the performance of an oil or gas reservoir accurately, we must have a thorough understanding of the fundamentals of reservoir material balance. This is true whether we use conventional tank-type methods of analysis or a reservoir computer model. Thus, this chapter is one of the most important; it presents the fundamentals of material balance and many of its important applications.

Additional Problems

7.13 A reservoir has thicknesses as indicated in the field map in Fig. 7–14. The grid on the map represents 40-acre squares. The gas composition is methane, 78%; ethane, 15%; normal butane, 5%; and carbon dioxide, 2%. The porosity is 21% and connate water is 19%. Determine the acre-feet of bulk volume of the reservoir by drawing an isopach map, estimating the average thickness of each square or portion of a square, estimating the area of the square portions, and calculating the sum of the area—average thickness products. If the initial pressure is 3,150 psia, how many thousand cubic feet of gas will be produced when the reservoir pressure has declined to 2,900 psia. The reservoir temperature is 130°F. (Note that the reservoir volume is normally based on planimetered areas within the individual isopach contours.[13])

7.14 Assume the gas reservoir in problem 7.13 is attached to an oil reservoir to form a gas-cap reservoir. If production from the oil-bearing portion of the reservoir causes the gas-cap pressure to decline to 2,900 psia with no gas-cap production, how many reservoir barrels of gas will encroach into the original oil zone? How many standard cubic feet of gas does this represent?

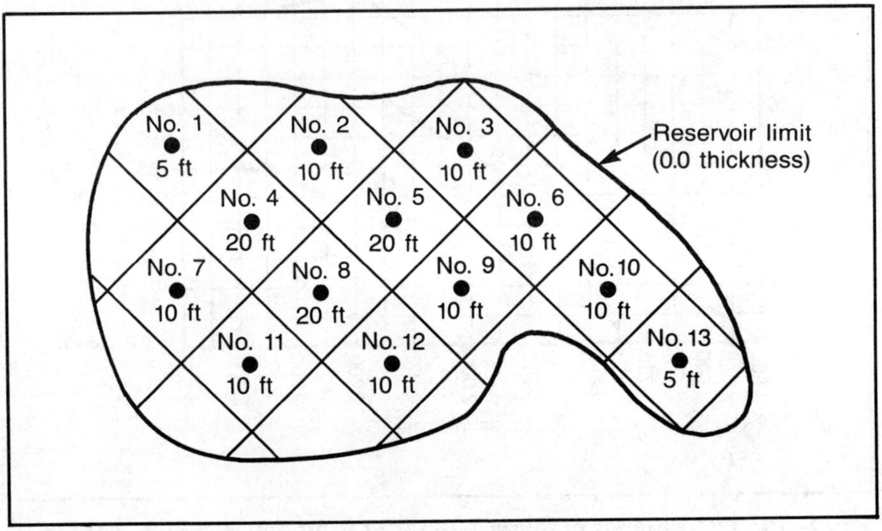

Fig. 7–14 Field map for problem 7.13

7.15 To demonstrate the effect of the cumulative produced gas-oil ratio on the oil recovery, rework problem 7.5 using a cumulative gas-oil ratio of 800 scf/stb.

7.16 Rework problem 7.6 using the actual production as 80,000 stb of oil and 48 MMscf of gas. Explain the results.

7.17 To illustrate the error resulting from an unknown water drive, assume that the reservoir in problem 7.7 has a water drive that results in water encroachment of 30,000 bbl by the time the reservoir pressure reaches 2,900 psia. What is the error in the material-balance calculated stock-tank barrels of oil originally in place if the answer to problem 7.7 is 4 MMstb?

7.18 Given the PVT data in Fig. 7–15 and a cell volume of 500 cc at 2,500 psia and 150°F, find the cubic centimeters of oil and gas in the cell at pressures of 2,350, 2,200, 1,900, and 1,500 psia and a cell temperature of 150°F. Also, find the oil and gas volumes at a cell pressure of atmospheric and a temperature of 60°F.

7.19 Use the empirical data in appendix B to determine the oil formation volume factor, gas in solution, oil viscosity, and gas viscosity at pressures of 2,000, 1,500, 1,000, and 500 psia. The following data apply:

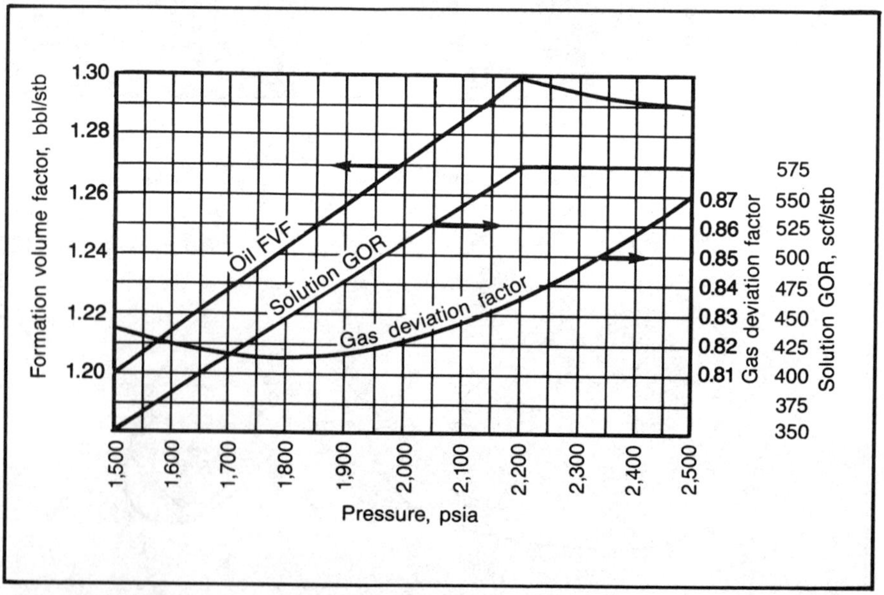

Fig. 7–15 PVT data for problem 7.18 (after Craft and Hawkins, *Applied Petroleum Reservoir Engineering*, courtesy Prentice-Hall, 1959)

Specific gravity of gas = 0.65
Stock-tank oil gravity = 40°API
Reservoir temperature = 195°F
Saturation pressure = 2,000 psia

7.20 Given the following data on a solution-gas-drive reservoir, find the stock-tank barrels of original oil in place.

Pressure, psia	*Cumulative Production, stb*	*Oil FVF*	*Solution Gas, scf/stb*	*Cumulative GOR, scf/stb*	*Gas FVF, bbl/scf*
3,112	0	1.42	885	—	0.00078
1,725	60,000	1.45	885	885	0.00141
1,400	242,000	1.40	772	975	0.00174

Bubble-point pressure = 1,725 psia
Reservoir temperature = 125°F
Average porosity = 7.7%
Average connate water = 20%

7.21 The following additional production and gas-injection data are available for the reservoir in problem 7.4:[14]

Cumulative Oil Production, N_P, MM stb	*Producing Gas-Oil Ratio, R, scf/stb*	*Cumulative Gas Injected, MMscf*
8	1,640	1,440
9	1,700	2,104
10	1,640	2,743

A. Calculate the average producing GOR during the production interval from 6–8 MMstb.
B. What is the cumulative produced GOR when 10 MMstb have been produced?
C. Calculate the net average producing GOR during the production interval from 6–8 MMstb.
D. Calculate the net cumulative produced GOR when 10 MMstb have been produced.

7.22 In problem 7.9 the average oil saturation and the gas-oil relative permeability are calculated for a reservoir pressure of 1,302 psia. Repeat this calculation for the reservoir pressure of 1,498 psia when the cumulative oil production is 948 Mstb and the cumulative gas production is 726 MMscf. At this time the producing gas-oil ratio in the reservoir is 1,360 scf/stb and PVT data are as indicated in problem 7.9. Compare the calculated values with those in Fig. 7–6.

7.23 The following data apply to the solution-gas-drive reservoir in problem 7.9:

$N = 10.025$ MMstb
$S_{wc} = 22\%$
$P_i = 3{,}013$ psia
$p_s = 2{,}496$ psia

P, psia	*B_o*	*R_s, scf/stb*	*B_g, Res bbl/scf*	$\frac{\mu_o}{\mu_g}$
3,013	1.315	650	0.000726	53.9
1,100	1.215	412	0.001998	115.2
900	1.195	369	0.002626	129.96

Gas-oil relative permeability data are as shown in the field curve in Fig. 7–6. When the reservoir pressure has declined to 1,100 psia, $N_p = 1.370$ MMstb, $G_p = 1{,}608$ MMscf, and the producing GOR = 3,019 scf/stb. Use the Schilthuis method to determine the cumulative oil and gas production and the producing gas-oil ratio when the reservoir pressure reaches 900 psia. The cumulative gas production is in the area of 2,200 MMcf.

7.24 Appendix C shows the solution to problem 7.9, which calculates the reservoir behavior to a pressure of 1,100 psia. In problem 7.23 we calculate the cumulative oil and gas production and the producing gas-oil ratio when the reservoir pressure has declined to 900 psia. Use the Tarner method and the results of problems 7.9 and 7.23 to calculate N_p, G_p, and R when the reservoir pressure has declined to 700 psia. The following reservoir characteristics are applicable when the reservoir pressure is 700 psia:

$B_o = 1.172$
R_s, scf/stb = 320
B_g, res bbl/scf = 0.003481
$\mu_o/\mu_g = 148.89$

7.25 Repeat the calculation of N_p, G_p, and R at 700 psia in problem 7.24 using the Muskat method. Make a plot of the R_s, B_o, and B_g data that are given in problems 7.9, 7.23, and 7.24 versus the pressure in order to apply the Muskat equation. We should not expect exact agreement since we are applying a finite-difference equation to a too-large pressure drop (200 psia).

7.26 Extend problem 7.10 by calculating the water encroachment after 300 days of production or until the saturation pressure is reached.

7.27 Extend problem 7.11 by calculating the cumulative oil and gas production after a total of 700 days of production.

7.28 Determine the drive indices after 700 days of production from the reservoir in problems 7.11 and 7.27. Compare these indices with those calculated for problem 7.12.

Notes

1. H.N. Hall, "Compressibility of Reservoir Rocks," *Trans.*, AIME (1953).
2. W. van der Knapp, "Non-Linear Behavior of Elastic Porous Media," *Trans.*, AIME (1959) volume 216.
3. van der Knapp, 1959.
4. J.W. Amyx, D.M. Bass Jr., and R.L. Whiting, *Petroleum Reservoir Engineering* (New York: McGraw-Hill, 1960), pp. 91–93.
5. Amyx, Bass, and Whiting, 1960.
6. D.L. Katz et al., *Handbook of Natural Gas Engineering* (New York: McGraw-Hill, 1959).
7. Ralph J. Schilthuis, "Active Oil and Reservoir Energy," *Trans.*, AIME (1936), p. 33.
8. J. Tarner, "How Different Size Gas Caps and Pressure Maintenance Programs Affect Amount of Recoverable Oil," *Oil Weekly* (June 12, 1944).
9. M. Muskat, "The Production Histories of Oil Producing Gas-Drive Reservoirs," *Journal of Applied Physics* (1945).
10. B.C. Craft and M.F. Hawkins, *Applied Petroleum Reservoir Engineering*, (Englewood Cliffs: Prentice-Hall, 1959).
11. Stewart, Callaway, and Gladfelter, "Comparison of Methods for Analyzing a Water Drive Field, Torchlight Tensleep Reservoir, Wyoming," *Trans.*, AIME (1955), p. 197.
12. A.F. van Everdingen and W. Hurst, "The Application of the Laplace Transformation of Flow Problems in Reservoir," *Trans.*, AIME (1949).
13. Craft and Hawkins, 1959.
14. Craft and Hawkins, 1959.

8

Decline-Curve Analysis

Many years ago it was discovered that a plot of the rate of oil production versus time for many wells could be extrapolated into the future to provide an estimate of the future rates of production for a well.[1] With the future rates known, it was of course possible to determine the future total production or reserves of the well. This represented the beginning of the art that has since become more of a science known as *decline-curve analysis.*

Decline-curve analysis may be one of the most misused reservoir engineering techniques and at the same time it appears to be one of the most neglected reservoir engineering techniques.[2] Decline-curve analysis can only be used as long as the mechanical conditions and reservoir drainage remain constant in a well and the well is produced at capacity. These limitations lead to much misuse. However, by combining several approximations Fetkovich provides examples of decline-curve analysis that include a change in the well pressure and a change in drainage.[3]

On the other hand the more theoretically inclined petroleum engineer may not appreciate decline-curve analysis and may fail to augment theoretical predictions. The science of reservoir engineering is such that the engineer should use every technique available in arriving at predictions for a reservoir. Just as it is foolish for a log analyst to ignore the core data available for a well when determining the characteristic porosity, thickness, and saturations, it is likewise foolish for the reservoir engineer to make reservoir predictions without using the results of decline-curve analysis.

Typical decline-curve analysis consists of plotting well production versus time on semilog paper and attempting to fit these data with a straight line, which is then extrapolated into the future. The reserves are calculated based on some average production rate per year for the extrapolated production rates. The typical engineer probably realizes that a hyperbolic decline-curve analysis gives a more accurate prediction. However, the complexity of the hyperbolic analysis techniques and the excuse that the late-life differences between a constant percentage—a straight line on semilog paper—and hyperbolic decline analysis often affect the present worth value very little leads the engineer to conclude that he should use the easiest method, the *constant-percentage decline*. Equations, rather than reading points, can be used more easily and more accurately in determining reserves, and hyperbolic decline-curve analysis is little more difficult than constant-percentage decline once the engineer is familiar with the methods.*

Decline-Rate Definition

The various methods of decline-curve analysis are based on the manner in which the rate of decline varies with time, rate, etc.[4,5,6] Consequently, it is important that we carefully define the decline rate. When the production rate is plotted versus time, we observe that the rate declines with time as indicated in Fig. 8–1. The decline rate can be defined as the fractional change in the rate with time:

$$a = -\,(\Delta q/q)/\Delta t \qquad (8.1)$$

The graphical interpretation of this definition is illustrated in Fig. 8–1. Consequently, the decline rate at any particular time can be determined graphically by determining the slope of the plot of rate versus time at the time of interest and dividing the slope by the rate at that particular time. For the decline rate, a, to be constant, the slope must decline at the same rate that the rate, q, declines.

The mathematician immediately recognizes that the expression of Eq. 8.1 is equal to the change in the natural log of q ($\Delta\ \ell n\ q$) with respect to the time:

$$a = -\,\Delta\ \ell n\ q/\Delta t \qquad (8.2)$$

Thus, the slope from a plot of the natural log of the rate versus time is the decline rate, a. However, as noted, natural log paper is not in common use so it is convenient to state the decline rate as a function of the slope of a plot of the log to the base 10 of the rate (log q) versus time. Since the natural log is equal to 2.3 times the log to the base 10, Eq. 8.2 becomes:

$$a = -\,2.3\ \Delta \log q/\Delta t \qquad (8.3)$$

The graphical interpretation of Eq. 8.3 is illustrated in Fig. 8–2. It is also convenient to recognize that the slope of a semilog plot can be

*No effort is made to justify theoretically the hyperbolic or constant-percentage decline of a well's productivity. Those readers who find this subject of interest are referred to the Brons and Fetkovich works referenced in this chapter (see references 2 and 3).

obtained by determining the change in the linear scale that occurs over a one-cycle change in the log scale, e.g., 1,000 to 100 and 0.1 to 0.01. This is true because the change in the log that occurs over a one-cycle change in the rate is equal to one. Consequently, the slope of a plot of the natural log of the rate versus time (Eq. 8.2) can be obtained by determining the change in time per cycle from a log to base 10 plot and dividing it into 2.3:

$$a = 2.3/(\Delta t/\text{cycle}) \tag{8.4}$$

Note that, in all of these equations for the decline rate, the units are the reciprocal of time. Physically, it may be helpful to consider the units to be a fractional change in the rate per unit of time. Also, note that the rate units used are stock-tank barrels per day. This is the only chapter in which we use the symbol q in stock-tank barrels without some appropriate subscript.

Some confusion exists in the industry in the evaluation of the decline rate. Some engineers evaluate the decline rate using the straight line representing the decline rate on the semilog plot and reading the rate of this straight line where it crosses the bottom of a cycle (e.g., at 1.0, 10.0 or 100.0) and one year later. The fractional decline rate is then taken to be the rate at the bottom of the cycle minus the rate one year later, divided by the rate at the bottom of the cycle. This is

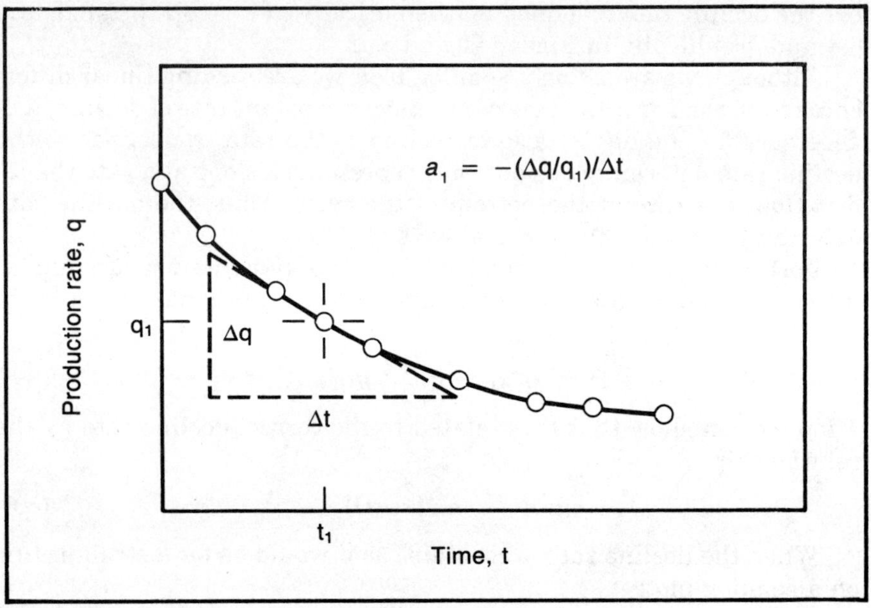

Fig. 8–1 Rate of decline definition—linear plot

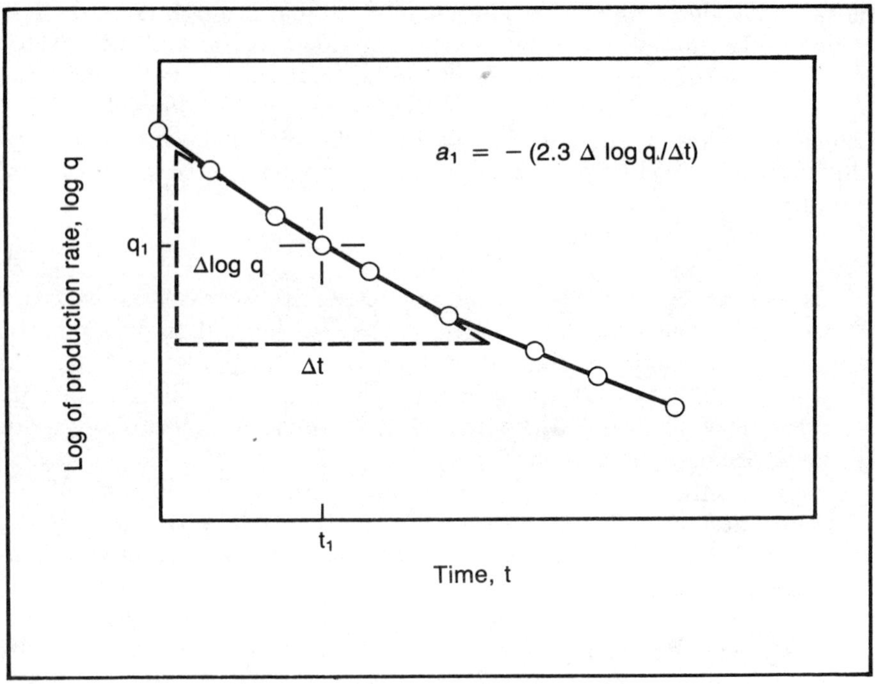

Fig. 8–2 Rate of decline definition—semilog plot

not the decline rate a defined mathematically in Eqs. 8.1, 8.2, 8.3, and 8.4 and graphically in Figs. 8–1 and 8–2.

Although we are using a semilog plot, we are reading linear differences from the log scale. In order to have a constant rate of decline, it is necessary for the $\Delta q/\Delta t$ ratio to decline as the rate, q, declines so the decline rate determined as outlined represents a slope at a rate that is less than the rate at the bottom of the cycle. Thus, the decline rate determined in this manner is always too small.

Fortunately, the error involved in this approximation is minimal for small decline rates. We can define the conventional decline rate as:

$$D = (q_{t=0} - q_{t=1.0})/q_{t=0} \tag{8.5}$$

Then we can show that it is related to the correct decline rate by the relationship:

$$a = -\ \ell n\ (1 - D) \tag{8.6}$$

When the decline rate is constant, as it would be for a straight line on a semilog plot:

$$\frac{q_i}{q} = e^{at} \tag{8.7}$$

Then using q_1/q for $q_{t=0}/q_{t=1}$ and $t = 1.0$ in Eq. 8.7, we can show that Eq. 8.6 is correct.

Eq. 8.6 shows that, when the conventional decline rate is 6%/year, an error of about 3% of the 6% is introduced by assuming $D = a$. When D is 12%, the error is 6.5%; when D is 20%, the error is 11%. To be safe, the engineer should avoid the conventional decline rate unless it is converted to a through Eq. 8.6.

Constant-Percentage Decline

Constant-percentage decline is also known as *exponential decline* since the mathematical expression that defines this type of decline is an exponential equation.[7,8] Constant-percentage decline appears to be more widely used than the other mathematical forms, even though engineers generally agree that hyperbolic decline more nearly describes the decline characteristics of most wells and it can be shown that there is some theoretical basis for such well behavior.[9,10] The simplicity of the constant-percentage decline technology makes it attractive to the engineer. Also, when decline rates are small, the accuracy added using the hyperbolic decline may not be significant for many purposes.

The constant-percentage decline rate equation. As the name implies, the constant-percentage decline is based on the assumption that the decline rate, a, does not change with time. We can then use the basic definition of the decline rate, a, as in Eq. 8.1 to derive the mathematical expression for constant-percentage decline. We can rearrange Eq. 8.1 as:

$$a\Delta t = -(\Delta q/q) \tag{8.8}$$

In this form it can be applied to very small increments of time. If it is then applied to all of the time increments from one rate, q_1, when the time is taken as zero to another rate, q, corresponding to a time, t, and all of these equations are added together, we obtain:

$$a \sum_{o}^{t} \Delta t = -\sum_{q_i}^{q} (\Delta q/q) \tag{8.9}$$

$$a\,t = \ell n\, q_i - \ell n\, q \tag{8.10}$$

Eq. 8.10 shows that a plot of the natural log of the rate, $\ell n\, q$, versus the time, t, gives a straight line. Differentiation of this equation yields Eq. 8.2 and shows that the slope of $\ell n\, q$ versus t is the decline rate, a. Eq. 8.10 can also be stated in log to base 10 values:

$$a\,t = 2.3 \log q_i - 2.3 \log q \tag{8.11}$$

When this expression is differentiated with respect to time, Eq. 8.3 is obtained. Thus, a plot of log q versus t gives a straight line whose slope is $-a/2.3$.

Conversely, we recognize that a well whose plot of the log of rate versus time yields a straight line is experiencing constant-percentage decline. Rates at any future time can be calculated from Eq. 8.10 or Eq. 8.11, or a time can be calculated that corresponds to the time necessary for the rate to decline to some particular value, for example, the economic limit rate for the well.

Many engineers prefer to use the exponential form of Eq. 8.10 for calculations:

$$a\,t = \ell n\,(q_i/q) \tag{8.12}$$

$$(q_i/q) = e^{at} \tag{8.13}$$

$$q = q_i\,e^{-at} \tag{8.14}$$

Eq. 8.14, Eq. 8.6, or Eq. 8.8 may be used to calculate future production rates or times. A decline curve generally does not immediately follow a constant-percentage decline. A plot of rate versus time on semilog paper does not generally approach a straight line immediately. However, it continues to curve at a lesser rate as the production life continues, until it is possible to approximate the plot with a straight line. An example of such a decline curve after the data have been smoothed is shown in Fig. 8–3.

Decline curves are generally maintained on a month-to-month basis by calculating a daily average production rate for the month and plotting this rate or by simply plotting the month's production versus time so the rate scale is in barrels per month. Generally, no effort is made to adjust the time scale for the difference in the lengths of the months. Often, the barrels of oil per month are not adjusted for the difference in the number of days in each month.

Regardless of the method used to plot the raw data, this plot normally exhibits a wide fluctuation around the trend of the data. This fluctuation may be caused by a variation in such periods as downtime, weather difficulties, and pipeline runs. To make the data easier to interpret, the slope can generally be smoothed by calculating averages for periods of time and plotting the averages at the middle of the time increment used for averaging. It may be necessary to determine averages on the basis of 3-month, 6-month, or even 1-year periods in order to obtain the smoothest production decline trend.

To understand better the principles of constant-percentage decline, work problem 8.1 and compare the solution with the one in appendix C.

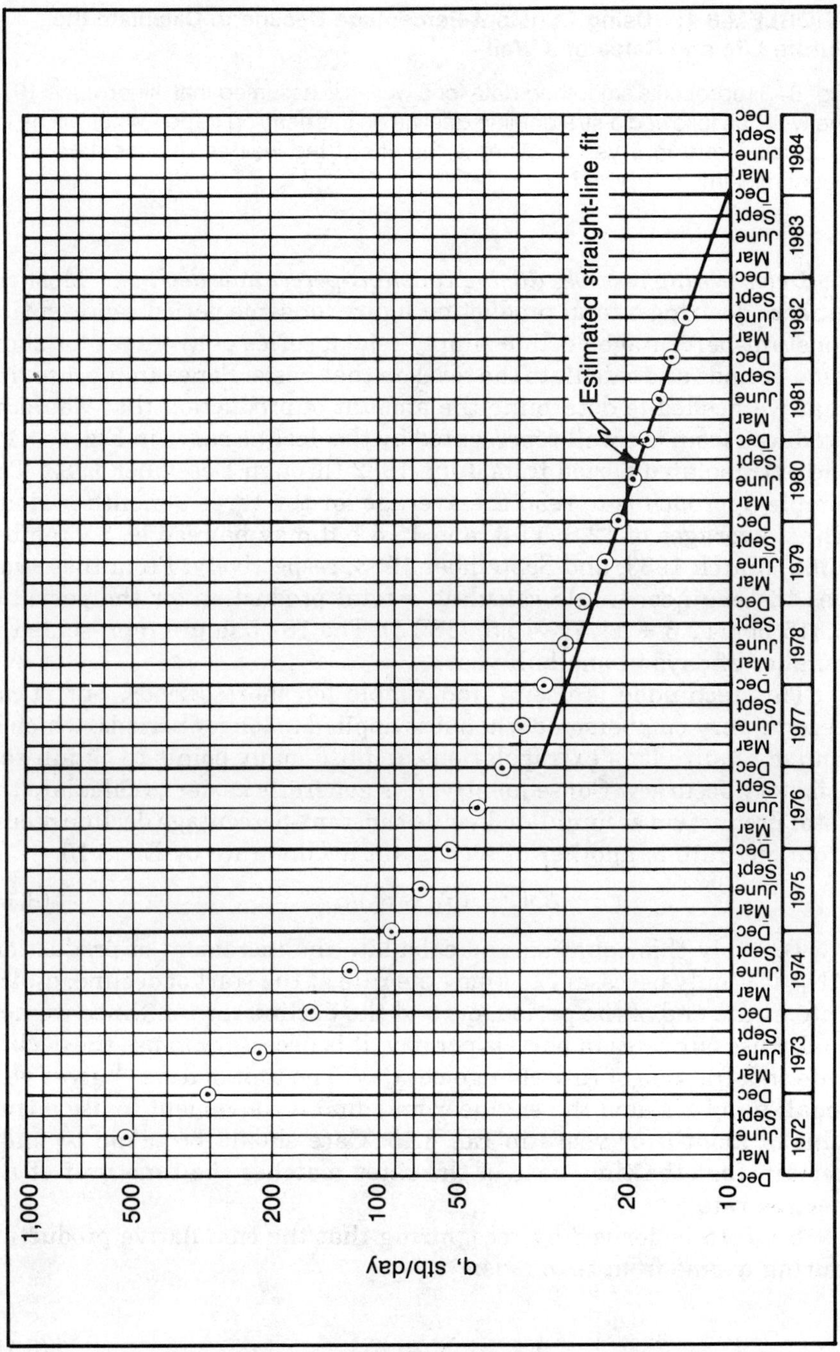

Fig. 8—3 Smoothed decline-curve data

PROBLEM 8.1: Using Constant-Percentage Decline to Calculate the Future Life and Rates of a Well

Fig. 8–3 represents smoothed data for a well. It is assumed that, since June 1979, the well has followed a straight-line decline on this semilog paper. What will be the producing rate in 5 years? What is the life of the well as of June 1982 if the economic limit is 1 bo/d?

Determining reserves during constant-percentage decline. Most engineers desiring a total production figure for some period representing constant-percentage decline simply read a series of averages for short time periods and calculate the total on that basis. For example, assume that we needed to determine the amount of production that would be produced from the well represented by the decline curve in Fig. 8–3 for the extrapolated period from June 1982 through December 1983. The simple approach is to read the average for the three 6-month periods. Thus, averages of 12.6, 11.5, and 10.5 b/d may be read at September 1982, March 1983, and September 1983, respectively. From these values the engineer would calculate a total production for the period of 6,315 bbl (12.6 + 11.5 + 10.5) (182.5). The 182.5 figure represents the number of days in one-half year.

This technique is useful and simple for short periods, but it can become very cumbersome when it is applied to longer periods, which of course require long extrapolations and use many points to obtain satisfactory accuracy. Consequently, it is generally easier to calculate the total production accumulated while constant-percentage decline occurs from one rate to another at a constant decline rate by Eq. 8.15:

$$\Delta N_p = (q_1 - q_2)/a \tag{8.15}$$

To apply this equation and calculate the increment of production, ΔN_p, it is only necessary to know the rate at the start of decline, q_1, the rate at the end of the period, q_2, and the decline rate, a. Since we normally use our rates in barrels per day, it is necessary to use the decline rate in a fraction of rate change per day. The unit of days^{-1} gives very small numbers, and the engineer may find it convenient to use a time unit of months or years in Eq. 8.15. Care should be taken to make certain that the time base of the rates matches the time unit of the decline rate.

Eq. 8.15 is derived by recognizing that the cumulative production during a time from t_1 to t_2 is:

$$\Delta N_p = \sum_{t_1}^{t_2} (q\,\Delta t) \tag{8.16}$$

By rearranging Eq. 8.1, we can derive an expression for $q\Delta t$ that can be substituted into Eq. 8.16 to obtain:

$$\Delta N_p = -\sum_{q_1}^{q_2} (\Delta q/a) \tag{8.17}$$

When the Δq terms are summed from q_1 to q_2, we obtain Eq. 8.15.

If we consider the initial rate, q_1, in Eq. 8.15 as a constant, the ΔN_p value is inversely proportional to the rate, q_2:

$$\Delta N_p = (q_1/a) - (q_2/a) \tag{8.18}$$

Consequently, a plot of N_p versus the rate, q_2, gives a straight line whose slope is $-(1/a)$, as long as the decline follows the constant-percentage decline. This type of plot can be very useful, especially in evaluating or understanding remedial workovers.

Fig. 8–4 shows a plot of the rate versus cumulative production for a well that is assumed to have entered constant-percentage decline prior to the performance of remedial work and following this work. In this case the reservoir drainage is unaffected by the workover because the primary mobile oil appears to have remained unchanged. The primary mobile oil is assumed to be all of the oil that would be produced if it were possible to produce the well to a rate of zero. For constant-percentage decline the mobile oil remaining in the reservoir at any time can be calculated using Eq. 8.15 with q_1 as the current rate and q_2 equal to zero:

$$\text{Remaining primary mobile oil} = q_1/a \tag{8.19}$$

Note that when mobile oil is defined in this way, the capacity of the wells in the reservoir does not affect the total mobile oil in the reservoir. The actual producible oil is influenced by the increased capacity of the wells because the time at which the economic limit is reached is pushed further into the future. However, if we assume the economic production rate is zero, this is no longer true. Thus, the mobile oil of a reservoir is fixed by the physical limits of the reservoir.

If there were only one well in a reservoir, its mobile oil would be independent of its producing capacity. However, with more than one well in a reservoir, the mobile oil of the individual well is a function of the well's producing capacity. Any increase in the mobile oil of one well results in a corresponding decrease in the mobile oil of one or more of the other wells in that reservoir. The adjustment should be minimal considering that the drainage in these adjusted volumes is very poor. Thus, for all practical purposes there is no significant change in the

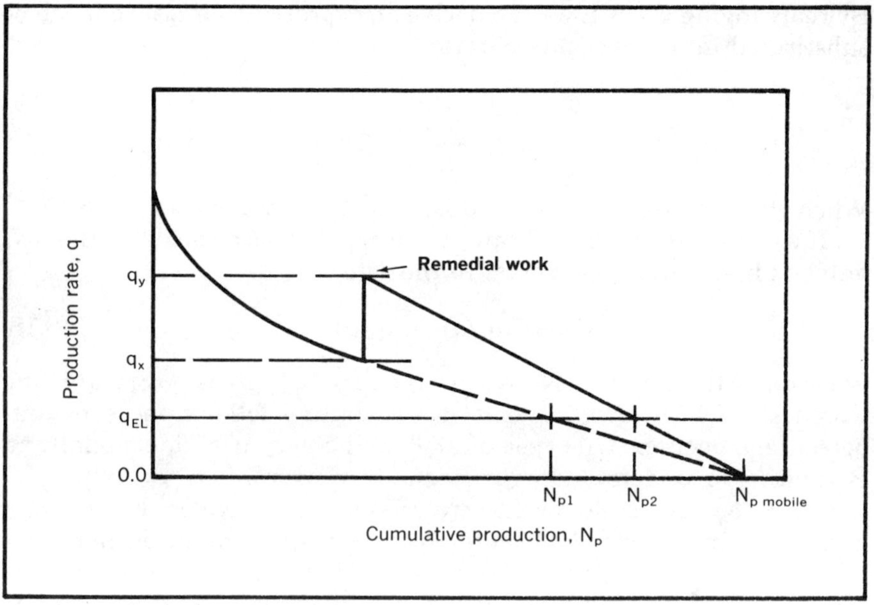

Fig. 8–4 Rate versus cumulative production for constant-percentage decline

mobile oil of a well whose production capacity has been changed by such methods as fracturing, and acidizing.

This discussion may not appear consistent with the general ideas of drainage. These ideas imply, for example, that one well in a 5,000-acre, 1.0-md reservoir would drain only some small portion of that reservoir, so many wells would be necessary to drain the reservoir. However, remember that, given enough time, one well would affect all of the 5,000 acres and some flow would take place from all portions of the 5,000 acres toward the single well. Drainage then is a relative term. When we say one well does not drain the 5,000 acres, we should say that one well does not effectively drain the 5,000 acres. Many more wells are necessary to drain the entire area effectively.

The remaining primary mobile oil in a reservoir is virtually unchanged by a workover that increases the producing rate from q_x to q_y. Therefore, the decline rate also is increased in the same ratio in order for the remaining primary mobile oil to be the same, according to Eq. 8.19. Conversely, if a well in constant-percentage decline before and after a workover has a coordinate plot of rate versus time that extrapolates to the same cumulative production for a rate of zero before and after the workover, we can conclude that the workover has not increased the drainage of the well. This generally seems to be the case for fracturing, acidizing, and similar methods of stimulating a well production rate. In such a case the increase in the ultimate production is the difference in the cumulative production that would be achieved at the economic limit, as indicated by the difference in N_{p1} and N_{p2} in Fig. 8–4.

This relatively small increase in the ultimate primary recovery that is indicated does not mean that the workover economics are unattractive. When corresponding lives are calculated using Eq. 8.14 and the appropriate rates and decline rates, the remaining life of the field is drastically shortened. Therefore, all of the operating expense difference as a result of the difference in the life is saved, in addition to the savings in the present-worth value of the production.

Work problem 8.2 and compare the solution with the one in appendix C.

PROBLEM 8.2: Using a Plot of Rate versus Cumulative Production during Constant-Percentage Decline

In problem 8.1 a straight-line extrapolation of Fig. 8–3 is used to calculate the remaining life as of June 1982. Now assume that in June 1982 this well is fractured and the producing rate is increased to 53 b/d without increasing the drainage volume of the well. Assume that the well again entered constant-percentage decline. What is the remaining primary mobile oil in June 1982? What will be the increase in the ultimate primary production and the change in the well life if the economic limit is 1.0 b/d? Note that it is not necessary to prepare a rate versus cumulative plot to solve this problem.

Hyperbolic Decline

An engineer who has been faced with the task of determining reserves in an area where solution-gas-drive reservoirs exist is aware of the shortcomings of the constant-percentage decline analysis. Such an engineer has undoubtedly tried to use a straight-line extrapolation of a semilog rate-time plot and has had to increase the reserves on these leases because the decline continued to flatten. If the engineer used hyperbolic methods to predict reserves from decline curves, he probably found that the best-known hyperbolic methods were too time consuming or too insensitive. The engineer probably reached a situation in which he simply extrapolated the decline curves using a favorite French curve in a manner that is more of an art than a science. The discussion of hyperbolic decline-curve analysis methods provides the subject engineer with methods that are relatively simple to use and much less arbitrary than the French-curve method.

The nature of the loss ratio, the conventional log-log plot, and the Gentry and Fetkovich methods of analyzing hyperbolic decline are reviewed with some of the difficulties associated with the practical application of these methods.

The hyperbolic decline equations. In working with the constant-percentage decline curve, we found the decline rate, a, to be constant. However, the decline rate in hyperbolic decline varies according to Eq. 8.20:

$$(a/a_i) = (q^n/q_i^n) \tag{8.20}$$

Where:

Constant, n = a number between, but not including, zero and 1.0

Note that when n is zero, q^n and q_i^n are 1.0 and the decline rate, a, is equal to a_i, which indicates that the decline rate is constant. Thus, when the decline rate, a, is constant, n is zero and the decline is the constant-percentage type.

When n is 1.0, a special type of decline known as *harmonic decline* prevails. In this case we can see that the decline rate, a, is proportional to the rate. In the more general case of hyperbolic decline, we see in Eq. 8.20 that the decline rate, a, is proportional to some fractional power of the rate.

To obtain the rate-time hyperbolic decline equation, we can first substitute for the decline rate, a, in Eq. 8.20 according to the definition of the decline rate as presented in Eq. 8.1:

$$-(\Delta q/q\Delta t)/a_i = (q^n/q_i^n) \tag{8.21}$$

Now, a_i and q_i can be treated as constants, and the variables can be separated and summed between limits:

$$a_i \sum_{0}^{t} \Delta t = -\, q_i^n \sum_{q_i}^{q} q^{-(n+1)}\, \Delta q \tag{8.22}$$

In this expression note that q_i is taken as the rate at the time when t is zero. Therefore, a_i is the decline rate at the time when t is zero. When the expression is integrated between limits, we obtain:

$$a_i\, t = -q_i^n \left[\frac{q^{-n}}{-n} - \frac{q_i^{-n}}{-n}\right] \tag{8.23}$$

This can be rearranged to obtain:

$$n\, a_i\, t = (q_i^n/q^n) - 1 \tag{8.24}$$

When Eq. 8.24 is solved for the rate, we obtain the hyperbolic rate-time decline-curve equation:

$$q = q_i/(1 + n a_i\, t)^{1/n} \tag{8.25}$$

In using Eq. 8.25 remember that a_i is the decline rate when the rate, q_i, prevails, which is the time equivalent of zero. Also, note that the time, t, is the time required for the rate to decline from q_i to q. The constant, n, is defined in the definition of hyperbolic decline in Eq. 8.20. Thus, in applying Eq. 8.25 to calculate the rate at some particular time or the time required to reach a particular rate, it is necessary to know one more parameter than is needed in applying the similar constant-percentage rate equation. The difficulty of determining the hyperbolic decline constant, n, from past rate-time data is the major difficulty of hyperbolic decline-curve analysis. Once this constant is determined, it is relatively simple to determine graphically the decline rate, a_i, corresponding to a q_i and calculate the rate, q, corresponding to any time, t. These same parameters can also be used to calculate the production accumulating during the time, t, when the production rate is declining from q_i to q.

The hyperbolic decline cumulative production equation is derived in the same manner that the constant-percentage decline cumulative production equation, Eq. 8.15, is derived. Eqs. 8.16 and 8.17 apply to any type of decline curve:

$$\Delta N_p = \sum_{t_1}^{t_2} q\Delta t \tag{8.16}$$

$$\Delta N_p = -\sum_{q_1}^{q_2} (\Delta q/a) \tag{8.17}$$

To apply Eq. 8.17 to hyperbolic decline, we substitute for the decline, a, according to Eq. 8.20:

$$\Delta N_p = \sum_{q_1}^{q_2} (-q_i^n/a_i)\, q^{-n}\, \Delta q \tag{8.26}$$

When this expression is integrated between q_1 and q_2, we obtain the hyperbolic cumulative production equation:

$$\Delta N_p = [q_i^n/a_i(1-n)][q_i^{(1-n)} - q_2^{(1-n)}] \tag{8.27}$$

It is convenient to note that for a particular hyperbolic decline curve, the term $[q_i^n/a_i(1-n)]$ is constant because, according to Eq. 8.20, q^n/a is constant. Thus, Eq. 8.27 can be written as a function of:

$$H = q_i^n/a_i\,(1-n) \tag{8.28}$$

$$\Delta N_p = H\,[q_1^{1-n} - q_2^{1-n}] \tag{8.29}$$

Thus, once the hyperbolic constant, n, has been determined and a decline rate, a_i, corresponding to any rate, q_i, has been determined, the cumulative production between any two rates, q_1 and q_2, can be calculated.

The curve-fitting hyperbolic decline-curve analysis method. It is believed that the simplest, most useful means of extrapolating hyperbolic decline performance and evaluating the future life and reserves is by comparing the actual decline-curve data with a series of hyperbolic semilog-type curves. These curves represent various combinations of n and a_i. The engineer should note that these curves are basically different from those devised to analyze transient pressure data. All of the transient pressure analyses use log-log data matches, whereas these decline-curve analysis-type curves use semilog data plots. Unique fits may still be difficult to obtain, but they present nothing like the difficulties encountered with a log-lot plot matching.

Once the engineer has determined which of the hyperbolic-type curves most closely fit the data, he has determined the n, a_i, and q_i that can be used to calculate future rates, reserves, etc., using Eqs. 8.25 and 8.27. Alternatively, he may be able to read the future rates directly from the type curves.

Examples of the hyperbolic type curves are shown in Figs. 8–5, 8–6, and 8–7. These curves have some peculiarities that should be noted. They are basically rate versus time curves for various combinations of n and a_i. In all cases q_i, the rate when the time, t, is zero, is 1.0. Thus, the Y axis for the rate-time curves can be considered to be the rate if a_i is 1.0, or it can be taken as the ratio of the rates, q/q_i. Since the rate is plotted on a log scale, the rate units do not affect the shape of the curve. Consequently, we can consider the hyperbolic-type rate curves to be plots of the rate stated in q_i units. Eq. 8.25 shows that the slope of the log of q plotted versus the time, t, is the same as the slope of a plot of the log of q/q_i plotted versus time.

Since an extremely wide range of values is associated with the various n and a_i combinations, it simplifies the presentation of the type curves to place the zero time of the type curve in the center of the plot rather than at the left-hand side of the plot as is customary. This feature makes it possible to obtain the best fit of the actual data for the latest times, which greatly enhances the accuracy of the graphical or mathematical extrapolations. The use of negative times does not mathematically interfere with the application of Eqs. 8.25 and 8.26.

It would be awkward to attempt to use Figs. 8–5, 8–6, and 8–7 in the scale presented for predicting hyperbolic decline behavior since normal graph paper is not this size. Also, a more extensive suite of curves should be used to obtain the greatest accuracy. Consequently, tabular data for plotting curves of n of 0.1, 0.2, 0.3, 0.4, 0.5, 0.6, 0.7, 0.8, and 0.9 are presented in appendix B, Tables B6 through B14. It is suggested that for the greatest utility the hyperbolic type curves be plotted on the particular size of semilog graph paper that is customarily used for decline-curve plots. The data plots on Codex No. 4272, twenty years

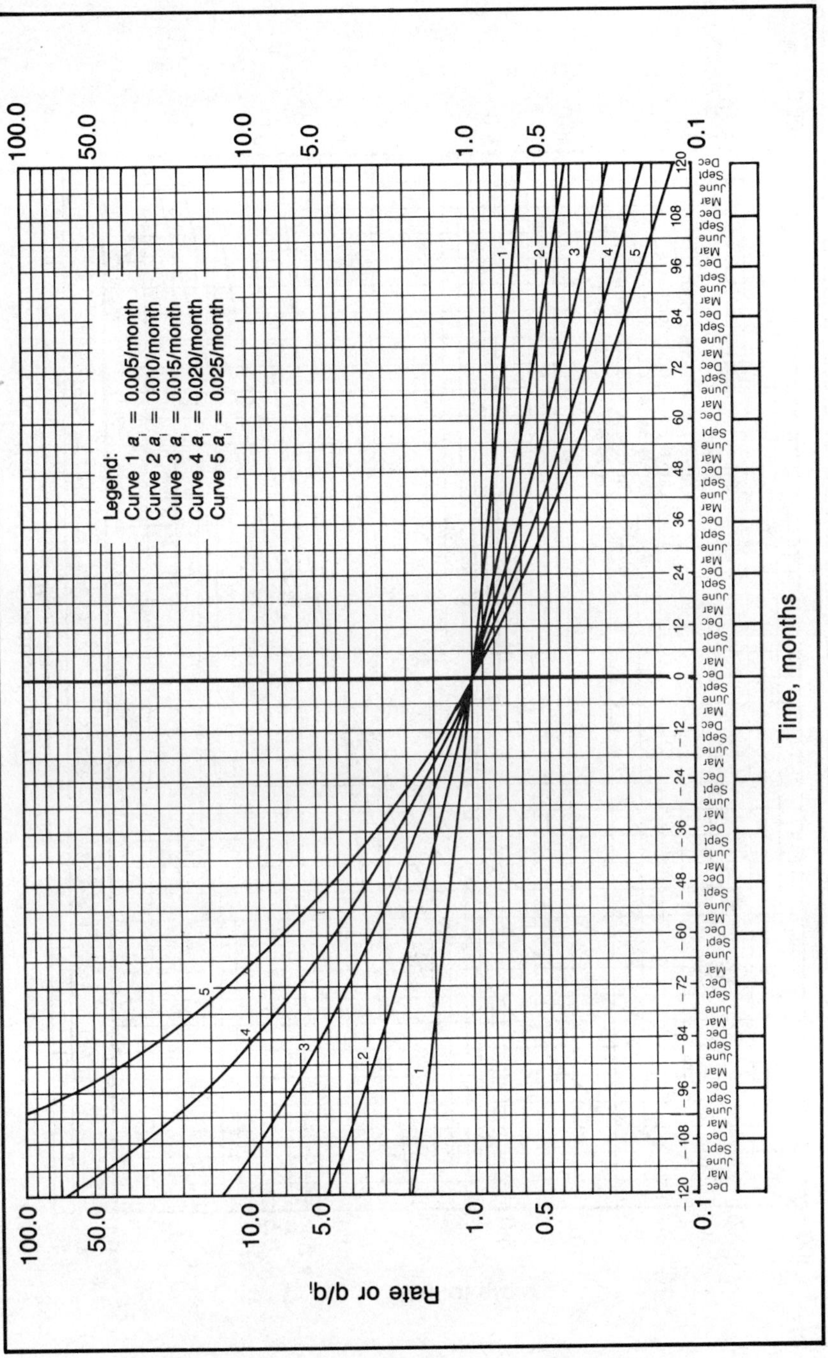

Fig. 8–5 Hyperbolic decline curves for n = 0.3

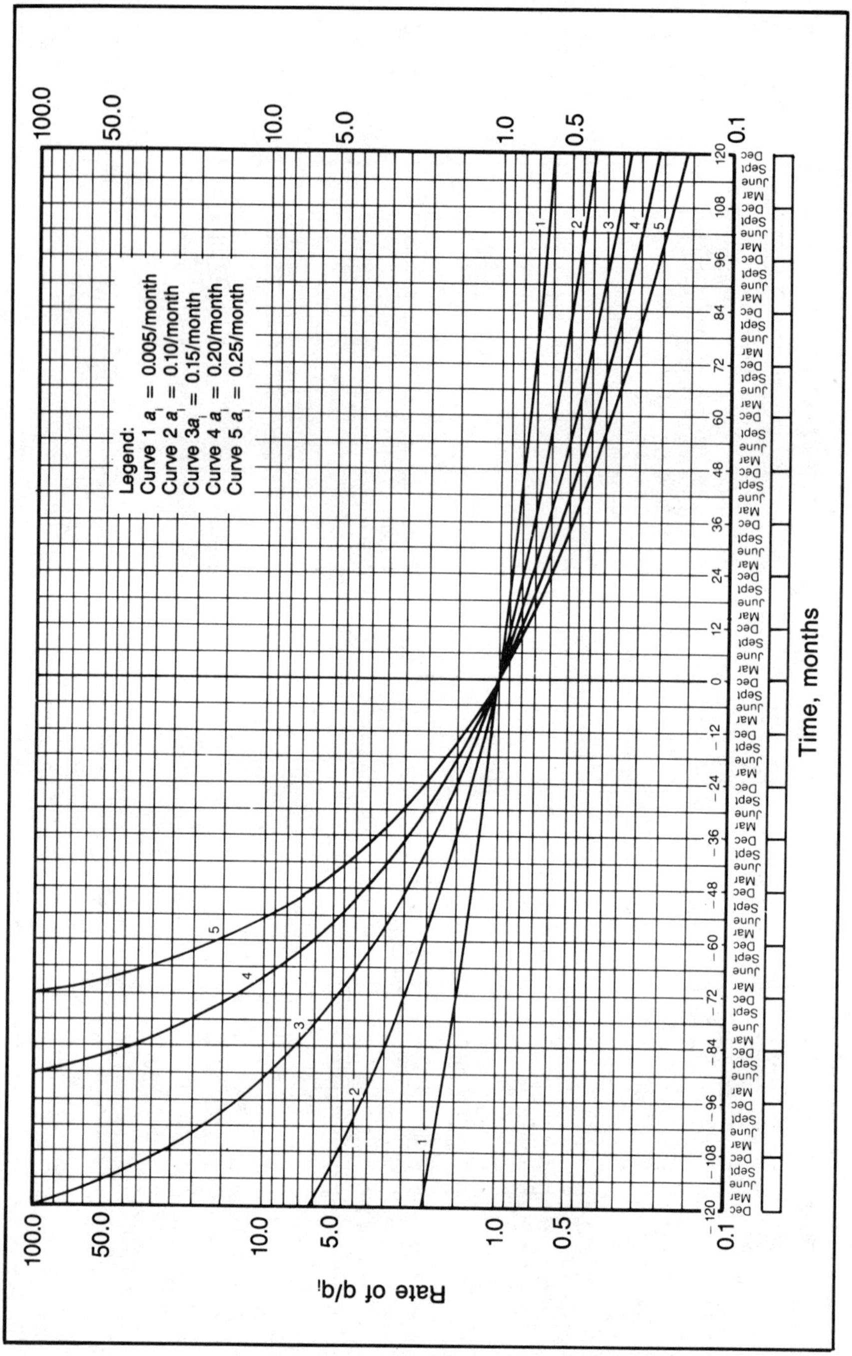

Fig. 8—6 Hyperbolic decline type curves for n = 0.5

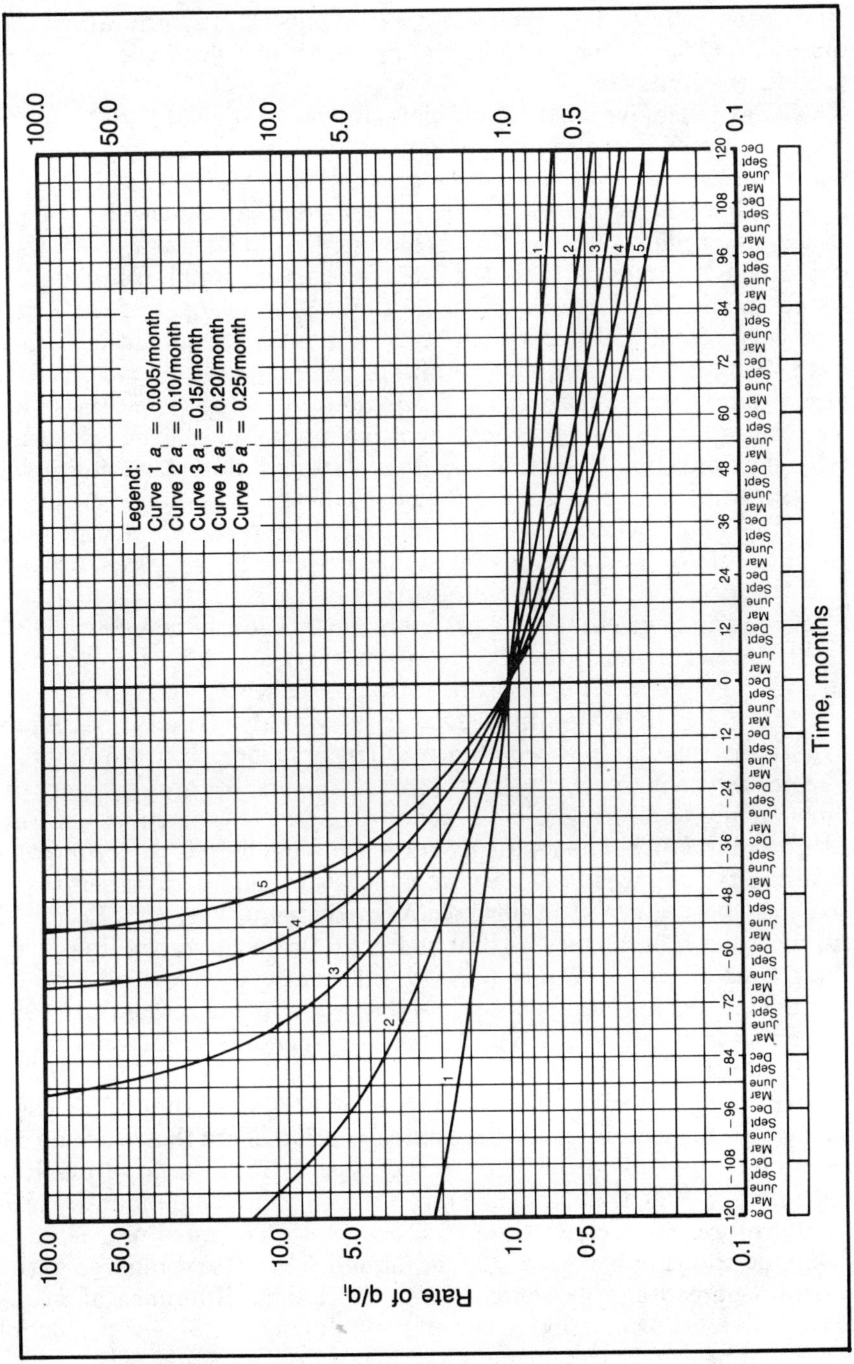

Fig. 8–7 Hyperbolic decline type curves for n = 0.7

by months × three 3-in. cycles (1 × 17-in. paper), can be obtained from the author. This appears to be the most common decline-curve paper used in the industry.

To use these hyperbolic data plots, it is necessary to plot the actual decline-curve data on semilog paper with the same linear-log cycle size and time scale as that of the type curves. Any rate units (e.g., barrels per day, barrels per month, or barrels per year) can be used for the plot of the actual data, but the time scale must be the same as that of the hyperbolic type curves. Once the plot of the actual data is completed, its shape should be compared with the shape of the family of hyperbolic type curves to determine which one best fits the actual data. To facilitate such a comparison, it is helpful to have the type curves on transparent paper. In comparing the actual data with the type curves, it is only necessary to keep the vertical or horizontal axis of the actual and type plots parallel. Once the best fit is obtained, the actual chronological time and producing rate (q_i) corresponding to the chart time of zero should be noted, as well as the a_i and n for the type curve that gives the best fit. With q_i, a_i, and n fixed, it is then possible to apply Eqs. 8.25 and 8.29 to calculate rates and cumulatives as desired.

An example calculation is probably most useful in determining the exact method of applying the type curves. In problem 8.1 the smoothed decline curve data of Fig. 8–3 are extrapolated with a straight line as a constant-percentage decline. However, if we analyze the data using the hyperbolic type curves, we obtain different, more accurate answers. The procedure is to plot the data on semitransparent graph paper with the same cycle size and time scale as that of the type curves of Figs. 8–5, 8–6, and 8–7. The resulting curve of actual data is then compared with the type curves to determine which curve fits best. When this is done, the best curve fit is obtained as indicated in Fig. 8–8. From this curve fit we should observe that the actual rate corresponding to the chart time of zero is 15.6 b/d or 474 bbl/month (15.6 × 30.4), and the chronological time corresponding to the chart time of zero is July 1981. The curve fit has an n of 0.5 and an a_i corresponding to the zero time of 0.015/month.

In problem 8.1 we evaluated the rate 5 years after June 1982, based on a constant-percentage decline extrapolation. Note that we can read a rate for this time directly from the hyperbolic curve fit. June 1982 represents a +11 time on the curve fit, so 5 years or 60 months later should represent a chart time of 71. At this chart time q/q_i is about 0.42, and the rate is 0.42 × 15.6 or 6.5 b/d. This rate compares with a constant-percentage extrapolation of 5.6 b/d, a difference of nearly 20%. The total production accumulating during the 5-year period following June 1982 can be calculated using Eqs. 8.28 and 8.29:

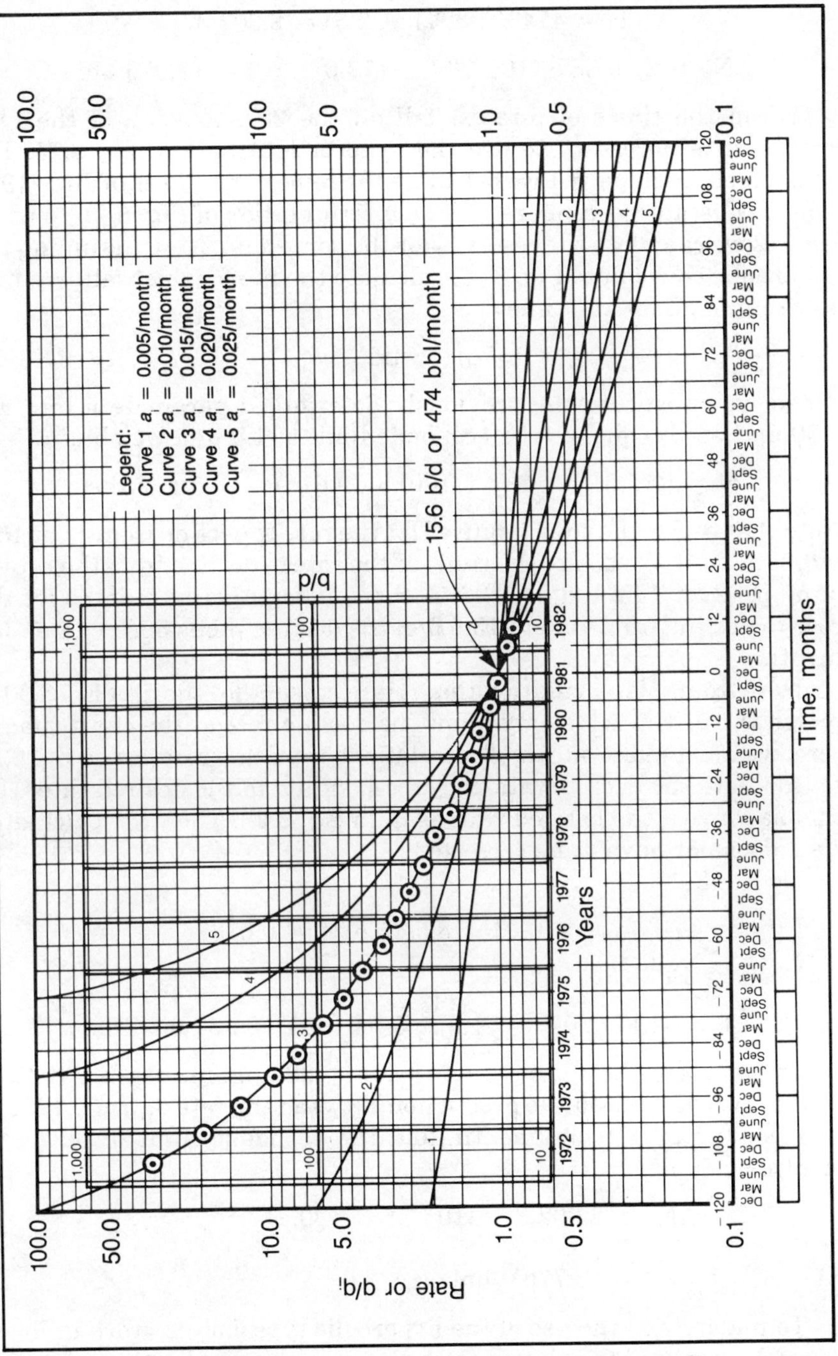

Fig. 8–8 Hyperbolic curve-fitting technique

$$H = 474^{0.5/0.015}(1 - 0.5) = 2{,}902.87$$

$$N_p = 2{,}902.87(410^{1-0.5} - 177.6^{1-0.5}) = 17{,}973 \text{ bbl}$$

When the times of interest fall off the 20-year span of the type curves, it is necessary to use the hyperbolic equation, Eq. 8.25. For example, in problem 8.1 we calculate the remaining life as of June 1982 from a constant-percentage decline interpretation of Fig. 8–3. In problem 8.2 we calculate the reserves for the same date. Now, assuming the hyperbolic curve fit of Fig. 8–8, calculate the life for this well. We first solve Eq. 8.25 for the time:

$$t = [(q_i/q)^n - 1.0]/n\, a_i \tag{8.30}$$

When we substitute the previously determined parameters into Eq. 8.30 and use q equal to the economic limit of 1.0 b/d, we obtain:

$$t = [(15.6/1.0)^{0.5} - 1.0]/(0.5)\,(0.015) = 395 \text{ months}$$

Note that this is the time required for the rate to decline from q_i, or 15.6 b/d, to 1.0 b/d. We desire the time for the rate to decline from the rate in June 1982, or 13.5 b/d, to 1.0 b/d. Since it requires 11 months for the rate to decline from 15.6 to 13.5 b/d, the desired life is 395 − 11 or 384 months.

Note from this calculation that it is necessary for the q_i of Eq. 8.30 to match the a_i. Since we only know the decline rate at the chart time of zero, we must make our time calculations from this rate or use Eq. 8.20 to calculate the decline rate, a, corresponding to the desired rate. Use this as a_i with the desired rate as q_i in Eq. 8.28. Then the calculated time does not have to be corrected.

By Eq. 8.20:

$$a_{13.5} = 0.015(13.5/15.6)^{0.5} = 0.01395$$

Then by Eq. 8.30:

$$t = [(13.5/1.0)^{0.5} - 1.0]/0.5(0.01395) = 384 \text{ months}$$

To calculate the reserves as of June 1982 again, use Eq. 8.29 to determine the cumulative production between the rate on June 1982 of 13.5 × 30.4 or 410 bbl/month, and the economic limit of 30.4 bbl/month:

$$\Delta N_p = 2{,}902.87\,[410^{1-0.5} - 30.4^{1-0.5}] \tag{8.29}$$

$$= 42{,}770.0 \text{ bbl}$$

To understand the use of the hyperbolic type curves, work problem 8.3 and compare the solution with the one in appendix C.

PROBLEM 8.3: Application of the Hyperbolic Decline Curves

The following production history of a well is given:

Month	*Production, stb*
1	2,580
2	2,100
3	2,090
4	1,780
5	1,860
6	1,470
7	1,510
8	1,250
9	1,330
10	1,220
11	1,090
12	1,150
13	982
14	940
15	883
16	850
17	713
18	700
19	743

Overlay Fig. 8–4, Fig. 8–5 or Fig. 8–6 with a sheet of transparent paper and plot a decline curve using these production data. It may be necessary to replot averages to smooth the data sufficiently for a good analysis. Find the remaining life and reserves if the economic limit is 100 bbl/month. Find the remaining life and reserves if the economic limit is 10 bbl/month.

Some additional guidance in the use of the hyperbolic type curves may prove helpful. For a particular hyperbolic constant, n, we generally can fit more than one curve. A figure representing a family of curves for one n is actually a plot of various segments of one mathematical curve. Consequently, for most plots the data curves overlap. In some cases we may be able to fit three different curves on the same figure. In such cases the life and reserves calculated are still the same. When there is a choice, we should always fit the curve that places the last data point nearest the chart time of zero. Since calculations are made using a match of q_i and a_i at a chart time of 1.0, extrapolations are made graphically or mathematically from this point.

When hyperbolic data plots are prepared from the appendix B data, remember that any reproduction of these plots must be by a method that gives exact-size copies. Otherwise, the basic graph size is different from the graph size used for the production data plot. Then curve fitting is inaccurate. Engineers familiar with the author's previous reservoir engineering book, *Practical Petroleum Reservoir Engineering Methods,* may recall that cumulative production-type curves were used.[11] It has been found that using equations for these evaluations of cumulative production is easier and less confusing than using the curves. Consequently, these curves are not included herein.

Other useful hyperbolic decline-curve methods. In 1972 Gentry published some curves for the purpose of simplifying hyperbolic decline-curve analysis.[12] He showed that by solving Eq. 8.25 for a_i:

$$a_i = \frac{(q_i/q)^n - 1}{nt} \tag{8.31}$$

Substituting this expression for a_i in Eq. 8.27 and rearranging, we obtain:

$$(\Delta N_p/q_i t) = \frac{1 - (q_i/q)^{n-1}}{(q_i/q)^n - 1}[n/(1 - n)] \tag{8.32}$$

Thus, the parameter a is shown to be a function of q_i, the intervening cumulative production, q, and time. Then using this equation Gentry prepared a plot of q_i/q versus $\Delta N_p/tq_i$ for different values of n (Fig. 8–9). With this set of curves available, n for a particular hyperbolic decline curve can be easily obtained using two rates and the intervening time and cumulative production for a hyperbolic decline curve. Once n has been evaluated, a_i can be calculated by Eq. 8.31. Gentry also provided another group of curves that can be used for evaluating a_i.[13] With a knowledge of these parameters, cumulative production and rates can be determined for any times.

This is an easy mathematical approach that many engineers are apparently using. The one drawback is that a curve of production data on a semilog plot of rate versus time does not necessarily represent one single hyperbolic curve. Some caution should be used in applying this method if we are not certain that the portion of the curve from which the two rates and intervening production are read is a hyperbolic curve. Normally, it appears that efforts to ensure that we are working with a hyperbolic curve by comparing the plotted data to the hyperbolic type curves leads us to the n and a_i characteristics without the Gentry data. We may often be certain of the nature of the hyperbolic decline data when a comparison with the type curves is made. However, we still have difficulty finding an exact fit of the data with a particular type curve. In such a case it is very helpful to use the Gentry data in Fig. 8–9 for the purpose of determining n and a_i.

Another approach to decline-curve analysis that appears to be receiving much practical use is the log-log type-curve analysis pro-

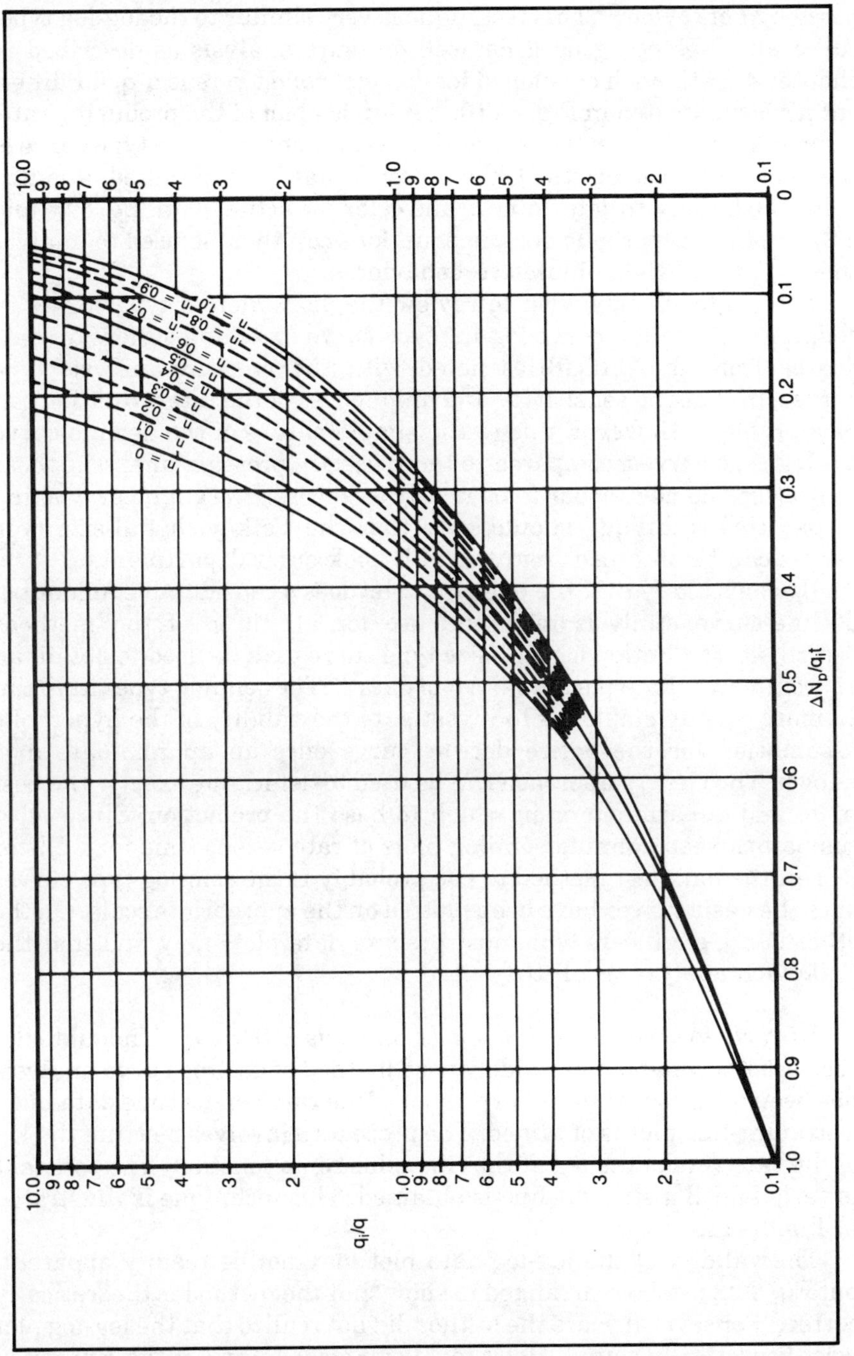

Fig. 8–9 Decline-curve analysis chart relating production rate to cumulative production (after Gentry, "Decline Curve Analysis," courtesy *Petroleum Technology*, January 1972), © SPE-AIME)

posed by Fetkovich.[14] This technique is very similar to the log-log type-curve analyses being used for well-pressure analysis as described in chapter 4. Fetkovich developed log-log plots of $a_i t$ versus q/q_i for different n's, as indicated in Fig. 8–10.[15] A log-log plot of the producing rate, q, versus time, t, gives a curve with the same shape as the type curve of the same n characteristic. Once a best fit has been obtained, a match point can be used to determine a_i and q_i for the actual data. Eq. 8.25 and 8.27 or other hyperbolic decline equations can then be used to analyze the rate, time, and cumulative behavior.

The engineer may wish to review the discussion of type curves in chapter 4. A decline-curve log-log type-curve matching suffers to some degree from the difficulties noted with the pressure-analysis type curves, in that the sensitivity and uniqueness of the interpretation can be a problem. However, due to the simple nature of the decline-curve log-log type curves, compared to those used in pressure analysis, these limitations do not appear to be nearly as severe. The engineer wishing to use this technique should note that the Fetkovich full-size type curves can be obtained from the SPE book order department.

It seems likely that the engineer who does a considerable amount of decline-curve analysis may find use for all three of the methods described. The Fetkovich full-size type curves can be used to obtain an approximate n for a particular set of data.[16] The semilog type curve can be more readily employed to investigate the validity of the hyperbolic assumption for the entire decline curve once an approximate n is known. The Gentry approach can be used to determine exactly the best n, a_i, and q_i parameters on which to base the prediction. Since most companies keep running semilog plots of rate versus time for wells or leases, the handiest method to use probably is the semilog type curves once the basic curves have been plotted on the appropriate scale. On the other hand, engineers who must prepare data plots may find that the Fetkovich analysis is all they need.

Historic hyperbolic decline-curve analysis methods. Undoubtedly, the best-known and most widely used method of decline-curve analysis has been a log-log trial-and-error plot of the rate versus time data until a straight-line plot is obtained. The procedure involves plotting the log of the rate versus the log of the time plus some constant. The constant is varied until a straight line is obtained. This technique is illustrated in Fig. 8–11.

The validity of the log-log data plot may not be readily apparent, but Eq. 8.22 can be rearranged to show that the method is theoretically correct. For several years the author did not realize that the log-log plot was theoretically sound. However, Leo Shrider (then of the Bureau of Mines) and a few of his colleagues showed the author that Eq. 8.25 can be rearranged and the log of the equation taken to give:

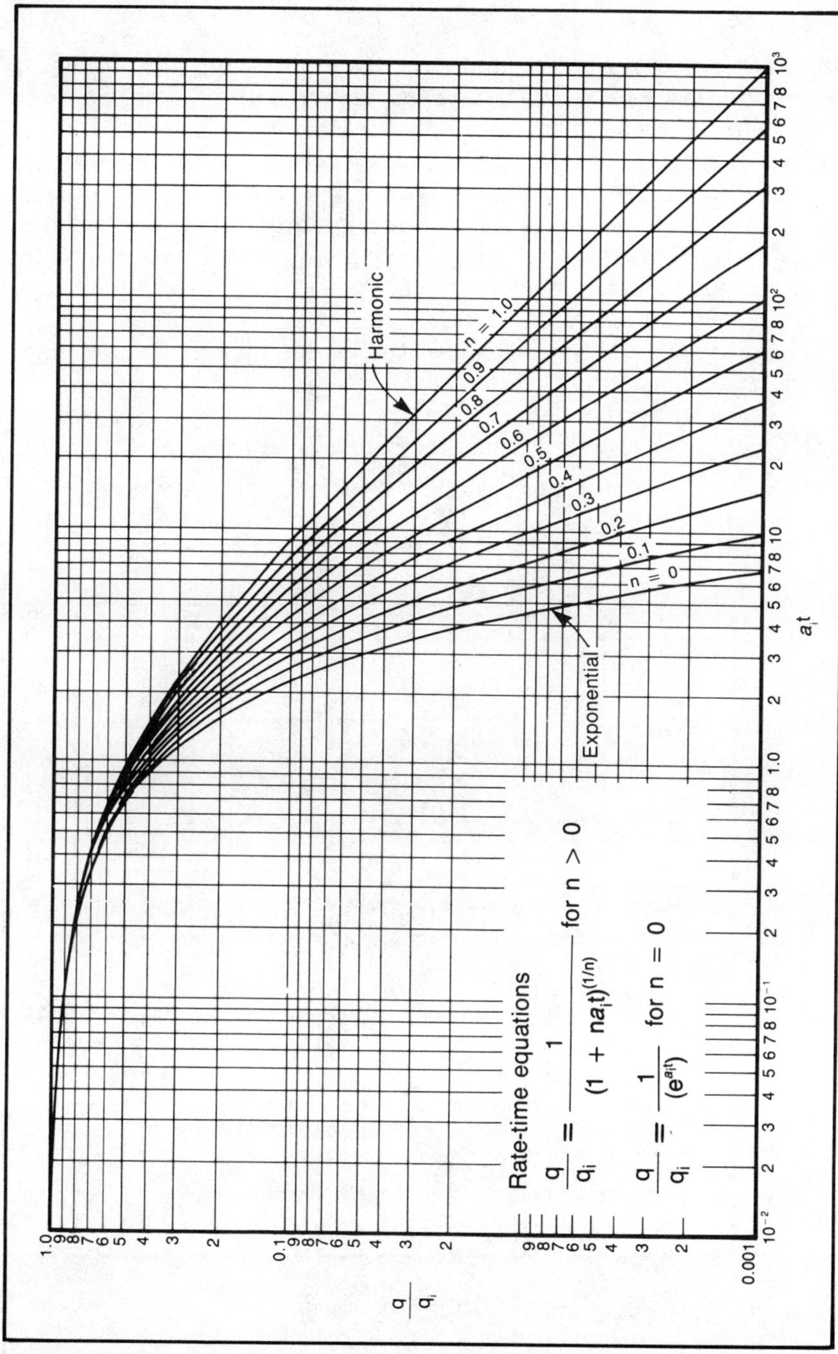

Fig. 8–10 Type curves for Arps empirical rate-time decline equations unit solutions—a_i = 1 (after Fetkovich, "Decline Curve Analysis Using Type Curves," courtesy *Petroleum Technology*, June 1980, © SPE-AIME)

$$\log q = \log [q_i(n\,a_i)^{-1/n}] + (-1/n)\log(n\,a_i + t) \qquad (8.33)$$

This transformation is accomplished by multiplying the numerator and denominator of Eq. 8.25 by the expression $(na_i)^{-1/n}$. Then when the equation is rearranged and the log is taken of both sides, Eq. 8.33 results. Thus, the log of the rate, q, plotted versus the log $(na_i + t)$ gives a straight line whose slope is $-1/n$. This is true because the first term of Eq. 8.33 contains only constants. Consequently, when plots of log q versus log (C + t) are made using various values for C, a straight-line plot is obtained when C is equal to na_i. The slope of this straight line is $-1/n$.

Normally, the engineer who uses the log-log hyperbolic plot simply extrapolates the straight line to obtain the rates, times, and cumulative production figures desired. He can of course evaluate n from the slope. Once n is evaluated, a_i can be determined from the C, na_i, that resulted in the straight line. Then Eqs. 8.25 and 8.27 can be used to

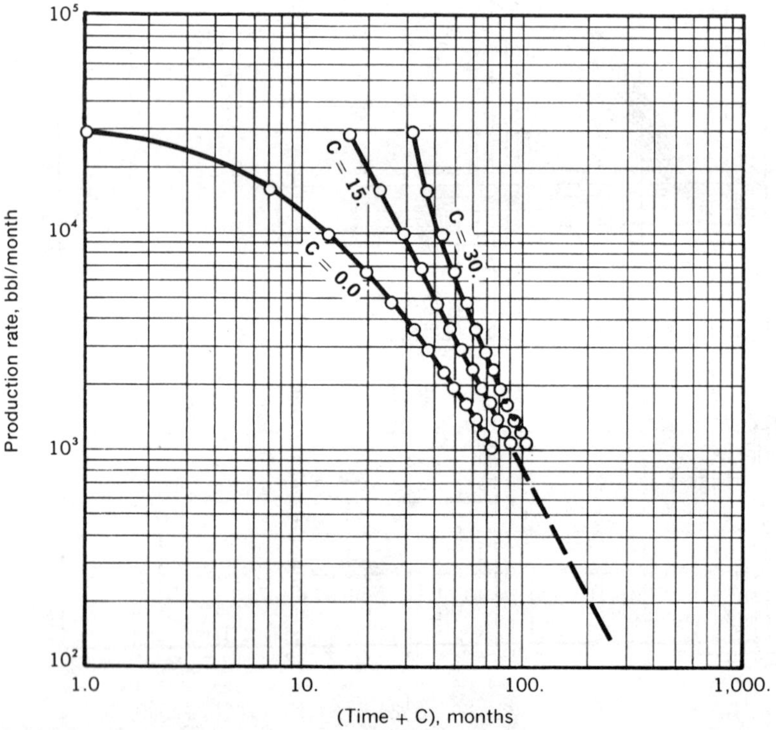

Fig. 8–11 Log-log extrapolation of hyperbolic decline

calculate the future rates and cumulatives. Generally, engineers who have attempted to use the log-log analysis method have found it to be time consuming and insensitive. By using a large C, it is generally possible to force a straight line from the production data even when it does not fit a hyperbolic equation.

Another common method for analyzing hyperbolic decline-curve data is the loss-ratio method. This technique uses the relationship between the decline rate and the hyperbolic constant:

$$n = \Delta\,(1/a)\,/\,\Delta t \tag{8.34}$$

Eq. 8.34 can be derived from Eqs. 8.20 and 8.25. First, raise both sides of Eq. 8.25 to a power n to obtain a value for q^n. Now, substitute this expression for q^n in Eq. 8.20 and solve for $1/a$. When the resulting expression is differentiated with respect to time, Eq. 8.34 results.

As proposed for use in the Campbell evaluation, production rates at the end of equal time intervals are tabulated.[17] The reciprocal of the decline rate for each period is calculated using the production rate at the end of the period as q in Eq. 8.1. The differences in successive reciprocals are also calculated. The n can then be calculated by Eq. 8.34. Constant reciprocals, $1/a$, imply exponential decline. Constant differences in reciprocals for the same length of time indicate hyperbolic decline and give a constant, calculated n for each time period.

The use of the rate at the end of a period, Δt, in calculating the decline rate, a, by Eq. 8.1 provides an accurate value of a only if Δq is very small compared to the rate, q. Consequently, a constant n can be obtained by this method only when the decline rates are small. At high decline rates hyperbolic data that exactly fit Eq. 8.25 do not yield a constant hyperbolic constant, n, by the loss-ratio method. The calculation of n by this method is also very sensitive to small variations of the actual data from hyperbolic decline and may be too involved when many hyperbolic declines must be evaluated, for example, in the case of an annual company reserve report. One big advantage of the loss-ratio method is that a hyperbolic decline can be forced by averaging the calculated n values and using this average n for the rate-time extrapolation.

Several computer methods for obtaining hyperbolic extrapolations of rate-time data are available. However, it does not appear that any of these methods can consistently obtain the best fit desired by most engineers.

Mobile oil by hyperbolic decline. As noted, we define primary mobile oil for decline-curve purposes as the oil that would be produced if it were possible to continue producing a well to a rate of zero. In the case of hyperbolic decline, this would mean that q_2 of Eq. 8.27 would be zero. Then if we let q_1 equal q_i, ΔN_p of Eq. 8.27 would be the remaining primary mobile oil:

$$\text{Remaining primary mobile oil} = q_i/a_i(1 - n) \qquad (8.35)$$

Eq. 8.35 then indicates a result similar to that noted for constant-percentage decline. If remedial work does not increase the drainage volume of a well, the remaining primary mobile oil volume is unchanged and any increase in the current rate of production must be matched by a corresponding increase in the decline rate. Note that the same conclusion can be reached based on hyperbolic decline if we further specify that the hyperbolic decline constant, n, is the same before and after the remedial work.

Work problem 8.4 and compare the solution with the one in appendix C.

PROBLEM 8.4: The Change in Reserves and Life following a Workover

Fig. 8–8 shows that the data of Fig. 8–3 fit a hyperbolic curve with an n of 0.5, a q_i of 15.6 b/d and an a_i of 0.015/month. If this well is worked over to obtain a rate of 53 b/d and the hyperbolic decline constant, n, remains unchanged, calculate the remaining life and reserves if the economic limit is 1.0 b/d. Compare answers with those to problem 8.2.

Harmonic Decline

In decline-curve analysis a third type of decline is generally recognized. This is the decline that occurs if the n of Eq. 8.20 is 1.0. In this case the decline rate, a, is proportional to the rate, q. This form of decline is sometimes encountered where production is controlled predominantly by gravity drainage, and it is known as *harmonic decline*. Eq. 8.25 shows that the rate-time equation for this type of decline is:

$$q = q_i/(1 + n\, a_i\, t) \qquad (8.36)$$

The cumulative production equation cannot, however, be obtained from the cumulative production equation for hyperbolic decline, Eq. 8.27, because the value of the second bracket in the equation is zero when n is 1.0. Thus, we must return to Eqs. 8.16 and 8.17 and derive a specific equation for harmonic decline. Note that when n is 1.0 and the decline rate, a, is proportional to the rate, q, the decline rate, a, can be stated as a function of the rates and the initial decline rate, a_i, as $(q/q_i)a_i$. This expression can be substituted for the decline rate, a, in Eq. 8.17, and the equation can be solved to obtain the cumulative production equation for harmonic decline:

$$\Delta N_p = -\sum_{q_1}^{q_2} (\Delta q/a) \qquad (8.17)$$

$$\Delta N_p = -\sum_{q_1}^{q_2} \Delta q/(q/q_i)a_i \quad (8.37)$$

$$\Delta N_p = (q_i/a_i)\,\ell n\,(q_1/q_2) \quad (8.38)$$

The harmonic decline analysis can be performed in much the same way that the hyperbolic analyses are performed, although no semilog type curves are available in the text for this purpose since harmonic decline is encountered so seldom.

If the engineer works in an area where harmonic decline is encountered often, it is a simple matter to prepare data plots similar to the hyperbolic type curves for these curves using Eqs. 8.37 and 8.38. These plots have the same general semilog shape as the hyperbolic type curves. However, note from Eq. 8.37 that a plot of the cumulative production versus the rate on semilog paper gives a straight line. This should simplify the application of the rate-cumulative curves analysis as compared to a similar hyperbolic analysis.

The data of Gentry and Fetkovich in Figs. 8–9 and 8–10, respectively, do include harmonic decline. Therefore, analyses using these techniques are the same as those discussed previously, except the rate-time and rate-cumulative curves are Eqs. 8.36 and 8.38.

Decline-curve analysis is perhaps the most used and at the same time the most misused reservoir engineering technique. It should be used when possible as a verification of the theoretical analyses. However, there are times when only decline-curve analysis can be used to predict the behavior of a reservoir. Most practicing engineers appear simply to use a constant decline-rate type of analysis, but this chapter presents a curve-fitting procedure for analyzing hyperbolic decline that is almost as simple to use as the constant-rate decline method. It is shown that other hyperbolic decline methods may also be useful in some situations.

Additional Problems

8.5 A. The following production data are given for a well in a solution-gas-drive reservoir. Assume the well is in constant-percentage decline and prepare a semilog plot of rate versus time. What will the producing rate be in December 1988? In December 2008? What is the remaining life of the well as of December 1983 if the economic limit is 1.4 stb/d? What are the reserves?

B. If the well is acidized in December 1983 and the producing rate is increased to 72 stb/d without increasing the mobile oil, find the remaining life and reserves as of December 1983.

	Average q_o, stb/d
December 1973	694
June 1974	422
December 1974	296
June 1975	209
December 1975	163
June 1976	126
December 1976	104
June 1977	84
December 1977	70
June 1978	61
December 1978	53
June 1979	46
December 1979	41
June 1980	35
December 1980	31
June 1981	28
December 1981	26
June 1982	24
December 1982	22
June 1983	20
December 1983	18

8.6 Rework problem 8.5 assuming hyperbolic decline and using the semilog type curves. The rate-time plot must be made on the same scale as the hyperbolic type curves. Assuming hyperbolic decline from December 1973 to December 1983, determine the total production during this period.

8.7 Using the cumulative production and the rates from December 1973 to December 1983, determine n and the a_i value in problem 8.5. Use the Fig. 8–9 data of Gentry.

8.8 Rework problem 8.5, assuming hyperbolic decline and using the Fetkovich log-log type curves in Fig. 8–10.

Notes

1. W.W. Cutler Jr., "Estimation of Underground Oil Reserves by Well Production Curves," *Bulletin,* USBM 91 (1924), p. 228.

2. Folkert Brons, "On the Use and Misuse of Production Decline Curves," API paper 801-39E (1963).

3. J.J. Fetkovich, "Decline Curve Analysis Using Type Curves," *Petroleum Technology* (June 1980).

4. J.J.Arps, "Analysis of Decline Curves," *Trans.,* AIME (1940), volume 160, 228–247.

5. J.J.Arps, "Estimation of Primary Oil Reserves," *Trans.,* AIME (1956), pp. 182–191.

6. A.T. Chatas and W.W. Yanhee, "Applications of Statistics to the Analysis of Decline Curves," *Trans.,* AIME (1958), pp. 399–401.

7. K.E. Gray, "Constant Percentage Decline Curve," *OGJ* (August 20, 1960).

8. K.E. Gray, "How to Analyze Yearly Production Data for Constant Percent Decline," *OGJ* (January 1962).

9. Brons, 1963.

10. Fetkovich, 1980.

11. H.C. Slider, *Practical Petroleum Reservoir Engineering Methods* (Tulsa: PennWell, 1976).

12. R.W. Gentry, "Decline Curve Analysis," *Petroleum Technology* (January 1972).

13. Gentry, 1972.

14. Fetkovich, 1980.

15. Fetkovich, 1980.

16. Fetkovich, 1980.

17. John M. Campbell, *Oil Property Evaluation* (Englewood Cliffs: Prentice-Hall, 1959), pp. 177–208.

Additional References

Slider, H.C. "A Simplified Method of Hyperbolic Decline Curve Analysis." JPT Forum, *Petroleum Technology,* March 1968.

9

Waterflooding and Its Variations

This chapter may well be the most valuable one in this book. There appears to be more opportunity for the engineer to conserve petroleum energy and make profits for a company through effective engineering of a waterflood than in any other single method.

Exactly what constitutes a waterflood or a variation of a waterflood is difficult to define. Polymers may be used in conjunction with water injection to control the effective mobility ratio. Some micellar solution, caustic, or other chemical may be used ahead of the injected water to reduce the residual oil saturation. Also, miscible displacements using carbon dioxide, LPG, or high-pressure gas may be followed by water injection. It is unlikely that all of these enhanced oil recovery systems should be called waterflood variations, but the ideas covered principally for water injection only can also be used in predicting the behavior of any of the aforementioned systems. Although many of the ideas presented apply theoretically only to immiscible systems, the miscible systems behave in a very similar fashion.

Note that none of the systems mentioned involve the effects of heat. Thermal effects involve the movement of heat between reservoir beds, and they require different treatment than the nonthermal effects. With the exception of steamflooding, it is the author's opinion that there is currently very little possibility for economic applicaton of any enhanced oil recovery technique outside the area of waterflooding and its variations. This is not meant to imply that there have been no economic successes using in situ combustion or other exotic recovery techniques. However, it is meant to suggest that such opportunities occur so infrequently that they are presently more in the area of research and development than in the realm of engineering.

Most of the fundamentals of waterflooding are useful in all types of frontal secondary recovery methods and in many cases in natural water-drive or gas-cap-drive applications. Many of the concepts covered are also fundamental to all enhanced oil recovery operations. For example, the effect of permeability distribution is important in all methods, including such nonfrontal techniques as steaming. Consequently, the concepts discussed are important, even to engineers who are never faced with reservoir problems in waterflooding.

As throughout this book, the author does not claim that the methods presented are the only useful, practical methods of waterflooding analysis. However, the objective is to present methods that can be completely understood by the engineer and that have proven reasonably accurate over the years. By using methods that are understood, the engineer can intelligently apply these methods and to a large extent can avoid misapplication of the calculating techniques. Consequently, we do not discuss in detail such well-known but highly complicated techniques as the Leighton and Higgins (Bureau of Mines) method. Only two prediction techniques are presented in sufficient detail to permit their practical use. One of the methods is useful for theoretical predictions, and the other has proven very useful and surprisingly accurate for the prediction of waterflood behavior by analogy.

However, before concerning ourselves with the rate-time behavior prediction, we will consider the problem of predicting the total flood recovery, which is where every waterflood begins. Discussion of the total waterflood recovery prediction requires consideration of many fundamental ideas involved in flooding, such as pattern, vertical, or stratification sweep, the evaluation of residual oil, and the determination of bulk swept volume.

This discussion of predicting total flood recovery is followed by a discussion of recommended methods for predicting oil and water production rate versus time. A comparison of recommended methods with some of the more common published methods is presented. Then the very important subject of flooding variations using mobility-ratio-control additives and residual-oil-reducing additives is considered so the engineer can determine which procedure or combination of procedures will result in the most profit.

Before we consider detailed flood calculations, let us first look at the flood displacement mechanism to provide a good physical picture of the displacement taking place in the reservoir.

The Waterflood Displacement Mechanism

A waterflood displacement is governed by the same fractional flow and Buckley-Leverett equations that describe all immiscible displacement processes. If there is no free gas present in the reservoir at the time the flood is initiated, the saturation distribution along any stream line, or flow line, in the reservoir can be calculated using the equations in chapter 6. However, if the reservoir contains a free-gas saturation at the time the flood is initiated, the analysis presents some additional problems.

The engineer's first conclusion concerning the displacement of oil and free gas concurrently in the reservoir is that three-phase relative permeability data are required. Fortunately, this is not generally the case. What does generally occur is depicted in Fig. 9–1. The lower portion of this figure shows a plot of water and total liquid (oil + water) saturation along a stream line before the displaced oil reaches the producing face. The basis for the lower diagram may be better understood by considering the upper diagram. This considers various classifications of reservoir pores grouped together. At the bottom of the diagram is shown the pore space originally occupied by immobile water, called *connate water*. The next horizontal strip of the diagram (not a horizontal strip of reservoir) represents the pores originally occupied by immobile oil, termed *residual oil*. The top strip represents the comparable gas pore volume; that is, the pores originally occupied by immobile gas is shown as being unrealistically high. For all practical purposes these

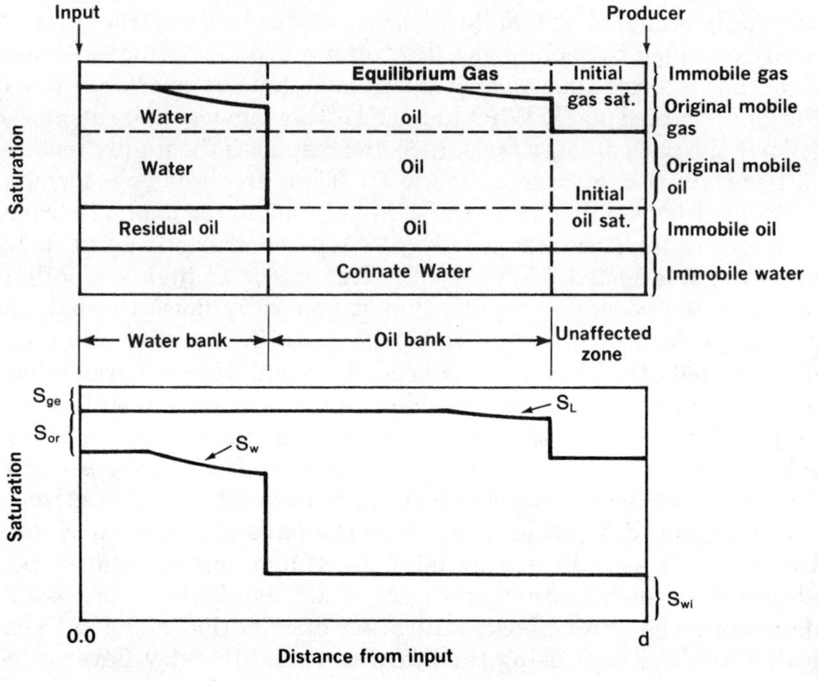

Fig. 9–1 Saturation distribution in a waterflood

three groups of pores can be considered as inactive or dead pores. If their fluids cannot be moved, they can play no important part in the displacement process.

However, the pores initially containing the mobile oil and mobile gas do play an active part in the displacement. The right-hand side of the diagram indicates the unaffected portion of the reservoir. Thus, this represents the original saturations in the reservoir. On the left is the water bank, which indicates that the water entering the formation displaces the original mobile oil and oil in the original mobile gas pores. Between the unaffected zone and the water bank we find a zone where the pore space originally occupied by mobile gas is almost completely filled with oil, so there is very little free gas in the oil bank except for that which is immobile. The question is how does this occur?

This question can be answered physically by considering what happens to the oil displaced by water. This displaced oil must choose a path to follow. It is obvious that it cannot enter pores occupied by immobile fluids. Consequently, it has a choice of traveling through pores occupied by mobile oil or pores occupied by mobile gas. Therefore, oil displaced from the water bank can either displace oil from the pore space originally occupied by mobile oil, or it can displace gas from pores originally occupied by mobile gas. The reservoir oil is normally about 100 times as viscous as gas and occupies pores that are smaller in size than the gas-occupied pores. What happens is then obvious. The displaced oil follows the path of least resistance and displaces the highly mobile gas rather than the much less mobile oil. Then an oil bank is formed.

Since free gas displaced from the oil and water banks is traveling through the unaffected zone in Fig. 9–1, it is not strictly accurate to say that it is unaffected. However, the free gas is so highly mobile compared to the reservoir liquids that it generally flows through to the producing wells with a negligible increase in the reservoir pressure that existed at the start of the flood. This point is further emphasized when we discuss bypassing outside producers by the oil bank. It should be specifically noted that a substantial quantity of equilibrium gas seldom remains in the reservoir after the flood front passes because the increase in pressure causes free gas to go back into solution. Also, equilibrium gas is difficult to measure in the lab and is seldom as large in the reservoir as indicated by lab data. If appropriate relative permeability data such as that in Fig. 9–2 are available, it is possible to determine the three-phase saturation distribution along a linear or radial displacement using the linear or radial Buckley-Leverett equations derived in chapter 6:

$$\Delta X_{sdj} = \frac{5.615\, q_t\, \Delta t}{\phi A} \left(\frac{\Delta f_d}{\Delta S_d}\right)_{sdj} \tag{6.14}$$

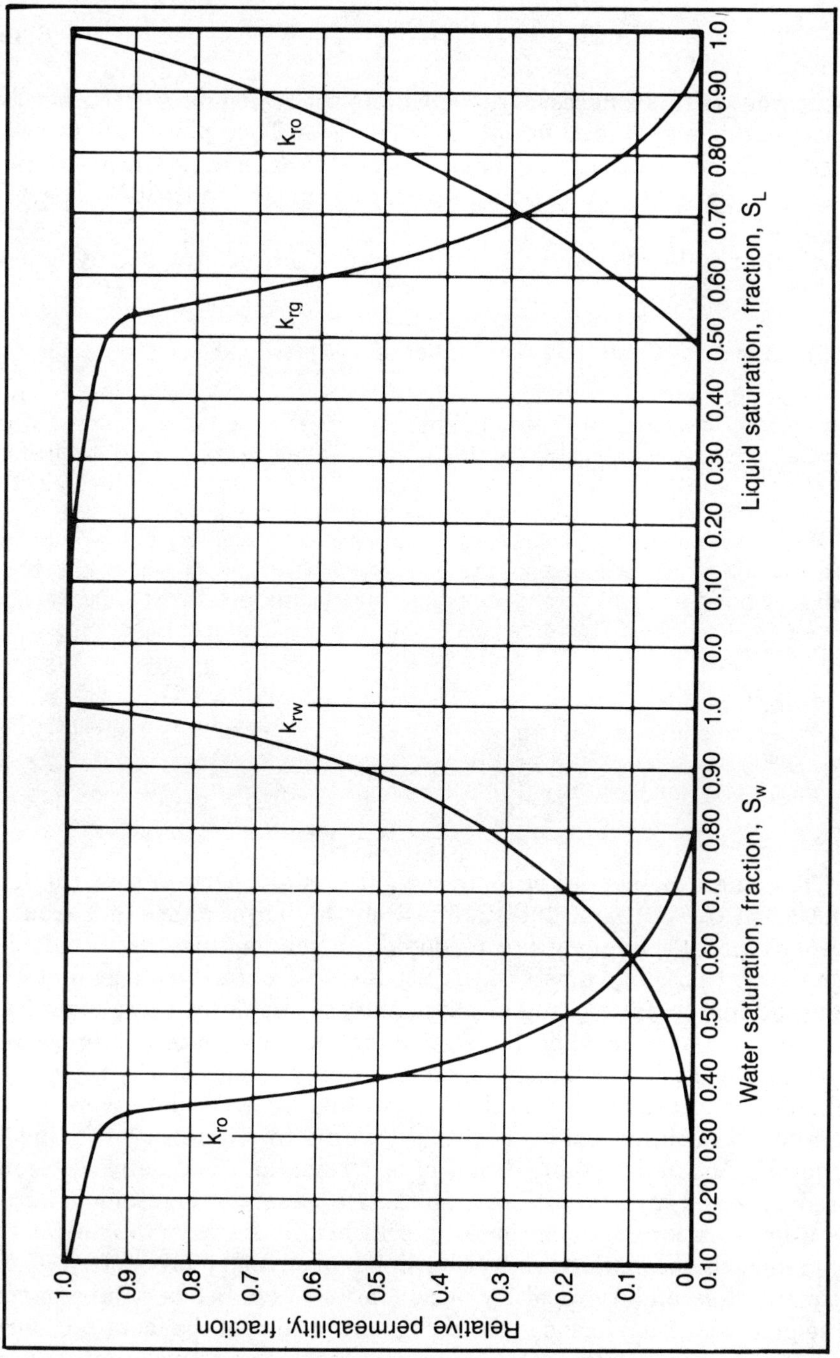

Fig. 9–2 Relative permeability data

$$\Delta(r^2)_{sdj} = \frac{1.79q_t\,\Delta t}{\phi h}\left(\frac{\Delta f_d}{\Delta S_d}\right)_{sdj} \tag{6.36}$$

It would first be necessary to calculate the fractional flow curves for water displacing oil and for oil displacing gas. Then plot the data as in Figs. 9–3 and 9–4 so the slopes can be used to determine the position of the water saturation and liquid saturation lines as in the lower part of Fig. 9–1.

Solve problem 9.1 and compare the solution with the one in appendix C.

PROBLEM 9.1: Saturation Distribution in a Waterflood

A large peripheral flood contains a ring of producers 1,000 ft within the surrounding injectors. If we treat this section of the reservoir as a linear displacement, what will be the saturation distribution in the most permeable zone when the first oil production increase occurs? The cross-sectional area of the most permeable section is 150,000 sq ft, and it has an effective injection rate of 3,000 b/d. Assume no free gas goes back into solution. Note that this assumption is only for the purposes of this problem. It is, of course, possible to calculate the amount of gas going back into solution as a result of the reservoir pressure increase. Use the fractional flow curves in Figs. 9–3 and 9–4 calculated from relative permeabilities. Other data are as follows:

Porosity = 0.2
Initial oil saturation, % = 55
Initial water saturation, % = 30
Initial gas saturation, % = 15

The engineer should be very careful to make certain that the formation of the oil bank and its effect on the performance of a frontal displacement are always considered in the engineering analysis. Many, and possibly most, of the studies of frontal displacements in general, and waterflooding in particular, assume that there is no free gas in the reservoir when the secondary recovery process is initiated. This can lead to erroneous conclusions. It may be justified by the investigator, engineer, or researcher by stating that secondary recovery should be initiated before a free gas forms in the reservoir. This is theoretically indisputable, but the fact remains that many different producers and royalty owners in the typical reservoir can seldom agree on a formula for participation in a unit before the reservoir begins to approach the economic production limit for primary production. Consequently, secondary recovery is seldom initiated before a substantial free-gas saturation forms in the reservoir, in spite of the fact that many states have compulsory unitization laws that permit pursuit of secondary recovery operations with less than 100% agreement among the involved parties.

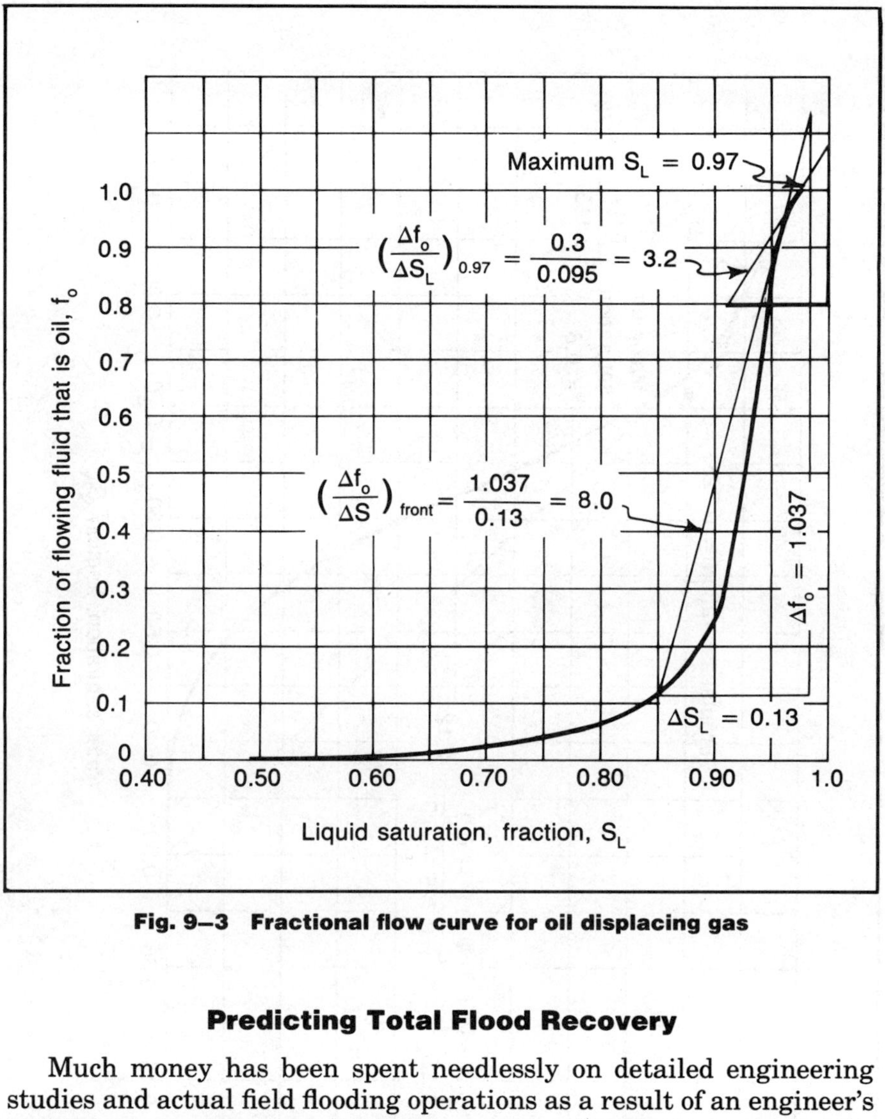

Fig. 9–3 Fractional flow curve for oil displacing gas

Predicting Total Flood Recovery

Much money has been spent needlessly on detailed engineering studies and actual field flooding operations as a result of an engineer's inability to predict the total flood recovery of a reservoir accurately. Conversely, many flooding opportunities have been missed for the same reason. The basic calculation need not be elaborate to obtain a reasonably accurate estimation of recovery. The following simple equation should yield accurate results if the reservoir parameters in the equation are evaluated with care:

$$N_{pf} = 7{,}758\ \phi\ E_t\ V_{sw} \left[\frac{S_{oP}}{B_{oP}} - \frac{S_{or}}{B_{or}}\right] \tag{9.1}$$

Eq. 9.1 is based on the difference in the oil in the gross swept vol-

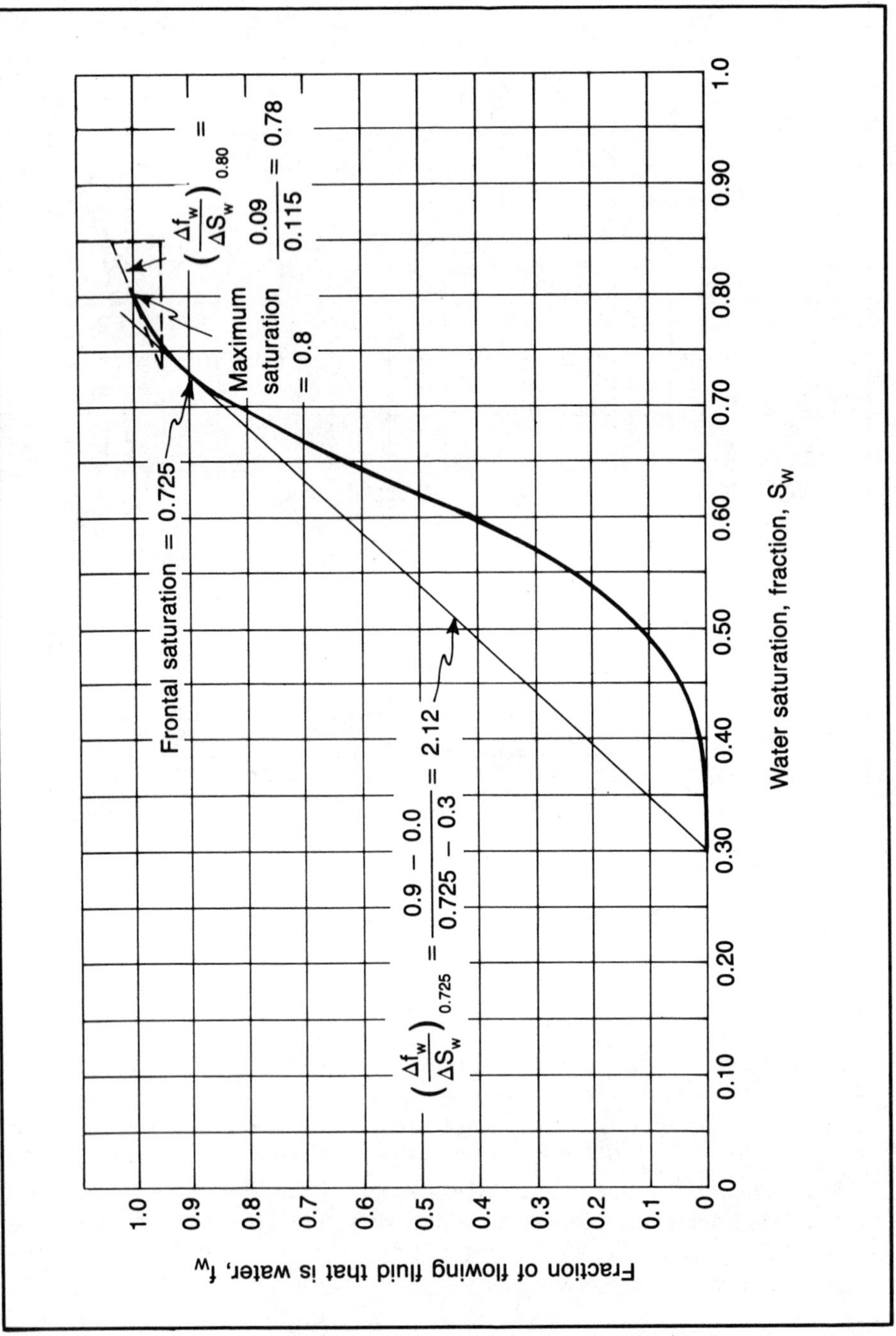

Fig. 9–4 Fractional flow curve for water displacing oil

ume, V_{sw}, at the start of the flood and the oil in the gross swept volume at the abandonment of the flood. In Eq. 9.1 the residual oil, S_{or}, represents the oil saturation that would remain at abandonment in a part of the reservoir that has been flushed by water. The saturation at the start of the flood or after primary recovery is denoted by the symbol, S_{oP}. Much of the reservoir volume included in the gross swept volume, V_{sw}, is not flushed by water so an overall efficiency, E_t (a fraction), is introduced to account for this difference.

Thus, the expression, $7{,}758\phi E_t V_{sw}$, represents the volume of the reservoir flushed by water. The constant 7,758 converts the acre-foot units of the gross swept volume, V_{sw}, to barrels. Thus, when multiplied by (S_{oP}/B_{oP}), this group of terms gives the stock-tank barrels of oil in the reservoir at the start of the flood. The same group of terms when multiplied by (S_{oP}/B_{oP}) gives the stock-tank barrels of oil in the reservoir after flooding. The difference is Eq. 9.1, the flood recovery.

All of the factors in Eq. 9.1 are interrelated. The efficiency, E_t, must take into account the method of evaluating the gross swept volume, V_{sw}, and the residual oil saturation, S_{or}, which in turn must be correlated with the basis for the sweep efficiency, the gross swept volume, and the porosity. Consequently, Eq. 9.1 appears to be a simple volumetric balance equation, but it is useless or even dangerously misleading unless the engineer considers the basis and interrelationship of its reservoir parameters very carefully.

It does not appear that the evaluation of porosity or formation volume factors should present many new problems. However, engineers do not always limit the basis for their statistical evaluation of porosity to the area included in the gross swept volume. Many times, the gross swept volume considered eliminates the edge areas of a reservoir that cannot be swept and, thus, eliminates some of the areas of marginal porosity. Also, in evaluating the oil formation volume factor at abandonment, remember that the oil probably does not have much more gas in solution than it contained at the start of the flood. The trapped free gas may go back into solution as the reservoir pressure is increased by water injection, but this amount is generally negligible. Most of the free gas is displaced from the reservoir ahead of the front of the oil bank and, thus, is displaced before there is an increase in the reservoir pressure. In any event the residual oil is far from saturated with gas after flooding has increased the reservoir pressure. Any sizable error is seldom introduced by assuming the oil formation volume factor at the start of the flood to be equal to the formation volume factor at abandonment.

Determining oil saturation, S_{oP}, at the start. Determining oil saturation, S_{oP}, represents one of the major problems in evaluating enhanced oil recovery prospects. The reservoir engineer generally assumes he has a reasonably accurate knowledge of the original oil in the reservoir prior to the initiation of primary production and, thus, can accurately determine the oil in place in the reservoir at the time an enhanced oil recovery project is initiated. However, experience has shown that the determination of the original oil in place is not always as accurate as we assume. Consequently, other methods have been explored for determining the oil saturation at the start of an enhanced oil recovery project.

One approach is to core and log carefully the development wells drilled for the purpose of an EOR project. The difficulty of determining reservoir saturations from conventional cores that are flushed by drilling mud filtrate or solution gas is well known. However, by using a pressure core barrel and tracers, the accuracy of the calculated reservoir oil saturations can be greatly improved. Also, in older reservoirs the logging technology may have changed considerably and would now permit a more accurate evaluation of reservoir saturations from logs.

The oil saturation at the start of the flood, S_{oP}, is generally evaluated by a material-balance equation based on the entire reservoir instead of just the gross swept volume of the flood. We assume a uniform saturation throughout the reservoir at the start of the flood. A material-balance equation subscripted specifically for our purposes is convenient for this calculation:

$$S_{oP} = \frac{(N - N_{pP})\, B_{oP}\, (1 - S_{wc})}{NB_{oi}} \tag{9.2}$$

Where:

N_{pP} = Cumulative stock-tank barrels of oil production at the start of the flood or after primary, denoted by the subscript P

Eq. 9.2 represents the remaining reservoir oil in the reservoir, $(N - N_{pP})B_{oP}$, divided by the pore volume of the reservoir determined from the initial oil saturation and volume of oil in place, $NB_{oi}/(1 - S_{wc})$.

The original stock-tank volume of oil in the reservoir before primary production, N, is calculated by material balance or volumetrically in the same way it would be evaluated for any reservoir. The following equation is very similar to the one derived in chapter 6 on material balance:

$$N = \frac{7{,}758\phi(1 - S_{wc})V_{bP}}{B_{oi}} \tag{9.3}$$

The only thing new about Eq. 9.3 is the symbol, V_{bP}, for the bulk volume. The subscript P is used to make certain the engineer recognizes the difference between the reservoir volume drained under primary production, V_{bP}, and the reservoir volume drained under secondary, V_{sw}. We now consider the specific difference between these two volumes.

Determining gross swept volume, V_{sw}. One of the common errors made by engineers in calculating total flood recovery is to assume that the swept volume is equal to the total reservoir volume without adjusting overall sweep efficiency or residual oil. Since all of these parameters are interrelated, there can be no right or wrong value for one without a discussion of the others. Consequently, the engineer should recognize that there can be no correct value for any one of these parameters without first fixing the others. However, the methods discussed give reasonably accurate results if applied conscientiously and consistently.

The necessity for differentiating between primary drainage volume and swept volume is shown in Fig. 9–5. This idealized reservoir is bounded by a shale out at the zero thickness isopach as indicated. Originally, it was developed with five wells and produced to near depletion

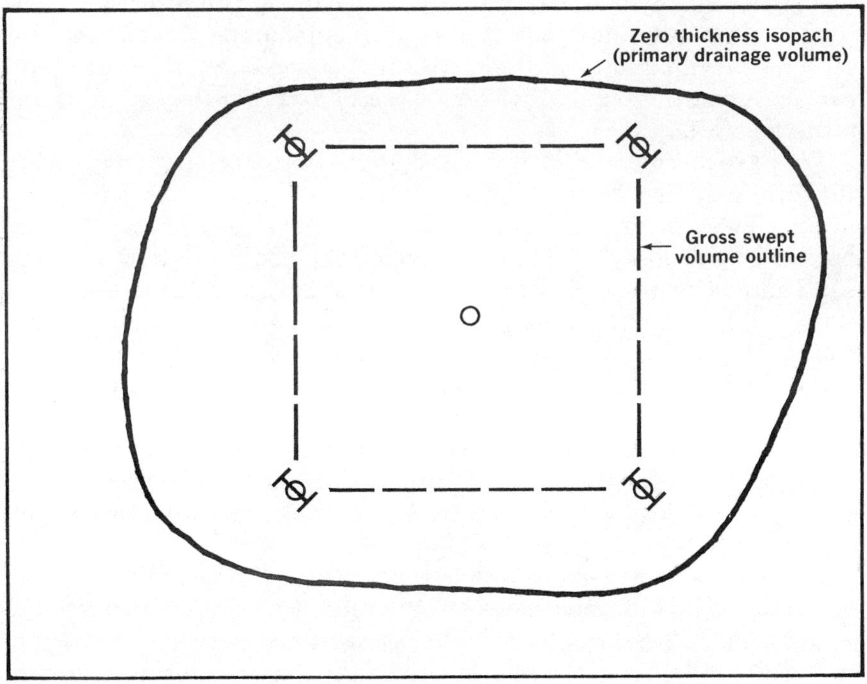

Fig. 9–5 Comparison of primary drainage volume and gross swept volume

with natural solution-gas-drive energy. During this phase of production, it is clear that the entire reservoir would be drained by the five pressure sinks represented by the five producing wells. However, when the four outside wells are converted to injection wells and the fifth well continues to produce, it is clear that during the economic life of this flood—before the water cut becomes uneconomical—little of the oil between the gross swept volume outline and the zero thickness isopach will be produced. Initially, the water injection into the four wells does move out radially to form an oil bank ahead of the water bank until the adjacent oil banks meet. The oil is squeezed from between the two water banks. About half of it goes toward the producing well and is produced, and the other half moves toward the reservoir boundary to be trapped there. Once all of the gas has been displaced from the gross swept volume, either by being produced or by being moved toward the reservoir boundary, most of the injected fluid moves along stream lines in the gross swept volume. This occurs because a pressure sink exists only at the producing well.

Late in the flood life after the reservoir has substantially reached steady state from a pressure standpoint, some flow takes place from outside the gross swept volume toward the producing well. It can be shown that steady-state stream lines cover the entire reservoir. However, the velocities along the stream lines outside the gross swept area are so low compared to the velocities in the flood area that they have little significance, especially when we recognize that they occur so late in the flood life.

We have discussed Fig. 9–5 based on the assumption that there is a substantial gas saturation in the reservoir at the time the flood is initiated. However, note that essentially the same situation results when there is no free gas in the reservoir when the flood is initiated. In this case the reservoir quickly approximates steady-state conditions, but substantial flow still takes place only within the flood area. Consequently, the gross swept volume remains the same.

Some published articles on flood sweep conclude that flow does take place outside the gross swept area.[1] However, these studies do not attach quantitative values to the stream-line flow outside the flood area. Therefore, for all practical purposes we can consider recovery to occur only in the area marked in Fig. 9–5 as the gross swept volume.

Fig. 9–5 is considered only for the purpose of demonstrating why there must be a difference between the recovery volume considered in primary and secondary recovery. In most situations the delineation of gross volume is more complicated than the simple case described in Fig. 9–5. Consider the more complex flood geometry depicted in Fig. 9–6. The solid line indicates the recommended gross swept volume if

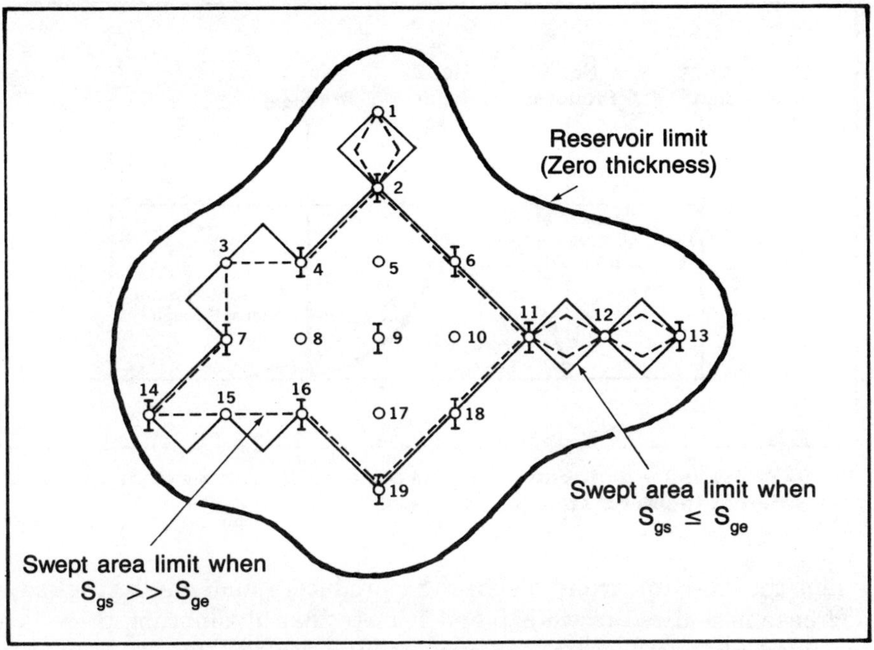

Fig. 9–6 Estimation of gross swept volume in a flood

the gas saturation at the start of the flood, S_{gs}, is less than the equilibrium gas saturation, S_{ge}. Therefore, gas would not flow and an oil bank would not form. In this case the recommended method for delineating the gross swept volume is simply to consider parts or elements of the flood geometry. In this five-spot pattern we have included one-fourth of a five spot between wells 1 and 2, one-half of a five spot between wells 3, 4, and 7, three-fourths of a five spot between wells 7, 14, 15, and 16, etc.

When a substantial gas saturation exists in the reservoir at the start of the flood, delineation of the gross swept area is more difficult—the edge producers cause the problem. Since the oil bank is pushed past the outside producers before the reservoir pressure in that vicinity is increased, the oil producing rate is increased very little prior to when the water bank reaches the outside producer. A look at the pressure distribution associated with outside producers during the early flood life helps provide an understanding of this phenomenon.

Consider Fig. 9–7, which is a pressure cross section of a portion of Fig. 9–6 through wells 9, 5, 2, and 1, and on to the reservoir limit. This can represent the pressure distribution in one strata (uniform permeability) of a reservoir that has a substantial gas saturation at the start of the flood, such that an oil bank would form. Now, consider the time

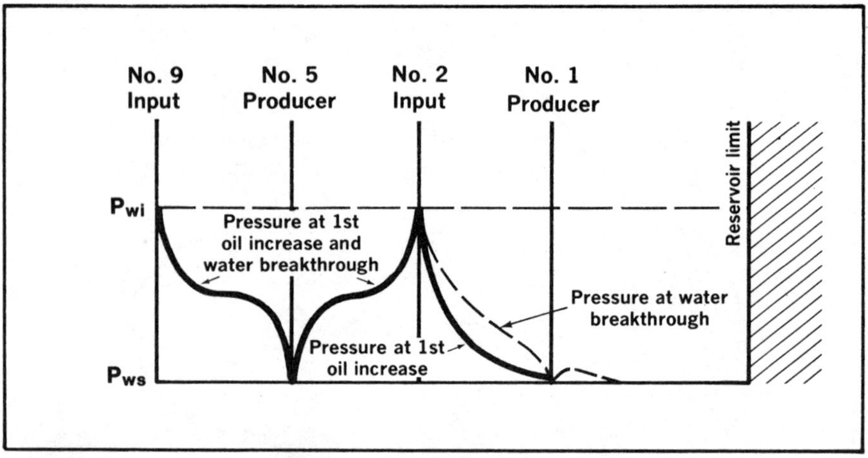

Fig. 9–7 Pressure distribution around edge wells with a substantial gas saturation at the flood start (see Fig. 9–6)

when the oil-bank front reaches the producers and the first oil-rate increase is realized in wells 5 and 1. Note that at this time there is a considerable increase in pressure around well 5 and a negligible increase around well 1. The pore volume between well 1 and the reservoir limit contains a large saturation of gas that has been compressed only a small amount. All of the gas around well 5 has been produced. The radial pressure gradient in the gas zone is negligible compared with the pressure gradient in the oil and water banks because the gas viscosity is only about 1/100 of the liquid viscosities at low pressures.

Since the oil production rate is a function of oil saturation and pressure drop in the drainage area, the rate increases in well 5 at this time is quite large, while the rate increase in well 1 is relatively small. Both wells have the same oil saturation, but well 1 has a very small pressure drop available for production.

At water breakthrough the pressure in the well 1 drainage area has still increased only a small amount. Therefore, most of the oil is simply pushed past well 1 and is trapped in the original gas-occupied pore volume between the flood area and the reservoir limit. Remember that the pressure shown is the result of radial flow from the injection well with the pressure drops caused by the production from well 1 superimposed on it. This results in the pressure hump between the producer and the reservoir limit. To the left of the hump, fluid is flowing toward the producing well. To the right of the hump, oil and gas continue to migrate toward the reservoir boundary. Eventually, the hump reaches the reservoir boundary, and all flow is toward the producer.

Consequently, there is a considerable difference in the gross swept volume of a producing well that is surrounded by injection wells, as is well 5 in Fig. 9–6; a producing well affected directly by three injection wells, for example, well 15; a producing well affected by two injection wells, wells 3 and 12; and a producing well affected by only one injection well, well 1. A reasonable gross swept volume is obtained by delineating the gross swept volume with a line constructed by joining adjacent edge production and injection wells and adjoining edge injection wells but not adjoining edge production wells. Then add one-half quadrant for isolated producers such as well 1 or well 2.

This technique is illustrated in Fig. 9–6. Adjoining producers and injection wells are connected with the line delineating the gross swept area between wells 4 and 3, 3 and 7, 14 and 15, 15 and 16, 19 and 20, 20 and 18, 18 and 21, and 21 and 11. Adjoining edge injection wells are joined by the delineating line between 2 and 4, 7 and 14, 16 and 19, 11 and 6, and 6 and 2. Note that adjoining edge producers 20 and 21 are not joined. In the pattern that exists in Fig. 9–6, note that one-half quadrant has been included in the gross swept volume between wells 1 and 2, 11 and 12, and 12 and 13. These rules may be varied with the initial gas saturation and the area between the outside wells and the zero thickness isopach. They are simply cited as an example of a procedure that gives reasonable values most of the time.

This discussion emphasizes the difficulty involved in evaluating a pilot flood in a reservoir. In such a case only one or two elements of a pattern are normally used. If the outside pattern wells are producing wells, much of the oil can be pushed past the wells without a corresponding increase in pressure and oil production. Furthermore, note that if the injection wells are placed outside of the pattern, the oil within the pattern is contained. However, most of the injected water tends to go into the unpressured reservoir outside of the pattern. Both of these effects can make a pilot operation result in little additional oil recovery, while a full-scale flood of the reservoir can be extremely successful.

Engineers should resist the temptation to misuse these rules for delineating the gross swept volume to increase the volume by using some irregular pattern around the edge of the flood. For example, we may be tempted to make well 18 a producer and convert wells 20 and 21 to injection wells because, according to the rules, we can then join the outside injection wells 20 and 21 in delineating the gross swept volume. This would apparently increase the gross swept volume by the amount of reservoir in the triangle between wells 20, 21, and 18. However, note that, if this were done, the displacement efficiency in the area drained by producers 10, 17, and 18 would be poor. The oil in the triangle formed by these producers would be virtually undrained. Oil would be trapped in this area by the pressure and injected fluid from the surrounding injection wells 16, 9, 6, 11, 21, 20, and 19. Also, if a substantial gas saturation existed at the start of the flood, oil from outside the subject triangle would be forced into this triangle before the reservoir pressure in the area would be substantially increased, resulting in trapping the oil. This effect is similar to the previously noted migration of the oil bank past outside producers. The rules cited are applicable to a five-spot pattern, not some variation thereof.

To test our knowledge of this section, work problem 9.2 and check the solution against the one in appendix C.

PROBLEM 9.2: Evaluating Gross Swept Volume

In subsequent problems we will predict the total flood recovery for the reservoir in Fig. 9–8. However, in this part the engineer is asked to determine the original (before primary) stock-tank volume of oil in place, the reservoir saturations at the start of the flood, and the acre-feet of gross swept volume based on the best 20-acre five-spot flood pattern. Use only the wells that presently exist as either injection or production wells. Note that a 20-acre five-spot is based on 10-acre well spacing. Use the 10-acre grid in Fig. 9–8 to predict the reservoir volumes by estimating areas based on the grid and average thicknesses for the individual areas. The reservoir data are as follows:

Porosity = 20%
Connate water = 20%
Original B_o = 1.2
B_o at the flood start = 1.1
Cumulative production at the flood start = 364,500 stb

Determining residual oil saturation. The residual oil saturation should represent the oil that cannot be moved regardless of the amount of displacing phase moved through the formation. It does not include the inability to make the injected fluid contact all of the formation. In this respect it may be thought of as a laboratory residual oil saturation. In water-oil relative permeability data it represents the oil saturation below which oil will not flow. The residual oil is caused by unbalanced capillary forces acting on the discontinuous nonwetting oil phase. This phenomenon is explained in chapter 6.

Historically, most engineers have based residual oil saturations, S_{or}, on relative permeability tests or other laboratory tests on actual reservoir cores designed to determine the residual oil saturation. Theoretically, such tests can be run on old cores that have been in storage for years using a restored-state method. Many companies have warehouses full of old cores because conventional core analysis only required about half of the core and then did not use all of that. Most companies stored the remainder.

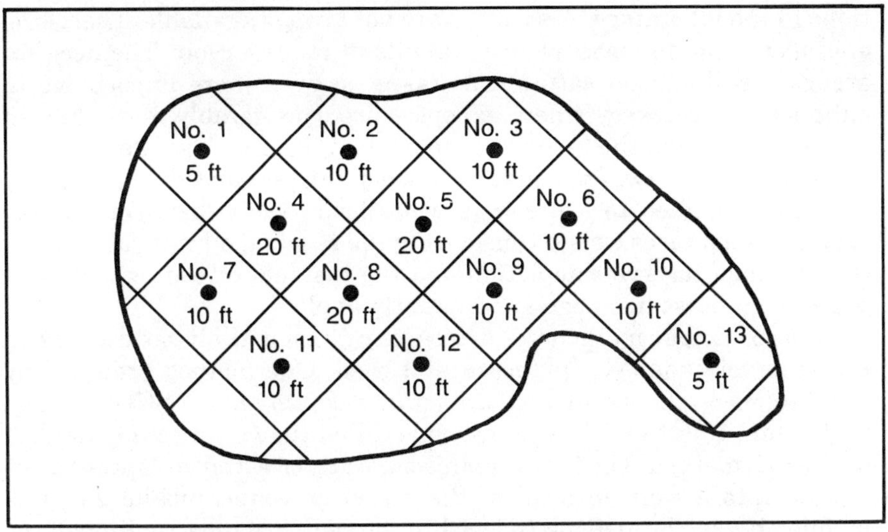

Fig. 9–8 Reservoir for problem 9.2 with superimposed 10-acre grid

The core is restored to its initial reservoir saturation by first cleaning all of the old oil and water from the cores, restoring the wettability (generally destroyed in the cleaning process), and running displacements with first water and then oil. The core is again displaced with water to reach a residual oil saturation. In some cases the residual oil saturation can be determined by analogy. The reported oil saturation from conventional core analysis has often been used as S_{or}.

In cutting a conventional core, the core is subjected to mud-filtrate invasion that may compare with flooding.* In addition, the core is subjected to a solution-gas drive when it is removed from the well and the pressure is reduced to atmospheric, causing the liberation of the solution gas and the corresponding displacement of oil from the core. Thus, conventional core-analysis oil saturations often closely approximate the residual oil obtained after solution-gas drive and water displacement.

When available, core-analysis oil saturations should always be compared with laboratory-measured residual oil values. So much difficulty is involved in determining residual oil saturations in the laboratory, especially control of the core wettability, that it has not been unusual for an engineer to use the core-analysis oil saturation in preference to the saturations resulting from laboratory analysis.

Laboratory-measured residual oil saturations can be inaccurate in some cases, and it is difficult to match wettability and reservoir condi-

*Discussed in chapter 6 in the section "Determining Original WOC and GOC."

tions in the laboratory. Also, cores are not always available. Therefore, engineers tend to measure residual oils in the reservoir. The need for accurate residual oil saturations takes on even more importance in enhanced oil recovery where expenses are considerably more than in waterflooding and the results are much less certain. Also, we are often concerned with two residual oils when an enhanced oil recovery method is applied to a previous waterflood, water drive, or gas-cap drive. In such cases we are considering the residual oil saturation after the previous reservoir depletion and the residual oil saturation after the EOR process has been subsequently applied.

Two principal methods of determining residual oil saturations in the reservoir has been proposed and used. One method proposed by Shell relies on the use of a *pulsed neutron-capture log* (PNC) combined with water injection.[2,3] The method is called the *log-inject-log* method or the LIL method. The idea is to inject a water of a known capture cross section into a well until all of the reservoir water, mobile gas, and mobile oil have been displaced within the radius of investigation of the logging device. Then the well is logged with PNC. The hydrocarbon saturation can be calculated from this log if the porosity, capture cross section of the formation matrix, and capture cross section of the hydrocarbons are known. However, the need for the two additional capture cross sections can be avoided by injecting a water with a different capture cross section to displace all of the first injected water within the radius of investigation. The PNC log is run again. The residual hydrocarbon saturation can then be calculated simply from the two log values and a knowledge of the porosity, which can be determined by running some type of porosity log. This method has been used extensively for determining residual hydrocarbon saturations after waterfloods and after subsequent EOR operations.

A second method proposed by Exxon Production Research relies on a chemical tracer, ethyl acetate, that has different solubilities in oil and water and the generation in the formation of ethyl alcohol as a result of hydrolyzing the ethyl acetate.[4] For all practical purposes the alcohol is soluble only in water. Before this test is run, the oil saturation must be at the residual level so no oil flows when water containing the ethyl acetate is injected. Following the injection of the acetate dissolved in formation water, the well is permitted to rest for a period to allow the ethyl acetate to hydrolyze and form the second tracer, ethyl alcohol. Then the well is produced, and the concentrations of the ethyl acetate and ethyl alcohol are measured continuously. Since ethyl acetate continues to dissolve in the residual oil but the alcohol does not dissolve, the profile of the concentrations versus cumulative produced fluid can be interpreted to determine the residual oil saturation.

This interpretation must be done with a computer model of the process that includes the equilibrium distribution of the ethyl acetate, the pore velocity of the fluids, the rate of ethyl acetate hydrolysis, and the radial dispersion. A match of the computer model and the observed data is obtained by varying the residual oil saturation, the hydrolysis rate, and the radial dispersion. The interpretation is also complicated in many cases by the drift of the injected bank during the shutin period. This method of determining residual oil saturations has also received wide application.

Determining ultimate horizontal sweep efficiency. The total displacement efficiency combines several different efficiencies, including the horizontal or sweep efficiency, the vertical or stratigraphic efficiency, and the conformance efficiency. The total efficiency is then the product of these three:

$$E_t = (E_H)(E_V)(E_C) \tag{9.4}$$

Most waterfloods are developed on patterns such as those illustrated in Fig. 9–9. By far the most used pattern is the five-spot, so most of our discussions are concerned with this. However, the same general methods can be applied to most patterns. Results from peripheral flooding, where the injection wells are along the perimeter of the reservoir, are more difficult to predict. Generally, peripheral floods are not recommended. They push the oil bank past the inside producers and, thus, are difficult to produce. Also, we cannot determine at what water cut to shut in the producers, and such floods are difficult to operate. Furthermore, injection rates generally are a problem because the injection wells continue to push the water greater distances.

Even if piston-like displacement efficiency could be obtained in the reservoir, it would be impossible to displace all of the mobile oil from a patterned element before any of the displacing fluid is produced because it is necessary to inject into a point (well) in the reservoir and produce from a point (well) in the reservoir. This is demonstrated in Fig. 9–10, which depicts the water-bank front at various stages of development in a five-spot element. This five-spot is in a theoretically infinite pattern with wells 1, 2, 3, and 4 producing and well 5 being a water-injection well.

The water bank tends to cusp into the producing wells as a result of the greater mobility of the water (k_w/μ_w) as compared with the mobility of the oil in the oil bank (k_o/μ_o). Also, the most direct flow path in the five-spot is much shorter than the least direct flow path along the edges of the five-spot quadrant, or the quarter of a five-spot. Thus, water tends to bypass oil because it moves more readily, it has a shorter distance to travel along the most direct stream line, and the pressure gradient is greater along the most direct stream line. Fig. 2–17 more

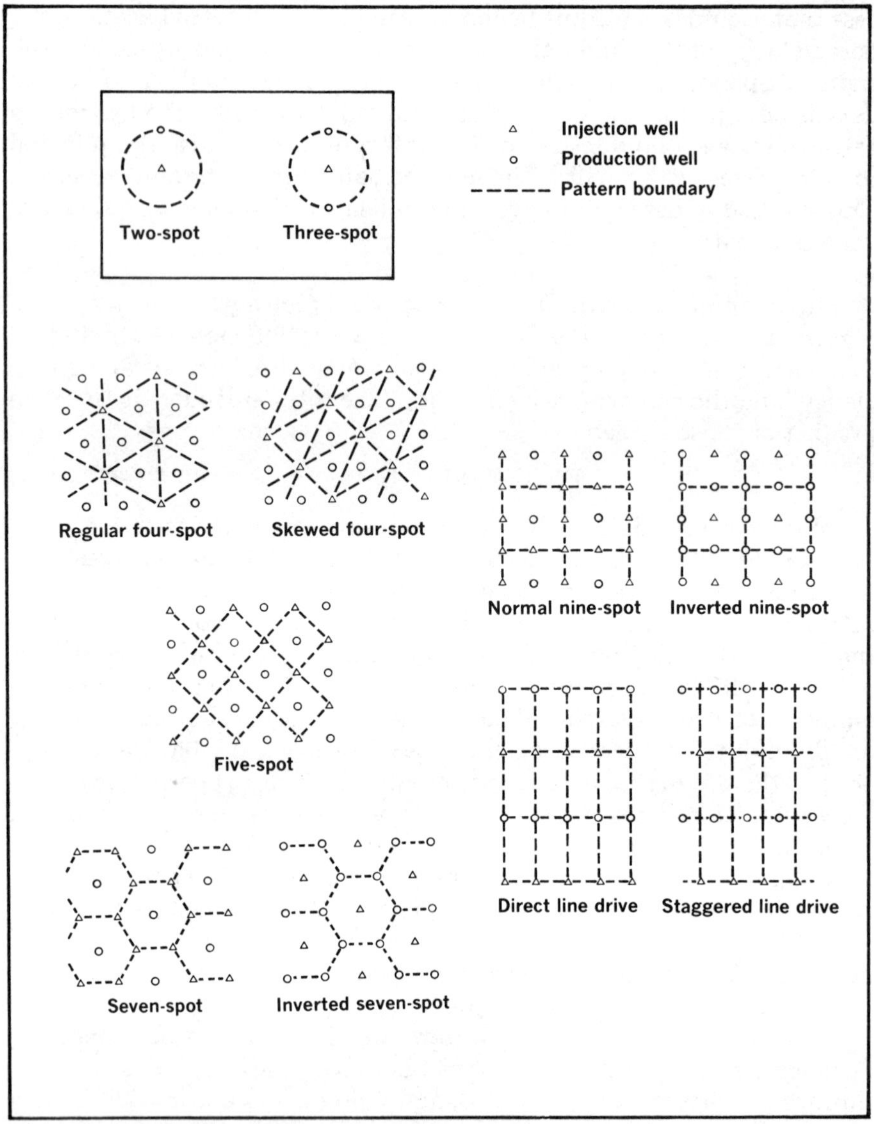

Fig. 9–9 Flooding patterns (after Craig, *Reservoir Engineering Aspects of Waterflooding*, Monograph No. 3, © 1971, SPE-AIME)

fully indicates the conditions in a five-spot quadrant by showing the stream lines and pressure distribution.

After the water front breaks into the producing well, there is a continued advance of the water front throughout the five-spot. However, the advance is retarded considerably because of the mobility ratio. Many studies have been made to relate the percentage of a five spot

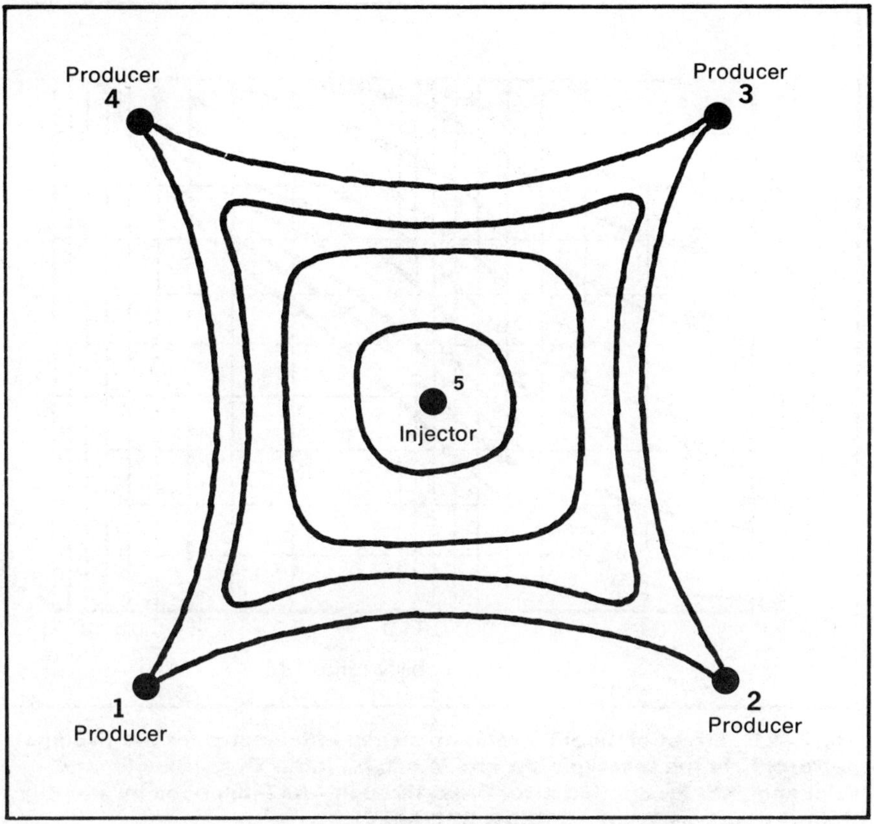

Fig. 9–10 Water-front cusping in a five-spot

swept by the water bank at different stages of depletion for different mobilities. The results of one such investigation are shown in Figs. 9–11 and 9–12.

The appendix of SPE Monograph No. 3 contains more data of this type covering many other flood patterns. These data are for homogeneous sands, and most sands perform like a series of parallel homogeneous sands stacked on top of each other, each with its own permeability and porosity. Consequently, by the time the economic produced water-oil ratio is reached in a flood, most of the strata either have had sufficient water throughout to give them a 100% horizontal sweep, or the water bank has not reached the producing well and no sweep efficiency correction is necessary. The incomplete sweep of the strata before water breakthrough at the producing well is included in the vertical sweep efficiency. This will be clearer after we consider the vertical sweep efficiency.

Experience has shown that the average horizontal sweep that should be applied in calculating the overall sweep should be high,

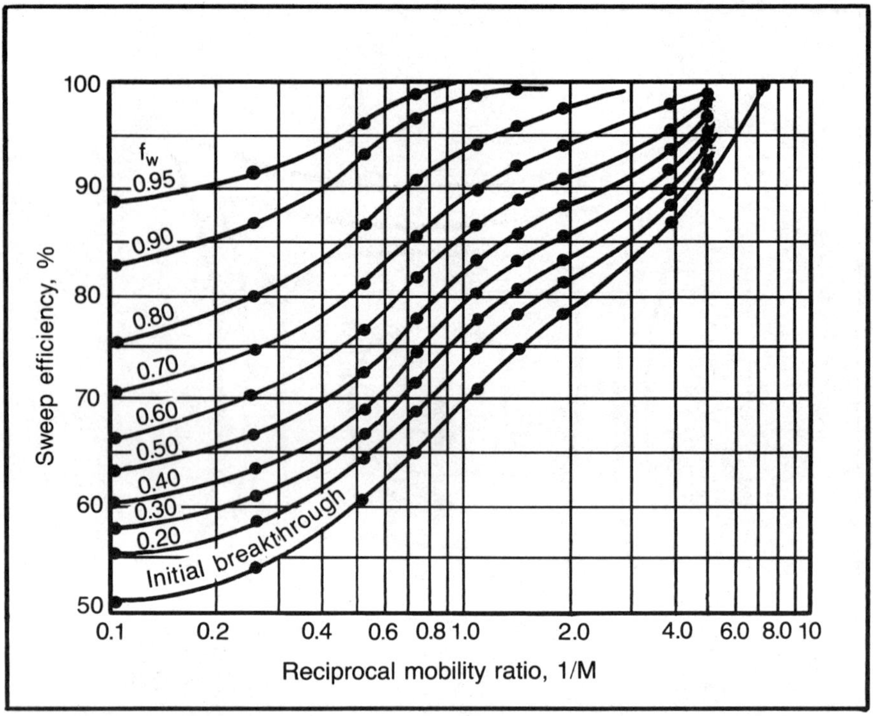

Fig. 9–11 Effect of mobility ratio on sweep efficiencies for the five-spot pattern; f_w is the reservoir cut and M = λ_w/λ_o (after Dyes, Caudle, and Erickson, "Oil Production after Breakthrough—As Influenced by Mobility Ratio," courtesy *Trans.*, © 1954, SPE-AIME)

about 0.9 or 0.95, depending on the adversity of the permeability distribution, which determines the percentage of the entire thickness that is producing both oil and water at abandonment. A very diverse permeability distribution means that many of the strata are producing both oil and water at abandonment and, thus, require a sweep efficiency correction. Therefore, the total sweep correction may be 0.9, whereas a more uniform permeability distribution may require only a 0.95 correction. Horizontal sweep data such as in Figs. 9–11 and 9–12 are very valuable and necessary when predicting the rate versus time behavior of a waterflood rather than just the ultimate recovery.

The conformance efficiency is used by some engineers to account for lateral variations in permeability caused by shale or low-permeability lenses. Other engineers use it to cover any displacement inefficiencies that are not included in the horizontal or vertical sweep. We consider the conformance efficiency to be 1.0 at all times. It is included in Eq. 9.4 for completeness.

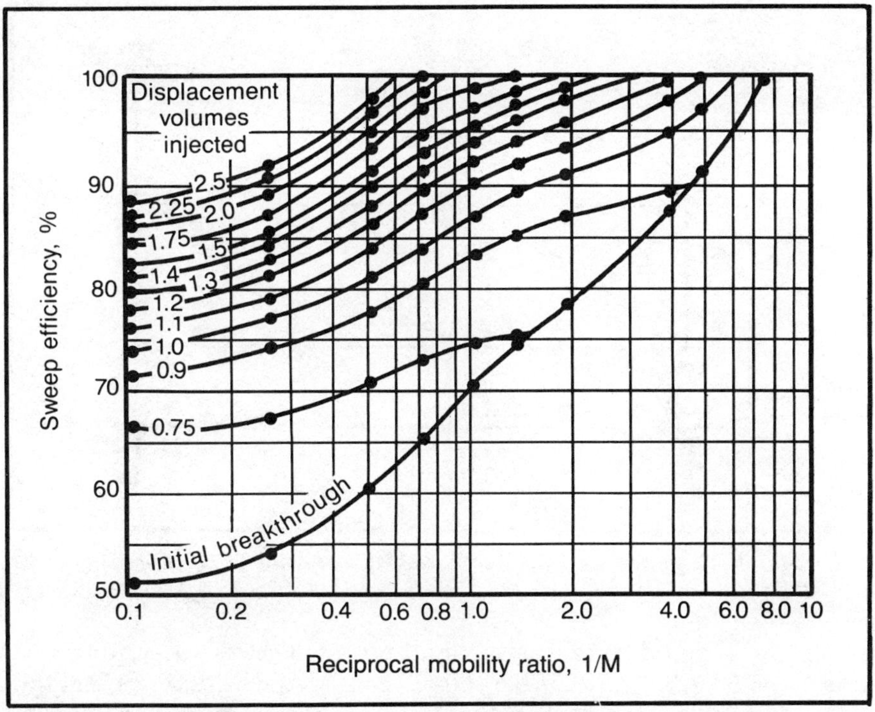

Fig. 9–12 Effect of mobility ratio on the displaceable volumes injected for the five-spot pattern; M = λ_w/λ_o (after Dyes, Caudle, and Erickson, "Oil Production after Breakthrough—As Influenced by Mobility Ratio, courtesy *Trans.*, © 1954, SPE-AIME)

Determining ultimate vertical sweep efficiency. Most reservoirs appear to perform as though they are comprised of a series of parallel reservoirs of different permeabilities. This is to be expected since the permeability tends to vary vertically or at right angles to the bedding plane. A thickness increment represents an increment of geologic time. Thus, we expect grain size and the type of minerals being deposited to vary with the geologic time. Therefore, we expect the nature of the reservoir to vary vertically or at right angles to the bedding plane.

This stratification of reservoirs results in a variation in permeability that causes considerable difficulty during a frontal displacement of the reservoir oil. Consider Fig. 9–13, which schematically represents a reservoir that contains three strata of different permeabilities. Suppose water injection is initiated in this reservoir with a water that has the same mobility as the oil being displaced. No free gas is in the reservoir. Then the fraction of the total injection going into each zone would be proportional to the permeability thickness product. The

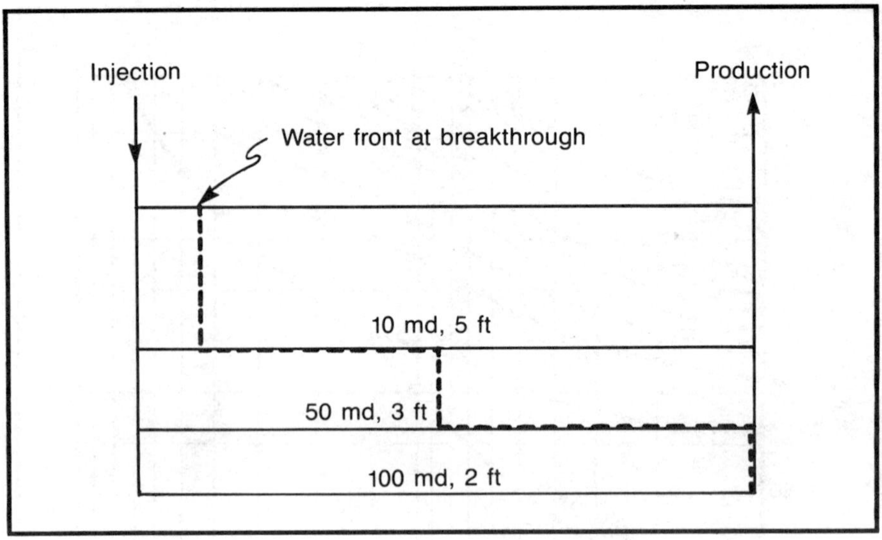

Fig. 9–13 Schematic of permeability variation in a reservoir

100-md zone would be taking one-half of the injected water, 100 × 2/(100 × 2 + 50 × 3 + 10 × 5). The 50-md zone would be taking 1/8 of the water. Also, the velocity of the fluid would be proportional to the permeability, so velocity of the water in the 100-md zone would be twice that in the 50-md zone and 10 times that in the 10-md zone.

Then at breakthrough of water into the producing well in the 100-md zone, the 50-md zone would only be one-half filled with water and the 10-md zone would only be one-tenth filled with water. At this time only 30% of the mobile oil has been produced, but from that time on only one-half of the produced fluid would be oil. The other half would be cycling through the 100-md zone. Therefore, at this time only half of the injected water is displacing oil, and the other half is circulating through the 100-md zone. After all of the mobile oil in the 50-md zone has been displaced, 60% of the mobile hydrocarbons would have been produced but only one-eighth of the injected water would then be displacing oil.

Unless a mobility control agent is being used in the injected water, such a situation is much worse because the water is more mobile than the oil. As the less mobile oil is displaced by the more mobile water, the percentage of water entering the more permeable zones becomes even higher. Stratification behavior such as this persists, even when there is little evidence of actual vertical stratification. In the latter case it is probably caused more by a microscopic permeability distribution than it is by actual stratification. Most of the early methods proposed for predicting the behavior of a waterflood concerned themselves almost exclusively with the effect of permeability stratification.

Probably the best-known method of waterflood prediction is the Stiles method that accounts only for permeability stratification in predicting the behavior history of a waterflood.[5] Consequently, the Stiles method usually does not accurately predict the early flood life when the sweep efficiencies in the most permeable strata control the behavior. The Stiles method predicts the oil producing rate prior to water breakthrough into the producing well as equal to the effective injection. Most floods experience a peak oil producing rate that is one-third to one-half of the effective injection rate. However, by the time a flood is approaching its economic limit, as noted under horizontal sweep, the horizontal sweep is relatively unimportant in controlling the behavior of the flood. Therefore, the Stiles method provides a good, easy means of estimating the ultimate vertical sweep efficiency, R. Consequently, it is recommended that the Stiles method with one modification be used to predict the ultimate vertical sweep efficiency.

The forms of the Stiles equations presented in the Craft and Hawkins text are probably the easiest to understand and the simplest to derive.[6] Basically, the results of a Stiles analysis provide a relationship between the fractional water cut, f_w, of the total (oil plus water) producing rate and the fraction of the mobile oil that has been produced, R. These two performance parameters are governed by the following equations:

$$f_w = \frac{Ac_j}{[Ac_j + (c_t - c_j)]} \tag{9.5}$$

$$R = \frac{[h_j k_j + (c_t - c_j)]}{h_t k_j} \tag{9.6}$$

Where:

$$A = \frac{B_o(k_r/\mu)_w}{(k_r/\mu)_o} \tag{9.7}$$

The meaning of the capacity symbols, c, and the thickness symbols, h, can best be explained by example. Examine Fig. 9–14. The permeability distribution curve is a plot of permeability versus the thickness of the formation that has a permeability equal to or greater than the subject permeability. For example, 15 ft (h_j) of the total reservoir thickness of 29 ft (h_t) has a permeability equal to or greater than 127 md (k_j). The capacity curve, c, is obtained from the permeability distribution curve by summing the products of the individual permeabilities and the thickness they represent from a zero thickness to the subject thickness. Mathematically, the capacity is the integral of the permeability

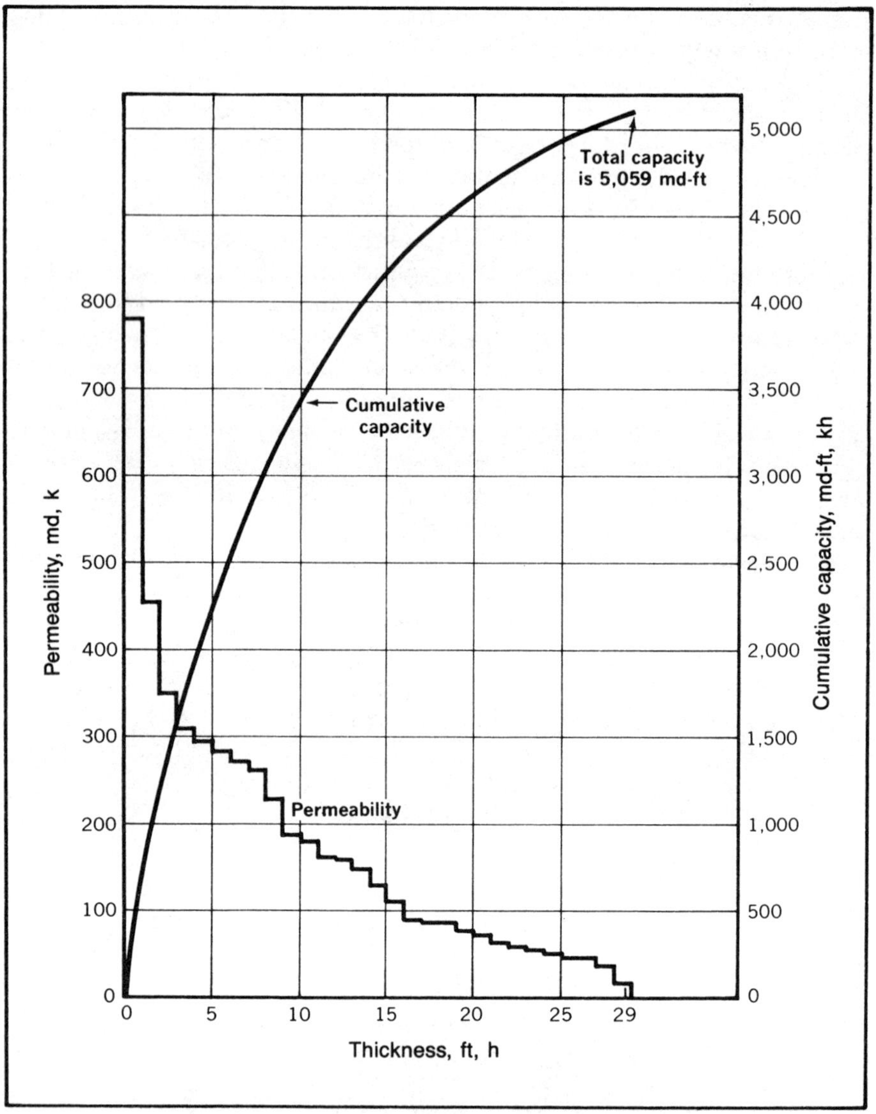

Fig. 9–14 Permeability and capacity distribution

distribution curve from a thickness of zero to the thickness of interest. Thus, the capacity curve represents the permeability thickness of the most permeable part of the reservoir being considered. For example, the most permeable 15 ft of the reservoir thickness represented in Fig. 9–14 has a capacity of 4,187 md-ft compared with the total reservoir capacity of 5,059 md-ft.

Eq. 9.5 is based on the assumption that at any particular time the thickness producing oil is the least permeable portion of the reservoir and the thickness producing the water is the most permeable portion of the reservoir. Thus, if the mobility of water and oil are the same, the ratio of oil to water production is the ratio of the capacity of the reservoir thickness producing oil to the capacity of that producing water. However, since water normally flows more readily than oil, the difference in mobilities of the two fluids must be taken into account by weighting the capacities with the mobility ratio.

An example may help in understanding Eq. 9.5. Assume that 15 ft of the formation represented by Fig. 9–14 has been flooded so 15 ft of the formation produces water. The remaining 14 ft produces oil. There is no free-gas saturation in the reservoir at the start of the flood. If mobilities of oil and water are the same, the rate of production of each is proportional to the capacity of the thickness from which each is producing. Then the water cut would be 4,187/[4,187 + (5,059 − 4,187)]. However, suppose the fluid in the water-producing thickness flows two times as easily as the fluid in the oil-producing thickness. Then the water cut would be (2)(4,187)/[(2)(4,187) + (5,059 − 4,187)]. These figures represent the water cut at reservoir conditions. To correct to stock-tank conditions, the oil production rate must be divided by the oil formation volume factor, B_o. In Eq. 9.5 this is done by including the oil formation volume factor in the parameter, A.

The justification of Eq. 9.5 may be simpler mathematically. The f_w is defined as $q_w/(q_w + q_{o\ stb})$. Remember that the Darcy equation is:

$$q = \frac{1.127kA}{\mu}(\Delta p/\Delta x) \tag{2.3}$$

Eq. 2.3 can be applied at the wellbore to the cross section of flowing water, A_w, to obtain q_w and to the cross section of flowing oil, A_o, to get q_o. Subsequently, $q_{o\ stb} = q_o/B_o$, so we can write the fractional water cut in stock-tank barrels as:

$$f_w = \frac{1.127k_{rw}k_{zw}\,A_w\,(\Delta p/\Delta x)_w/\mu_w}{1.127k_{rw}\dfrac{k_{zw}}{\mu_w}A_w\,(\Delta p/\Delta x)_w + \dfrac{1.127k_{ro}}{B_o\mu_o}k_{zo}A_o\,(\Delta p/\Delta x)_w} \tag{9.8}$$

Where:

k_{zw} and k_{zo} = Average permeabilities of the zones producing water and oil, respectively

$(\Delta p/\Delta x)_w$ = Pressure gradient at the wellbore

Since the cross section at the well is $2\pi r_w h$, note that the $2\pi r_w$ in A_w and A_o cancel, leaving h_j and h_o. When Eq. 9.8 is simplified, we obtain:

$$f_w = \frac{k_{rw}k_{zw}\,h_j/\mu_w}{(k_{rw}k_{zw}h_j/\mu_w) + (k_{ro}k_{zo}h_o/B_o\mu_o)} \tag{9.9}$$

Dividing the numerator and denominator by $(k_{ro}/B_o\mu_o)$ and substituting the capacity of the zone producing water, c_j, for $(k_{zw}h_j)$ and the capacity of the zone producing oil $(c_T - c_j)$ for $(k_{zo}h_o)$, we obtain:

$$f_w = \frac{\dfrac{B_o(k_{rw}/\mu_w)}{(k_{ro}/\mu_o)}\,c_j}{\dfrac{B_o(k_{rw}/\mu_w)}{(k_{ro}/\mu_o)}\,c_j + (c_T - c_j)} \tag{9.10}$$

Then substituting A according to Eq. 9.7 into Eq. 9.10, we obtain Eq. 9.5.

The basis for Eq. 9.6, the fractional recovery equation, may be best understood by again referring to Fig. 9–14. At the instant the production in the 15th most permeable foot of the formation goes from oil to water, notice that all of the oil has been produced from the 15 most permeable feet (h_j), which are 15/29 (h_j/h_t). Thus, we can conclude that the fraction of the total mobile oil that has been recovered from the h_j thickness after it begins producing water is:

$$(R)_{hj} = \left(\frac{h_j}{h_t}\right) \tag{9.11}$$

Also, note that some oil has been produced from all of the thicknesses with lesser permeabilities, those still producing oil, $h_t - h_j$. For example, we can assume that fluid velocity in the individual zones is proportional to the permeability prior to water breakthrough in that zone. Then at the instant the 15th foot with a permeability of 127 md (k_j) is flooded out—the water front just reaches the producer in this zone—the 16th foot with a permeability of 109 md would have had 109/127 of the oil in that foot displaced. This fraction (k_i/k_j) would be filled with water. Thus, we can say that the section of the reservoir that is less permeable than the zone that has just been flooded—the thickness producing oil $(h_t - h_j)$—has a fractional oil recovery of the total mobile oil of:

$$(R)_{h_t - h_j} = \sum_{i=j}^{i=t} [(k_i/k_j)h_i/h_t] \tag{9.12}$$

Note that k_j is a constant at any particular time, and thus, k_j and h_t can be removed from the summation:

$$(R)_{h_t - h_j} = \frac{1}{k_j\,h_t} \sum_{i=j}^{i=t} (k_i h_i) \tag{9.13}$$

Then the sums of the k_ih_i products represent the capacity of the (h_t − h_j) thickness. This capacity is the difference between the capacity of the total thickness and the capacity of the h_j thickness, or (c_t − c_j). Eq. 9.13 then becomes:

$$(R)_{h_t - h_j} = \frac{(c_t - c_j)}{k_j h_t} \tag{9.14}$$

If we add this expression to the fraction of total mobile oil recovered from the completely flooded portion of the reservoir at this time, as shown in Eq. 9.11, we obtain an expression for the total fraction of recovery, R, when a thickness h_j has been flooded:

$$R = \frac{h_j}{h_t} + \frac{(c_t - c_j)}{k_j h_t} \tag{9.15}$$

When the first term of Eq. 9.15 is multiplied by k_j/k_j, it is possible to rewrite the equation as Eq. 9.6.

Stiles assumes that Eq. 9.5 and 9.6 can be used to obtain a curve of the fractional recovery, R, versus the water cut, f_w. This curve can then be interpreted in terms of oil and water producing rates versus time. The Stiles method has no provision for an initial gas saturation, except to assume that the first water injected displaces all of the free gas from all of the permeability zones before oil is produced from any zone. This is, of course, inconsistent with the basic permeability distribution concept. Also, the Stiles method does not permit the variation of pattern or horizontal sweep with cumulative injection into a zone. The Stiles publication uses a constant sweep efficiency correction. Neither of these assumptions should introduce much prediction inaccuracy late in the flood life. All of the free gas will have been displaced from most zones that will ultimately affect performance. Also, as noted, most of the zones will be producing 100% water or will not have had water breakthrough and, thus, will require no horizontal sweep correction. Therefore, the average horizontal sweep will be very large, perhaps 0.9 or 0.95.

The engineer should note that the method also omits consideration of many other phenomena that influence the behavior of a flood, such as the variation in saturation along stream lines—the Stiles method assumes piston-like displacement—the variation in average mobility along a stream line with time, and the variation in saturations, residuals, and porosities from zone to zone. However, for most floods these errors are relatively unimportant in determining the recovery at abandonment.

In using the Stiles method to determine the flood efficiency at flood abandonment, it is not necessary to generate the entire recovery versus water cut curve or even all of the permeability and capacity distribution curves. With the water-cut fraction at abandonment assumed, the corresponding c_j can be calculated from Eq. 9.5. The total capacity, c_t, can be evaluated from the average permeability and total thickness if it is not more accurately known. The thickness corresponding to c_j can then be evaluated by summing the capacities from the least permeable bed in an increasing permeability direction until a capacity of $c_t - c_j$ is reached. The thickness corresponding to this capacity is $h_t - h_j$, from which the thickness, h_j, can be evaluated. Once h_j and the corresponding k_j are known, the corresponding fraction recovery, R, can be determined from Eq. 9.6.

When using Eq. 9.5 and 9.7, care should be taken in evaluating mobility for the zone producing oil. Mobility for the zone producing water can be taken as the mobility of water in the water bank without introducing significant error. However, note that the zone producing oil is flowing oil at the producing well, but is is flowing water at the injection well. This should be considered when evaluating fluid mobility in the oil producing zone. For example, for a five-spot pattern mobility of the zone producing oil is affected by the mobility of oil in the oil bank that exists at the producing well and the mobility of water in the water bank that exists at the injection well.

The effective mobility in a zone can generally be closely estimated by considering only the resistance to flow at the injection and production wells. The resistance to flow is the reciprocal of the mobility, γ. Since resistances in series are additive, like pressure drops in series, we can write an expression for the resistance to flow in the oil producing zone:

$$\frac{1}{\gamma_{oz}} = \frac{1}{\gamma_w} + \frac{1}{\gamma_o} \tag{9.16}$$

Then the mobility in the oil zone is the reciprocal of this, or:

$$\gamma_{oz} = \frac{1}{[(1/\gamma_w) + (1/\gamma_o)]} \tag{9.17}$$

If we develop a similar expression for the resistance to flow and the corresponding mobility in the water producing zone, we obtain:

$$\frac{1}{\gamma_{wz}} = \frac{1}{\gamma_w} + \frac{1}{\gamma_w} \tag{9.18}$$

$$\gamma_{wz} = \frac{\gamma_w}{2} \tag{9.19}$$

Then the expression for the mobility ratio is obtained by dividing the expression for γ_{wz} by the expression for γ_{oz}, to obtain:

$$\frac{\gamma_{wz}}{\gamma_{oz}} = \frac{[1 + (\gamma_w/\gamma_o)]}{2} \tag{9.20}$$

Since the Stiles A is $(\gamma_{wz}/\gamma_{oz})B_o$, we can write an expression similar to Eq. 9.7 for the effective A as:

$$A_{eff} = \frac{B_o[1 + ((k_r/\mu)_w/(k_r/\mu)_o)]}{2} \tag{9.21}$$

Evaluation in this manner amounts to a minor modification of the Stiles method. The example analysis provided by Stiles simply bases A on mobility of water in the water bank and mobility of oil in the oil bank.

Work problem 9.3 and check the solution against the one in appendix C to test knowledge of this section and to provide a background for the discussions that follow. In this problem note that the best estimates available on the connate water and residual oil saturations are different than those indicated by the relative permeability. This is a common occurence apparently caused by the difficulty of measuring the irreducible saturations during a relative permeability measurment.

To overcome this inconsistency, remember that the relative permeability to the nonwetting phase in the presence of an irreducible wetting phase should be close to 1.0 since the wetting phase is in the smaller pores. These pores flow insignificant amounts of oil if they contain oil, since the oil viscosity is normally greater than that of the water. Thus, the relative permeability of oil in the oil bank should be read at the irreducible water saturation indicated by the relative permeability data.

However, the relative permeability to water in the water bank should be read at the best estimate of the residual oil saturation, not at the residual oil saturation indicated by the relative permeability data, since the nonwetting phase tends to stop flow in the largest pores. Thus, the relative permeability to the water becomes sensitive to the residual oil saturation.

PROBLEM 9.3: Calculating the Total Displacement Efficiency and Recovery from a Waterflood

In problem 9.2 the saturations at the start of the flood and the gross swept volume are evaluated for a proposed flood. Complete this problem by calculating the ultimate flood recovery and stating it as (1) stock-tank barrels, (2) percentage of oil originally in the reservoir, and (3) percentage of oil originally in the swept area. Assume the economic water cut as 95% and S_{wc} = 0.2.

When performing this calculation, it will be necessary to determine the total displacement efficiency based on the Stiles method. Assume that the oil formation volume factor does not change significantly during the flood, that the average horizontal sweep at the economic limit is 0.9, that the average conventional core-analysis oil saturation is 19%, and that the permeability distribution of the reservoir—listed in descending order of magnitude—is as follows:

Permeability, md	*Thickness, ft*
780	1
455	1
348	1
309	1
295	1
282	1
271	1
260	1
227	1
185	1
179	1
160	1
158	1
147	1
127	1
109	1
87	1
85	2
76	1
70	1
61	1
57	1
54	1
50	1
46	2
35	1
15	1
Average 174	Total 29

Use Fig. 9–15 for relative-permeability data. The reservoir data are as follows:

Oil formation volume factor = 1.1
Oil viscosity = 6.38 cp
Water viscosity = 0.7 cp
Oil saturation after primary = 0.561
Original oil in place before primary = 155.2×10^4stb
Gross swept volume under flood = 900 acre-ft.

Problem 9.3 emphasizes that the fractional recovery from a reservoir depends on the basis for the fraction. The problem illustrates the difference caused by using the swept reservoir volume or the total reservoir volume. Obviously, there also is a difference depending on whether the basis is mobile oil or total oil. If the swept volume is used, the method of delineating the swept volume is very important. The point is that when we are comparing recovery efficiencies of floods, we must be careful that all of the bases are the same.

Permeability distribution data used in problem 9.3 imply that some previous study of the reservoir has led to the conclusion that a particular strata in the reservoir is the most permeable strata and has a particular average permeability. For example, the 1-ft interval between 5–6 ft from the top of the reservoir may have the peak permeability in each well and an average permeability of 780 md. Similarly, the strata 3 ft from the bottom of the pay zone may be the least permeable and may have an average permeability of 15 md. A strict interpretation of this type would then mean that reservoir thickness in each well would have to be 29 ft. However, we learned in problem 9.3 that this permeability distribution can be used as a general distribution that applies to all of the reservoir, regardless of the thickness variation throughout the reservoir.

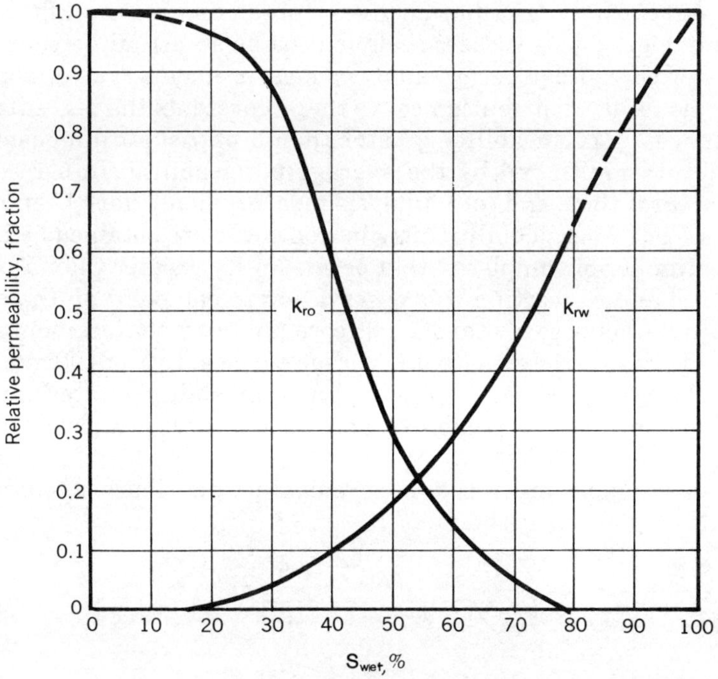

Fig. 9–15 Example relative permeability data

Generally, reservoirs are not sufficiently uniform from well to well to permit assumption that the 5th foot from the top is the most permeable or the 3rd foot from the bottom is the least permeable. When a reservoir can be so characterized, permeability distribution should be based on such an analysis. However, in most cases we find that it is impossible to characterize a reservoir in such a manner. Nevertheless, flooded reservoirs behave like stratified reservoirs. Thus, we must approach the stratification problem from a statistical rather than a geological standpoint to obtain realistic predictions.

Permeability distribution can be put on a statistical basis by dividing thickness values by total thickness and permeability values by average permeability. Then the thickness is the fraction of total thickness, and the permeability is a multiple of average permeability. Since the average permeability on this scale is 1.0 and the total thickness is 1.0, the total capacity is 1.0 × 1.0, or 1.0. The individual values of capacity are a fraction of the total capacity. This dimensionless capacity curve can be obtained mathematically by dividing individual capacity values by total capacity and, as before, dividing individual thickness values by total thickness.

In most reservoirs it is found that the dimensionless form of permeability and capacity distribution gives a prediction that best fits actual flood behavior. The thickness scale is used as an actual percentage of samples or percentage of core-analysis measurements. The dimensionless permeability distribution curve then represents the percentage of samples with a permeability greater than a particular dimensionless permeability multiplied by the average permeability. In a reservoir where several thousand core-analysis measurements have been made, the data for the permeability distribution curve are obtained by counting the number of samples with a permeability greater than a particular value and converting this observation to one point on the curve. For example, suppose a total of 5,000 core-analysis measurements have been made in a reservoir with an average permeability of 100 md. Then if we find that 500 of the samples have a permeability greater than 300 md, one point on the permeability distribution curve would be 3.0 (300/100) versus 0.1 (500/5,000).

When a dimensionless type of permeability distribution is used, the two Stiles equations become:

$$f_w = \frac{(A\ c_j')}{[A\ c_j' + (1 - c_j')]} \tag{9.5a}$$

$$R = \frac{[h_j' k_j' + (1 - c_j')]}{k_j'} \tag{9.6a}$$

In Eqs. 9.5a and 9.6a the prime (′) denotes the dimensionless thickness, permeability, and capacity. These dimensionless forms of the Stiles equations are derived from Eqs. 9.5 and 9.6 by dividing the numerator and denominator of both equations by $h_t k_{avg}$ where k_{avg} is the average permeability. When Fig. 9–14 is converted to a dimensionless permeability and capacity distribution, we obtain Fig. 9–16.

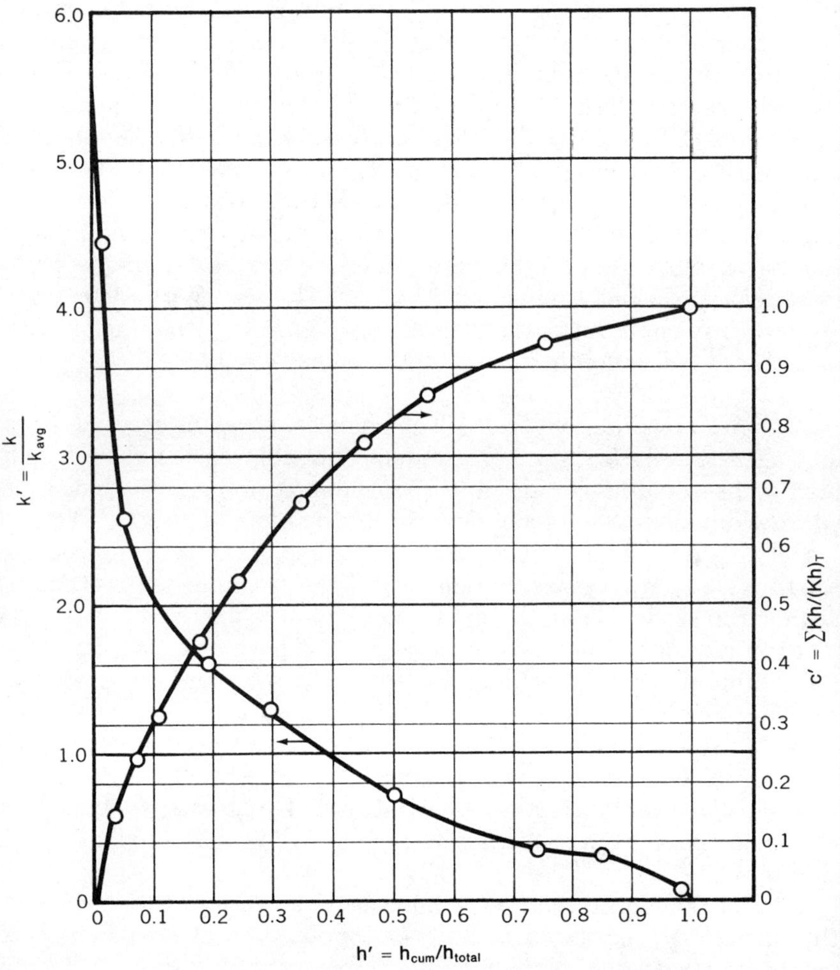

Fig. 9–16 Dimensionless permeability and capacity distribution

Predicting Rate Versus Time Performance

Most practicing engineers can do a creditable job of predicting total flood recovery. However, the prediction of the rate at which oil will be recovered is a more difficult task. From a standpoint of profitability, this estimate is almost as important as predicting total ultimate recovery. Due to the time value of money, the rate at which production is obtained can be the difference between a profitable flood and an unprofitable flood, both of which reach the same ultimate recovery. Much has been written on the prediction of waterflood behavior. No effort is made here to review all of the various contributions of the engineers and researchers. Generally, each method seems to concentrate on one particular difficulty in predicting the behavior of a flood. As noted, Stiles emphasizes the need to account for the permeability distribution.[7] Suder and Calhoun account for the oil bank formation; Prats et al., work with the variation of the injectivity into a particular strata with time; Higgins and Leighton account for the variation of saturation along a particular stream line; Dyes, Caudle, and Erickson are mainly concerned with sweep efficiency; and Dykstra and Parsons predict residual saturations empirically.[8,9,10,11,12] These authors and researchers certainly point out that they also consider many other factors. However, these items dominate the methods discussed.

The recommended method. The basic *bookkeeping method* for predicting rate versus time performance probably comes closest to the method presented by Craig in the SPE Monograph on waterflooding.[13] Others may believe it is more similar to the method used by Prats and his associates.[14] At any rate the idea is not original. It may be of interest that, for many years in teaching this recommended method, the author referred to it as the *modified Prats method.* The bookkeeping method assumes a knowledge of injectivity of each permeability zone as a function of cumulative injection into that zone and a knowledge of recovery of each zone as a function of cumulative injection into that zone. Many engineers feel that, once they have the data necessary to perform the bookkeeping calculations, they have practically solved the problem. This conclusion is somewhat true. The problem of generating the basic data for the bookkeeping system will be discussed later in the chapter.

The bookkeeping method is necessary because of the variation in the injectivity, resistance to flow, or conductance of a particular permeability strata during its flood life. The strata may start its flood life essentially with only gas flowing. Therefore, the fluid flowing has a high mobility. Later, all of the gas may have been displaced from the strata, and much less mobile oil and water are flowing in the strata.

Still later, all of the mobile oil may have been displaced from the strata, and the less mobile oil has been displaced by the normally more mobile water. We learned that various strata do not take water at the same rate because of permeability differences. Consequently, no two strata of different permeabilities are at the same state of depletion at the same time. Therefore, the relation between the injectivity or resistance to the flow in the various strata is continually changing as the flood develops.

For example, a particular strata in a flood may originally take 10% of the total injected water, but later, it may be taking only 4% of the water. As production continues, the take may increase to 8–9%. We cannot determine recovery from each zone, unless we know how much water is in each zone at a particular time.

The injectivity rate into a particular strata may be approximated by a history such as that shown in Fig. 9–17, if minor variations are omitted and the reservoir contains a substantial gas saturation at the start of the flood. These data are more useful to us in a dimensionless form where the injection rate is stated as the ratio of the initial rate divided by the existing rate into this strata. Cumulative injection into the strata is stated as a function of strata pore volume. Replotting the data in Fig. 9–17 gives us a plot such as Fig. 9–18.

As noted, before our bookkeeping procedure can be initiated, we must also have a recovery curve for each of the permeability strata. The recovery data should be in the form presented in Fig. 9–19, with cumulative oil production from the strata and cumulative injection stated as a function of pore volume. The oil formation volume factor is included in the recovery term so it represents a fractional recovery in the reservoir, with recovery in barrels at stock-tank conditions being N_{pf}.

By using the injectivity data such as that given in Fig. 9–18, the permeability distribution, and other reservoir parameters, it is possible to construct a plot of (W_i/V_p) versus a function of time, t_r, for each of the permeability zones of the reservoir. Such a set of data is illustrated in Fig. 9–20 for a reservoir where only three zones are considered. The reduced time, t_r, is the injection time divided by a group of reservoir constants that do not vary from zone to zone. Consequently, the relationship between the cumulative injection into each zone at a particular time can be determined from a plot such as Fig. 9–20 when zone 1 has a cumulative injection of 1.0 bbl of water/bbl of pore volume. Zone 2 has a cumulative injection of 0.45, and zone 3 has a cumulative injection of 0.2.

Since our basic data include recovery versus cumulative injection for each zone, we can determine oil production for each zone at this particular time represented by this t_r value. Since we know the pore volume for each zone, we can convert recovery and cumulative injection

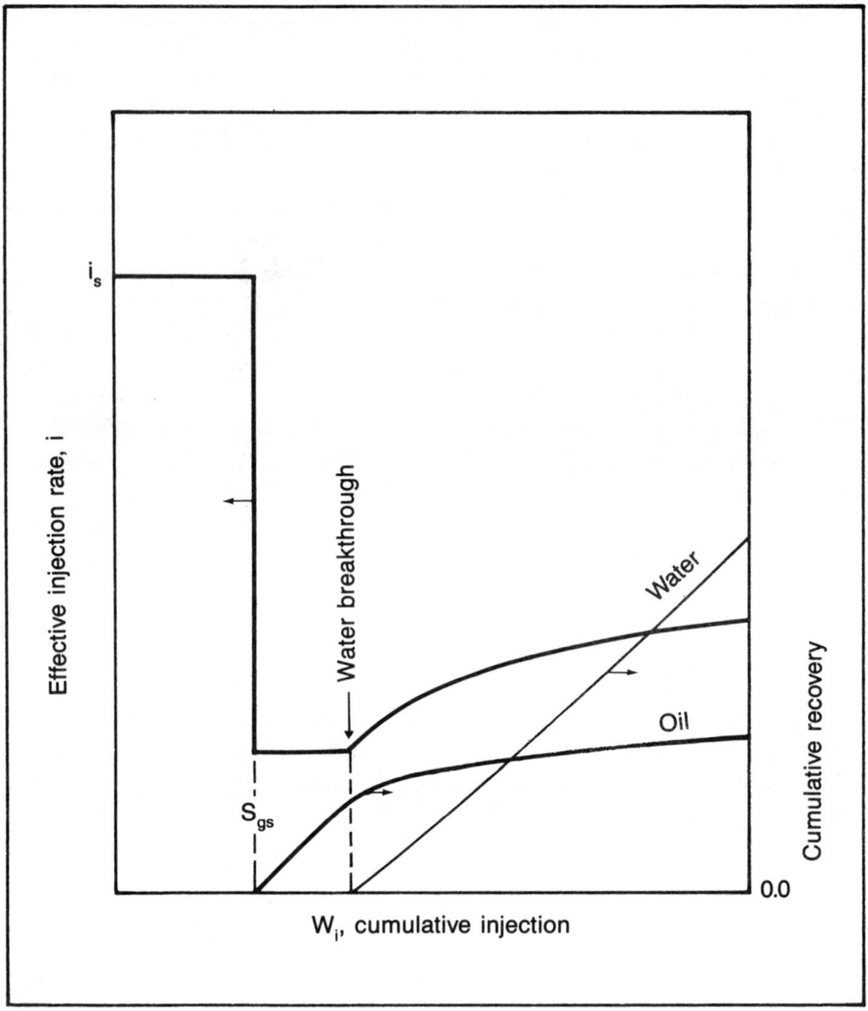

Fig. 9–17 Approximate variation of the injection rate in a strata of a five-spot ($S_{gs} > S_{ge}$)

data to barrels. The cumulative injection and injection rate can then be used to calculate injection time in days.

To determine the basis for the plot of cumulative injection versus t_r for a particular zone, note that the increment of time, Δt, necessary for the injection of an increment of water, ΔW_{in}, into a particular strata at an average rate, i_n, is $\left(\frac{\Delta W_i}{i}\right)_n$. The total time required for the injection of W_{in} cumulative barrels of water into a particular strata is the summation of the time increments:

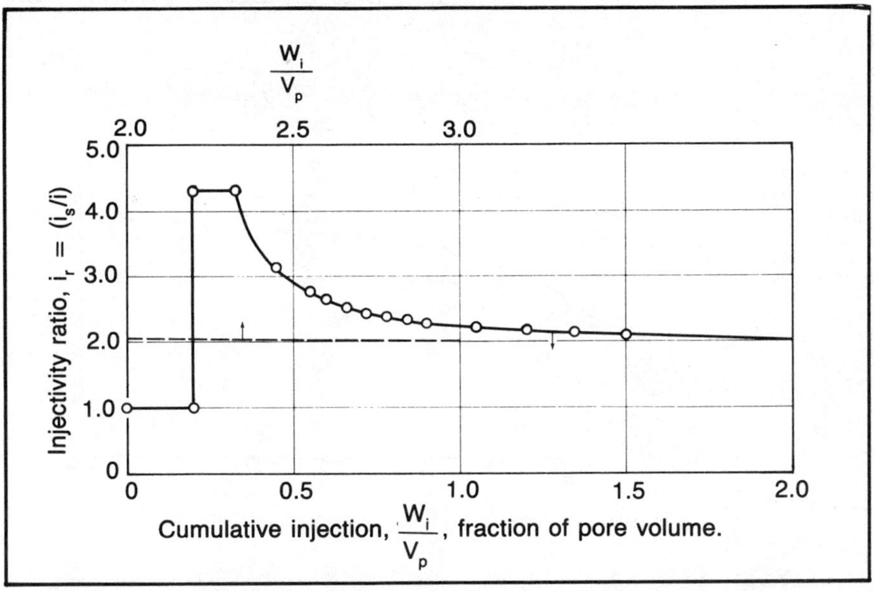

Fig. 9–18 Example injectivity ratio data

$$t = \sum \left(\frac{\Delta W_i}{i} \right)_n \tag{9.22}$$

Our given data are not in barrels, W_{in}, and barrels per day, i_n, but are stated as W_i/V_p and i_s/i. Therefore, we must substitute functions of these values for W_i and i in Eq. 9.22:

$$i_n = \frac{i_{sn}}{(i_s/i)_n} \tag{9.23}$$

$$(W_i)_n = \left(\frac{W_i}{V_p} \right)_n \left(\frac{\phi A h}{5.615} \right)_n \tag{9.24}$$

The second set of parentheses in Eq. 9.24 is the pore volume stated as a function of the horizontal area, A, affected by each injection well and the thickness, h. Both are stated in feet.

When Eqs. 9.23 and 9.24 are substituted into Eq. 9.22, we obtain:

$$t = \sum \frac{\Delta (W_i/V_p)_n (\phi A h/5.615)_n (i_s/i)_n}{i_{sn}} \tag{9.25}$$

This expression can be further simplified when we substitute a function of the capacity, kh, for the initial injection rate into the *n*th strata, i_{sn}. Any flow rate can be equated to the transmissibility, kh/μ, multiplied by the pressure drop, a geometric constant, and a numerical constant. If we group all of the parameters together that are the same

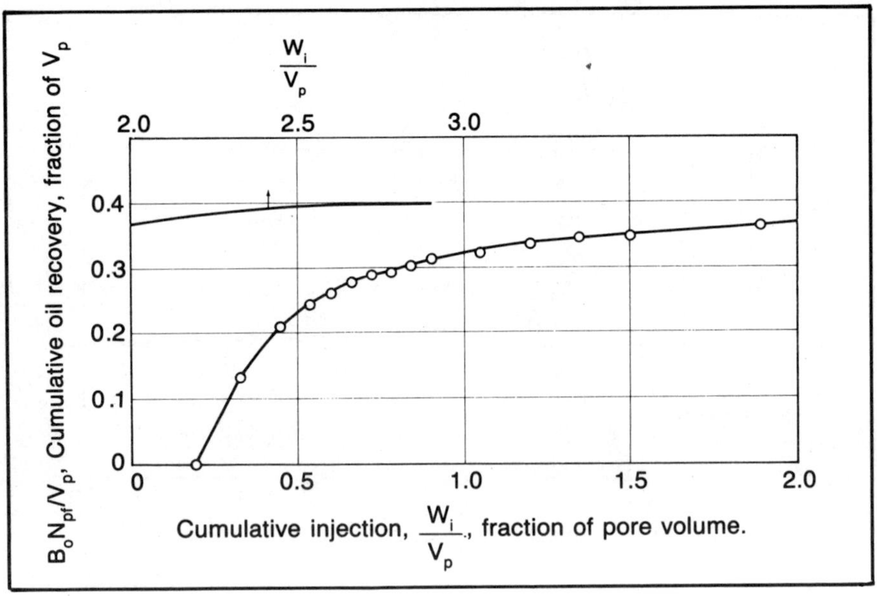

Fig. 9–19 Example recovery data

for all of the strata, we can show that the injection rate into each strata initially is proportional to the capacity of the strata $(kh)_n$:

$$i_{sn} = (\text{numerical constant})(\text{geometric constant})\Delta p(kh/\mu)_n \qquad (9.26)$$

$$i_{sn} = (\text{constant})(kh)_n \qquad (9.27)$$

Note that the effective viscosity of the strata, μ_n, is included in the group of constants. This is done because the effective viscosity in the reservoir initially is based on essentially the same saturation throughout the reservoir. Thus, μ_{sn} is the same for all strata. At any other time in the flood history, the saturations differ from strata to strata. However, at the start of the flood, they differ only to the extent that the primary drainage differs from strata to strata. This difference has a negligible effect on the difference in the initial injection rates from strata to strata.

When Eq. 9.27 is substituted for the initial injection rate of a strata, i_{sn}, in Eq. 9.25, we note that the thickness, h_n, cancels. Then the equation can be arranged to obtain:

$$t = \left[\frac{A}{(\text{constant})(5.615)}\right]\left(\frac{\phi}{k}\right)_n \sum\left[(i_s/i)_n \Delta(W_i/V_p)_n\right] \qquad (9.28)$$

Remember that the constant includes only reservoir parameters that are the same for all strata. Note that the drainage area, A, and the

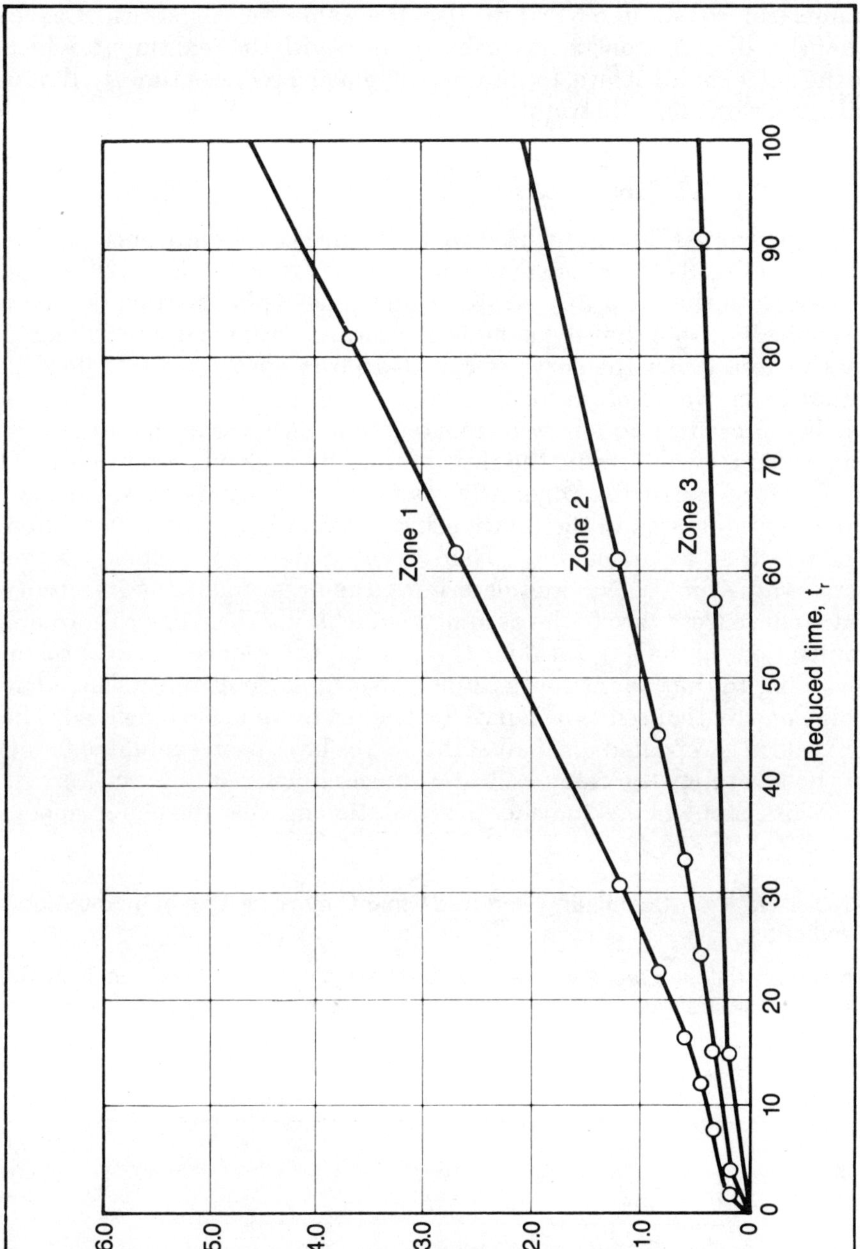

Fig. 9–20 Reduced-time, t_r, curves

numerical constant 5.615 are also the same for all strata. Consequently, all of the constants can be grouped with the real time, t, which is the same for all strata, to obtain a correlating reduced time, t_r, that is also the same for all strata:

$$t_r = \frac{t}{A/5.615(\text{constant})} = (\phi/k)_n \sum (i_s/i)_n \Delta(W_i/V_p)_n \tag{9.29}$$

This expression can be used with the injectivity data, such as that shown in Fig. 9–18, to construct curves as in Fig. 9–20. Eq. 9.29 is used to calculate the t_r for any strata at any time. This equation is solved graphically. Note that the summation term is graphically represented by the area under the curve of Fig. 9–16 from $(W_i/V_p) = 0$ to (W_i/V_p) equal to the value of interest.

The procedure to follow in constructing t_r curves such as those in Fig. 9–20 is to determine the ratio of the porosity and absolute permeability for a particular zone. Any consistent unit can be used since we are only interested in the relationship of the (W_i/V_p) values for different strata at a particular t_r. Then to calculate the t_r values, assume some value for (W_i/V_p) and determine the area under the injectivity ratio curve from zero to the assumed value of W_i/V_p. When this area is multiplied by the ϕ/k ratio for that strata, the reduce time, t_r, corresponding to that particular assumed (W_i/V_p) is determined. Then one point on the reduced time curve for that strata has been defined. The procedure is repeated until all of the curves have been evaluated for all of the strata for the range of water injection desired.

Work problem 9.4 and check the solution against the one in appendix C.

PROBLEM 9.4: Calculating Reduced-Time Curves for Use in a Waterflood Prediction

To demonstrate the recommended method, assume that a 10-acre five spot in a reservoir can be treated as having three zones with properties as follows:

Zone No.	*h, ft*	*k, md*	ϕ, %	V_p, *bbl*
1	5	25	25	97,000
2	10	10	20	155,000
3	15	2	15	175,000

The following reservoir data apply for all zones:

S_{gi} = 20%
S_{os} = 65%
S_{wc} = 15%
S_{or} = 25%

μ_w = 0.7 cp
μ_o = 4.15 cp

Assume i_r and recovery curves are as shown in Figs. 9–18 and 9–19, respectively, and relative permeabilities are as in Fig. 9–15. Note that the connate water and residual oil values differ from those indicated by the relative permeability data.

In other problems we determine strata injectivity curves, strata recovery curves, the five-spot recovery in barrels of oil, and water versus time. At this time the engineer is asked to verify the reduced-time curves of Fig. 9–20.

Converting reduced-time data to production versus injection. Once reduced-time curves have been generated, they can be interpreted in terms of cumulative total injection and production of both oil and water. The reduced-time curves show the relationship between cumulative injection into each zone at any particular reduced time. Consequently, even though we do not know directly the calendar time represented by any particular time, we can conclude that at an unknown real time the cumulative total injection will be some particular value. This becomes meaningful when we recognize that a particular cumulative injection into a particular zone represents a particular cumulative production from that zone. Thus, for some unknown time we can determine the cumulative total injection and the corresponding total oil and water recovery. When such numbers are calculated for enough reduced times, we have data representing cumulative oil and water production versus cumulative injection. By assuming some value for the injection rate, these data can in turn be used to calculate the cumulative oil and water production versus time.

To determine the cumulative injection representing a particular reduced time, we plot reduced-time data as in Fig. 9–20 and read the pore volumes of cumulative injection into each zone for that reduced time. Each of the pore volumes of cumulative injection can then be converted to barrels by multiplying by the pore volume of that zone. The total injection then is the sum of injections into each zone:

$$(W_i)_{total} = \sum_{n=1}^{n=y}\left[(W_i/V_p)_n\, V_{p_n}\right] \qquad (9.30)$$

The cumulative oil recovery corresponding to this cumulative injection can be determined by finding the pore volumes of oil that correspond to the pore volumes of cumulative injection for each zone. This calculation is accomplished using basic recovery data such as that in Fig. 9–19. Once the pore volumes of reservoir oil recovered from each zone have been determined, these figures can be converted to stock-tank barrels by dividing by the oil formation volume factor:

$$(N_{pf})_{total} = (1/B_o)\sum_{n=1}^{n=y}\left[(B_oN_{pf}/V_p)_nV_{pn}\right] \qquad (9.31)$$

In Eq. 9.31 the oil formation volume factor, B_o, is taken outside the summation sign because it is the same for all zones in most cases. In some cases where reservoirs are very thick and the pressure and oil formation volume factor vary from zone to zone, B_o for each zone should be left inside the summation sign as B_{on}.

If we assume that the free-gas saturation in an oil-producing zone is negligible, we can determine the amount of water production from each zone by material balance based on our knowledge of oil recovery, cumulative injection, and the saturations at the start of the flood. All of the water injected into a particular zone either is in that zone or has been produced. Furthermore, all of the injected water that remains in the zone has displaced either free gas or oil. We have shown that we can evaluate the cumulative oil production at any particular time. We also know how much free gas is originally in the zone. If we assume that the gas saturation in an oil-producing zone is zero, we then know that all of the original free gas has been produced.

When we put these observations together, we find that the pore volumes of injection less the pore volumes of gas produced, less the pore volumes of oil produced give the pore volumes of water produced. When this is multiplied by the appropriate zonal pore volume, we have the barrels of water that have been produced from a particular zone. The summation of water produced from each zone results in the total water recovery from the flood:

$$(W_p)_{total} = \sum_{n=1}^{n=y}\left[(W_i/V_p)_n - S_{gin} - (B_oN_{pf}/V_p)_n\right]V_{pn} \qquad (9.32)$$

When applying Eq. 9.32, the engineer should ensure that no negative water production figures are included for any zone. The nature of the terms is such that, if the oil production is zero, the gas saturation in the zone would not be zero. Then the volume of free gas produced would not be the initial gas saturation. Therefore, the injected water volume would be greater than the initial gas saturation, and a negative number would be obtained for the volume of water produced. Thus, Eq. 9.32 should only be applied to those zones producing oil.

Work problem 9.5 and check the solution against the one in appendix C to check understanding of this section.

PROBLEM 9.5: Calculating Oil and Water Production Rates from Reduced-Time Curves

In problem 9.4 the reduced-time curves are calculated for a flood. Using these reduced-time curves (Fig. 9–20) and the recovery curves of Fig. 9–19, calculate and plot the oil production and water production versus cumulative injection to a cumulative injection value of about 320,000 bbl, or a reduced time of 40. Assume $B_o = 1.0$.

Recovery curves for individual strata. Much data are available that are usable for determining recovery curves for individual strata. We previously referred to one such set of data for a five-spot flood, Figs. 9–11 and 9–12. Dyes, Caudle, and Erickson published similar data for other patterns, as did Kimbler, Caudle, and Cooper.[15,16] All of these data can be found in appendix D of the SPE Monograph on waterflooding.[17] Since the problem lends itself to an interesting and not too difficult analysis either by use of flow models or computer models, there is much other data that can be used for the same purpose.

Remember that data must provide cumulative recovery versus time for a single layer of a homogeneous flood pattern element. Most of the data assume piston-like displacement or a minimal drag zone, a zone containing mobile oil behind the flood front, characteristic of a very permeable sand. However, the error introduced using the piston-like concept seems minimal compared to the effect of sweep efficiency. The Higgins and Leighton method can be used to obtain the strata recovery curve and can include the variation in saturation along each stream line if sufficient reservoir data and computer time are available.[18]

Fig. 9–12 is used to illustrate the method of determining a strata recovery curve from published data. Note that the recovery must be obtained from the percentage of the five-spot area swept and the corresponding cumulative injection.

The percentage of the five-spot area occupied by water must be equal to the percentage of mobile hydrocarbons (oil and gas) displaced from the reservoir by the injected water. Thus, the percentage of the five-spot area occupied by injected water is equal to the oil and gas production caused by water injection. Note in Fig. 9–12 that, if the reciprocal of the mobility ratio is greater than about 7.5, 100% of the reservoir is swept before breakthrough. Therefore, the mobility ratio needs to be less than 1/7.5, or 0.133. Since the viscosity ratio of oil to gas approaches 0.01, it appears safe to assume that the mobility ratio for oil displacing free gas generally is less than the critical 0.133.

Consequently, we can assume that mobile free gas in a strata is produced before displaced oil is produced. Thus, for a particular strata we can assume that before gas fillup the injected water volume is equal to the reservoir volume of free gas produced. After gas fillup the oil produced is equal to the injected water in the reservoir, less the reservoir volume of mobile free gas in the strata at the start of the flood. When the swept area is 100%, all of the mobile oil and gas have been displaced from the reservoir. Thus, when the swept area is less than 100%, we can state the oil and gas production as:

$$\text{Oil and gas produced} = E_H\,(S_{os} - S_{or} + S_{gi} - S_{ge})V_p \qquad (9.33)$$

If the initial mobile gas volume is subtracted from the oil and gas production and the resulting expression is solved for the reservoir barrels of oil production stated as a fraction of pore volume, we obtain:

$$B_oN_{pf}/V_p = E_H(S_{os} - S_{or} + S_{gi} - S_{ge}) - (S_{gi} - S_{ge}) \qquad (9.34)$$

To obtain the corresponding cumulative water injected from the displacement volumes injected, it is only necessary to recognize that one displacement volume is equal to the mobile oil saturation at the start of the flood, plus the mobile gas saturation at the start multiplied by the pore volume. Then the cumulative water injection as a fraction of the pore volume is:

$$\left(\frac{W_i}{V_p}\right) = (DVI)(S_{os} - S_{or} + S_{gi} - S_{ge}) \qquad (9.35)$$

Using Fig. 9–12, it is then possible to read corresponding values of the percentage of the five-spot swept and the displacement volumes injected for a particular reciprocal of the mobility ratio. These corresponding values can then be converted to reservoir barrels of oil production, as a fraction of pore volume, and water injection, as a fraction of the pore volume, using Eqs. 9.34 and 9.35.

When evaluating the residual gas saturation, S_{ge}, the engineer should remember that this value should represent the saturation in the oil bank and the water bank. The gas saturation just behind the oil-bank front is undoubtedly greater than the gas saturation further behind the oil-bank front and the water bank. The pressure in these latter positions is considerably larger, which means the gas volume and saturation are reduced because of the compression of the gas and because some or all of the gas goes back into solution in the oil as a result of the increase in pressure. Thus, for all practical purposes the residual gas saturation is often 0.0.

Work problem 9.6 to understand this conversion procedure. The solution is given in appendix C.

Problem 9.6: Calculating Individual Strata Recovery Curves

In problems 9.4 and 9.5 calculations are made predicting oil recovery and water production versus cumulative effective injection based on the individual strata injectivity and recovery curves in Figs. 9–18 and 9–19, respectively. Verify Fig. 9–19 by calculating the points representing gas fillup, water breakthrough, and injected displacement volumes of 0.75 and 1.5. Fig. 9–15 represents the relative permeability for this reservoir, and the oil and water viscosities are 4.15 and 0.7 cp, respectively. The initial gas saturation, connate water, and residual oil are 20%, 15%, and 25%, respectively. Assume the residual gas saturation is 0.0.

Injectivity curves for individual strata. As mentioned, much data can be used to generate the desired recovery data for a particular strata. However, little data are available for determining the injectivity curves for an individual strata. Fortunately, we can approximate an injectivity curve of acceptable accuracy in most cases by observing a few fundamental concepts.

To approximate the injectivity curve, we must first recognize that most pressure drops in a pattern occur at the wells, whether they are injection or production wells. The pressure distribution in a five-spot pattern quadrant emphasizes this point. Fig. 2–17 verifies that about 80% of the total pressure drop between the injection and production wells occurs in about 4% of the pattern around the injection well and 4% of the pattern surrounding the production well. Thus, accounting for the pressure drop at the injection and production wells provides a reasonably accurate estimate of the individual strata injectivity variations in the flood history.

Physically, the cross-sectional area between the injection and production wells is very large compared with the small cross section at the wells. Therefore, the wells provide most of the resistance to flow. This is true in a five spot as well as in virtually any waterflood pattern. Thus, we model the flood pattern by assuming it represents two radial systems back to back, as shown in Fig. 9–21. To illustrate the use of simple geometry to approximate the behavior of a more complex geometry, we show in chapter 2 that the model of Fig. 9–21 gives a flow equation almost the same as the exact analytical equation. We further expand this idea by assuming that saturations on the injection side are the

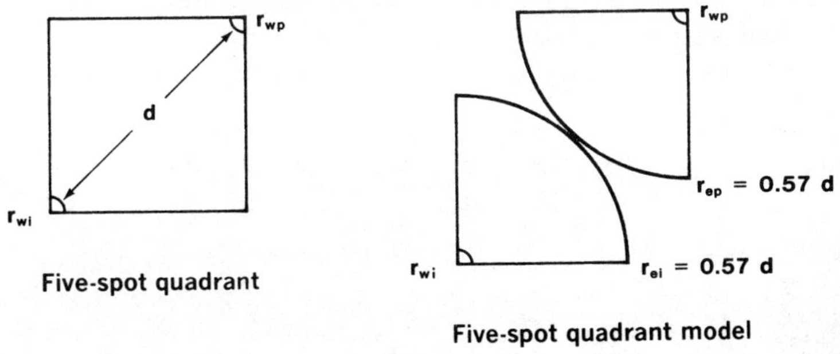

Fig. 9–21 Modeling a five spot

same as saturations at the injection well and that saturations on the production side are the same as saturations at the production well.

Using the model of Fig. 9–21, we can say that the total pressure drop between the injection and production wells is the sum of the pressure drops in the injection and production sides of the pattern:

$$p_{wi} - p_{wp} = \Delta p_i + \Delta p_p \tag{9.36}$$

We can determine an expression for the two pressure drops from the steady-state radial flow equation as a function of the injection rate, i:

$$\begin{aligned} p_{wi} - p_{wp} &= 0.141i\ \ell n(r_{ei}/r_{wi})(\mu_i/k_i)/h \\ &\quad + 0.141i\ \ell n(r_{ep}/r_{wp})(\mu_p/k_p)/h \end{aligned} \tag{9.37}$$

When we solve Eq. 9.37 for the injection rate and assume the radius ratio on the injection and production sides is equal, we obtain:

$$i = \frac{(p_{wi} - p_{wp})h/0.141\ \ell n(r_e/r_w)}{(\mu_i/k_i) + (\mu_p/k_p)} \tag{9.38}$$

Note that the numerator of Eq. 9.38 is the same for all zones at any particular time. Thus, the ratio of injection rates into two zones at a particular time is the ratio of the denominators of Eq. 9.38. By writing the numerator of Eq. 9.38 as a constant, we obtain:

$$i = \frac{\text{constant}}{[(\mu_i/k_i) + (\mu_p/k_p)]} \tag{9.39}$$

To obtain the initial injection rate into each zone, i_s, we consider only water to be flowing at the injection well and free gas to be flowing at the production well. This represents a typical situation when the reservoir has been substantially depleted by solution-gas drive prior to the start of the flood. However, if the gas saturation is very low at the start of the flood, the initial injection rate should be based on the assumption that oil is being produced.

If we assume that production is substantially free gas, note that the initial injection based on Eq. 9.39 is:

$$i_s = \frac{\text{constant}}{[(\mu_w/k_w) + (\mu_g/k_g)]} \tag{9.40}$$

When Eq. 9.40 is applied to a substantially depleted solution-gas-drive reservoir, note that the reciprocal of gas mobility is negligible compared to the reciprocal of water mobility in the water bank. Thus, we can consider the reciprocal gas mobility to be zero, and Eq. 9.40 becomes:

$$i_s = \frac{\text{constant}}{(\mu_w/k_w)} \tag{9.41}$$

Based on our assumption that injectivity does not change until a substantial change in saturations occurs at the injection or production well, and that all of the free gas is produced from a particular strata before any oil rate increase occurs, the injection rate remains constant until gas fillup has occurred in the strata. Thus, the injectivity ratio, i_s/i, is 1.0 until W_i/V_p equals S_{gi}. At that time the oil bank reaches the producing well and oil starts flowing. Then during the time between gas fillup and water breakthrough (BT) into the producing well, the injectivity ratio according to Eq. 9.39 is:

$$i \text{ for fillup to BT} = \text{constant}/[(\mu_w/k_w) + (\mu_o/k_o)] \quad (9.42)$$

In Eq. 9.42 water mobility should still be evaluated at the saturations existing in the water bank, and oil mobility should be evaluated at the saturations existing in the oil bank. A ratio of Eqs. 9.41 and 9.42 provides the injectivity ratio for the period from gas fillup to first water breakthrough:

$$(i_s/i) \text{ for fillup to BT} = 1 + (\mu_o/k_o)(k_w/\mu_w) \quad (9.43)$$

Recognizing the last term as the mobility ratio, M, we obtain:

$$(i_s/i) \text{ for fillup to BT} = 1 + M \quad (9.44)$$

During the period when water and oil are both being produced from a particular strata, the evaluation of the injectivity ratio is more difficult. It is necessary to modify the five-spot model, Fig. 9–21, so both oil and water are flowing in parallel stream lines in the producing side of the model as shown in Fig. 9–22. Using this model, the pressure drop on the production side of the model must be based on the oil or water flow rate, but the flow rate base is not the same as the flow or injection rate base used to calculate the pressure drop on the injection side. Thus, Eq. 9.39 cannot be used to determine the injectivity ratio when both oil and water are being produced from the strata.

It is convenient to state the pressure drop on the producing side of the five spot as a function of the water flow rate, q_w, and θ_w, the fraction of the well radius flowing water:*

$$p_{wi} - p_{wp} = \Delta p_p + \Delta p_i \quad (9.36)$$

$$p_{wi} - p_{wp} = 0.141i \ \ell n(r_e/r_w)(\mu_w/k_w)/h + 0.141q_w \ \ell n(r_e/r_w)(\mu_w/k_w)/h\theta_w \quad (9.45)$$

If the mobility ratio were 1.0, θ_w would be equal to the water cut. For example, if the water cut is 0.667 with a mobility ratio of 1.0, it appears to be clear that the fraction of the well radius flowing water would be 0.667, i.e, $\theta_w = 0.667$. However, the mobility ratio is seldom

*See chapter 2, the section on "General Problems in Fluid Flow Applications."

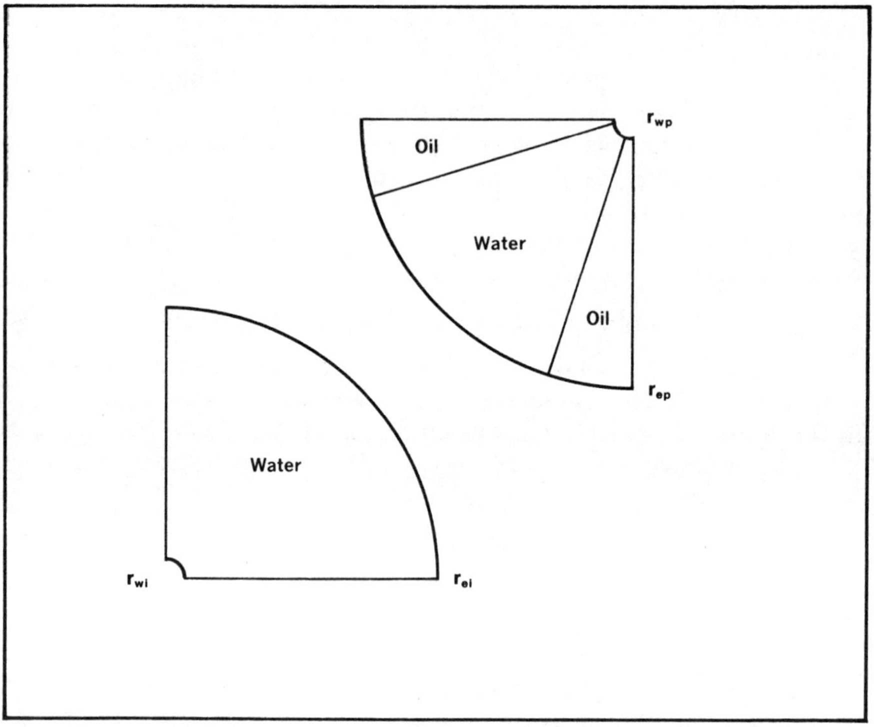

Fig. 9–22 Five-spot quadrant model after water breakthrough

1.0, so we must adjust the actual water cut on the basis of the mobility ratio to obtain the water cut that would exist if the mobility ratio were 1.0. This adjusted f_w then equals θ_w.

Modifying our previous example, suppose that at a particular state of depletion a strata has the same water cut of 0.667 but the mobility ratio is 2.0. To get the water cut that would exist at this time if the mobility ratio were 1.0, the actual oil cut of 0.33 must be weighed by 2.0 so the water cut with mobility of 1.0 would be:

$$0.667/[0.667 + (0.333)(2.0)] = 0.5$$

Therefore, in this example θ_w would equal 0.5. In equation form we can then write θ_w as:

$$\theta_w = f_w/[f_w + (1 - f_w)M] \tag{9.46}$$

Substituting for θ_w in Eq. 9.45 according to Eq. 9.46, substituting $f_w i$ for q_w, and solving the equation for the injection rate, we obtain:

$$\text{i after fillup} = \frac{(p_{wi} - p_{wp})h/0.141\ \ell n(r_e/r_w)}{(\mu_w/k_w)[1 + f_w + M(1 - f_w)]} \tag{9.47}$$

Note that the numerator is the same as the numerator of Eq. 9.38, which is the constant of Eq. 9.41. Thus, the injection ratio is obtained by dividing Eq. 9.41 by Eq. 9.47.

$$(i_s/i) \text{ after fillup} = (1 + M) - (M - 1)f_w \qquad (9.48)$$

Eq. 9.48 actually applies to any time after gas fillup, as indicated, which includes the period of time during which the water cut is zero. We discussed the period from fillup to breakthrough and showed that the injectivity ratio is as in Eq. 9.44. Note that Eq. 9.48 gives the same expression when f_w is zero. Thus, Eq. 9.48 and the realization that the injectivity ratio is 1.0 prior to the time of gas fillup provides the entire history of the injectivity ratio for a particular strata.

When using Figs. 9–11 and 9–12 to evaluate the injectivity-ratio history for a strata, it is necessary to determine the displacement volumes injected (DVI) at a particular water cut to calculate the injectivity ratio at that particular stage of injection. Since the water cut and DVI injected do not appear on the same graph, it is necessary to relate them on the basis of the swept area, which appears on both graphs. For example, for a reciprocal mobility ratio of 0.3, note in Fig. 9–12 that the sweep efficiency is 0.8 when 1.1 DVI have been injected. In Fig. 9–11 note that, when the sweep efficiency is 0.8 with a mobility ratio of 0.3, f_w is about 0.78. Thus, when 1.1 DVI have been injected, f_w is 0.78. When these values are converted to W_i/V_p and i_s/i using Eqs. 9.35 and 9.48, respectively, one point for the injectivity-ratio plot is determined.

To understand this procedure better, work problem 9.7 and compare the solution with the one in appendix C.

Problem 9.7: Calculating Individual Strata Injectivity Curves

In problems 9.4 and 9.5 calculations predict the oil recovery and water production versus cumulative effective injection, based on the individual strata injectivity and recovery curves of Figs. 9–18 and 9–19, respectively. Verify Fig. 9–18 by calculating the points representing gas fillup, water breakthrough, and displacement volumes injected of 0.75 and 1.5. Fig. 9–15 represents the oil and water relative permeability for this reservoir, and the oil and water viscosities are 4.15 and 0.7 cp, respectively.

Stratifying the reservoir for waterflood predictions. It has been noted that it is preferable to use geologic stratification (e.g., the middle of the formation is most permeable or the top of the formation is least permeable), but it is generally impossible to do so. Yet, in all cases the reservoirs behave during a frontal displacement as though they are stratified. Thus, when geologic stratification is impossible, we choose to use a statistical stratification based on a permeability and capacity distribution such as that in Fig. 9–16 if the x axis is the total number of samples rather than total feet. Then the y axis is microdarcy samples rather than microdarcy feet.

Conversion of such data into a series of strata for the purposes of predicting the waterflood behavior can be based on several ideas. Since permeabilities placed in a geometric progression tend to give a normal histogram, as noted in chapter 2, it seems logical to stratify the reservoir in such a fashion. For example, suppose that we decide we will base our prediction on 10 permeability zones. Then we would choose the ratio of the lowest permeability to the highest permeability in each zone so to provide 10 zones. Therefore, the ratio we would choose must be such that the ratio raised to a power of 10 would give the ratio of the highest permeability to the lowest permeability in the total reservoir data. In the case of the permeability data in problem 9.3, which ranges from 15–780 md, the ratio would have to satisfy the relationship, $(\text{ratio})^{10} = (780/15)$. When this relationship is solved, it is found that a ratio of 1.48 would satisfy the equation.

Therefore, the most permeable zone would be 780/1.48, or 527 to 780 md. The least permeable zone would be 15 md to 15 × 1.48, or 22 md. Using this technique in whole feet and arbitrarily assigning one foot to the 22–33-md zone—which does not contain an actual sample in that permeability range—we would have strata thicknesses from most to least permeable of 1, 1, 6, 3, 5, 4, 5, 1, 1, 1 ft, with corresponding average permeabilities of 780, 455, 294, 197, 140, 83, 58, 46, 35, 15 md.

Some engineers prefer to stratify the reservoir on the basis of equal thicknesses, which tends to make the late-life prediction most accurate. In problem 9.3, which is represented by Fig. 9–14 and in the dimensionless form by Fig. 9–16, if 10 zones are used, each zone would be about 3 ft and the average permeability for each zone would be the average of the 3-ft interval (3 samples). Therefore, the peak permeability zone would be about 527 md, and the least permeable zone would be 32 md, which would make the early life prediction less accurate than the late-life prediction.

Another possibility is to make each zone have about the same capacity. In this way all of the zones have similar amounts of water entering them initially. Equal-capacity stratification can be accomplished most readily using a dimensionless capacity curve like that in Fig. 9–16. Then dimensionless thickness increments corresponding to the dimensionless capacity increments can be read from the curve. The dimensionless thickness increments can be converted to feet of thickness by multiplying by the total thickness, and the dimensionless capacity increment can be converted to microdarcy-feet by multiplying by the total formation capacity in md-ft. The averge permeability in microdarcies for a particular increment is then calculated by dividing the capacity increment in md-ft by the thickness increment in feet.

For example, the most permeable 0.1-capacity increment has a corresponding dimensionless thickness of about 0.02. When applied to the total thickness of 29 ft and total capacity of 5,059 md-ft as indicated in Fig. 9–14 (the basis for Fig. 9–16), we find that the most permeable zone of the formation has a thickness of 0.58 ft and a capacity of 505.9 md-ft. The average permeability of this zone would then be 505.9/0.58 or 872 md, which is higher than the highest permeability measured. Therefore, if we had taken more samples of the permeability, we would eventually have obtained a permeability equal to or greater than 872 md. Furthermore, the equal capacity method of stratification would result in a more accurate early-life prediction as a result of the greater detail in the high permeabilities.

It is helpful to have considerable detail in the most permeable portion of the data so considerable accuracy can be obtained for the early-flood peak oil rates that are necessary for sizing equipment such as tanks, pumps, and treating facilities. It is also helpful to have details on the least permeable portion of the data so the economic life and ultimate recovery can be determined as accurately as possible. However, it does not appear that adding detail to the model in the intermediate permeability range adds much value to the prediction.

Stratification of the reservoir using the geometric progression ranges of permeability appears generally to give the detail at the very high and very low permeability range as illustrated. Alternatively, the engineer may wish to use equal capacity stratification in the most permeable range and equal thickness in the least permeable range with some arbitrary treatment of the intermediate permeability range. This procedure assures the desired detail in the high- and low-permeability ranges. These procedures are recommended for the statistical stratification of a reservoir for the purposes of frontal displacement analyses.

Permeability distribution classification. Dykstra and Parsons made popular a method that permitted determining recovery at different water cuts for floods of different permeability distribution classifications.[19] The classification is called the *coefficient of permeability variation,* V. Statistically, it can be shown that if permeability is plotted versus the portion of the total sample having higher permeabilities on log-probability paper and the plot gives a straight line, the coefficient of permeability variation can be calculated using the permeability of the straight line at 50% and at 84.1% in the expression:

$$V = (\log k \text{ at } 50\% - \log k \text{ at } 84.1\%)/\log k \text{ at } 50\%$$

Reservoirs exhibiting the same V of course have the same permeability distribution efficiency regardless of the absolute permeability level; i.e., the effect of the permeability distribution on the displacement efficiency is the same for all reservoirs with the same V characteristic.

Unfortunately, it has been the author's experience that most reservoirs do not exhibit the permeability-percentage sample straight line on log-probability paper over the entire range of permeabilities. For this to occur a permeability histogram such as Fig. 2–2 based on geometric progression for the permeability ranges must be a perfect bell shape. Typically, based on the author's experience, the most permeable data and the least permeable data fall off of the best log-probability straight line for the data. Consequently, a prediction based on a permeability distribution forced to fit a particular V characteristic is inaccurate in the very areas that are most important in the flood life—the early life and the late life. Thus, the engineer is cautioned to make certain all of the permeability data fit the permeability-percentage sample log-probability straight line before relying on a prediction method that uses this parameter or bases a statistical stratification on data smoothed to fit a particular V.

In spite of the possible inadvisability of using the coefficient of permeability variation, V, in a prediction method, the term is very valuable as a general classification of permeability distribution and is widely used for this purpose.

Predicting Waterflood Performance by Analogy or from Pilot-Flood Results

It is not always obvious how the results of a previous flood project can be used to predict the results in a proposed project. Ideally, the model and a proposed flood should be exactly the same; then results would be exactly as predicted. However, this is never the case. When the results of a previous flood are used to predict the results of a proposed flood, we are invariably faced with many differences in the two projects. These differences include such values as size, percentage of injection lost outside the swept area, injection pressure, reservoir capacity, the state of depletion, mobility ratio, and permeability profile.

It is impossible to account for variations in all of these factors. However, if the permeability distribution and the mobility ratios are roughly the same and the states of depletion are similar, prediction by analogy probably results in a more accurate prediction than does prediction by straight theoretical considerations. The most accurate predictions are obtained by using both methods. The methods presented are especially valuable in expanding the results of a pilot flood to a full-scale flood prediction. Many profitable floods have been missed as a result of misinterpreting pilot-flood results.

The engineer should carefully note that the method presented is not meant to replace the theoretical methods of predicting flood performance. The method is proposed as a supplement to theoretical methods and as a means of obtaining a prediction when sufficient reservoir data are not available for a theoretical prediction or when time does not permit a theoretical prediction. Even under these circumstances prediction by analogy may be impossible if the reservoirs are not similar. The proposed method supplements theoretical techniques the same way that decline-curve analysis should supplement theoretical primary production predictions.

As noted, flood recovery preditions fall into two general categories: prediction of total ultimate flood recovery and prediction of rate versus time. Prediction of total ultimate recovery is not generally as difficult as the prediction of rate versus time since it is, to a large extent, independent of injection rates, percentage of injection that is ineffective, and many other parameters that affect displacement efficiency but do not significantly alter ultimate recovery.

Predicting total flood recovery. The most logical basis for predicting total flood recovery is the total secondary plus primary recovery from the swept area as a fraction of total pore volume. We assume that the total fractional recovery in the swept area is the same in similar floods:

$$\text{Fractional recovery in swept area} = \frac{(\text{total recovery, stb/acre-ft})B_{os}}{7{,}758\phi} \tag{9.49}$$

The total recovery to be used in Eq. 9.49 should be determined very carefully:

$$\text{Total recovery, stb/acre-ft} = \frac{\text{stb of primary production}}{\text{acre-ft of primary drainage}} + \frac{\text{stb of waterflood production}}{\text{acre-ft of gross swept volume}} \tag{9.50}$$

When using Eq. 9.50, primary production is defined as the cumulative production up to the start of the flood. Also, note that the primary recovery and secondary recovery per acre-foot are based on different drainage volumes. Care should be exercised in making certain that all of the primary drainage volume is included in this term. A good rule of thumb is to assume that original oil in the reservoir is produced by the nearest producing well if it is produced during the solution-gas-drive phase of production. Unequal production rates, completion dates, and other factors should be used to adjust this rule.

The acre-feet in the gross swept volume can be determined by any reasonable technique since the same technique is applied to both the model and proposed flood. Nevertheless, care should be exercised that the gross swept volume is not excessive. The techniques recommended and discussed in the section, "Determining Gross Swept Volume," can be used for this purpose.

Once the necessary parameters have been determined for the application of Eqs. 9.49 and 9.50 to the model and proposed flood, the prediction of total recovery is a matter of calculating the fractional total recovery in the swept area for the model. Then reverse this procedure by using the fractional total recovery in the swept area for the model as the basis for calculating the volume of waterflood production for the proposed flood.

To check knowledge of this section, work problem 9.9 and check solution against the one in appendix C.

Problem 9.8: Predicting Total Flood Recovery by Analogy

In subsequent problems effective injection rates and the performance in terms of rate versus time are determined. At this time it is desired to calculate the total flood recovery for the proposed flood using the data obtained for the old flood.

Data on old flood:

- Formation thickness, ft = 20
- Average porosity, % = 18
- Estimated connate water, % = 25
- Original B_o = 1.25
- B_o at flood start = 1.01
- Primary drainage volume, acre-ft = 10,000
- Gross swept flood volume, acre-ft = 8,000
- Oil production to flood start, MM stb = 2.410
- Waterflood recovery, MM stb = 1.570

Data on proposed flood:

- Average porosity, % = 20
- Estimated connate water, % = 20
- Original B_o = 1.20
- B_o at flood start = 1.00
- Primary drainage volume, acre-ft = 15,000
- Swept flood volume, acre-ft = 15,000
- Cumulative oil production at flood start, MM stb = 3.645

Predicting effective injection rates. Prediction of effective injection rates for a waterflood involves two important considerations. First, it is necessary to account for the difference in the total injection rates that result from a difference in the transmissibility, kh/μ, the pressure drop $(p_{wi} - p_{wp})$, and geometry. Just as important is the necessity for estimating how much of the total injection rate enters the gross swept volume of the reservoir and how much is lost outside the gross swept volume. For example, the total injection rate in a single five spot consisting of four injection wells and one production well, as in Fig. 9–5, may be 400 b/d. However, the effective injection rate may initially be only one-fourth of this amount. In other words, with radial flow into all four injection wells, only one-fourth would enter the swept volume.

If there is an insignificant gas saturation at the start of the flood, radial flow does not prevail and virtually 100% of the injected fluid enters the gross swept volume since there are no other paths available. However, when a substantial free-gas saturation exists at the start of the flood, estimating the percentage of total injection that enters the gross swept volume is much more difficult. Referring to Fig. 9–6, note that if the initial gas saturation is substantial, radial flow initially exists around all injection wells. Therefore, well 20 would have about 25% of its water entering the gross swept volume, while well 6 would have 50% entering the swept volume. Well 9 would have 100% entering the gross swept volume.

It can be argued that when the gross swept volume is pressured to a point higher than that of the reservoir outside the gross swept volume, a larger percentage of the injection from the edge wells enters the region outside of the flood as a result of the differences in reservoir pressure. However, if the zero thickness contour is near the injection wells, it seems likely that, eventually, the reservoir outside the gross swept volume is filled and attains a pressure greater than the average pressure in the gross swept volume. Then virtually all of the injected fluid enters the gross swept volume since pressure sinks exist at the producing wells.

Thus, as a first estimate it is suggested that edge wells in a flood with a substantial initial gas saturation be treated as though radial flow were taking place throughout the flood life so the percentage of each well's injection that is entering the gross swept volume can be calculated as illustrated. This assumption can be checked for the old flood during its late life by comparing the total injection rate with the total fluid production rate. By late life virtually all of the gas has been displaced from all of the zones with significant permeability. Therefore, if 100% of the injected water is entering the gross swept volume, there is a one-to-one ratio between the injection rate and the total (oil plus water) producing rates.

Study of the steady-state injection rate representing a particular flood geometry results in methods for calculating the injection rate for the proposed flood based on experience gained from the old flood. For example, the steady-state flow equation for a five spot flowing one fluid of viscosity μ is shown to be:

$$i = \frac{3.54 k_{avg} h (p_{wi} - p_{wp})}{\mu \left(\ell n \frac{d}{r_w} - 0.619 \right)} \tag{2.48}$$

From Eq. 2.48 we note that the injection rate is proportional to the transmissibility, kh/μ, the pressure drop between the injection and producing wells $(p_{wi} - p_{wp})$, and the reciprocal of the geometry factor $[(\ell n(d/r_w) - 0.619)]$. Consequently, when these terms are different for the old and proposed floods, the total injection rate for the old flood can be corrected to the estimated total injection rate for the proposed flood by ratio:

$$i_{proposed} = i_{old} \frac{(kh/\mu)_{proposed} (p_{wi} - p_{wp})_{proposed} \left(\ell n \frac{d}{r_w} - 0.619 \right)_{old}}{(kh/\mu)_{old} (p_{wi} - p_{wp})_{old} \left(\ell n \frac{d}{r_w} - 0.619 \right)_{proposed}} \tag{9.51}$$

The viscosity values can probably be depicted accurately using an arithmetic average of the water and oil viscosity. The pressure at the water-injection well, p_{wi}, can be approximated from the well depth, wellhead pressure, and water density:

$$p_w \approx 0.433 \gamma D + p_{wh} \tag{9.52}$$

Eq. 9.52 assumes that the pressure loss caused by friction in the tubing is negligible and simply equates the bottom-hole pressure, p_{wi}, to the static bottom-hole pressure. Tubing flow equations are available for making more accurate approximations.

Another analogy technique often overlooked in predicting total injectivity for a new flood is the use of initial oil-production test data for the well at the time of completion or peak primary production. Except for highly depleted reservoirs, it should always be safe to assume that an injection pressure can be used that is at least as high as the initial reservoir pressure. Under these conditions the injection rate for most five-spot pattern floods would be one-half or less of the initial productivity of the well if k/μ for injected water is equal to the initial k/μ for oil. This is true because the injection pressure must force water into an injection well and force the fluids out of the producing well in a five spot. The initial reservoir pressure in the reservoir must only force oil out of the producing well. That is, the pressure drop during flooding results in flow into the radial injection-well system and out of the radial production-well system. This flow requires about twice as much energy as the initial production test where the pressure drop from the reservoir to the production well causes flow only out of the radial production-well system.

Note that the injection rate is one-half or less of the initial productivity of the well. The "or less" is added to account for differences that may exist in the relative permeability of water in the water bank and oil initially produced under primary conditions. Relative permeability characteristics are such that this ratio may be as much as 3–4 in some cases. A comparison of the relative permeability to the wetting phase at an irreducible nonwetting-phase saturation and the relative permeability of the nonwetting phase at the irreducible wetting-phase saturation illustrates this phenomenon.

Thus, we can simply correct the initial oil-production rate for the differences in viscosity and relative permeability and show that a well converted to injection with a pressure equal to the initial reservoir pressure has an injection rate related to the well's initial oil producing rate as:

$$i \approx q_{oi}B_{oi}\,(\mu_{oi}k_{rw}/\mu_{w}k_{roi})/2 \tag{9.53}$$

Predicting oil production rate versus time. To predict production rate versus time, we must account for several major differences between the model and prototype that affect the rate versus time curve. These include effective injection rate, cumulative effective injection at a particular time, difference in size of the floods, difference in the patterns, and difference in oil saturation at the start of the flood. The rate versus time curve can be normalized for the effects of these differences by plotting production rate as a fraction of current effective injection rate versus cumulative effective injection plotted as a multiple of total ultimate flood recovery. The assumption then is that such a plot is the same for similar floods. This plot is illustrated in Fig. 9–23, which shows curves for three different floods.

The rate is stated as a ratio of effective injection rate since most floods, after they have had an initial increase in the oil producing rate, are very sensitive to injection rate. A change in injection rate causes a proportional change in the oil producing rate in a few hours or at most a few days. Using this ratio to calculate the oil producing rate in the proposed project also gives an adjustment for the differences in injection rates between the two floods. Some engineers are confused by the fact that they have observed wells where the oil producing rate is increasing when there is a decrease in the injection rate. The engineer should remember that in such a situation the oil producing rate would have continued increasing at a greater rate if there had not been a reduction in the injection rate. We must compare what would have happened without the injection rate decrease with the actual rate.

The cumulative effective injection must also be adjusted for the differences in flood size. If the same saturations exist at the start of the

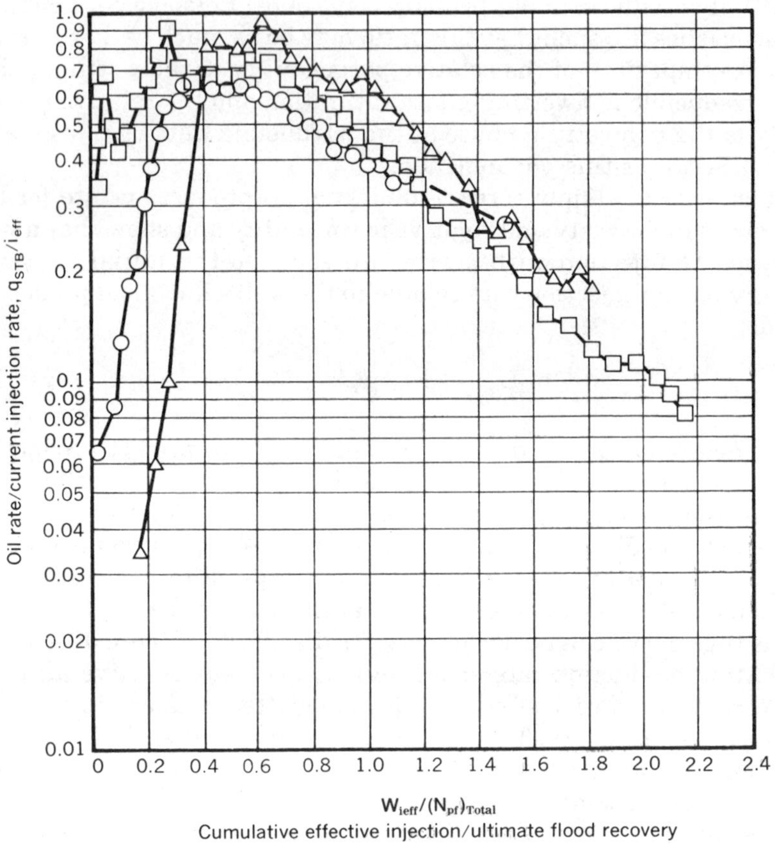

Fig. 9–23 Normalized performance data for three Illinois waterfloods (reservoir characteristics are shown in Table 9–1)

flood in both reservoirs and the same residual oil is to be experienced, the pore volume can be used to normalize the cumulative effective water injection. However, under these conditions normalizing with the total recoverable oil provides exactly the same degree of normalization. Use of ultimate total flood recovery for this purpose accounts for mobile oil differences between the two reservoirs. Since the correction for the difference in mobile oil is so inexact, it is suggested that this analogy technique of predicting rate versus time for a waterflood be limited to those reservoirs that are in about the same state of depletion at the time of flood initiation. A detailed modification of the rate versus time curve can be accomplished where mobile oil differences are considerable, but such a detailed correction is beyond the scope of this chapter.

To indicate the validity of the normalization approach, a normalized plot has been prepared for three floods with substantial production history. Since floods normally experience injection problems and are thus difficult to keep in balance, it is difficult to find a number of similar old floods. The author used data on only three similar floods in sufficient detail to prepare the subject plots. Consequently, the data fit as indicated in Fig. 9–23 is not as good as would normally be expected. However, a study of the reservoir data for the three floods indicated in Table 9–1 reveals that the similarity of the plots in Fig. 9–23 is certainly very good under the circumstances. Note first that while two of the reservoirs are in the Benoist formation, the third is in an entirely different formation 1,000 ft deeper. Consequently, we may expect the best fit with the two Benoist floods. The parameters of the two Benoist reservoirs are represented by squares and circles in Fig. 9–23.

TABLE 9–1 Comparison of Reservoir Parameters for Illinois Floods Shown in Fig. 9–23

	Benoist Sand	*Benoist Sand*	*Tar Springs Sand*
Depth, ft	1,400	1,300	2,300
Porosity, %	19.0	20.5	19.0
Permeability, md	110	250	450
Area, acres	500	300	230
Thickness, ft	27	14	?
Flood pattern	10-acre 5-spot	20-acre 5-spot	20-acre 5-spot
Average injection rate, b/d	8,500	1,800	1,500
Cumulative oil Production at flood start, stb/acre-ft	184	284	?
Symbol in Fig. 9–23	□	○	△

Note that these two reservoirs represent vastly different permeabilities, flood areas, thicknesses, injection rates, and five-spot spacings. Yet, we would expect the normalization to overcome these differences and provide a good curve fit. However, the cumulative primaries at the start of the flood are very different. Therefore, greatly different gas saturations result in the reservoir at the start of the floods and affect the time—effective cumulative injection/ultimate flood recovery—of the first production increase. The plots reflect this situation in much the same way that we would anticipate with a good fit between the two normalized plots during most of the life. However, the Benoist flood with the lower gas saturation responds to injection somewhat earlier than the other that has a higher gas saturation. This comparison may suggest a method of adjusting an analogy prediction for differences in initial saturations.

Actually, the fit of the two Benoist plots could be much better in spite of the differences in gas saturations at the start. However, one flood is initiated on a staggered-time basis, resulting in the early oil-rate peaks.

The plot for the Tar Springs flood (triangles) is included to indicate that reservoirs from different geologic formations are sometimes suitable for predicting flood behavior by analogy. The plot fit with the Benoist reservoirs is not too inaccurate, and the availability of detailed information on flood balance, spacing irregularities, and other performance can conceivably result in an excellent fit. Reliable reservoir thickness data are not available on this reservoir since the wells were drilled with cable tools. Consequently, gas saturation at the start of the flood is not known.

Thus, the proposed method for predicting rate versus time by analogy for reservoirs with similar characteristics and saturations at the start of the flood is to use the model data to determine the relationship between (q_o/i_{eff}) versus $(W_{i\,eff}/N_{pf})$. This relationship is then assumed to hold equally for the proposed flood. The data are converted to q_o versus $W_{i\,eff}$ for the proposed project using the i_{eff} and N_{pf}, respectively, to convert the data determined for the old flood. Note that it is not necessary to make a plot like Fig. 9–23 for either the old or proposed flood. The old data can be used to calculate corresponding values of (q_o/i_{eff}) and $(W_{i\,eff}/N_{pf})$. These values can then be converted to corresponding values of q_o and $W_{i\,eff}$ or time by applying the values of i_{eff} and N_{pf} for the proposed flood. To check knowledge of this section, work problem 9.9 and check the solution against the one in appendix C.

PROBLEM 9.9: Predicting the Rate versus Time Performance by Analogy

In problem 9.8 the engineer is asked to predict the total flood recovery by analogy. For this same reservoir predict the oil production rate versus time. Fig. 9–24 represents the rate versus time history of the old flood, including the oil rate, the injection rate, and the cumulative injection versus time. The following reservoir data are also necessary for this prediction:

Data on old flood:
- Formation depth, ft = 3,000
- Average permeability, md = 100
- Total injection-well thickness, ft = 401
- Average injection rate at 600 psi wellhead pressure (fresh water), b/d = 1,700
- Estimated effective injection rate, % = 90

Data on proposed flood:

Formation depth, ft = 2,500

Average permeability, md = 75

Total injection-well thickness, ft = 780

Estimated wellhead pressure, psi = 800

Estimated effective injection rate, % = 95

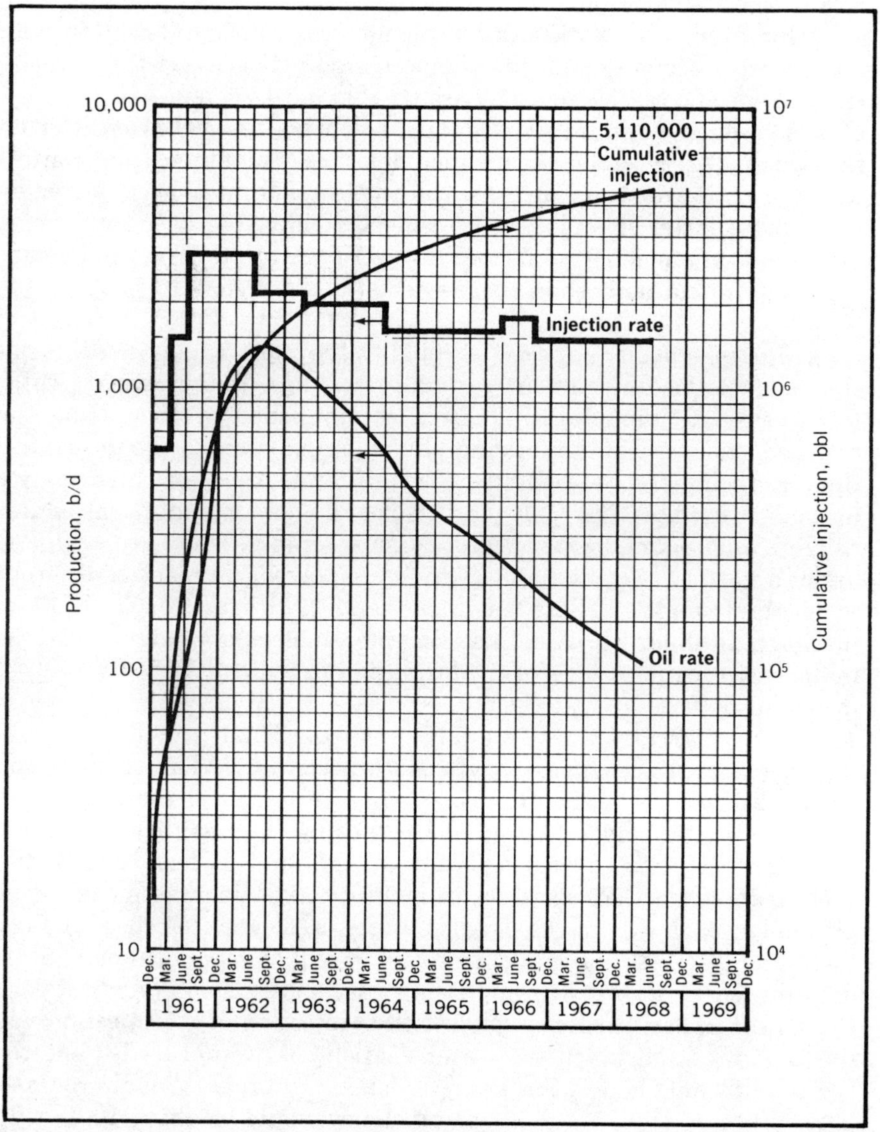

Fig. 9–24 Production and injection data for a flood; data for problem 9.9

Waterflooding Variations

Much effort and money have been expended to find more efficient, economical means of recovering oil from subsurface reservoirs. Research and development continue on in situ combustion and various forms of miscible displacement. However, the most useful ideas that have come from modern research on recovery methods have been variations of waterflooding.

A long list of waterflooding variations can be compiled. However, for the purposes of evaluating the possible application of these variations to specific reservoirs, they are treated in two categories. The first of these reduces residual oil in that portion of the reservoir contacted by the water. The second category includes those techniques that control mobility ratio and, thus, increase the amount of formation contacted by water. This latter effect is accomplished by increasing sweep efficiency, stratification efficiency, and frontal water saturation. We also consider means of including both of these effects in our predictions.

Adjusting mobility ratios. Probably the most widely used variation of waterflooding currently is adding polymer to the injected water to increase its immobility. This is accomplished by increasing the apparent viscosity of the fluid and by plugging pores in the formation, thus, reducing the permeability. Two different types of polymers are normally used.[20] The polyacrylamides have very high molecular weights and reduce the permeability considerably. These are available under a variety of trade names which use polyacrylamides of different molecular weights. These polymers tend to degrade as a result of mechanical shearing. Also, the polymer molecular weight must be matched to the pore sizes of the formation in which it is to be used for maximum effect.

Polysaccharrides work in a different way. Their principal effect is the increase in the apparent viscosity of the fluid. The reduction in permeability associated with this polymer is much less than that associated with the polyacrylamides. Decreasing the mobility of the displacing water has several desirable effects and at least one highly undesirable effect. A decrease in mobility ratio increases the sweep efficiency as shown in the Dyes, Caudle, and Erickson data of Figs. 9–11 and 9–12. For example, using an M of 4 or reciprocal of 0.25, we determine that at breakthrough the strata sweep is 53%. At f_w of 0.95, the strata sweep efficiency is 92% after injection of 2.5 injection volumes. If we quadrupled the water viscosity, M would be 1.0 and the reciprocal would be 1.0. The sweep efficiency at breakthrough would be 70%. When f_w is 0.95, the sweep efficiency would be about 100% with about 1.8 displacement volumes injected.

A decrease in mobility ratio also increases vertical efficiency. To put this effect in some perspective, consider the ultimate sweep efficiency at the economic limit as indicated by the Stiles method. As shown, the strata sweep efficiencies play a minor role in the ultimate overall efficiency, which is affected mainly by the vertical or stratification sweep.

In problem 9.3 we found that using an effective mobility of about 4.0 for A_{eff}, we obtained an efficiency of 0.76. If we quadrupled the water viscosity from 0.7 to 2.8 cp, the comparable efficiency would be 0.93. The engineer is again reminded that these efficiencies are too high because a high lower limit on the permeability distribution is considered. Nevertheless, improving (reducing) the mobility ratio greatly improves the vertical sweep efficiency.

The decrease in mobility ratio also affects the microscopic displacement efficiency as seen in the Buckley-Leverett calculations. Decreasing the mobility ratio by simply changing the viscosity ratio changes the f_d versus S_d relationship and, thus, increases the frontal S_d saturation. This effect is illustrated in Fig. 9–25. It generally shifts the displacing phase saturation behind the front by decreasing all S_d values greater than the S_d equivalent of residual oil saturation. However, if the change in mobility ratio is caused also by a change in relative permeability characteristics, the change in saturation distribution is more complex.

The obvious disadvantage of decreasing water mobility is the adverse effect on injection rate obtained for a particular injection pressure. This effect seems to limit economic applications to high-permeability reservoirs and high-viscosity oils. The high-viscosity oil is necessary because the percentage of decrease in the injection rate compared with a normal flood resulting from the use of viscous water to displace a high-viscosity oil is much less than the percentage of decrease associated with the use of viscous water to displace low-viscosity oil. For example, consider the situation that exists for a five-spot flood. We noted that for all practical purposes we can consider all of the resistance to flow to exist at the wells. Thus, over most of the life of a flood, only water is flowing at the injection wells and mainly oil is flowing at the production wells. In this situation we may be justified in considering the arithmetic average of the oil and water as being characteristic of the effective viscosity in the five-spot flood. If this is the case, note that a water viscosity of 1.0 combined with a low-viscosity reservoir oil of say 0.7 would give an effective viscosity of 0.85. If a water thickener that increases the water viscosity to 4.0 is employed, the effective viscosity would be increased to 2.35 and the average injection rate may be reduced by a factor of about 2.75, 2.35/0.85.

Compare this reaction with the effect of the same thickened water

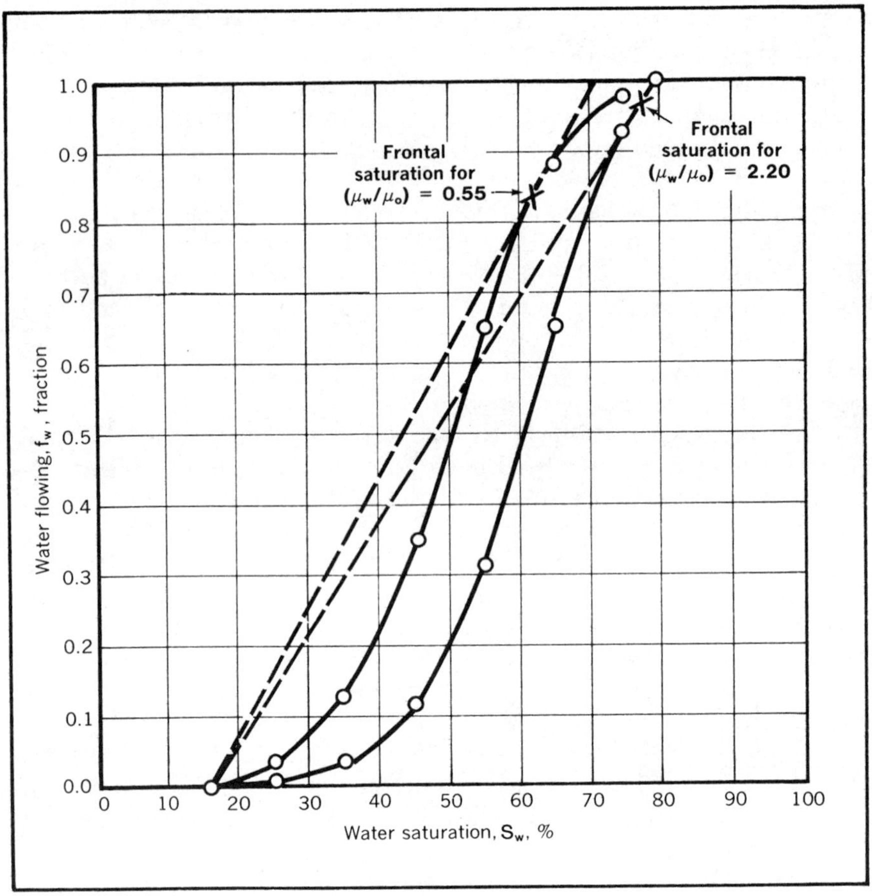

Fig. 9–25 Comparison of f_w curves for different viscosity ratios

on the injection rate where the reservoir oil viscosity is 7.0. Without the thickener, average viscosity would be 4.0. With the thickener the average would be 5.5, and the injection rate would be reduced by a factor of only 1.38. Thus, it is clear that the thickening of the water has a much less pronounced effect on the injection rate in reservoirs containing high-viscosity oils.

This effect of the apparent viscosity of the polymer solution on the injection rates is not quite as bad as is indicated in the discussion because most polymers have a shear, thinning nature. Fig. 9–26 demonstrates this shear, thinning property. This, of course, is advantageous from an injection rate standpoint because the greatest flow velocities in the injection system are near the wellbore in the formation and in the tubing in the well. The pressure drop in the formation around the

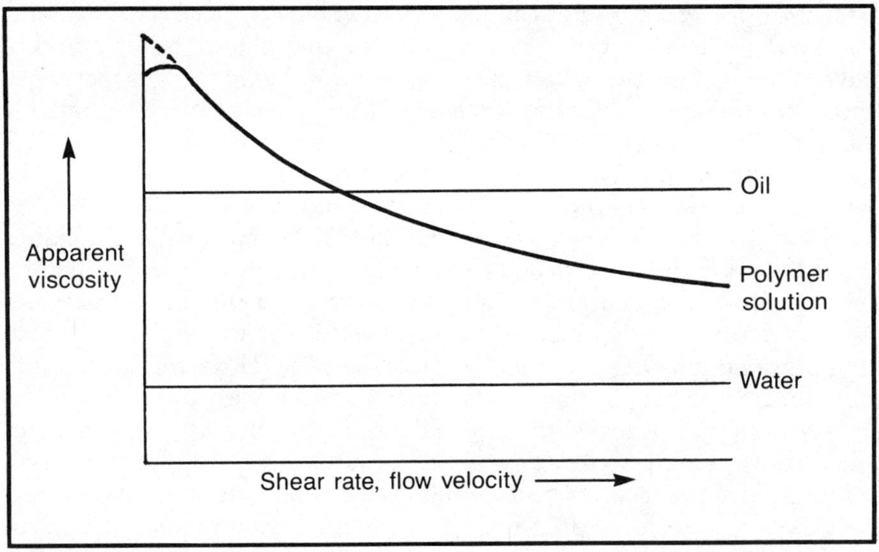

Fig. 9–26 Apparent viscosity versus shear rate or flow velocity (after Skelland, from van Poollen & Associates, *Fundamentals of Enhanced Oil Recovery*, PennWell, 1980)

wellbore represents the largest resistance to flow in the system. This area is where the apparent viscosity is least. Once the fluid travels away from the wellbore, the velocity is drastically reduced. This point is where we need to develop the greatest apparent viscosity. Consequently, the injection rate ratios as stated are overly pessimistic but still illustrate the desirability of applying the polymer-displacement mechanism to high-viscosity oil reservoirs. There is, of course, no point in applying a polymer displacement to reservoirs containing oil so viscous that it cannot be moved. The U.S. Department of Energy at one point suggested limiting polymer-displacement applications to reservoir oils with viscosities less than 200 cp. They also suggested the oil saturation at the start should be greater than 50%, the average permeability greater than 20 md, and the reservoir temperature less than 200°F.

At the same time to move very viscous oil at economic rates, it is necessary to have relatively high-permeability reservoirs. Hence, the statement that economic applications of thickened water seem to be limited to reservoirs of relatively high permeability and high-viscosity oils. It should be noted that shear, thinning polymers are not ideal in all respects. They do aid in getting the polymer into the reservoir at a maximum rate, but once in the reservoir, they tend to reduce the displacement efficiency by increasing the velocity along the high-velocity stream lines and through the more permeable strata. Examination of the sweep efficiency and stratification efficiency shows that it would be advantageous for the velocities to be reduced along the characteristically high-velocity stream lines and through the more permeable strata. Thus, shear, thickening polymers are advantageous at this point. A shear, thickening polymer that does not develop, or become viscous, until it gets into the reservoir would be ideal.

The same recommended method of predicting flood behavior can be used to predict the results of a polymer flood. The easiest way to accomplish such an application is simply to assume that the thickened water will be used throughout the flood so the application is a matter of making a prediction with an adjusted mobility ratio. However, such a prediction represents the maximum result obtainable from a thickened water flood of that mobility ratio. This estimate indicates whether further study is advisable. However, it certainly does not give a realistic prediction of the most economic application of polymer flooding, since it is uneconomic to thicken all of the water in a flooding operation. The normal procedure is to use a slug of thickened water amounting to 10–40% of the pore volume and then displace this with water. Lab and field tests show that the polymers reduce the permeability to the polymer solution and that they permanently damage the permeability so the effective permeability to the water displacing the polymer solution is also reduced.

Consequently, it is necessary to assume that displacement of the slug is by a water with an abnormally high viscosity, which gives the same effect on the performance as would a reduced effective permeability. To make an accurate prediction of the performance of a polymer flood, it is thus necessary to know the cost of thickening water to various effective mobility ratios and to know the residual effect of the polymer on the permeability to normal injection water. Care should be exercised to make certain that the supplier of the polymer also provides the polymer absorption data. The absorption is generally substantial.

When slug-type displacement is used, the generation of injectivity and recovery curves for use in the recommended method is somewhat more difficult than the process previously described for a regular waterflood. An acceptable recovery curve can be obtained using the Dyes, Caudle, and Erickson data with the mobility ratio existing between the mobile oil and the fluid directly displacing it, i.e., the slug. This approach would not be sufficiently accurate if a very small slug size, say less than 15% of the pore volume, is used, but generally, such a small slug is impractical because dispersion takes place between the slug and the water displacing it.

Determining the slug distribution in the various zones. To determine the injectivity curves to be employed in the recommended method it is necessary to determine the portion of the slug that will enter each permeability zone. This is most readily obtained by applying the prediction method to the reservoir as though the thickened water would be continuously applied.

In so doing, reduced-time curves such as those in Fig. 9–20 are obtained. With these curves the cumulative injection corresponding to any particular reduced time can be determined, assuming continued injection of the thickened water. This technique is used to solve problem 9.7, and the solution is shown in appendix C.

Once we have determined the amount of slug we wish to inject into the formation, we can determine the reduced time that corresponds to this particular cumulative effective injection. Then a plot such as Fig. 9–20 can be used to determine the pore volumes of injection into each zone that correspond to this reduced time. These values can be converted to cumulative injection of slugs into each zone by applying the appropriate pore volume to each of the values.

For example, in problem 9.7 we are planning to inject a slug of fluid until the injected fluid breaks into the producing wells. The solution to problem 9.7 in appendix C shows that this would occur when 86,700 bbl of slug have been injected. The solution also shows that this would occur at a reduced time of 7.63 and that at this particular time the pore volumes of injection into zones I, II, and III would be 0.33, 0.24, and 0.1, respectively.

Since the pore volumes of the three zones are given as 97,000, 155,000, and 175,000 bbl, respectively, we can calculate the water injected into each zone at this time as 0.33 × 97,000, 0.24 × 15,000, and 0.1 × 175,000 bbl, respectively, or about 20% of the total pore volume. After the amount of slug going into each zone has been determined, the principles previously discussed for determining an injectivity curve for each zone can be applied and the prediction calculations can be repeated using these new injectivity curves.

The engineer may be concerned about the accuracy of determining how much of the slug enters each zone in this way since the injectivity curves on which the method is based assume that the only resistance to flow is at the injection and production wells. This assumption can be unacceptably inaccurate when applied to very high or very low mobility ratios. However, the mobility ratio between the oil displaced and the polymer solution is generally designed to approach 1.0 where the assumption should be most accurate.

The recovery data of Dyes, Caudle, and Erickson, found in Figs. 9–11 and 9–12 apply theoretically only to two-phase flow; however, the error introduced by the third phase generally has a minimum effect on the recovery curve when a slug of sufficient size is used to keep the water displacing the slug from the oil being displaced by the slug.

Determining injectivity curves for a slug-type displacement. Once the amount of slug that will enter each zone has been determined, it is possible to calculate the curve of an injectivity ratio versus cumulative injection in pore volumes for each zone. This, of course, is required for the recommended method of performance prediction. Each zone now has an injectivity curve that is unique since all of the zones have different amounts of slug injected into them.

The determination of injectivity curves for a slug-type operation is necessary to predict many different enhanced oil recovery mechanisms. Thus, it is not just peculiar to a polymer flood. It is difficult to generalize the procedure for determining the injectivity curves for a slug-type process, but it is hoped that the development of a series of equations for one or two sets of circumstances permits the engineer to formulate equations for other circumstances.

If the saturations in the reservoir at the start of the flood include a substantial gas saturation, the expression for the initial injection rate, Eq. 9.41, can be adapted to our slug injection:

$$i_s = \text{constant}/(\mu_{sl}/k_{sl}) \tag{9.54}$$

Thus, until the injected slug is equal to the initial mobile gas saturation or until water injection starts, the injectivity ratio, i_s/i, is 1.0. The other equations for the injectivity ratio also apply until the slug injection is discontinued and water injection or injection of another displacing phase is initiated.

Using the same principles employed in Eq. 9.48, we can derive the following injectivity ratio equations. When water is being injected and oil and slug are being produced:

$$i_r = \frac{1}{M_{w/s1}} + f_{s1} + M_{s1/o}\,(1 - f_{s1}) \tag{9.55}$$

When water is being injected and water and slug are being produced:

$$i_r = \frac{1}{M_{w/s1}}\,(1 + f_w) + (1 - f_w) \tag{9.56}$$

There will be a period when water, slug, and oil are being produced after the injected water breaks through the slug. This situation is not reflected in Eq. 9.56, which is based on only water and slug production. However, this production does not introduce much of an error in most cases for two reasons. First, most of the oil is produced from a particular zone before water breakthrough, and second, every effort is made to have the slug-oil mobility ratio near one for recovery and economic reasons.

If the mobility ratio of the slug to the oil is 1.0, Eq. 9.56 does not need to be modified to account for the concurrent production of water, slug, and oil. Consequently, it is suggested that Eq. 9.55 be used until water breakthrough. Then Eq. 9.56 can be used to obtain the injection ratio. The point of water breakthrough can be determined from Figs. 9–11 and 9–12, based on water injection alone and using the water-slug mobility ratio.

As noted, a thickened waterflood is usually applied to a reservoir containing a high-viscosity oil that has experienced very little primary production. Therefore, little or no free gas is present in the reservoir at the time the flood is initiated. In such cases initial injection takes place with oil flowing at the production well and the pusher slug flowing at the injection well. Applying Eq. 9.39 to obtain the initial injection rate, we obtain:

$$i_s = (\text{constant})(k_{slug}/\mu_{slug})/(1 + M_{slug/o}) \tag{9.57}$$

Where:

$M_{slug/o}$ = Mobility ratio between the slug and the oil

We can then use Eq. 9.39 repeatedly to obtain the injection rate at various stages of displacement stated as a function of the geometrical constant and the slug mobility in the denominator. This permits the formation of i_r for these various stages of displacement. For the slug flowing at the injection well and oil at the producer, $i_r = 1.0$. For the slug flowing at the injection well and oil and slug flowing at the producer:

$$i_r = \frac{1 + f_{slug} + M_{slug/o}(1 - f_{slug})}{1 + M_{slug/o}} \tag{9.58}$$

For water flowing at the injection well and oil and slug flowing at the producer:

$$i_r = \frac{M_{slug/w} + f_{slug}(1 - M_{slug/o}) + M_{slug/o}}{1 + M_{slug/o}} \tag{9.59}$$

For water flowing at the injection well and water and slug at the producer:

$$i_r = \frac{M_{slug/w}(1 + f_w) + (1 - f_w)}{1 + M_{slug/o}} \tag{9.60}$$

Eqs. 9.58, 9.59, and 9.60 can be used for flowing fractions of 1.0 or 0.0. Consequently, they describe the entire range of i_r values based on the assumption that all of the resistance to flow exists only at the wells. Once the strata injectivity and recovery curves have been generated, the same recommended procedure can be employed in predicting the behavior of a thickened waterflood.

Work problem 9.10 and compare the solution with the one in appendix C.

PROBLEM 9.10: Calculating Injectivity Curves When using a Slug

To demonstrate the application of the modified Prats method to a performance prediction when a slug or pusher is used, we can consider the use of a viscous slug in waterflooding the hypothetical reservoir in problems 9.6, 9.7, 9.8, and 9.9. In this application assume the pusher is designed to give a slug/oil mobility ratio of about 1.0. This is accomplished using a slug with a viscosity of 2.37. We arbitrarily assume that the pusher is injected until it breaks through at the producing well and that the pusher causes a reduction in the relative permeability to water of 13%. Determine the injectivity ratio curve for each zone.

Reducing residual oil. As noted, it is difficult to define what constitutes a modification of a waterflood and what should be considered as a separate enhanced oil recovery method. All enhanced oil recovery methods except the use of polymers alone affect the residual oil. We will consider the use of micellar solutions and other surfactants and caustic flooding here. Other thermal and miscible methods will be considered in chapter 10.

Most industry engineers appear to give credit to Marathon Oil for the development of a micellar solution of hydrocarbons, water, and petroleum sulfonates capable of displacing all of the oil from the reservoir contacted.[21] However, much effort has been put forth by many others both prior to and since Marathon obtained its patents on the process that they gave the trade name of Maraflood. Union Oil developed a similar product with the trade name Uniflood. In spite of this work, there still appears to be very few if any economic successes to which the industry can point. H.K. van Poollen and Associates point out that: "Although much laboratory work has been done on this process, no field project has as yet been reported as economic."[22] However, it is this author's opinion that widespread economic application of the process is eminent, and the practicing reservoir engineer should be prepared to predict the effect of the use of a micellar or surfactant slug on the behavior of a proposed waterflood. The principal application of this process has been attempting to increase the recovery in conventional waterfloods that are at or approaching their economic limit.

A similar process that has been tried widely is the use of caustic (sodium hydroxide) in the flood water. The reasons for the increase in oil recovery that accompanies the use of caustic in the injection water range from the effect on surface tension to the reversal of wettability and emulsification of the crude oil being displaced. Although the results from caustic flooding have not been as spectacular as that of some of the other enhanced oil recovery methods, it appears that some economic success has been realized with the method largely resulting from the modest additional investment that is necessary to implement a caustic flood as compared with the cost of a conventional flood.

If the micellar or caustic flood has been designed so frontal displacement is uniform throughout the reservoir—i.e., the residual oil is the same throughout all of the reservoir contacted—then modification of the prediction method to account for the effects of the reduced residual oil saturation is simple. First reduce the residual oil saturation used to predict calculations of S_{or} and proceed with the prediction calculations as described.

If a mobility control slug is to follow the micellar or caustic slug used to reduce the residual oil, it is then necessary to use the methods developed previously for handling the mobility control slug in addition to reducing the residual oil. If conditions at the displacing front cannot be maintained so the residual oil actually varies throughout the reservoir, it is then necessary to use a more detailed analysis along the lines of reservoir modeling.

The U.S. Department of Energy at one time suggested that chemical processes for reducing the residual oil be used only if the oil saturation is greater than 25%, the average permeability is greater than 20 md, the reservoir temperature is less than 250°F, the oil viscosity is less than 30 cp, the reservoir water salinity is less than 200,000 ppm, and the water hardness is less than 1,000 ppm. The engineer will also want to consider the prospect of other enhanced recovery methods discussed in chapter 10.

Additional Problems

9.11 Rework problem 9.1 to find the saturation distribution when the average liquid saturation at the first ring of producers is 0.96 instead of at the first oil increase. Assume $\Delta f_o/\Delta S_L$ at $S_L = 0.96$ is 3.77.

9.12 The portion of the reservoir shown in Fig. 9–27 has net thickness as indicated on the well spacing diagram (Fig. 9–27). Each dotted square is 10 acres, and the outer solid line is the lease line. During primary production the lease produced 260,000 stb of oil. We now desire to flood this reservoir by converting some of the wells on the lease to injection wells to form a five-spot pattern. Core-analysis data for two wells are shown in Table 9–2. To simplify

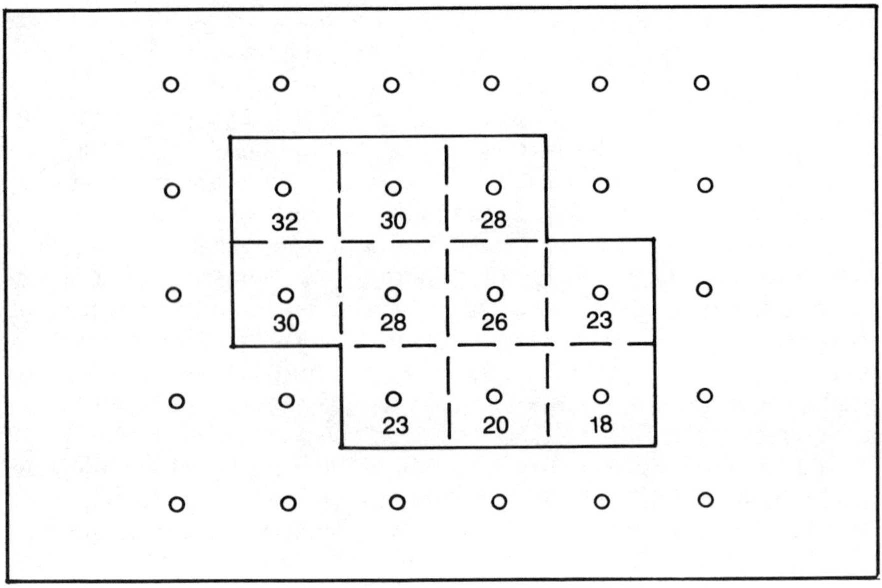

Fig. 9–27 Well spacing for problem 9.12

calculations, assume the thickness for each well's drainage square is uniform. Values for k_{ro} and k_{rw} are given in Fig. 9–28. The reservoir characteristics are as follows:

$\mu_o = 2.0$ cp $\quad B_{oi} = 1.2$
$\mu_w = 0.63$ cp $\quad B_{op} = 1.05$
$S_{wc} = 0.2$ $\quad B_{or} = 1.05$

A. Find the oil saturation at the start of the flood.
B. Find the gross swept volume based on the best five-spot conversion pattern.
C. Find the ultimate flood recovery based on the Stiles method if the economic limit of the water cut is 0.95 and the horizontal sweep at this time is 90%.

9.13 In problems 9.4, 9.5, 9.6, and 9.7, the prediction of the behavior of a flood of a hypothetical 10-acre five-spot is achieved. To demonstrate the effect of the mobility ratio on the flood behavior, predict the oil and water cumulative production versus time if the water viscosity can be increased by a factor of 3 to 2.1 cp and all other data remains the same. Fig. 9–15 is used as the relative permeability. The prediction should be accomplished in the following manner:

A. Determine a recovery curve like Fig. 9–19 for each zone.
B. Determine an injectivity ratio curve like Fig. 9–18 for each zone.
C. Calculate reduced-time curves like Fig. 9–20 for each zone.
D. Calculate and plot cumulative oil and water production versus cumulative injection for the flood.
E. Prepare an oil and water rate versus time plot if the effective injection rate is 300 b/d.

The following data apply:

S_{gi}, % = 20
S_{os}, % = 65
S_{wc}, % = 15

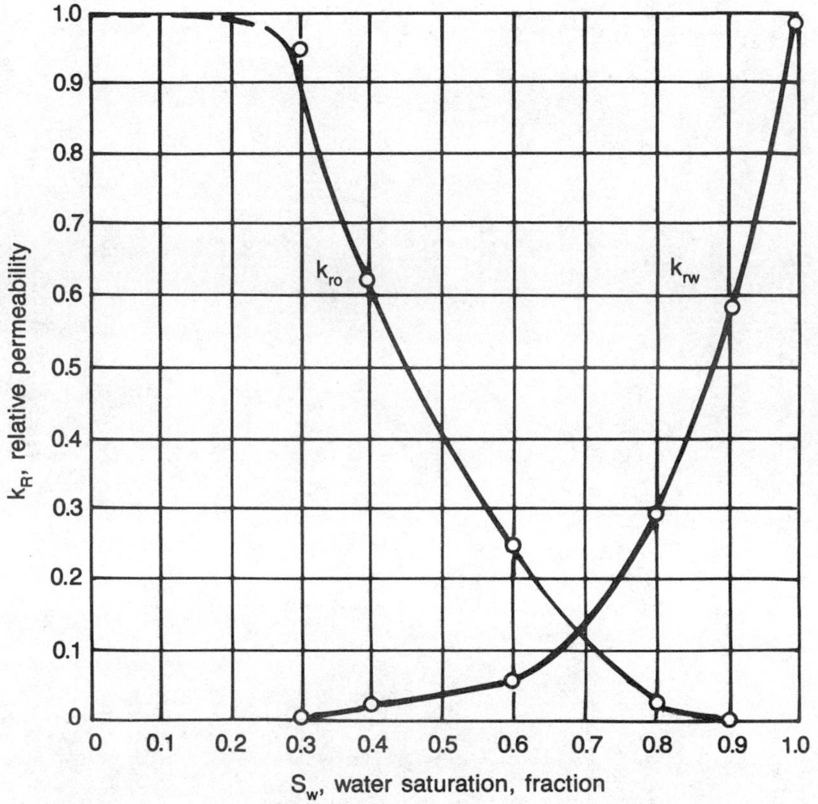

Fig. 9–28 Relative permeability data for problem 9.12

TABLE 9–2 Proposed Flood Core-Analysis Data for Problem 9.12

Depth, ft, From–To	*Horizontal k, md*	*Effective Porosity, %*	*Total Porosity, %*	*Oil Saturation, %*	*Water Saturation, %*
Well 23:					
2,488–2,489	143.0	17.6	18.2	14.6	46
2,489–2,490	127.0	18.4	18.6	16.1	40
2,490–2,491	100.0	19.8	21.6	6.0	61
2,491–2,492	118.0	20.6	19.2	12.8	44
2,492–2,493	60.0	17.9	18.0	17.1	45
2,493–2,494	141.0	19.1	18.6	13.4	46
2,494–2,495	97.0	19.1	19.8	15.2	49
2,495–2,496	100.0	19.5	24.1	14.5	48
2,496–2,497	36.7	16.3	16.9	18.3	43
2,497–2,498	299.0	22.6	22.2	15.8	49
2,498–2,499	97.0	18.2	24.0	13.7	36
2,499–2,500	24.6	14.0	19.7	18.0	29.6
2,500–2,501	407.0	20.6	20.4	16.5	52
2,501–2,502	98.0	20.2	25.4	11.8	43
2,502–2,503	154.0	21.0	23.4	15.1	48
2,503–2,504	191.0	22.7	23.2	14.7	51
2,504–2,505	61.0	19.4	19.5	18.6	50
2,505–2,506	103.0	19.2	20.8	15.9	48
2,506–2,507	21.0	15.8	18.2	19.4	42
2,507–2,508	31.5	16.5	18.8	20.4	42
2,508–2,509	33.3	18.3	17.2	18.7	50
2,509–2,510	81.0	18.6	19.9	18.7	45
2,510–2,511	14.1	15.3	18.0	20.9	39
2,511–2,512	24.7	15.0	16.3	17.9	38
2,512–2,513	14.7	14.3	16.7	20.7	44
2,513–2,514	8.7	14.4	15.3	25.0	46
2,514–2,515	8.7	14.9	18.2	23.4	42

2,515–2,516	13.2	17.4	17.7	23.3	45
2,516–2,517	8.1	17.0	17.8	22.4	38
2,517–2,518	5.7	15.9	18.1	24.2	46
2,518–2,519	9.5	15.9	17.0	22.1	38
2,519–2,520	1.9	8.9	12.7	16.3	45
Well 12:					
2,509–2,510	110.0	20.1	22.0	15.4	52
2,510–2,511	347.0	21.8	20.0	17.9	56
2,511–2,512	158.0	20.5	21.8	17.6	46
2,512–2,513	267.0	21.6	22.9	17.2	46
2,513–2,514	166.0	21.0	24.4	15.0	45
2,514–2,515	290.0	21.6	21.2	16.3	46
2,515–2,516	80.0	17.9	19.4	18.8	38
2,516–2,517	139.0	19.9	20.7	18.0	50
2,517–2,518	232.0	20.2	19.4	21.2	44
2,518–2,519	402.0	16.7	18.9	21.2	37
2,519–2,520	146.0	19.9	23.2	15.7	47
2,520–2,521	49.0	18.1	20.1	17.7	53
2,521–2,522	21.6	15.2	17.5	23.2	42
2,522–2,524	185.0	19.5	21.4	15.9	39
2,524–2,525	5.7	13.2	18.7	21.4	39
2,525–2,526	216.0	19.6	21.2	17.4	47
2,526–2,527	10.9	17.3	18.1	20.9	36
2,527–2,528	600.0	21.3	22.2	17.6	44
2,528–2,529	390.0	20.6	22.4	20.0	36
Averages (both wells included)	124.0	18.2	19.8	17.8	44.5

S_{or}, % = 25
μ_w, cp = 2.1
μ_o, cp = 4.15

Zone No.	h, ft	k, darcies	ϕ	V_p, bbl
1	5	0.025	0.25	97,000
2	10	0.010	0.20	155,000
3	15	0.002	0.15	175,000

Note: A practical application of this recommended method should not be made with less than 10 zones.

9.14 Prepare a computer or programmable calculator program for the recommended method of predicting flood behavior. Generate the injectivity curve data in the program or provide an option for direct input of the injectivity curves. Input the recovery curves. Provide for the use of 25 or less separate injectivity and recovery curves.

9.15 In problem 9.12 we calculate the ultimate recovery for a flood using the Stiles method and an assumed pattern sweep to determine the total sweep efficiency at the economic limit. Now predict the rate versus time behavior using the recommended waterflood prediction method. Use a total injection rate of 10 b/d/ft and an average thickness of 26.1 ft in determining the gross swept volume. Limit the zonation so the following conditions apply:

$$(k_{max}/k_{min}) < 1.1 \text{ in any zone}$$

Or:

$$\Delta h < 0.1(h_{total}) \text{ and } \Delta c < 0.1(c_{total})$$

9.16 Use the injectivity ratio curves generated in problem 9.10 (see appendix C solution) together with the data of problems 9.4, 9.5, 9.6, and 9.7. Predict the rate versus time behavior of this hypothetical flood.

9.17 Rework problem 9.8 with the oil production to flood start and waterflood recovery changed to 2 MM stb each. All other data remain the same. This problem demonstrates the effect of having the flood start at a later point in the primary life.

9.18 Rework problem 9.9 using the curve with circles in Fig. 9–23 as the performance of the old flood. Do not change any other characteristics for the old flood.

Notes

1. R.L. Slobod and B.H. Caudle, "X-Ray Shadowgraph Studies of Areal Sweepout Efficiencies," *Trans.,* AIME (1954), volume 204, 81.

2. J.E. Richardson et al., "Methods for Determining Residual Oil with Pulsed Neutron Capture Logs," *JPT* (May 1973), pp. 593–606.

3. C.W. Kidwell and A.J. Guillory, "A Recipe for Residual Oil Saturation Determination," *JPT* (November 1980), pp. 1999–2008.

4. J.F. Tomich et al., "Single-Well Tracer Method to Measure Residual Oil Saturation," *JPT* (February 1973), pp. 211–218.

5. W.E. Stiles, "Use of Permeability Distribution in Water Flood Calculations," *Trans.,* AIME (1949), volume 186, 9–13.

6. B.C. Craft and M.F. Hawkins, *Applied Petroleum Reservoir Engineering* (Englewood Cliffs: Prentice-Hall, 1959).

7. Stiles, 1949.

8. E. Suder and J.C. Calhoun Jr., "Water-Flood Calculations," *Drilling and Production Practice,* API (1949), p. 260.

9. M. Prats et al., "Prediction of Injection Rate and Production History for Multifluid Five-Spot Floods," *Trans.,* AIME (1959), volume 216, 98–105.

10. R.V. Higgins and A.J. Leighton, "Computer Method to Calculate Two Phase Flow in Any Irregularly Bounded Porous Medium," *JPT* (June 1962), pp. 679–683.

11. A.B. Dyes, B.H. Caudle, and R.A. Erickson, "Oil Production After Breakthrough—As Influenced by Mobility Ratio," *Trans.,* AIME (1954), volume 201, 81.

12. H. Dykstra and H.L. Parsons, "The Prediction of Oil Recovery by Waterflooding," *Secondary Recovery of Oil in the United States,* 2nd edition (New York: API, 1950), pp. 160–174.

13. H.C. Slider, "New Method Simplifies Prediction of Water Flood Behavior," *Petroleum Engineer* (February 1961).

14. Prats et al., 1959.

15. Dyes, Caudle, and Erickson, 1954.

16. Slider, 1961.

17. F. Craig, *Reservoir Engineering Aspects of Waterflooding,* Monograph No. 3, SPE (1971).

18. Higgins and Leighton, 1962.

19. Dykstra and Parsons, 1950.

20. H.K. van Poollen and Associates, *Fundamental of Enhanced Oil Recovery* (Tulsa: PennWell, 1980).

21. W.B. Gogarty and W.C. Tosch, "Miscible-Type Waterflooding: Oil Recovery with Micellar Solutions," *JPT* (December 1968).

22. van Poollen and Associates, 1980.

Additional References

Slider, H.C. "Predicting Waterflood Performance by Analogy." unpublished SPE paper 3442, 1971, presented at Regional SPE meetings in Los Angeles and in Evansville.

10

Enhanced Oil Recovery

According to Webster the term "enhance" means "to advance, augment, or elevate; to make or become larger."[1] Thus, grammatically, the term *enhanced oil recovery* (EOR) can be applied to any situation where some action has been taken to increase the recovery of oil or gas from a reservoir. However, to most engineers and scientists EOR probably refers to any recovery operation more sophisticated than the injection of water or gas into a hydrocarbon reservoir. To others the term probably is used to include waterflooding and repressuring with gas injection.

The confusion arises historically. Originally, the term *secondary recovery* was applied to waterflooding and gas flooding although some engineers included pressure maintenance through water and gas injection in secondary recovery. Later, when there was an effort to increase the recovery from reservoirs that had previously been subjected to water or gas injection, the term *tertiary* (third) recovery was used. However, this became confusing when these tertiary processes were applied to reservoirs that had not previously been subjected to secondary recovery. Hence, the term enhanced oil recovery became popular and was generally applied to the same processes that had previously been called tertiary recovery.

For the purpose of clarifying what will be covered in this chapter, we define enhanced oil recovery as any processes used to increase the ultimate recovery from a reservoir except the injection of plain water or low-pressure gas into a reservoir. Variations of waterflooding or gas injection are included.

The objectives are different from the general objectives of the first nine chapters of this book. In the first chapters the objective is to teach the engineer to solve problems using proven methods. The objective of the last chapters is to teach the engineer to recognize when help is needed. Most practicing reservoir engineers require specialists in designing EOR projects and predicting their behavior. Engineers in major oil companies should seek the help of research and development groups. Engineers working for smaller companies that do not have such specialists should contact consultants with proven track records in this area of expertise. The purpose of this chapter is to teach the reservoir engineer to recognize when EOR prospects warrant further study.

We start with a listing of the reservoir characteristics that indicate when a particular EOR process should be considered for application in that reservoir. This subject is presented so practicing engineers can judge how much time to invest in obtaining more detailed knowledge of the EOR processes. If the engineer does not have reservoirs that are good EOR prospects, he may wish to skip this particular chapter or spend a minimum amount of time on it. However, engineers who recognize some EOR prospects among the reservoirs for which they are responsible may want to study this chapter to obtain an understanding of the EOR mechanisms before discussing their problems with experts.

The list of reservoir characteristics conducive to various EOR processes is followed by an explanation of the forces and conditions that limit the recovery of oil from a subsurface reservoir. Then a description of the various EOR methods and an explanation of the mechanisms that are encompassed by each method are given. Methods for conducting a feasibility analysis of a prospect also are presented.

Recognizing an EOR prospect. The classification of EOR processes is difficult. They can be classified as thermal or nonthermal, mobility control or residual oil control, and polymer, miscible, micellar, or surfactant. However, for the purpose of grouping processes that seem to be generally successful in the same type of reservoir, we consider EOR processes in the categories of steam and hot waterflooding, in situ combustion, micellar and surfactant floods, miscible displacement, and mobility control polymers. Table 10–1 lists some of the reservoir characteristics that give the most promise of successful application of each group of EOR processes. These classifications overlap. Miscible displacement as used here refers to the use of the miscible mechanism alone as in the use of carbon dioxide, high-pressure natural gas, and very high-pressure inert gas. Miscibility also plays a big part in steamflooding, in situ combustion, and micellar and surfactant floods. Also, micellar and surfactant floods are generally combined with the use of mobility control polymers.

Basic forces and difficulties limiting oil recovery. Two basic problems limit the recovery of oil from a reservoir. The most obvious problem is the difficulty of making all of the oil mobile (movable) in a

TABLE 10–1 Reservoir Characteristics Amenable to Enhanced Oil Recovery

	Steamflooding and Hot Waterflooding	*In Situ Combustion*	*Micellar and Surfactant Floods*	*Mobility Control Polymers*	*Miscible Displacement*
Oil saturation, %	> 65	> 50	> 25	>50	>25**
Average permeability, md	100–5,000	High	> 20	>20	5–100†
Oil viscosity or gravity	1–45°API	>10°API	< 30 cp	< 200 cp	< 5 cp**
Reservoir temperature, °F	—	—	< 250°F	< 200	< Critical temperature
Thickness, ft	10–300	< 10	—	—	< 25**
Formation depth, ft	< 3,000 Ft*	—	—	—	> 2,500 ft for CO_2**
Miscellaneous	Reservoir pressure < 2,000 psia	Coking < 1 lb/cu ft	Water salinity < 200,000 ppm	—	pressure 1,000–4,000 psia

*After "DOE Report Examines EOR Constraints," *OGJ* (April 28, 1980).
**After van Poollen and Associates, *Fundamentals of Enhanced Oil Recovery* (Tulsa: PennWell, 1980).
†After Timmerman, *Practical Reservoir Engineering,* volume 2 (Tulsa: PennWell, 1982) and van Poollen and Associates, *Fundamentals of Enhanced Oil Recovery* (Tulsa: PennWell, 1980).

particular process. Most of the processes used on a practical or experimental basis employ a frontal displacement of oil, and it is necessary to provide a mechanism that makes as much of the oil mobile as possible. The one process that does not use a frontal displacement is the huff-and-puff steam process where steam is injected into a well and later the same well is used as a producing well. However, even here, the basic problem is making the oil mobile by reducing the viscosity.

Even when 100% of the oil is made mobile by a particular process, we are still faced with the problem of making the process—frontal displacement in most cases—contact all of the reservoir (see chapter 9).

The problem of making the oil mobile by overcoming capillary forces in the reservoir has been discussed briefly in chapter 6. Fig. 6–10 illustrates a pore with a changing radius that contains an isolated drop of oil so one end of the drop has a radius that is smaller than the other end. Therefore, one end of the drop is subjected to a capillary pressure that is greater than the capillary pressure at the other end. If the direction of the imbalanced capillary forces is opposite to the direction of flow, the drop of oil does not flow unless the applied pressure differential is greater than the imbalanced capillary forces. This mechanism results in a large amount of oil being immobile in an immiscible system. It may be worthwhile to note that an immiscible system is one in which the fluids do not mix, i.e., they are unmixable. We also may say it is a system in which the two fluids exhibit some finite interfacial tension.

Since the capillary forces are proportional to the interfacial tension, σ, $p_c = 2\sigma \cos \theta / r$, it is clear that any reduction in the interfacial tension reduces the capillary pressures or the difference in capillary pressures and, thus, reduces the residual oil. If the interfacial tension can be reduced to zero, e.g., liquid petroleum gas displacing oil, the two fluids are miscible (mixable). All of the oil can be produced because the capillary forces that result in the residual oil are zero.

The engineer may wonder why the oil breaks up into discontinuous drops during displacement when the water is normally the wetting phase. As explained in chapter 6, this occurs because the porous media is made of a variety of pore sizes and fluids traveling through these different size pores at different velocities. Since the pores are interconnected, oil and water may travel through different size pores at different velocities. When they reach the intersection of the pores, there is a mixing of the oil and water, which forms a discontinuous oil phase since water is the wetting phase.

Many companies feel that some reservoirs are oil wet rather than water wet. If this is true, the oil is in the smallest pores relative to those occupied by water. Then the displacement process should be different. When engineers and researchers first began looking at the water-wet/

oil-wet phenomena, many concluded that reservoirs were oil wet without realizing that they were examining a core whose wettability had been changed by the process used to clean the oil from the core prior to the wettability tests. Most companies have found it difficult to core a reservoir without changing the wettability of the core. The important point to be made is that an engineer should be very careful in concluding that a reservoir is oil wet based on lab tests. The relationship between the magnitude of the irreducible water saturation and the residual oil saturation as observed in the reservoir together with the observed relative permeability characteristics as determined from reservoir saturations and produced water-oil ratios should provide a more reliable conclusion on the wettability characteristics of a reservoir.

Steam and Hot-Water Injection

The principle effect of steam and hot-water injection that results in increased oil production is the reduction in the oil viscosity that accompanies the increase in temperature. Inspection of the empirical relationship between the oil viscosity with no gas in solution and the temperature can be seen in the top graph of Fig. B14. It is not unusual for a heavy oil of 13°API to have a viscosity of 2,000 cp at a reservoir temperature of 110°F and have a viscosity of only 60 cp at 220°F. This potential 33-fold reduction in viscosity means there is a corresponding 33-fold increase in the production rate. In addition, there appears to be some effect on the wettability of the formation making it more strongly water wet, as indicated by the fact that the irreducible water saturation appears to be increased. This is evidenced by the fact that it is not unusual in a huff-and-puff operation for no more than 30% of the injected water (steam) to be produced with the oil. This could mean that the increase in the irreducible water saturation moves the oil into larger pores where it can flow more readily.

Steam also tends to strip the light hydrocarbons from the residual oil over which it passes and, thus, becomes a mixture of steam and light hydrocarbons that tend to be miscible with the oil.

Steamflooding and steam huff and puff represent perhaps 80% of the total oil produced by EOR methods in 1980.[2] Virtually all of this production is from heavy crude-oil reservoirs in California, Venezuela, and Alberta. A huge steamflood costing $2.7 billion involving nearly 4,500 wells is under construction in the Duri field in Sumatra.[3] Because of the steam stripping and wettability phenomena, there also appears to be some potential for the application of the steam processes to reservoirs containing lighter oils.

The biggest problems with steam EOR processes may be more mechanical than reservoir. Only a small fraction (5–10%) of the energy generated in the steam generator actually reaches the reservoir as a result of heat losses.[4] Also, the completion of the wells with the potential thermal expansion and contraction of the casing and tubing presents difficult problems. The heat loss during steam movement down the well generally limits the depth of application to less than 3,000 ft.

Huff-and-puff operations. The huff-and-puff process is also called *steam stimulation* or *steam soak* (Fig. 10–1). This process was born when a major oil company in Venezuela was attempting a steamflood, and the injection pressure had become so high that steam broke through to the surface. When the injection well was back flowed to relieve the steam pressure, substantial amounts of oil were produced with the steam.[5] The result of this high-powered research was the huff-and-puff steam process. Steam is injected for a short period of time, perhaps 5 days, the well is permitted to rest (soak) for 2–3 days, and the same well is put on production. Typically, the oil rate increases dramatically and declines until in 3–6 months it is back to the rate at which it produced prior to the steam injection. The steam injection and soak may then be repeated two to four times, probably with less success. Eventually, the increase in rate becomes uneconomic.

A typical huff-and-puff result may be that indicated in Fig. 10–2, which has been reported as the average results obtained in the Huntington Beach Field in California.[6] This 2.3-darcy, 100-ft sandstone contained about 700-cp oil at a depth of 2,000 ft. In many cases huff-and-puff operations are conducted on all of the wells in a proposed steamflood to improve the injectivity and productivity of the wells and to stimulate the production from the producing wells until the frontal displacement of the steamflood affects the producing wells.

Since most of the pressure drop in a radial system occurs near the producing well, an estimation of the initial oil producing rate can be made on the basis of the viscosity of the oil before and after steaming the well:[7]

$$\frac{q_{hot}}{q_{cold}} = \frac{\text{oil viscosity before steaming}}{\text{oil viscosity after steaming}} \tag{10.1}$$

Determining how long and in what manner the increased rates are maintained is a more difficult problem. In addition to the problems of permeability stratification and mobility ratio difficulties discussed in chapter 9, it is necessary to account for the flow of heat between strata and from the reservoir to the formations above and below the oil reservoir. The conduction of heat during the soak period is illustrated in Fig. 10–3. The radial conduction of the heat to increase the reservoir

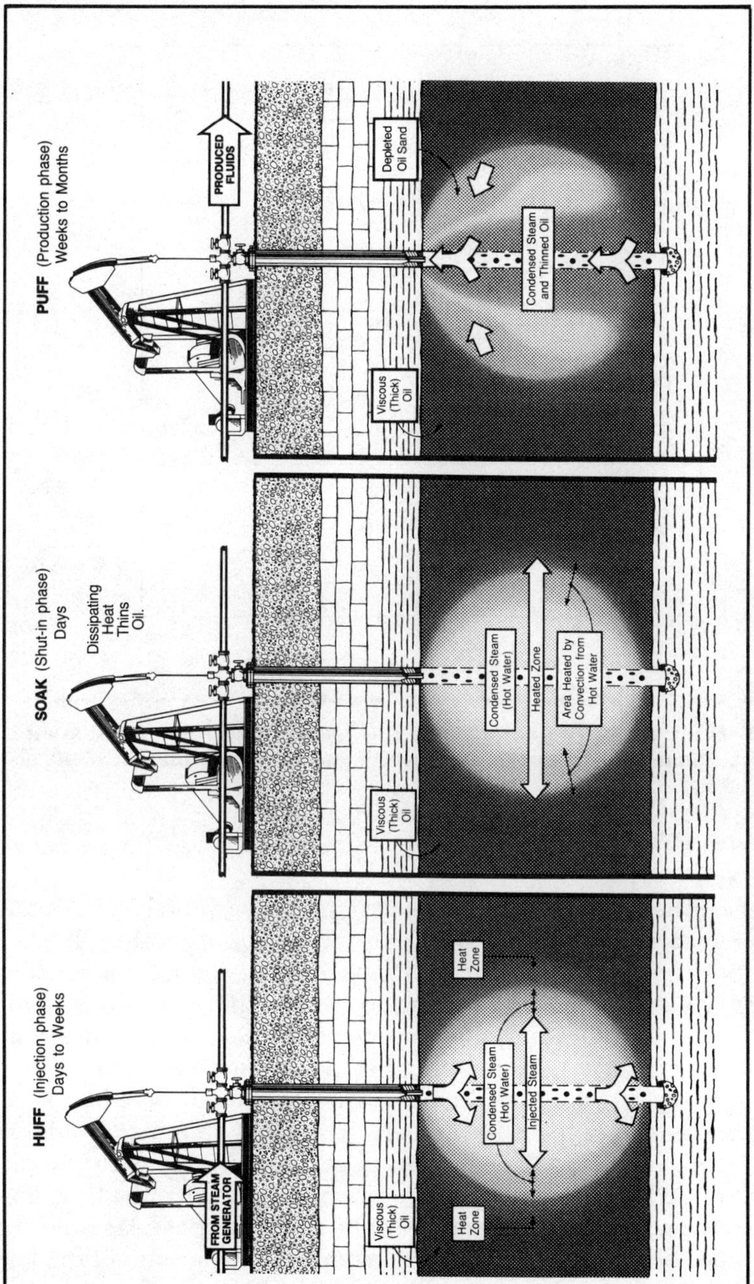

Fig. 10–1 Cyclic steam stimulation (reprinted from *Enhanced Oil Recovery and Improved Drilling Technology*, Report No. DOE/BETC-82/1, courtesy Bartlesville Energy Technology Center, DOE). Steam injected into a well in a heavy-oil reservoir introduces heat that, coupled with alternate soak periods, thins the oil, allowing it to be produced through the same well. This process may be repeated until production falls below a profitable level.

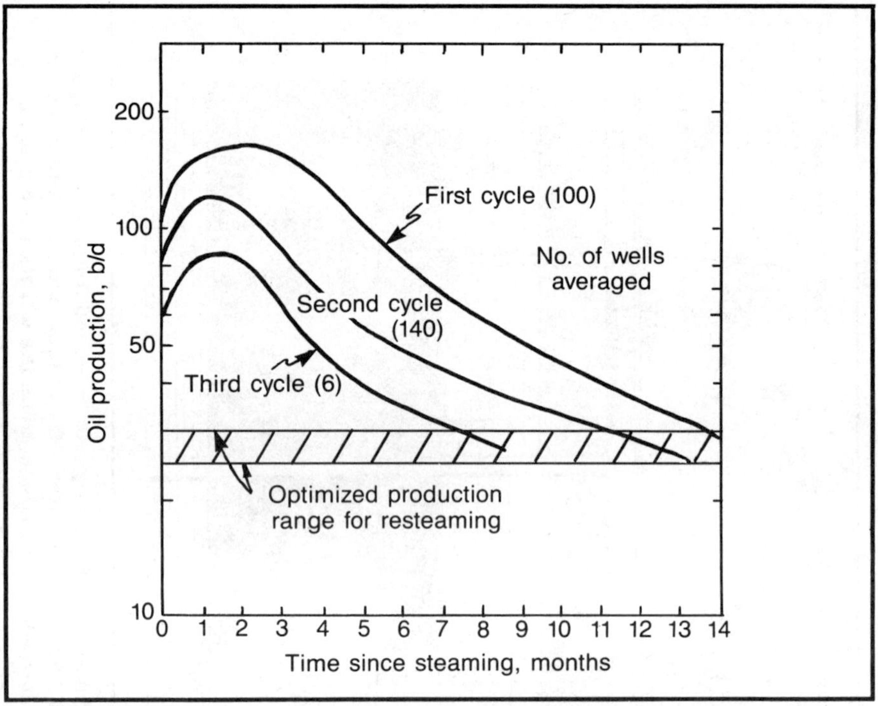

Fig. 10–2 Huntington Beach huff-and-puff results (after van Poollen and Associates, *Fundamentals of Enhanced Oil Recovery*, PennWell, 1980, courtesy *JPT*)

temperature as far as possible radially makes a longer soak period desirable.

However, the loss of heat to the surrounding formations above and below the reservoir makes a shorter soak period desirable. It then appears that there may be some optimum soak period for a particular volume of injected steam, stratification, conductivity of surrounding formations, etc. Determining the optimum soak period is further complicated by the choice between starting production before all of the steam condenses in order to make use of the high compressibility of the steam to increase productivity or permitting the steam to condense completely to avoid the adverse mobility ratio associated with the displacement of oil by a gas, steam. Once production begins, the heat flow becomes even more complex because of the movement of heat from the heated zone to the steam zone in the flowing oil and because of the loss of heat from the formation through the hot oil being produced.

The complexities of the heat flow in huff-and-puff operations are offset to some degree by the fact that in most cases flow is caused almost entirely by gravity drainage, which is relatively simple to ana-

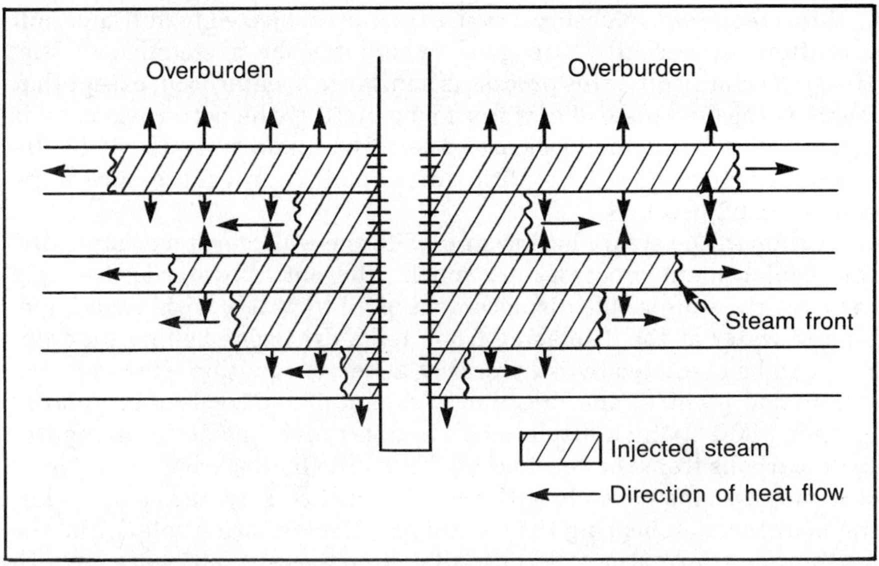

Fig. 10–3 Conduction of heat during huff-and-puff soak period in stratified reservoir

lyze. Since very viscous oils seldom have little if any gas in solution, it appears that most of the production results from gravity drainage. Viscous oil reservoirs that have the misfortune of having strong water drives do not generally produce much oil because of the extremely adverse mobility ratio between water and the highly viscous oil. This causes the water to break through into the producing well after a minimum amount of oil has been produced, at which time the water cut quickly approaches 100%.

The huff-and-puff operation has been modeled by many investigators based on a wide range of assumptions. Van Poollen and Associates discuss the Boberg and Lantz model along with the modifications of Jones.[8,9,10] Timmerman presents the Seba and Perry method in some detail.[11,12] Reservoir response to huff-and-puff operations have generally been better than that predicted by model studies. Production as high as 20 times that predicted by model studies has been reported.[13] Consequently, it is recommended that the engineer back up any model study predictions with a prediction by analogy when possible. Experience with huff and puff in very viscous reservoirs has been so extensive that an engineer should be able to find performance data for a huff-and-puff operation from a similar depth, thickness, oil viscosity, stratigraphy, etc., on which to base huff-and-puff design and performance prediction.

Steamflooding. Most reservoirs that are subjected to huff-and-puff operations successfully are good candidates for a steamflood (Fig. 10–4). Mechanically, this process is similar to a waterflood, except that steam is injected instead of water and generally the pattern spacing is much less. Also, steamfloods are generally limited to depths of 4,000–5,000 ft because of heat-loss difficulties and the critical pressure of the steam (3,202 psia).

Although the steamflood is similar to the waterflood mechanically, the displacement processes are much different. A waterflood simply involves the immiscible displacement of oil by water with a mixing of oil and water at the displacing front and a drag zone behind the front that can be predicted by the Buckley-Leverett equation. However, the steamflood involves the injection of steam of relatively high quality (perhaps 80%), which displaces oil and gas and may strip the lighter hydrocarbons from the residual oil. Thus, the hydrocarbon-steam mixture may become miscible with the oil. Concurrently, the steam is losing heat through heating the oil and gas, the formation solids, and the surrounding formations. Eventually, it condenses into water, which advances ahead of the steam and displaces oil ahead of it. A simplified fluid and temperature distribution is shown in Fig. 10–5.

Willman shows that the hot-water front displaces more oil than a regular waterflood and that additional oil displacement occurs when the reservoir is contacted by the steam (Fig. 10–6).[14] This is based on a Buckley-Leverett type of calculation for a radial system equivalent of a 0.625-acre five-spot containing 900-cp oil.

The displacement process is further complicated by the tendency of the steam to override the oil as a result of the extreme density and viscosity differences between the steam and the oil. Most reservoir engineers are familiar with the difficulty associated with the displacement of a liquid by a gas. The engineer should remember that steam is a gas, and the same mobility and density contrasts exist between the steam and oil as exist between a gas and oil. However, the steam-oil relationship is probably more severe because of the high temperature at which the steam exists. The fact that the steam is not 100% gas—the quality is less than 100%—helps some in this regard. Nevertheless, there is a severe tendency for the steam to override the oil being displaced and to finger through the oil.

The injection of steam near the bottom of the oil zone may minimize the overriding tendency of the steam. An initial gas cap or a water zone in the reservoir further complicates the displacement mechanism.

Several reports of successful steamfloods can be found. One such report is for the Schoonebeck Field in the Netherlands as reported by Dijk.[15] This steamflood consists of about four inverted five spots involving four injection wells and eight producing wells with a gross swept

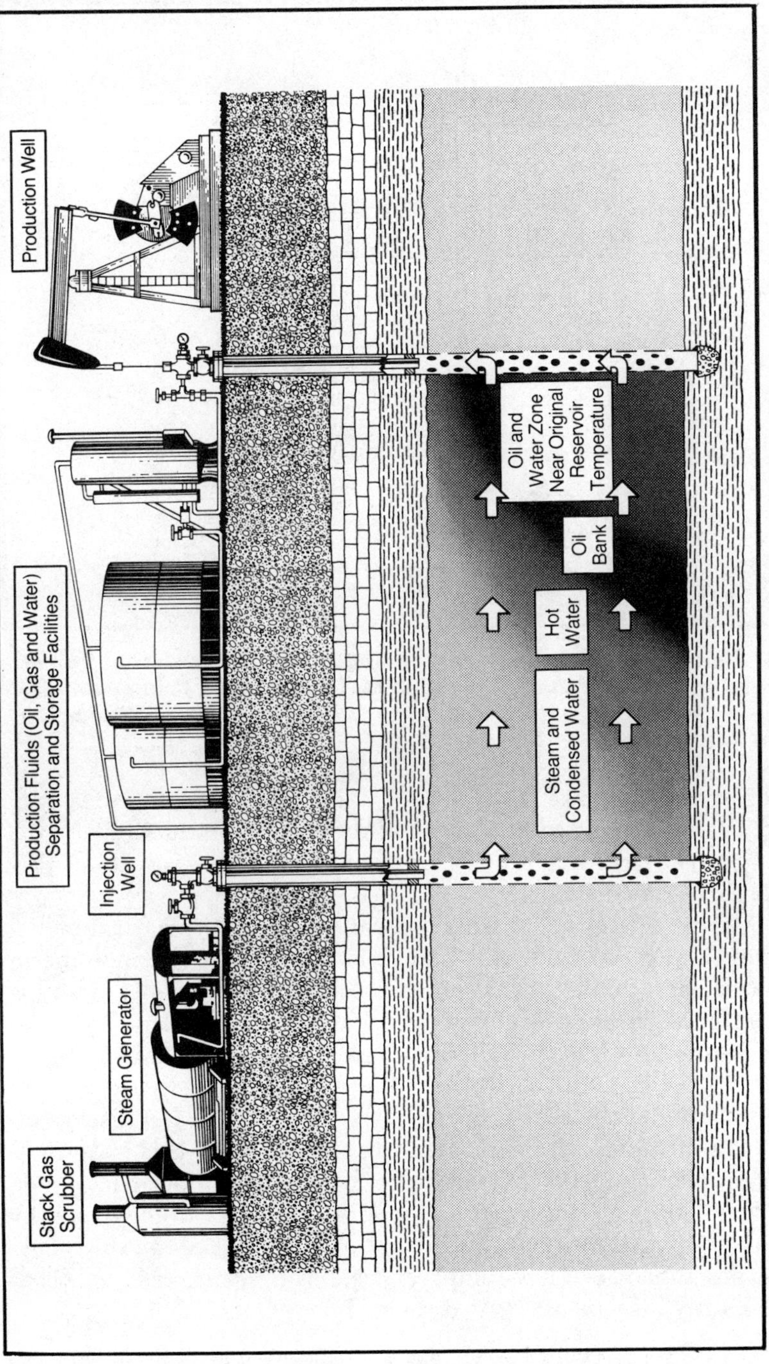

Fig. 10–4 Steamflooding (reprinted from *Enhanced Oil Recovery and Improved Drilling Technology*, Report No. DOE/BETC-82/1, courtesy Bartlesville Energy Technology Center, DOE). Heat from steam injected into a heavy-oil reservoir thins the oil, making it easier for the steam to push the oil through the formation toward production wells. Heat reduces viscosity of oil and increases its mobility.

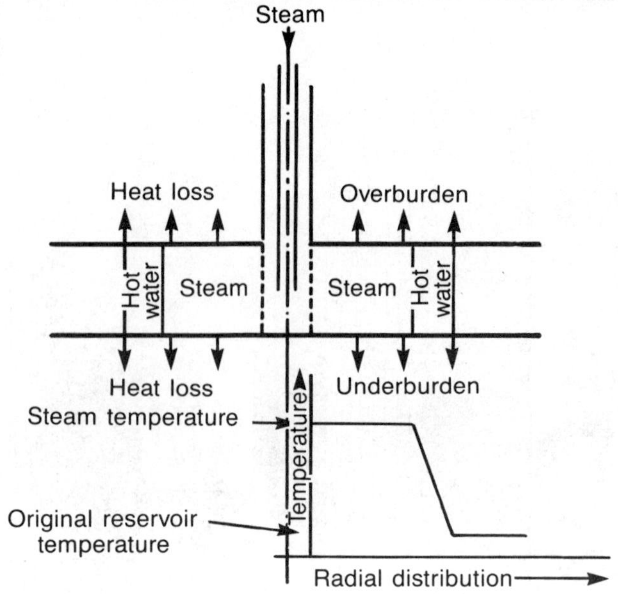

Fig. 10–5 Schematic of steam injection and approximate distribution of formation temperature (after van Poollen and Associates, *Fundamentals of Enhanced Oil Recovery*, PennWell, 1980, courtesy *Journal of Canadian Petroleum Technology*)

volume of about 65 acres. Injection of about 10.5 MM bbl of steam (water equivalent) resulted in about 4 MM bbl of oil production and 9.2 MM bbl of water production over a 6-year period. The results of this steamflood are shown in Fig. 10–7.

Many attempts have been made to formulate a mathematical model of a steamflood. Perhaps the best known is the Marx and Langenheim model that ignores radial heat conduction, that assumes only steam is displacing the oil without formation of a water bank, and that assumes fluids are incompressible.[16] Willman et al., Lauwerier, Malofeev, Rubinshtein, Spillette, and many others have contributed to the mathematical modeling of the steam flood.[17,18,19,20,21] As noted, a discussion of steamflood modeling is not a part of the objective of this chapter.*

A means of estimating the recovery from a steamflood has been presented by Gomaa.[22] He used a computer simulation developed by Coats et al., to investigate the effect of several parameters on the recovery from a steamflood of a low-temperature, low-pressure reservoir containing a highly viscous oil.[23] As a result he recommended the follow-

*Such methods are discussed in *Fundamentals of Enhanced Oil Recovery,* by H.K. van Poollen & Associates (Tulsa: PennWell, 1980).

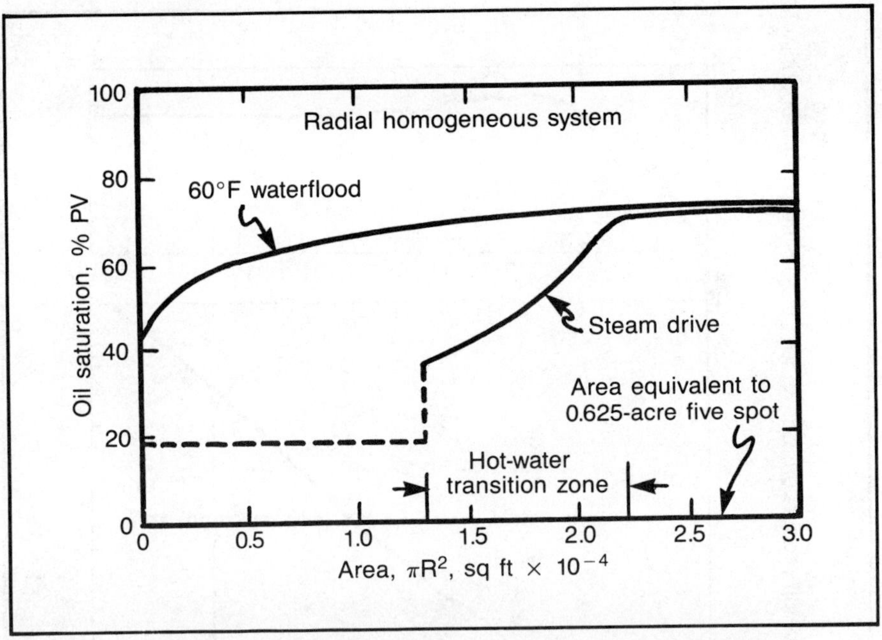

Fig. 10–6 Saturation distribution after 1 year of injection (after van Poollen and Associates, *Fundamentals of Enhanced Oil Recovery*, PennWell, 1980, courtesy *Trans.*, AIME)

ing procedure for estimating the recovery versus cumulative steam injection for a proposed flood, with the figure numbers changed to conform with the numbering system of this book:

1. Read the vertical heat loss (f_{hv}) as fraction of input from Fig. 10–8.
2. Read the heat utilization factor (Y) from Fig. 10–9.
3. Calculate the net heat injected (Q_{inj}) in MMBTU/gross acre-ft from:

$$Q_{inj} = 0.128\ \Sigma\ [I\ h(1 - f_{hv})\ \Delta t]_i,$$

Where:

I = Injection rate, b/d/gross acre-ft
h = Enthalpy, BTU/lb_m, from Fig. 10–10
Δt = Time increment, years
i = Index of time increments

4. Calculate the effective heat injected (Q_e) in MMBTU/gross acre-ft from:

$$Q_e = YQ_{inj}$$

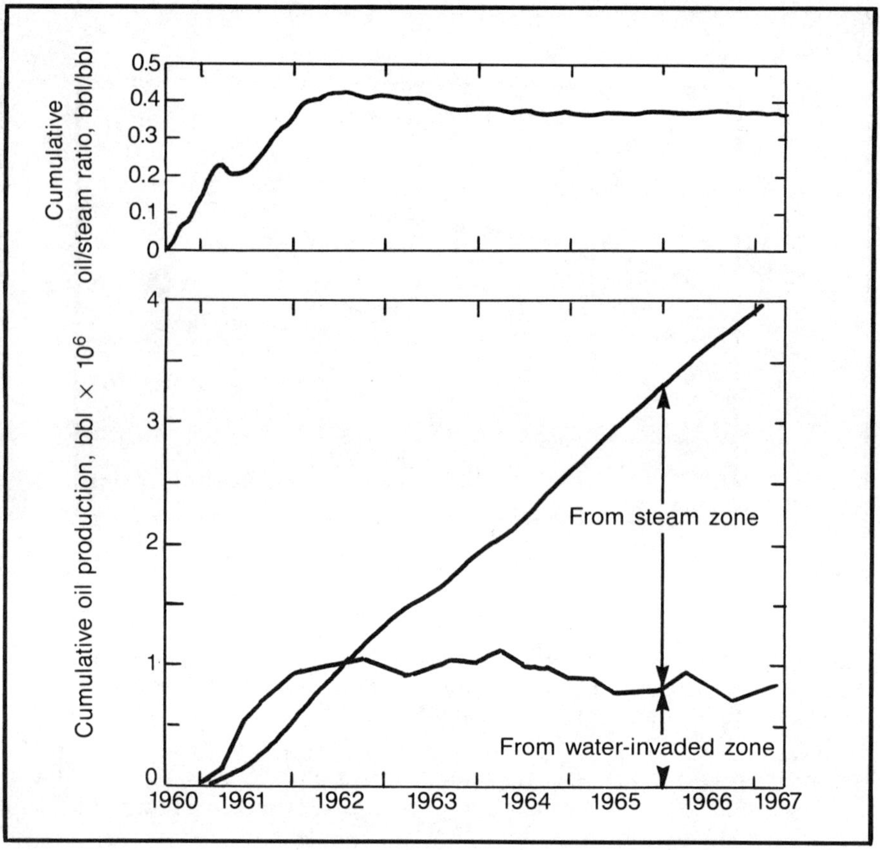

Fig. 10–7 Cumulative production and oil-steam ratio (after van Poollen and Associates, *Fundamentals of Enhanced Oil Recovery*, PennWell, 1980, courtesy *Trans.*, AIME)

Where:

Q_e = Effective heat injected, MMBTU/acre-ft
Q_{inj} = Net heat injected, MMBTU/acre-ft
Y = Heat utilization factor

5. Read the oil recovery from Fig. 10–11.
6. Repeat steps 3, 4, and 5 as many times as necessary until the ultimate recovery is reached.

If the results of such an empirical study are promising, the engineer should proceed to a more detailed, explicit study of the prospect.

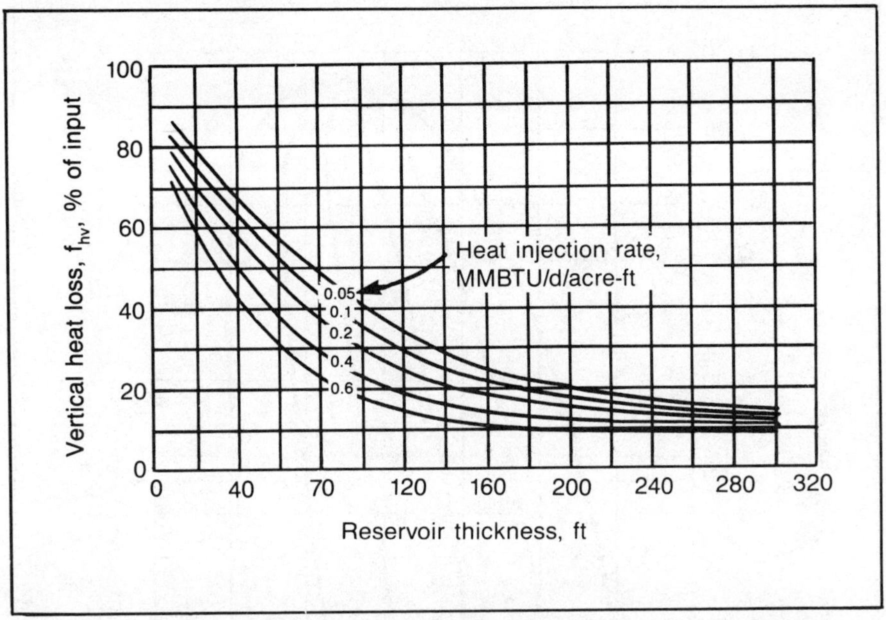

Fig. 10–8 Heat loss to overlying and underlying strata (after Gomaa, "Corrections for Predicting Oil Recovery by Steamflood," courtesy *JPT*, February 1980, © SPE-AIME)

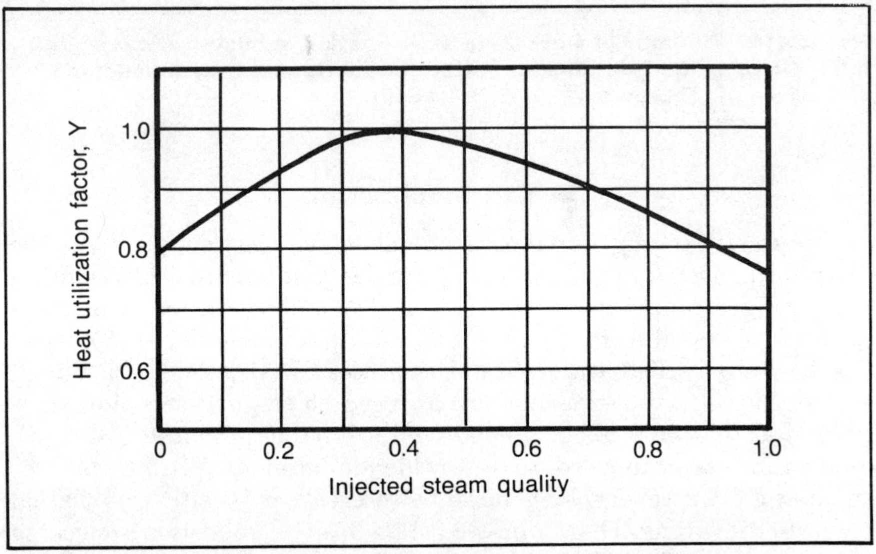

Fig. 10–9 Heat utilization factor as a function of steam quality (after Gomaa, "Correlations for Predicting Oil Recovery by Steamflood," courtesy *JPT*, February 1980, © SPE-AIME)

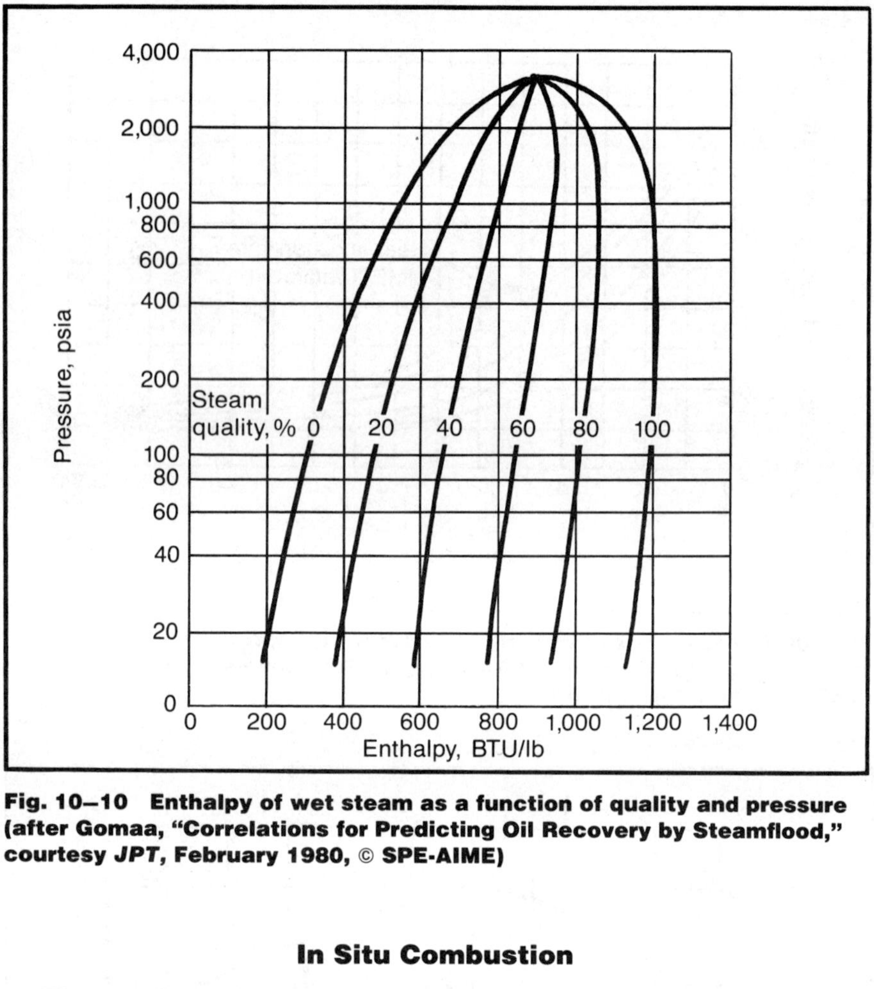

Fig. 10–10 Enthalpy of wet steam as a function of quality and pressure (after Gomaa, "Correlations for Predicting Oil Recovery by Steamflood," courtesy *JPT*, February 1980, © SPE-AIME)

In Situ Combustion

The in situ combustion process involves the injection of air into the reservoir in such a way that a portion of the oil is burned. The resulting heat and gases that are the product of the combustion result in 100% removal of the oil from the reservoir that is contacted by the process. Fig. 10–12 is a schematic of the in situ process. In situ combustion must be considered in the experimental or research stage at this time. It is estimated that only about 55,000 bo/d are being produced by in situ combustion as compared to a worldwide total of 681,000 bbl for enhanced oil recovery in general.[24] Interest in in situ combustion seems to be waning. There were 38 active in situ combustion projects in 1971 and only 17 in 1980.[25]

In situ combustion can be carried out in several different ways. A conventional forward combustion is depicted in the saturation and tem-

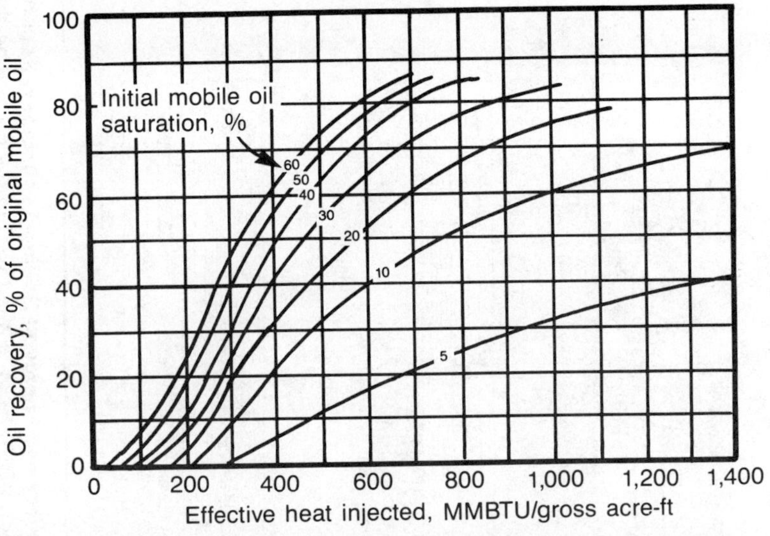

Fig. 10–11 Steamflood oil recovery as a function of effective heat injected and mobile oil saturation (after Gomaa, "Correlations for Predicting Oil Recovery by Steamflood," courtesy *JPT*, February 1980, © SPE-AIME)

perature diagrams of Fig. 10–13. A reverse combustion process can be used where the air injected travels in one direction and the flame front travels in the opposite direction—like blowing air through a cigarette. Forward combustion may also be combined with water injection resulting in wet combustion or partially quenched combustion.

When air is injected into a reservoir containing oil, the air tends to oxidize a portion of the oil. This oxidation raises the temperature of the air and combustion gas, and the rate of oxidation is increased resulting in a temperature rise until the temperature becomes so high that instantaneous combustion or burning takes place. Burning of the oil produces much heat and results in heating the oil ahead of the burning front. This heating greatly reduces the viscosity of the oil and makes it easy for the combustion gases to displace the oil toward the producing well.

This heating of the oil also results in the formation of coke ahead of the burning front. The high temperatures also cause flashing of the formation water into steam, which in turn picks up light hydrocarbons from the hot oil and forms a gas with the products of combustion that may tend to be miscible with the oil that the steam is displacing.

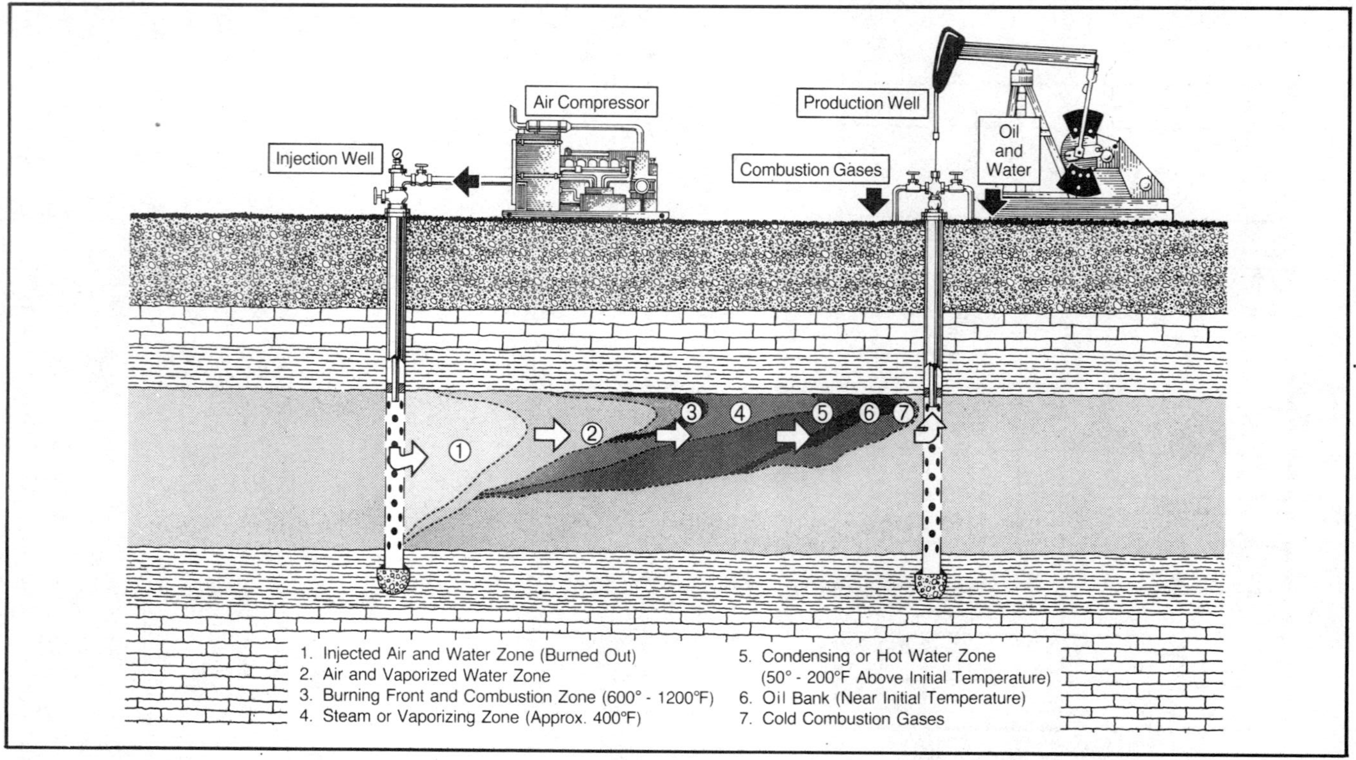

Fig. 10–12 In situ combustion (reprinted from *Enhanced Oil Recovery and Improved Drilling Technology*, Report No. DOE/BETC-82/1, courtesy Bartlesville Energy Technology Center, DOE). Heat is used to thin the oil and permit it to flow more easily toward production wells. In a fireflood the formation is ignited, and by continued injection of air, a fire front is advanced through the reservoir. Mobility of oil is increased by reduced viscosity caused by heat and solution of combustion gases.

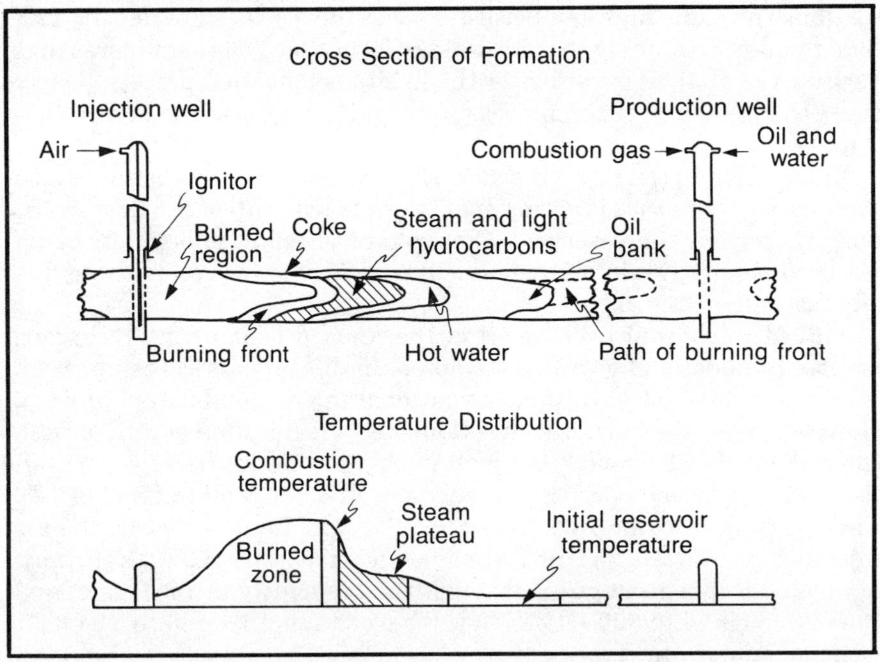

Fig. 10–13 Schematic of in situ combustion process (after van Poollen and Associates, *Fundamentals of Enhanced Oil Recovery*, PennWell, 1980, courtesy Nelson and McNeil)

When the steam has been displaced a sufficient distance ahead of the burning front, it is cooled by the transfer of its heat to the oil and rock matrix. It is then condensed and forms a water bank ahead of the steam bank. The advancing water bank reduces the oil saturation to a value comparable to that in a conventional waterflood. That residual is further reduced by the zone of steam and light hydrocarbons so only a coke remains, which in turn is burned by the flame front. This process can completely remove all of the oil from the reservoir rock contacted by the process. This has been verified by coring formations that have been subjected to in situ combustion.

In spite of the fact that all of the hydrocarbons can be removed from the reservoir contacted by the process, many factors make the in situ combustion process of questionable economic worth at the present time. Perhaps the biggest difficulty is the inability to make the process contact a significant portion of the reservoir. The combustion gases and steam at extremely high temperatures provide a minimal viscosity for the mixture that in turn provides a horribly adverse mobility ratio and a very strong tendency for the displacement to finger through the formation. The same high-temperature gases have a very low density and tend to migrate quickly to the top of the formation. The same cores that verify a residual oil of zero after the in situ combustion process clearly show the tendency of the process to burn through the top of a formation.

In addition to the terrible sweep efficiency that characterizes in situ combustion, note that it is very inefficient to leave all of the heat in the burned zone in the reservoir. Consequently, many efforts are being made to modify the in situ combustion process to overcome these disadvantages.

In order to avoid leaving all of the unused heat in the reservoir, reverse combustion has been attempted. In this process the combustion front is established as in the conventional in situ combustion process. However, after the burn is well established, air injection is shifted from the regular injection well (left well in Fig. 10–13) to the adjacent well that previously had been the producing well (right well in Fig. 10–13). Then the injection well becomes the producer. In other words the air injection is reversed. In this way the oil is forced to move with the flow of air and combustion gases through the previously burned zone, and the oil is heated to 500–700°F in the burned zone. Therefore, the heat does not remain in the reservoir.

The theory of reverse in situ combustion works well in the laboratory, but it does not appear to live up to its expectations in the field. It is believed that the failure of the process in the field results from the spontaneous combustion of the oil near the new (reversed) injection well that results in a conventional forward combustion rather than the reversed in situ combustion desired.[25] In most cases spontaneous combustion results from the injection of air into an oil reservoir.[26,27]

Other efforts have been made to scavenge the heat from the burned zone in the reservoir and to improve the effective mobility ratio by injecting water into the reservoir with the air. Due to the heat in the burned zone, the water injected flashes to steam within a few feet of the injection well and simply helps to transport the heat from the burned zone to the oil ahead of the burned zone. Physically, this process appears to enlarge the hot-water and steam zones in a profile such as Fig. 10–13 and to reduce the temperature in the burned zone. Such a process is known as *wet combustion* or *partially quenched combustion,* depending on the amount of water used. Injection and production data for a wet in situ combustion project operated by Cities Service and supported by the Department of Energy are shown in Figs. 10–14 and 10–15.

Another problem that exists in the design of an in situ combustion project is maintaining a proper coke supply during the process. The flame front can only move forward after it has burned all of the coke in

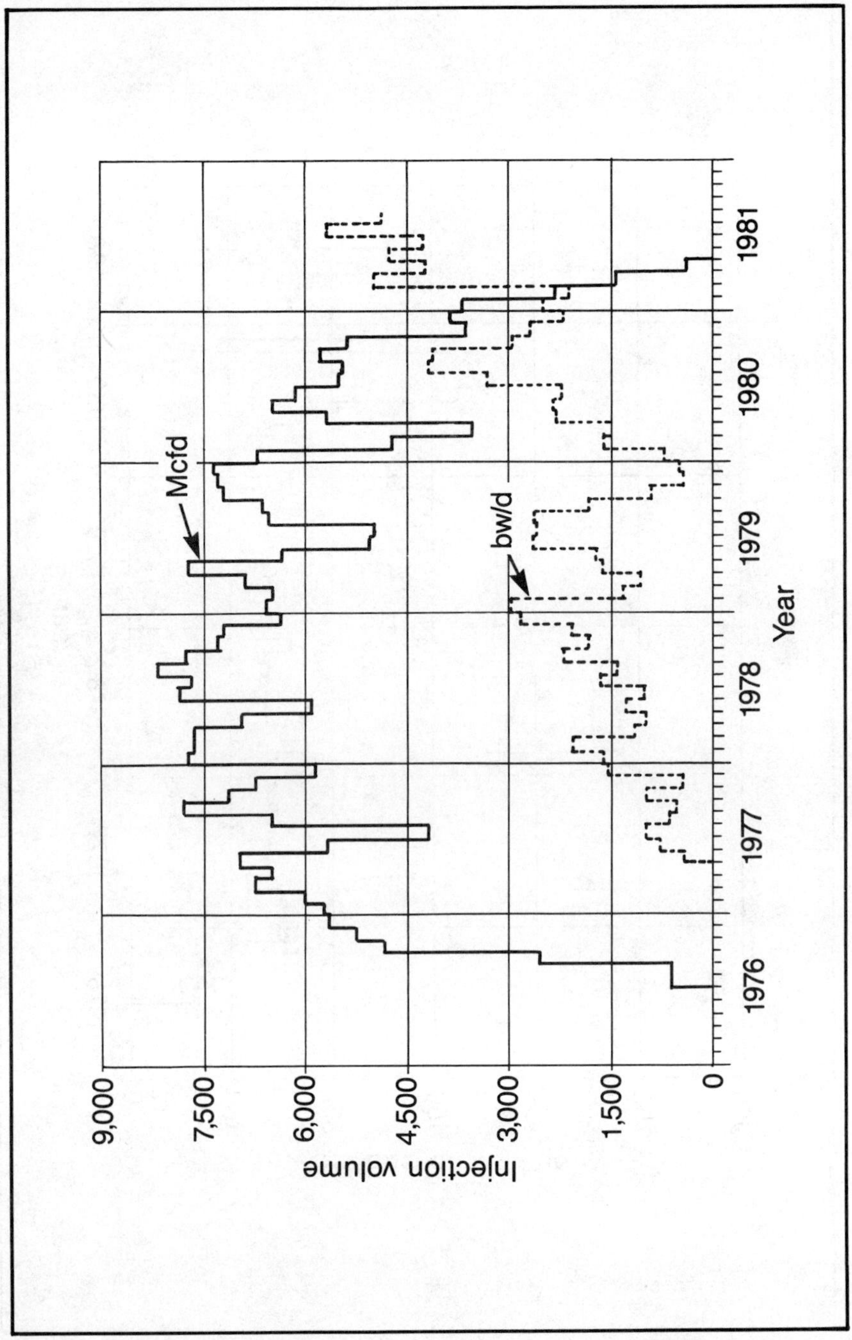

Fig. 10–14 Injection rates for Bodcau in situ project, July 1976 to August 1981

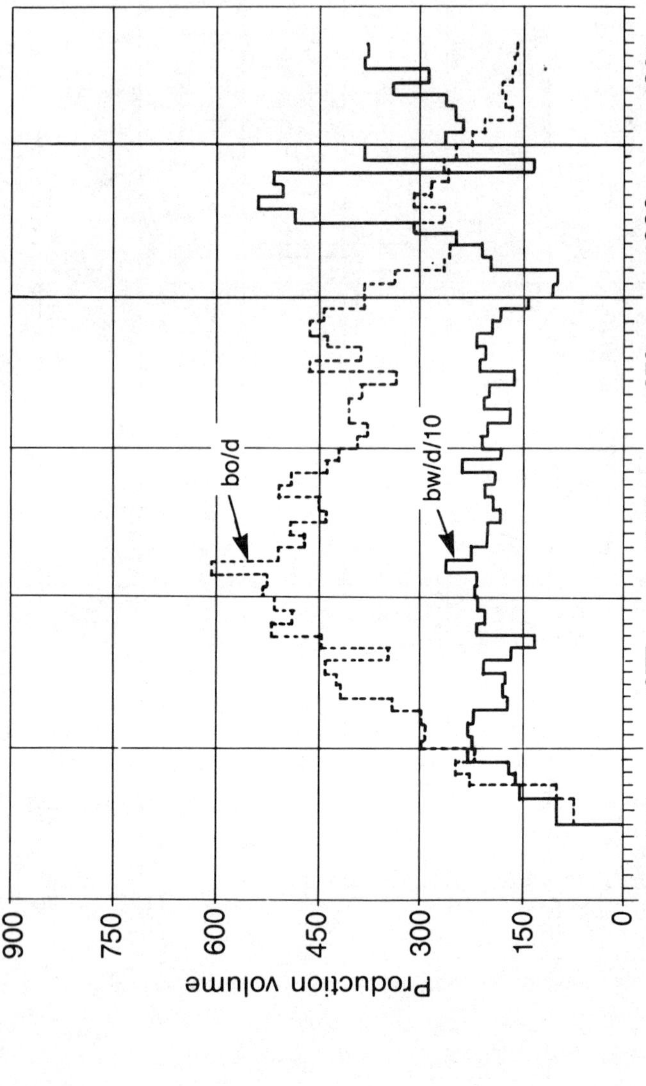

Fig. 10–15 Production rates for Bodcau in situ project, July 1976 to August 1981

its path. Consequently, crude oils that form excessive coke make it necessary to use large amounts of air, and the combustion process progresses slowly. However, if the oil in the reservoir has a low viscosity, all of the oil may be displaced ahead of the advancing flame front and no coke forms, in which case the flame front may be extinguished.[28]

These difficulties in the design and operation of an in situ combustion process have apparently made the successful computer modeling of the process impossible.

Micellar and Surfactant Floods

Marathon Oil appears to have pioneered the development of a mixture of oil and water termed a *micellar solution* that can displace all of the oil from the contacted portion of a reservoir under the right set of conditions. A micellar solution is a microemulsion of oil and water preferably with the oil being the continuous phase. With the oil as the continuous phase, the displacement of oil by a micellar solution puts oil in contact with oil. Thus, a displacement of oil by oil results. At the same time the oil in the micellar solution may be only 5–10%, which keeps the cost low.

The primary chemicals used in a micellar solution are petroleum sulfonates. The chemistry of petroleum sulfonates is very complex. They vary widely in their molecular weight from 350–550. One of the problems in using the micellar solution is the cost of the sulfonates. The commercial market for sulfonates at present is relatively small, but a flood may require a million pounds of sulfonate per year. It has been theorized that if the use of sulfonates in large amounts were stabilized it would be possible to manufacture them at a much cheaper price. However, there will be no big market until the price comes down.

One of the biggest problems in any chemical flood process is the absorption of the chemical on the surface of the porous media. The typical sandstone reservoir may have a porosity of 20–25%, but it may contain pore diameters measured in microns. This combination means that the porous media must contain enormous amounts of solid surface, which in turn means that tremendous quantities of chemical can be absorbed on these surfaces in the porous media. Of course, the excessive absorption of the chemicals from a micellar solution causes the micellar solution to break down and lose its ability to displace all of the oil. Some investigators have shown that a preflush of selected chemicals satisfies the absorption needs of the reservoir surfaces, and the subsequent travel of a micellar solution through the porous media results in greatly reduced absorption.[29,30]

The original formulation of micellar solutions seems to suggest that the viscosity of the solution can be controlled, but this quality has apparently been dropped. It is now common to combine a micellar or surfactant flood with a polymer flood for mobility control, i.e., a slug of micellar solution is followed by a polymer in order to control the effective mobility ratio. The polymer is subsequently displaced by water. Thus, the process is one of displacing the oil with a micellar solution or surfactant to reduce the residual oil to zero. Then displace the slug of micellar solution or surfactant with a slug of polymer that controls the mobility ratio and makes it possible to maximize the amount of reservoir contacted by the residual oil-reducing slug. Since it would be uneconomical to continue injection with a polymer, the polymer slug is displaced with water. Fig. 10–16 is a schematic of a micellar-polymer flood. It is obvious that optimizing the design of such a system represents a considerable task.

Some companies use microscopic methods of qualitatively investigating the effect of specific micellar solutions and surfactants on a drop of oil under static conditions or with a slight movement of the micellar solution or surfactant.[31,32] In addition, actual displacements are conducted in Berea sandstone, a common, easily obtained sandstone used in the laboratory. Finally, the promising systems are used for displacements in actual reservoir sandstones or in a reservoir media made from pulverized reservoir material.

Fig. 10–17 shows the result of a micellar solution-polymer flood conducted by Cities Service and supported by the Department of Energy. This test was conducted in a reservoir that had been previously waterflooded successfully. It consists of four 3.2-acre five-spots in the El Dorado, Kansas, field. The polymer injection was designed to provide a graded decrease in viscosity, but the decrease in viscosity was accelerated as a result of injection-rate difficulties.

Often, adding sodium hydroxide to the injected water in a waterflood results in added recovery of oil. This process may be referred to as an alkaline flood (Fig. 10–18). The recovery increase is not as spectacular as the results of some of the other EOR processes, but it is much less costly to implement.

The mechanism that results in the increased recovery is not fully understood.[33] It is clear that the addition of sodium hydroxide reduces the interfacial tension between the water and the oil and, thus, results in lower capillary forces trapping the oil as previously described to reduce the residual oil. It is also speculated that if a reservoir is preferentially oil wet, the sodium hydroxide may change the wettability to water wet and increase the oil that can be displaced. However, if the wettability is changed from water wet to oil wet after the displacement of oil by water has resulted in a discontinuous oil phase, this wettability change can result in the oil again becoming a continuous phase. Then additional oil can be displaced.[34]

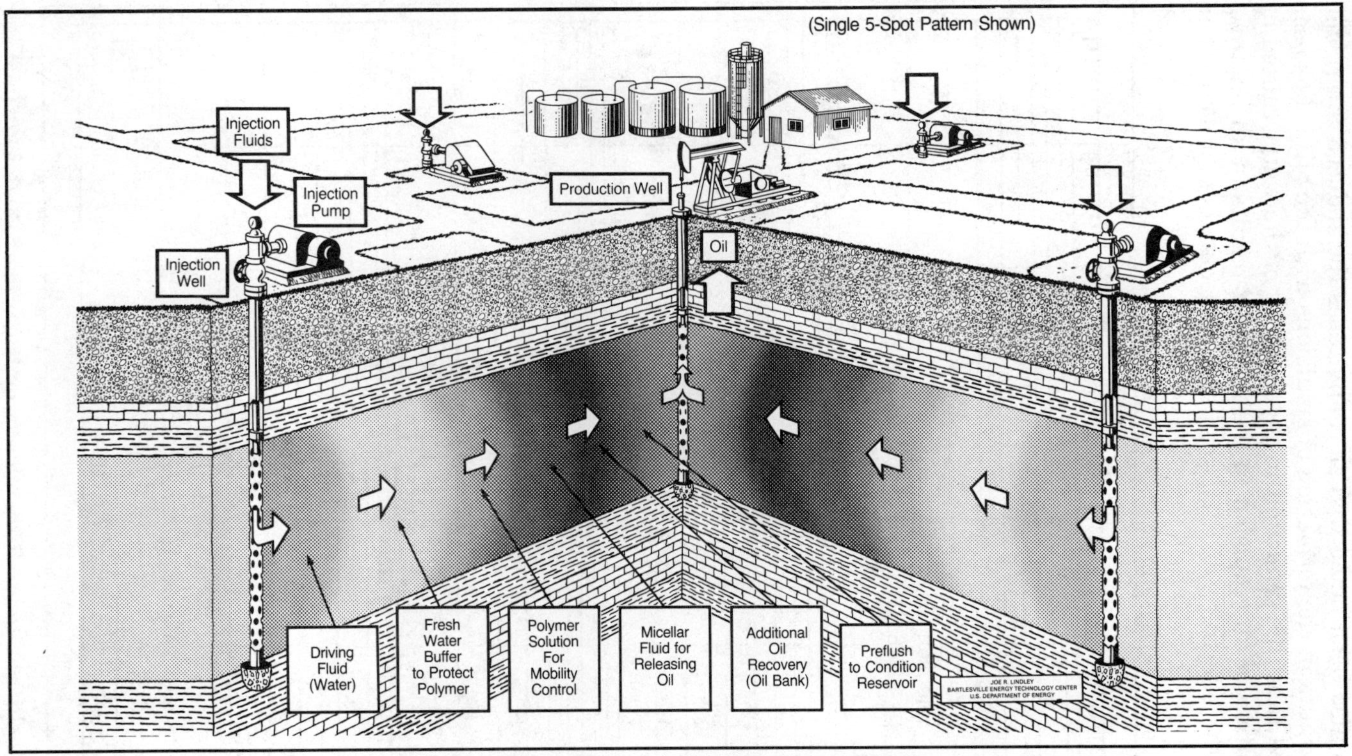

Fig. 10–16 Chemical flooding, micellar-polymer (reprinted from *Enhanced Oil Recovery and Improved Drilling Technology*, Report No. DOE/BETC-82/1, courtesy Bartlesville Energy Technology Center, DOE). The method shown requires a preflush to condition the reservoir, the injection of a micellar fluid for releasing oil, followed by a polymer solution for mobility control to minimize channeling, and a driving fluid (water) to move the chemicals and resulting oil bank to production wells.

It has also been noted that in many cases the reduced interfacial tension results in the formation of emulsions with the oil. It has been speculated that this reduction in the mobility of the water caused by the formation of an emulsion in the reservoir results in an improvement in the effective mobility ratio and subsequent improvement in

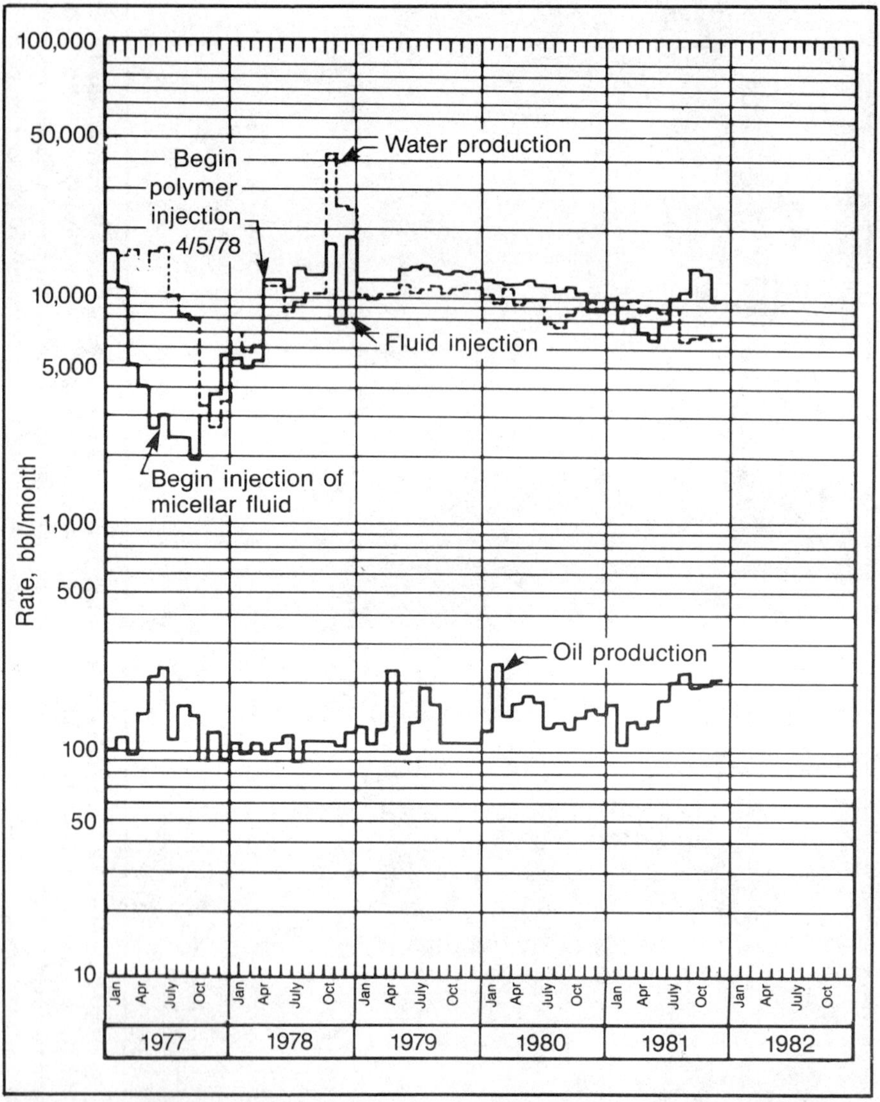

Fig. 10–17 Cities Service-DOE, micellar-polymer flood, El Dorado field, Hegberg pattern—four 3.2-acre five spots (after *Enhanced Oil Recovery and Improved Drilling Technology*, Report No. DOE/BETC-82/1, courtesy Bartlesville Energy Technology Center, DOE)

the displacement efficiency. Some field results have indicated an initial reduction in water production without a corresponding increase in the oil producing rate that tends to support this conclusion.[35]

Several situations should be avoided in the application of alkaline flooding. Some reservoirs are so acidic that too much caustic is required to obtain an effective alkaline flood. Excessive reservoir temperature may also result in the excessive consumption of caustic chemicals. Remember that an alkaline flood is simply a modification of a conventional flood. Consequently, a reservoir that is not a good waterflood prospect because of such problems as adverse permeability distribution or the presence of a gas cap should not be considered for alkaline flooding.

Miscible Displacement

The miscible displacement method of oil recovery involves the injection of a fluid into the reservoir to displace the oil that is miscible (mixable) with the oil. Physically, it works like cleaning grease out of the concrete pores of a garage floor. We can flush a grease spill with water forever without removing all of the grease, but it can be removed with relative ease with gasoline—if we do not burn the garage down—or with alcohol. These materials dissolve or mix with the grease. In a more scientific fashion we note that the gasoline or alcohol displaces the grease from the concrete pores because the interfacial tension between the grease and these liquids is zero. Therefore, there is no capillary attraction tending to hold the grease in the pores.

Miscible displacement of oil from a reservoir was originally obtained by injecting liquid petroleum gas (LPG) into the reservoir in rare instances where a relatively cheap source of LPG was available. Several attempts have been made to achieve miscibility using high-pressure gas, and some studies have tried to use inert gases such as nitrogen to achieve miscibility. In recent years carbon dioxide (CO_2) has been used with increasing frequency to achieve a miscible displacement.

Some of these miscible systems are more difficult to achieve and maintain than others because of the limitations on the temperatures and pressures and mixtures of the oil and displacing materials under which miscibility is obtained. There is also the question of how much or how large of a slug of the displacing material must be used to displace the oil. This is generally described as being some specified number of pore volumes of displacing material that is required, and it may be necessary to run displacements in the laboratory that measure many feet in length in order to obtain a realistic evaluation of the size of the slug required.

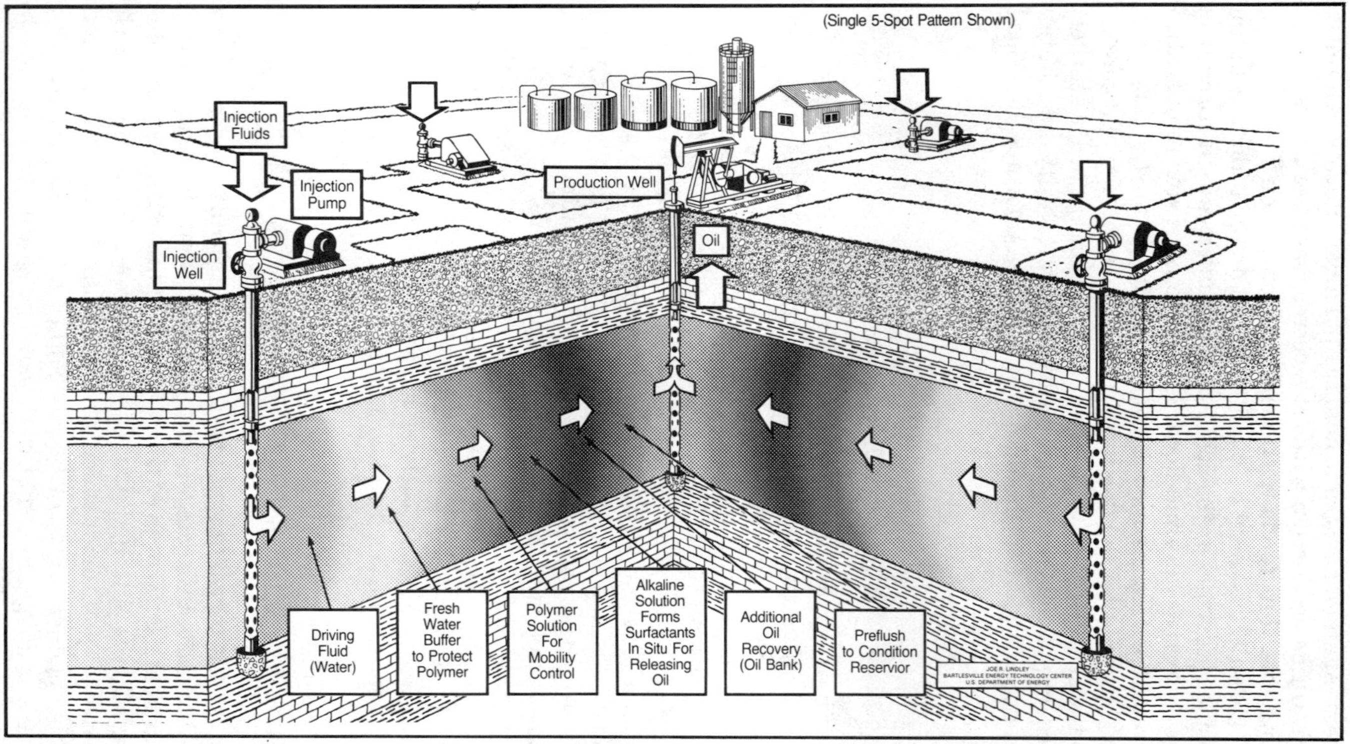

Fig. 10–18 Chemical flooding, alkaline (reprinted from *Enhanced Oil Recovery and Improved Drilling Technology*, Report No. DOE/BETC-82/1, courtesy Bartlesville Energy Technology Center, DOE). The method shown requires a preflush to condition the reservoir and injection of an alkaline or alkaline/polymer solution that forms surfactants in situ for releasing oil. This is followed by a polymer solution for mobility control and a driving fluid (water) to move the chemicals and resulting oil bank to production wells.

In miscible displacement relative permeabilities lose their significance since there is no interface between the fluids. This may lead the engineer to believe that the mobility ratio is then unimportant; however, this is not the case. The mobility ratio is simply the ratio between the viscosities of the oil and the displacing material, but it still is very important in controlling the amount of the reservoir that can be contacted by the displacement process and the efficiency with which the displacement can be achieved.

Next, we discuss LPG and high-pressure natural gas displacement, carbon dioxide displacement, and displacement with high-pressure inert gases. As noted, many other EOR methods of different names such as steamflooding, in situ combustion, and surfactant flooding achieve some degree of miscibility with the oil they are displacing.

LPG and high-pressure natural gas displacement. LPG is comprised of ethane, propane, and butane. These materials are miscible with oil only when they are in the liquid state. Even then they take on gaseous characteristics when the temperature exceeds the critical temperature of the hydrocarbon, which is 90°F for ethane, 206°F for propane, and 305°F for normal butane. The pressures required for the hydrocarbon to be a liquid varies with the temperature. At 50°F the minimum pressure required for the liquid state is 460, 92, and 22 psia for the ethane, propane, and normal butane, respectively. At the critical temperatures noted, this pressure is 709, 617, and 550 psia, respectively. Thus, the miscibility between the crude oil and LPG is not too difficult to achieve. However, in order to make the process economical, the LPG must be miscibly displaced with some cheaper material such as high-pressure natural gas, nitrogen, or a flue gas. Fig. 10–19 indicates the typical pressures needed to obtain miscibility at different temperatures for different LPG-displacement-gas combinations. Such a system as this may typically use a 5% pore volume LPG slug and supplement the displacement gas with water injection in order to minimize the cost of the displacement gas and to maximize the sweep efficiency.

The high cost of LPG in most areas has led to other miscible hydrocarbon displacement systems. One of these involves the use of an enriched natural gas that is in turn displaced with a lean natural gas. This process uses a slug of 10–20% pore volume of natural gas that has been enriched with ethane through hexane. This slug achieves miscibility with the oil by enriching the oil with ethane through hexane and thus making the oil-ethane mixture miscible with the enriched-gas slug. Lean gas or lean gas and water are then used to displace the enriched-gas slug. The design of such a system obviously requires a great deal of laboratory and phase-behavior design work. For example,

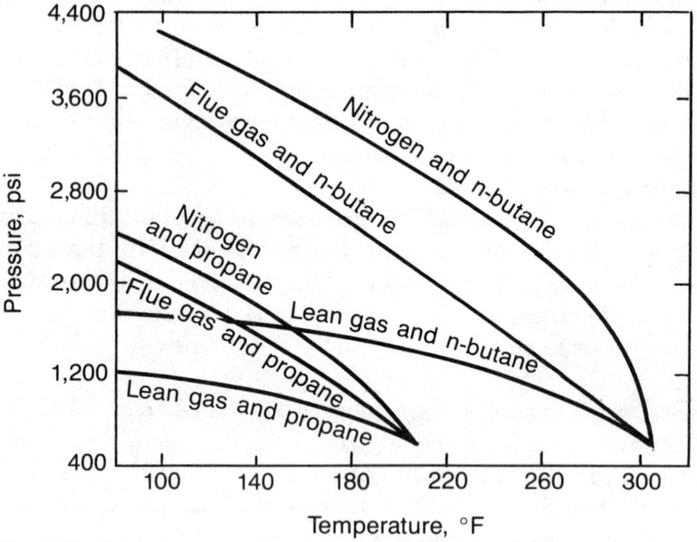

Fig. 10–19 Minimum pressure for miscibility, miscible slug-driving gas combinations (after van Poollen and Associates, *Fundamentals of Enhanced Oil Recovery*, PennWell, 1980, courtesy *Petroleum Engineer*)

using about 23 mol% of propane in natural gas may require a pressure of 3,000 psia to achieve miscibility. However, using 93% ethane and 7% propane as the enriching materials and enriching the natural gas with about 50 mol% of this mixture requires only a pressure of 1,500 psia for miscibility.[36]

A third miscible hydrocarbon displacement process may be usable in reservoirs containing undersaturated oils with gravities of more than 40°API and depths greater than 5,000 ft.[37] This involves the displacement of the reservoir oil at a pressure and temperature that make it miscible with the oil. In this case the lean gas picks up ethane through hexane components from the reservoir oil and becomes miscible with the reservoir oil. Thus, the application of this process is restricted to oils that have a high percentage of ethane through hexane—hence, the restriction to undersaturated oil with a gravity greater than 40°API. The considerable reservoir depth is necessary because relatively high pressures must be maintained to ensure miscibility. The high pressures and corresponding large amounts of gas that must be injected make the economics highly questionable with the current wellhead natural gas price.

An alternative to high-pressure lean gas injection to achieve miscibility is to use an inert gas such as nitrogen, which is much less expensive than natural gas. The mechanism of achieving miscibility appears to be very similar to that of achieving miscibility by lean natural gas injection. The high-pressure nitrogen strips the light hydrocarbons from the reservoir oil and eventually becomes miscible with the reservoir oil. Fig. 10–20 is a schematic of a miscible nitrogen flood.

Laboratory work has shown that nitrogen at a pressure of 5,000 psia can be used to displace 54°API crude with 700 scf/bbl of natural gas in solution to a residual of about 10%.[38] Such results make it appear that the nature of the injected gas is relatively unimportant. Any injected gas seems to obtain miscibility with a relatively volatile reservoir oil by stripping the light components from the oil and, eventually, becoming miscible with the reservoir oil. Air or flue gas could be used except for the problems of corrosion, explosive gas mixtures, and spontaneous ignition of the oil in the reservoir caused by the free oxygen in the gas.

Carbon dioxide injection. Perhaps the hottest EOR method, including the thermal processes, in the late 1970s and early 80s has been miscible carbon dioxide injection. Fig. 10–21 is a schematic of a carbon dioxide miscible flood.

At one time the Department of Energy recognized CO_2 injection and steamflooding as the most promising EOR processes. When carbon dioxide injection into a reservoir was initiated in the 1950s, it was simply injected with or in water to provide a carbonated waterflood. Although it was pointed out at that time that the CO_2 went into solution in the oil and, thus, caused the oil to swell, most of the engineers, including the author, seemed to think that the most beneficial effect of the carbon dioxide was to improve water-injection rates by making the water acidic. Increased oil production rates associated with the initiation of CO_2 use in a waterflood was usually associated with a corresponding increase in the injection rates. Patents were also sought and perhaps awarded for use of carbon dioxide as a well-treatment chemical for the purpose of increasing injectivity or productivity.

The use of carbon dioxide as a miscible-displacement material received a huge boost with the initiation of the Kelly Snyder/SACROC Unit in Texas that involved about 50,000 acres and 273 injection wells. Carbon dioxide has caught on as a miscible-displacement material largely because of the high cost of LPG and natural gas. In 1971 there was only one active CO_2 miscible project, but in 1980 there were 17 such projects in operation.[39]

The critical pressure and temperature of carbon dioxide are 1,073 psia and 88°F, respectively. Thus, on the surface it appears that it would be difficult to find applications where the reservoir pressure and temperature could be less than the critical values and a miscible dis-

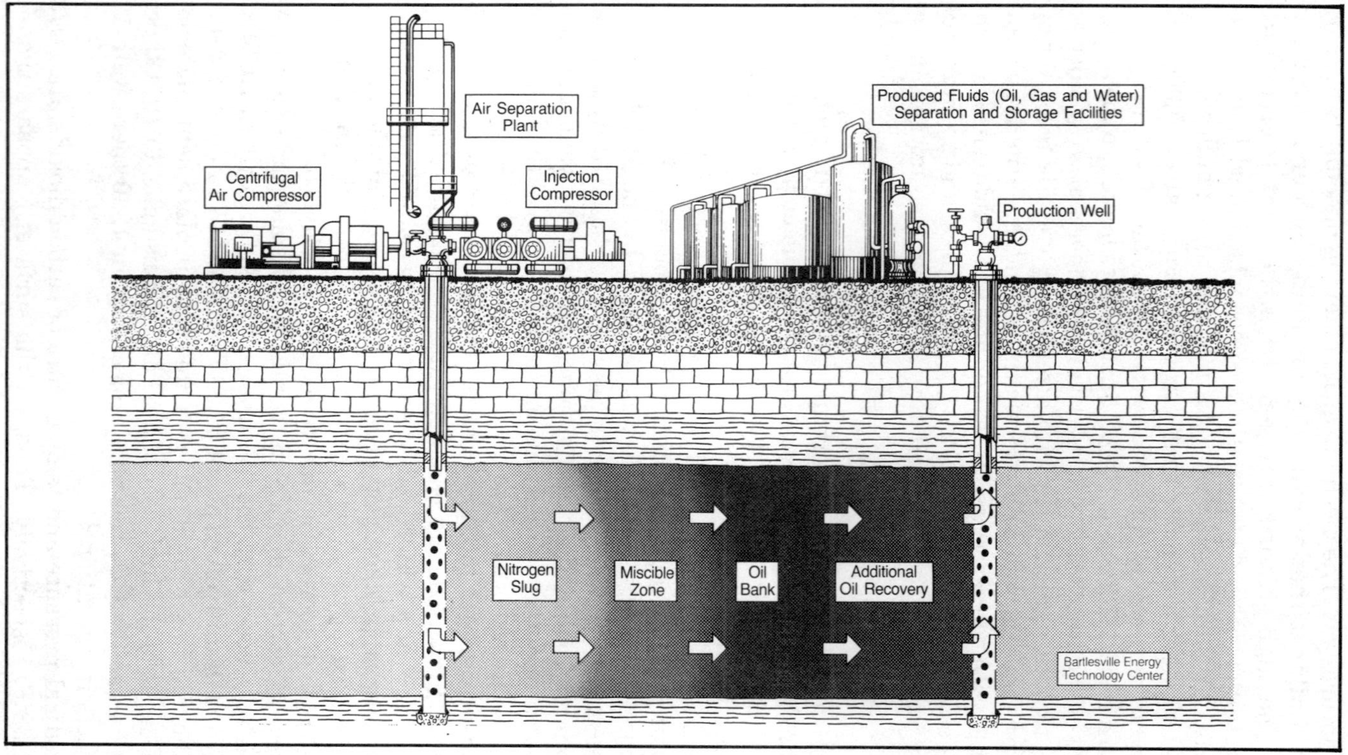

Fig. 10–20 Nitrogen flooding (reprinted from *Enhanced Oil Recovery and Improved Drilling Technology*, Report No. DOE/BETC-82/1, courtesy Bartlesville Energy Technology Center, DOE). This method can be used as a substitute for CO_2 in deep reservoirs with high API gravity oil. When injected at high pressure, nitrogen can form a miscible slug that aids in freeing the oil from the reservoir rock. Viscosity of oil is reduced providing more efficient miscible displacement.

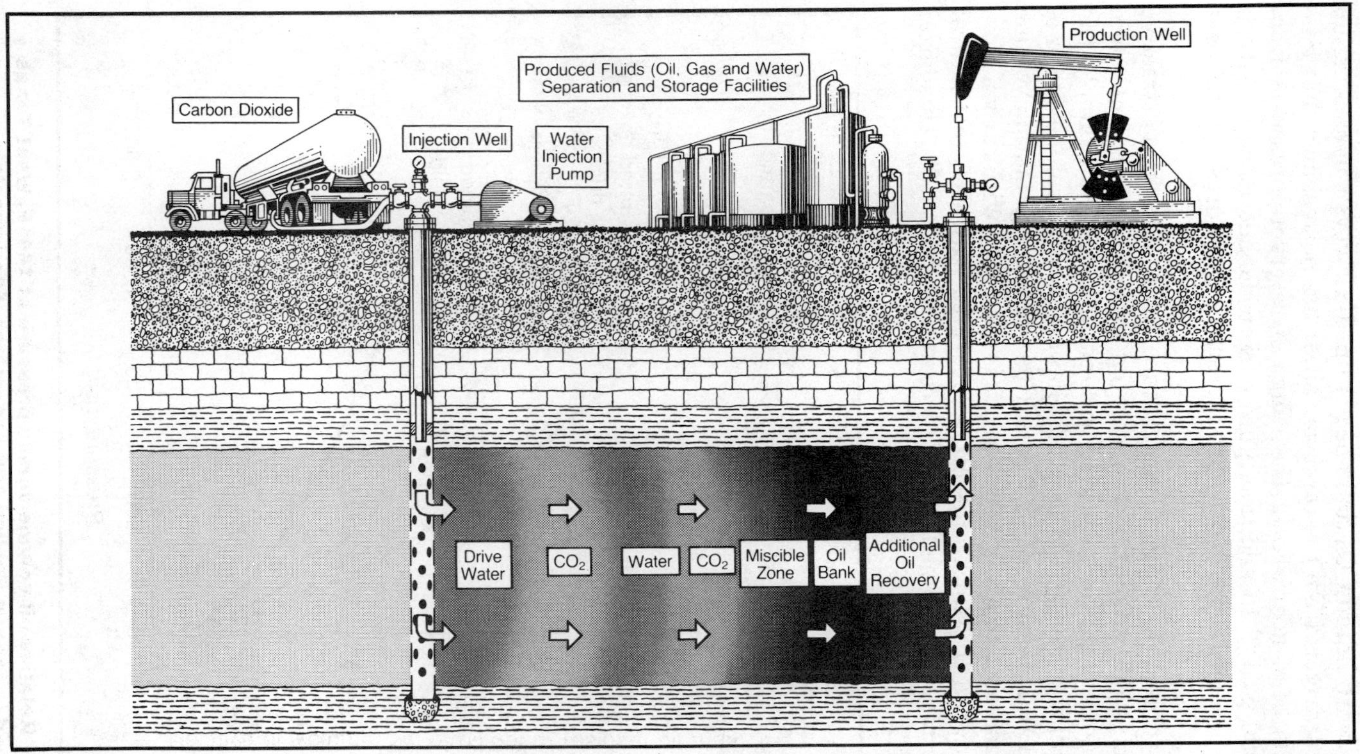

Fig. 10–21 Carbon dioxide flooding (reprinted from *Enhanced Oil Recovery and Improved Drilling Technology,* Report No. DOE/BETC-82/1, courtesy Bartlesville Energy Technology Center, DOE). This method is a miscible displacement process applicable to many reservoirs. A CO_2 slug followed by alternate water and CO_2 injections (WAG) is usually the most feasible method. Viscosity of oil is reduced providing more efficient miscible displacement.

placement with liquid CO_2 could be obtained. However, by picking up light hydrocarbons from the reservoir oil, it has been shown that miscibility has been achieved in the range of reservoir pressures from 1,000–4,000 psia.[40] In addition, pure carbon dioxide at supercritical pressures becomes more dense with increasing pressures and behaves more like a liquid.

Carbon dioxide has other physical properties that make it attractive for displacing oil. An oil that is saturated with CO_2 swells more than does the same oil that is saturated with natural gas. For example, the reservoir oil represented in Fig. 10–22 has a formation volume factor of about 1.28 at a pressure of 1,500 psia. However, this formation volume factor is increased to about 1.45 when it is saturated with carbon dioxide. Also, when the stock-tank oil is saturated with CO_2 (no natural gas), its formation volume factor is increased to about 1.63. This latter

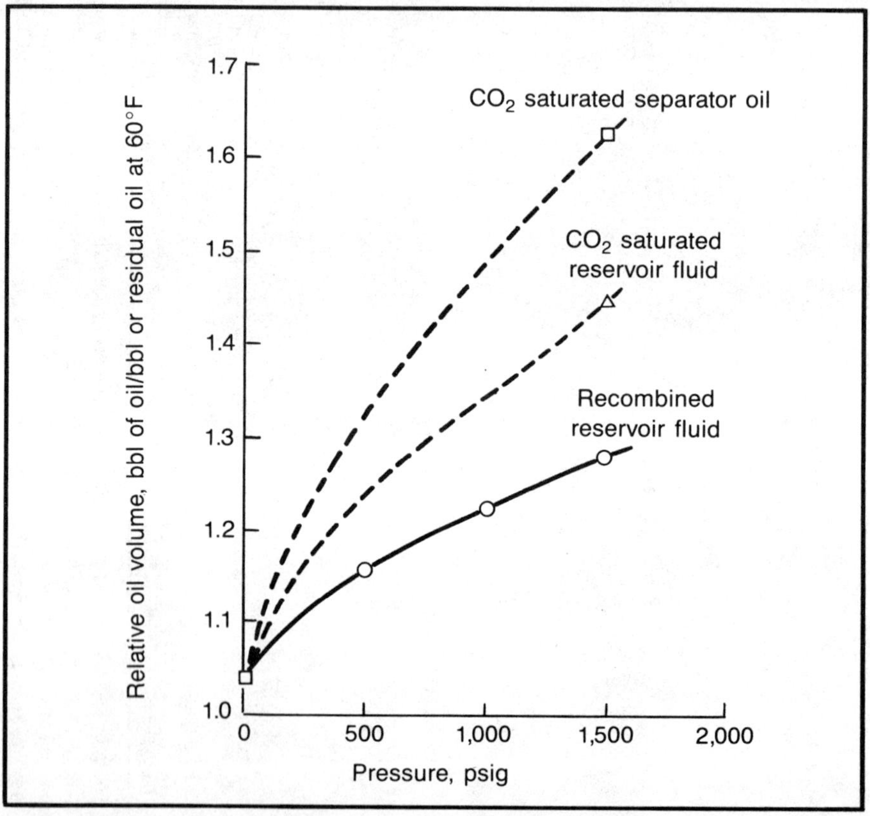

Fig. 10–22 Relative oil volume versus pressure at 144° F, West Texas reservoir fluid (after van Poollen and Associates, *Fundamentals of Enhanced Oil Recovery*, PennWell, 1980, courtesy *JPT*)

case represents the swelling of the oil that occurs if the CO_2 miscible displacement is initiated when the reservoir pressure approaches atmospheric. The increase in the reservoir volume of oil of course means that for a particular residual-oil saturation, fewer stock-tank barrels of oil are left in the reservoir. Also, when the reservoir pressure is reduced at the end of the displacement a solution-gas drive results from the CO_2 in solution.

In addition, carbon dioxide has the advantages of swelling the water it contacts, reducing the oil viscosity, being nonhazardous as compared with hydrocarbons, and having a density under some reservoir conditions that closely approximates that of the reservoir oil. Thus, it largely avoids gravity segregation difficulties.

These advantages are offset to some degree by the difficulty of finding an economic source of carbon dioxide and the tendency of the CO_2 stripping the light hydrocarbons from the oil to form an asphaltic residue that has a high viscosity and causes a corresponding increase in the injection pressures.[41] To reduce costs and improve the effectiveness of a given amount of carbon dioxide, water and carbon dioxide are sometimes injected alternately as in Fig. 10–21.

Results of a CO_2 miscible-displacement project are shown in Figs. 10–23 and 10–24. These data through 1980 are for two five spots in the Rock Creek field in West Virginia that were isolated from the rest of the reservoir by injecting water into 12 injection wells that surround the two five spots. The water-injection figures include the water injected into these backup wells. When CO_2 injection was discontinued in June 1980, water was injected into the injection wells to continue the displacement. It appears that in January 1981 injection into a smaller four spot was initiated in a project supported by DOE.

Polymer Flooding

Polymer flooding involves the use of polymers dispersed in water as an injection fluid (Fig. 10–25). The polymer content reduces the mobility of the water by increasing the effective viscosity and plugging some of the pores in the formation. This reduction in water mobility in turn improves (reduces) the mobility ratio and, thus, improves sweep efficiencies, stratification efficiencies, and frontal saturations.

This subject has been discussed in some detail in chapter 9. In that chapter the characteristics of polyacrylamides and polysaccharides, the plugging of the formation by polymers, the effect of the reduced water mobility on the flood recovery, the shear thickening and shear thinning polymers, and the methods of predicting the rate versus time behavior when a slug of polymer is used are presented.

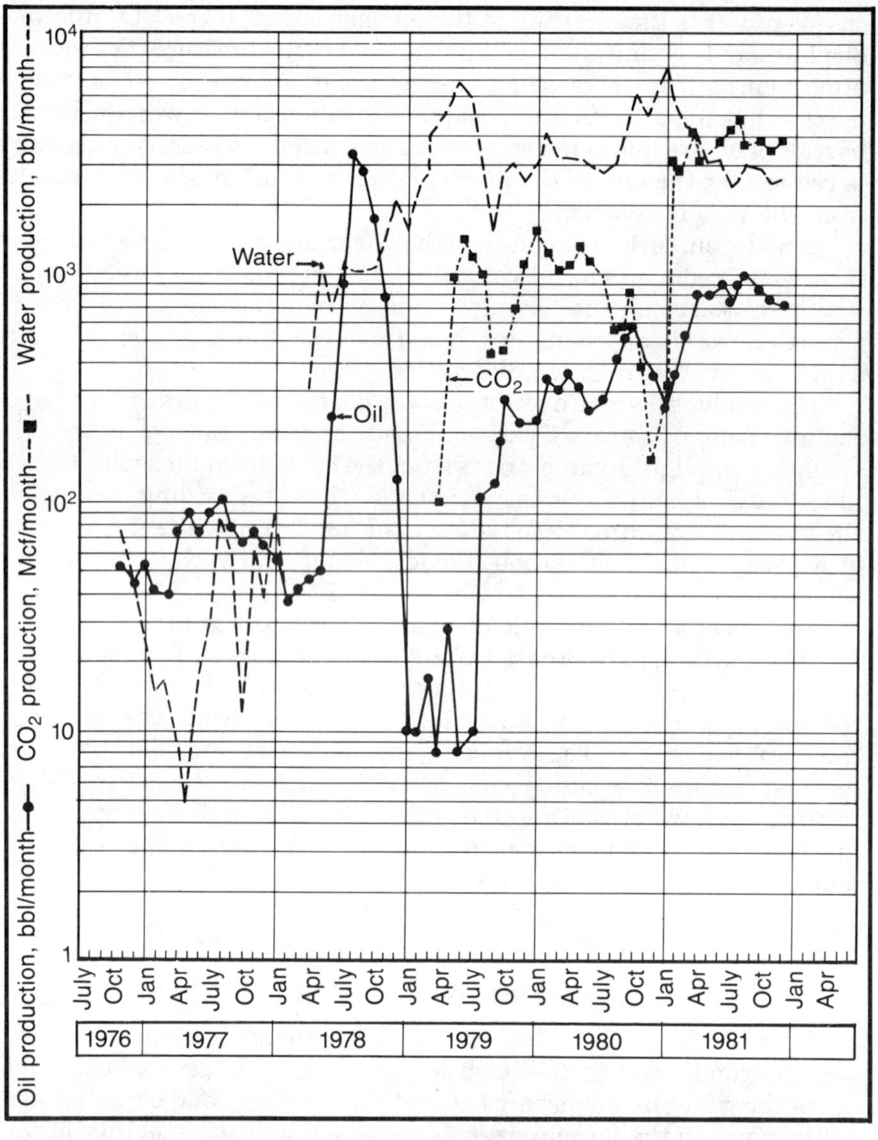

Fig. 10–23 Rock Creek pilot CO_2 project production history—two five spots (after *Enhanced Oil Recovery and Improved Drilling Technology*, Report No. DOE/BETC-82/1, courtesy Bartlesville Energy Technology Center, DOE)

Polymers are also used in connection with many other EOR processes. Most surfactant and micellar floods use a slug of polymer behind the chemicals to control vertical and horizontal sweep efficiencies.

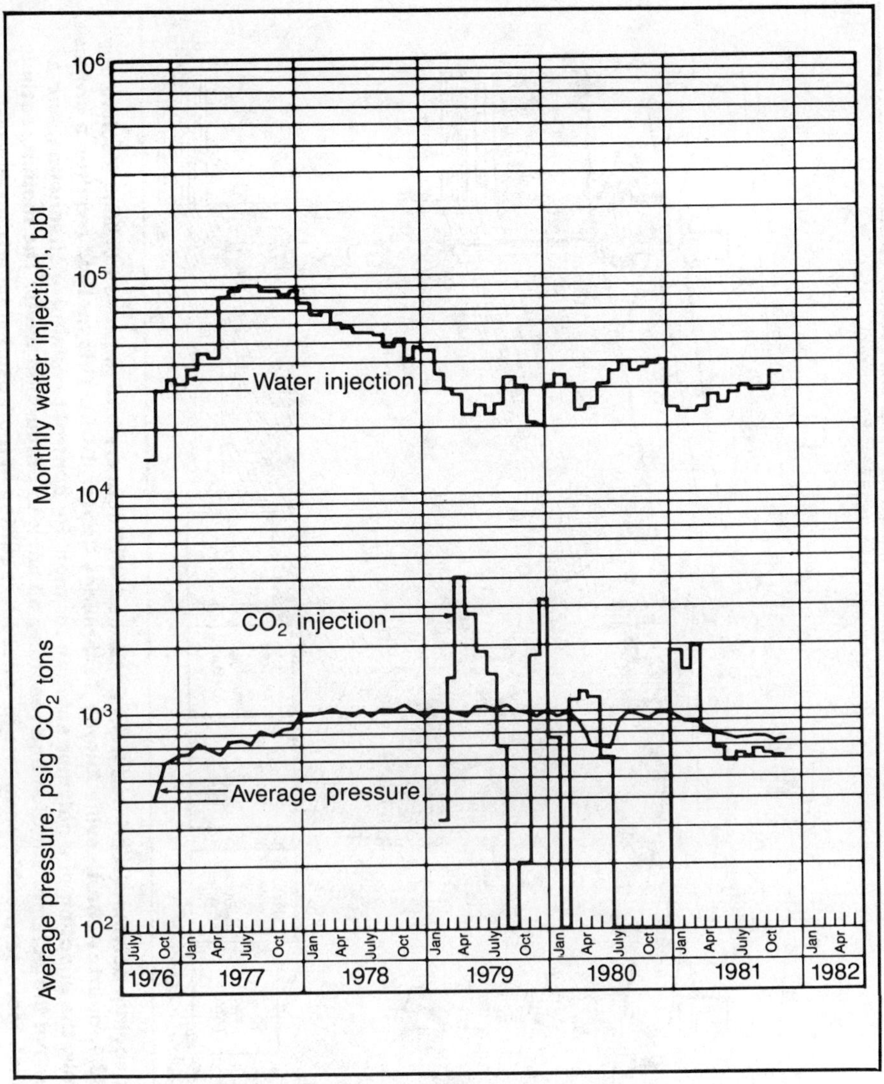

Fig. 10–24 Rock Creek pilot CO_2 project injection history—two five spots (after *Enhanced Oil Recovery and Improved Drilling Technology*, Report No. DOE/BETC-82/1, courtesy Bartlesville Energy Technology Center, DOE)

Feasibility Analysis of EOR Processes

As noted, it is the objective of this chapter to give the engineer enough background in EOR to recognize the prospects for EOR. Then the engineer can seek assistance from company or consulting special-

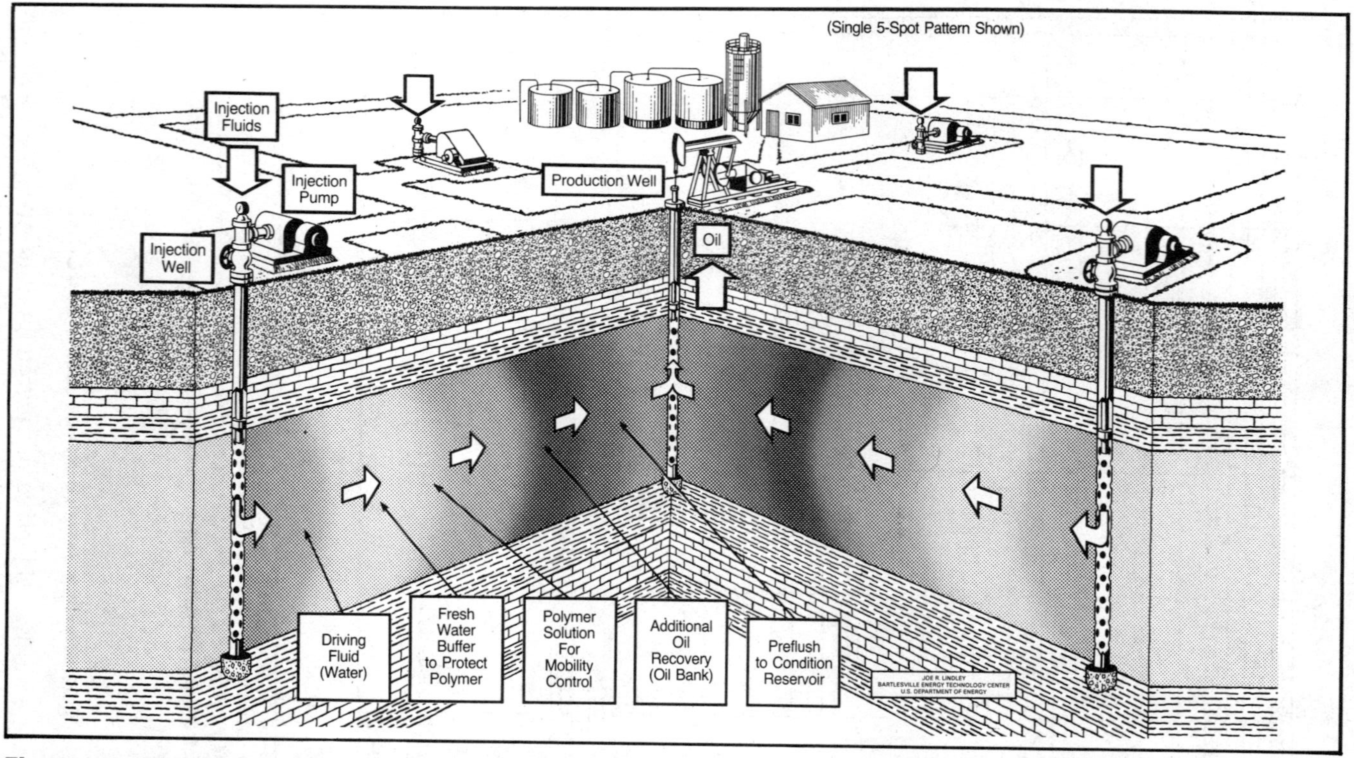

Fig. 10–25 Chemical flooding, polymer (reprinted from *Enhanced Oil Recovery and Improved Drilling Technology*, Report No. DOE/BETC-82/1, courtesy Bartlesville Energy Technology Center, DOE). This method requires a preflush to condition the reservoir, the injection of a polymer solution for mobility control to minimize channeling, and a driving fluid (water) to move the polymer solution and resulting oil bank to production wells. The mobility ratio is improved and flow through more permeable channels is reduced, resulting in increased volumetric sweep.

ists that can assist in a further evaluation or design of a suitable process.

It should be possible for the engineer to eliminate most prospects by relatively simple analyses that would provide the most optimistic prediction of the results to be obtained by the application of a particular EOR process. Most performance evaluations result in the evaluated process being unattractive, and only a few such analyses result in predictions that warrant further study.

Perhaps the simplest process for feasibility analysis is polymer flooding. As detailed in chapter 9, the first analysis of a polymer flood can simply use the improved mobility ratio and the recommended method of waterflood analysis to complete a rate-time prediction, assuming continued injection of the polymer to the economic limit. In this analysis assume that the residual oil is unaffected by the use of the polymer. If this analysis makes the proposed application look attractive, the engineer can then proceed to analyze a slug process as outlined. If the process continues to look attractive, the engineer should contact a service company that sells polymers to obtain details of polymer costs, absorption phenomena, residual permeability, and assistance in predicting the behavior of the proposed flood under a variety of conditions and slug sizes.

In the feasibility analysis of chemical floods, the same general procedure is recommended. However, in this case the engineer must first estimate what the residual oil will be as a result of the injected chemicals. Then the prediction can be made using the recommended method and incorporating the residual oil that is anticipated and the mobility ratio that would result if a polymer displacement follows the slug of chemical. In this first analysis no effort needs to be made to account for the mobility difference within the chemical slug. If this result is encouraging, it is recommended that the analysis be repeated using a finite chemical-and-polymer slug size and corresponding chemical-slug/oil bank and polymer/chemical-slug mobilities. If this analysis still is encouraging, the engineer can then seek assistance for further study.

In the feasibility analysis of steamfloods, in situ combustion, and miscible-displacement processes, the same approach can be used as outlined for chemical floods. However, the mobility ratio assumed is simply the result of the process rather than the result of a following polymer slug. The results of these analyses do not account for cross flow of heat or gravity effects. Nevertheless, they should be useful in providing the most optimistic performance prediction of a process and, thus, eliminate the further study of most prospects. The Gomaa technique for the feasibility analysis of a steamflood as outlined is also recommended.[42]

The engineer should bear in mind that the recommended waterflood methods for predicting the behavior of an EOR process can be based on a recovery versus cumulative injection relationship and an injection rate versus cumulative injection relationship for one homogeneous zone. The bookkeeping procedure of the recommended waterflood method then combines the performance of the individual strata.

Huff-and-puff steam stimulation does not lend itself to analysis using the recommended waterflood prediction method. It is recommended that, when possible, feasibility analyses of prospective huff-and-puff projects be analyzed by analogy with other projects.

Notes

1. D.B. Guralnik, ed., *Webster's New World Dictionary* (New York: Simon & Schuster, 1980).

2. Matheny, "EOR Methods Help Ultimate Recovery," *OGJ* (March 31, 1980).

3. "Duri Full Steam Ahead," *Chevron World* (Winter 1982).

4. E.H. Timmerman, *Practical Reservoir Engineering*, volume 2 (Tulsa: PennWell, 1982).

5. H.J. de Haan and J. van Lookern, "Early Results of the First Large-Scale Steam Soak Project in the Tia Juana Field, Western Venezuela," *JPT* (January 1969), p. 246.

6. B.H. Adams and A.M. Khan, "Cyclic Steam Injection Project Performance Analysis and Some Results of a Continuous Steam Displacement Pilot," *JPT* (January 1969).

7. H.K. van Poollen and Associates, *Fundamentals of Enhanced Oil Recovery* (Tulsa: Pennwell, 1980).

8. van Poollen and Associates, 1980.

9. T.C. Boberg and R.B. Lantz, "Calculation of the Production Rate of a Thermally Stimulated Well," *JPT* (December 1966).

10. J. Jones, "Cyclic Steam Reservoir Model for Viscous Oil, Pressure Depleted, Gravity Drainage Reservoir," SPE paper 6544 presented at the SPE-AIME 47th Annual California Regional Meeting (Bakersfield: April 1977).

11. Timmerman, 1982.

12. R.D. Seba and G.E. Perry, "A Mathematical Model of Repeated Steam Soaks of Thick Gravity Drainage Reservoir," *JPT* (January 1969).

13. van Poollen and Associates, 1980.

14. B.T. Willman et al., "Laboratory Studies of Oil Recovery by Steam Injection," *Trans.*, AIME (1961), p. 222.

15. C. van Dijk, "Steam-Drive Project in the Schoonebeek Field, The Netherlands," *JPT* (March 1968).

16. van Poollen and Associates, 1980.

17. Willman, 1961.

18. H.A. Lauwerier, "The Transport of Heat in an Oil Layer Caused by the Injection of Hot Fluid," *Applied Science Res.*, section A (1955).

19. G.E. Malofeev, "Calculation of the Temperature Distribution in a Formation when Pumping Hot Fluid into a Well," *Neft'i Gaz*, volume 3, No. 7 (1960).

20. L.I. Rubinshtein, "An Asymptotic Solution of an Axially Symmetric Contact Problem in Thermal Convection for High Values of the Convection Parameter," *Dan SSSR,* volume 146, No. 5 (1962).

21. A.G. Spillette, "Heat Transfer during Hot Fluid Injection into an Oil Reservoir," *Journal of Canadian Petroleum Technology* (October–December 1965).

22. E.E. Gomaa, "Correlations for Predicting Oil Recovery by Steamfood," *JPT* (February 1980).

23. K.H. Coats, W.D. George, and B.E. Marcum, "Three Dimensional Simulation of Steamflooding," *SPEJ* (December 1974), pp. 573–592.

24. van Poollen and Associates, 1980.

25. van Poollen and Associates, 1980.

26. D.M. Dietz and J. Weijdema, "Reverse Combustion Seldom Feasible," *Producers Monthly* (May 1968).

27. van Poollen and Associates, 1980.

28. van Poollen and Associates, 1980.

29. L.W. Holm and V.A. Josendal, "Reservoir Brines Influence Soluble-Oil Flooding Process," *OGJ* (November 13, 1972).

30. J.H. Bae and C.B. Petrick, "Adsorption/Retention of Petroleum Sulfonates in Berea Cores," *SPEJ* (October 1977).

31. H.R. Froning and L.E. Treiber, "Development and Selection of Chemical Systems for Miscible Waterflooding," SPE paper 5816 presented at the SPE-AIME 4th Symposium on Improved Methods for Oil Recovery, (Tulsa: March 1976).

32. H.J. Hill, J. Reisberg, and G.L. Stegemeier, "Aqueous Surfactant Systems for Oil Recovery," *JPT* (February 1973).

33. van Poollen and Associates, 1980.

34. van Poollen and Associates, 1980.

35. "EOR and Improved Drilling Technology," No. 28 (Bartlesville: DOE/BETC, September 30, 1981).

36. E.F. Herbeck, R.C. Heintz, and J.R. Hastings, "Fundamentals of Tertiary Oil Recovery," parts 2, 3, and 4, *Petroleum Engineer* (February, March, and April 1976).

37. van Poollen and Associates, 1980.

38. Crawford et al., *Petroleum Engineer* (November 1977).

39. "DOE Report Examines EOR Constraints," *OGJ* (April 28, 1980).

40. "DOE Report Examines EOR Constraints," 1980.

41. van Poollen and Associates, 1980.

Additional References

Closmann, P.J. "Steam Zone Growth during Multiple-Layer Steam Injection." *SPEJ,* March 1967.

"EOR and Improved Drilling Technology." Report No. 29, Bartlesville: DOE/BETC, December 31, 1981.

Farouq Ali, S.M. "Current Status of Steam Injection as a Heavy Oil Recovery Method." *Journal of Canadian Petroleum Technology,* January–March 1974.

Hicks and Foster. "Evaluation of Target Oil in 50 Major Reservoirs in the Texas Gulf Coast for EOR." 5th Annual DOE Symposium on EOR, Tulsa, 1979.

Holm, L.W.; and Josendal, V.A. "Mechanism of Oil Displacement by Carbon Dioxide." *JPT,* December 1974.

Kuo, C.H.; Shain, S.A.; and Phocas, D.M. "A Gravity Drainage Model for the Steam-Soak Process." *SPEJ,* June 1970.

Martin, J.C. "A Theoretical Analysis of Steam Stimulation." SPE paper 1579 presented at the SPE-AIME 41st Annual Fall Meeting, Dallas: October 1966, revised 1967, SPE Reprint Series No. 10.

Marx, J.W.; and Langenheim, R.H. "Reservoir Heating by Hot Fluid Injection." *Trans.*, AIME, 1959.

Nelson, T.W.; and McNeil, J.S. "How to Engineer an In-Situ Combustion Project." *OGJ*, June 5, 1961.

Williamson, A.S.; Drake, L.P.; and Chappelear, J.E. "A Steam Soak Well Model for an Isothermal Reservoir Simulator." SPE paper 5739 presented at the 4th Annual Symposium on Numerical Simulation of Reservoir Performance, Los Angeles: February 1976.

11

Computer Modeling of a Reservoir

Throughout this book references have been made to solving problems using some sort of a computer reservoir model. The reader may have obtained the idea that without question any reservoir problem can be solved by reservoir modeling. This is a misleading and dangerous impression. The author has seen many types of computer-modeling results that were being accepted as fact by engineers and managers who believed that the computer model was infallible. Reservoir behavior predictions for solution-gas-drive reservoir have indicated a constant and unchanging producing gas-oil ratio. Water-drive reservoirs have exhibited constant water-encroachment rates. Also, an aquifer has been modeled by using one huge cell to represent the aquifer with the same pressure throughout the huge cell at any particular time. Engineers who have used the previous chapters of this book should recognize such results and methods as ridiculous.

Admittedly, most of these errors were observed several years ago before reservoir models had reached their present state of refinement. However, considerable care should still be exercised in the use of reservoir models. Although many errors will be made in the application of computer modeling, it represents the only possible solution to many of our reservoir engineering problems.

In chapter 2 it is noted that some reservoir flow geometries are so complex that analysis of the flow can only be obtained by reservoir modeling. In chapter 3 we analyze unsteady-state systems that treat compressibilities, viscosities, and permeabilities as constants. However, when these parameters vary significantly in the reservoir, it may be necessary to turn to the reservoir computer model to solve the problem. In coning and fingering problems we can only evaluate the time factors involved by reservoir modeling; all but the simplest material-balance studies are presently made with a reservoir computer model; and the more involved enhanced oil recovery applications can only be made by using reservoir computer modeling.

The purpose of this chapter is to familiarize the practicing reservoir engineer with the fundamental ideas involved in computer reservoir modeling so he will recognize the reservoir engineering problems requiring computer modeling and will be able to discuss the problems with the reservoir modeling specialists. Very few practicing reservoir engineers are faced with the task of designing a reservoir computer model since most major petroleum companies already have sophisticated programs. For engineers in smaller companies, many programs are available commercially from software companies such as Scientific Software, Intercomp, Garrett Computing Systems, and Watterbarger and Associates. These programs can be purchased or leased for sums that probably make it financially unattractive for most companies to develop their own reservoir modeling software. Many of these programs can be accessed via commercial computer networks.

Repeated use of the most popular reservoir modeling programs available commercially has led to their continued debugging and refinement so development of a comparable program by any individual company would require a major effort of questionable economic worth. Engineers who must design a reservoir modeling computer program are referred to Crichlow's book, *Modern Reservoir Engineering—A Simulation Approach* or Aziz and Settari's *Petroleum Reservoir Stimulation*.[1,2]

Preparing a Reservoir Modeling Study

Computer reservoir models do not always provide a unique solution to most reservoir problems. If reservoir parameters and conditions are allowed to vary without reasonable limits, a computer solution can be obtained that is completely misleading in many or even most cases. Reasonable, useful, and accurate solutions to reservoir modeling problems can only be reached if the engineer has a thorough knowledge of the reservoir. Such an understanding can only be obtained using conventional analytical methods to analyze the reservoir problem as far as possible.

If the problem is one concerned with a single well such as a coning, fingering, or similar problem, the engineer should attempt a solution using the idealized analytical equations of chapter 6, plus all of the data available from flow tests, pressure buildups, etc. Such an analysis limits the parameters to be used and the results that are realistic. If the problem involves several wells or the whole reservoir, the engineer should attempt to analyze the problem using material-balance methods such as those in chapter 7. Only then will he know the best possible limits to be placed on the original oil in place, the gas-cap size, the productivity, and the water-drive strength. If the problem involves waterflooding or enhanced oil recovery, the methods proposed in chapters 9 and 10 should be used as far as possible, so the engineer understands what is known and not known about the problem before a model is used.

In addition, it is of course necessary to perform the more obvious tasks of studying reservoir and fluid characteristics in as much detail as possible.

An Incompressible Flow Model

In order to provide as much basic understanding as possible of the computer reservoir model, we first consider models that treat only incompressible fluids. These are not, strictly speaking, steady-state models because the change in saturations changes the transmissibility and, thus, changes the pressure drops or flow rates.

In chapter 2 it is shown that flow geometries more complex than radial, linear, or spherical can often be analyzed by breaking the complex geometry into a combination of simple geometries that would approximate the flow pattern. Then an equation can be written for (1) the sum of the pressure drops if the simple geometries are in series or (2) the sum of the flow rates if the simple geometries are in parallel. It is further noted that when the simple geometries became too numerous, it is necessary to use a computer to solve the equations that can be generated.

To illustrate the principles involved, suppose we desire to determine the pressure distribution in a five-spot flood pattern by this technique, since we already have an analytical solution that gives us the exact answers for comparison. We find that all of the flood pattern is made up of units such as those illustrated in Fig. 11–1, which is one-half of a quadrant of a five spot. In this figure there is no flow across any of the outside figure boundaries except at the injection well, r_{wi}, and the production well, r_{wp}.

We can write an equation for flow into and out of each of the numbered segments shown in such a way that the equations are stated as functions of the pressures only. By assuming pressures at the injection and production wells, we have 15 equations, one for each numbered segment, and 15 unknowns, the pressures for the 15 points shown. These simultaneous equations can then be solved by various numerical procedures for the 15 pressures that define the pressure distribution. Once the pressure distribution is known, the rate distribution can also be determined. To illustrate this technique consider the steady-state flow equations governing flow into and out of segment 8 of Fig. 11–1. We assume that flow is upward and to the right. (The validity of this

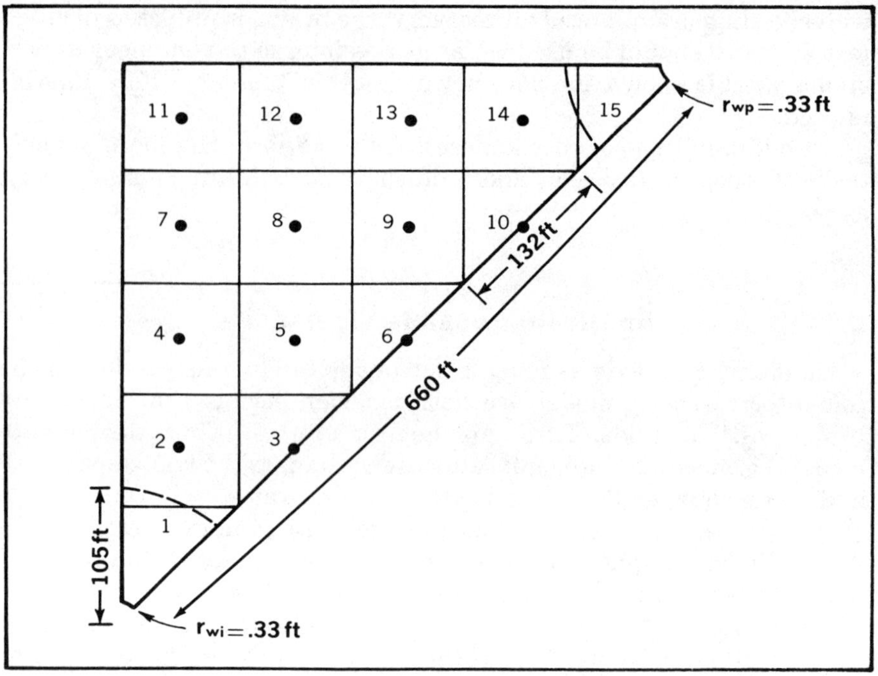

Fig. 11–1 Segmenting a five-spot for computer analysis

assumption does not affect the results.) Based on this assumption, we can say:

$$q_{7-8} + q_{5-8} = q_{8-12} + q_{8-9} \tag{11.1}$$

The subscript numbers are the segment numbers between which flow is represented by that particular rate. Each rate can be found using the Darcy equation:

$$q = -\frac{1.127kA}{\mu}\frac{\Delta p}{\Delta x} \tag{2.3}$$

Now note that for the square grid, Δx and A are the same for all of the rates in Eq. 11.1. If we assume that only one fluid is flowing and that the permeability is the same in all directions, k/μ is the same for all rates in Eq. 11.1. Under these conditions the rates are simply proportional to the respective pressure drops, so we can write Eq. 11.1 as:

$$(p_7 - p_8) + (p_5 - p_8) = (p_8 - p_{12}) + (p_8 - p_9) \tag{11.2}$$

Eq. 11.2 can be rearranged to:

$$p_5 + p_7 - 4p_8 + p_9 + p_{12} = 0 \tag{11.3}$$

Note that such an equation can be written for each of the 15 segments. Many of the segment equations have only two or three terms.

Segments 1 and 15 pose special problems. To approximate the pressure distribution into the very small wellbore with acceptable accuracy requires excessive linear grid segments. Consequently, these segments are treated as radial flow from the well radius to a radius of 105 ft or vice versa. Thus, flow across these segments is stated as flow from one radial boundary to the opposite radial boundary, rather than from the center of one segment to the center of another as is the case with most of the segment flow equations.

Consequently, the pressure at grid points 1 and 15 actually are the pressures along the dotted lines near the numbers in Fig. 11–1. The dotted line represents the radial and the solid line the linear extremity of segment 2. A similar situation exists for segment 14.

When we assume the injection pressure at r_{wi} as 100 and the producing pressure at r_{wp} as 0.001, we can show that the equations for segments 1 and 15 are:

$$110.2\ p_1 - 103.11\ p_2 = 708 \tag{11.4}$$

$$110.2\ p_{15} - 103.11\ p_{14} = 0.00708 \tag{11.5}$$

Note that in writing these two equations, the Δx for the flow rate from point 1 to 2 and from point 14 to 15 is one-half of the Δx from one point to the other between the other segments. Once all of the equations have been written, the coefficients of the pressures and the constant terms can be fitted into an augmented matrix such that each column of the matrix represents the coefficient of a pressure point or the equation constant term. Each matrix row represents an equation for a particular segment. For example, −103.11 in Eq. 11.4 may be equal to A (1, 2) where A is the matrix value for row 1 and column 2. Row 1 represents the coefficients for the pressures representing the equation for segment 1, and column 2 means that the coefficient is for pressure point 2. Similarly, from Eq. 11.3 the coefficient for p_5, which is 1.0, would be designated A (8, 5).

When a complete matrix has been generated, a digital computer subroutine such as the Gauss-Jordan or Gauss-Seidel methods can be used to reduce the matrix or in effect solve for the pressures.[3] Such subroutines can be obtained from IBM or other digital computer companies. The results for this example, Fig. 11–1, are shown in Table 11–1. Comparison with the analytically obtained data of Fig. 2–17 indicates close agreement. Accuracy of course is greatly improved by using a finer grid.

A reservoir flow problem for incompressible flow with all segments having the same unchanging k/μ values is seldom encountered in prac-

TABLE 11–1 Pressure Distribution in a Half-Quadrant of a 5-Spot during Steady-State Flow

				Well pressure in = 100.0 psi
50.0	51.0	53.0	56.3	59.1
49.0	50.0	51.8	54.1	
47.0	48.2	50.0		
43.7	45.9			
40.9				
0.001 psi = the well pressure out				

After Rusnak, undergraduate chemical engineering report, Ohio State University, 1977.

tice. However, by keeping a running material balance on each segment and changing the k/μ for each segment as the saturations change, this type of program can be used frequently to study many enhanced oil recovery displacement problems. Similar programs written in radial coordinates with segments as illustrated in Fig. 11–2 can often be used to study coning, fingering, and other well problems.

The General Reservoir Model

In the study of total reservoir behavior, the expansion of the reservoir liquids and gases is important and cannot be neglected. Thus, the assumption used in the incompressible flow model that the volumetric flow rates in and out of a segment are equal is no longer possible. Consequently, we must base our equations for each segment on an equation that accounts for the expansion of the fluids in the segment as the pressure declines. For radial flow we can use the radial diffusivity equation derived in chapter 3 as the basis for the flow equations for each segment. However, most reservoir models are based on linear flow equations so we must examine the basis for a similar linear diffusivity equation.

The linear diffusivity equation. Consider the one-dimensional flow of fluid through a segment as shown in Fig. 11–3. The pressure is declining in the direction of flow at any particular time as a result of the pressure drop caused by the flow. This in turn causes an expansion of the flowing fluids. If the reservoir pressure is also declining with time, the change in the average pressure in the Δx segment causes the expansion of the fluid in the pores. Thus, we can equate the difference in the fluid that moves into the segment and the fluid that leaves the

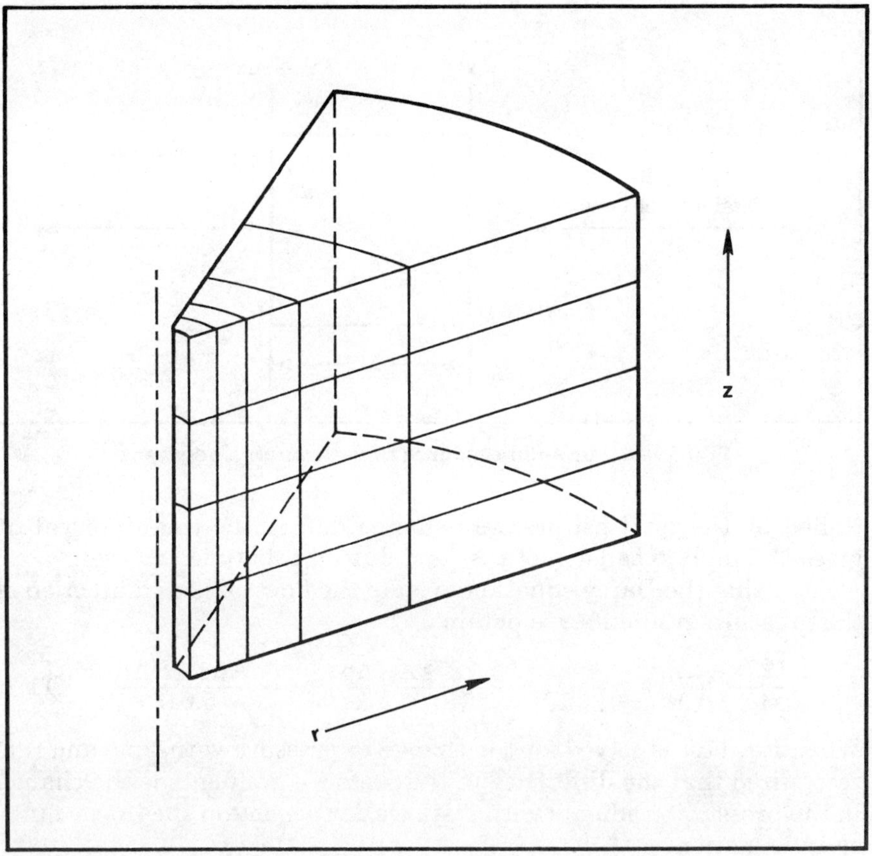

Fig. 11–2 A computer-coning model with linear (vertical) and radial (horizontal) segments

segment during a time interval, Δt, to the expansion of the fluid in the segment caused by the drop in the average pressure in the segment, Δp, during the time interval, Δt:

$$q_{x+\Delta x}\,\Delta t - q_x\,\Delta t = A\Delta x\,\phi c\,\Delta p/5.615 \tag{11.6}$$

In Eq. 11.6 the left-hand side is the difference between the total fluid flow out and into the segment during the time interval, Δt, based on the average flow rates during the time interval. The right-hand side represents the expansion of the fluid in the segment during the same time interval. The $A\Delta x$ is the bulk volume of the segment, $A\Delta x\,\phi$ is the pore volume of the segment, and $\frac{A\Delta x\,\phi c}{5.615}$ is the barrels of expansion of the fluid for every psi of pressure change. Then this expression mul-

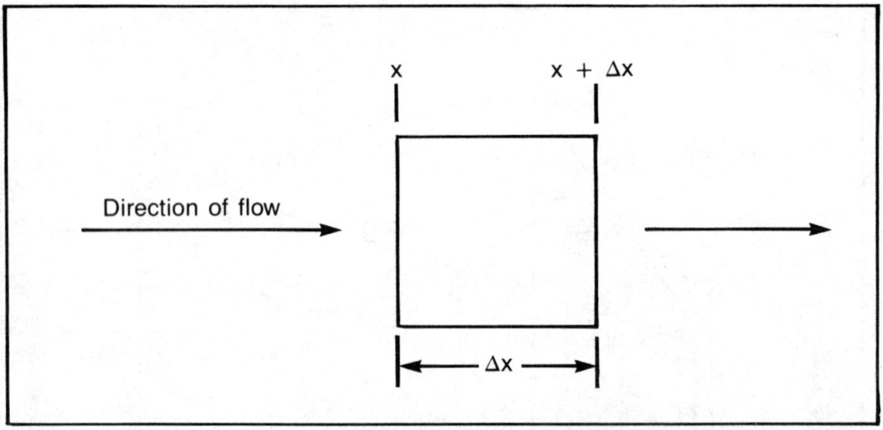

Fig. 11–3 One-dimensional flow through a segment

tiplied by the total psi pressure change during the time interval Δt gives the total expansion of the fluid during this time interval.

By using the Darcy equation to state the flow rates as a function of the pressure gradients, we obtain:

$$\frac{1.127kA}{\mu}\left(\frac{\Delta p}{\Delta x}\right)_{x+\Delta x}\Delta t - \frac{1.127kA}{\mu}\left(\frac{\Delta p}{\Delta x}\right)\Delta t = \frac{A\Delta x\ \phi c\ \Delta p}{5.615} \tag{11.7}$$

When Eq. 11.7 is solved for the change in pressure with time and it is recognized that the difference in the pressure gradients is the change in the pressure gradients with distance Δx, we obtain the linear diffusivity equation:

$$\frac{\Delta p}{\Delta t} = \left(\frac{6.33k}{\phi\mu c}\right)\frac{\Delta(\Delta p/\Delta x)}{\Delta x} \tag{11.8}$$

Eq. 11.8 can then be used to write a flow equation for each of the segments in the reservoir model. Note that the constant $(6.33k/\phi\mu c)$ is the same as the diffusivity constant η derived with the radial diffusivity equation in chapter 3.

Application of the linear diffusivity equation to reservoir modeling. To demonstrate how Eq. 11.8 is used to write a flow equation for each reservoir segment, first note that we can assume the pressure gradient, $\Delta p/\Delta x$, to be the difference in the pressures between successive segments divided by the distance between the points at which the pressures apply. Then the pressure gradient between segment $j-1$ and segment j in Fig. 11–4 at some particular time is $(p_j - p_{j-1})/\Delta x$. Similarly, the pressure gradient between segments j and $j+1$ is $(p_{j+1} - p_j)/\Delta x$. The change in the pressure gradients with distance can then be represented by the difference between the two pressure gradients divided by the distance between the midpoints of the pressure gradients, Δx:

$$\left[\frac{\Delta(\Delta p/\Delta x)}{\Delta x}\right]_j = \frac{[(p_{j+1} - p_j)/\Delta x] - [(p_j - p_{j-1})/\Delta x]}{\Delta x}$$

$$= \frac{p_{j+1} - 2p_j + p_{j-1}}{(\Delta x)^2} \qquad (11.9)$$

Eq. 11.9 of course assumes that the Δx terms are equal.

The change in the pressure with time at any particular point in the reservoir can be similarly described as the difference in the pressure at two successive times, n and n + 1, divided by the time interval separating the two pressures, $(p_n - p_{n+1})/\Delta t$. Note that when we apply Eq. 11.8, we must apply it to the change in pressure gradients with distance at a particular point in the reservoir at a particular time and equate it to the change in the pressure with time in the future or in the past.

To be exact, we identify all pressures with two subscripts, the first to identify the position in the reservoir and the second to identify the time. Then the equation for a particular change in pressure with time at some point j in the reservoir can be written in two ways. If we use the change in the pressure gradient with distance at the start of the time interval, we obtain:

$$\frac{(p_{j,n} - p_{j,n+1})}{(t_{n+1} - t_n)} = \left(\frac{6.33k}{\phi\mu c}\right)\frac{(p_{j+1,n} - 2p_{j,n} + p_{j-1,n})}{(\Delta x)^2} \qquad (11.10)$$

If we use the change in the pressure gradient with distance at the end of the time interval, we obtain:

$$\frac{(p_{j,n} - p_{j,n+1})}{(t_{n+1} - t_n)} = \left(\frac{6.33k}{\phi\mu c}\right)\frac{(p_{j+1,n+1} - 2p_{j,n+1} + p_{j-1,n+1})}{(\Delta x)^2} \qquad (11.11)$$

Fig. 11–4 may help in understanding the difference between Eqs. 11.10 and 11.11. In both cases we are calculating the same change in pressure with time. However, in one case we are basing this calculation on the pressure gradient at the start of the time interval, t_n. In the second case we are basing the calculation on the pressure distribution at the end of the time interval, t_{n+1}.

Explicit versus implicit programming. Examination of the Fig. 11–4 plot of pressure versus distance at times n and n + 1 indicates that Eqs. 11.10 and 11.11 can only give the same results if the pressure gradients at the two times are the same, which can only occur if the time interval separating the two approaches 0.0. Application of the two

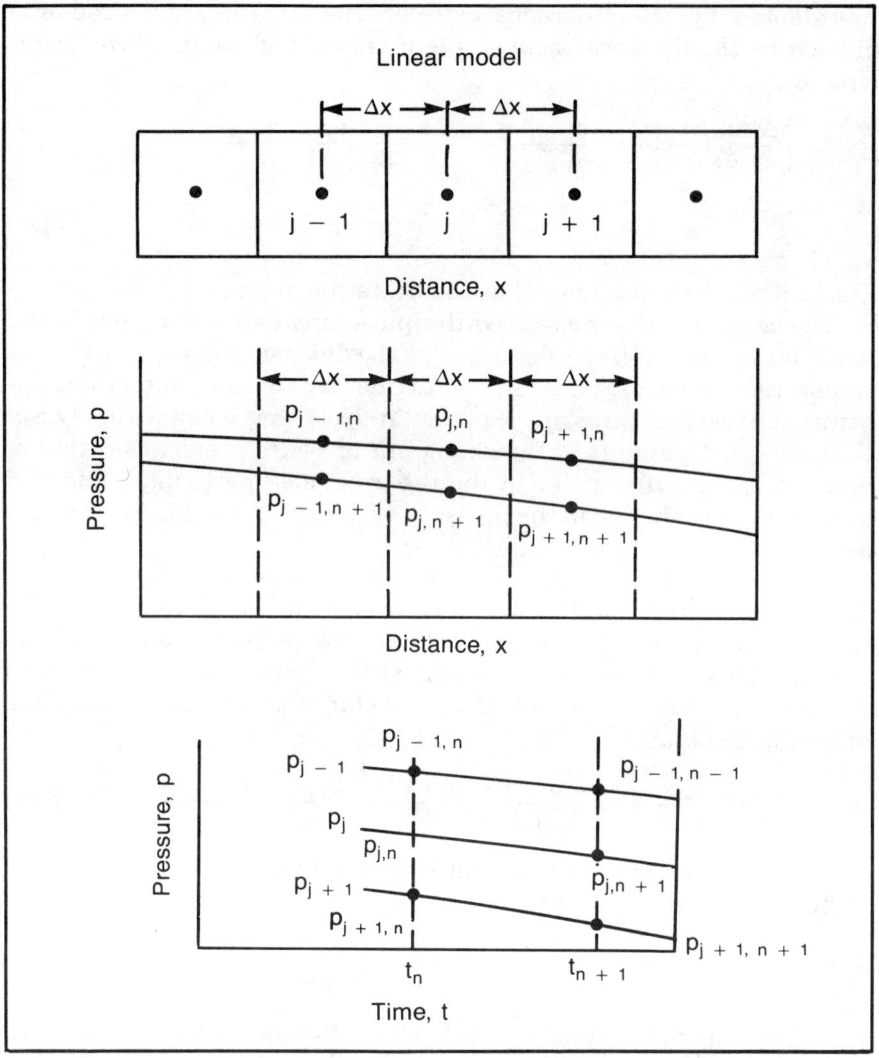

Fig. 11–4 Pressure-time-distance relationship for linear unsteady-state flow

equations presents different problems. If we assume that at any particular time n all of the pressures in the model have been given or have been predicted, we can then calculate all of the pressures at the n + 1 time by Eq. 11.10. With all of the n values known, there is only one unknown, $p_{j,n+1}$. This method is known as the *explicit calculating technique*.

The explicit calculating technique has the advantage of being very easy to program. However, to get accurate results, it is necessary to use such small time increments that the process is impractical for most applications.

Eq. 11.11 used as a calculating procedure is more difficult to program, but accurate results can be obtained with much larger time steps. If we again assume that we know all of the model pressures at time n, then Eq. 11.11 contains three unknowns: the pressures at $j + 1$, j, and $j - 1$ at time $n + 1$. However, if we write the equation for each segment in the linear model, we end up with the same number of equations as we have unknowns in the same manner previously demonstrated for incompressible fluid flow. This set of simultaneous equations can then be solved using the Gauss-Seidel, Gauss-Jordan, or some other mathematical process. The use of Eq. 11.11 as a calculating procedure is referred to as an *implicit method.*

Until now we have talked only about the flow of one fluid. However, it should be clear that a similar set of equations can be generated for both oil and water. The equations for the simultaneous flow of gas are somewhat more difficult because the gas is moving as both free gas and gas in solution. Nevertheless, it should be clear that three sets of equations can be determined for the flow of oil, gas, and water. Solving these equations results in a pressure matrix for the reservoir, which can then be used to calculate the saturations in each segment of the reservoir.

We also have only considered one-dimensional flow, but again it should be clear that similar equations can be written for two- or three-dimensional flow, both linear and radial. When the vertical flow of fluid is calculated, it is necessary to account for static pressure differences. In many cases it may also be necessary to include capillary pressures in the model.

History Matching

The foregoing should indicate that a reservoir simulator is an extremely complex and detailed computer program. This program requires that the engineer come up with very detailed reservoir and fluid characteristics. Such data are normally uncertain to say the least. It is difficult to know how the permeability and porosity varies, both vertically and laterally in the reservoir when we only have core data potentially from the wells and only a portion of these are normally cored. Even more difficult is knowing whether there is an initial gas cap and what its size is and whether the reservoir is connected to an aquifer and what the characteristics of the aquifer are. Aquifer data are obtained only by accident from dry holes, which are normally quickly plugged.

With so much uncertain data it is generally impossible to input the engineer's best estimates of the reservoir parameters and calculate a prediction of the future reservoir behavior. The normal procedure is to use the past reservoir history to prove the validity of the reservoir data. The past history amounts to the oil, gas, and water production from the various wells in the reservoir, together with the average reservoir pressure measurements throughout the reservoir. At this point the practical reservoir engineer should exert his best judgment as to the accuracy of the data that are being matched. He should recognize the unreliability of unsold gas production, water production, and reservoir pressure measurements. The engineer should recognize the inherent inaccuracy of the gas and water production, but he may not be aware of the inaccuracy of the average reservoir pressures he is using. It is strongly recommended that the careful review of the methods used to obtain the average reservoir pressure data be a routine part of reservoir modeling. The use of grossly inaccurate methods in determining average reservoir pressures is so common as to be a major problem in the history-matching portion of a reservoir simulation. The engineer should be especially careful in non-U.S. areas where reservoir simulation is badly needed but where average reservoir pressures may be obtained by using pseudosteady-state methods in steady-state reservoirs.

The engineer should also resist the temptation to change reservoir data to unreasonable data in the interest of obtaining a past reservoir-behavior fit. This simply substitutes one problem for another. Justification of data only because they obtain a past history fit when they do not fit a logical range of values is no justification at all. It simply guarantees an inaccurate prediction.

The use of a single cell or reservoir segment to simulate the behavior of a water drive is an excellent example of an assumption that can be used to obtain a past reservoir match that cannot hope to result in a long-range accurate prediction because a water drive does not act in that manner. Early or short time behavior can be approximated in this way. However, the longer the reservoir is produced, the more inaccurate the water-drive model becomes.

When the engineer is unable to solve reservoir engineering problems using analytical methods, he should then consider their solution by reservoir simulation. By first making every effort to solve problems analytically, he guarantees the best possible knowledge of the reservoir to guide computer simulator data input and modification to obtain a past history match.

Computer reservoir simulators are available within most major petroleum companies or can be bought or leased from several software companies such as Scientific Software, Intercomp, Garrett Computing Systems, and Wattenbarger and Associates. Engineers should give preference to weather-worn programs that have received much use resulting in extensive debugging.[4] Engineers should resist the temptation to use unreasonable data in history matching attempts. Engineers who are or will be involved with computer simulation are referred to Crichlow's *Modern Reservoir Engineering—A Simulation Approach* or Aziz and Settari's *Petroleum Reservoir Simulation* for a more thorough knowledge of the subject.[5,6]

Notes

1. Henry B. Crichlow, *Modern Reservoir Engineering—A Simulation Approach* (Englewood Cliffs: Prentice-Hall, 1977).
2. K. Aziz and A. Settari, *Petroleum Reservoir Simulation* (London: Applied Science Publishers, 1979).
3. S.D. Conte, *Elementary Numerical Analysis* (New York: McGraw-Hill, 1965), pp. 191–197.
4. Larry Baggett, Texaco Inc., from personal communication in March 1982.
5. Crichlow, 1977.
6. Aziz and Settari, 1979.

Appendix A

Reservoir Engineering Symbols

Symbol	Description	Dimension
a	Fractional change in the rate per day (decline rate)	$(\text{day})^{-1}$
a_i	Decline rate at q_i	$(\text{day})^{-1}$
A'	Area	sq cm
A	Stiles mobility ratio	dimensionless
A	Area	sq ft
A_{eff}	Modified Stiles mobility ratio	dimensionless
A_r	Cross-sectional area at radius r	sq ft
b	Flowing thickness at the well	ft
B	Angle between interface and bed	degrees
B_{oabd}	B_o at abandonment	dimensionless
B_o	Oil formation volume factor	dimensionless
B_g	Gas formation volume factor	res bbl/scf
B_{gi}	Gas formation volume factor at p_i	res bbl/scf
B_{oi}	Oil formation volume factor at p_i	dimensionless
B_{os} or B_{op}	B at flood start	dimensionless
B_t	Total formation volume factor	dimensionless
B_w	Water formation volume factor	dimensionless
c	Capacity; also a constant	md-ft
c'	Capacity as fraction of total, c_T	fraction
c	Capillary length/core length	dimensionless
c	Compressibility	volume/volume/psi
c_f	Formation (rock) compressibility	volume/volume/psi
c_g	Gas compressibility	volume/volume/psi
c_R	Reduced gas compressibility, $c_g p_c$	dimensionless
c_o	Oil compressibility	volume/volume/psi
c_{sta}	Gas compressibility at static pressure	psi^{-1}
c_w	Water compressibility	volume/volume/psi
c_e	Effective compressibility	volume/volume/psi
C	Constant in gas flow equation	—
d	Distance between input and producer in five-spot	ft

Symbol	Description	Dimension
d	Distance to boundary	ft
D	Depth or distance	ft
D	Conventional decline rate	$years^{-1}$
DR	Damage ratio	dimensionless
DVI	Displaceable volumes injected	dimensionless
E	Efficiency	dimensionless
E_V	Vertical efficiency: hydrocarbon pore space invaded (affected, contacted) by the injected fluid divided by the hydrocarbon pore space enclosed in all layers behind the injected fluid when $E_H = 1.0$	dimensionless
E_H	Pattern sweep efficiency: hydrocarbon pore space enclosed behind the injected fluid front divided by total hydrocarbon pore space when $E_V = 1.0$	dimensionless
E_c	Conformance efficiency: accounts for bypassed volumes due to lateral permeability variations, lenses, and other difficulties	dimensionless
E_T	Volumetric efficiency: product of pattern, conformance, and vertical efficiencies	dimensionless
Ei	Exponential integral	dimensionless
f	Friction factor	—
f_d	Fraction of phase in the total flow rate, q_t	dimensionless
f_{dfr}	f_d at the front of the displacing phase	dimensionless
f_{dPf}	f_d at the producting face (well)	dimensionless
f_{dno}	f_d with no gravity	dimensionless
f_w or F_w	Fraction of water in q_t	dimensionless
f_g	Fraction of gas in q_t	dimensionless
G	Total initial gas in place in reservoir	scf
GCDI	Gas-cap-drive index	fraction
G_p	Cumulative gas produced	scf
G_{pn}	Cumulative gas production at time n	scf
h'	Thickness as fraction of total, h_T	dimensionless
h	Net thickness (general and individual bed)	ft
h_a	Thickness of zone a	ft

Symbol	*Description*	*Dimension*
h_g or h_o	Net thickness of gas or oil zone	ft
h_T	Total thickness	ft
H	Hyperbolic decline constant	—
i	Injection rate	res b/d
i_{eff}	Effective injection rate	res b/d
i_r	Injection rate ratio i_s/i	dimensionless
i_s	Initial injection rate (at start)	res b/d
j	Mathematical counter	—
J	J function	dimensionless
J	Productivity index	stb/d/psi
J_a	Productivity index at condition a	stb/d/psi
J_s	Specific productivity index	stb/d/psi/ft
k	Absolute permeability (fluid flow)	darcy
k′	Permeability as fraction of average permeability	dimensionless
k_a	Permeability of a vertical or lateral zone	darcy
k_{actual}	Existing permeability	darcy
k_{avg}	Overall effective permeability of a zone where permeability varies laterally or at right angles to the bedding plane	darcy
k_d	Permeability to displacing phase	darcy
k_e	Effective permeability	darcy
k_g	Effective permeability to gas	darcy
k_o	Effective permeability to oil	darcy
k_r	Relative permeability	dimensionless
k_{rg}	Relative permeability to gas	dimensionless
k_{ro}	Relative permeability to oil	dimensionless
k_{rw}	Relative permeability to water	dimensionless
k_w	Effective permeability to water	darcy
k_i	Effective permeability at input	darcy
k_p	Effective permeability at producer	darcy
$k_{undamaged}$	Permeability with no damage	darcy
ℓn	Natural logarithm, base e	—
log	Common logarithm, base 10	—
L	Length of system	ft
L'_{cap}	Capillary length	cm
L'_{core}	Capillary length in core	cm
L_{mi}	Pipeline length	miles

Symbol	*Description*	*Dimension*
m	Ratio of initial free-gas volume to oil volume	dimensionless
m	Slope	various
m′	Slope of p_w^2 versus log of time plot	—
m_L	Slope of well pressure versus square root of time in days	psi/days$^{0.5}$
m(p)	Pseudo gas potential, real gas potential, real gas pseudo pressure, or modified pressure squared	psia^2/cp
M	Mobility ratio (λ displacing/λ displaced)	dimensionless
$M_{x/y}$	Mobility ratio (mobility of x/ mobility of y)	dimensionless
MW	Molecular weight	lb
MW_j	Molecular weight of component j	lb
n	Exponent of back-pressure curve, gas well	—
n	Number of mols of gas	mols
n	Math counter	—
n	Exponent of hyperbolic decline	—
N	Initial oil in place in reservoir	stb
N_p	Cumulative oil produced	stb
N_{pf}	Oil produced during flood	stb
$(N_{pf})_{total}$	Oil produced from all zones during flood (performance prediction)	stb
N_{pn}	Cumulative oil production at time n	stb
N_{pP}	Cumulative oil produced during primary	stb
N_{ps}	N_p at flood start	stb
p	Pressure	psia
p′	Pressure	atm
$p_{1\ hr}$	Well pressure after one hour of shut in	psia
p_a	Pressure at point *a*	psia
p_{avg}	Average gas pressure	psia
p_b	Bubble-point (saturation) pressure	psia
p_c	Critical pressure	psia
$_pp_c$	Pseudo critical pressure	psia
p_{cj}	Critical pressure of component j	psia
Δp_D	Dimensionless pressure function	dimensionless

Symbol	*Description*	*Dimension*
$(\Delta p_D)_a$	Dimensionless pressure function based on time or radius *a*	dimensionless
Δp_{Dpj}	Dimensionless pressure drop due to pulse j	dimensionless
p_e	External boundary pressure	psia
Δp_{flow}	Pressure drop due to flow	psi
p_i	Initial pressure	psia
(piez)	Piezometric pressure drop	ft of fluid
Δp_j	Changes in pressure at time j	psi
$p_{j,n}$	Pressure at point j and time n	psia
p_m	Arithmetic average pressure	psia
p_{new}	Pressure after change	psia
p_{old}	Pressure before change	psia
Δp_{pj}	Change in pressure associated with pulse j	psi
Δp_q	Pressure change caused by rate change to 0.0	psi
$\Delta p_{\Delta q}$	Pressure change caused by rate change Δq	psia
p_{ref}	Reference pressure	psia
$p_{r,t}$	Pressure at radius r and time t	psia
p_R	Reduced pressure	dimensionless
p^*	Initial pressure necessary for a well to be infinite acting	psia
$(\Delta p_q)_{ts}$	Pressure change caused by rate change when time is t_s	psi
Δp_{total}	Total pressure drop	psi
p_r	Pressure at r	psia
p_s	Static reservoir pressure	psia
Δp_{skin}	Additional pressure drop in damaged zone	psia
p_{sc}	Pressure, standard conditions	psia
p_{tD}	Dimensionless pressure function at dimensionless time t_D	dimensionless
p_w	Bottom-hole pressure, general	psia
p_{wf}	Bottom-hole pressure, flowing	psia
p_{wi}	Injection well bottom-hole pressure, flowing	psia
p_{wp}	Producing well bottom-hole pressure	psia
$(\Delta p/\Delta r)_r$	Pressure gradient at radius r	psi/ft

Symbol	Description	Dimension
p_w'	Well pressure without effect of last disturbance	psia
$\Delta(p^2)_{visc}$	Change in pressure squared due to viscous flow	psi^2
$(\Delta p/\Delta t)_{pseudo}$	Rate of pressure change in pseudosteady state	psi/day
P_c	Capillary pressure	psi
q'	Flow rate	cc/sec
q	Production rate or flow rate	res b/d
q_a	Producing rate for zone or thickness *a*	res b/d
q_{actual}	Oil flow rate as is	res b/d
q_{cC}	Minimum rate at which a cone reaches a well	res b/d
q_{cF}	Minimum rate at which a finger forms	res b/d
q_d	Displacing phase flow rate	res b/d
q_{DST}	DST flow rate	res b/d
q_{EL}	Rate at the economic limit	res b/d or stb/d
q_g	Gas production rate at well	Mscfd
q_{gr}	Gas production rate at r	Mscfd
q_{gn}	Last gas flow rate	Mscfd
q_i	Reference rate in decline curve analysis	stb/d
q_m	Gas volumetric flow rate at p_m	res b/d
q_o	Oil production rate	res b/d
q_{oi}	Initial oil producing rate	res b/d
q_r	Flow rate at r	res b/d
$q_{scf/hr}$	Pipeline flow rate	scf/hr
q_{stb}	Flow rate at the well	stb
q_t	Total flow rate of all fluids	res b/d
$q_{undamaged}$	Oil flow rate if no damage	res b/d
q_w	Water production rate	res b/d
Q	Cumulative fluid flow	res bbl
Q_{Mscf}	Cumulative gas production	Mscf
Q_i	Cumulative influx of the displacing phase	res bbl
Q_{tD}	Dimensionless fluid influx function at dimensionless time t_D	—
r	Radial distance	ft

Symbol	*Description*	*Dimension*
r'	Radius	cm
r_D	Dimensionless reservoir size	dimensionless
r_{De}	Dimensionless reservoir size, r_e/r_w	dimensionless
r_e	External boundary radius	ft
r_e^*	External radius equivalent of the isochronal producing time, t*	ft
r_{ew}	Effective well radius in spherical flow	ft
r_i	Inner radius of radial segment	ft
r_o	Outer radius of radial segment	ft
r_{rw}	Actual radial well radius	ft
r_w	Well radius	ft
r_{we}	Effective well radius	ft
r_{wp}	Radius of the producing well	ft
R	Gas constant	varies
R	Producing gas-oil ratio	scf/stb
R	Stile's percentage of recovery from a flood pattern	dimensionless
R_p	Cumulative gas-oil ratio	scf/stb
R_s	Solution gas-oil ratio (gas solubility in oil)	scf/stb
R_{si}	Initial solution gas-oil ratio	scf/stb
S	Skin factor	dimensionless
s	Exponent in Smith tubing or casing flow equation	—
S	Saturation	dimensionless
$\overline{S}_d$	Average S_d	dimensionless
S_d	Displacing phase saturation	dimensionless
S_{di}	Initial displacing phase saturation	dimensionless
$(S_d)_{pf}$	S_d at producing face	dimensionless
S_{df}	Displacing phase saturation at front	dimensionless
S_g	Gas saturation	dimensionless
SGDI	Solution-gas-drive index	fraction
S_{ge}	Equilibrium gas saturation	dimensionless
S_{gi}	Gas saturation at flood start	dimensionless
S_L	Total (combined) liquid saturation	dimensionless
S_o	Oil saturation	dimensionless
S_{oP}	Oil saturation after primary	dimensionless
S_{or}	Residual oil saturation	dimensionless

Symbol	*Description*	*Dimension*
S_{os}	Oil saturation at flood start	dimensionless
S_s	Spherical skin factor	dimensionless
S_w	Water saturation	dimensionless
S_{wc}	Connate water or irreducible water saturation	dimensionless
t	Time	days
Δt	Change in time or shutin time	days
t^*	Isochronal producing time	days
t_D	Dimensionless time	dimensionless
$t_D{}^*$	Dimensionless time equivalent of the isochronal producing time, t^*	dimensionless
t_{Di}	Dimensionless time for image well	dimensionless
$t_{D\ image}$	Dimensionless time for image well	dimensionless
t_{DLj}	Dimensionless time lag associated with pulse j	dimensionless
t_{Dp}	Dimensionless time based on producing time	dimensionless
t_{DS}	Dimensionless stabilization time	dimensionless
$t_{D\ real}$	Dimensionless time for real well	dimensionless
t_{Dw}	Dimensionless time based on r_w	dimensionless
t_{Dxf}	Dimensionless time based on fracture length	dimensionless
t_p	Producing time	days
Δt_{pj}	Time lag associated with pulse j	days
t_r	Reduced time (for flood prediction method)	—
t_s	Time to reach pseudosteady state (stabilize)	days
T	Temperature	varies
T_{avg}	Average temperature	°R
T_c	Critical temperature	°R
T_{cj}	Critical temperature of component j	°R
T_f	Formation temperature	°R or °F + 460
${}_pT_c$	Pseudo critical temperature	°R
T_R	Reduced temperature	dimensionless
T_{sc}	Temperature, standard conditions	°R
T_w	Temperature at the well	°R
v	Mathematical expression	—
v′	Velocity	cm/sec
V	Volume	varies
V_b	Bulk volume	acre-ft or cu ft

Symbol	*Description*	*Dimension*
VF_j	Volume fraction of component j	dimensionless
V_p	Pore volume	cu ft or bbl
V_{pT}	Total formation pore volume	cu ft or bbl
V_{BP}	Bulk volume under primary drainage	acre-ft
V_{sw}	Gross swept volume	acre-ft
V_o	Volume of oil	varies
$(V_o)_{st}$	Volume of oil at stock-tank conditions	varies
V_t	Total volume	varies
$(V_t)_{st}$	Total volume at stock-tank conditions	varies
w	Width	ft
WDI	Water-drive index	fraction
W_e	Cumulative water influx (encroachment)	res bbl
W_i	Cumulative water injected	res bbl
$(W_i)_{total}$	W_i sum for all zones (waterflood prediction method)	res bbl
W_p	Cumulative water produced	res bbl
ΔW_i	Water injected during Δt	res bbl
$(W_p)_{total}$	Water produced during flood from all zones (prediction method)	res bbl
x'	Distance	cm
x_f	Fracture length measured in one direction from well	ft
ΔX	Tubing or casing length	ft
X_f	Distance to front	ft
X_{sdj}	Position of displacing phase saturation s_{dj}	ft
y	Math counter	—
z	Gas deviation factor (compressibility factor)	dimensionless
z_{avg}	Gas deviation factor at average pressure	dimensionless
z_i	Gas deviation factor at original reservoir conditions	dimensionless
z_r	Gas deviation factor at reservoir conditions	dimensionless
z_w	Gas deviation factor at well conditions	dimensionless
W_{ieff}	Effective cumulative water injected	res bbl

Symbol	*Description*	*Dimension*
α (alpha)	Angle of formation dip	degrees
β (beta)	Acute angle between interface and bed	degrees
β	Turbulence constant	ft^{-1}
β'	Turbulence constant	cm^{-1}
σ (sigma)	Interfacial tension	dynes/cm
γ (gamma)	Specific gravity	dimensionless
γ_g	Gas specific gravity	dimensionless
γ_L	Liquid specific gravity	dimensionless
γ_o	Oil specific gravity	dimensionless
γ_w	Water specific gravity	dimensionless
η (eta)	Hydraulic diffusivity ($6.33k/\phi\mu c$)	—
λ (lambda)	Mobility (k/μ)	darcies/cp
λ_g	Gas mobility	darcies/cp
λ_o	Oil mobility	darcies/cp
λ_w	Water mobility	darcies/cp
μ (mu)	Viscosity	cp
μ_{avg}	Average viscosity	cp
μ_d	Viscosity displacing phase	cp
μ_g	Gas viscosity	cp
μ_i	Fluid viscosity at input	cp
μ_o	Oil viscosity	cp
μ_p	Fluid viscosity at producer	cp
μ_w	Water viscosity	cp
ϕ (phi)	Porosity	fraction
ϕ_w	Fraction of circle flowing water	fraction
ρ (rho)	Density in gm/cc or specific gravity relative to water	gm/cc
ρ_g	Gas density	lb/cu ft
θ (theta)	Fraction of circle	fraction
θ	Wetting angle	degrees

Appendix B

Empirical Reservoir Engineering Data

TABLE B1 Empirical Relative Permeability Equations

I. Two-Phase Equations

Wyllie's equations for oil-gas relative permeabilities:*

$$S^* = S_o/(1 - S_{wi})$$

Where:

S_{wi} = Irreducible water saturation

	k_{ro}	k_{rg}
Unconsolidated sand—well sorted	$(S^*)^{3.0}$	$(1 - S^*)^3$
Unconsolidated sand—poorly sorted	$(S^*)^{3.5}$	$(1 - S^*)^2(1 - S^{*1.5})$
Cemented sand, oolitic lime, and vugular lime	$(S^*)^{4.0}$	$(1 - S^*)^{2.0}(1 - S^{*2.0})$

Wyllie's equations for water-oil relative permeabilities:*

$$S^* = \left[\frac{S_w - S_{wi}}{1 - S_{wi}}\right]$$

Where:

S_{wi} = Irreducible water saturation

	k_{ro}	k_{rg}
Unconsolidated sand—well sorted	$(1 - S^*)^{3.0}$	$(S^*)^{3.0}$
Unconsolidated sand—poorly sorted	$(1 - S^*)^2(1 - S^{*1.5})$	$(S^*)^{3.5}$
Cemented sand, oolitic lime, vugular lime	$(1 - S^*)^2(1 - S^{*2.0})$	$(S^*)^{4.0}$

Corey's equations for a drainage system (gas drive)**

$$k_{ro} = (1 - S)^4$$
$$k_{rg} = S^3(2 - S)$$

Where:

$$S = S_g/(1 - S_{wc})$$

Naar-Henderson equations for a water-drive reservoir**

$$k_{ro} = \frac{(1 - 2S)^{1.5}}{[2 - (1 - 2S)^{0.5}]}$$
$$k_{rg} = S^4$$

Where:

$$S = \frac{(S_w - S_{wc})}{(1 - S_{wc})}$$

*After Frick, *Petroleum Production Handbook* (New York: McGraw-Hill, 1962).
**After Crichlow, *Modern Reservoir Engineering—A Simulation Approach* (Englewood Cliffs: Prentice-Hall, 1977)

TABLE B1 *continued*

II. Three-Phase—Stone's equations**

k_{rw} is determined from water-oil relative permeability data and is used in three-phase flow as a function of only the water saturation. Can be calculated from two-phase equations.

k_{rg} is determined from oil-gas relative permeability data and is used in three-phase flow as a function of only the gas saturation. Can be calculated from two-phase equations.

$$k_{ro} = (k_{row} + k_{rw})(k_{rog} + k_{rg}) - (k_{rw} + k_{rg})$$

Where:

k_{row} = k_{ro} from two-phase oil-water relative permeability data
k_{rw} = k_{rw} from two-phase oil-water relative permeability data
k_{rog} = k_{ro} from oil-gas relative permeability data
k_{rg} = k_{rg} from oil-gas relative permeability data

**After Crichlow, *Modern Reservoir Engineering—A Simulation Approach* (Englewood Cliffs: Prentice-Hall, 1977)

TABLE B2 Sandstone Relative Permeability Ratio Curves*

Curve No.	*Formation Name*	*Permeability, md*	*Porosity, %*	*Comments*	*Reference*
1	Tensleep Sandstone				e
2	Torpedo Sandstone				e
3	Berea Sandstone				e
4	Velma Springer Sandstone				e
5	Cromwell Sand	8			e
6	Cromwell Sand	2,300			e
7	Wilcox Sand	Oklahoma		Largely unconsolidated	a
8	Simpson Bromide Sandstone	Oklahoma		Consolidated	a
9	Unconsolidated Sandstone			Unconsolidated	a
10	Consolidated Sands	500	21.8	Consolidated	a
11	Unconsolidated Sands			Unconsolidated	b
12	Nichols Buff Sand	500	21.8	Consolidated	c
13	Berea Sand	115	?		d
	Maximum			Typical of unconsolidated	
	Minimum			Highly cemented sandstones	
	Average			Average consolidated sand or sandstone	

*See Fig. B1

TABLE B2 *continued*

Reference:

[a]L.E. Elkins, "The Importance of Injected Gas as a Driving Medium in Limestone Reservoirs as Indicated by Recent Gas-Injection Experiments and Reservoir-Performance History," *Drilling and Production Practices,* API (1946), p. 160.

[b]M.C. Leverett and W.B. Lewis, "Steady Flow of Gas-Oil-Water Mixtures through Unconsolidated Sands," *Trans.,* AIME (1941), volume 142, 107.

[c]H.G. Botset, "Flow of Gas-Liquid Mixtures through Consolidated Sand," *Trans.,* AIME (1940), volume 136, 91.

[d]J.G. Richardson et al., "Laboratory Determination of Relative Permeability," *Trans.,* AIME (1942), volume 195, 187.

[e]J. Arps and T.G. Roberts, "The Effect of the Relative Permeability Ratio, the Oil Gravity, and the Solution Gas-Oil Ratio on the Primary Recovery from a Depletion Type Reservoir," *Trans.,* AIME (1955), volume 204, 120.

TABLE B3 Limestone and Dolomite Relative Permeability Ratio Curves*

Curve No.	*Formation Name*	*Permeability, md*	*Porosity, %*	*Comments*	*Reference*
1	San Andreas Lime				d
2	Pettit Lime				d
3	Oolithic Lime				d
4	Devonian Chert A				d
5	Devonian Chert B				d
6	Penn Reef A				d
7	Penn Reef B				d
8	Strawn Reef				d
9	Palo Pinto Reef				d
10	Intergranular Louisiana A	78	18.3	Basically intergranular porosity	a
11	Intergranular Louisiana B	3.0	17.6	Basically intergranular porosity	a
12	Reef Limestone C	99.0	17.0	Very vuggy pores—some calcite crystals in them	a
13	Reef Limestone D-1	36.3	28.4	Large vugs, fossils, very fine pores	a
14	Reef Limestone D-2	47.2	35.6	Large vugs, fossils, very fine pores	a
15	Fractured Louisiana E	55	12.8	Large cavities connected with hairline fracs	a
16	Fractured Louisiana F	2.4	30.3	Large fractures connected together	a
17	Fractured Louisiana F	2.4	30.3	Large fractures connected together	a
18	San Andres Dolomite A			West Texas pool	b
19	San Andres Dolomite B			Limestone pool West Texas	b

*See Fig. B2

TABLE B3 *continued*

20	Panhandle Dolomite			Panhandle Region Texas	b
21	Pettit Lime			Louisiana	b
22	Hunton Lime			Oklahoma	b
23	Wasson Dolomite	78	23.6		b
24	Permian Dolomite			An average of 26 samples	c
	Average six Limestone fields				c
	Average			Vugular-type limestone	
	Minimum			Fractured chert	
	Maximum			West Texas Permian dolomite	

Reference:

[a]C.R. Stewart, F.F. Craig Jr., and R.A. Morse, "Determination of Limestone Performance Characteristics by Model Flow Tests," *Trans.*, AIME (1953), volume 198, 93.

[b]L.E. Elkins, "The Importance of Injected Gas as a Driving Medium in Limestone Reservoirs as Indicated by Recent Gas-Injection Experiments and Reservoir-Performance History," *Drilling and Production Practices,* API (1946), p. 160.

[c]A.C. Balnes and R.U. Fitting Jr., "An Introductory Discussion of the Reservoir Performance of Limestone Formations," *Trans.*, AIME (1945), volume 160, 179.

[d]J. Arps and T.G. Roberts, "The Effect of the Relative Permeability Ratio, the Oil Gravity, and the Solution Gas-Oil Ratio on the Primary Recovery from a Depletion Type Reservoir," *Trans.*, AIME (1955), volume 204, 120.

TABLE B4 Empirical Determination of Oil Compressibility

Example Determination.

A reservoir oil has a solution-gas-oil ratio of 600 scf/stb. The gas has a specific gravity of 0.7, and the stock-tank oil is 40° API (0.825 sp gr). The reservoir pressure is 3,500 psia, the reservoir temperature is 200°F, the oil formation volume factor is 1.3, and the original reservoir bubble-point pressure is 4,000 psia. Find the oil compressibility.

Solution.

First calculate the density of the reservoir oil at reservoir conditions:

$$\begin{aligned}\text{Res oil density} &= [(\text{wt of 1.0 stb of oil} + \text{solution gas wt})]/B_o\\ &= [(0.825 \times 350) + 0.7(29/379)(600)]/1.3\\ &= 247\ \text{lb/bbl}\\ \text{Res oil sp gr (or g/cc)} &= 247/350 = 0.706\end{aligned}$$

To determine the reservoir oil specific gravity at 60°F, enter Fig. B5 at the reservoir temperature of 200°F and specific gravity of 0.706. Follow the correlation lines to 60°F and read a specific gravity at 60°F as 0.77.

From Fig. B4(b) read the pseudocritical pressure and temperature as 470 psia and 880°R. Reduced temperatures and pressure can then be calculated:

$$T_r = (200 + 460)/880 = 0.75$$
$$p_r = 3{,}500/470 = 7.447$$

TABLE B4 ***continued***

Now the reduced oil compressibility can be read from Fig. B4(a) as 0.0055, and the oil compressibility can be calculated as:

$$c_o = c_r/p_c$$
$$= 0.0055/470 = 1.17 \times 10^{-5}/\text{psi}$$

Note that any oil that does not contain free gas can be considered undersaturated.

TABLE B5 Physical Properties of Petroleum Components

Compound	*Molecular Weight*	*Boiling point at 14.7 psia, °F*	*Critical Constants* Pressure, p_c, psia	*Critical Constants* Temperature, T_c, °R
Methane	16.04	−258.7	673.1	343.2
Ethane	30.07	−127.5	708.3	549.9
Propane	44.09	−43.7	617.4	666.0
Isobutane	58.12	10.9	529.1	734.6
n-butane	58.12	31.1	550.1	765.7
Isopentane	72.15	82.1	483.5	829.6
n-pentane	72.15	96.9	489.8	846.2
n-hexane	86.77	155.7	440.1	914.2
n-heptane	100.20	209.2	395.9	972.4
n-octane	114.20	258.2	362.2	1,024.9
n-nonane	128.30	303.4	334.0	1,073.0
n-decane	142.30	345.4	312.0	1,115.0
Air	28.97	−317.7	547.0	239.0
Carbon dioxide	44.01	−109.3	1,070.2	547.5
Helium	4.003	−452.1	33.2	9.5
Hydrogen	2.016	−423.0	189.0	59.8
Hydrogen sulfide	34.08	−76.6	1,306.5	672.4
Nitrogen	28.02	−320.4	492.2	227.0
Oxygen	32.00	−297.4	736.9	278.6
Water	18.02	212.0	2,109.5	1,165.2

Source: Eilerts and Others, *Phase Relations of Gas-Condensate Fluids,* U.S. Bureau of Mines Monograph 10, volume 1 (New York: American Gas Association, 1957), pp. 427–434.

TABLE B6 Hyperbolic Decline-Curve Functions (q/q_i) for n = 0.1*

Time, months	*a = 0.005*	*a = 0.010*	*a = 0.015*	*a = 0.020*	*a = 0.025*
−120.000	1.857	3.591	7.275	15.555	35.401
−117.000	1.827	3.470	6.888	14.378	31.823
−114.000	1.798	3.355	6.523	13.299	28.638
−111.000	1.770	3.243	6.179	12.308	25.800
−108.000	1.742	3.136	5.855	11.398	23.269

*Decline rates listed *a* are at time = 0 and in months^{-1}.

TABLE B6 *continued*

Time, months	*a* = 0.005	*a* = 0.010	*a* = 0.015	*a* = 0.020	*a* = 0.025
−105.000	1.715	3.032	5.550	10.562	21.008
−102.000	1.688	2.932	5.262	9.792	18.986
−99.000	1.661	2.836	4.991	9.084	17.176
−96.000	1.835	2.744	4.734	8.431	15.555
−93.000	1.610	2.654	4.493	7.830	14.100
−90.000	1.585	2.568	4.264	7.275	12.793
−87.000	1.560	2.485	4.049	6.764	11.619
−84.000	1.536	2.405	3.845	6.292	10.562
−81.000	1.512	2.327	3.652	5.855	9.609
−78.000	1.489	2.253	3.471	5.452	8.751
−75.000	1.466	2.181	3.298	5.079	7.976
−72.000	1.443	2.111	3.136	4.734	7.275
−69.000	1.421	2.044	2.982	4.415	6.642
−66.000	1.399	1.979	2.836	4.119	6.069
−63.000	1.377	1.917	2.698	3.845	5.550
−60.000	1.356	1.857	2.568	3.591	5.079
−57.000	1.335	1.798	2.444	3.355	4.652
−54.000	1.315	1.742	2.327	3.136	4.264
−51.000	1.295	1.688	2.216	2.932	3.912
−48.000	1.275	1.635	2.111	2.744	3.591
−45.000	1.256	1.585	2.011	2.568	3.298
−42.000	1.236	1.536	1.917	2.405	3.032
−39.000	1.218	1.489	1.827	2.253	2.790
−36.000	1.199	1.443	1.742	2.111	2.568
−33.000	1.181	1.399	1.661	1.979	2.366
−30.000	1.163	1.356	1.585	1.857	2.181
−27.000	1.146	1.315	1.512	1.742	2.011
−24.000	1.128	1.275	1.443	1.635	1.857
−21.000	1.111	1.236	1.377	1.536	1.715
−18.000	1.095	1.199	1.315	1.443	1.585
−15.000	1.078	1.163	1.256	1.356	1.466
−12.000	1.062	1.128	1.199	1.275	1.356
−9.000	1.046	1.095	1.146	1.199	1.256
−5.000	1.031	1.062	1.095	1.128	1.163
−3.000	1.015	1.031	1.046	1.062	1.078
0.0	1.000	1.000	1.000	1.000	1.000
3.000	0.985	0.970	0.956	0.942	0.928
15.000	0.928	0.862	0.801	0.744	0.692
27.000	0.875	0.766	0.672	0.591	0.520

TABLE B6 continued

Time, months	a = 0.005	a = 0.010	a = 0.015	a = 0.020	a = 0.025
39.000	0.824	0.682	0.566	0.472	0.394
51.000	0.777	0.608	0.478	0.379	0.301
63.000	0.733	0.543	0.405	0.305	0.232
75.000	0.692	0.485	0.344	0.247	0.179
87.000	0.653	0.434	0.293	0.201	0.140
99.000	0.617	0.389	0.250	0.164	0.110
111.000	0.583	0.349	0.214	0.135	0.086

TABLE B7 Hyperbolic Decline-Curve Functions (q/q_i) for n = 0.2*

Time, months	a = 0.005	a = 0.010	a = 0.015	a = 0.020	a = 0.025
−120.000	1.895	3.944	9.313	26.302	97.656
−117.000	1.863	3.792	8.685	23.466	81.238
−114.000	1.832	3.647	8.107	20.990	68.023
−111.000	1.801	3.508	7.575	18.820	57.306
−108.000	1.771	3.376	7.084	16.914	48.552
−105.000	1.741	3.250	6.631	15.236	41.355
−102.000	1.712	3.129	6.212	13.753	35.401
−99.000	1.684	3.014	5.824	12.440	30.447
−96.000	1.656	2.904	5.465	11.274	26.302
−93.000	1.629	2.798	5.132	10.238	22.815
−90.000	1.602	2.697	4.824	9.313	19.869
−87.000	1.576	2.601	4.537	8.487	17.368
−84.000	1.551	2.508	4.271	7.747	15.236
−81.000	1.526	2.420	4.023	7.084	13.410
−78.000	1.501	2.335	3.792	6.487	11.840
−75.000	1.477	2.254	3.577	5.950	10.486
−72.000	1.453	2.176	3.376	5.465	9.313
−69.000	1.430	2.101	3.189	5.027	8.295
−66.000	1.407	2.030	3.014	4.630	7.407
−63.000	1.385	1.961	2.850	4.271	6.631
−60.000	1.363	1.895	2.697	3.944	5.950
−57.000	1.341	1.832	2.554	3.647	5.351
−54.000	1.320	1.771	2.420	3.376	4.824
−51.000	1.299	1.712	2.294	3.129	4.357
−48.000	1.279	1.656	2.176	2.904	3.944
−45.000	1.259	1.602	2.065	2.697	3.577

Decline rates listed *a* are at time = 0 and in months^{-1}.

TABLE B7 *continued*

Time, months	*a* = 0.005	*a* = 0.010	*a* = 0.015	*a* = 0.020	*a* = 0.025
−42.000	1.239	1.551	1.961	2.508	3.250
−39.000	1.220	1.501	1.863	2.335	2.958
−36.000	1.201	1.453	1.771	2.176	2.697
−33.000	1.183	1.407	1.684	2.030	2.464
−30.000	1.165	1.363	1.602	1.895	2.254
−27.000	1.147	1.320	1.526	1.771	2.065
−24.000	1.129	1.279	1.453	1.656	1.895
−21.000	1.112	1.239	1.385	1.551	1.741
−18.000	1.095	1.201	1.320	1.453	1.602
−15.000	1.078	1.165	1.259	1.363	1.477
−12.000	1.062	1.129	1.201	1.279	1.363
−9.000	1.046	1.095	1.147	1.201	1.259
−6.000	1.031	1.062	1.095	1.129	1.165
−3.000	1.015	1.031	1.046	1.062	1.078
0.0	1.000	1.000	1.000	1.000	1.000
3.000	0.985	0.971	0.956	0.942	0.928
15.000	0.928	0.863	0.802	0.747	0.697
27.000	0.875	0.769	0.677	0.599	0.531
39.000	0.826	0.687	0.575	0.484	0.410
51.000	0.780	0.615	0.491	0.395	0.321
63.000	0.737	0.552	0.421	0.325	0.254
75.000	0.697	0.497	0.363	0.269	0.203
87.000	0.659	0.448	0.314	0.225	0.164
99.000	0.624	0.405	0.272	0.189	0.134
111.000	0.591	0.367	0.238	0.159	0.110

TABLE B8 Hyperbolic Decline-Curve Functions (q/q_i) for n = 0.3*

Time, months	*a* = 0.005	*a* = 0.010	*a* = 0.015	*a* = 0.020	*a* = 0.025
−120.000	1.938	4.427	13.309	69.631	2,154.372
−117.000	1.903	4.225	12.086	56.573	1,095.308
−114.000	1.868	4.036	11.004	46.527	624.354
−111.000	1.835	3.857	10.046	38.682	386.021
−108.000	1.802	3.688	9.193	32.473	253.599
−105.000	1.771	3.529	8.432	27.500	174.635
−102.000	1.739	3.379	7.751	23.472	124.863
−99.000	1.709	3.237	7.139	20.180	92.060

*Decline rates listed *a* are at time = 0 and in months^{-1}.

TABLE B8 *continued*

Time, months	*a* = 0.005	*a* = 0.010	*a* = 0.015	*a* = 0.020	*a* = 0.025
−96.000	1.679	3.103	6.589	17.463	69.631
−93.000	1.650	2.975	6.093	15.203	53.816
−90.000	1.622	2.855	5.644	13.309	42.369
−87.000	1.594	2.741	5.238	11.710	33.896
−84.000	1.567	2.632	4.868	10.353	27.500
−81.000	1.540	2.529	4.532	9.193	22.587
−78.000	1.514	2.432	4.225	8.196	18.757
−75.000	1.489	2.339	3.945	7.336	15.730
−72.000	1.464	2.251	3.688	6.589	13.309
−69.000	1.439	2.166	3.453	5.938	11.350
−66.000	1.416	2.086	3.237	5.369	9.751
−63.000	1.392	2.010	3.038	4.868	8.432
−60.000	1.369	1.938	2.855	4.427	7.336
−57.000	1.347	1.868	2.686	4.036	6.418
−54.000	1.325	1.802	2.529	3.688	5.644
−51.000	1.304	1.739	2.385	3.379	4.987
−48.000	1.283	1.679	2.251	3.103	4.427
−45.000	1.262	1.622	2.126	2.855	3.945
−42.000	1.242	1.567	2.010	2.632	3.529
−39.000	1.223	1.514	1.903	2.432	3.169
−36.000	1.203	1.464	1.802	2.251	2.855
−33.000	1.184	1.416	1.709	2.086	2.580
−30.000	1.166	1.369	1.622	1.938	2.339
−27.000	1.148	1.325	1.540	1.802	2.126
−24.000	1.130	1.283	1.464	1.679	1.938
−21.000	1.113	1.242	1.392	1.567	1.771
−18.000	1.096	1.203	1.325	1.464	1.622
−15.000	1.079	1.166	1.262	1.369	1.489
−12.000	1.062	1.130	1.203	1.283	1.369
−9.000	1.046	1.096	1.148	1.203	1.262
−6.000	1.031	1.062	1.096	1.130	1.166
−3.000	1.015	1.031	1.046	1.062	1.079
0.0	1.000	1.000	1.000	1.000	1.000
3.000	0.985	0.971	0.956	0.942	0.929
15.000	0.929	0.864	0.804	0.750	0.701
27.000	0.876	0.771	0.682	0.606	0.541
39.000	0.827	0.692	0.583	0.496	0.425
51.000	0.782	0.622	0.502	0.411	0.340
63.000	0.740	0.562	0.435	0.343	0.275

TABLE B8 continued

Time, months	a = 0.005	a = 0.010	a = 0.015	a = 0.020	a = 0.025
75.000	0.701	0.508	0.379	0.290	0.226
87.000	0.664	0.462	0.332	0.247	0.187
99.000	0.630	0.420	0.293	0.211	0.157
111.000	0.598	0.384	0.259	0.182	0.133

TABLE B9 Hyperbolic Decline-Curve Functions (q/q_i) for n = 0.4*

Time, months	a = 0.005	a = 0.010	a = 0.015	a = 0.020	a = 0.025
−120.000	1.986	5.129	24.105	3,124.851	
−117.000	1.947	4.844	20.628	965.021	
−114.000	1.910	4.581	17.815	435.296	
−111.000	1.873	4.338	15.511	238.203	
−108.000	1.837	4.113	13.603	146.604	
−105.000	1.803	3.903	12.009	97.655	
−102.000	1.769	3.708	10.664	68.858	
−99.000	1.736	3.527	9.521	50.680	99,985.187
−96.000	1.704	3.358	8.542	38.572	3,124.887
−93.000	1.673	3.200	7.699	30.158	771.339
−90.000	1.642	3.052	6.968	24.105	316.223
−87.000	1.613	2.913	6.330	19.625	164.111
−84.000	1.584	2.783	5.772	16.230	97.655
−81.000	1.556	2.662	5.279	13.603	63.550
−78.000	1.528	2.547	4.844	11.535	44.049
−75.000	1.501	2.439	4.458	9.882	32.000
−72.000	1.475	2.338	4.113	8.542	24.105
−69.000	1.450	2.242	3.804	7.444	18.689
−66.000	1.425	2.152	3.527	6.533	14.835
−63.000	1.400	2.067	3.277	5.772	12.009
−60.000	1.377	1.986	3.052	5.129	9.882
−57.000	1.353	1.910	2.847	4.581	8.248
−54.000	1.331	1.837	2.662	4.113	6.968
−51.000	1.309	1.769	2.492	3.708	5.950
−48.000	1.287	1.704	2.338	3.358	5.129
−45.000	1.266	1.642	2.196	3.052	4.458
−42.000	1.245	1.584	2.067	2.783	3.903
−39.000	1.225	1.528	1.947	2.547	3.441
−36.000	1.205	1.475	1.837	2.338	3.052

*Decline rates listed *a* are at time = 0 and in months^{-1}.

TABLE B9 *continued*

Time, months	*a* = 0.005	*a* = 0.010	*a* = 0.015	*a* = 0.020	*a* = 0.025
−33.000	1.186	1.425	1.736	2.152	2.722
−30.000	1.167	1.377	1.642	1.986	2.439
−27.000	1.149	1.331	1.556	1.837	2.196
−24.000	1.131	1.287	1.475	1.704	1.986
−21.000	1.113	1.245	1.400	1.584	1.803
−18.000	1.096	1.205	1.331	1.475	1.642
−15.000	1.079	1.167	1.266	1.377	1.501
−12.000	1.063	1.131	1.205	1.287	1.377
−9.000	1.046	1.096	1.149	1.205	1.266
−6.000	1.031	1.063	1.096	1.131	1.167
−3.000	1.015	1.031	1.046	1.063	1.079
0.0	1.000	1.000	1.000	1.000	1.000
3.000	0.985	0.971	0.956	0.942	0.929
15.000	0.929	0.864	0.806	0.753	0.705
27.000	0.877	0.774	0.687	0.613	0.550
39.000	0.829	0.696	0.591	0.507	0.439
51.000	0.784	0.629	0.513	0.425	0.357
63.000	0.743	0.570	0.449	0.360	0.295
75.000	0.705	0.519	0.395	0.309	0.247
87.000	0.670	0.474	0.350	0.267	0.209
99.000	0.637	0.434	0.312	0.233	0.179
111.000	0.606	0.399	0.279	0.204	0.155

TABLE B10 Hyperbolic Decline-Curve Functions (q/q_i) for n = 0.5*

Time, months	*a* = 0.005	*a* = 0.010	*a* = 0.015	*a* = 0.020	*a* = 0.025
−120.000	2.041	6.250	99.998		
−117.000	1.998	5.806	66.638		
−114.000	1.956	5.408	47.562		
−111.000	1.916	5.050	35.642		
−108.000	1.877	4.726	27.701		
−105.000	1.839	4.432	22.145		
−102.000	1.802	4.165	18.108		
−99.000	1.766	3.921	15.081	9,998.816	
−96.000	1.731	3.698	12.755	624.982	
−93.000	1.698	3.494	10.928	204.078	
−90.000	1.665	3.306	9.467	99.999	
−87.000	1.633	3.133	8.281	59.171	

*Decline rates listed *a* are at time = 0 and in months^{-1}.

TABLE B10 continued

Time, months	*a* = 0.005	*a* = 0.010	*a* = 0.015	*a* = 0.020	*a* = 0.025
−84.000	1.602	2.973	7.305	39.062	
−81.000	1.572	2.825	6.491	27.701	
−78.000	1.543	2.687	5.806	20.661	1,599.910
−75.000	1.515	2.560	5.224	16.000	255.995
−72.000	1.487	2.441	4.726	12.755	99.999
−69.000	1.460	2.331	4.295	10.406	52.892
−66.000	1.434	2.228	3.921	8.650	32.653
−63.000	1.409	2.131	3.594	7.305	22.145
−60.000	1.384	2.041	3.306	6.250	16.000
−57.000	1.360	1.956	3.051	5.408	12.098
−54.000	1.336	1.877	2.825	4.726	9.467
−51.000	1.314	1.802	2.623	4.165	7.610
−48.000	1.291	1.731	2.441	3.698	6.250
−45.000	1.270	1.665	2.278	3.306	5.224
−42.000	1.248	1.602	2.131	2.973	4.432
−39.000	1.228	1.543	1.998	2.687	3.807
−36.000	1.208	1.487	1.877	2.441	3.306
−33.000	1.188	1.434	1.766	2.228	2.897
−30.000	1.169	1.384	1.665	2.041	2.560
−27.000	1.150	1.336	1.572	1.877	2.278
−24.000	1.132	1.291	1.487	1.731	2.041
−21.000	1.114	1.248	1.409	1.602	1.839
−18.000	1.096	1.208	1.336	1.487	1.665
−15.000	1.079	1.169	1.270	1.384	1.515
−12.000	1.063	1.132	1.208	1.291	1.384
−9.000	1.047	1.096	1.150	1.208	1.270
−6.000	1.031	1.063	1.096	1.132	1.169
−3.000	1.015	1.031	1.047	1.063	1.079
0.0	1.000	1.000	1.000	1.000	1.000
3.000	0.985	0.971	0.956	0.943	0.929
15.000	0.929	0.865	0.808	0.756	0.709
27.000	0.878	0.776	0.692	0.620	0.559
39.000	0.830	0.700	0.599	0.518	0.452
51.000	0.787	0.635	0.523	0.439	0.373
63.000	0.746	0.578	0.461	0.376	0.313
75.000	0.709	0.529	0.410	0.327	0.266
87.000	0.675	0.486	0.366	0.286	0.229
99.000	0.643	0.447	0.329	0.253	0.200
111.000	0.613	0.414	0.298	0.225	0.175

TABLE B11 Hyperbolic Decline-Curve Functions (q/q_i) for n = 0.6*

Time, months	*a* = 0.005	*a* = 0.010	*a* = 0.015	*a* = 0.020	*a* = 0.025
−120.000	2.104	8.345			
−117.000	2.056	7.521			
−114.000	2.009	6.821			
−111.000	1.964	6.219	99,853.437		
−108.000	1.921	5.698	387.299		
−105.000	1.879	5.244	125.714		
−102.000	1.838	4.845	64.610		
−99.000	1.799	4.492	40.205		
−96.000	1.761	4.179	27.803		
−93.000	1.725	3.899	20.560		
−90.000	1.690	3.648	15.925		
−87.000	1.655	3.422	12.761		
−84.000	1.622	3.218	10.496		
−81.000	1.590	3.032	8.811	387.302	
−78.000	1.559	2.863	7.521	97.654	
−75.000	1.529	2.706	6.509	46.415	
−72.000	1.500	2.567	5.698	27.803	
−69.000	1.472	2.437	5.038	18.798	
−66.000	1.444	2.317	4.492	13.695	2,154.183
−63.000	1.418	2.206	4.035	10.496	125.715
−60.000	1.392	2.104	3.648	8.345	46.415
−57.000	1.367	2.009	3.317	6.821	24.987
−54.000	1.343	1.921	3.032	5.699	15.925
−51.000	1.319	1.838	2.784	4.845	11.174
−48.000	1.296	1.761	2.567	4.179	8.345
−45.000	1.273	1.690	2.376	3.648	6.509
−42.000	1.252	1.622	2.206	3.218	5.244
−39.000	1.230	1.559	2.056	2.863	4.331
−36.000	1.210	1.500	1.921	2.567	3.648
−33.000	1.190	1.444	1.799	2.317	3.123
−30.000	1.170	1.392	1.690	2.104	2.708
−27.000	1.151	1.343	1.590	1.921	2.376
−24.000	1.133	1.296	1.500	1.761	2.104
−21.000	1.115	1.252	1.418	1.622	1.879
−18.000	1.097	1.210	1.343	1.500	1.690
−15.000	1.080	1.170	1.273	1.392	1.529
−12.000	1.063	1.133	1.210	1.296	1.392

*Decline rates listed *a* are at time = 0 and in months^{-1}.

TABLE B11 *continued*

Time, months	*a* = 0.005	*a* = 0.010	*a* = 0.015	*a* = 0.020	*a* = 0.025
−9.000	1.047	1.097	1.151	1.210	1.273
−6.000	1.031	1.063	1.097	1.133	1.170
−3.000	1.015	1.031	1.047	1.063	1.080
0.0	1.000	1.000	1.000	1.000	1.000
3.000	0.985	0.971	0.957	0.943	0.929
15.000	0.929	0.866	0.810	0.759	0.713
27.000	0.878	0.779	0.696	0.626	0.567
39.000	0.832	0.704	0.606	0.527	0.464
51.000	0.789	0.641	0.533	0.451	0.388
63.000	0.749	0.586	0.473	0.391	0.330
75.000	0.713	0.538	0.423	0.343	0.285
87.000	0.679	0.497	0.381	0.304	0.249
99.000	0.648	0.460	0.346	0.271	0.219
111.000	0.619	0.427	0.315	0.244	0.195

TABLE B12 Hyperbolic Decline-Curve Functions (q/q_i) for n = 0.7*

Time, months	*a* = 0.005	*a* = 0.010	*a* = 0.015	*a* = 0.020	*a* = 0.025
−120.000	2.178	13.708			
−117.000	2.122	11.494			
−114.000	2.070	9.825			
−111.000	2.019	8.531			
−108.000	1.971	7.502			
−105.000	1.924	6.667			
−102.000	1.879	5.979			
−99.000	1.836	5.403			
−96.000	1.795	4.916			
−93.000	1.755	4.499	212.337		
−90.000	1.717	4.139	63.019		
−87.000	1.680	3.825	33.002		
−84.000	1.644	3.549	21.178		
−81.000	1.610	3.306	15.104		
−78.000	1.577	3.090	11.494		
−75.000	1.545	2.896	9.139		
−72.000	1.514	2.723	7.502		
−69.000	1.484	2.566	6.307	125.278	
−66.000	1.455	2.424	5.403	39.704	

*Decline rates listed *a* are at time = 0 and in months^{-1}.

TABLE B12 continued

Time, months	a = 0.005	a = 0.010	a = 0.015	a = 0.020	a = 0.025
−63.000	1.427	2.295	4.700	21.178	
−60.000	1.400	2.178	4.139	13.708	
−57.000	1.374	2.070	3.683	9.825	5,212.395
−54.000	1.349	1.971	3.306	7.502	63.019
−51.000	1.324	1.879	2.990	5.979	24.193
−48.000	1.300	1.795	2.723	4.916	13.708
−45.000	1.277	1.717	2.494	4.139	9.139
−42.000	1.255	1.644	2.295	3.549	6.667
−39.000	1.233	1.577	2.122	3.090	5.150
−36.000	1.212	1.514	1.971	2.723	4.139
−33.000	1.192	1.455	1.836	2.424	3.424
−30.000	1.172	1.400	1.717	2.178	2.896
−27.000	1.152	1.349	1.610	1.971	2.494
−24.000	1.134	1.300	1.514	1.795	2.178
−21.000	1.115	1.255	1.427	1.644	1.924
−18.000	1.097	1.212	1.349	1.514	1.717
−15.000	1.080	1.172	1.277	1.400	1.545
−12.000	1.063	1.134	1.212	1.300	1.400
−9.000	1.047	1.097	1.152	1.212	1.277
−6.000	1.031	1.063	1.097	1.134	1.172
−3.000	1.015	1.031	1.047	1.063	1.080
0.0	1.000	1.000	1.000	1.000	1.000
3.000	0.985	0.971	0.957	0.943	0.930
15.000	0.930	0.867	0.811	0.762	0.717
27.000	0.879	0.781	0.700	0.633	0.575
39.000	0.833	0.708	0.612	0.537	0.476
51.000	0.791	0.647	0.542	0.463	0.402
63.000	0.752	0.593	0.484	0.405	0.346
75.000	0.717	0.547	0.436	0.359	0.302
87.000	0.684	0.507	0.396	0.320	0.267
99.000	0.654	0.471	0.361	0.289	0.238
111.000	0.626	0.440			

TABLE B13 Hyperbolic Decline-Curve Functions (q/q_i) for n = 0.8*

Time, months	a = 0.005	a = 0.010	a = 0.015	a = 0.020	a = 0.025
−120.000	2.265	55.900			
−117.000	2.201	31.065			
−114.000	2.140	20.864			

*Decline rates listed *a* are at time = 0 and in months^{-1}.

TABLE B13 *continued*

Time, months	*a* = 0.005	*a* = 0.010	*a* = 0.015	*a* = 0.020	*a* = 0.025
−111.000	2.083	15.434			
−108.000	2.028	12.108			
−105.000	1.976	9.882			
−102.000	1.926	8.298			
−99.000	1.878	7.119			
−96.000	1.832	6.211			
−93.000	1.789	5.492			
−90.000	1.747	4.910			
−87.000	1.707	4.430			
−84.000	1.668	4.029			
−81.000	1.631	3.688	87.305		
−78.000	1.596	3.396	31.065		
−75.000	1.562	3.144	17.783		
−72.000	1.529	2.923	12.108		
−69.000	1.497	2.728	9.028		
−66.000	1.467	2.556	7.119		
−63.000	1.438	2.402	5.831		
−60.000	1.409	2.265	4.910	55.900	
−57.000	1.382	2.140	4.221	20.864	
−54.000	1.356	2.028	3.688	12.108	
−51.000	1.330	1.926	3.266	8.298	
−48.000	1.305	1.832	2.923	6.211	55.901
−45.000	1.282	1.747	2.640	4.910	17.783
−42.000	1.258	1.668	2.402	4.029	9.882
−39.000	1.236	1.596	2.201	3.396	6.637
−36.000	1.215	1.529	2.028	2.923	4.910
−33.000	1.194	1.467	1.878	2.556	3.852
−30.000	1.173	1.409	1.747	2.265	3.144
−27.000	1.154	1.356	1.631	2.028	2.640
−24.000	1.134	1.305	1.529	1.832	2.265
−21.000	1.116	1.258	1.438	1.668	1.976
−18.000	1.098	1.215	1.356	1.529	1.747
−15.000	1.080	1.173	1.282	1.409	1.562
−12.000	1.063	1.134	1.215	1.305	1.409
−9.000	1.047	1.098	1.154	1.215	1.282
−6.000	1.031	1.063	1.098	1.134	1.173
−3.000	1.015	1.051	1.047	1.063	1.080
0.0	1.000	1.000	1.000	1.000	1.000
3.000	0.985	0.971	0.957	0.943	0.930

TABLE B13 continued

Time, months	*a* = 0.005	*a* = 0.010	*a* = 0.015	*a* = 0.020	*a* = 0.025
15.000	0.930	0.868	0.813	0.764	0.720
27.000	0.880	0.783	0.704	0.638	0.583
39.000	0.834	0.712	0.619	0.545	0.486
51.000	0.793	0.652	0.551	0.474	0.415
63.000	0.755	0.600	0.495	0.418	0.361
75.000	0.720	0.556	0.448	0.373	0.318
87.000	0.688	0.517	0.409	0.336	0.284
99.000	0.659	0.482	0.376	0.305	0.255
111.000	0.632	0.452	0.347	0.279	

TABLE B14 Hyperbolic Decline-Curve Functions (q/q_i) for n = 0.9*

Time, months	*a* = 0.005	*a* = 0.010	*a* = 0.015	*a* = 0.020	*a* = 0.025
−120.000	2.370				
−117.000	2.295				
−114.000	2.224				
−111.000	2.158	2,152.321			
−108.000	2.095	53.133			
−105.000	2.035	25.095			
−102.000	1.979	16.102			
−99.000	1.926	11.736			
−96.000	1.875	9.178			
−93.000	1.826	7.505			
−90.000	1.780	6.330			
−87.000	1.737	5.461			
−84.000	1.695	4.794			
−81.000	1.655	4.266			
−78.000	1.617	3.839			
−75.000	1.580	3.486			
−72.000	1.545	3.190	53.133		
−69.000	1.511	2.939	19.664		
−66.000	1.479	2.723	11.736		
−63.000	1.448	2.535	8.262		
−60.000	1.419	2.370	6.330		
−57.000	1.390	2.224	5.107	53.133	
−54.000	1.363	2.095	4.266	16.102	
−51.000	1.336	1.979	3.654		

*Decline rates listed *a* are at time = 0 and in $months^{-1}$.

TABLE B14 ***continued***

Time, months	a = 0.005	a = 0.010	a = 0.015	a = 0.020	a = 0.025
−43.000	1.310	1.875	3.190	9.178	
−45.000	1.286	1.780	2.827	6.330	
−42.000	1.262	1.695	2.535	4.794	25.095
−39.000	1.239	1.617	2.295	3.839	10.308
−36.000	1.217	1.545	2.095	3.190	6.330
−33.000	1.196	1.479	1.926	2.723	4.515
−30.000	1.175	1.419	1.760	2.370	3.486
−27.000	1.155	1.363	1.655	2.095	2.827
−24.000	1.135	1.310	1.545	1.875	2.370
−21.000	1.117	1.262	1.448	1.695	2.035
−18.000	1.098	1.217	1.363	1.545	1.760
−15.000	1.081	1.175	1.286	1.419	1.580
−12.000	1.064	1.135	1.217	1.310	1.419
−9.000	1.047	1.098	1.155	1.217	1.286
−6.000	1.031	1.064	1.098	1.135	1.175
−3.000	1.015	1.031	1.047	1.064	1.081
0.0	1.000	1.000	1.000	1.000	1.000
3.000	0.985	0.971	0.957	0.943	0.930
15.000	0.930	0.869	0.815	0.767	0.724
27.000	0.880	0.785	0.708	0.644	0.590
39.000	0.836	0.716	0.625	0.554	0.497
51.000	0.795	0.657	0.559	0.485	0.428
63.000	0.758	0.607	0.505	0.431	0.375
75.000	0.724	0.564	0.460	0.387	0.333
87.000	0.693	0.526	0.422	0.351	0.300
99.000	0.664	0.493	0.389	0.321	0.272
111.000	0.638	0.463	0.362	0.295	0.249

TABLE B15 Gas Equivalent of Stock-Tank Condensate

(1) Oil Gravity, °API	(2) Oil Specific Gravity	(3) Molecular Weight	(4) Gas Equivalent of Stock-Tank Condensate, scf/stb
45	0.802	156	684
50	0.780	138	752
55	0.759	124	814
60	0.739	113	870
65	0.720	103	930

(2) From $°\text{API} = \dfrac{141.5}{\lambda_L} - 131.5$

(3) From mol wt (MW) $= 44.29\,\lambda_L/(1.03 - \lambda_L)$*

(4) From gas equivalent $= 133{,}000\,\lambda_L/\text{MW}$

*Molecular weight equation developed by C.S. Cragoe, "Thermodynamic Properties of Petroleum Products," Bureau of Standards (1929), Miscellaneous Publication No. 97.

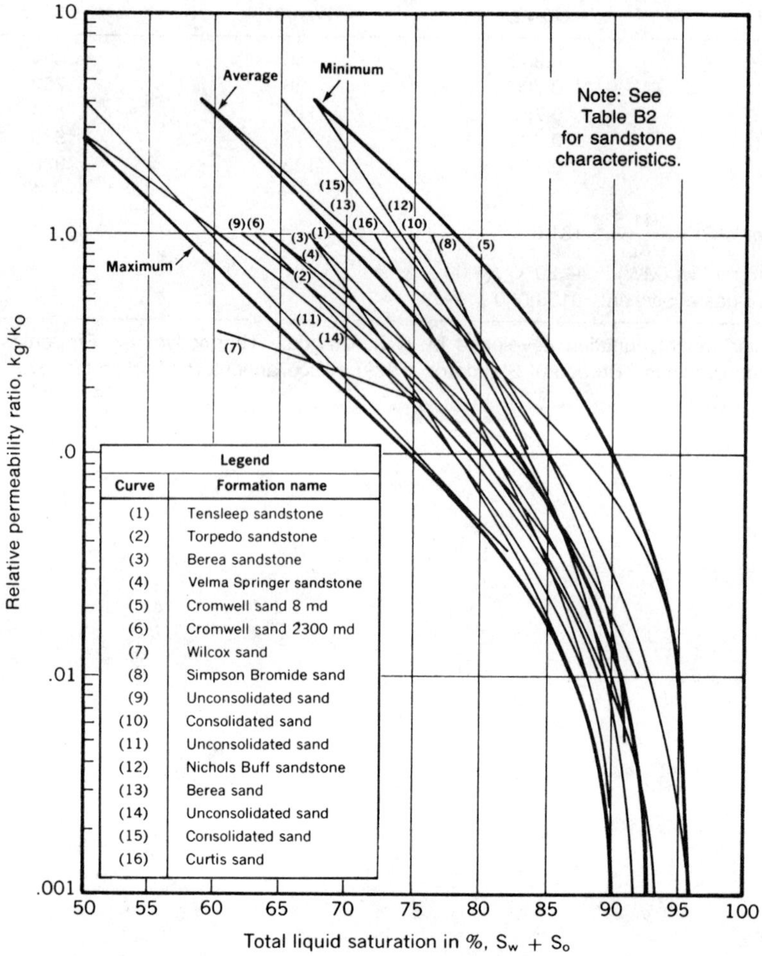

Fig. B1 Relative permeability data for typical reservoirs (after Arps and Roberts, "The Effect of the Relative Permeability Ratio, the Oil Gravity, and the Solution Gas-Oil Ratio on the Primary Recovery from a Depletion Type Reservoir," courtesy *Trans.*, © 1955, SPE-AIME)

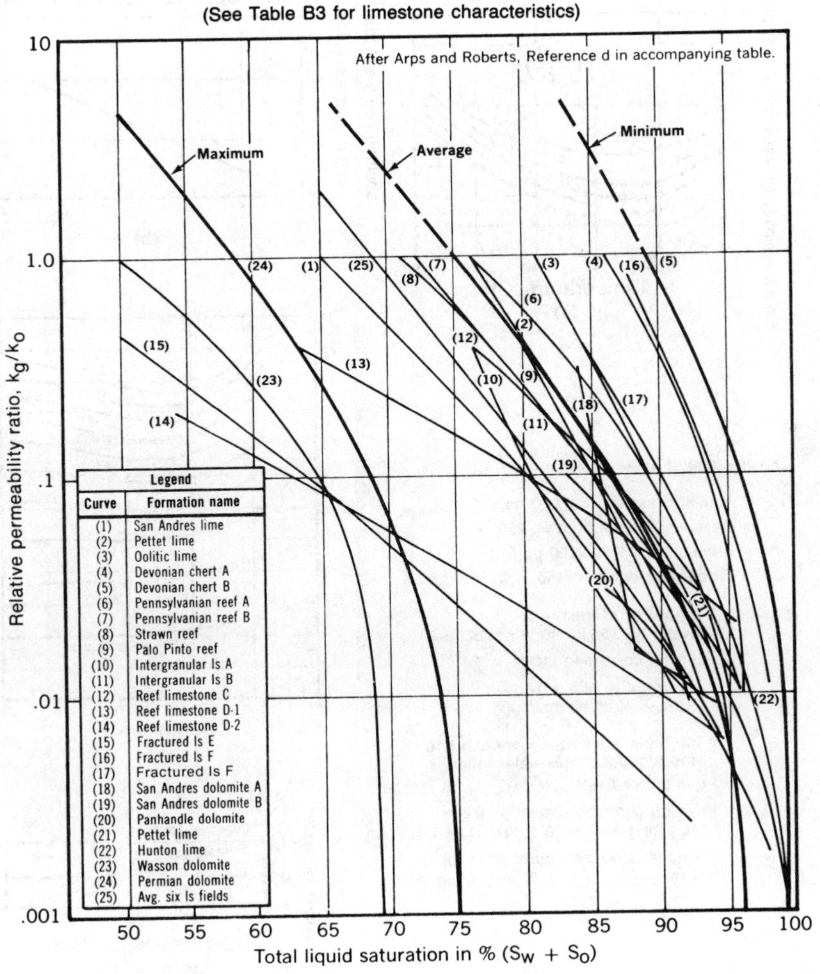

Fig. B2 Relative permeability data for typical reservoirs (after Arps and Roberts, "The Effect of the Relative Permeability Ratio, the Oil Gravity, and the Solution Gas-Oil Ratio on the Primary Recovery from a Depletion Type Reservoir," courtesy *Trans.*, © 1955, SPE-AIME)

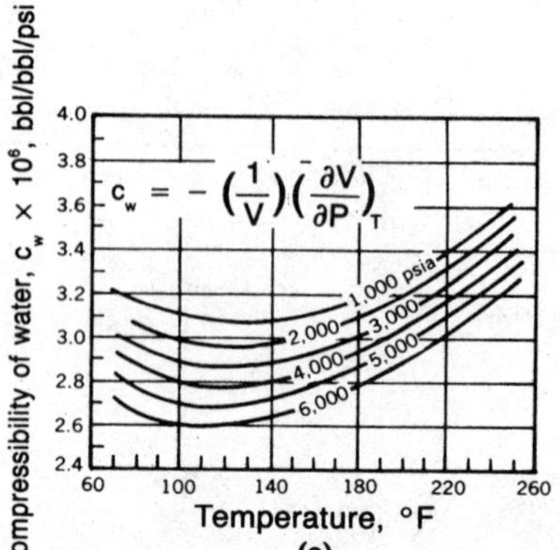

(a)

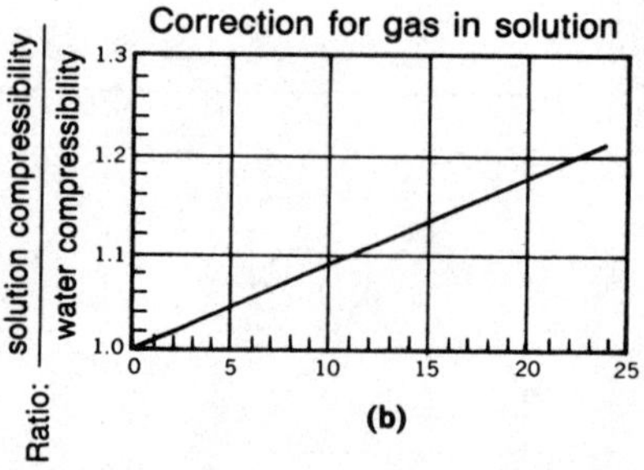

(b)

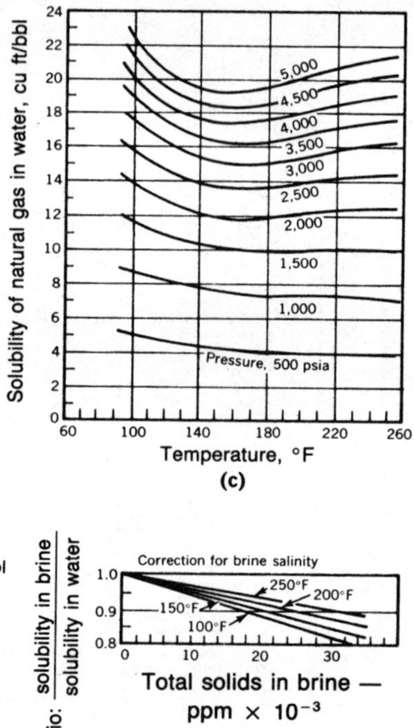

(c)

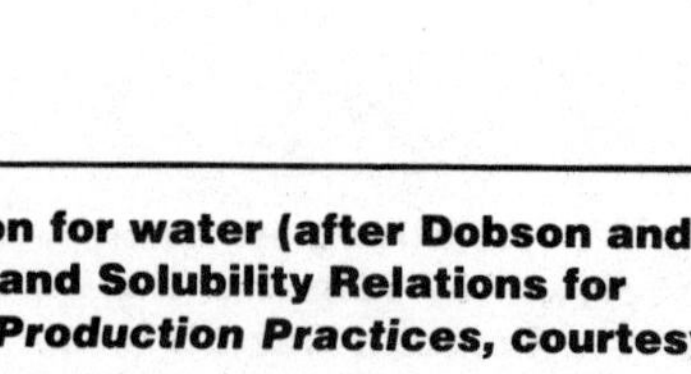

(d)

Example use of data

Given: Aquifer pressure = 4,000 psia
Reservoir temperature = 200 °F
Brine salinity = 25,000 ppm
Produced gas-water ratio = 11 scf/bbl

Analysis: From (a) c_w of fresh water = 3.05×10^{-6} bbl/bbl/psi
From (b) correction factor = 1.1
c_w of brine = 3.05×1.1 = 3.36×10^{-6} bbl/bbl/psi

If the brine is saturated, we can find the production gas-water ratio:
From (c) for fresh water, R_{sw} = 18 scf/bbl
From (d) correction factor = 0.9
R_{sw} for brine = 18×0.9 = 16.2 scf/bbl
Then the correction factor from (b) is 1.15 and $c_w = 1.15 \times 3.05 \times 10^{-6}$ = 3.5×10^{-6}/psi

Fig. B3 Compressibility and gas in solution for water (after Dobson and Standing, "Pressure-Volume-Temperature and Solubility Relations for Natural Gas-Water Mixtures," ***Drilling and Production Practices,*** **courtesy API, 1944)**

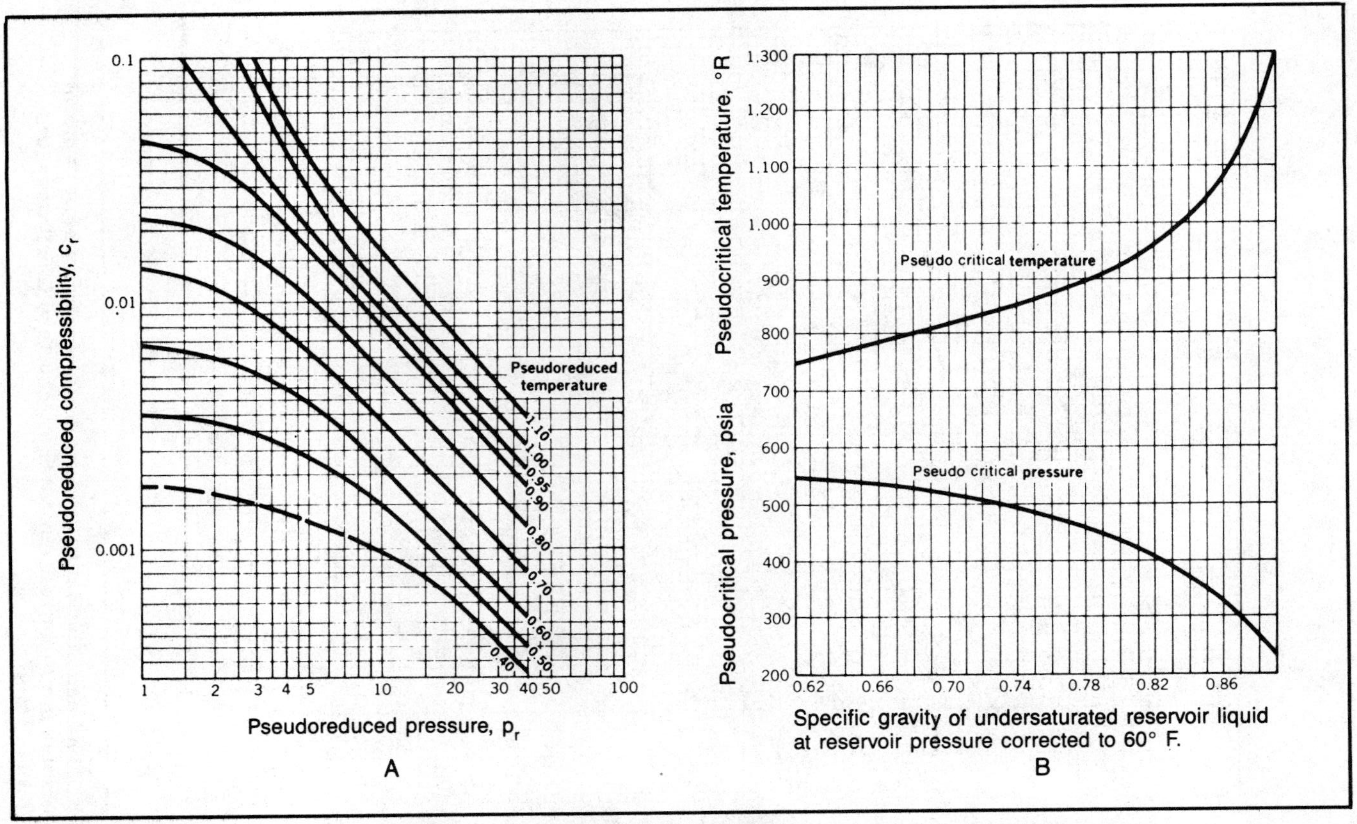

Fig. B4 Empirical determination of oil compressibility (after Trube, "Compressibility of Undersaturated Hydrocarbon Reservoir Fluids," courtesy *Trans.*, © 1957, SPE-AIME)

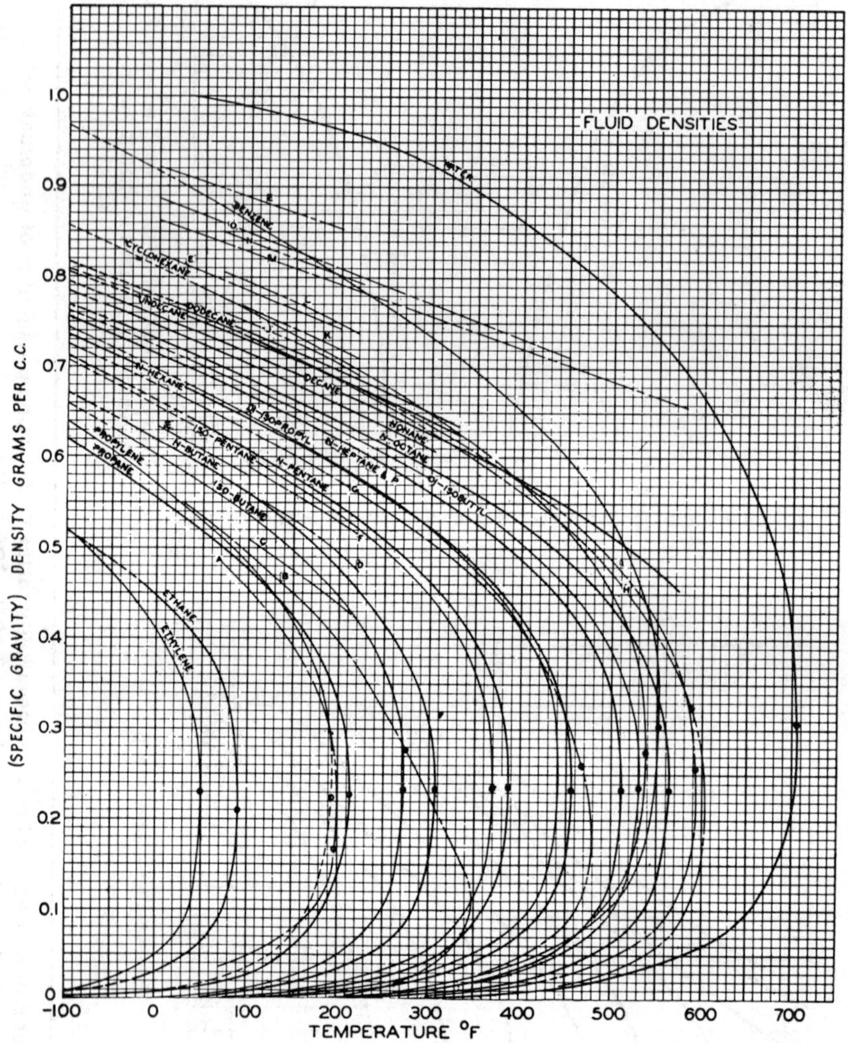

Fig. B5 Fluid densities for hydrocarbons (after Brown et al., *Natural Gasoline and the Volatile Hydrocarbons*, courtesy Gas Processors Association, 1948)

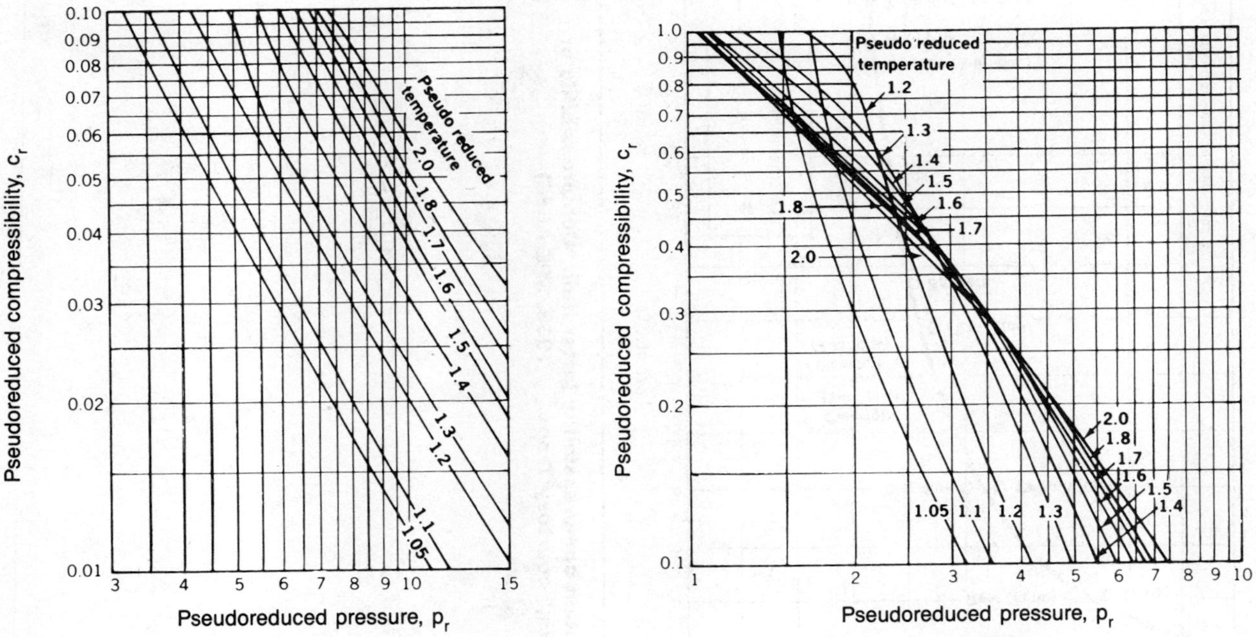

$$c_g = (\Delta V/V)\Delta_p$$

$$c_g = c_r r/p_c$$

Fig. B6 Compressibility of natural gases (after Trube, "Compressibility of Natural Gases," courtesy *JPT*, © 1957, SPE-AIME)

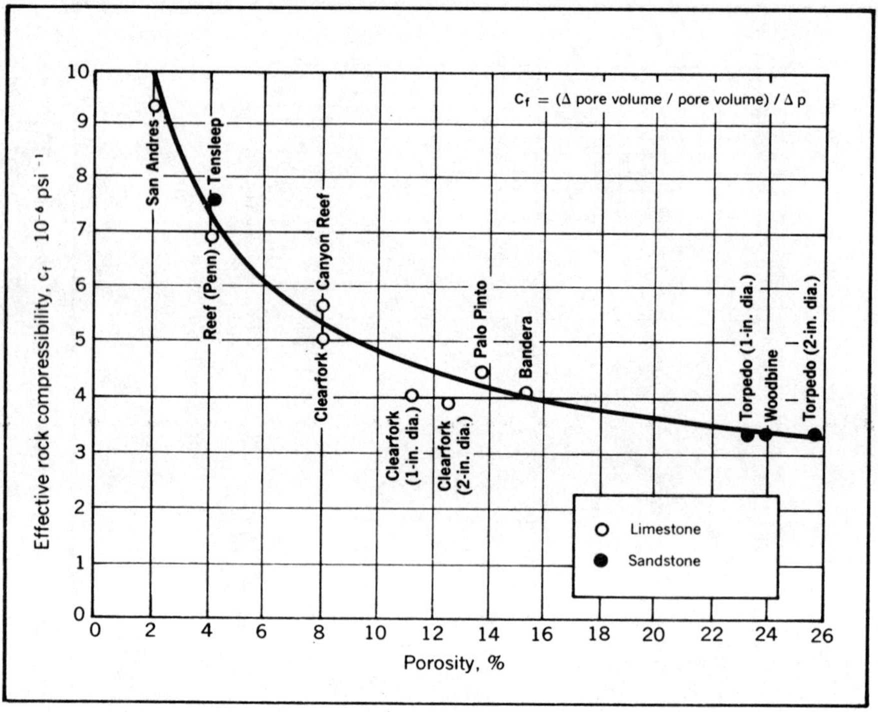

Fig. B7 Formation compressibility (after Hall, "Compressibility of Reservoir Rocks," courtesy *Trans.*, © 1953, SPE-AIME)

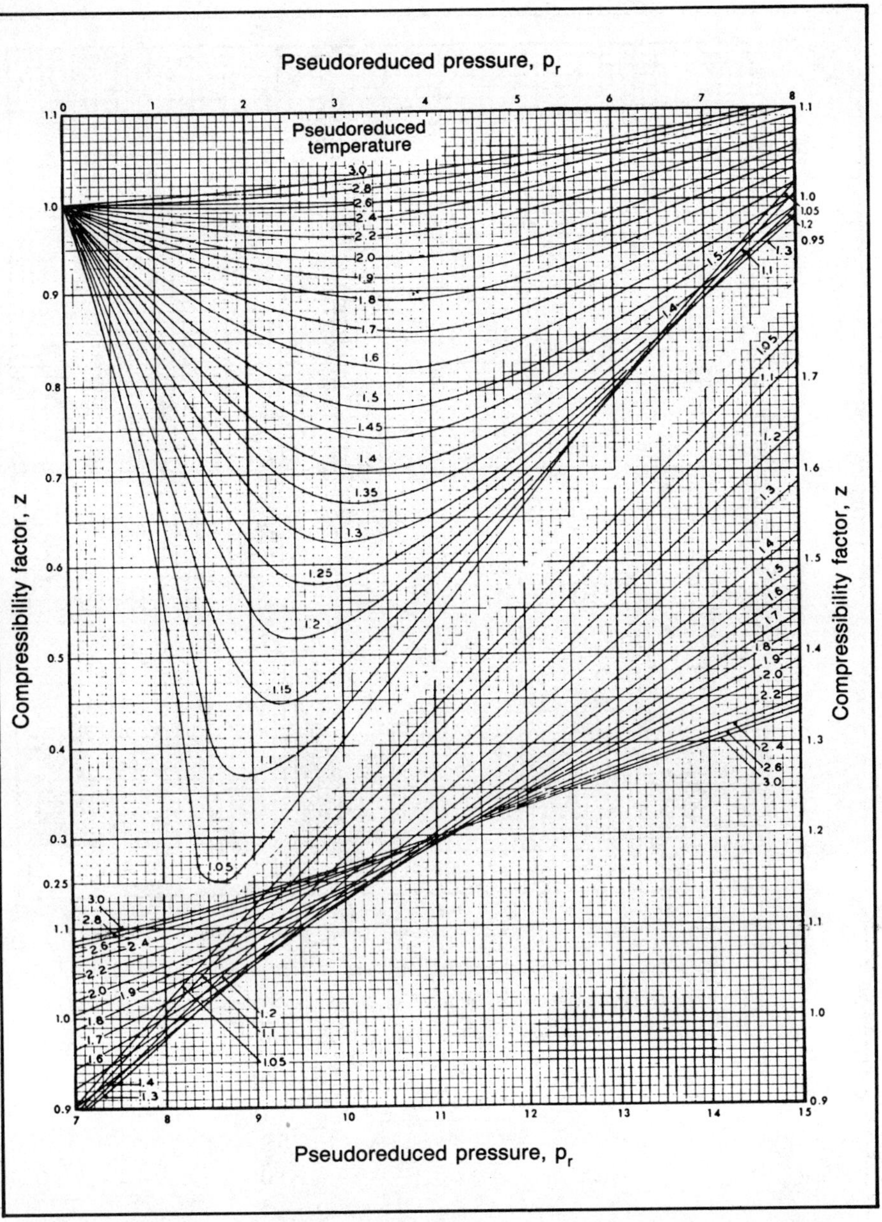

Fig. B8 Gas deviation (compressibility) factors (after Standing and Katz, "Density of Natural Gases," courtesy *Trans.*, © 1942, SPE-AIME). Note: T_c and p_c should be calculated from the gas composition and the critical values in Table B5 or alternatively obtained from the empirical data in

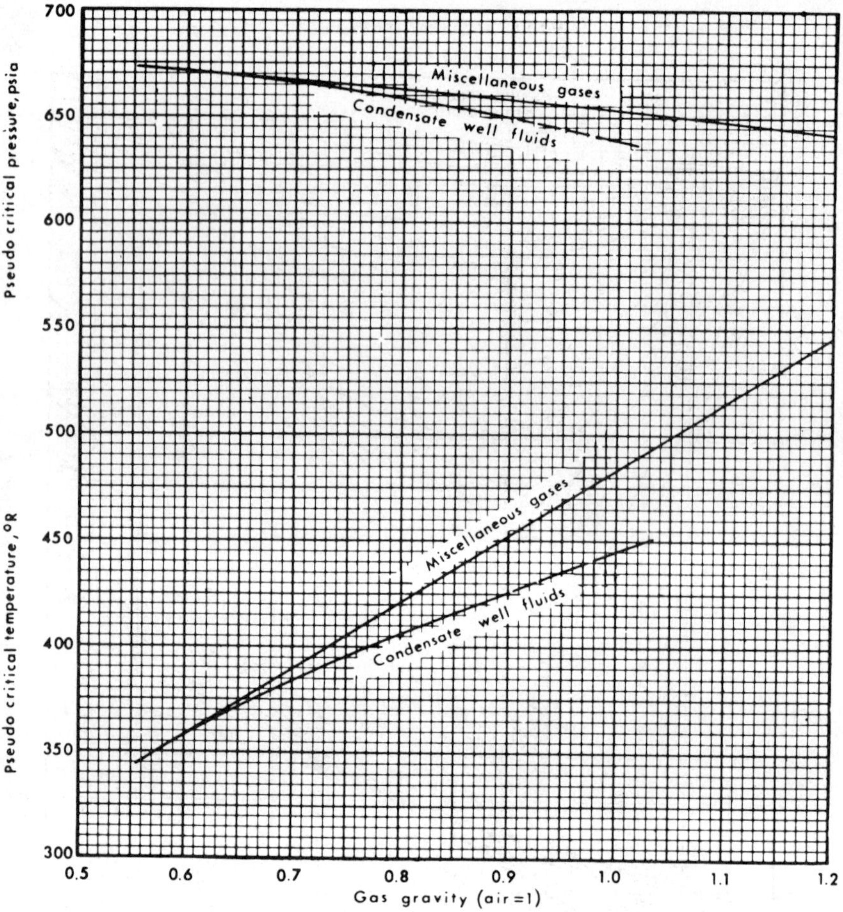

Miscellaneous gases should be used for gases with low-condensate content.
Condensate well fluids should be used for gases with high-condensate content.

Corrections for Nonhydrocarbons*

	$_pP_c$, psi/mol, %	$_pT_c$, °R/mol, %
CO_2	+4.4	−0.8
H_2S	+6.0	+1.3
N_2	−2.7	−2.5

*From Carr, Kobayashi, and Burroughs, "Viscosity of Hydrocarbon Gases under Pressure," *Trans.* AIME, (1954).

Fig. B9 Pseudo critical properties of natural gas (after Brown et al., *Natural Gasoline and the Volatile Hydrocarbons*, courtesy Gas Processors Association 1948)

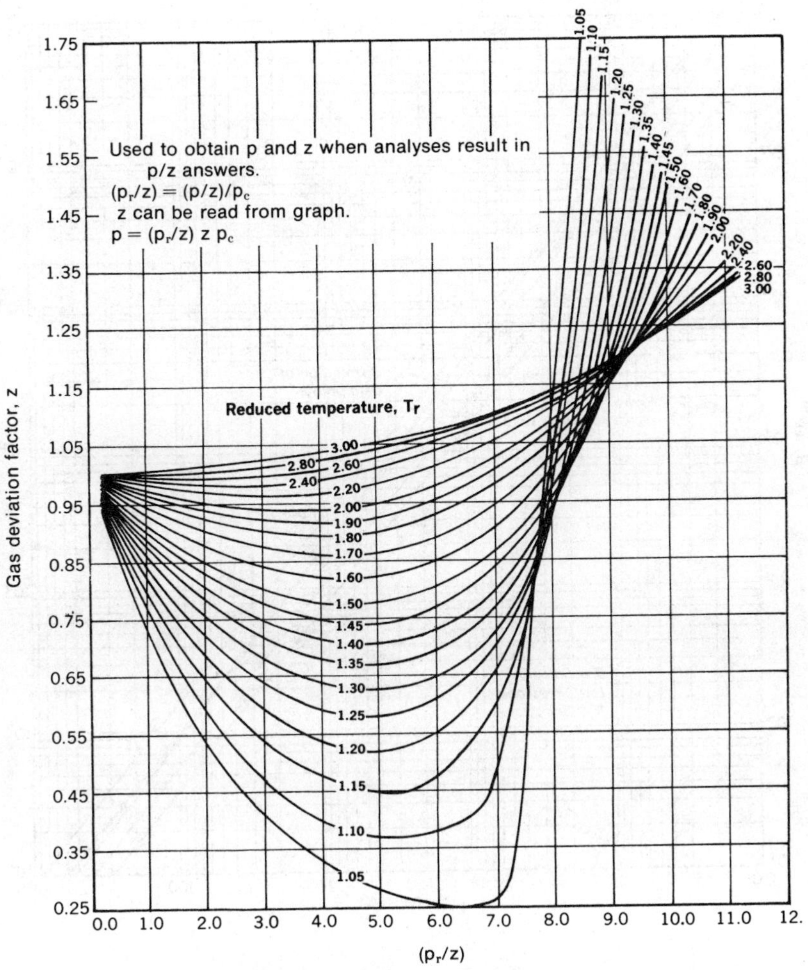

Fig. B10 Gas deviation factor, z versus p_r/z (unpublished data from Slider, *Gas Reservoir Engineering Short Course*)

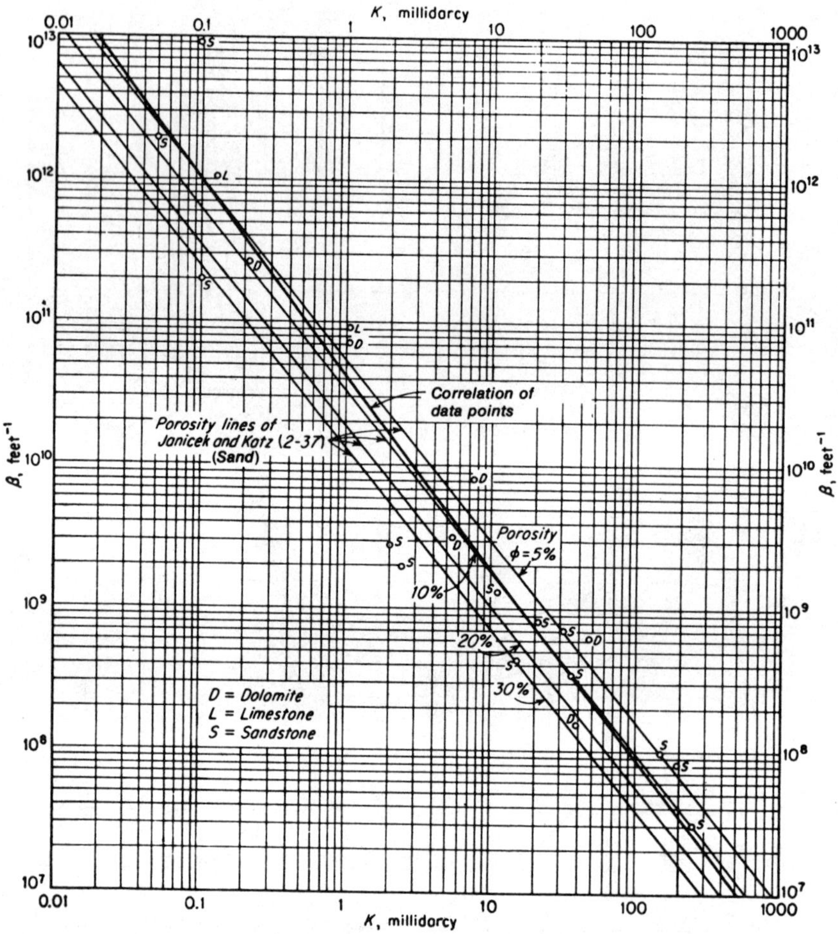

Fig. B11 Gas turbulence factor (after *Handbook of Natural Gas Engineering*, courtesy McGraw-Hill, 1959). Porosity lines of Janicek and Katz are not a correlation of plotted data points (after Janicek and Katz, "Application of Unsteady State Calculations," courtesy University of Michigan, 1955). Correlation of data points (after Katz and Correli, "Flow of Natural Gas from Reservoirs," course notes, University of Michigan, 1955).

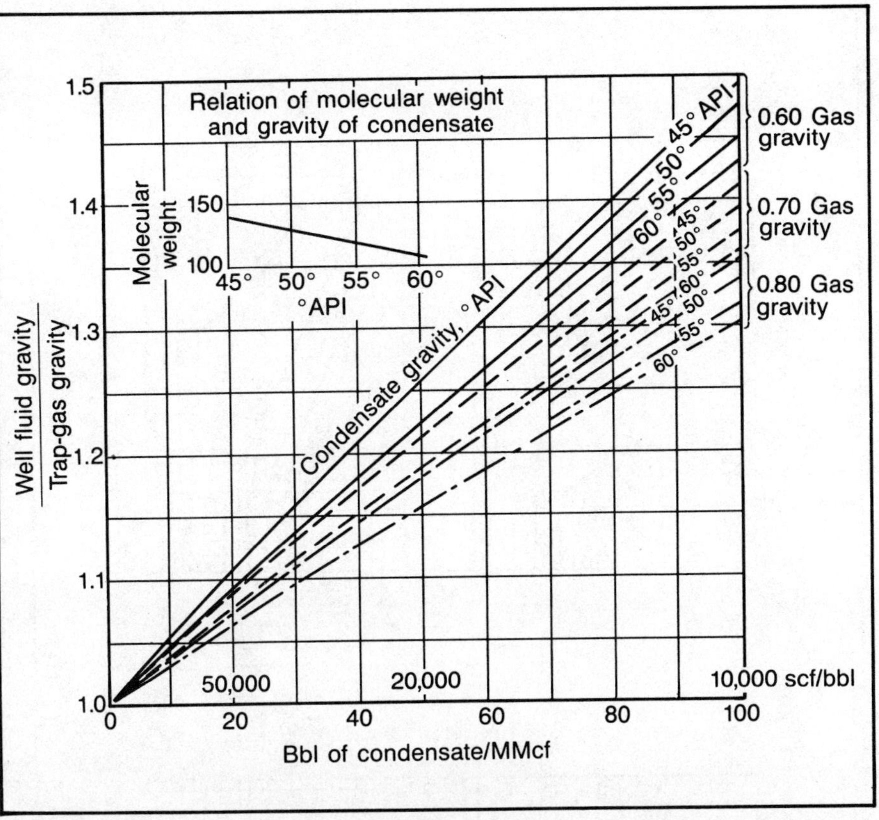

Fig. B12 Effect of condensate volume on the ratio of surface gas gravity to well fluid gravity (after Standing, *Volumetric and Phase Behavior of Oil Field Hydrocarbon Systems*, Reinhold, 1952)

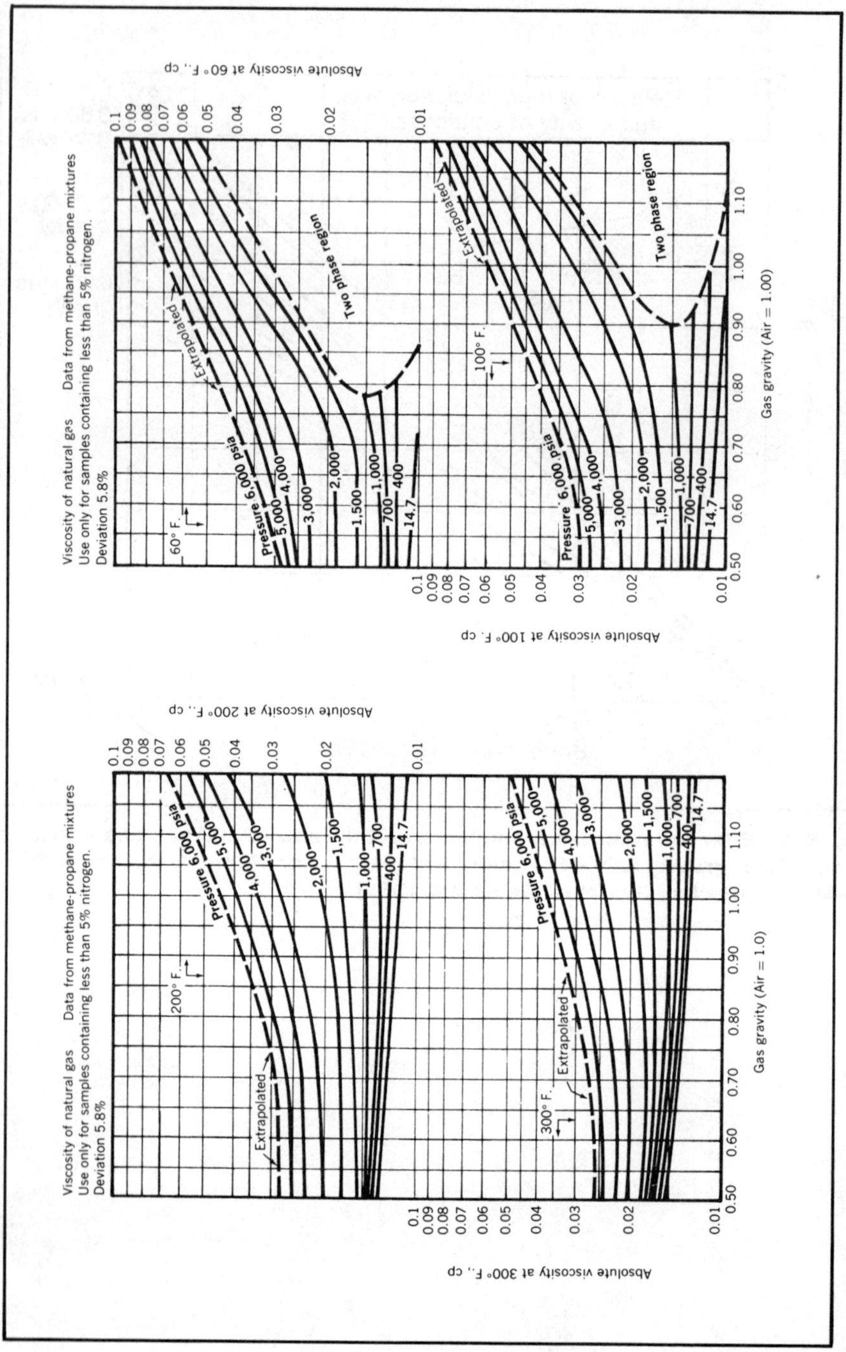

Fig. B13 Viscosity of natural gas (after Bicher and Katz, "Viscosity of Natural Gases," courtesy *Trans.*, © 1944, SPE-AIME)

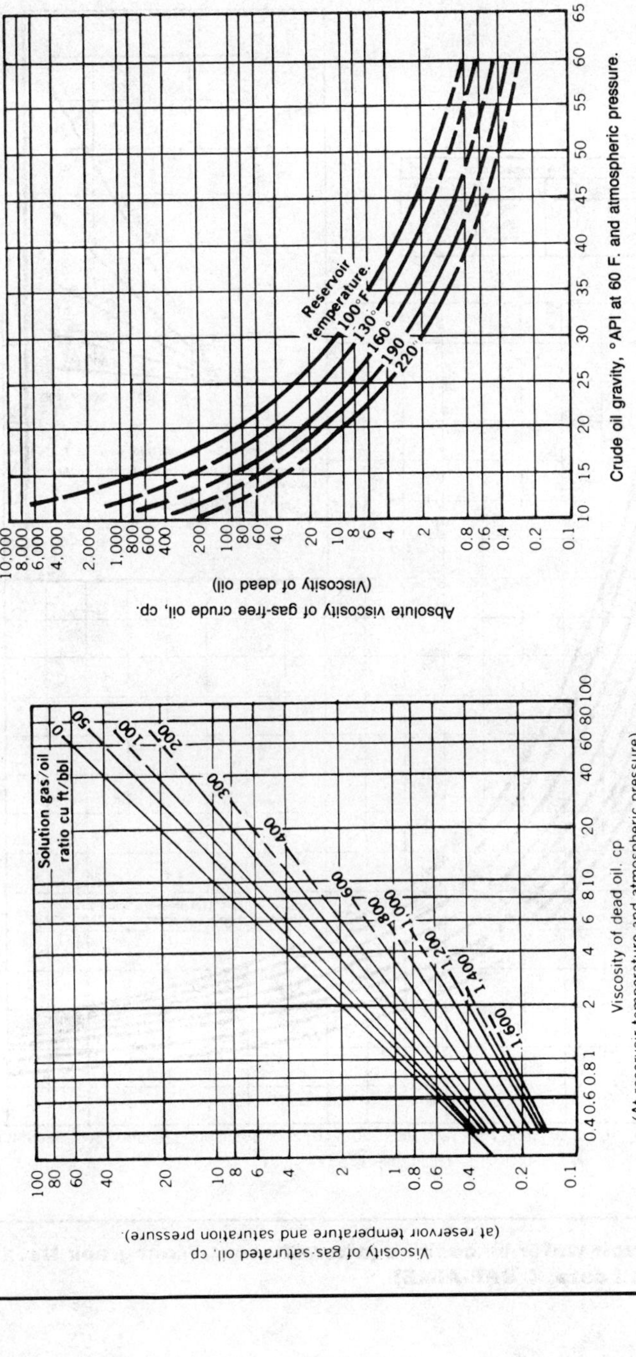

Viscosity of gas-free crude oil at oil-field temperatures. From Beal. Trans., AIME (1946) 165.94.

Viscosity of gas-saturated crude oils. From Chew and Connally. Trans., AIME (1959) 216.23.

Fig. B14 Reservoir oil viscosities

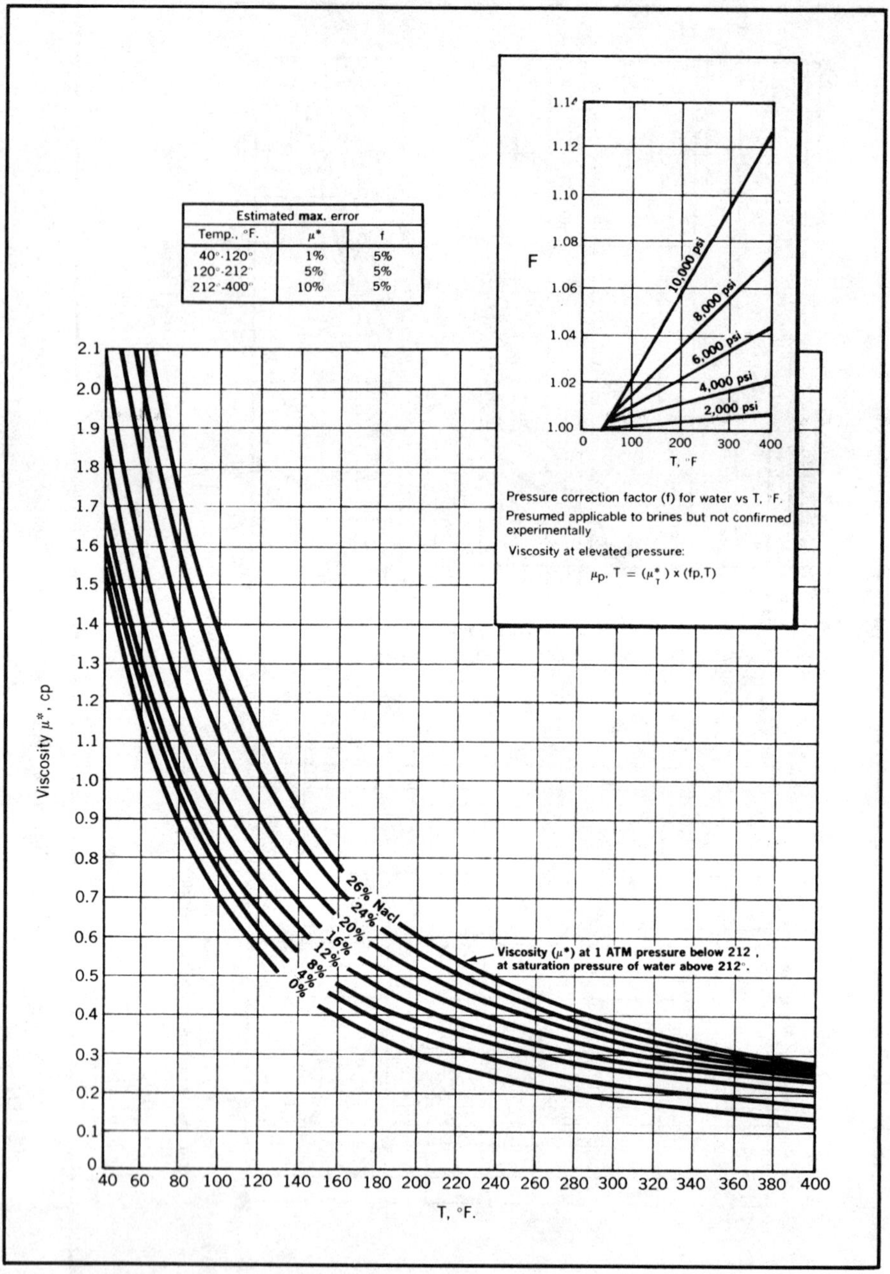

Fig. B15 Reservoir water viscosities (after Chesnut, Monograph No. 1, unpublished Shell data, © SPE-AIME)

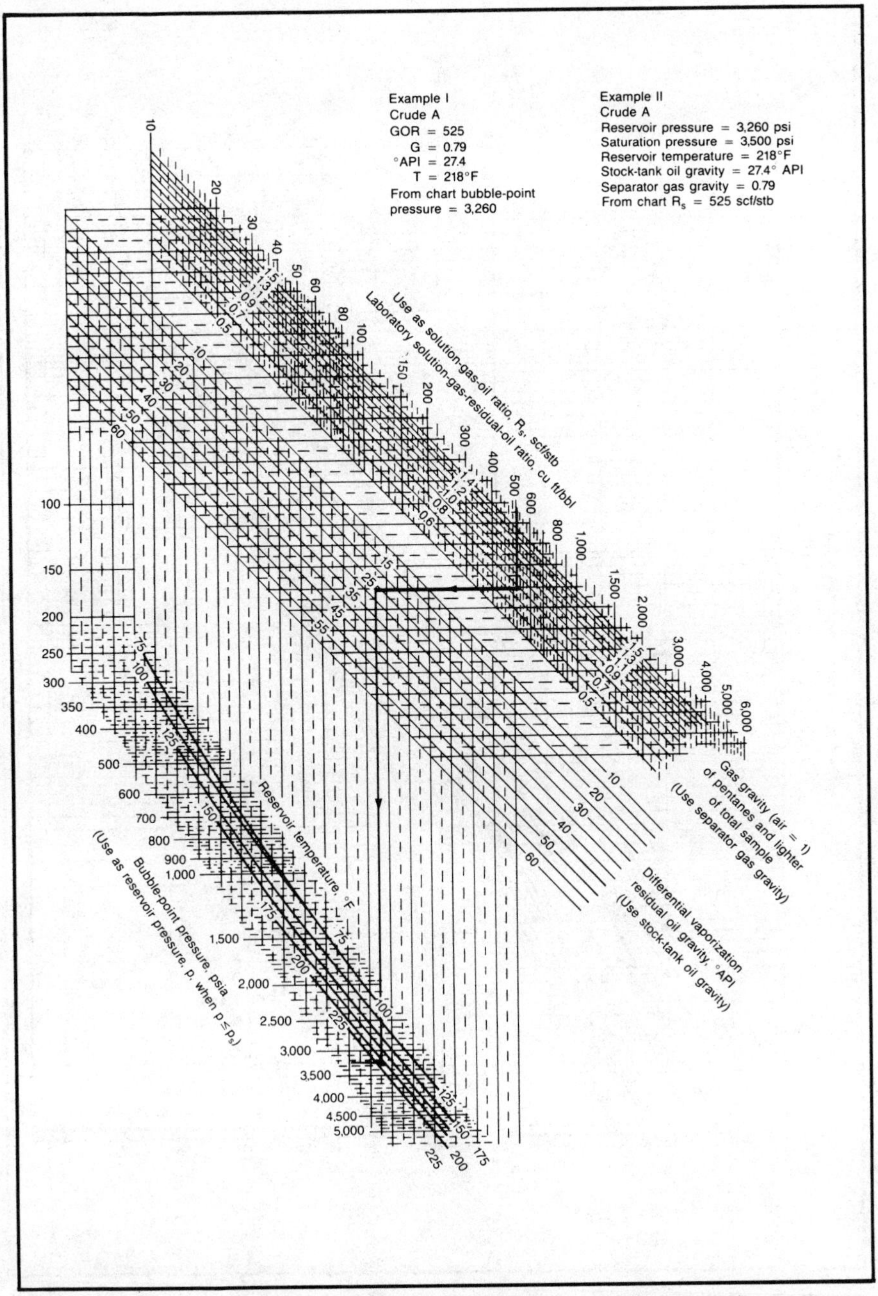

Fig. B16 Gas in solution or bubble-point pressure (after Borden and Rzasa, "Correlation of Bottom Hole Sample Data," courtesy *Trans.*, AIME, 1950). Note: For West Coast crudes use California research data (after Katz et al., *Handbook of Natural Gas Engineering*, McGraw-Hill, 1959).

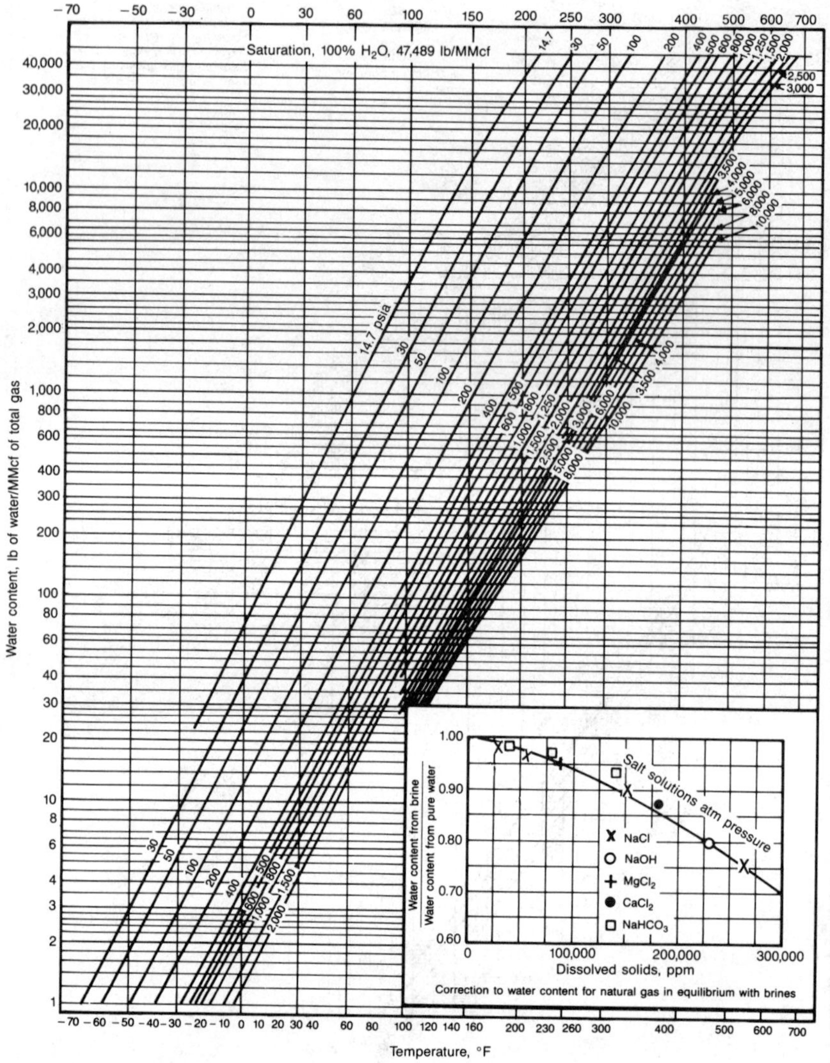

Fig. B17 Water content of natural gas in the reservoir (after Katz et al., *Handbook of Natural Gas Engineering*, courtesy McGraw-Hill, 1959)

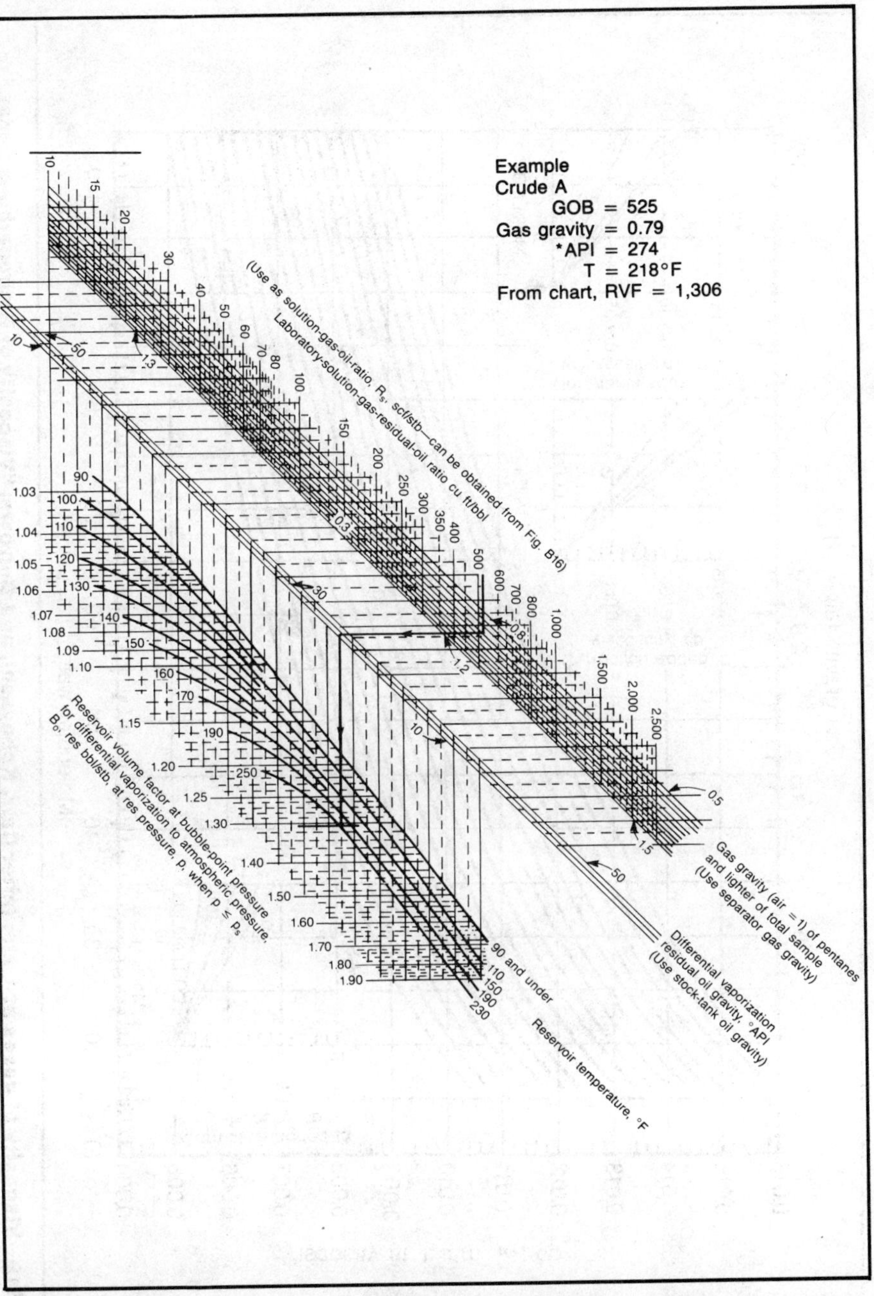

Fig. B18 Oil formation volume factors (after Borden and Rzasa, "Correlation of Bottom Hole Sample Data," courtesy *Trans.*, AIME, 1950). Note: For West Coast crudes use California research data (after Katz et al., *Handbook of Natural Gas Engineering*, McGraw-Hill, 1959).

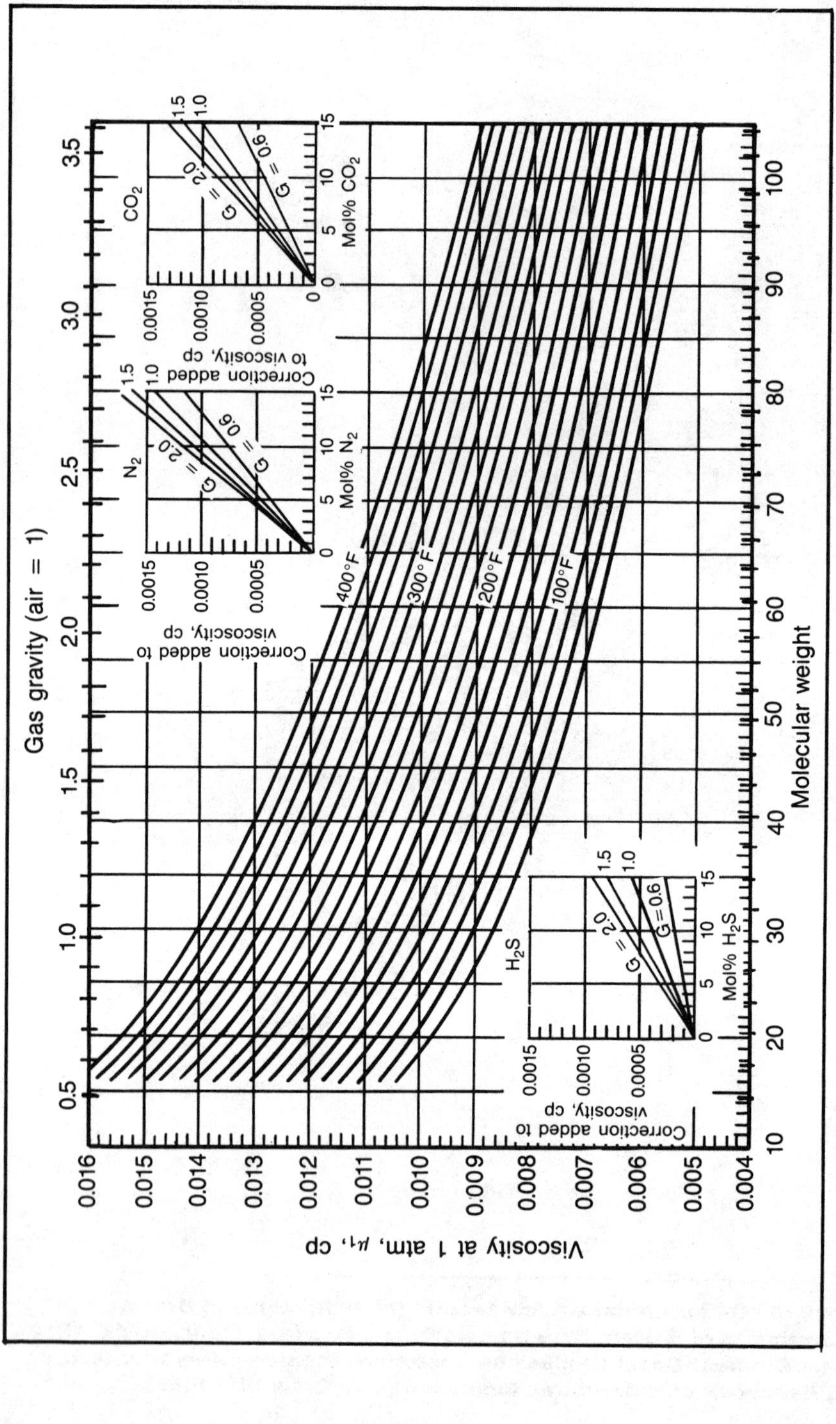

Fig. B19 Viscosity of gases at 1 atm (after Carr, Kobayashi, and Burrows, "Viscosity of Hydrocarbons under Pressure," *Trans.*, AIME, 1954, courtesy McGraw-Hill)

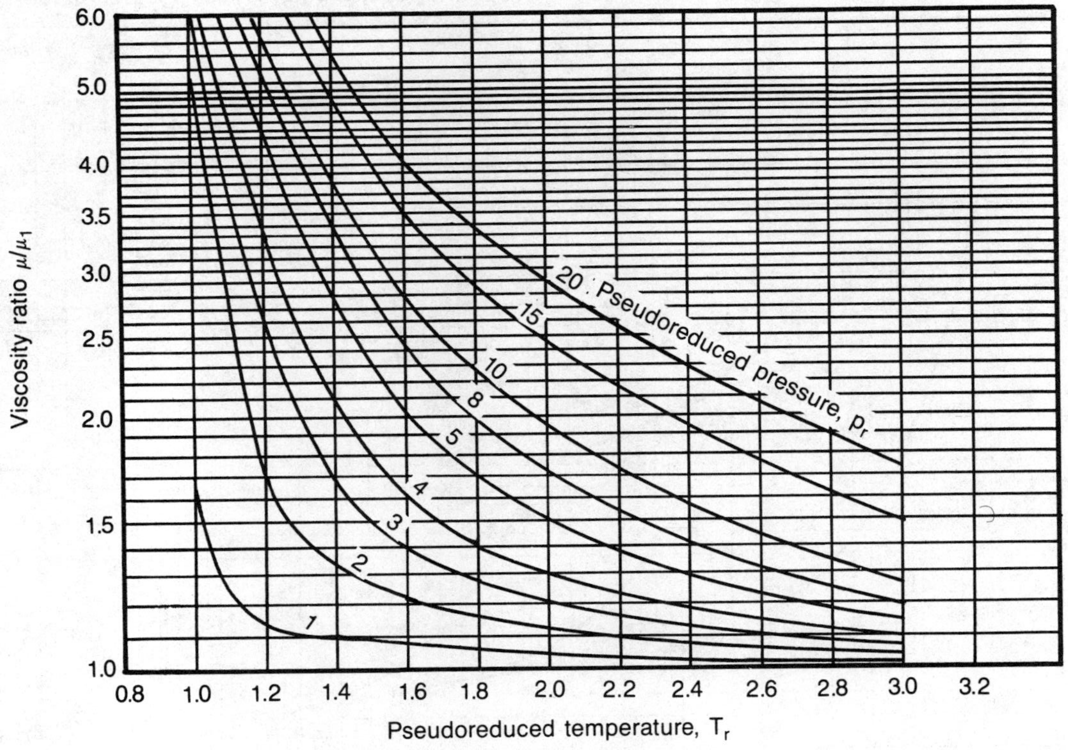

Fig. B20 Correction of viscosity ratio with reduced temperature (after Carr, Kobayashi, and Burrows, "Viscosity of Hydrocarbons under Pressure," *Trans.*, AIME, courtesy McGraw-Hill, 1954)

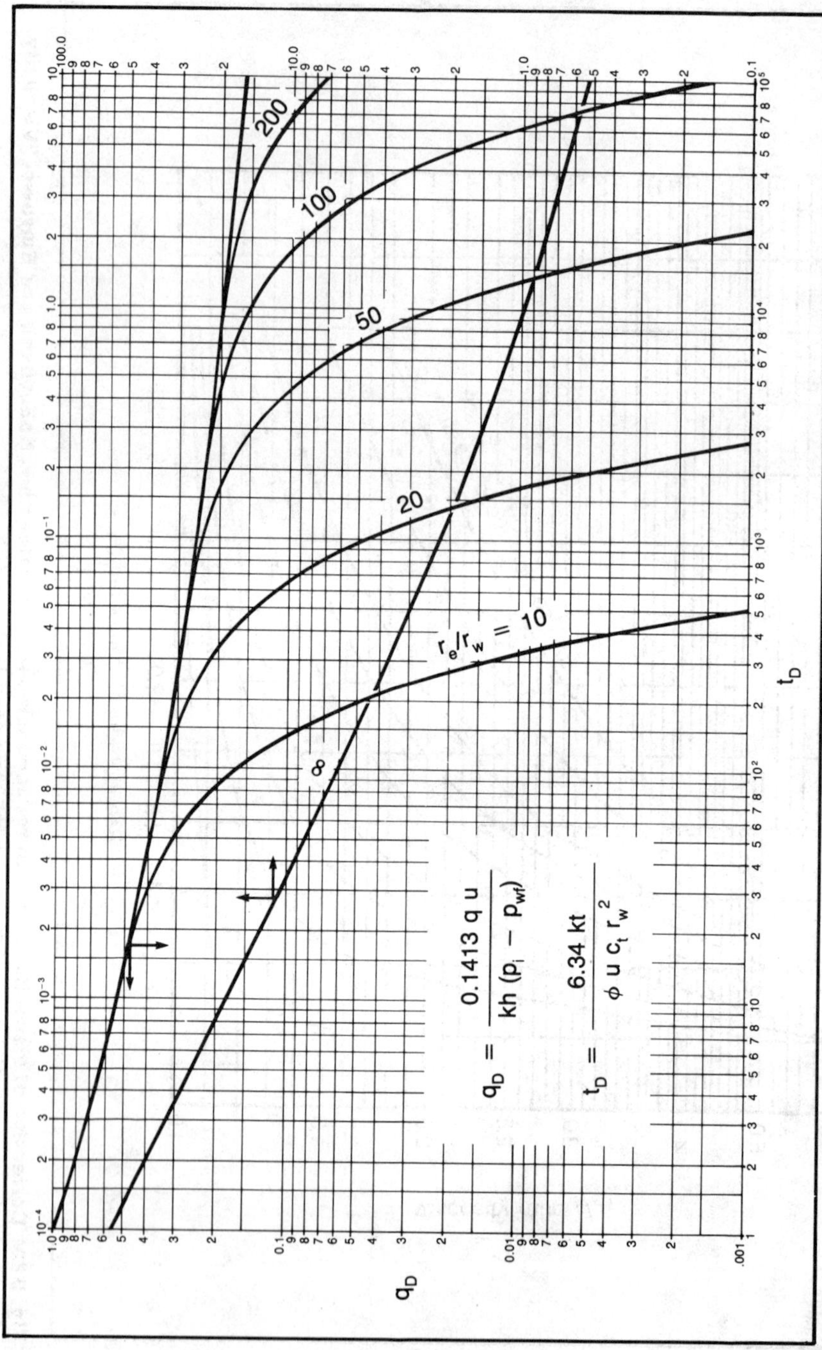

Fig. B21a Dimensionless flow rate functions for a plane radial system, infinite and finite outer boundary, constant pressure at inner boundary (after Fetkovich, "Decline Curve Analysis Using Type Curves," courtesy *JPT*, © June 1980, SPE-AIME)

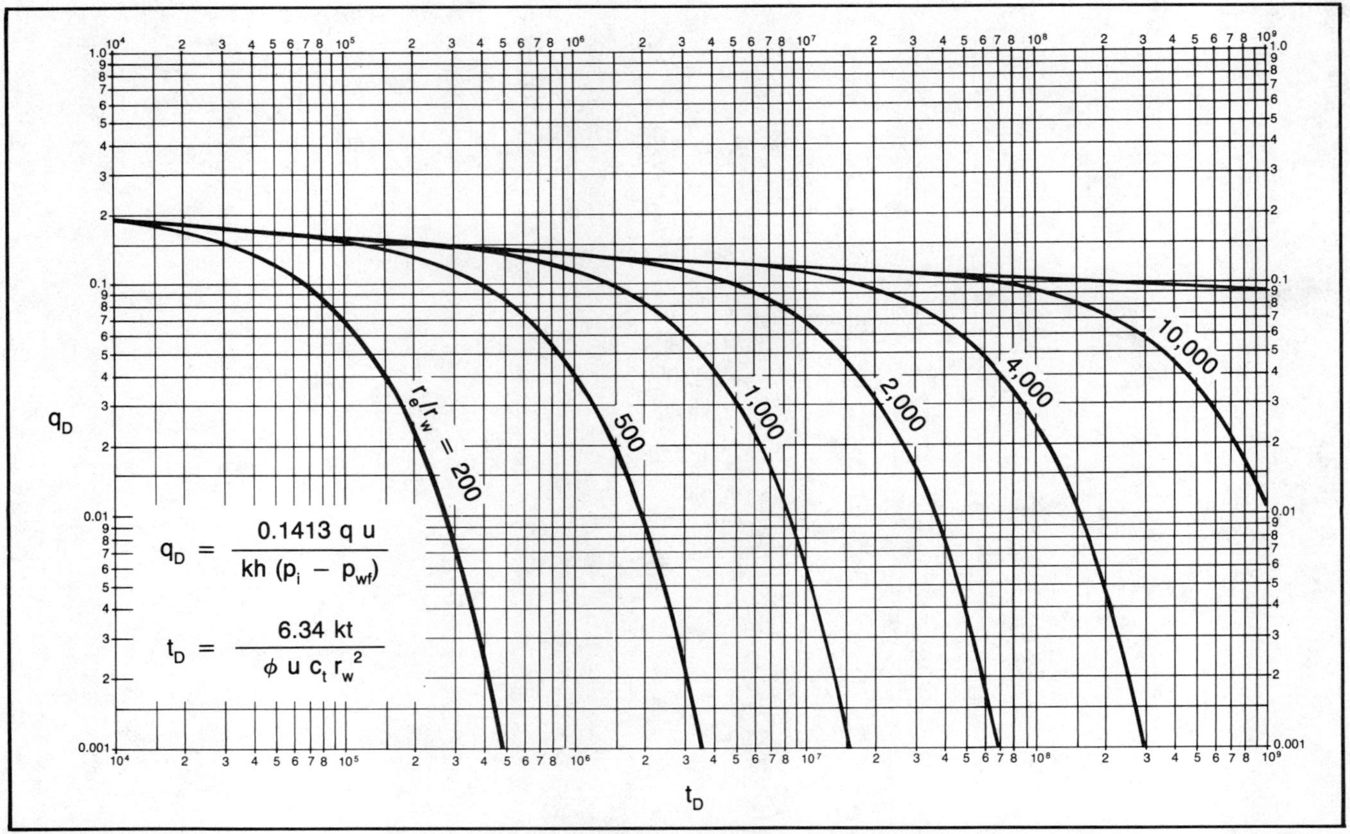

Fig. B21b Dimensionless flow rate functions for a plane radial system, infinite and finite outer boundary, constant pressure at inner boundary (after Fetkovich, "Decline Curve Analysis Using Type Curves," courtesy *JPT*, © June 1980, SPE-AIME)

Appendix C

Problem Solutions

PROBLEM 2.1 Solution: Calculating Permeability Data from Lab Tests

The absolute permeability can be calculated only when the core is 100% saturated with one fluid, so use data when $S_w = 100\%$:

$$q_w' = \frac{k_w A'}{\mu_w} \frac{\Delta p'}{\Delta x'} \qquad \text{(2.2 applied to water flow)}$$

$$k_w = \frac{q_w' \mu_w}{A'} \frac{\Delta x'}{\Delta p'}$$

$$k_w = \frac{(0.5)(1.0)}{(5)} \frac{(3)}{(2-1)}$$

$$= 0.3 \text{ darcies or } 300 \text{ md, absolute permeability}$$

At $S_w = 30\%$:

$$K_o = \frac{q_o' \mu_o}{A'} \frac{\Delta x'}{\Delta p'} \qquad \text{(2.2 applied to oil flow)}$$

$$= \frac{(0.38)(1.25)(3)}{(5)(2-1)}$$

$$= 0.285 \text{ darcies or } 285 \text{ md}$$

Calculation of Relative Permeabilities

(1) S_w	*(2)* k_w	*(3)* k_o	*(4)* k_{rw}	*(5)* k_{ro}
100	0.300	0.0000	1.00	0.000
90	0.180	0.0000 (critical)	0.60	0.000 (critical)
80	0.090	0.0075	0.30	0.025
60	0.018	0.0750	0.06	0.250
40	0.006	0.1875	0.02	0.625
30	0.000 (critical)	0.285	0.00 (critical)	0.950

(1) Given in problem
(2) From k_w equation above
(3) From k_o equation above
(4) $k_w/0.300$
(5) $k_o/0.300$
Critical zeros refer to point where data first begin to deviate from zero.

PROBLEM 2.2 Solution: Radial Steady-State Flow from a Damaged Well

A. To calculate Δp_{skin}, use Eq. 2.20:

$$q = \frac{7.08 k_{undamaged} h (p_e - p_w - \Delta p_{skin})}{\mu \, \ell n \, (r_e/r_w)} \tag{2.20}$$

$$(80)(1.25) = \frac{(7.08)(0.01)(6.9)(2{,}000 - 900 - \Delta p_{skin})}{0.708 \, \ell n \left(\dfrac{700}{0.7}\right)}$$

$$\Delta p_{skin} = 100 \text{ psia}$$

B. If the Δp skin is reduced to zero, the production rate is:

$$q = \frac{7.08 kh (p_e - p_w)}{\mu \, \ell n (r_e/r_w)} \tag{2.14}$$

$$= \frac{(7.08)(0.01)(6.9)(2{,}000 - p_w)}{(0.708) \, \ell n \dfrac{(700)}{(0.7)}}$$

$$= 110 \text{ b/d}$$

$$\frac{110}{1.25} = 88 \text{ stb/d}$$

C. Letting $r_1 = 7.0$ and $r_2 = 700$ and substituting into Eq. 2.15:

$$(80)(1.25) = \frac{(7.08)(0.010)(6.9)(2{,}000 - p_1)}{(7.08) \, \ell n (700/7)}$$

$$p_1 = 1{,}333 \text{ psia}$$

$$\Delta p \text{ from 7–0.7 ft} = 1{,}333 - 900 = 433 \text{ psia}$$

PROBLEM 2.3 Solution: Determining Average Permeabilities

A.

$$k_{avg} = \frac{\Sigma k_j h_j}{h} \quad \frac{(2)(0.2) + (3)(0.1) + (5)(0.01)}{2 + 3 + 5} \tag{2.53}$$

$$= 0.075 \text{ darcies or 75 md}$$

$$q = \frac{7.08 kh}{\mu} \, \frac{(p_e - p_w)}{\ell n \dfrac{r_e}{r_w}} \tag{2.14}$$

$$= \frac{(7.08)(0.075)(10)(600 - 500)}{1.0 \, \ell n \dfrac{100}{1/3}}$$

$$= 93 \text{ b/d}$$

B. $$k_{avg} = \frac{\ell n\left(\frac{r_e}{r_w}\right)_{total}}{\sum \frac{\ell n\left(\frac{r_o}{r_j}\right)_j}{k_j}} \quad (2.57)$$

$$= \frac{2.303 \log \frac{1{,}000}{1.0}}{\left(2.303 \times \frac{\log \frac{1{,}000}{10}}{0.075}\right) + \left(2.303 \times \frac{\log \frac{10}{1}}{0.0075}\right)}$$

$$= 0.019 \text{ darcies or } 19 \text{ md}$$

PROBLEM 2.4 Solution: Correcting for Static Pressure Differences

A. $$\Delta p_{Flow} = \Delta p_{Total} \pm 0.433\, \gamma\, \Delta D \quad (2.58)$$

$$= (3{,}000 - 2{,}250) - (0.433)(1.0)(1{,}000)$$

$$= 750 - 433 = 317 \text{ psi}$$

$$q = \frac{7.08kh(p_e - p_w)}{\mu\, \ell n \frac{r_e}{r_w}} \quad (2.14)$$

$$= \frac{(7.08)(0.092)(10)(317)}{(0.708)\, \ell n \frac{(4{,}000)}{0.4}}$$

$$= 317 \text{ res b/d}$$

B. All parameters are the same except $\Delta p = 750$. Thus, by ratio:

$$q_{res\ bbl} = 317 \frac{(750)}{(317)} = 750 \text{ res b/d}$$

PROBLEM 3.1 Solution: Application of the p_{tD} Function

Calculate the diffusivity constant:

$$\eta = \frac{6.33k}{\phi \mu c} = \frac{(6.33)(10^{-4})}{(6.95)(10^{-2})(0.4)(6.33)(10^{-6})} = 3{,}600 \quad (3.9)$$

Calculate the reduced time:

$$t_{Dw} = \frac{\eta t}{r_w^2} = \frac{\eta t_{min}}{r_w^2\ 1{,}440} = \frac{3{,}600\ t_{min}}{(0.25)(1{,}440)} \quad (3.11)$$

$$= 10t_{min}$$

$$t_{Dw} = (10)(6) = 60$$

Read p_{tD} from Fig. 3–7 using the $(r_e/r_w) = \infty$ curve since it is obvious that after 6 min of production the well is still infinite acting. Since $\Delta p_{skin} = 0.0$:

$$p_i - p_{wt} = \frac{0.141q\mu}{kh}(p_{tD}) = \left[\frac{(0.141)(18.0)(1.39)(0.4)}{(10^{-4})(141)}\right] p_{tD} \qquad (3.17)$$

$$= (100)\,p_{tD} = (100)\,2.48 = 248 \text{ psia}$$

$p_{w,t} = 3{,}000 - 248 = 2{,}752$ psia, which is the flowing pressure after 6 min of production if there is no well damage.

PROBLEM 3.2 Solution: Application of the Ei Function

To find the pressure at any radius where the rate is not known, we must use the Ei function. Since we only know the rate at the well radius in this example, we must use the Ei function.

Pressure after 6 min at 5-ft radius:

We must first test to see if we can use the Ei solution. From problem 3.1 we know $\frac{\eta t}{r_w^2} = 60 < 100$; therefore, we cannot use the Ei function and there is no solution.

Pressure after 1 hr at 5-ft radius:

$$\frac{\eta t}{r_w^2} = 10\,(60) = 600 > 100 \longrightarrow \text{OK}$$

$$\frac{r_e^2}{4\eta} = \frac{(3{,}000)^2}{(4)(3{,}600)} = 625 > t = 1/24 \text{ days} \longrightarrow \text{OK}$$

Therefore, we can use the Ei solution:

$$t_D = \frac{\eta t}{r^2} = \frac{(3{,}600)(1/24)}{(5)^2} = 6.0 \qquad (3.11a)$$

$$½\left[-\text{Ei}\,\frac{1}{-4t_D}\right] = 1.33 \text{ (from function graph Fig. 3–9)}$$

From problem 3.1:

$$(0.141q\mu/kh) = 100$$

$$p_{r,t} = p_i - \frac{0.141q\mu}{kh}(½)\left(-\text{Ei}\,\frac{-1}{4t_D}\right) \qquad (3.20a)$$

$$p_{r,t} = 3{,}000 - (100)(1.33)$$

$$= 3{,}000 - 133 = 2{,}867 \text{ psia}$$

Pressure after 1 hr at 50-ft radius:

From previous part $(\eta t/r_w^2) > 100$ and $(r_e^2/4\eta) > t$. Therefore, we can use the Ei solution:

$$t_D = \frac{\eta t}{r^2} = \frac{(3{,}600)(1/24)}{(50)^2} = 0.06 \text{ and} \qquad (3.11a)$$

$$\tfrac{1}{2}\left[-\text{Ei}\,\frac{-1}{4t_D}\right] \approx 0.0 \text{ (from function graph)}$$

$$p_{r,t} = 3{,}000 - (100)(0)$$

$$\text{p at 50 ft and 1 hr} = 3{,}000 \text{ psia}$$

The pressure influence has not yet significantly affected the pressure at 50 ft.

PROBLEM 3.3 Solution: Calculating the Time Necessary to Obtain a Particular Pressure Drop in the Reservoir

$$p_{r,t} = p_i - \frac{0.141q\mu}{kh}\left[\tfrac{1}{2}\left(-\text{Ei}\,\frac{-1}{4t_D}\right)\right] \quad (3.20a)$$

$$7.5 = \frac{(0.141)(540)(1.39)(0.5)}{(0.1)(14.1)}\left[\tfrac{1}{2}\left(-\text{Ei}\,\frac{-1}{4t_D}\right)\right]$$

$$\left[\tfrac{1}{2}\left(-\text{Ei}\,\frac{-1}{4t_D}\right)\right] = 0.2$$

From Fig. 3–9 the t_D corresponding to this value is 0.38:

$$t_D = \frac{\eta t}{r^2}$$

$$0.38 = \frac{(10^5)(t)}{(2{,}500)^2}$$

$$t = 24 \text{ days}$$

PROBLEM 3.4 Solution: Choosing a Pressure-Function Solution

Since we want a well pressure, we can use the p_{tD} function. In problem 3.1 we find that t_D at the well is 60 when t is 6 min. Thus, when t is 100 min, t_D at the well is 1,000. Fig. 3–7 only goes to $t_D = 100$, so we must use the extension equation:

Is $t > r_e^2/4\eta$?

$$t = \left(\frac{100}{1{,}440}\right)_{\text{day}} < r_e^2/4\eta = \frac{(3{,}000)^2}{(4)(3{,}600)} = 625$$

Therefore, use the natural log equation for the pressure function:

$$p_{tD} = \tfrac{1}{2}(\ell n\, t_D + 0.809) = \tfrac{1}{2}(\ell n\, 1{,}000 + 0.809) = 3.85$$

$$p_{w,t} = p_i - \frac{0.141q_r\mu}{kh}\,p_{tD} + \Delta p_{skin} \quad (3.17)$$

From problem 3.1 $(0.141q_r\mu/kh) = 100$ and $\Delta p_{skin} = 0.0$, so p_w at 100 min = $3{,}000 - (100)(3.85) = 2{,}615$ psia. Find the best pressure function for the well pressure after 1,000 days of production.

In problem 3.1 we find that t_D at the well is 60 when t is 6 min, so when t is 1,000 days × 1,440 min/day or 1.44×10^6 min:

$$t_D = 1.44 \times 10^7 > 100$$

From $r_e^2/4\eta = 625$, $t = 1{,}000 > 625$. Therefore, we must use p_{tD} by Eq. 3.24:

$$p_{tD} = \frac{2t_D}{r_{De}^2} + \ell n\, r_{De} - 3/4 \tag{3.24}$$

$$r_{De} = \frac{3{,}000}{0.5} = 6{,}000$$

$$p_{tD} = \frac{(2)(1.44 \times 10^7)}{(6{,}000)^2} + \ell n\, 6{,}000 - 3/4$$

$$= 8.74$$

$$p_{w,t} = p_i - \frac{0.141 q_r \mu}{kh} p_{tD} + \Delta p_{skin} \tag{3.17}$$

From problem 3.1 $(0.141q_r\mu/kh) = 100$ and $\Delta p_{skin} = 0.0$, so p_w after 1,000 days = 3,000 − (100)(8.74) = 2,126 psia.

PROBLEM 3.5 Solution: A Pseudosteady-State Flow-Test Analysis

A. Note constant producing rate and constant rate of pressure decline. Therefore, use pseudosteady-state equations:

$$\left(\frac{\Delta p}{\Delta t}\right)_{pseudo} = \frac{1.79q}{r_e^2 \phi ch} \tag{3.33}$$

$$\therefore \phi h = \frac{1.8q}{r^2 c} \frac{\Delta t}{\Delta p_{avg}}$$

$$= \frac{(1.8)(850)(1.1)}{(4 \times 10^6)(10^{-5})} \left(\frac{1}{2}\right)$$

$$= 21\ \phi\ \text{ft}$$

B.

$$kh = \frac{q\mu\, \ell n[(r_e/r_w) - 1/2 + S]}{7.08(p_e - p_w)} \tag{3.43a}$$

$$\ell n\, r_e/r_w = \ell n \frac{2{,}000}{0.5} = 8.298$$

$$kh = \frac{(850)(1.1)(0.5)[(8.298) - 1/2 + 0.0]}{(7.08)(3)(10^2)}$$

$$kh = 1.720 \text{ darcy-ft or } 1{,}720 \text{ md-ft}$$

C.

$$q = 7.08 \frac{kh}{\mu} \frac{(p_e - p_w)}{\ell n(r_e/r_w)} \tag{2.14}$$

$$kh = \frac{q\mu\,(\ell n\, r_e/r_w)}{7.08(p_e - p_w)}$$

$$= \frac{(850)(1.1)(0.5)(2.303 \times 3.603)}{7.08(3 \times 10^2)}$$

$$= 1.830 \text{ darcy-ft or } 1{,}830 \text{ md-ft}$$

PROBLEM 3.6 Solution: Water Disposal in an Aquifer

A. For reservoir to be infinite acting, $(r_e^2/4\eta) >$ injection time of 100 days:

$$\eta = \frac{6.33k}{\phi\mu c} = \frac{(6.33)(0.01)}{(0.2)(0.5)(6.33 \times 10^{-6})} = 10^5 \quad (3.9)$$

$$\frac{r_e^2}{4\eta} = \frac{(10{,}000)^2}{(4)(10^5)} = 250$$

$$250 > 100$$

Reservoir is infinite acting.

B. To find the total water injected:

$$r_{wa} = r_w e^{-S} \quad (2.26)$$
$$= 0.5\ e^{-1.7}$$
$$= 0.09134$$

$$t_D = \frac{\eta t}{r_w^2}$$

$$= \frac{(10^5)(100)}{(0.09134)^2} = 1.2 \times 10^9$$

From Table 3–1:

$$Q_{tD} = 1.1551 \times 10^8 \quad (3.57)$$

$$Q_r = 1.12\ \phi c r^2 h\ \Delta p_r\ Q_{tD}$$

$$Q_r = (1.12)(0.2)(6.33 \times 10^{-6})(14.1)(0.09134^2)(2{,}200 - 1{,}000) \times (1.1551 \times 10^8)$$

$$= 23{,}000 \text{ bbl}$$

PROBLEM 3.7 Solution: Pressure Resulting from Multiple Wells in a Reservoir

$$\eta = \frac{6.33k}{\phi\mu c} = \frac{(6.33)(0.1)}{(0.2)(0.5)(6.33 \times 10^{-5})} = 10^5 \quad (3.9)$$

To find Δp caused by a rate of 540 stb/d. For 62.5 days at a radius of 2,500 ft. Note that 62.5 days is less than $r_e^2/4\eta = 10{,}000^2/(4 \times 10^5)$. Therefore:

$$\Delta p_{2500,\,62.5} = \frac{0.141 q_w \mu}{kh}\left[(½)\left(-\text{Ei}\,\frac{-1}{4t_{D1}}\right)\right] \quad (\text{3.20a solved for } p_i - p_{r,t})$$

$$t_{D1} = \frac{\eta t}{r^2} = \frac{(10^5)(62.5)}{(2{,}500)^2} = 1.0$$

From Fig. 3–9:

$$½\left(-\text{Ei}\,\frac{1}{4t_{D1}}\right) = 0.5$$

$$\Delta p_{2500,\,62.5} = \frac{0.141(540)(1.39)(0.5)}{(0.1)(14.1)}(0.5) = (37.5)(0.5) = 18.7 \text{ psi}$$

To find Δp caused by a rate of 1,080 stb/d for 10 days at a radius of 0.5 ft:

$$\Delta p_{0.5,\,10} = \frac{0.141 q_r \mu}{kh}[(½)(\ell n\, t_{D2} + 0.809) + S] \qquad \text{(3.17a and 3.25 combined)}$$

$$t_{D2} = \frac{\eta t}{r^2} = \frac{(10^5)(10)}{(0.5)^2} = 4 \times 10^6$$

$$½\,[\ell n(4 \times 10^6) + 0.809] = 8.0$$

$$\Delta p_{0.5,\,10} = \frac{(0.141)(1{,}080)(1.39)(0.5)}{(0.1)(14.1)}(8) = 600 \text{ psia}$$

p_w at well 2 = 3,000 − 19 − 600 = 2,381 psi

PROBLEM 3.8 Solution: Accounting for Variations in Producing Rates

Δp at well 2 = (Δp for r = 2,500, Δq = 750, and t = 72.5)
− (Δp for r = 2,500, Δq = (540 − 180) 1.39, and t = 10)
+ (Δp for r = 0.5, q = 1,500, and t = 20)

Δp for r = 2,500, Δq = 750, and t = 72.5:

$$t_D = \frac{\eta t}{r^2} = \frac{(10^5)(72.5)}{(2{,}500)^2} = 1.16$$

From Fig. 3–9:

$$(½)\left(-\text{Ei}\,\frac{-1}{(4)(1.16)}\right) = 0.57$$

$$\Delta p = \frac{0.141\,\mu q}{kh}\,(½)\left(-\text{Ei}\,\frac{-1}{4t_D}\right)$$

$$= \frac{(0.141)(0.5)}{(0.1)(14.1)}(750)(0.57)$$

$$= (0.05)(750)(0.57) = 21.4 \text{ psi}$$

Δp for r = 2,500, Δq = 500, and t = 10:

$$t_D = \frac{(10^5)(10)}{(2{,}500)^2} = 0.16$$

From Fig. 3–9:

$$(½)\left(-\text{Ei}\,\frac{-1}{(4)(0.16)}\right) = 0.04$$

$$\Delta p = \frac{0.141\,\mu}{kh}q\left[½\left(-\text{Ei}\,\frac{-1}{(4)(0.16)}\right)\right]$$

$$= (0.05)(500)(0.04) = 1.0 \text{ psi}$$

Δp for r = 0.5, q = 1,500, and t = 20:

$$t_D = \frac{(10^5)(20)}{(0.5)^2} = 8 \times 10^6$$

$$\tfrac{1}{2}[\ell n\, t_D + 0.809] = \tfrac{1}{2}[\ell n\,(8 \times 10^6) + 0.809] = 8.35$$

$$\Delta p = (0.05)(1{,}500)(8.35) = 626.5 \text{ psi}$$

Substituting into the first equation, Δp at well 2 = 21.4 − 1.0 + 626.5 = 647 psi. Then p_w at well 2 = 3,000 − 647 = 2,353 psia.

PROBLEM 3.9 Solution: Simulating Boundary Effects

$$\Delta p \text{ at well 1} = (\Delta p \text{ for } q = 750,\ t = 52.5, \text{ and } r = 0.5) + [\Delta p \text{ for } q = 750,\ t = 52.5, \text{ and } r = (2)(1{,}250)]$$

Δp for q = 750, t = 52.5, and r = 0.5:

$$t_D = \frac{\eta t}{r^2} = \frac{(10^5)(52.5)}{(0.5)^2} = 21 \times 10^6$$

$$\tfrac{1}{2}[\ell n\, t_D + 0.809] = \tfrac{1}{2}[\ell n\,(21 \times 10^6) + 0.809]$$

$$= 8.83$$

$$\Delta p = \frac{0.141\mu q}{kh} (\tfrac{1}{2})\, [\ell n\, t_D + 0.809]$$

$$= \frac{(0.141)(0.5)}{(0.1)(14.1)} (750)(8.83)$$

$$= (0.05)(750)(8.83)$$

$$= 331 \text{ psi}$$

Δp for q = 750, t = 52.5, and r = 2,500:

$$t_D = \frac{(10^5)(52.5)}{(2{,}500)^2} = 0.84$$

From Fig. 3–9:

$$(\tfrac{1}{2}) \left[-\text{Ei}\, \frac{-1}{(4)(0.84)}\right] = 0.45$$

$$\Delta p = (0.05)(750)(0.45)$$

$$= 17 \text{ psi}$$

Substituting in the first equation, Δp_w at well 1 = 331 + 17 = 348 psi and p_w = 3,000 − 348 = 2,652 psia.

PROBLEM 4.1 Solution: Productivity-Index Evaluation

A.

(1)	*(2)*	*(3)*	*(4)*	*(5)*
	Initial		*When N_p = 35,000 stb*	
p_w, psig	*q_o, stb/d*	*PI, stb/d/psig*	*PI, stb/d/psig*	*q_o, stb/d*
3,000	0	—	—	—
2,660	90	0.265	0.157	—
2,380	165	0.266	0.158	—
1,980	270	0.265	0.157	—
1,600	330	0.236	0.140	—
1,195	373	0.207	0.123	197 maximum flow rate
0	455**	0.152	0.09	252 maximum pump rate

(1) From Fig. 4–2.
(2) From Fig. 4–2; **Maximum pumping rate obtained by extrapolation.
(3) $q/(p_e - p_w) = q/(3{,}000 - p_w)$ where q is in stb. Based on Eq. 4.1. Where 3,000 is obtained from Fig. 4–2 extrapolation.
(4) (PI when N_p is 35,000) $=$ Initial PI $(k_o/B_o\mu_o)_{35{,}000}/(k_o/B_o\mu_o)_{\text{Intitial}}$ (4.6)

$$= \text{Initial PI}(0.5/(1.15 \times 1.1))]/[(.0/(1.25 \times 1.2)]$$
$$= \text{Initial PI} \times 0.5925$$

(5) $(q_o)_{35{,}000} = \text{PI} \times (p_e - p_w) = (\text{PI})_{35{,}000} \times (2{,}800 - p_w)$ (4.1)

B. If only rates less than 270 b/d are used in the test, the indicated maximum rate is based on the PI of 0.265 stb/d/psi or (3,000) × (0.265) = 795 stb/d.

$$DR = k_{undamaged}/k_{actual} \quad (4.7)$$

$$J = \frac{7.08k_{act.}h}{B_o\mu_o[\ell n(r_e/r_w) - 0.5]} \quad (4.4 \text{ where } k_{actual} = k_{avg})$$

$$0.265 = \frac{7.08k_{act.}\,14}{1.25 \times 1.2 \times [\ell n(745/0.333) - 0.5]}$$

$$k_{actual} = 0.0288 \text{ darcies or } 28.8 \text{ md}$$

$$DR = \frac{20.5}{28.8} = 0.712 - \text{permeability around wellbore has been improved}$$

C.

$$\left(\frac{\Delta p}{\Delta t}\right)_{pseudo} = \frac{1.79q}{\phi chr_e^2} \quad (3.33)$$

$$3.0 = \frac{1.79 \times 75 \times 1.15}{(\phi cr_e^2)\ 14}$$

$$\phi cr_e^2 = \frac{3.7\ \text{ft}^2}{\text{psi}}$$

$$t_S = \frac{0.04(\phi cr_e^2)\mu}{k} \quad (3.50)$$

$$= \frac{0.04 \times 3.7 \times 1.1}{(0.5 \times 0.0205)}$$

$$= 15.9 \text{ days}$$

PROBLEM 4.2 Solution: Using the Vogel IPR Curve

From Fig. 4–2:

p_e = 3,000 psig (Curve extrapolation to a 0.0 oil rate.)

From Fig. 4–1:		*From Fig. 4–3:*		
p_w, psia	q_{ostb}	p_w/p_e	q_o/q_{omax}	q_{omax}
2,660	90	0.887	0.190	473
2,380	165	0.793	0.337	489
1,980	270	0.660	0.520	519
1,600	330	0.533	0.664	497
1,195	373	0.398	0.795	469

Maximum producing rate estimated as 469 (minimum extrapolation).

If p_w = 500 psia:

$$p_w/p_e = 500/3{,}000 = 0.1667$$

From Fig. 4–3:

$$q_o/q_{o\,max} = 0.93$$

$$q_o = (0.93)(469) = 436 \text{ stb/d}$$

PROBLEM 4.3 Solution: Constant-Rate Drawdown Tests

From graph:

$$m = 80 \text{ psi/cycle}$$

$$m = \frac{0.1625q\mu}{kh}, \text{ i.e., } \frac{kh}{\mu} = \frac{(0.1625)(80)(1.25)}{80} = 0.20$$

Therefore, $\frac{k}{\mu} = 0.02$ when k is in darcies and k = (0.02) 1.2 = 0.024 darcies

Using the pressure at 10 days from the extrapolation:

$$p_w = p_i - 0.867m[0.5(\ell n\, t_D + 0.809) + S] \quad (4.14)$$

$$0.5(\ell n\, t_D + 0.809) = 0.5\left(\ell n \frac{(6.33)(0.02)(10)}{(0.2)(10^{-5})(1/16)} + 0.809\right) = 8.47$$

$$2{,}337 = 2{,}800 - (0.867)(80)[(8.47) + S]$$

$$S = -2.76$$

PROBLEM 4.4 Solution: Determining the Distance to a Reservoir Boundary from a Drawdown Test

$$\Delta p'_{(10\ days)} = 67\ psi$$

$$\tfrac{1}{2}\left[-Ei\left(\frac{-1}{4t_D}\right)\right] = \frac{\Delta p'}{0.867m} = \frac{(1.151)(67)}{(80)} = 0.97 \quad (4.20\ \text{where}\ p_w' - p_w = \Delta p')$$

Then from Fig. 3–9:

$$t_D = 3.0$$

However:

$$t_{D\,image} = \frac{\eta t}{(2d)^2}\ \text{and}\ t_{D\,image} = \frac{6.33kt_{days}}{\phi\mu c(2d)^2} \qquad (4.19)$$

$$(2d)^2 = \frac{\eta t}{t_{D\,image}} = \frac{(6.33)(0.02)(10)}{(0.2)(10 \times 10^{-6})(3.0)} = 2.11(10^5)$$

$$d^2 = 52{,}750$$

$$d = 229\ ft$$

PROBLEM 4.5 Solution: Pressure Buildup from an Unchanging Well Pressure

Determine the undamaged k from the slope of the buildup plot (Fig. 4–9):

$$m\ \text{from the plot} = \frac{1{,}484 - 1{,}424}{1.0 - 0.3} = 86$$

$$m = \frac{0.1625q\mu}{kh} \qquad (4.13)$$

$$k = \frac{0.1625q\mu}{mh} = \frac{(0.1625)(199)(1.15)(2.0)}{(86)(22)}$$

$$= 0.04\ \text{darcy or 40 md}$$

ϕc can be determined from the time-travel equation and the observation that the nearest boundary is 300 ft away:

$$t_s = \frac{0.04\phi\mu c r_e^2}{k} \qquad (3.50)$$

$$\phi c = \frac{t_s k}{0.04\mu r_e^2} = \frac{(7.5/24)(0.04)}{(0.04)(2.0)(300)^2}$$

$$= 1.737 \times 10^{-6}$$

For convenience, the Δp_{skin} calculation is based on the shutin pressure on the straight-line extrapolation at 10 hr (log = 1.0):

$$\Delta t_D\ \text{for 10 hr} = \frac{6.33k\Delta t}{\phi c\mu r_w^2} \qquad (4.25)$$

$$= \frac{(6.33)(0.04)(10/24)}{(1.737)(10^{-6})(2.0)(1/3)^2} = 2.73 \times 10^5$$

$$\ell n\ \Delta t_D = (5)(2.3) + 1.0 = 12.5$$

$$p_w - p_{wf} = 0.867m[(½)(\ell n\ \Delta t_D + 0.809) + S] \quad (4.29)$$

$$\Delta p_{skin} = p_w - p_{wf} - 0.867m(½)(\ell n\ \Delta t_D + 0.809)$$

$$1{,}484 - 702 = (0.867)(86)[(½)(12.5 + 0.8) + S]$$

$$S = 3.85$$

The rate after skin removal can be determined by ratio since, with all other parameters equal, the rate is proportional to the effective pressure drop:

$$\Delta p_{skin} = 0.867m\ S \quad (2.22a)$$
$$= 0.867(86)(3.85) = 287 \text{ psi}$$

$$q_{zero\ skin} = q_{stb} \frac{(p_{unchanging} - p_{wf})}{(p_{unchanging} - p_{wf} - \Delta p_{skin})} \quad \text{(similar to 2.28)}$$

$$= \frac{(199)(1{,}482 - 702)}{(1{,}482 - 702 - 287)}$$

$$= 315 \text{ stb/d}$$

PROBLEM 4.6 Solution: Pressure-Buildup Analysis of an Old Well

Two points for Δp_q versus log Δt plot check Fig. 4–11.

At t = 24 hr:

$$\left(\frac{\Delta p}{\Delta t}\right)_{pseudo} = \frac{1 \text{ psi}}{\text{hr}} \text{ or } \frac{24 \text{ psi}}{\text{day}}$$

$$\Delta p_q = p_w - p_{wf} + \Delta t\left(\frac{\Delta p}{\Delta t}\right)_{pseudo} \quad (4.38)$$

$$\Delta p_q = 2{,}675 - 1{,}123 + 24 = 1{,}576 \text{ psia at 24 hr}$$

At t = 30 hr:

$$\Delta t\left(\frac{\Delta p_w}{\Delta t}\right) = \frac{1 \text{ psi}}{\text{hr}}(30 \text{ hr}) = 30 \text{ psia}$$

$$\Delta p_q = 2{,}687 - 1{,}123 + 30 = 1{,}594 \text{ psia}$$

Calculate (ϕc) product:

$$\frac{\Delta p}{\Delta t} = \frac{1.8q\ B_o}{r_e^2 \phi hc} \quad (3.33)$$

Where:

$$q = \text{stb}$$

$$r_e = [(43{,}560)(40)/\pi]^{0.5} = 745 \text{ ft}$$

$$(\phi c) = \frac{1.8q\,B_o}{r_e^2 h}\,\frac{dt}{dp}$$

$$= \frac{(1.8)(280)(1.31)}{(7.45)^2(10^4)(40)}\,\frac{1}{24}$$

$$(\phi c) = 1.24 \times 10^{-6}$$

Calculate permeability, k:

$$m = \frac{0.1625 q_{stb}\,B_o \mu}{kh} \qquad \text{(4.13 modified)}$$

$$k = \frac{0.1625 q_{stb}\,B_o \mu}{mh}$$

Where:

$$q = \text{stb}$$

$$m = \text{slope} = \frac{1{,}594 - 1{,}304}{\log 31 - \log 1} = \frac{290}{1.491} = 196.3$$

$$k = \frac{0.1625(280)(1.31)(2.0)}{(196.3)(40)} = 0.0151 \text{ darcies or } 15.1 \text{ md}$$

Beginning of pseudosteady state:

$$t_s = \frac{0.04\phi c \mu r_e^2}{k} \qquad (3.50)$$

$$r_e^2 = 40 \times 43{,}560/\pi = 5.55 \times 10^5$$

$$t_s = \frac{(4)(10^{-2})(1.24)(10^{-6})(2)(5.55)(10^5)}{0.0151}$$

$$t_s = 3.65 \text{ days}$$

Calculate Δp_{skin} and S based on Δp_q of 1,576 psia at a shutin time of 24 hr:

$$\Delta p_q = 0.867m(0.5)(\ell n\,\Delta t_D + 0.809) + \Delta p_{skin} \qquad (4.37)$$

$$\ell n\,\Delta t_D + 0.809 = \ell n(6.33k\Delta t/\phi\mu c r_w^2) + 0.809$$

$$= \ell n(6.33 \times 0.0151 \times 1.0/1.24 \times 10^{-6} \times 2 \times 0.333^2) + 0.809$$

$$= 13.6$$

$$1{,}576 = 0.867 \times 196.3 \times 0.5 \times 13.6 + \Delta p_{skin}$$

$$\Delta p_{skin} = 415 \text{ psi}$$

$$S = \frac{\Delta p_{skin}}{0.867m} \qquad (2.22a)$$

$$= \frac{415}{0.867 \times 196.3} = 2.43$$

PROBLEM 4.7 Solution: Determining the Average Pressure in the Drainage Area of a Pseudosteady-State Well

To find p_s we must first evaluate p^* using pressure after 8 hr of shutin:

$$p^* = p_w + m \log \frac{t_p t \Delta t}{\Delta t} \tag{4.63}$$

$$p^* = 2{,}592 + 196.3 \log \frac{240 + 8}{8}$$

$$= \ell n\,(3.47 \times 10^6) = 15.064$$

$$p^* = 2{,}886$$

$$\frac{(p_i - p^*)}{m} = \frac{(2{,}960 - 2{,}886)}{196.3} = 0.376$$

From Fig. 4–15:

$$\frac{(p_1 - p_s)}{m} = 1.25$$

$$\frac{(2{,}960 - p_s)}{196.3} = 1.25; \quad p_s = 2{,}713 \text{ psia}$$

If we assume a radial drainage system, we can calculate p_s as:

$$p_s = p_{wf} - 0.439m + (\Delta p_q)_{ts} \tag{4.67}$$

$$= 1{,}123 - (0.439)(196.3) + 1{,}685$$

$= 2{,}722$ psia compared with the p_s of 2,713 psia calculated from the Odeh data

To find the damage ratio, we must first evaluate p_e:

$$p_e = 0.217m + p_s = 0.217 \times 196.3 + 2{,}713 \tag{4.41}$$

$$= 2{,}756 \text{ psia}$$

$$DR = \frac{(p_e - p_{wf})}{(p_e \times p_{wf} - \Delta p_{skin})} \tag{2.28}$$

$$= \frac{(2{,}756 - 1{,}123)}{(2{,}756 - 1{,}123 - 415)}$$

$$= 1.34$$

PROBLEM 4.8 Solution: Analysis of a Two-Rate Pressure Buildup

$$m = \frac{0.1625 \Delta q \mu}{kh} \tag{4.68}$$

$$k_o = \frac{0.1625 \Delta q \mu}{mh}$$

$$m = 8.75 \text{ at shutin times greater than 4 hr}$$

$$= \frac{(0.1625)(14.)(1.322)(0.39)}{(8.75)(10)}$$

$$k_o = 0.0134 \text{ darcies or } 13.4 \text{ md}$$

Base the skin calculations on the $\Delta p_{\Delta q}$ value after 1 day of shutin:

$$\Delta t_D \text{ for 1 day} = \frac{6.33k\Delta t}{\phi\mu c r_w^2} \quad (4.25)$$

$$= \frac{(6.33)(0.0134)(1)}{(0.2)(0.39)(1.379)(10^{-4})(0.265)^2}$$

$$= 1.12(10^5)$$

$$\ell n\ \Delta t_D = 11.6$$

$$\Delta p_{\Delta q} = 0.867m(\tfrac{1}{2})(\ell n\ \Delta t_D + 0.809) + \Delta p_{skin} \quad (4.69)$$

$$\Delta p_{skin} = (2{,}211 - 1{,}963) - 0.867(8.75)(\tfrac{1}{2})(11.6 + 0.8)$$

$$\Delta p_{skin} = 201 \text{ psi}$$

$$S = \frac{\Delta p_{skin}}{0.867m} = \frac{201}{(0.867)(8.75)} = 26.5 \quad (2.22a)$$

To adjust m and Δp_{skin} from a Δq to a q base:

$$m \text{ for } q = (78)(1.322) = \frac{78}{14}(8.75) = 48.75$$

$$\Delta p_{skin} \text{ at } q = (78)(1.322) = 0.867(48.75)(26.5)$$

$$= 1{,}123 \text{ psi}$$

$$p_s = p_{wf} + \Delta p_{skin} + 0.867m\left(\ell n\ \frac{r_e}{r_w} - \tfrac{3}{4}\right) \quad (4.44a)$$

$$= 1{,}963 + 1{,}123 + (0.867)(48.75)\left(\ell n\ \frac{1{,}490}{0.265} - 0.75\right)$$

$$= 3{,}419 \text{ psia}$$

PROBLEM 4.9 Solution: Pressure Falloff Analysis

Plot p_w versus Δt.

To find the undamaged permeability:

$$\text{Slope } m = \frac{0.1625\mu i}{kh} \quad (4.73)$$

Where:

$$k = \text{Undamaged permeability}$$

From plot m = 82.0:

$$82.0 = \frac{(0.1625)(250)(1.0)}{k(50)}$$

$$k = 0.00995 \text{ darcies or } 10 \text{ md}$$

To find Δp skin:

$$p_w = p_{wf} - \frac{0.141\ \mu i}{kh}(\tfrac{1}{2})\ (\ell n\ \Delta t_D + 0.809) - \Delta p_{skin} \qquad (4.71)$$

At t = 15 min:

$$t_d = \frac{6.33kt \text{ days}}{\phi\mu c r_w^2} = \frac{6.33(0.01)t_{min}}{(0.2)(1.0)(5 \times 10^{-6})(1/16)\left(1{,}440\ \frac{min}{day}\right)}$$

$$= 703 \times t_{min}$$

$$4 = 500 - \frac{(0.141)(1.0)(250)}{(0.01)(50)}\ \frac{\ell n(703 \times 15) + 0.809}{2} - \Delta p_{skin}$$

$$\Delta p_{skin} = 141 \text{ psi}$$

$$\Delta p_{skin} = \frac{0.141\ \mu i}{kh} S$$

$$141 = \frac{(0.141)(250)(1.0)}{(0.01)(50)} S$$

$$S = 2.0$$

To find r_e at t = 15 min, use Eq. 3.50:

$$t_s = \frac{0.04\phi\mu c r_e^2}{k} \qquad (3.50)$$

$$r_e^2 = \frac{kt_{days}}{0.04\phi\mu c} = \frac{0.01\ t_{days}}{(0.04)(0.2)(1.0)(5 \times 10^{-6})}$$

$$r_e = \frac{(t_{days})^{1/2}}{2(10^{-3})} = 13.2(t_{min})^{1/2}$$

$$r_e = 13.2(15)^{1/2} = 51.0 \text{ ft}$$

i. when $\Delta p_{skin} = 0.0$:

Bottom-hole pressure in injection well = 500 + (0.433)(3,000) = 1,300 psig
New rate = 250(1,300 − 0)/(1,300 − 0 − 141) = 271 b/d

PROBLEM 4.10 Solution: Pressure Buildup in a North Sea Well Demonstrating Spherical Flow

Fig. C1 indicates negligible afterflow after 0.6 hr of shutin. From Fig. C2, plot of Table C1 data:

$$m = 8.75 \text{ psi} \Big/ \frac{1}{\sqrt{\text{days}}}$$

$$m = .0158q\,\phi^{0.5}\mu^{1.5}c^{0.5}/k^{1.5} \tag{4.86}$$

$$8.75 = \frac{0.0158(4{,}510 \times 1.734)(0.15)^{0.5}(0.4)^{1.5}(20 \times 10^{-6})^{0.5}}{k^{1.5}}$$

$$k = 0.0336 \text{ darcies or } 33.6 \text{ md}$$

TABLE C1 Spherical Pressure Buildup Plot Data

t, hr	p_w*, psia*	$(1/\sqrt{t_{days}} - 1/\sqrt{\Delta t_{days} + t_{days}})$
0.000	5,264.6	
0.059	5,492.1	18.940
0.119	5,587.9	12.940
0.178	5,604.1	10.360
0.238	5,613.9	8.804
0.297	5,621.2	7.741
0.357	5,626.6	6.982
0.476	5,635.3	5.882
0.595	5,641.1	5.123
0.774	5,647.5	4.227
0.952	5,652.4	3.795
1.190	5,656.5	3.301
1.369	5,659.1	2.990
1.846	5,663.7	2.425
2.441	5,668.0	1.973
3.453	5,672.4	1.507
4.347	5,675.3	1.271
8.098	5,681.1	0.710
10.421	5,683.1	0.554
13.636	5,685.1	0.379
16.137	5,685.7	0.345

PROBLEM 4.11 Solution: Analyzing DST Data

A. For t = 60 min:

$$\frac{t + \Delta t}{\Delta t} = \frac{80 + 60}{60} = 2.33$$

t = 100 min:

$$\frac{t + \Delta t}{\Delta t} = \frac{80 + 100}{100} = 1.8$$

continued on p. 753

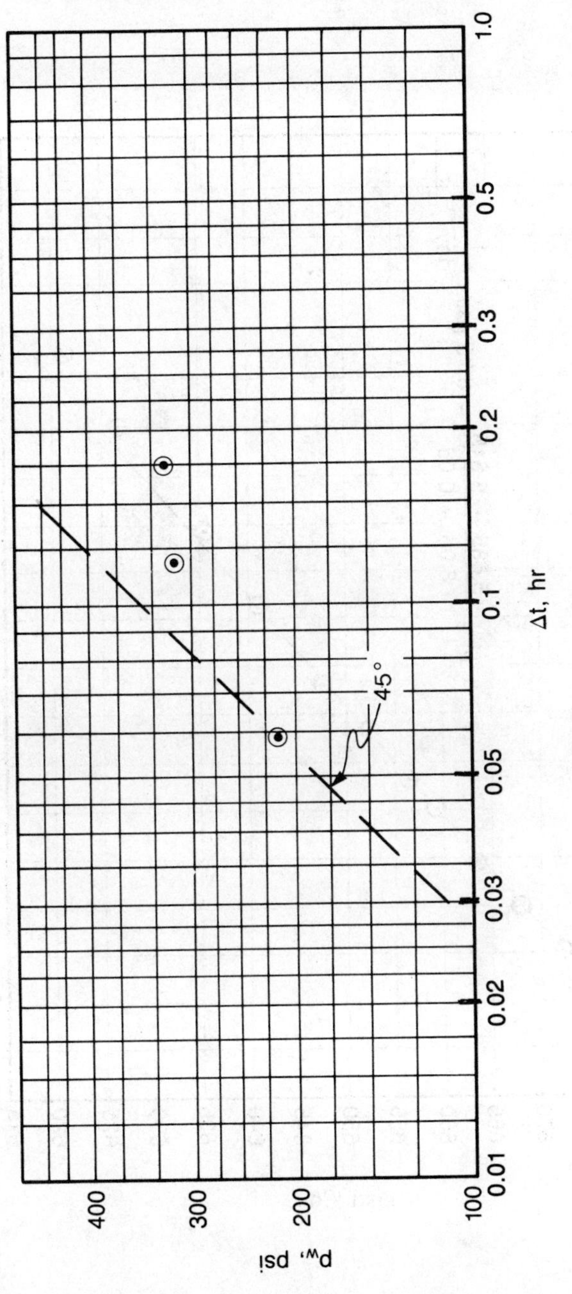

Fig. C1 Afterflow plot, problem 4.10 solution

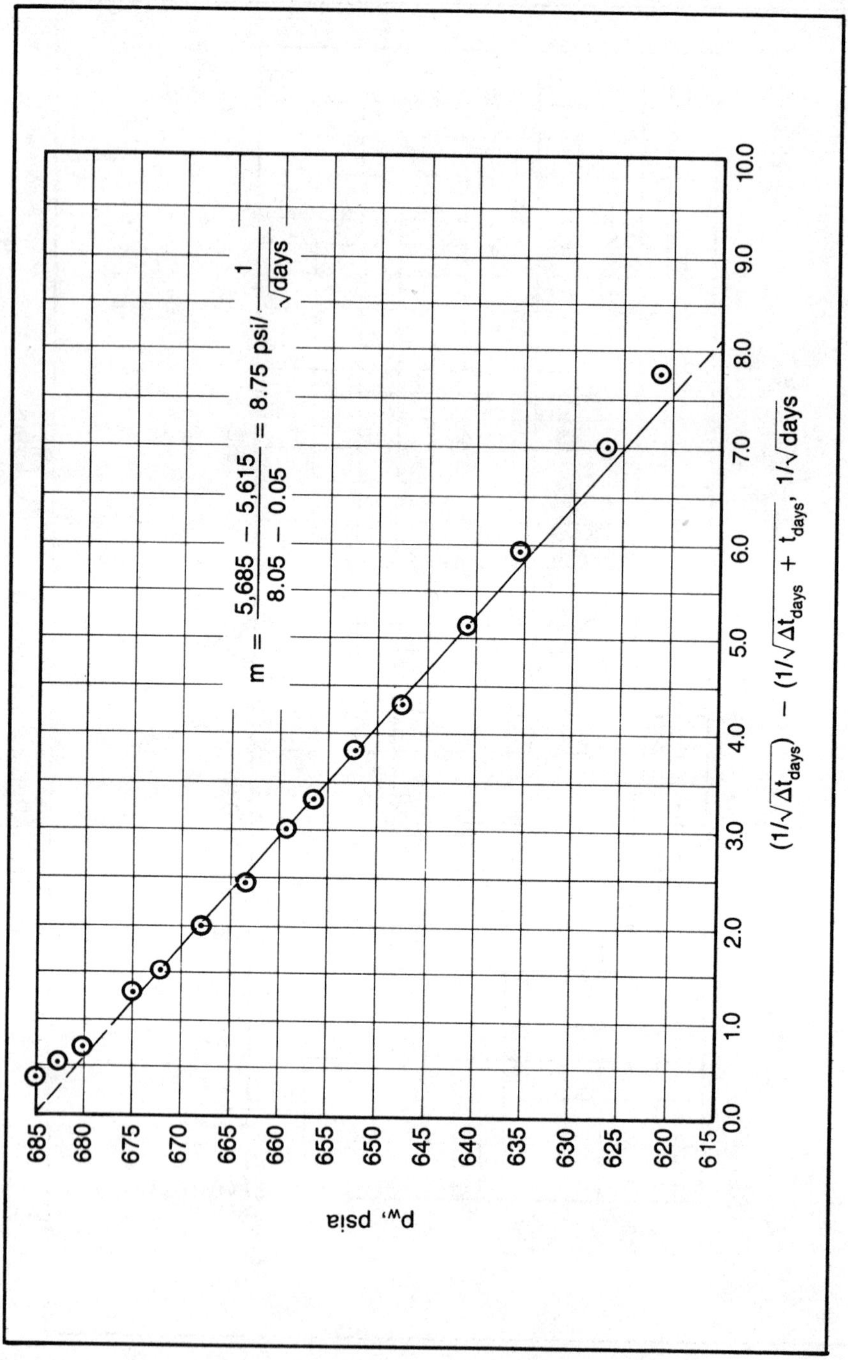

Fig. C2 Spherical pressure-buildup plot, problem 4.10 solution

t = min:

$$\frac{t + \Delta t}{\Delta t} = \frac{80 + 80}{80} = 2.0$$

p_i from plot at $(\Delta t + t)/\Delta t = 1.0$ (Fig. 4–13):

$$p_i = 4{,}500 \text{ psia}$$

$$m \text{ from plot} = 300 \text{ psi/cycle}$$

B. Hydrostatic mud pressure $= (0.433)\left(\frac{\text{mud wt}}{8.33}\right) \times \text{bomb depth}$ (4.96 modified)

$$= (0.433)\left(\frac{9.8}{8.33}\right)(10{,}000) = 5{,}100 \text{ psig}$$

$$\text{Drillpipe capacity} = \frac{(3.826)^2}{1{,}000} = 0.0147 \text{ bbl/ft}$$

$$\text{Fluid recovery} = (0.0147)(7{,}500) = 110 \text{ bbl}$$

$$q_{stb} = \frac{110}{80/1{,}440} = 1{,}980 \text{ b/d}$$

$$(k_o/\mu_o) = \frac{0.1625qB}{hm} = \frac{(0.1625)(1{,}980)(1.75)}{(20)(300)}$$

$$(k_o/\mu_o) = 0.0938 \text{ darcies/cp or } 93.8 \text{ md/cp}$$

$$k_{o\,md} = (93.8)(0.4) = 37.5 \text{ md}$$

C.

$$p_w = (0.825)(0.433)\,10{,}000 = 3{,}572 \text{ psig}$$

$$q_{undamaged} = \frac{7.08kh(p_e - p_w)}{B_o\mu\left[\left(\ell n\,\frac{r_e}{r_w} - \frac{1}{2}\right) + S\right]} \quad (4.93)$$

$$= \frac{(7.08)(0.0938)(20)(4{,}500 - 3{,}572)}{1.75(\ell n\ 2{,}980 - 1/2)}$$

$$= 939 \text{ stb}$$

D. $p_i - p_{wf} = 0.867m(0.5)(\ell n\ t_D + 0.809) + \Delta p_{skin}$ (4.51 rearranged)

$$\ell n\ t_D = \ell n\left[\frac{(6.33)(0.0938)(80/1{,}440)}{(0.2)(5 \times 10^{-5})(0.25)^2}\right] = 10.87$$

$$4{,}500 - 2{,}700 = (0.867)(300)(0.5)(10.87 + 0.809) + \Delta p_{skin}$$

$$\Delta p_{skin} = 280 \text{ psi}$$

$$S = \frac{\Delta p_{skin}}{0.867m} = \frac{280}{(0.867)(300)} = 1.08 \quad (2.22a)$$

E.
$$t_s = \frac{0.040\phi\mu c r_i^2}{k} \tag{4.97}$$

$$r_e = \sqrt{\frac{(80)(0.0938)}{(1{,}440)(40)(0.2)(5 \times 10^{-5})}} = \sqrt{1.30 \times 10^4}$$

$$= 114 \text{ ft}$$

PROBLEM 4.12 Solution: Pulse Test Analysis

$$\frac{t_{L2}}{t_c} = \frac{0.9}{6} = 0.15$$

$$F' = \frac{\Delta t_{odd}}{\Delta t_c} = \frac{3.0}{6.0} = 0.5$$

From Fig. 4–36:

$$p_D[t_L/\Delta t_c]^2 = 0.0025$$

$$\Delta p_D = \frac{0.0025}{(0.15)_2} = 0.111$$

$$\Delta p_{Dpj} = \frac{\Delta p_{pj}}{(0.141 q\mu/kh)} \tag{4.104}$$

From Fig. C3:

$$\Delta p_{p2} = 6.8 \text{ psia}$$

$$0.111 = \frac{6.8}{(0.141)(700)(0.86)/kh}$$

$$kh = \frac{(0.119)(0.141)(700)(0.86)}{6.8}$$

$$= 1.3879 \text{ darcy-ft}$$

$$= 1{,}387.9 \text{ md-ft}$$

From Fig. 4–32:

$$t_{LD} = 0.095$$

$$t_{pLj} = \eta t_{Lj}/r^2 \tag{4.105}$$

$$0.095 = \eta(0.9/24)/(330)^2$$

$$\eta = (0.095)(330)^2/(0.9/24)$$

$$\eta = 275{,}880 = \frac{6.33k}{\phi c}$$

$$275{,}880 = \frac{(6.33)k}{(0.16)(0.86)(9.6 \times 10^{-6})}$$

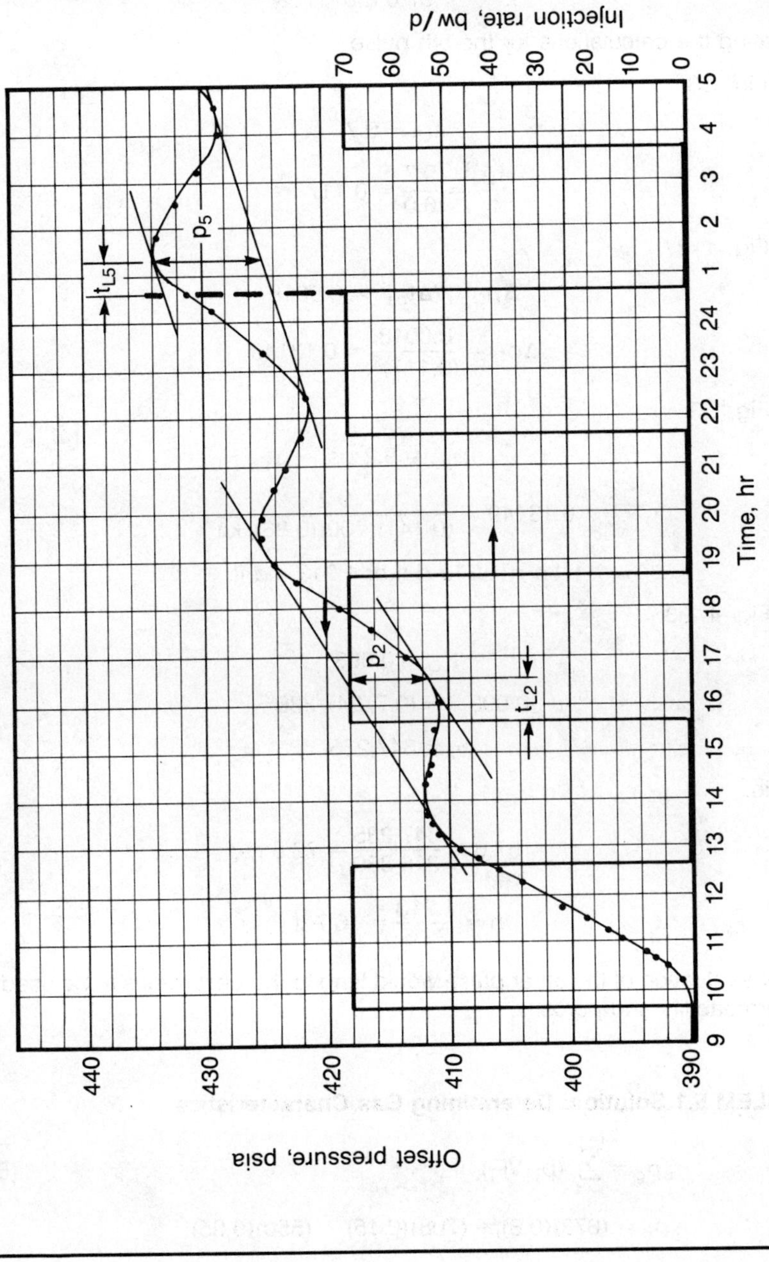

Fig. C3 Pulse pressure response for problem 4.12 solution

$$k = 0.0576 \text{ darcy or } 57.6 \text{ md}$$

$$h = \frac{kh}{k} = \frac{1{,}387.9 \text{ md-ft}}{57.6 \text{ md}} = 24.1 \text{ ft}$$

Repeating the calculations for the 5th pulse.

From Fig. C3:

$$t_{L5} = 0.7$$

$$\frac{t_{L5}}{t_c} = \frac{0.7}{6.0} = 0.117$$

From Fig. 4–37:

$$\Delta p_D[t_L/\Delta t_c]^2 = 0.0018$$

$$\Delta p_D = \frac{0.0018}{(0.117)^2} = 0.13149$$

From Fig. C3:

$$p_{p5} = 9.2 \text{ psi}$$

$$0.13149 = \frac{9.2}{(0.141)(700)(0.86)/kh}$$

$$kh = 1.213 \text{ d-ft or } 1{,}213 \text{ md-ft}$$

From Fig. 4–33:

$$t_{LD} = 0.093$$

$$0.093 = \eta(0.7/24)/330^2$$

$$\eta = 347{,}235$$

By ratio:

$$k = 57.6 \times \frac{347{,}235}{275{,}880} = 72.5 \text{ md}$$

$$h = \frac{1{,}213}{72.5} = 16.7 \text{ ft}$$

Further evaluation of the other pulse would lead to the best values to be used for the permeability and porosity.

PROBLEM 5.1 Solution: Determining Gas Characteristics

A. $$_p p_c = \sum_{j=1}^{j=n} (p_{cj} VF_j) \tag{5.8}$$

$$_p p_c = (673)(0.8) + (708)(0.15) + (550)(0.05)$$
$$= 672 \text{ psia}$$

$$_pT_c = \sum_{j=1}^{j=n} (T_{cj})(VF)_j \tag{5.9}$$

$$= (343)(0.8) + (549)(0.15) + (766)(0.05)$$

$$= 395°R$$

$$_pp_r = \frac{3,000}{672} = 4.46 \tag{5.7}$$

$$_pT_r = \frac{170 + 460}{395} = 1.59 \tag{5.7}$$

B. From Fig. B8:

$$z = 0.82$$

C.

$$MW = \sum_{j=1}^{j=n} (MF)_j(MW)_j \tag{5.10}$$

$$= (0.8)(16) + (0.15)(30) + (0.05)(58)$$

$$= 20.2$$

$$\gamma = \frac{20.2}{29} = 0.70$$

D. From Fig. B9:

$$_pp_c = 668;\ _pT_c = 390$$

$$_pp_r = \frac{3,000}{668} = 4.49 \tag{5 7}$$

$$_pT_r = \frac{170 + 460}{390} = 1.62 \tag{5.7}$$

$$z = 0.83$$

E.

$$B_g = \frac{0.00504zT}{p} \tag{5.12}$$

$$= \frac{(5.04)(10^{-3})(0.82)(6.3)(10^2)}{(3.0)(10^3)}$$

$$= (8.68)(10^{-4}) \text{ res bbl/scf}$$

F.

$$\rho = \frac{2.7\gamma p}{zT} \tag{5.14}$$

$$= \frac{(2.7)(0.7)(3.0)(10^3)}{(0.82)(6.3)(10^2)}$$

$$= 10.98 \text{ lb/ft}^3$$

G. $$c \approx \frac{1}{p} = \frac{1}{3{,}000} = \frac{(10)(10^{-1})}{(3)(10^{3})} = (3.3)(10^{-4})$$

From Fig. B6, $c_r = 0.2$:

$$c_g = 0.2/672 = (3.0)(10^{-4})$$

H. At 100°F, when $\gamma = 0.7$ and $p = 3{,}000$, $\mu_g = 0.023$:

$$\text{at } 200°\text{F},\ \mu_g = 0.021$$

$$\text{at } 170°\text{F},\ \mu_g = 0.021 + \frac{30}{100}(0.002) = 0.0216 \text{ cp}$$

I. From Fig. B12:

$$\frac{\text{Well fluid gravity}}{\text{Trap gas gravity}} = 1.16$$

$$\text{Well fluid gravity} = \text{Reservoir fluid gravity} = (1.16)(0.7) = 0.81$$

From Fig. B9, condensate data:

$$_pp_c = 657 \text{ psia}$$

$$p_r = \frac{3{,}000}{657} = 4.57 \qquad (5.7)$$

$$_pT_c = 408°\text{R}$$

$$T_r = \frac{170 + 460}{408} = 1.54 \qquad (5.7)$$

From Fig. B8:

$$z = 0.81$$

PROBLEM 5.2 Solution: Determining Gas Production for Use in Material Balance

G_p = scf of dry gas + scf of original reservoir water vapor + scf of condensate vapor + scf of vent gas. (5.25)

From Table B15:

$$GE = 814 \text{ scf/stb}$$

Condensate gas = (814)(150) = 122,000 scf

Water vapor in original gas (Fig. B17) = (350 lb/MMcf)(0.89) = 312 lb/MMcf

$$\frac{15 \text{ bbl} \times 350 \text{ lb/bbl}}{10.152 \text{ MMcf}} = 517 \text{ lb/MMcf} = \text{water production}$$

$$\text{scf of water vapor} = \frac{(312)}{(350)}(7{,}390)(10.152) = 66{,}900 \text{ scf}$$

$$G_p/\text{day} = 10 + 0.067 + 0.122 + 0.030 \text{ (vent gas)}$$
$$= 10.219 \text{ MMcfd}$$

$$W_p/\text{day} = \frac{(517 - 312)(10.152)}{350} = 5.95 \text{ bw/d}$$

PROBLEM 5.3 Solution: Calculating Static Bottom-Hole Pressure from Surface Pressure Measurements

$$p_2 = p_1\, e^{(0.01875\, \gamma_g D/Z_{avg} T_{avg})} \qquad (5.31)$$

$$= p_1\, e^{(0.01875 \times .68 \times 3676/Z_{avg} \times 560)} = p_1\, e^{(0.0838/z)}$$

z can be determined by trial and error using Figs. B8 and B9:

$$\text{Assume } p_2 = 3{,}035;\ p_{avg} = \frac{3{,}035 + 2{,}715}{2} = 2{,}875$$

$$p_r = 2{,}875/668 = 4.303;\ T_r = \frac{560}{380} = 1.474$$

$$z = 0.77$$

Checking the p_2 assumption:

$$p_2 = 2{,}715\, e^{(0.0838/0.77)} = 3{,}027$$

Now, assume $p_2 = 3{,}027$; $p_{avg} = (3{,}027 + 2{,}715)/2$; $p_r = \frac{2{,}871}{668} = 4.3$

$$z = 0.77$$

$$p_2 = 2{,}715\, e^{(0.0838/0.77)} = 3{,}027 \text{ psia}$$

The bottom-hole pressure calculated from the program in Table 5–3 is 3,034 psia.

PROBLEM 5.4 Solution: Application of Graphical Gas Material-Balance Techniques

To construct the graphical material-balance plot, we must first determine the p/z values. Using Figs. B8 and B9 for a gas gravity of 0.68:

$${}_pp_c = 667.5 \text{ psia};\ {}_pT_c = 385;\ T_R = \frac{560}{385} = 1.45$$

G_p, MMcf	p, psia	p_r	z	p/z
1,809	3,461	5.18	0.796	4,348
3,901	3,370	5.05	0.790	4,266
5,850	3,209	4.81	0.778	4.125
9,451	3,029	4.54	0.765	3,959

The p/z versus G_p plot is shown in Fig. C4.

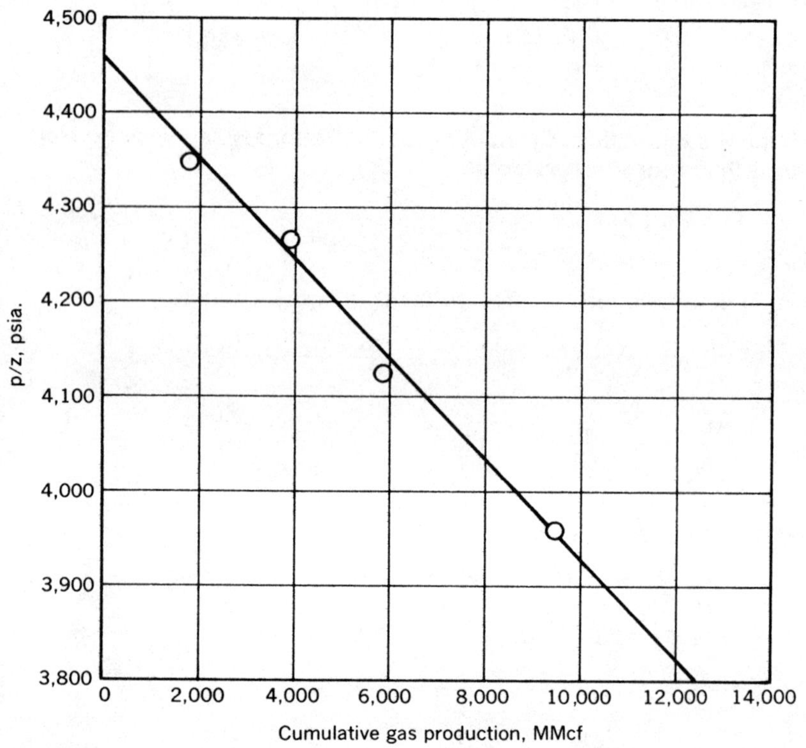

Fig. C4 Plot of p/z versus cumulative gas production, problem 5.4 solution

To determine the initial pressure, read:

$$\frac{p_i}{z_i} = 4{,}458 \text{ at } G_p = 0.0; \text{ So } \frac{(p_i/z_i)}{p_c} = \frac{p_{ri}}{z_i} = \frac{4{,}458}{667.5} = 6.678$$

From Fig. B10:

$$z = 0.81$$
$$p_i = 4{,}458 \times 081$$
$$p_i = 3{,}611 \text{ psia}$$

The best straight-line fit of the (p/z) versus G_p plot gives:

$$\frac{p}{z} = 4{,}458 - 0.053\ G_p$$

Where:

$$0.053 = (4{,}458 - 3{,}800)/12{,}400 \text{ MMcf}$$

When p/z = 0.0:

$$G = \frac{4{,}458}{0.053} = 84{,}113 \text{ MMcf (original gas in place)}$$

Cumulative production at the end of the contract = (20 MMcfd)(365 days/year) (5 years) + 9,450 MMcf = 45,950 MMcf.

When G_p = 45,950:

$$\begin{aligned}(p/z) &= 4{,}458 - (0.053)(45{,}950)\\ &= 2{,}023 \text{ psia}\\ (p_r/z) &= \frac{2{,}023}{667.5} = 3.031\end{aligned}$$

From Fig. B10:

$$\begin{aligned}z &= 0.775\\ p &= (2{,}023)(0.775) = 1{,}568 \text{ psia}\end{aligned}$$

PROBLEM 5.5 Solution: Calculating Water Encroachment

Production Time, months	*Water Influx Time, months*	t_D	Q_{tD}*	*Average Reservoir Pressure, psia*	*Pressure at Contact*	Δp
0	0	—	—	2,500	2,500	0
6	0	0	0	2,450	2,500	47
12	6	5	4.539	2,356	2,406	87
18	12	10	7.411	2,276	2,326	70
24	18	15	9.949	2,216	2,266	

$$\begin{aligned}Q \text{ at 12 months} &= 192{,}000 = B(47)(4.539)\\ B &= 192{,}000/(47)(4.539) = 900 \text{ bbl/psi}\end{aligned}$$

$$\begin{aligned}Q \text{ at 18 months} &= 1.12\phi c h r^2\, \Sigma\Delta p\, Q_{tD}\\ &= 900[(47)(7.411) + (87)(4.539)\\ &= 668{,}889 \text{ bbl}\end{aligned}$$

$$\begin{aligned}Q \text{ at 24 months} &= 900[(47)(9.949) + (87)(7.411) + (70)(4.539)]\\ &= 1{,}287{,}081 \text{ bbl}\end{aligned}$$

*Q_{tD} is read from Table 3–1 for each t_D.

PROBLEM 5.6 Solution: Simultaneous Solution of Reservoir Pressure and Water Encroachment

At 100 days:

$$B_{gi} = 0.00504\, Z_i T/p_i \qquad (5.12)$$

From Fig. 5–4 when p = 807 psia, (p/z) = 900 psia:

$$B_{gi} = (0.00504)(538)/900 = 0.003013$$

If $W_e = 0.0$:

$$G\,B_{gi} = (G - G_p)B_g \quad (5.21)$$

$$(1{,}118 \times 10^6)(0.003013) = [1{,}118 \times 10^6 - (1{,}000 \times 10^3)(100)]B_g$$

$$B_g = 3.309 \times 10^{-3} = 0.00504\ T/(p/z) \quad (5.12)$$

$$p/z = 819.4 \text{ psia}$$

From Fig. 5–4:

$$p = 740 \text{ psia}$$

If p = 740 psia:

$$W_e = 1.12\ \phi hcr^2 \sum_{j=1}^{j=n} \Delta p_j Q_{tD_j} \quad (3.64a)$$

$$t_D = (6.33\ k/\phi c)t/r^2$$

$$= [6.33(0.014)/(0.11)(1.0)(3.3 \times 10^{-6})]100/1{,}702^2$$

$$= 8.428$$

Interpolated from Table 3–1:

$$Q_{tD} = 6.5515$$

$$W_e = (1.12)(0.11)(28)(3.3 \times 10^{-6})(1{,}702)^2 \frac{(807 - 740)}{2} 6.5515$$

$$= 32.98 \frac{(807 - 740)}{2} 6.5515$$

$$= 7{,}237.5 \text{ bbl}$$

If W_e = 7,237.5 bbl:

$$GB_{gi} = (G - Gp)B_g + (W_e - W_p) \quad (5.22)$$

$$(1{,}118 \times 10^6)(0.003013) = [1{,}118 \times 10^6 - (1{,}000 \times 10^3)100]B_g + (7{,}237.5 - 0.0)$$

$$B_g = 3.302 \times 10^{-3} = 0.00504/(p/z)$$

$$p/z = 821.2$$

From Fig. 5–4:

$$p = 742 \text{ psia}$$

If p = 742 psia:

$$W_e = 32.98 \frac{(807 - 742)}{2} 6.5515$$

$$= 7{,}021 \text{ bbl}$$

No significant change with previously calculated W_e, so assume correct and that p = 742 psia.

At 300 days if $W_e = 7{,}021$ bbls:

$$(1{,}118 \times 10^6)(0.003013) = [1{,}118 \times 10^6 - (1{,}000 \times 10^3)300]B_g + 7{,}021$$

$$B_g = 4.109 \times 10^{-3} = 0.00504(538)/(p/z)$$

$$p/z = 659.8 \text{ psia}$$

From Fig. 5–4:

$$p = 608 \text{ psia}$$

If p = 608 psia:

$$t_{D300} = 3(t_{D100}) = 3(8.428) = 25.28$$

Interpolated from Table 3–1:

$$Q_{tD300} = 14.696$$

$$t_{D200} = 2(8.428) = 16.86$$

$$Q_{td} = 10.846$$

$$W_e = 32.98\left[\frac{(807 - 742)}{2}14.696 + \frac{(807 - 608)}{2}10.846\right]$$

$$= 51{,}337 \text{ bbl}$$

If $W_e = 51{,}337$:

$$(1{,}118 \times 10^6)(0.003013) = (1{,}118 \times 10^6 - 3{,}000 \times 10^5)B_g + 51{,}337$$

$$B_g = 4.055 \times 10^{-3} = 0.00504(538)/(p/z)$$

$$(p/z) = 668.6 \text{ psia}$$

From Fig. 5–4:

$$p = 615 \text{ psia}$$

If p = 615 psia:

$$W_e = 32.98\left[\frac{807 - 742}{2}14.696 + \frac{(807 - 614)}{2}10.846\right]$$

$$= 50.086 \text{ bbl}$$

Assume change with previously calculated W_e is negligible, so $W_e = 50{,}086$ bbls after 300 days and p = 615 psia.

PROBLEM 5.7 Solution: Turbulence Effects in Gas Wells

From Fig. B11:

$$B = 3.5 \times 10^{10}$$

$$\Delta(p^2)_{turbulence} = \frac{3.161(10^{-12})B\gamma q_g{}^2 zt\left(\frac{1}{r_2} - \frac{1}{r_1}\right)}{h^2}$$

(2nd term, 5.36)

$$= \frac{(3.161)(10^{-12})(3.5)(10^{10})(0.76)(3.9)^2(10^6)(0.97)(7.12)(10^2)\left(\frac{1}{1/3} - \frac{1}{550}\right)}{(30)^2}$$

$$= \frac{(3.161)(3.5)(0.76)(10^6)(3.9)^2(0.97)(7.12)(3 - 0)}{(9)(10)^2}$$

$$= (293)(10^4) = (2.93)(10^6)$$

$$\Delta p^2_{viscous} = \frac{1.424 \mu z_R t_f q_g \ell n(r_1/r_2)}{kh} \qquad \text{(1st term, 5.36)}$$

$$= \frac{(1.424)(2.7)(10^{-2})(0.97)(7.12)(10^2)(3.9)(10^3)\ell n\left(\frac{550}{1/3}\right)}{(1.5)(10^{-3})(3.0)(10)}$$

$$= 17.05 \times 10^6$$

A. $$(p_w)^2_{no\ turbulence} = (4{,}583)^2 - 17.05 \times 10^6$$

$$p_w = 1{,}988 \text{ psia}$$

B. $$(p_w)_{turbulence} = \sqrt{(4{,}583)^2 - (17.05)(10^6) - (2.93)(10^6)}$$

$$= 1{,}012 \text{ psia}$$

$$\Delta p_{w\ turbulence} = 1{,}988 - 1{,}012 = 976 \text{ psia}$$

C. $$(\Delta p_{skin}) = 1{,}400 - 976 = 424 \text{ psia}$$

PROBLEM 5.8 Solution: Conventional Gas-Well Back-Pressure Test

A.

Stabilized BHP, psia	*$(p_w)^2$*	*$\Delta(p^2)$*	*log $\Delta(p^2)$*	*q_g, Mscfd*	*log q_g*
2,798	7.84×10^6				
2,669	7.13×10^6	0.71×10^6	5.851	1,810	3.255
2,591	6.71×10^6	1.13×10^6	6.053	2,710	3.431
2,499	6.25×10^6	1.59×10^6	6.201	3,590	3.556
2,426	5.88×10^6	1.96×10^6	6.292	4,510	3.653

See Fig. 5–9 for the data plot:

$$\log 5{,}000 \text{ Mscfd} = 3.699$$
$$\log \Delta(p^2) = 6.349$$
$$\Delta p^2 = 2.23 \times 10^6$$
$$p_e^2 - p_w^2 = 2.23 \times 10^6$$
$$(4.0 \times 10^6) - (2.23 \times 10^6) = p_w^2$$
$$p_w^2 = 1.77 \times 10^6$$
$$p_w = 1{,}340 \text{ psia}$$

B. $$t_s = \frac{0.04\phi \mu c r_e^2}{k} \qquad r_e^2 = (2{,}250)(43{,}560)/3.14 \qquad (3.50)$$

$$t_s = \frac{(0.04)(0.15)(0.021)(0.000357)(2{,}250)(43{,}560)}{(0.074)(3.14)} = 18.9 \text{ days}$$

PROBLEM 5.9 Solution: Adjustment of a Gas Deliverability Curve for a Change in Well Spacing or Depletion

For 2 wells:

$$r_{e2} = \left[\frac{\frac{4{,}500}{2}(43{,}560)}{3.14}\right]^{1/2} = 5{,}580 \text{ ft}$$

For 5 wells:

$$r_{e5} = \left[\frac{\frac{4{,}500}{5}(43{,}560)}{3.14}\right]^{1/2} = 3{,}530 \text{ ft}$$

$$\log q_g = \log C + n \log (p_e^2 - p_w^2) \tag{5.46}$$

Where:

$$C = \frac{0.703 k_{avg} h}{u z_{avg} T_f \ln(0.606\, r_e/r_w)} \tag{5.47}$$

$$\frac{C_5}{C_2} = \frac{\ln(0.606\, r_{e2}/r_w)}{\ln(0.606\, r_{e5}/r_w)} \qquad C_5 = C_2 \frac{\ln(0.606\, r_{e2}/r_w)}{\ln(0.606\, r_{e5}/r_w)} \tag{5.48}$$

Calculate C_2. Fig. C5 is the plot of data from problem 5.8:

at $q_g = 1{,}800$ Mscfd, $\log \Delta(p^2) = 5.851$

$$\log q_g = \log C + n \log (\Delta(p)^2)$$

$$3.255 = \log C_2 + (0.900)(5.851)$$

$$C_2 = 0.00975$$

$$C_5 = (0.00975)\frac{\ln\left(0.606\,\frac{5{,}580}{0.25}\right)}{\ln\left(0.606\,\frac{3{,}530}{0.25}\right)}$$

$$C_5 = 0.01025$$

At contract completion $p_e = 1{,}568$ psia and the five wells are producing at a total rate of 20 MMscfd, so one well is producing 4 MMscfd:

$$\log q_g = \log C + 0.900 \log \Delta(p^2) \tag{5.46}$$

$$\log \Delta(p^2) = \frac{\log 4{,}000 - \log 0.01025}{0.900} = \frac{3.602 - (-1.989)}{0.900} = \frac{5.591}{0.900} = 6.21$$

$$\Delta(p^2) = 1.621 \times 10^6$$

$$(1{,}568)^2 - (p_w)^2 = 1.621 \times 10^6$$

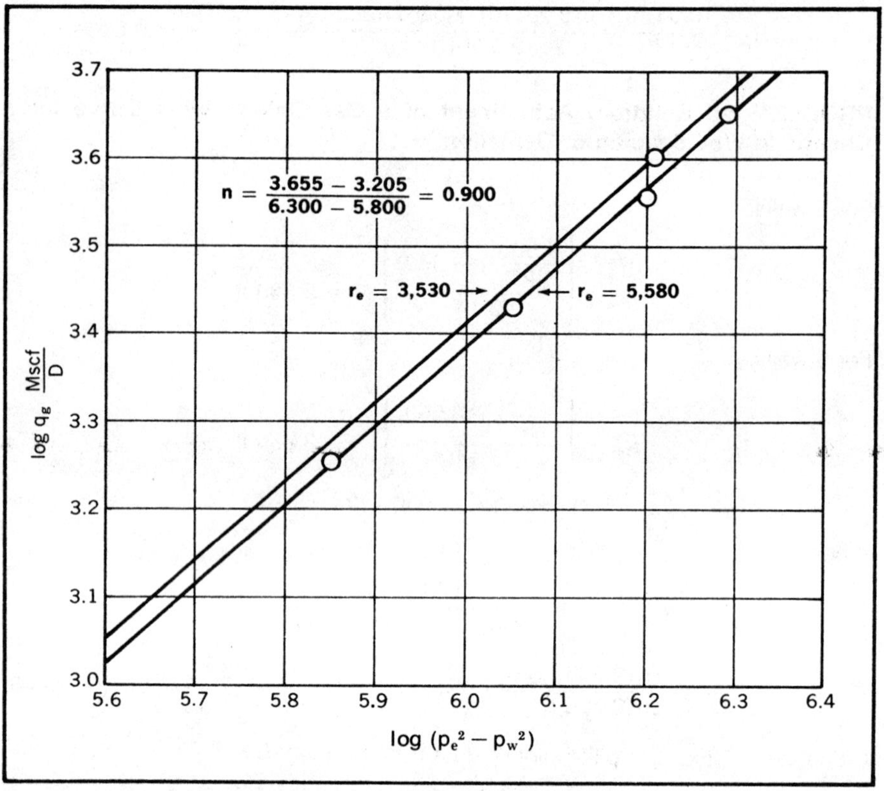

Fig. C5 Data from problem 5.8

$$(p_w)^2 = (1{,}568)^2 - 1.621 \times 10^6 = 2.459 \times 10^6 - 1.621 \times 10^6 = 0.838 \times 10^6$$

$$p_w = 915 \text{ psia}$$

Curve for 5 wells is parallel to the one for 2 wells and passes through $q_g = 4{,}000$ Mscfd, $\Delta(p^2) = 1.621 \times 10^6$. See Fig. C5:

$$\log 4{,}000 = 3.602; \ \log \Delta(p^2) = 6.21$$

PROBLEM 5.10 Solution: Using Isochronal Data

A.
$$r_e{}^* = \left(\frac{38.5kt^*}{\phi\mu c}\right)^{1/2} \tag{5.61}$$

$$r_e = \left(\frac{38.5 \times 0.074 \times \frac{1}{24}}{0.15 \times 0.021 \times 0.000357}\right)^{1/2}$$

$$r_e = (10.55 \times 10^4)^{1/2}$$

$$r_e = 325 \text{ ft}$$

B.

q_g, Mscfd	log q_g	BHP at q, psia	BHP^2	$\Delta(p^2)$	log $\Delta(p^2)$
		2,798	7.84×10^6	—	—
1,810	3.255	2,710	7.35×10^6	0.49×10^6	5.690
2,710	3.431	2,659	7.06×10^6	0.78×10^6	5.892
3,590	3.556	2,605	6.79×10^6	1.05×10^6	6.021
4,510	3.653	2,545	6.48×10^6	1.36×10^6	6.134

$$\log q_g = \log c + n \log \Delta(p^2) \qquad (5.46)$$

At 1,810 Msfd:

$$\log 1{,}800 = \log C + 0.900 \log (0.49 \times 10^6)$$

$$3.25 = \log C + (0.900)(5.690)$$

$$\log C = 3.25 - 5.12$$

$$\log C = -1.87$$

$$C = 0.0135$$

$$C_{60\ min} = 0.0135$$

For 2 wells:

$$r_e = \sqrt{\frac{(2{,}250)(43{,}560)}{\pi}} = 5{,}580 \text{ ft}$$

$$\frac{C_{5.580}}{C_{60\ min}} = \frac{\ell n(0.606 r_{60\ min}/r_w)}{\ell n(0.606 r_{5{,}580}/r_w)} \qquad (5.48 \text{ modified})$$

$$C_{5{,}580} = C_{60\ min} \frac{\ell n(0.606 r_{60\ min}/r_w)}{\ell n(0.606 r_{5{,}580}/r_w)}$$

$$C_{5{,}580} = 0.0135 \frac{\ell n(0.606 \times 325/0.25)}{\ell n(0.606 \times 5{,}580/0.25)} = 0.0135 \frac{\ell n(788)}{\ell n(13{,}510)} = 0.0135 \frac{6.66}{9.51}$$

$$C_{5{,}580} = 0.00945$$

At $q_g = 1{,}810$

$$\log 1{,}810 = \log 0.00945 + n \log \Delta(p^2)$$

$$3.255 = -2.024 + 0.900 \log \Delta(p^2)$$

$$\log \Delta(p^2) = \frac{5.279}{0.900}$$

$$\log \Delta(p^2) = 5.865$$

Therefore, the stabilized back-pressure curve for $r_e = 5{,}580$ ft passes through the point log 1,810 = 3.255, log $\Delta(p^2) = 5.865$ and is parallel to the curve for $r_e = 325$ ft.

C.

$$q_g = 4{,}500 \text{ Mscfd}$$

$$BHP = 2{,}425 \text{ psia}$$

$$\log q_g = \log 4{,}500 = 3.653$$

$$\log \Delta(p^2) = \log(p_e^2 - p_w^2) = \log(2{,}810^2 - 2{,}425^2)$$

$$= \log(7.84 \times 10^6 - 5.89 \times 10^6)$$

$$= \log(1.95 \times 10^6)$$

$$= 6.290$$

$$\therefore \log q_g = 3.653$$

$$\log \Delta(p^2) = 6.290$$

The point is plotted on the graph (Fig. C6) and does fall near the $r_e = 5{,}580$ curve.

PROBLEM 5.11 Solution: Determining Approximate Isochronal Data from Conventional Drawdown Data

$$t_{Dw} = \frac{6.33kt}{\phi\mu c r_w^2} = \frac{6.33(0.074)(t_{min})}{(0.15)(0.021)(0.000357)(0.25)^2(1{,}440)} \quad (3.11)$$

$$= 4.62(10^3)t_{min}$$

$$t_{60} = 27.7(10^4) \qquad \ell n\, t_{60} = 12.5$$

$$t_{120} = 55.5(10^4) \qquad \ell n\, t_{120} = 13.22$$

$$t_{180} = 83.2(10^4) \qquad \ell n\, t_{180} = 13.64$$

$$t_{240} = 111.0(10^4) \qquad \ell n\, t_{240} = 14.00$$

For 60 min:

$$\frac{\Delta(p^2)_{actual}}{\Delta(p^2)_{desired}} = \frac{1{,}800(\ell n\, t_{60} + 0.809)}{1{,}800(\ell n\, t_{60} + 0.809)} \quad \text{(Similar to 5.73)}$$

From test data:

$$\Delta(p^2)_{actual} = (2{,}800)^2 - (2{,}710)^2$$

$\Delta(p^2)_{desired} = 0.5(10^6)$; log $0.5(10^6) = 5.699$, which is plotted versus log $1{,}800 = 3.255$ in Fig. C6.

For 120 min:

$$\frac{\Delta(p^2)_{actual}}{\Delta(p^2)_{desired}} = \frac{1{,}800(\ell n\, t_{120} + 0.809) + 900(\ell n\, t_{60} + 0.809)}{2{,}700(\ell n\, t_{60} + 0.809)} \quad (5.73)$$

From test data:

$$\Delta(p^2)_{actual} = (2{,}800)^2 - (2{,}653)^2$$

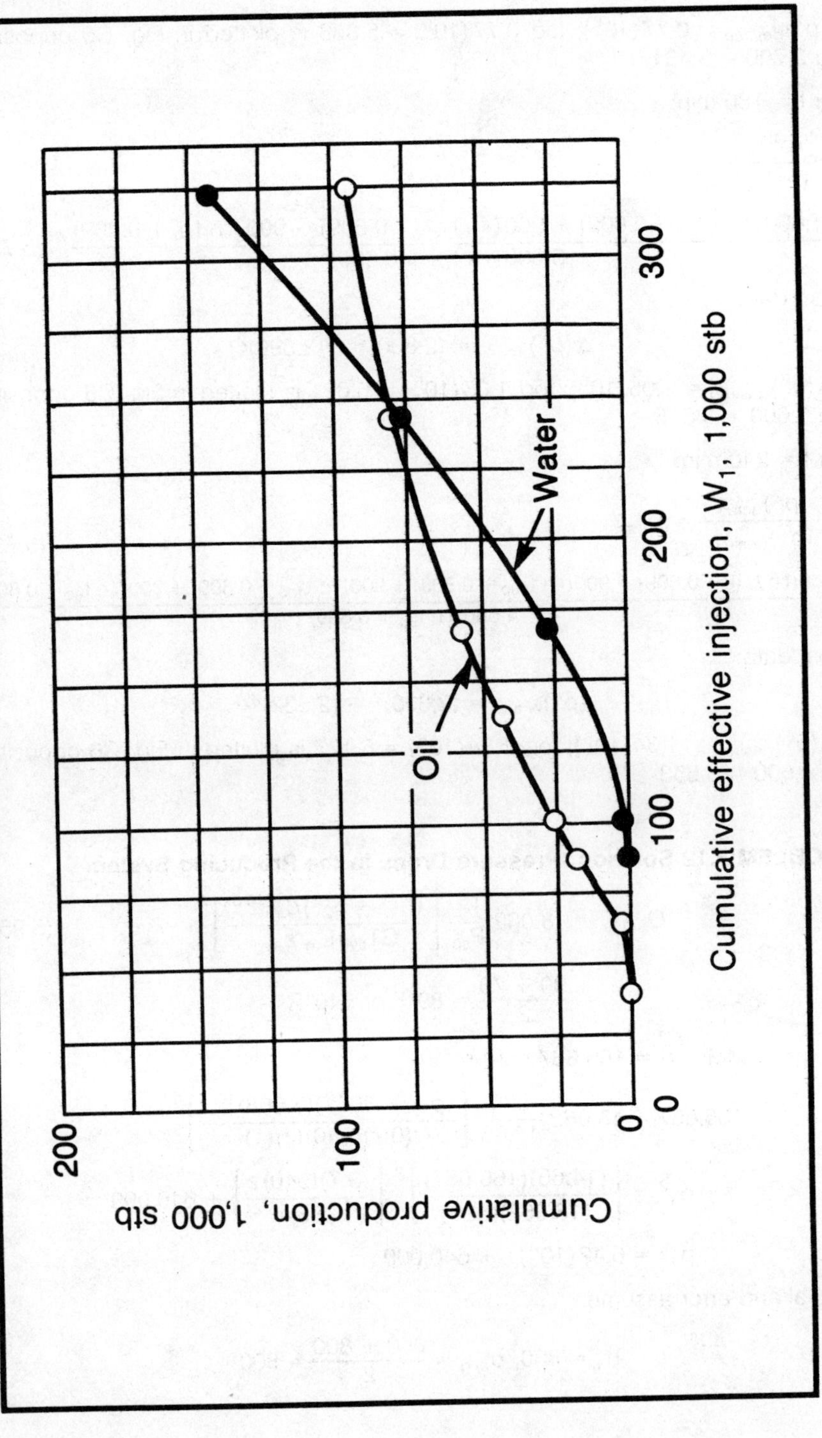

Fig. C6 Solution to problem 5.10

$\Delta(p^2)_{desired} = 0.77(10^6)$; log $0.77(10^6) = 5.886$ is plotted in Fig. C6 opposite log 2,700 = 3.431.

For t = 180 min:

$$\frac{\Delta \ (p^2)_{actual}}{\Delta \ (p^2)_{desired}} =$$

$$\frac{1{,}800(\ell n\ t_{180} + 0.809) + 900(\ell n\ t_{120} + 0.809) + 900(\ell n\ t_{60} + 0.809)}{3{,}600(\ell n\ t_{60} + 0.809)} \quad (5.73)$$

From data:

$$\Delta(p^2)_{actual} = (2.800)^2 - (2{,}596)^2$$

$\therefore \Delta(p^2)_{desired} = 1.05(10^6)$; log $1.05(10^6) = 6.021$ is plotted in Fig. C6 opposite log 3,600 = 3.556.

For t = 240 min:

$$\frac{\Delta \ (p^2)_{actual}}{\Delta \ (p^2)_{desired}} = \quad (5.73)$$

$$\frac{1{,}800(\ell n\ t_{240} + 0.809) + 900(\ell n\ t_{180} + 0.809) + 900(\ell n\ t_{120} + 0.809) + 900(\ell n\ t_{60} + 0.809)}{4{,}500(\ell n\ t_{60} + 0.809)}$$

From data:

$$(p^2)_{actual} = (2{,}800)^2 - (2{,}532)^2$$

$\therefore \Delta(p^2)_{desired} = 1.34(10^6)$; log $1.34(10^6) = 6.127$ is plotted in Fig. C6 opposite log 4,500 = 3.653

PROBLEM 5.12 Solution: Pressure Drops in the Producing System

$$Q_{scf/hr} = 18.062 \frac{T_{sc}}{P_{sc}} \left[\frac{(p_1^2 - p_2^2) d_{in.}^{16/3}}{GT_{avg} L_{mi} Z_{avg}} \right]^{0.5} \quad (5.95)$$

$$T_{avg} = \frac{90 + 70}{2} = 80°F \text{ or } 540°R$$

For q = 4 MMscfd = 166,667 scf/hr:

$$166{,}667 = 18.062 \frac{520}{14.65} \left[\frac{(p_1^2 - 800^2)(3.068)^{16/3}}{(0.7)(540)(z)(1)} \right]^{0.5}$$

$$p_1^2 = \left[\frac{(14.65)(166{,}667)}{(18.062)(520)} \right]^2 \left[\frac{(0.7)(540)z}{(3.068)^{16/2}} \right] + 640{,}000$$

$$p_1^2 = 6.42(10^4)z + 640{,}000$$

By trial and error assume:

$$p_1 = 800,\ p_{avg} = \frac{800 + 800}{2} = 800$$

Therefore:

$$p_r = \frac{800}{666} = 1.2$$

$$T_r = \frac{540}{390} = 1.38$$

$$z = 0.84$$

Calculating, we find:

$$p_1 = 833 \text{ psia}$$

So assume:

$$p_1 = 833 \text{ psia}$$

$$p_{avg} = \frac{833 + 800}{2} = 817$$

$$p_r = \frac{817}{666} = 1.225$$

$$z = 0.84$$

$$p_1 = 833 \text{ psia which matches assumption}$$

Using a wellhead pressure of 833 psia, we can find the bottom-hole pressure from the graphical method. Note that the available figures are for 2½-in. tubing, so the conversion chart (Fig. 5–26) must be used to find that 4 MMscfd in 3-in. tubing is equivalent to 2.35 MMscfd in 2.5-in. tubing.

An interpolation is necessary between charts for static bottom-hole pressure and 4 MMscfd flowing bottom-hole pressure.

From Fig. 5–7: 180 − 80 = 100 psia gas column pressure
From Fig. 5–22: 260 − 120 = 140 psia gas column pressure

Interpolating for the 2.35 MMscfd rate gives:

$$100 + \frac{2.35}{4.0}(140 - 100) = 123.5 \text{ psia weight pressure of gas column}$$

$$\text{BHP} = 833 + 123.5 = 956.5 \text{ psia}$$

Similarly, for 5 MMscfd in 3-in. pipe (same as 2.95 MMscfd in 2.5-in. pipe):

Wellhead pressure =	851 psia
Gas column pressure =	133 psia
Bottom-hole pressure =	984 psia

For 6.67 MMscfd in 3-in. pipe (same as 4.0 MMscfd in 2.5-in. pipe):

Wellhead presure =	887 psia
Gas column pressure =	147 psia
Bottom-hole pressure =	1,036 psia

For 10.0 MMscfd in 3-in. pipe (same as 5.9 MMscfd in 2.5-in. pipe):

Wellhead pressure =	986 psia
Gas column pressure =	196 psia
Bottom-hole pressure =	1,182 psia

PROBLEM 5.13 Solution: Determining Gas-Well Spacing for Completing a Pipeline Purchasing Contract

At contract completion, the following information is known:

$$q = 20 \text{ MMscfd}$$

$$p_e = 1{,}568 \text{ psia}$$

$$p_e^2 = 2.459(10^6)$$

If there are five wells at completion:

$$q = 4{,}000 \text{ Mscfd/well}$$

$$\log 4{,}000 = 3.602$$

From Fig. 5–29 for 5 wells:

$$\log \Delta(p^2) = 6.210$$

$$\Delta(p^2) = 1.622(10^6)$$

$$(p_e^2 - p_w^2) = 1.622(10^6)$$

$$p_w^2 = 2.459(10^6) - 1.622(10^6)$$

$$p_w^2 = 0.837(10^6)$$

$$p_w = 915 \text{ psia}$$

This well pressure is less than the 956 psia needed for a flow rate of 4.0 MMscfd through the tubing and flow line as shown in Fig. 5–28. Thus, we cannot fulfill the contract with five wells using this equipment. If there were six wells at completion: $q = 3{,}333$ Mscfd/well. From $\log 3{,}333 = 3.52$:

$$\log \Delta(p^2) \text{ for six wells} = 6.115$$

$$\Delta(p^2) = 1.303(10^6)$$

$$p_e^2 - p_w^2 = 1.303(10^6)$$

$$p_w^2 = 2.459(10^6) - 1.303(10^6)$$

$$p_w = 1.000(10^3)$$

$$p_w = 1{,}075 \text{ psia}$$

$p_w = 1{,}075$ psia, which is more than the psia needed for a flow rate of 3,333 MMscfd as indicated in Fig. 5–28. Therefore, six equally spaced wells are needed to meet this contract if the equipment investigated must be used. Use of larger tubing or flow lines would undoubtedly prove advisable.

PROBLEM 5.14 Solution: Predicting Rate versus Time Behavior for a Gas Reservoir

To find the time when the rate begins decline, i.e., the reservoir and equipment will no longer produce at 30 MMcfd, 5 MMcfd per well.

$$\log 5{,}000 \text{ Mcfd} = 3.7$$

From Fig. 5–29:

$$\log (p_e^2 - p_w^2) = 6.31$$

$$p_e^2 - p_w^2 = 2{,}041{,}738$$

From Fig. 5–28:

$$p_w = 983 \text{ psia}$$

$$p_e = 1{,}735$$

From Fig. 5–30:

$$(p/z)_{avg} = 2{,}270$$

$$G_p\text{, MMcf} = (4{,}558 - (p/z)_{avg})/0.053$$
$$= (4{,}558 - 2{,}270)/0.053$$
$$= 43{,}170$$

$$t = \frac{43{,}170 - 9{,}450}{80 \times 365} = 3.08 \text{ years}$$

When the rate = 4 MMcfd/well

From Fig. 5–29:

$$(p_e^2 - p_w^2) = 1{,}596{,}000$$

From Fig. 5–28:

$$p_w = 956 \text{ psia}$$
$$p_e = 1{,}584 \text{ psia}$$

From Fig. 5–30:

$$(p/z)_{avg} = 2{,}035$$

$$G_p\text{, MMcf} = (4{,}558 - 2{,}035)/0.053$$
$$= 45{,}717 \text{ MMcf}$$

$$t = \frac{45{,}717 - 43{,}170}{(4.5)(6)365} = 0.26 \text{ years}$$

Contract time corresponding to a rate of 4 MMcfd/well is:

$$3.08 + 0.26 = 3.34 \text{ years}$$

Similarly, the data in the following table are applicable. See Fig. C7 for rate-time plot.

q_g/well, MMcfd	p_w, psia	(p^2), psia^2	p_e psia	p/z, psia	G_p, MMcf	Δt, years	t, years
5	983	2,042,000	1,735	2,270	43,170	3.08	3.08
4	956	1,596,000	1,584	2,035	45,717	0.26	3.34
3	932	1,161,450	1,424	1,810	49,962	0.55	3.89
2	912	746,449	1,256	1,570	54,490	0.83	4.72
1	894	350,752	1,072	1,345	58,736	1.29	6.10

PROBLEM 5.15 Solution: Predicting the Pseudosteady-State Flow Rate for a Microdarcy Gas Reservoir Containing a Massive Hydraulic Fracture

The undamaged permeability can be calculated from the radial flow behavior of the constant-rate drawdown, Fig. 5–32:

$$m = 0.1625\, q\mu/kh \qquad (4.13)$$
$$542 = (0.1625)(342)(0.014)/k(10)$$
$$k = 0.00014 \text{ darcy or } 0.14 \text{ md}$$

The linear infinite-acting portion of the drawdown (Fig. 5–33) can be used to calculate the fracture length:

$$m_L = -\frac{1.259q}{hL}\left(\frac{\mu}{k\phi c}\right)^{0.5} \qquad (5.98)$$

The slope in Fig. 5–33 is:

$$8 \text{ psi}/\sqrt{\text{min}} \text{ or } 8\sqrt{1{,}440 \text{ min/day}} = 303.58 \text{ psi}/\sqrt{\text{day}}$$

$$-303.58 = -\frac{1.259(342)}{10\,L}\left[\frac{0.014}{(0.00014)(0.11)(0.0008787)}\right]^{0.5}$$

$$L = 144 \text{ ft}$$

When the well reaches pseudosteady state:

$$q_g = \frac{0.703\, kh\,(p_e^2 - p_w^2)^n}{\mu\, zT[\ell n(r_e/r_w) - 0.5 + S]} \qquad (5.41)$$

Assume the fracture fluid does not significantly damage the sand so S = 0.0. Also, the low flow rates and large flowing cross section at the well (due to the fracture) give low velocities. Thus, no turbulence exists, and n = 1. Furthermore, Prats showed that the flow capacity of a fracture well is about the same as the flow capacity of a well whose radius = (1/4) total fracture length:

$$q_g = \frac{(0.703)(0.00014)(10)(1{,}138^2 - 150^2)^{1.0}}{(0.014)(0.86)(560)\left[\text{ln}\,\dfrac{915}{0.25(144)} + 0.5 + 0.0\right]}$$

$$= 49.7 \text{ Mcfd}$$

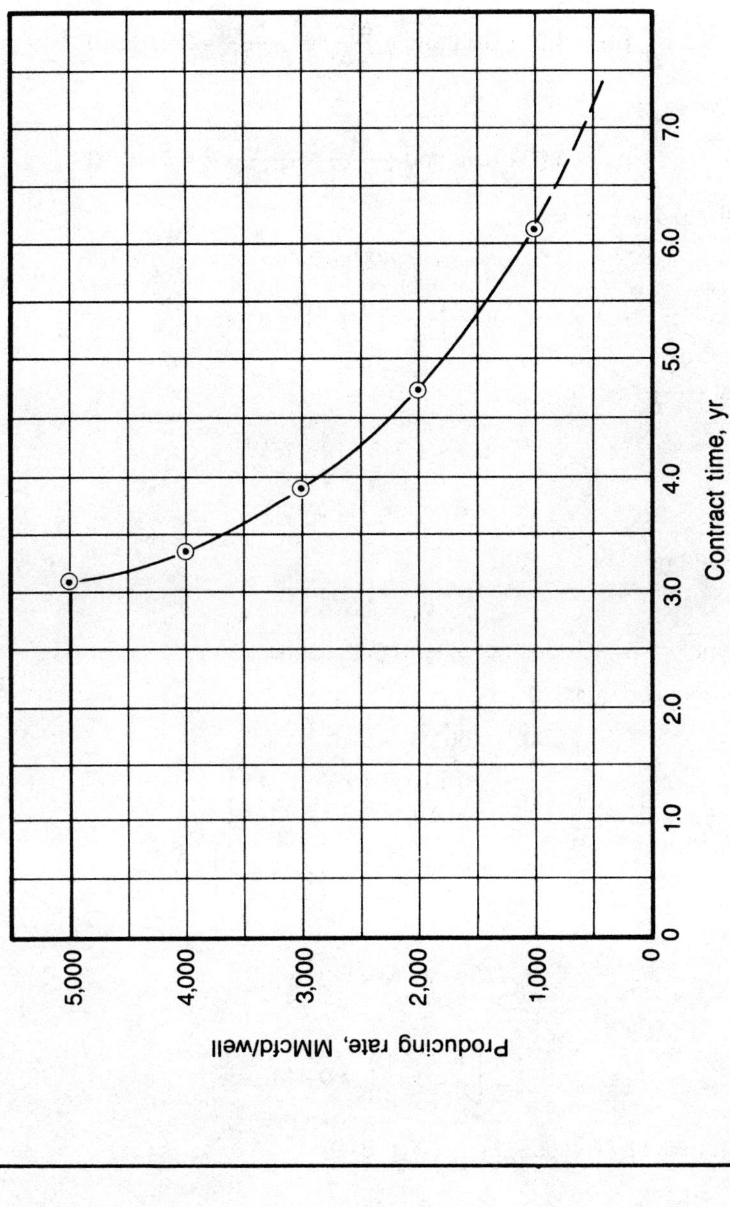

Fig. C7 Rate versus time for gas reservoir, problem 5.14 solution

PROBLEM 5.16 Solution: Evaluating a Pressure Buildup in a Gas Reservoir and Determining the Distance to a Reservoir Barrier

A. $t = \frac{248}{3.9} = 63.6$ days or 1,526 hr

At 5 hr:

$$p_w = 4{,}011 \text{ psia and } \frac{\Delta t}{t + \Delta t} = \frac{5}{1{,}531} = 3.27 \times 10^{-3}$$

At 10 hr:

$$p_w = 4{,}094 \text{ psia and } \frac{\Delta t}{t + \Delta t} = \frac{10}{1{,}536} = 6.51 \times 10^{-3}$$

From the plotted points:

$$m = 275/\text{cycle}$$

$$p_w \text{ at} \left(\frac{\Delta t}{t + \Delta t} = 10^{-2}\right) = 4{,}145$$

Then at $\frac{t}{t + \Delta t} = 1.0$:

$$p_w = 4{,}145 + 2(275)$$
$$= 4{,}695 < 4{,}990$$
$$\therefore \text{reservoir is finite acting}$$

B. At $\Delta t = 68$ hr:

$$p_w = 4{,}378$$

Note that Fig. 5–35 does not extend to this time.

p_w extrapolated = 4,320

$$\Delta p' = 58 \text{ psi}$$

$$\Delta p' = \frac{0.141 q \mu}{kh} \left[\frac{1}{2}\left(\text{Ei} \frac{-1}{4t_D}\right)\right]$$

$$58 = (0.867)(275) \left[\frac{1}{2} \text{Ei} \frac{-1}{4t_D}\right]$$

Where:

$$275 = m \text{ from Fig. 5–35}$$

Then:

$$\left[\frac{1}{2} \text{Ei} \frac{-1}{4t_D}\right] = 0.243$$

From graph of t_D versus $\frac{1}{2} \text{Ei} \frac{-1}{4t_D}$ (Fig. 3–9):

$$t_D = 0.43$$

$$t_D = \frac{6.33(k/\mu)t}{\phi c(2d)^2} \tag{4.19}$$

Must evaluate k/μ from m:

$$m = \frac{0.1625q\mu}{kh} \tag{4.13}$$

Must evaluate q at the average reservoir pressure in res b/d:

$$T_c = 410°F$$

From Fig. B9:

$$p_c = 665 \text{ psia}$$

$$T_R = \frac{253 + 460}{410} = 1.74$$

Assume:

$p_{avg} = (4{,}695 + 998)/2 = 2{,}847$ (using 4,695 as the estimated p_e)

$$p_R = \frac{2{,}847}{665} = 4.28$$

From Fig. B8:

$$z = 0.867$$

$$B_g = \frac{(5.04) \times (10^{-3})(0.867)(7.13 \times 10^2)}{2{,}847} = 0.00109 \tag{5.12}$$

$$q = (10.9 \times 10^{-4})(3.9 \times 10^6) = 4{,}251 \text{ b/d}$$

$$275 = \frac{(1.625 \times 10^{-1})(4.251 \times 10^3)}{(30)(k/\mu)}$$

From $p_{avg} = 2{,}847$:

$$\frac{k}{\mu} = 0.0836 \text{ darcy/cp}$$

From Fig. B6:

$$c_r = 0.22$$

$$c_g = \frac{0.22}{665} = 3.3 \times 10^{-4}$$

Substituting into Eq. 4.19:

$$0.43 = \frac{(6.33)(8.36)(10^{-2})\frac{68}{24}}{(5.0 \times 10^{-2})(4)(d^2)(3.3 \times 10^{-4})}$$

$$d^2 = 5.27 \times 10^4$$

$$d = 230 \text{ ft}$$

C.

$$(\Delta p/\Delta t)_{pseudo} = \frac{5.615q}{c\phi V_p} = \frac{q}{cV_{pbbl}} \tag{3.34}$$

For $p_s = 4{,}600$:

$$p_r = \frac{4{,}600}{665} = 6.92$$

From Fig. B6:

$$c_r = 0.1 \text{ and } c_g = \frac{0.1}{665} = 1.5 \times 10^{-4}$$

$$V_{p(bbl)} = \frac{3{,}025}{(6)(1.5 \times 10^{-4})} = 3.36 \times 10^6 \text{ bbl}$$

D.

$$GB_{gi} = 3.36 \times 10^6$$

For p = 4,990 psia:

$$p_r = \frac{4{,}990}{665} = 7.5$$

From Fig. B8:

$$z = 1.0$$

$$B_{gi} = \frac{(5.04)(10^{-3})(1.0)(7.12)(10^2)}{(4.990)(10^3)} = 7.19 \times 10^{-4} \tag{5.12}$$

$$G = \frac{3.36 \times 10^6}{7.19 \times 10^{-4}} = 4.67 \times 10^9$$

$$GB_{gi} = (G - G_p)B_g \tag{5.21}$$

$$\frac{p}{z} = \frac{(4.670 - 0.248)\,10^9}{(4.67)(10^9)(1.0)/4{,}990}$$

$$= \frac{4.422}{4.67}(4{,}990) = 4{,}725$$

$$\frac{p_r}{z} = \frac{p/z}{p_c} = \frac{4{,}725}{665} = 7.11$$

From Fig. B10:

$$z = 0.97$$

$$p = 4{,}725 \times 0.97 = 4{,}583 \text{ psia}$$

Note that the assumption of p_s in part C and the calculation in part D can be used as a trial-and-error method of determining p_s when $(\Delta p/\Delta t)_{pseudo}$ is known.

PROBLEM 6.1 Solution: Determining the Static Saturation Distribution from Reservoir Capillary Pressure Data

To find the free-water level, read the threshold pressure for the bottom zone, zone V, as 0.45 psi. This corresponds to the 100% saturation point in the reservoir:

$$P_c = 0.433\ \Delta\gamma h \tag{6.2}$$

$$0.45 = 0.433\left(\frac{65.3 - 56.2}{62.4}\right)h$$

$$h = \frac{(0.45)(62.4)}{(0.433)(9.1)} = 7.13 \text{ ft}$$

Free-water level = 4,053 + 7 = 4,060 ft
Find $\Delta p_{c\ res}$ equivalent of 1 ft:

$$\frac{0.45}{7.13} = (\Delta p_{c\ res})_{1\ ft}$$

$$(\Delta P_{c\ res})_{1\ ft} = \frac{0.45}{7.13} = 0.063 \text{ psi/ft}$$

(1) Depth, ft	*(2)* Height above Free-Water Level, h, ft	*(3)* Equivalent p_c	*(4)* Zone	*(5)* S_w, %
4,053	7.0	0.45	V	100 (critical)
4,049	11.0	0.69	V	55
4,047	13	0.82	V	47
4,045	15	0.94	IV	80
4,042	18	1.07	IV	67
4,039	21	1.33	IV	53
4,036	—	Shale		100
4,033	—	Shale		100
4,030	30	1.89	III	18
4,021	39	2.46	III	16
4,018	42	2.65	II	23
4,009	51	3.21	II	22
4,006	54	3.40	I	16
4,000	60	3.78	I	16

(2) h = 4,060 − depth
(3) Equivalent $p_c = h(0.063)$
(5) S_w is read from p_c curve

PROBLEM 6.2 Solution: Conversion of Lab Capillary Pressure Data to Reservoir Capillary Pressure Data

Zone I 3,998 − 4,007
Average permeability = 564 md; use Curve 4, permeability = 569 md

Zone II 4,007 – 4,019
Average permeability = 166 md; use Curve 3 with average permeability = 157 md

Zone III 4,019 – 4,031
Average permeability = 591 md; use Curve 4 with permeability = 569 md

Zone IV 4,037 – 4,046
Average permeability = 10.2 md; use Curve 1 with permeability = 11.2 md

Zone V 4,046 – 4,055
Average permeability = 72 md; No close curve available. Interpolate between curves 2 and 3; ⅔ of distance

72 – 34 = 38
157 – 72 = 85

$$\frac{85}{38 + 85} \approx 2/3$$

$$(P_c)_{res} = (P_c)_{lab} \frac{(\sigma \cos \theta)_{res}}{(\sigma \cos \theta)_{lab}} \quad (5.6)$$

$$= (P_c)_{lab} \frac{(28)(1)}{(70)(1)} = 0.4\,(P_c)_{lab}$$

	(1) $(P_c)_{lab}$ at 50% S_w	(2) $(P_c)_{res}$ at 50% S_w
Curve 1	3.65	1.46
2	2.53	1.01
3	1.35	0.54
4	0.80	0.32
Interpolated curve	2.00	0.80

(1) From Fig. 6–7
(2) = 0.4 × column 1

PROBLEM 6.3 Solution: Using the J Function to Average Capillary Pressure Data

(1) Core No.	(2) P_c for S_w = 50%	(3) k, md	(4) θ	(5) J for S_w = 50%
1	3.66	11.2	0.147	0.46
2	2.50	34.0	0.174	0.50
3	1.38	157.0	0.208	0.54
4	0.85	569.0	0.275	0.55

(1), (2), (3), (4) from Fig. 6–7

(5)

$$J = \frac{P_c(k_{md}\phi)^{0.5}}{\sigma \cos \theta} \tag{6.13}$$

$$= \frac{P_c(k_{md}\phi)^{0.5}}{(70)(1.0)}$$

Note that the average J value for $S_w = 50\%$ from Fig. 6–8 is about 0.5. To calculate the P_c curve for zone II of problem 6.1 from the J function:

$$P_c = \frac{J\,\sigma \cos \theta}{(k_{md}/\phi)^{0.5}} \quad \text{(6.13 modified)}$$

$$= \frac{J(28)(1.0)}{(166/0.208)^{0.5}}$$

$$= (0.99)\,J$$

(1) S_w, %	*(2)* J	*(3)* $P_C = 0.99\,J$
100	0.35	0.35
70	0.40	0.40
54	0.45	0.45
44	0.60	0.59
30	1.45	1.43
20	3.15	3.12

(1) Assumed
(2) From Fig. 6–8

PROBLEM 6.4 Solution: Calculating a Fractional Flow Curve

$$f_d = \frac{1 - (0.488k\,k_{ro}A\,|\Delta\gamma|\,\sin\alpha/\mu_o q_t)}{1 + (k_o/k_w)(\mu_w/\mu_o)} \quad \text{(6.18 modified)}$$

$$= \frac{1 - (0.488 \times 0.108 \times k_{ro} \times 240{,}000 \times 0.04 \times \sin 15.5)/(1.51 \times 2{,}830)}{1 + (k_o/k_w)(0.83/1.51)}$$

$$= \frac{1 - 0.0316k_{ro}}{1 + 0.5497\,(k_{ro}/k_{rw})}$$

S_w	k_{ro}	k_{rw}	(k_{ro}/k_{rw})	$(1 - 0.0316k_{ro})$	$1 + 0.5497\,(k_{ro}/k_{ro})$	f_w
79	0.00	0.63	0.000	1.000	1.000	1.000
75	0.02	0.54	0.037	0.9994	1.0203	0.980
65	0.09	0.37	0.243	0.9972	1.1336	0.880
55	0.23	0.23	1.000	0.9927	1.5497	0.641
45	0.44	0.13	3.385	0.9861	2.8607	0.345
35	0.73	0.06	12.167	0.9769	7.6882	0.127
25	0.94	0.02	47.000	0.9703	26.8359	0.036
16	0.98	0.00		0.9690		0.000

See Fig. 6–15 for a plot of f_w versus S_w.

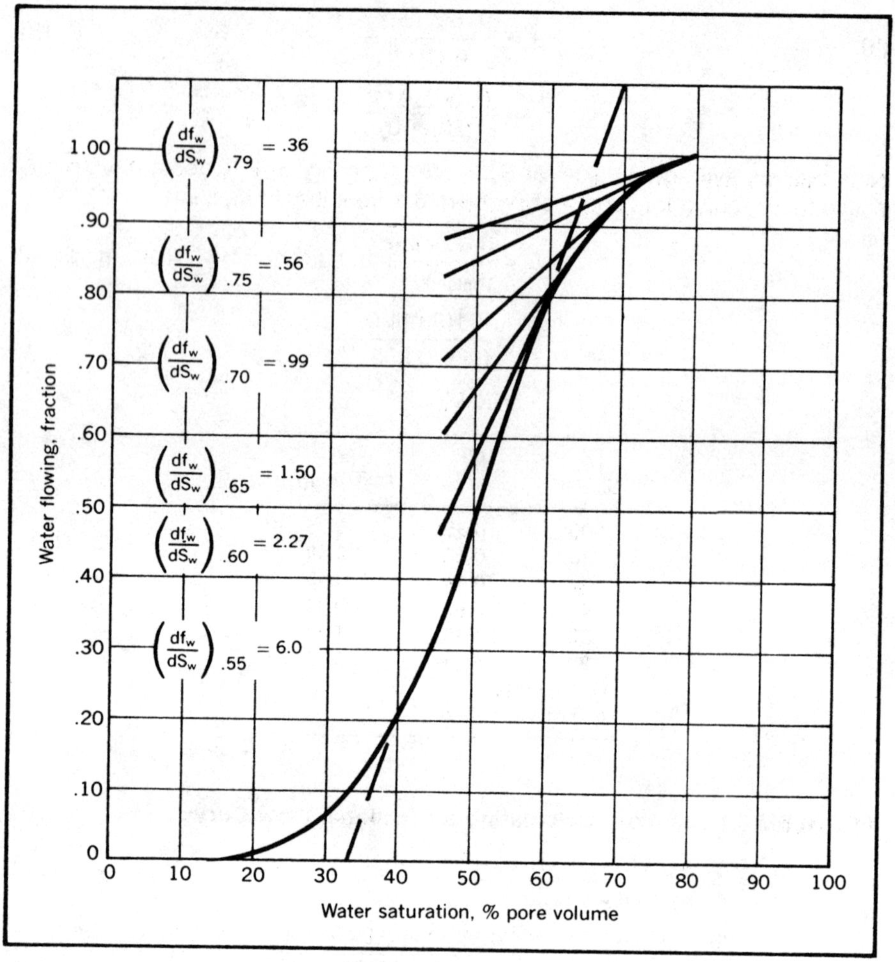

Fig. C8 Solution to problem 6.5

PROBLEM 6.5 Solution: Application of the Buckley-Leverett Equation

$$\Delta x_j = \left(\frac{5.615 q_t}{\phi A}\right) \Delta t \left(\frac{df_w}{dS_w}\right)_j \qquad \text{(6.14 modified)}$$

$$= \frac{(5.615 \times 2{,}830)}{(0.215 \times 240{,}000)} \Delta t \left(\frac{df_w}{dS_w}\right)_j$$

$$= 0.308\ \Delta t \left(\frac{df_w}{dS_w}\right)_j$$

Table of Δx Values, ft

S_w	$\left(\frac{df_w}{dS_w}\right)^*$	Δx at ½ year (182.5 days)	Δx at 1 year (365 days)	Δx at 2 years (730 days)	** $x_{init.}$	x at 2 years	x at ½ year
0.79	0.36	20	40	81	10	91	30
0.75	0.56	31	63	126	12	138	43
0.70	0.99	56	111	222	15	237	71
0.65	1.50	84	169	338	18	356	102
0.60	2.27	128	255	510	22	532	150
0.55	3.00	168	336	672	26	698	194

*Evaluated graphically on Fig. C8.
**From initial saturation curve of Fig. 6–17.

Fig. 6–17 is a plot of the calculated saturation profiles.

PROBLEM 6.6 Solution: Evaluating the Frontal Position by Material Balance

$$\frac{5.615\, q_t\, t}{\phi A} = \Sigma[(S_d - S_{di})\Delta X] \qquad (6.22)$$

When t = 182.5 days (½ year):

$$\frac{5.615\, q_t\, t}{\phi A} = 5.615 \times 2{,}830 \times 182.5/0.215 \times 240{,}000$$

$$= 56.2$$

Graphically integrating as in Fig. 6–18

ΔX	$\Sigma\Delta X$	$\overline{S_d}$	S_{di}	$\overline{S_{di}}$	$\Delta X(\overline{S_d} - \overline{S_{di}})$	$\Sigma\Delta X(\overline{S_d} - \overline{S_{di}})$
—	0	—	1.0	—		
10	10	0.895	0.79	0.895	0	0
20	30	0.790	0.50	0.620	3.4	3.4
20	50	0.765	0.36	0.405	6.7	10.1
20	70	0.720	0.280	0.35	8.0	18.1
20	90	0.686	0.250	0.265	8.4	26.5
20	110	0.654	0.220	0.235	8.4	34.9
20	130	0.628	0.209	0.216	8.2	43.1
20	150	0.605	0.203	0.206	8.0	51.1
20*	170	0.583	0.197	0.200	7.7*	58.8*

*Too large
$\overline{S_d} = S_d$ average from Fig. 6–18.
$\overline{S_{di}} = S_{di}$ average; S_{di} from Fig. 6–18.

To calculate size of last increment to satisfy Eq. 6.21:

$$56.2 = 51.1 + (0.583 - 0.200)\,\Delta X$$
$$\Delta X = 13$$
$$X_f = 150 + 13 = 163 \text{ ft}$$
$$S_{wf} = 58\%$$

PROBLEM 6.7 Solution: Using the Welge Graphical Method to Calculate Displacement in a Reservoir

A. Refer to Fig. C9. The tangent drawn through S_{di} = 35 touches the fractional flow curve at S_w = 0.56 pore volumes. Thus, the water saturation at the front prior to breakthrough is 56% pore volume.

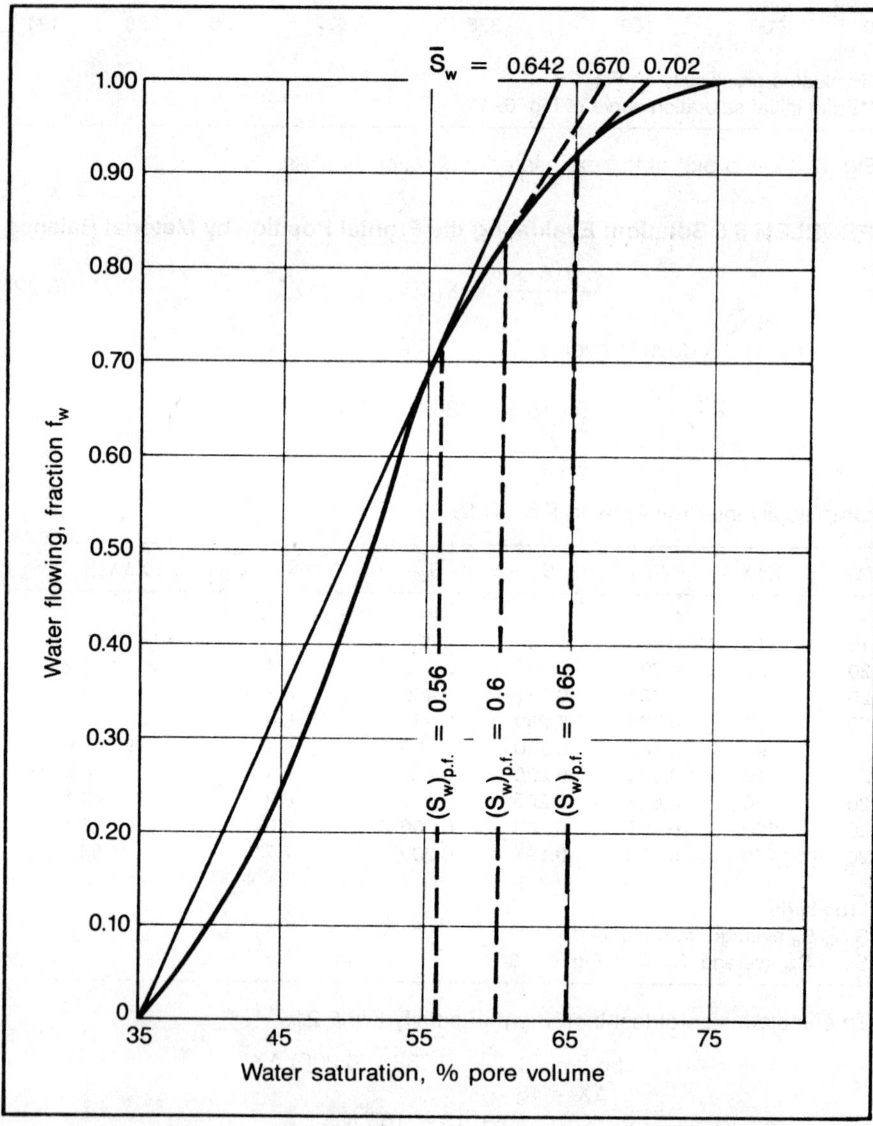

Fig. C9 Fractional flow curve for problem 6.7 solution

B. Prior to breakthrough of water at the producing face, oil recovery is equal to water injected less initial gas saturation (all expressed in pore volumes). After breakthrough the following method may be used.

The cumulative injection (pore volume) is related to the saturation at the producing face as:

$$\frac{W_i}{V_p} = \left(\frac{dS_w}{df_w}\right)_{p.f.} = \frac{1}{\left(\frac{df_w}{dS_w}\right)_{p.f.}} \qquad \text{(6.29 modified)}$$

Also, by the graphical method shown in Fig. C9, we can determine the average saturation of the invaded zone. The average change in the water saturation less the initial gas saturation is equal to the oil recovery expressed in pore volumes.

(1) S_w at Production Face	(2) $\left(\frac{df_w}{dS_w}\right)_{PF}$	(3) $\overline{S_w}$	(4) ΔS_w	(5) B_oN_{PF}/V_p	(6) W_i/V_p
0.560	3.42	0.642	0.292	0.242	0.292
0.600	2.36	0.670	0.320	0.270	0.424
0.650	1.45	0.702	0.352	0.302	0.690
0.700	0.73	0.728	0.378	0.328	1.370

B_oN_{PF}/V_p is plotted versus W_i/V_p in Fig. C10.

(1) Assumed

(2) Calculated slope of tangent in Fig. C9, e.g., when S_w at production face is 0.56: $(df_w/dS_w)_{p.f.} = (1 - 0)/(0.642 - 0.35) = 3.42$

(3) Read from Fig. C9.

(4) $\Delta S_w = \overline{S_w} - 0.35$.

(5) $B_oN_{PF}/V_p = \Delta S_w - 0.05$.

(6) Calculated from Eq. 6.29.

Column 5 plotted versus column 6 in Fig. C10.

PROBLEM 7.1 Solution: Gas Material Balance

$$G\,B_{gi} = (G - G_p)\,B_g \qquad (7.3)$$

$$(4 \times 10^8)(0.001) = [(4 \times 10^8) - G_p](0.0011)$$

$$G_p = 36.4 \text{ MMscf}$$

PROBLEM 7.2 Solution: Gas-Cap Expansion

$$\text{Gas expansion} = (G - G_{pc})\,B_g - G\,B_{gi} \qquad (7.5)$$

$$= [(4 \times 10^8) - 0](0.0011) - (4 \times 10^8)(0.001)$$

$$= 40{,}000 \text{ res bbl or } 40{,}000/0.0011 = 36.4 \text{ MMscf}$$

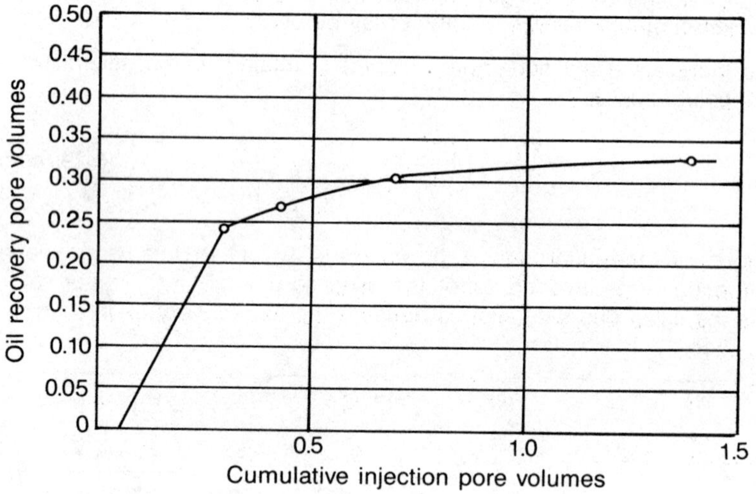

Fig. C10 Solution to problem 6.7—oil recovery versus cumulative injection

PROBLEM 7.3 Solution: PVT Lab Data Exercise

$$(V_o)_{ST} = 500 \text{ cc}$$

$$R_{si} = \frac{44{,}500}{500} = \text{cc/cc or bbl/bbl}$$

$$= \left(\frac{44{,}500}{500}\right) 5.615 \text{ cu ft/bbl} = 500 \text{ scf/bbl}$$

At 2,000 psia:

$$B_o = \frac{V_o}{(V_o)_{ST}} = \frac{650}{500} = 1.3$$

$$B_t = \frac{V_t}{(V_o)_{ST}} = \frac{650}{500} = 1.3$$

$$R_s = R_{si} = 500 \text{ scf/stb}$$

At 1,500 psia:

$$B_o = \frac{669}{500} = 1.34$$

$$B_t = \frac{669}{500} = 1.34$$

$$R_s = R_{si} = 500 \text{ scf/stb}$$

At 1,000 psia:

$$B_o = \frac{650}{500} = 1.3$$

$$B_t = \frac{650 + 150}{500} = 1.6$$

$$\text{Liberated gas} = (150)\left(\frac{1{,}000}{14.7}\right)\left(\frac{520}{655}\right)\left(\frac{1.0}{0.91}\right) = 8{,}902 \text{ cc}$$

$$R_s = (44{,}500 - 8{,}902)\ 5.615/500 = 400 \text{ scf/stb}$$

At 500 psia:

$$B_o = \frac{615}{500} = 1.23$$

$$B_t = \frac{615 + 700}{500} = 2.63$$

$$\text{Liberated gas} = (700)\left(\frac{500}{14.7}\right)\left(\frac{1.0}{0.95}\right) = 19{,}897 \text{ cc}$$

$$R_s = (44{,}500 - 19{,}897)\ 5.615/500 = 276 \text{ scf/stb}$$

Oil compressibility:

$$c = \frac{\Delta V/V}{\Delta p} = \frac{(699 - 650)/669}{2{,}000 - 1{,}500} = 5.7 \times 10^{-5}/\text{psi}$$

PROBLEM 7.4 Solution: Determining Gas-Oil Ratios*

(1) N_p, MM stb	(2) R, scf/stb	(3) ΔN_p, MM stb	(4) R_{avg}, scf/stb	(5) $R_{avg}\Delta N_p$, MMscf	(6) G_p, $\Sigma(R_{avg}\Delta N_p)$	(7) G_i, MMscf	(8) Net (G_p), MMscf
0	300	—	—	—	—	—	—
1	280	1	290	290	290	0	290
2	280	1	280	280	570	0	570
3	340	1	310	310	880	0	880
4	560	1	450	450	1,330	0	1,330
5	850	1	705	705	2,035	0	2,035
6	1,120	1	985	985	3,020	520	2,500
7	1,420	1	1,270	1,270	4,290	930	3,360

Columns 1, 2, and 7 data are given.
Column 3 = $(N_p)_n - (N_p)_{n-1}$
Column 4 = $(R_n + R_{n-1})/2$
Column 5 = (4) × (3)
Column 6 = Σ (5)
Column 8 = $G_p - G_i$

*See Fig. C11 for plots of R, G_p, $(G_p)_{net}$, and G_i versus N_p.

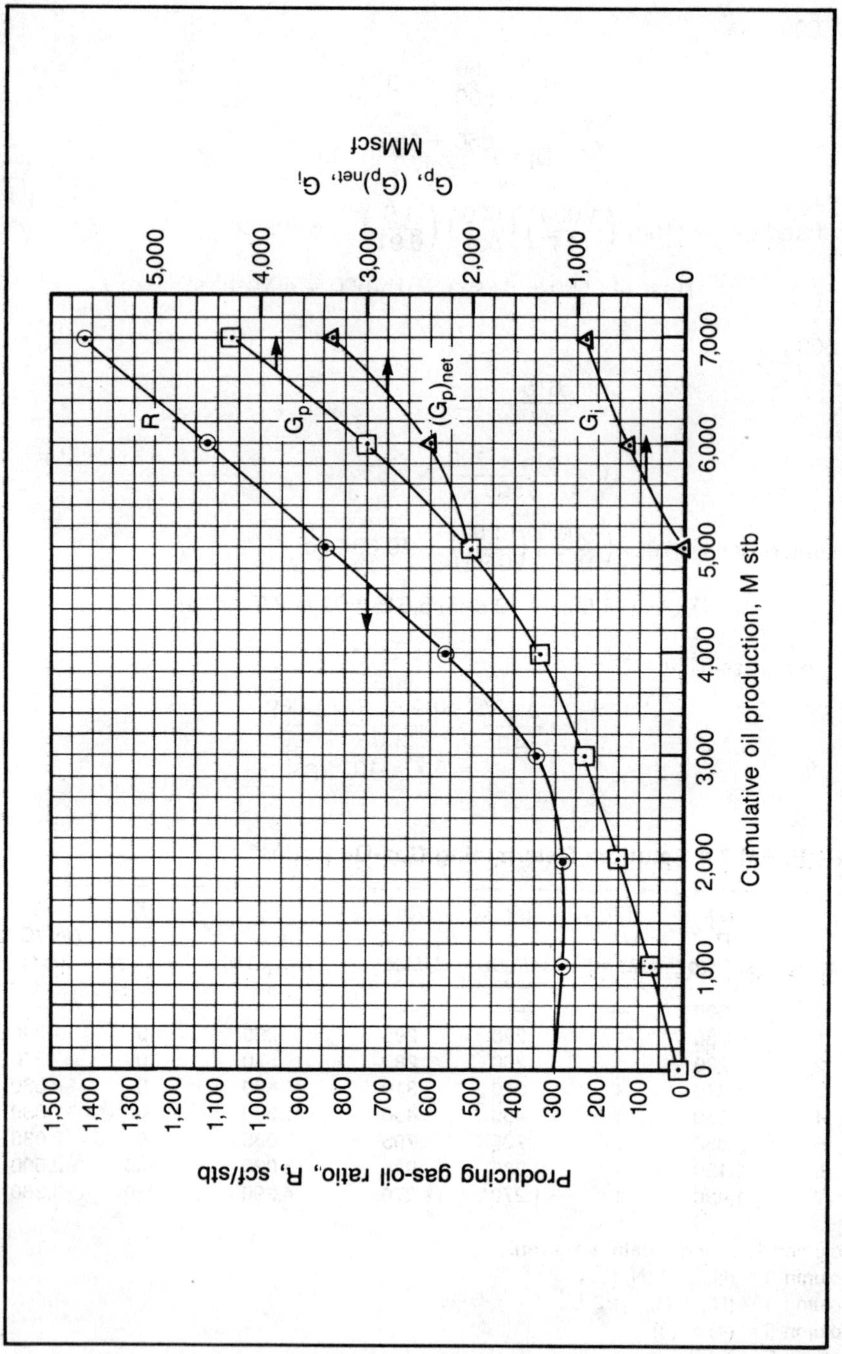

Fig. C11 Gas data plots, problem 7.4 solution

PROBLEM 7.5 Solution: Material Balance in a Solution-Gas-Drive Reservoir

$$N = \frac{N_p B_o + B_g (G_{ps} - N_p R_s)}{B_o - B_{oi} + (R_{si} - R_s) B_g} \qquad (7.17)$$

$$G_{ps} = R_p N_p = 600\ N_p$$

Then:

$$4 \times 10^6 = \frac{N_p\ 1.32 + 0.0011\ (600\ N_p - 550\ N_p)}{1.32 - 1.34 + (600 - 550)\ 0.0011}$$

$$N_p = 101{,}800 \text{ stb}$$

PROBLEM 7.6 Solution: Calculating Oil-Zone Shrinkage

$$\begin{aligned} \text{Oil zone shrinkage} &= (W_e - W_p) + \text{change in gas-cap volume} \\ &= N\,B_{oi} - [(N - N_p) B_o + [N R_{si} - (N - N_p) R_s - G_{ps}] B_g] \qquad (7.18) \\ &= (4 \times 10^6)\ 1.34 - [(4 \times 10^6) - 130{,}800]\ 1.32 \\ &\quad - [(4 \times 10^6)\ 600 - (4 \times 10^6 - 130{,}800)\ 550 \\ &\quad - (600)(130{,}800)] \times 0.0011 \\ &= 39{,}850 \text{ res bbl} \approx 40{,}000 \text{ res bbl} \end{aligned}$$

PROBLEM 7.7 Solution: Determining the Original Stock-Tank Barrels of Oil in a Reservoir

$$N = \frac{N_p B_o + B_g (G_p - N_p R_s) - G (B_g - B_{gi}) - (W_e - W_p)}{B_o - B_{oi} + (R_{si} - R_s)\ B_g} \qquad (7.20)$$

$$N = \frac{(130{,}800)(1.32) + 0.0011\ [(600 \times 130{,}800) - (130{,}800 \times 550)] - 4 \times 10^8 (0.0011 - 0.001)}{1.32 - 1.34 \times (600 - 550)\ 0.0011}$$

$$N = 4.0 \text{ MM stb}$$

PROBLEM 7.8 Solution: Determining PVT Data Empirically

The gas in solution at 500, 1,000, and 1,500 psia can be determined from Fig. B16. Using R_s and Fig. B18, the B_o values can be determined for $p \leqq p_s$:

p, psia	*R_s, scf/stb*	*B_o*
14.7	0	1.000
500	90	1.058
1,000	220	1.108
1,500	370	1.155
2,000	370*	1.149**

*Since 2,000 psia is above the saturation pressure, R_s is the same as R_s when $P = p_s$.

**To obtain B_o for $p > p_s$, it must be calculated from the oil compressibility, c_o. The oil compressibility is determined from Fig. B4 by determining the oil specific gravity at p_s; finding an equivalent specific gravity at 60°F, from Fig. B5 using this to determine the critical pressure and temperature from Fig. B4; calculating the corresponding reduced values; determining c_r from Fig. B4a; and calculating c_o from c_r.

$$\text{Res oil sp gr} = (\text{stock-tank oil wt} + \text{dissolved gas wt})/(B_{os})(\text{water density})$$
$$\text{Stock-tank oil wt} = (\text{sp gr oil})(350\ \text{lb/bbl})$$
$$\text{sp gr oil} = 141.5(131.5 + 40)$$
$$\text{Stock-tank oil wt} = [141.5/(131.5 + 40)]\ 350 = 289\ \text{lb}$$
$$\text{Dissolved gas wt} = [\text{MW air/scf/mol})]\,R_s(\text{gas sp gr})$$
$$\text{Dissolved gas wt} = \left(\frac{29}{379}\right)(370)(0.7) = 20$$
$$\text{Res oil sp gr} = (289 + 20)/(1.155 \times 350) = 0.764$$

Entering Fig. B5 with this specific gravity and the reservoir temperature, establish a density-temperature correlation line near cyclohexane. Read the established correlation line at 60°F as 0.795 sp gr.

From this specific gravity read from Fig. B4:

$$T_c = 920°R$$
$$p_c = 440\ \text{psia}$$

Then:

$$T_r = (125 + 460)/920 = 0.635$$
$$p_r = 1{,}500/440 = 3.41$$

From Fig. B4a:

$$c_r = 0.0059$$
$$c_o = 0.0059/440 = 1.3 \times 10^{-5}/\text{psia}$$
$$B_o \text{ at } 2{,}000\ \text{psia} = B_{os} - B_{os}c_o(2{,}000 - p_s)$$
$$= 1.155 - 1.155 \times 1.3 \times 10^{-5}(2{,}000 - 1{,}500)$$
$$= 1.147$$

PROBLEM 7.9 Solution: Predicting the Behavior of a Gas-Drive Reservoir

A.
$$S_o = \frac{(1 - S_{wc})(N - N_p)\,B_o}{N\,B_{oi}} \quad (7.13)$$
$$S_o = \left(1 - \frac{N_p}{N}\right)\frac{B_o(1 - S_{wc})}{B_{oi}}$$
$$S_o = \left(1 - \frac{1{,}179{,}000}{10{,}025{,}000}\right)\left(\frac{1.233}{1.315}\right)(1 - 0.22)$$
$$S_o = 0.646 \therefore S_L = 0.646 + 0.22 = 0.866$$
$$R = R_s\left(\frac{k_g}{k_o}\right)\left(\frac{\mu_o}{\mu_g}\right)\left(\frac{B_o}{B_g}\right) \quad (7.11)$$

$$\frac{k_g}{k_o} = (R - R_s)/\left[\left(\frac{\mu_o}{\mu_g}\right)\left(\frac{B_o}{B_g}\right)\right]$$

$$\frac{k_g}{k_o} = (2{,}080 - 450)/\left[(102.61)\left(\frac{1.233}{1.616 \times 10^{-3}}\right)\right]$$

$$\frac{k_g}{k_o} = 0.0207$$

B. $$N_{pn} = \frac{N[B_o - B_{oi} + (R_{si} - R_s)B_g] + G(B_g - B_{gi}) - B_g[G_{p(n-1)} - (R_n + R_{n-1})N_{p(n-1)}/2]}{B_o - B_g R_s + (R_n + R_{n-1})B_g/2} \quad (7.34)$$

For p = 1,200:

$$\text{Numerator} = 10{,}025{,}000[1.224 - 1.315 + (650 - 431)0.001807] - 0.001807 \times [1.123 \times 10^9 - (2{,}080 + R_{1,200})1{,}179{,}000/2]$$

$$= 3{,}241{,}299 + 1{,}065.23\, R_{1,200}$$

$$\text{Denominator} = 1.224 - 0.001807 \times 431 + (2{,}080 + R_{1,200})0.001807/2$$

$$= 2.324 \times 0.0009035\, R_{1,200}$$

$$N_{p\,1,200} = \frac{3{,}241{,}299 + 1{,}065.23\, R_{1,200}}{2.324 + 0.0009035\, R_{1,200}}$$

$R_{1,200\ est}$ = 2,520 from GOR versus pressure extrapolation, Fig. 7–9

$$N_{p\,1,200} = \frac{3{,}241{,}358 + 1{,}065.23(2{,}520)}{2.324 + 0.0009035(2{,}520)}$$

$$= 1{,}288{,}000 \text{ stb}$$

$$S_o = (1 - S_{wc})\left(\frac{B_o}{B_{oi}}\right)\left(1 - \frac{N_p}{N}\right) \quad (7.13)$$

$$S_o = (0.78)\left(\frac{1.224}{1.315}\right)\left(1 - \frac{N_p}{10{,}025{,}000}\right) = 0.726\left(1 - \frac{N_p}{10{,}025{,}000}\right)$$

$$= 0.726[1 - (1{,}288{,}000/10{,}025{,}000)]$$

$$S_o = 0.6327 \therefore S_L = 0.6327 + 0.22 = 0.8527$$

From S_L versus k_g/k_o curve for field data:

$$\frac{k_g}{k_o} = 0.0274$$

$$R_{1,200} = R_s + \left(\frac{k_g}{k_o}\right)\left(\frac{\mu_o}{\mu_g}\right)\left(\frac{B_o}{B_g}\right)$$

$$R_{1,200} = 431 + (k_g/k_o)(108.96)\left(\frac{1.224}{1.807 \times 10^{-3}}\right) = 431 + 73{,}806(k_g/k_o)$$

$$R_{1,200} = 431 + (73{,}806)(0.0274)$$

$R_{1,200}$ = 2,453 compared with an estimated value of 2,520

$R_{1,200}$ calculated $<$ $R_{1,200}$ estimated

C. Assume $R_{1,200\ est} = 2{,}453$ scf/stb. Then:

$$N_{p\ 1,200} = 1{,}289{,}000 \text{ stb}$$

$$S_L = 0.8526$$

$$\frac{k_g}{k_o} = 0.0274 \text{ (same)}$$

$$R_{1,200} = 2{,}453 \text{ scf/stb}$$

$$= R_{1,200\ est}$$

So $$R_{1,200} = 2{,}453 \text{ scf/stb}$$

$$N_{p\ 1,200} = 1{,}289{,}000 \text{ stb}$$

$$\Delta G_p = \left(\frac{R_{1,302} + R_{1,200}}{2}\right)(N_{p\ 1,200} - N_{p\ 1,302})$$

$$\Delta G_p = \left(\frac{2{,}080 + 2{,}453}{2}\right)(1{,}289{,}000 - 1{,}179{,}000)$$

$$= 2.49315 \times 10^8$$

$$G_{p\ 1,200} = 1.123 \times 10^9 + 2.5025 \times 10^8$$

$$= 1.37325 \times 10^9 \text{ scf}$$

D. After adjusting the GOR plot to coincide with $R_{1,200} = 2{,}453$ and extrapolating to 1,100, $R_{1,100}$ is estimated as 2,830:

$$N_{p\ 1,100} = \frac{10{,}025{,}000[1.215 - 1.315 + (650 - 412)0.001998] - 0.001998\left[1.37325 \times 10^9 - (2{,}453 + R_{1,100})\dfrac{1{,}289{,}000}{2}\right]}{1.215 - 0.001998(412) + (2{,}453 + R_{1,100})\dfrac{0.001998}{2}}$$

$$= \frac{4{,}181{,}499 + 1{,}287.711\ R_{1,100}}{2.842371 + 0.000999\ (R_{1,100})}$$

If $R_{1,100} = 2{,}830$ scf/stb:

$$N_p = 1{,}380{,}306$$

$$S_o = 0.78\left(\frac{1.215}{1.315}\right)\left(1 - \frac{N_{p\ 1,100}}{10{,}025{,}000}\right)$$

$$= 0.72068\left(1 - \frac{N_{p\ 1,100}}{10{,}025{,}000}\right)$$

$$S_o = 0.621$$

$$S_L = 0.841$$

$$k_g/k_o = 0.0385$$

$$R_{1,100} = 412 + \left(\frac{k_g}{k_o}\right)(115.2)\frac{1.215}{0.001998}$$

$$= 412 + 70{,}054\,(k_g/k_o)$$

$$(R_{1,100})_{calc} = 3{,}109$$

If $R_{1,100} = 3{,}109$:

$$N_p = 1{,}376{,}030$$

$$S_L = 0.842$$

$$k_g/k_o = 0.0375$$

$$(R_{1,100})_{calc} = 3{,}039$$

If $R_{1,100} = 3{,}039$:

$$(N_p) = 1{,}377{,}066$$

$$S_L = 0.842 \text{ (same)}$$

$$k_g/k_o = 0.0375$$

$$(R_{1,100})_{calc} = 3{,}039 \text{ OK}$$

$$G_{p\,1,100} = 1.37325 \times 10^9 + (1{,}377{,}066 - 1{,}289{,}000)\,\frac{(2{,}453 + 3{,}039)}{2}$$

$$= 1.614 \times 10^9 \text{ scf}$$

$$N_{p\,1,100} = 1{,}377{,}066 \text{ stb}$$

PROBLEM 7.10 Solution: Analyzing an Undersaturated Oil Reservoir with a Water Drive

$$B = 1.12\,\phi\,hc\,r^2\,\theta$$

$$= 1.12(0.22)(60)(4 + 3)10^{-6}(2{,}500)^2$$

$$= 646.8$$

$$\eta = \frac{6.33\,k}{\phi c} = \frac{6.33(0.1)}{(0.22)(0.3)(4 + 3)10^{-6}}$$

$$= 1.37 \times 10^6$$

To find W_e after 100 days:

$$t_{D100} = \frac{\eta t}{r^2} = \frac{(1.37 \times 10^6)(100)}{(2{,}500)^2} = 21.92$$

$$r_{De} = \frac{20{,}000}{2{,}500} = 8$$

Interpolating Table 3–2:

$$Q_{tD\,21.92} = 12.13 + (12.95 - 12.13)\left(\frac{21.92 - 20}{22 - 20}\right)$$

$$= 12.917$$

$$B\Sigma\Delta p\, Q_{tD} = N_p B_o + W_p - N(B_o - B_{oi}) - N\,\frac{[(c_f + c_w s_{wc})\Delta p\, B_{oi}]}{1 - S_{wc}} \tag{7.39}$$

$$\begin{aligned} B_{op} &= B_{oi} + c_o(p_i - p)B_{oi} \\ &= 1.34 + (7.7 \times 10^{-6})(3{,}500 - p)\,1.34 \\ &= 1.376 = 1.0318 \times 10^{-5}\,p \end{aligned}$$

$$N = \frac{\pi(2{,}500)^2(60)(0.22)(1 - 0.26)}{(5.615)(1.34)} = 2.54907 \times 10^7 \text{ stb}$$

Substituting into Eq. 7.39 and rearranging:

$$646.8\left(\frac{3{,}500 - p}{2}\right) 12.917 = [(1{,}500 \times 100) - (2.54907 \times 10^7)]$$

$$\times\,(1.376 - 1.032 \times 10^{-5}\,p_{100}) + (2.54907 \times 10^7)\,1.34 - (2.54907 \times 10^7)$$

$$\times\left[\frac{[4 + 3(0.26)]\,10^6(1.34)}{(1 - 0.26)}\right] \times (3{,}500 - p_{100})$$

$$p_{100} = 3{,}456 \text{ psia}$$

$$W_{e100} = 646.8(12.917)\left[\frac{3{,}500 - 3{,}456}{2}\right]$$

$$= 183{,}803 \text{ bbl}$$

To find W_e after 200 days:

$$t_{D200} = t_{D100} \times 2 = 21.92 \times 2 = 43.84$$

Q_{tD200} is interpolated from Table 3–2:

$$Q_{tD200} = 18.97 \times (20.26 - 18.97)\,\frac{(43.84 - 40)}{(45 - 40)} = 19.96$$

Substituting into Eq. 7.39:

$$646.8\left[\left(\frac{3{,}500 - 3{,}456}{2}\right) 19.96 + \frac{(3{,}500 - p_{200})}{2}\,12.917\right] =$$

$$(1{,}500 \times 200)(1.376 - 1.032 \times 10^{-5}\,p_{200}) - 2.54907 \times 10^7$$

$$\times\,[(1.376 - 1.032 \times 10^{-5}\,p_{200}) - 1.34] + 2.54907 \times 10^7$$

$$\times\,\frac{[(4 + (0.26 \times 3)\,10^{-6}(3{,}500 - p_{200})\,1.34]}{(1 - 0.26)}$$

$$p_{200} = 3{,}474 \text{ psia}$$

$$W_e = 646.8\left[\frac{(3{,}500 - 3{,}456)}{2}\,19.96 + \left(\frac{3{,}500 - 3{,}474}{2}\right) \times 12.917\right]$$

$$= 392{,}634 \text{ bbl}$$

PROBLEM 7.11 Solution: Analyzing a Combination Solution-Gas-Water-Drive Reservoir

Follow the previously outlined procedure for predicting the Behavior of a Water-Drive Reservoir.

1. We guess that $p_{500\ days} = 2{,}300$ psia. Since no past history is available for this initial prediction, a first guess of $p_{500\ days}$ could have been obtained by material balance by assuming W_e at 500 Days is 0.0. Since we are using an unrealistically large time increment for demonstration purposes, we simply assume $p_{500} = 2{,}300$ psia.

2. $$Q = B \sum_{j=1}^{j=n} \Delta p_j Q_{tDj} \qquad (3.64b)$$

$$t_D = 6.33Kt/\phi\mu c r^2 \qquad (3.11a)$$

$$= (6.33)(0.1)(500)/(0.25)(0.3)(7 \times 10^{-6})(1{,}860)^2$$

$$= 174.26$$

From Table 3–1:

$$Q_{tD} = 66.336 + \left(\frac{174.26 - 170}{175 - 170}\right)(67.928 - 66.336)$$

$$= 67.69$$

$$Q = (225)\left(\frac{2{,}500 - 2{,}300}{2}\right) 67.69$$

$$= 1{,}523{,}000 \text{ bbl}$$

3. $$\text{Water-invaded reservoir volume} = \frac{Q}{(1 - S_{wc} - S_{OBY})\phi}$$

$$= \frac{1{,}523{,}000}{(1 - 0.22 - 0.4)0.25}$$

$$= 16{,}032{,}000 \text{ bbl}$$

$$= 9.0 \times 10^7 \text{ cu ft}$$

$$\text{Uninvaded volume} = 6.44 \times 10^8 - 9 \times 10^7$$

$$= 5.54 \times 10^8 \text{ cu ft}$$

Fig. 7–13 shows that this uninvaded volume is well below the first line of producers. Assume all 6 wells produce during the first 500 days of production.

4. The initial producing rate averages 750 stb/well/d or 4,500 stb/d for the reservoir. No past rate-time is available for the history so we estimate the rate at the end of 500 days as 250 stb/well/d or $N_{p500} = [(750 + 250)/2]\ 6 \times 500 = 1{,}500{,}000$ stb.

5. $$S_{oun\,500} = \frac{(N - N_p)B_o - S_{OBY}(W_e - W_p)/(1 - S_{OBY} - S_{wc})}{[N\,B_{oi}/(1 - S_{wc})] - (W_e - W_p)/(1 - S_{OBY} - S_{wc})}$$

$$= \frac{(16.89 \times 10^6 - 1.5 \times 10^6)(1.311) - [0.4(1{,}523{,}000)/0.38]}{[(16.89 \times 10^6)\,1.325/0.78] - (1{,}523{,}000/0.38)}$$

$$= 0.752$$

6. and 7.

$$J_{2,500} = \frac{750}{2{,}500 - 2{,}150}$$

$$= 2.14 \text{ stb/psi}$$

$$\frac{J_{2,300}}{J_{2,500}} = \frac{(k_{ro}/B_o\mu)_{2,300}}{(k_{ro}/B_o\mu)_{2,500}}$$

$$(k_{ro})_{2,300} = \left(\frac{S_o}{1 - S_{wi}}\right)^{3.0} = \left(\frac{0.752}{1.0 - 0.22}\right)^{3.0}$$

$$= 0.8961$$

$$J_{2,300} = (2.14)\,\frac{0.8961/(1.311)(0.39)}{1/(1.325)(0.38)}$$

$$q_{res} = (1.888)(2{,}300 - 2{,}150)(6)$$

$$= 1{,}699.6 \text{ b/d}$$

$$(N_p)_{calc\,500} = [(4{,}500 + 1{,}699.6)/2]\,500$$

$$= 1{,}550{,}000 \text{ stb} > 1{,}500{,}000 \text{ assumed}$$

Return to step 4:

$N_{p\,500}$ assumed = 1,530,000
5. $S_{oun} = 0.751$
6. and 7. $k_{ro\,2,300} = 0.892$
$N_{p\,500} = 1{,}548{,}000 > 1{,}530{,}000$ assumed

Return to step 4:

$N_{p\,500}$ assumed = 1,545,000
5. $S_{oun} = 0.750$
6. and 7. $k_{ro\,2,300} = 0.8892$
$(N_p)_{calc} = 1{,}547{,}000 > 1{,}545{,}000$ assumed

8. $$G_{pn} = G_{p(n-1)} + [(R_n + R_{n-1})/2](N_{pn} - N_{pn-1}) \qquad (7.32)$$

$$R = R_s + \frac{k_g}{k_o}\,\frac{\mu_o}{\mu_g}\,\frac{B_o}{B_g} \qquad (7.11)$$

$$G_{p(n-1)} = 0.0$$

$$R_{2,300} = 618 + \frac{k_g}{k_o}(56.60)\,\frac{1{,}311}{0.000843}$$

$$k_{ro} = 0.8892$$

$$k_{rg} = \left[1 - \frac{S_o}{1 - S_{wi}}\right]^3 = \left[1 - \frac{0.750}{0.78}\right]^3$$

$$= 0.0000568$$

$$R_{2,300} = 618 + \frac{0.0000568}{0.8892}(56.6)\left(\frac{1.311}{0.000843}\right)$$

$$R_{2,300} = 624 \text{ scf/stb}$$

$$G_{p\,500} = 0 + \left(\frac{650 + 624}{2}\right)(1{,}546{,}000 - 0.0)$$

$$= 9.839 \times 10^8 \text{ scf}$$

9. $$W_e = N_p B_o + B_g(G_p - N_p R_s) + W_p - G(B_g - B_{gi}) - N\left[B_o - B_{oi} + (R_{si} - R_s)B_g + \frac{(c_f + c_w S_{wc})}{1 - S_{wc}}\Delta p B_{oi}\right]$$

$$W_e = 1{,}546{,}000(1.311) + 0.000843[9.839 \times 10^8 - (1{,}546{,}000)(618)] + 0 - 0 - 16.89 \times 10^6 \times \left[[B_o - 1.325 + (650 - R_s) \times B_g + \frac{(4 \times 10^{-6} + 3 \times 10^{-6} \times 0.22)(2{,}500 - 2{,}300)1.325}{0.78}\right]$$

$$W_e = 1.804 \times 10^6$$

10c. $W_e > Q$; go back to step 1:

1. Try Q of about 1,800,000

$$1{,}800{,}000 = 225\frac{(2{,}500 - p_{500})}{2}67.69$$

$$p_{500} \approx 2{,}264 \text{ psia}$$

$$Q = 1.797 \times 10^6 \text{ bbl}$$

2. Determine reservoir parameters for p = 2,264 psia

$$B_{o\,2,264} = 1.311 - (1.311 - 1.296)\frac{2{,}300 - 2{,}264}{200} = 1.3083$$

$$B_{g\,2,264} = 0.000843 = (0.000907 - 0.000843)0.18$$

$$= 0.00085452 \text{ res bbl/scf}$$

$$R_{s\,2,264} = 618 - (618 - 586)0.18 = 612.24$$

$$\mu_{o\,2,264} = 0.39 + (0.4 - 0.39)0.18 = 0.3918$$

$$\mu_o/\mu_g = 61.46 + (67.35 - 61.46)0.18 = 62.5202$$

3. All 6 wells producing
4. Estimate N_{p500} = 1,547,000 bbl

5. $S_{oun} = 0.7588$
6. and 7. $k_{ro} = 0.9205$
   $N_{p\,calc} = 1{,}560{,}373 > 1{,}547{,}000$ assumed in step 4
4. Estimate $N_{p500} = 1{,}560{,}373$ bbl
5. $S_{oun} = 0.7580$
6. and 7. $k_{ro} = 0.9179$
   $N_{p\,calc} = 1{,}559{,}118 \approx 1{,}560{,}373$
8. $k_{rg} = 0.00002233$ res bbl/scf
   $R_{2,264} = 614.6$ scf/bbl
   $G_{p500} = 9.866 \times 10^8$ scf
9. $W_e = 1{,}773{,}688$
10a. $W_e - Q = -23{,}481$

Assume $W_e = Q$, so at 500 days:

$$W_e = 1.773 \times 10^6 \text{ bbl}$$

$$N_p = 1{,}559{,}000 \text{ stb}$$

$$p = 2{,}264 \text{ psia}$$

$$G_p = 9.866 \times 10^8 \text{ scf}$$

PROBLEM 7.12 Solution: Calculating Drive Indices

The following data are taken from the problem 7.11 solution. After 500 days of production:

$$W_e = 1.773 \times 10^6 \text{ bbl}$$

$$N_p = 1{,}559{,}000 \text{ stb}$$

$$p = 2{,}264 \text{ psia}$$

$$G_p = 9.866 \times 10^8 \text{ scf}$$

Also, the following reservoir parameters apply when p = 2,264 psia:

$$B_o = 1.308$$

$$B_g = 0.0008545 \text{ res bbl/stb}$$

$$R_s = 612 \text{ scf/stb}$$

In calculating SGDI and WDI, we need $B_t = B_o + (R_{si} - R_s)B_g$

$$= 1.308 + (650 + 612)0.0008545$$

$$= 1.3405$$

$$R_p = G_p/N_p = 9.866 \times 10^8/1.559 \times 10^6 = 632.8 \text{ scf/stb}$$

$$\text{SGDI} = \frac{N(B_t - B_{ti})}{N_p[B_t + (R_p - R_s)B_g]} \tag{7.45}$$

$$= \frac{16.89 \times 10^6(1.3405 - 1.325)}{1.559 \times 10^6[1.3405 + (632.8 - 612)0.0008545]}$$

$$= 0.124 \text{ or } 12.4\%$$

$$\text{WDI} = \frac{(W_e - W_p)}{N_p[B_t + (R_p - R_s)B_g]} \tag{7.47}$$

$$= \frac{(1.773 \times 10^6) - 0.0}{1.559 \times 10^6[1.3405 + (632.8 - 612)0.0008545]}$$

$$= 0.837 \text{ or } 83.7\%$$

Note that with the gas saturation still very small (1.0 − 0.758 − 0.22 = 2.2%), the change in pore volume contributes considerably to the production:

$$\Delta V_p \text{ index} = 1 - 0.837 - 0.124 = 3.9\%$$

PROBLEM 8.1 Solution: Using Constant-Percentage Decline to Calculate the Future Life and Rates of a Well

The constant decline rate can be calculated from the straight-line slope of Fig. 8–3. From December 1976 to December 1983 (85 months), the rate changes from 35 to 10 stb/d.

$$a = (2.3)\frac{\log 35 - \log 10}{85} = 0.0147/\text{month} \tag{8.3}$$

In June 1982 the rate is 13.2 b/d or:

$$q_i = 13.2 \text{ b/d} \times 30.4 \text{ days/month}$$

$$= 401.3 \text{ bbl/month}$$

$$q = q_i\, e^{-at} \tag{8.14}$$

$$q_{5\text{ years}} = 401.3e^{-0.0147(5\times 12)}$$

$$= 166 \text{ bbl/month or } 6.5 \text{ b/d}$$

To find the remaining life use Eq. 8.14 to find the time of decline to the economic limit of 1.0 b/d or 30.4 bbl/month from 401.3 bbl/month:

$$30.4 = 401.3e^{-0.0147\,t}$$

$$t = 176.0 \text{ months}$$

PROBLEM 8.2 Solution: Using a Rate versus Cumulative Plot During Constant-Percentage Decline

Quadrupling the rate (53/13.2) also quadruples the decline rate if the mobile oil remains the same. After frac, q_i = 53 and a = 0.0147 × 4 = 0.0588/month. Then to find the remaining life:

$$q = q_i\, e^{-at} \tag{8.14}$$

$$30.4 = (53 \times 30.4)e^{-0.0588(t)}$$

$$t = 68 \text{ month}$$

By fracturing, the life is decreased 176 − 68 = 108 months. To see the effect of the fracture treatment on the reserves, apply Eq. 8.15 before and after the fracture treatment. Without the fracture treatment:

$$\Delta N_p = (q_1 - q_2)/a \qquad (8.15)$$

$$\text{Reserves} = (13.2 - 1.0)\ 30.4/0.0147 = 25{,}100 \text{ stb}$$

After the fracture treatment:

$$\text{Reserves} = (53 - 1)\ 30.4/0.0588 = 26{,}700 \text{ stb}$$

The reserves are increased only 26,700 − 25,100 or 1,600 stb, but the reduction in the life gives a much quicker return of the profits plus saving much of the operating expense.

$$\text{The remaining primary mobile oil} = q_1/a \qquad (8.19)$$

$$= (13.2)(30.4)/0.0147 \text{ or}$$

$$= (4)(13.2)(30.4)/(4)(0.0147) = 27{,}300 \text{ stb}$$

PROBLEM 8.3 Solution: Application of the Hyperbolic Decline Curves

The data are plotted on an overlay of Fig. 8–6 with the scales chosen as indicated in the solution illustration (Fig. C12). When this data plot is fit to curves of n = 0.3, 0.5, and 0.7 (Figs. 8–5, 8–6 and 8–7) it is found that a fit can be obtained for all three n's. Thus, the average, n = 0.5, is employed as shown in the illustration. In practice the plot would be compared with curves representing several additional n's to determine the best fit.

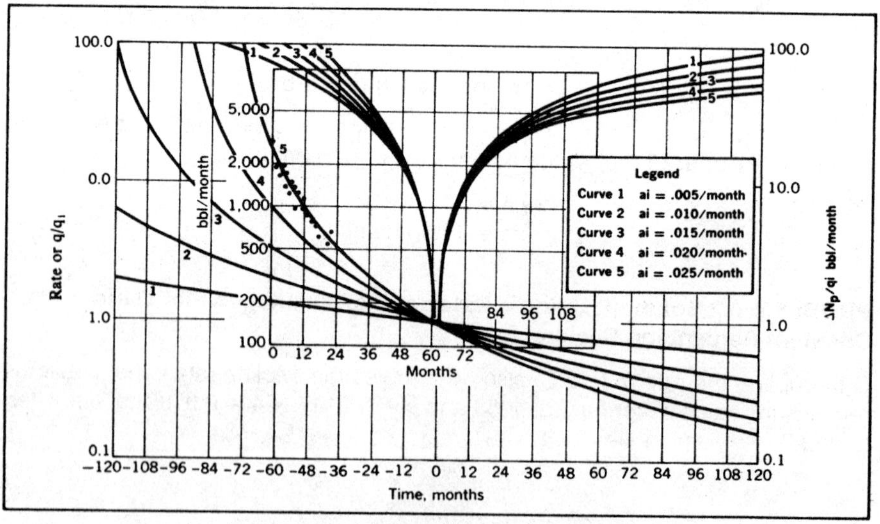

Fig. C12 Solution to problem 8.3

From the solution illustration fit, we note:

$$t(\text{chart}) = 0$$

This corresponds to a real time of about 62 months when the rate is:

$$q_i = 150 \text{ bbl/month}$$

$$a_i = 0.025\text{/month (characteristic of curve 5)}$$

If the economic limit is 100 bbl/month, the remaining life (after 19 months) = (62 − 19) + 18 = 61 months, where 18 months is the chart time where the fit rate curve (curve 5) intersects the economic limit rate of 100 bbl/month.

To find the remaining reserves (as of 19 months) if the economic limit is 100 bbl/month:

$$\Delta N_p = H[q_1^{1-n} - q_2^{1-n}] \quad (8.29)$$

Where:

$$H = q_i^n/a_i (1 - n) \quad (8.28)$$

$$H = (150)^{0.5}/0.025 (1 - 0.5)$$

$$= 979.8$$

$$\Delta N_p = 979.8[700^{0.5} - 100^{0.5}]$$

Reserves as of 19 months = 16,100 stb

If the economic limit is 10 bbl/month reserve as of 19 months:

$$= 979.8[700^{0.5} - 10^{0.5}]$$

$$= 22{,}800 \text{ stb}$$

The remaining life from the 0.0 type-curve time is calculated from:

$$q = \frac{q_i}{(1 + na_i t)^{1/n}} \quad (8.25)$$

$$10 = \frac{150}{[1 + (0.5)(0.025)t]^{1/0.5}}$$

$$t = 230 \text{ months}$$

The life as of "19 months" = 230 + 62 − 19 = 273 months

This life can also be calculated by first determining the *a* corresponding to the 19-month curve rate of 700 bo/month:

$$\frac{a}{a_i} = \frac{q^n}{q_i^n} \quad (8.20)$$

$$\frac{a}{0.025} = \frac{(700)^{0.5}}{(150)^{0.5}}$$

$$a_{700} = 0.054$$

Then using Eq. 8.25, we can calculate the time directly:

$$10 = \frac{700}{[1 + (0.5)(0.054)t]^{1/0.5}} \tag{8.25}$$

$$t = \text{life as of 19 months} = 273 \text{ months}$$

PROBLEM 8.4 Solution: The Change in Reserves and Life Following a Workover

$$\text{Remaining mobile oil} = q_i/a_i\,(1 - n) \tag{8.35}$$

The rate is increased in June 1982, when the rate (before the increase) is 13.5 b/d. Then a can be calculated:

$$\frac{a_{13.5}}{0.015} = \frac{13.5^{0.5}}{15.6^{0.5}} \tag{8.20}$$

$$a_{13.5} = 0.01395$$

$$\text{The mobile oil is then} = \frac{q_i}{a_i(1 - n)} \tag{8.35}$$

$$= \frac{13.5 \times 30.4}{0.01395(1 - 0.5)} = 59{,}000 \text{ stb}$$

When the rate is increased to 53 stb/d the decline factor a can be calculated:

$$59{,}000 = \frac{53 \times 30.5}{0.01395(1 - 0.5)} \tag{8.35}$$

$$a_{53} = 0.0546$$

Then:

$$1.0 = \frac{53}{[1 + (0.5)(0.01395)t]^{1/0.5}} \tag{8.25}$$

t = remaining life = 900 months, which compares with a life of 384 months without a workover. Then reserves would be:

$$N_p = [q_i^n/a_i(1 - n)][q_i^{1-n} - q_2^{1-n}] \tag{8.27}$$

$$= [(53 \times 30.4)^{0.5}/0.0546(1 - 0.5)][53 \times 30.4)^{1-0.5} - (1.0 \times 30.4)^{1-0.5}]$$

= 50,900 stb, which compares with reserves of 42,800 stb without the workover

PROBLEM 9.1 Solution: Saturation Distribution in a Waterflood

The total liquid saturation may be found at any location at any time by applying Eq. 6.14 to the displacement of the oil by water:

$$\Delta X_{Swj} = \frac{5.615 \Delta t\, q_t}{\phi A}(df_w/dS_w)_{Swj} \tag{1}$$

Then applying Eq. 6.14 to the displacement of free gas by the oil bank:

$$\Delta X_{SLj} = \frac{5.615 \Delta t\, q_t}{\phi A} (df_o/dS_L)_{SLj} \tag{2}$$

$(df_w/dS_w)_{Swj}$ = slope of f_w versus S_w curve at a particular saturation
$(df_o/dS_L)_{SLj}$ = slope of f_o versus S_L curve at a particular saturation
X = distance from injectors in feet

Now, find the time for the first production increase to occur. This will be when the oil bank reaches the producing well. The initial liquid saturation is:

$$S_L = S_o + S_w = 0.55 + 0.30 = 0.85$$

Then from Fig. C13 $(df_o/ds_L) = 8.0$ at the producing well when the oil bank reaches the well.

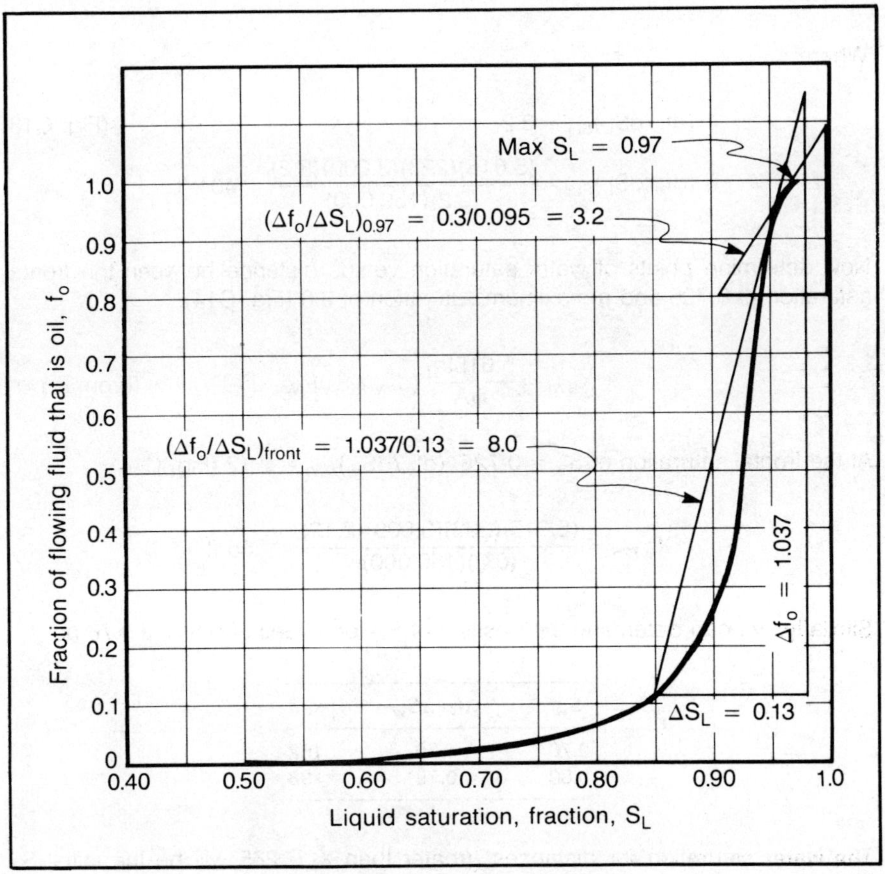

Fig. C13 Solution to problem 9.1

From Eq. (2):

$$t = \frac{x_{SLj}\phi A}{5.615q_t}\left(\frac{dS_L}{df_o}\right) = \frac{(1{,}000)(0.2)(150{,}000)}{(5.615)(3{,}000)}\ \frac{1}{8.0}$$

$$t = 223 \text{ days}$$

The liquid saturations must be between the frontal saturation (0.95, the intersection of the curve and the tangent line that passes through the 0.85 initial saturation) and the 0.97, which is the maximum saturation. (Fig. C13).

To find the position where $S_L = 0.97$ (maximum):

$$X_{SLj} = \frac{5.615tq_t}{\Phi A}(df_o/dS_L)_{SLj} \qquad \text{(From Eq. 2)}$$

Where:

$$(df_o/dS_L)_{SLj} = 3.2 \qquad \text{(Fig. C13)}$$

$$(df_o/dS_L)_{SLj} = \frac{(5.615)(223)(3{,}000)(3.2)}{(0.2)(150{,}000)} = 401 \text{ ft}$$

Now determine points of water saturation versus distance between the frontal saturation of 0.725 and a maximum saturation of 0.8 (Fig. C14):

$$X_{SWj} = \frac{5.615tq_t}{\phi A}(df_w/dS_w)_{SWj} \qquad \text{(From Eq. 1)}$$

At the frontal saturation of $S_w = 0.725$, $(df_w/dS_w)_{SWj} = 2.12$ (Fig. C14)

$$X_{0.725} = \frac{(5.615)(223)(3{,}000)(2.12)}{(0.2)(150{,}000)} = 265 \text{ ft}$$

Similarly, we can determine the position of S_w for values of 0.8 and 0.76 as:

S_w	$\Delta f_w/\Delta S_w$	x, ft
0.76	1.30	162
0.80	0.78	98

The water saturation for distances greater than $X_f = 265$ will be the initial S_w of 0.30. Saturations can then be plotted as in Fig. C15.

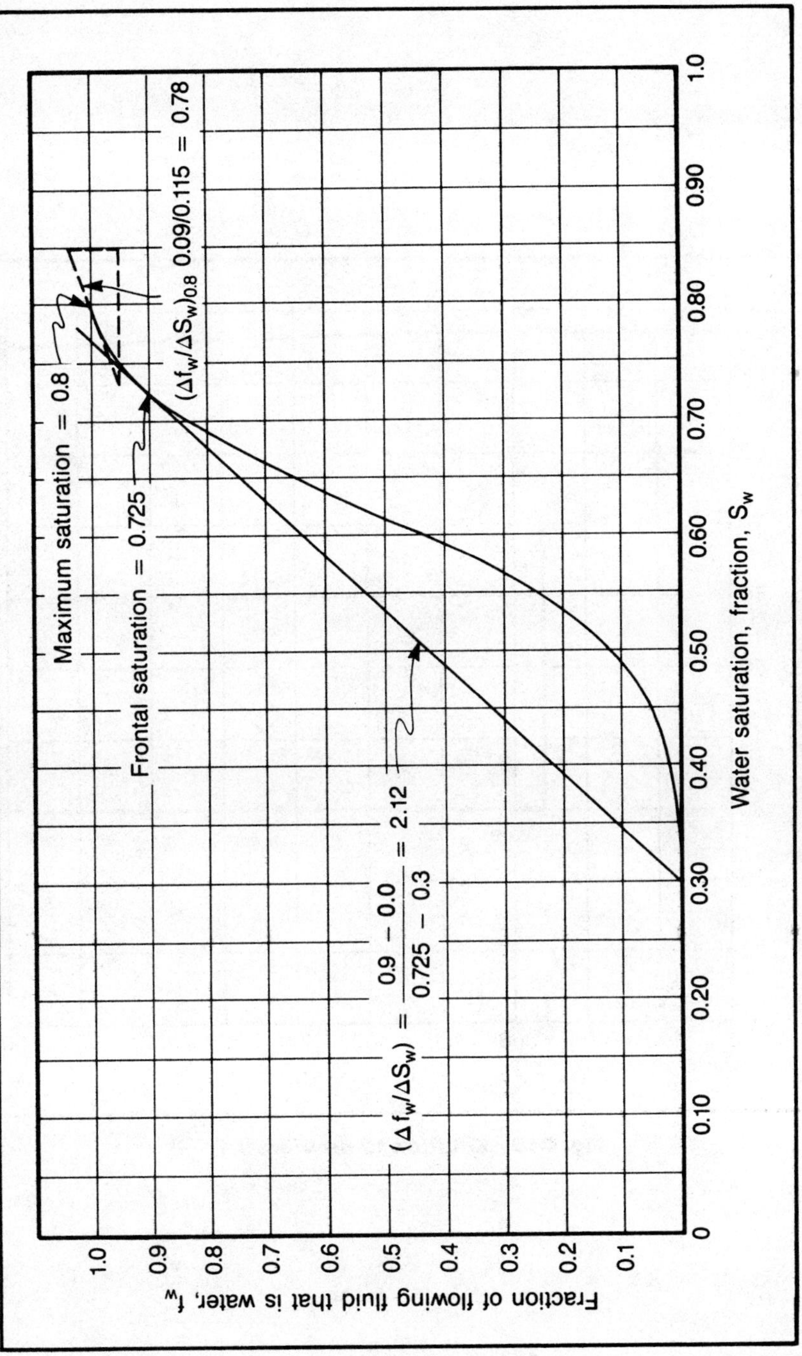

Fig. C14 Solution to problem 9.1

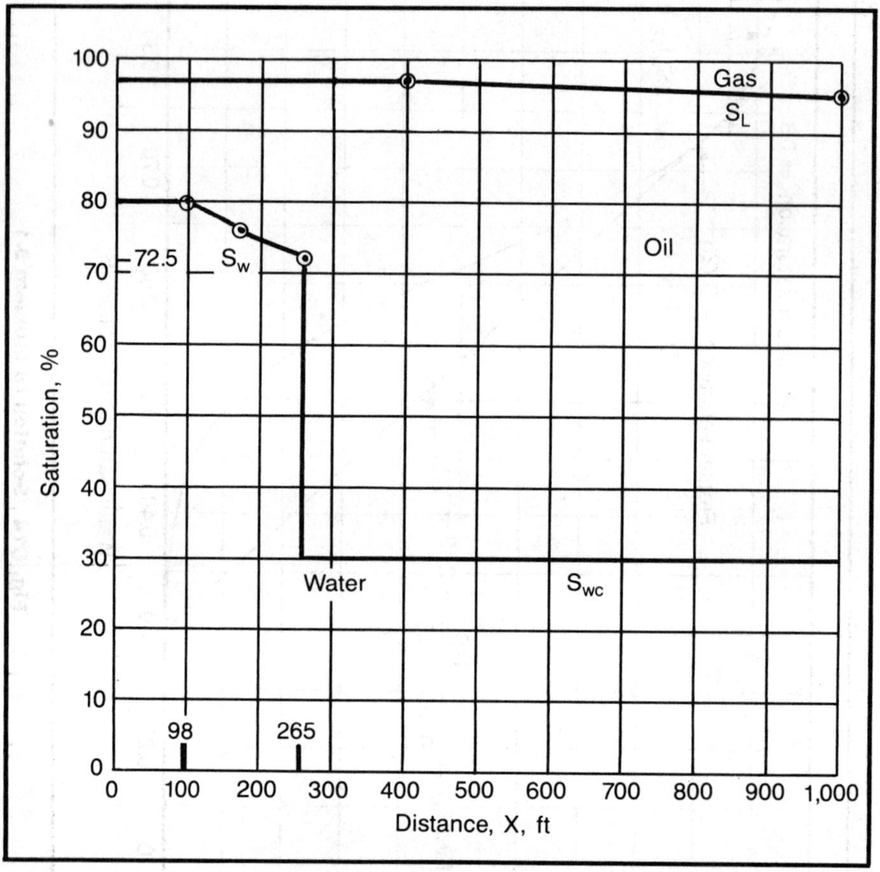

Fig. C15 Solution to problem 9.1

PROBLEM 9.2 Solution: Evaluating Gross Swept Volume*

Plot	*Estimated Average Thickness, ft*	*Estimated Area, acres*	*Acre-ft*
NE of 1	3	3	9
NE of 2	3	3	9
NE of 3	2	2	4
1	5	9	45
2	10	9	90
3	9	10	90
NE of 6	2	3	6
SW of 1	5	5	25
4	16	10	160
5	18	10	180
6	10	10	100
NE of 10	2	3	6
7	10	10	100
8	19	10	190
9	10	10	100
10	8	10	80
NE of 13	1	2	2
SW of 7	2	2	4
11	10	10	100
12	10	10	100
SE of 9	2	5	10
13	4	10	40
SW of 11	3	4	12
SE of 11	5	6	30
SE of 12	2	3	6
SE of 13	1	3	3
		Total volume =	1,501 acre-ft
		$V_{BP} \approx$	1,500 acre-ft

*Refer to Fig. C16

$$N = 7{,}758\,\phi\,(1 - S_{wc})\,V_{BP}/B_{oi} \tag{9.3}$$

$$N = (7{,}758)(0.2)(1 - 0.2)\,1{,}500/1.2$$

$$N = 15.516 \times 10^5 \text{ stb}$$

$$S_{oP} = (N - N_{pP})B_{os}(1 - S_{wc})/NB_{oi} \tag{9.2}$$

$$= (15.516 - 3.645) \times 10^5 \times 1.1 \times (1 - 0.2)/(15.516 \times 10^5 \times 1.2)$$

$$= 0.561$$

The best five-spot conversion pattern is as shown in Fig. C16. It totals 62½ acres. The alternative pattern totals 52½ acres.

The gross swept volume is estimated as:

$$= (20 \times 15) + (20 \times 17) + (5 \times 15) + (5 \times 14) + (10 \times 11) + (2.5 \times 8)$$

$$= 915 \text{ acre-ft} \approx 900 \text{ acre-ft}$$

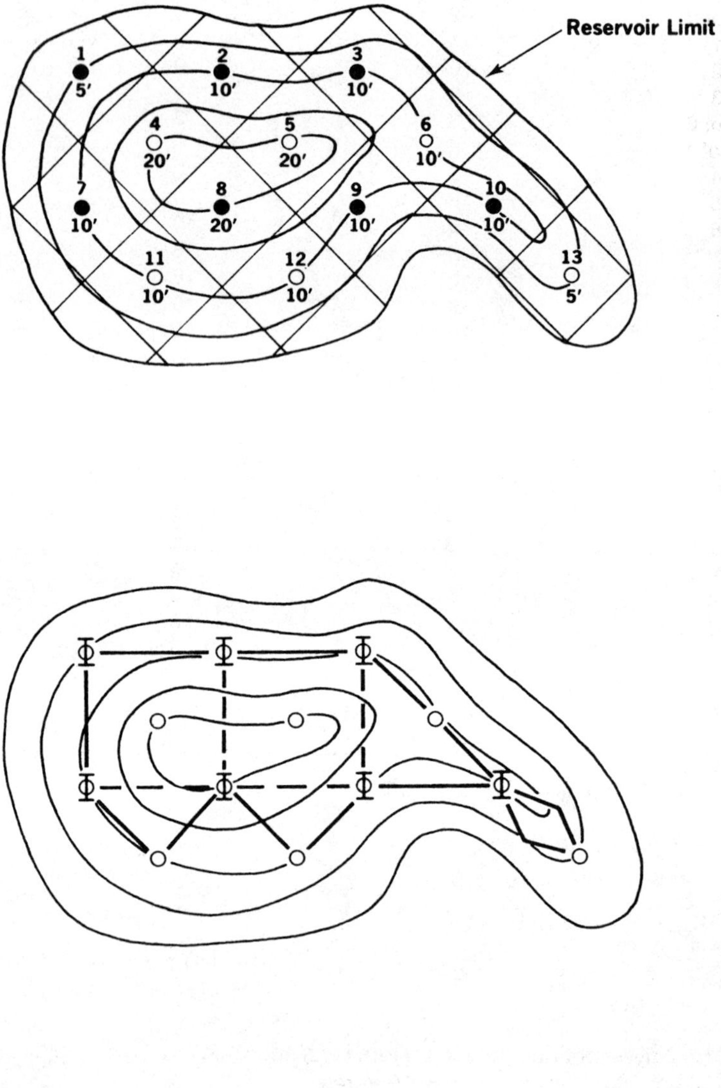

Fig. C16 Solution to problem 9.2

PROBLEM 9.3 Solution: Calculating Total Displacement Efficiency and Recovery from a Waterflood

Assume the total displacement efficiency at the economic limit is Stiles R at the economic water cut:

$$f_w = \frac{A\,c_j}{[A\,c_j + (c_t - c_j)]} \qquad (9.5)$$

$$c_t = \Sigma(kh) = 5{,}059 \text{ md-ft from the given data}$$

$$A_{eff} = \frac{B_o\left[1 + \left(\left(\frac{k_r}{\mu}\right)_w \Big/ \left(\frac{k_r}{\mu}\right)_o\right)\right]}{2} \qquad (9.21)$$

Reading k_{rw} and k_{ro} from Fig. 9–15

$$A_{eff} = \frac{1.1\left[1 + \left(\frac{0.67}{0.7} \Big/ \frac{0.98}{6.38}\right)\right]}{2} = 3.98$$

$$0.95 = \frac{3.98\,c_j}{[3.98\,c_j + 5{,}059 - c_j]}$$

$$c_j = 4{,}183 \text{ md-ft}$$

$$c_t - c_j = 5{,}059 - 4{,}183 = 876 \text{ md-ft}$$

To find h_j corresponding to c_j, sum the kh values from the least permeable in an increasing permeability direction until 876 md-ft is reached. The corresponding thickness will be h_j:

$$15 + 35 + (46 \times 2) + 50 + 54 + 57 + 61 + 70 + 76 + (85 \times 2) + 87 + 109 = 876 \text{ md-ft}$$

This corresponds to a thickness of 14 ft:

$$R = \frac{h_t k_j + (c_t - c_j)}{h_j k_j} \qquad (9.6)$$

$$h_j = 29 - 14 = 15 \text{ ft}$$

$$= \frac{(15 \times 127) + 876}{(29 \times 127)}$$

$$= 0.76$$

This efficiency is high because of the lower permeability limit being so high. If the formation down to a permeability of 1.0 md is considered, the efficiency would be more normal:

$$N_{pf} = 7{,}758\,\phi\,[(S_{op}/B_{op}) - (S_{or} - B_{or})]\,E_t V_{sw} \qquad (9.1)$$

From problem 9.2 solution we know $S_{op} = 0.561$; $V_{sw} = 900$ acre-ft and $N = 155.2 \times 10^4$ stb:

$$N_{pf} = 7.758 \times 0.2 \times \left[\left(\frac{0.561}{1.1}\right) - \left(\frac{0.19}{1.1}\right)\right] \times 0.76 \times 900$$

$$N_{pf} = 357{,}900 \text{ stb}$$

$$N_{pf} + N_{pP} = 357{,}900 + 364{,}500$$

$$= 722{,}400 \text{ stb}$$

$$\left(\frac{N_{pf}}{N}\right) = \frac{357{,}900}{1{,}552{,}000} = 0.23$$

$$\frac{(N_{pf} + N_{pP})}{N} = \frac{722{,}400}{1{,}552{,}000} = 0.47$$

Oil originally in swept area = (900/1,500) 1,552,000
= 931,200 stb

$$\frac{N_{pf}}{931{,}200} = \frac{357{,}900}{931{,}200} = 0.38$$

N_{pP} from gross swept flood volume = (900/1,500) × 364,500 = 218,700

$$\frac{(N_{pf} + N_{pP} \text{ from swept volume})}{931{,}200} = \frac{(357{,}900 + 218{,}700)}{931{,}200} = 0.62$$

PROBLEM 9.4 Solution: Calculating Reduced-Time Curves for Use in a Waterflood Prediction

(1) (W_i/V_p)	(2) $\Delta(W_i/V_p)$	(3) i_r	(4) $(i_r)_{avg}$	(5) $\Sigma i_r \Delta(W_i/V_p)$	(6) t_{r1} $\frac{0.25}{0.025}\Sigma$	(7) t_{r2} $\frac{0.2}{0.01}\Sigma$	(8) t_{r3} $\frac{0.15}{0.002}\Sigma$
0	0	1.0	—	0	0	0	0
0.2	0.2	1.0	1.0	0.2	2.0	4.0	15.0
0.33	0.13	4.33	4.33	0.763	7.63	15.26	57.23
0.45	0.12	3.165	3.75	1.213	12.13	24.26	90.98
0.60	0.15	2.63	2.90	1.648	16.48	32.96	123.60
0.84	0.24	2.33	2.48	2.243	22.43	44.86	168.23
1.20	0.36	2.18	2.26	3.057	30.57	61.14	229.28
2.68	1.48	2.00	2.09	6.150	61.50	123.00	461.25
3.68	1.00	2.00	2.0	8.150	81.50	163.00	611.25

(1) Assumed
(2) Change in column 1
(3) Read from Fig. 9–18 for column 1 value
(4) Average of column 3 over interval of column 2
(5) Sum of products of column 2 × column 4
(6), (7), and (8) Application of Eq. 9.29 to the individual zones. The summation is column 5

PROBLEM 9.5 Solution: Calculating Oil and Water Production Rates from Reduced-Time Curves

(1)* t_r: Zone	(2)* W_i/V_p	(3)* $N_{pf}B_o/V_p$	(4)* V_p, Mstb	(5)* W_i Mstb	(6)* N_p, Mstb	(7)* W_p, Mstb
2:1	0.20	0.0 (Crit)	97			0.0
:2	0.10	0.0	155			0.0
:3	0.027	0.0	175			0.0
2: Total				39.6	0 (Crit)	0.0
4:1	0.24	0.04	97			0.0
:2	0.20	0.0 (Crit)	155			0.0
:3	0.05	0.00	175			0.0
4: Total				63.0	3.9	0.0
7.63:1	0.33	0.13	97			0.0 (Crit)
:2	0.24	0.04	155			0.0
:3	0.10	0.00	175			0.0
7.63: Total				86.7	18.8	0.0 (Crit)
10:1	0.39	0.173	97			1.6
:2	0.26	0.060	155			0.0
:3	0.13	0.000	175			0.0
10: Total				100.9	26.1	1.6
15:1	0.55	0.249	97			9.8
:2	0.32	0.120	155			0.0
:3	0.20	0.000	175			0.0
15: Total				138	43	9.8
20:1	0.73	0.290	97			23.3
:2	0.39	0.173	155			2.6
:3	0.21	0.010	175			0.0
20: Total				168	57	25.9
30:1	1.17	0.338	97			61.3
:2	0.55	0.249	155			15.7
:3	0.25	0.050	175			0.0
30: Total				243	80	77.0
40:1	1.65	0.360	97			105.7
:2	0.74	0.291	155			38.6
:3	0.28	0.080	175			0.0
40: Total				324	94	144.3

(1)* t_r is assumed
(2)* From Fig. 9–20
(3)* From Fig. 9–19
(4)* Given data
(5)* $\Sigma[(W_i/V_p)V_p]$ (From Eq. 9.30)
(6)* $\Sigma[N_{pf}B_o/V_p)(V_p/B_o)$ (From Eq. 9.31)
(7)* $\Sigma[(W_i/V_p) - (N_{pf}B_o/V_p) - S_{gi}]V_p$: Eq. 9.32; set negative terms to 0.0
N_p and W_p are plotted versus W_i in Fig. C17

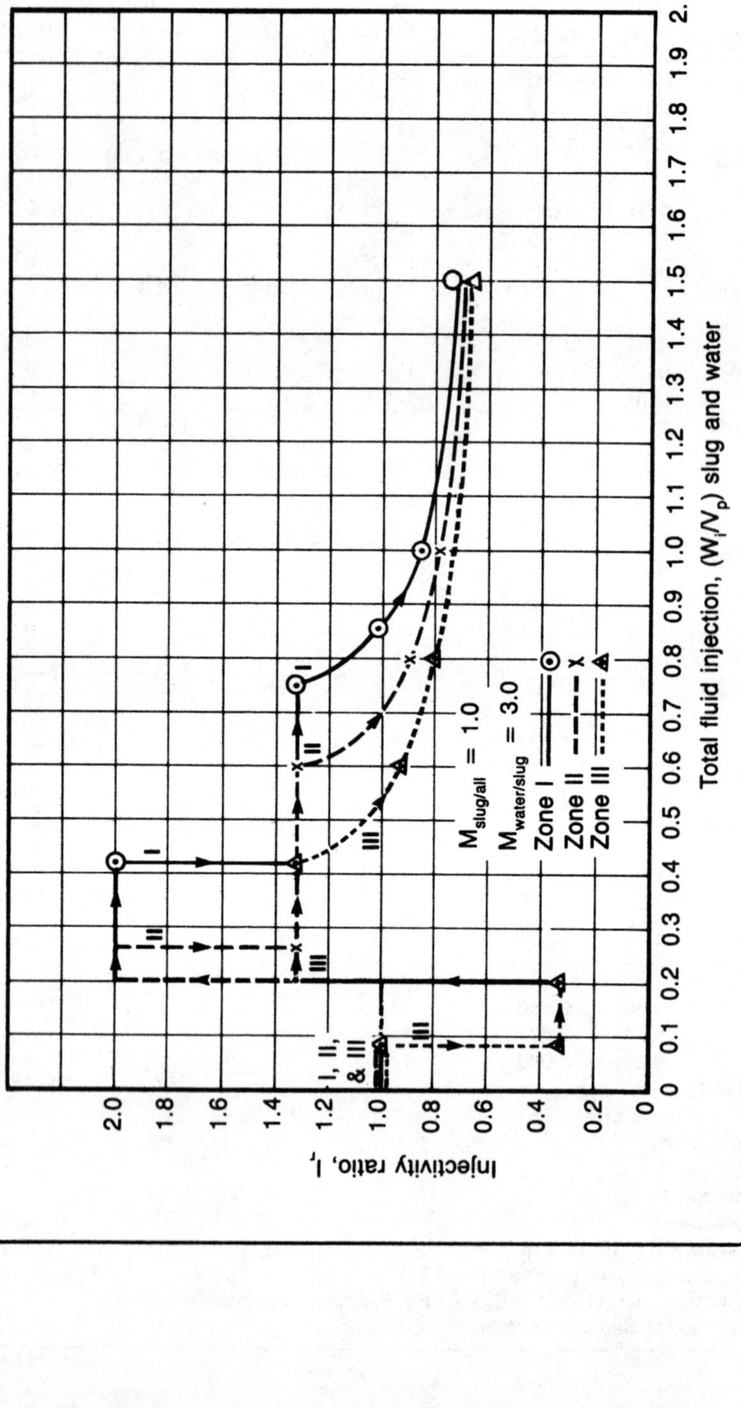

Fig. C17 Solution to problem 9.5

PROBLEM 9.6 Solution: Calculating Individual Strata Recovery Curves

$M = (k_{rw}/k_{ro})(\mu_o/\mu_w) = (0.55/0.98)(4.15/0.7) = 3.33$ and $1/M = 1/3.33 = 0.3$

At gas fillup:

$$(W_i/V_p) = 0.2 = S_{gi}$$

At water BT:

$$DVI = 0.55 \text{ at } 1/M \text{ of } 0.3 \text{ from Fig. 9–12}$$

$$\text{Displacement volume} = (1.0 - S_{wc} - S_{or})V_p = (1.0 - 0.15 - 0.25)V_p = 0.6\ V_p$$

(1) Displacement Volumes Injected	*(2) (W_i/V_p)*	*(3) E_p*	*(4) (B_oN_{pf}/V_p)*
0	0	0	0.0
—	0.2	—	0.0 (critical)
0.55	0.33	0.55	0.130
0.75	0.45	0.68	0.208
0.90	0.55	0.74	0.244
1.00	0.60	0.77	0.262
1.10	0.66	0.80	0.280
1.20	0.72	0.82	0.292
1.30	0.78	0.83	0.298
1.40	0.84	0.85	0.310
1.50	0.90	0.86	0.316
1.75	1.05	0.88	0.328
2.00	1.20	0.90	0.340
2.25	1.35	0.91	0.346
2.50	1.50	0.92	0.352
	2.68*	1.00	0.400

(1) : Assumed
(2) = (0.6)(column 1)
(3) : Read from Fig. 9–12 at DVI of column 1 and 1/M of 0.3
(4) = (0.6)(column 3) – (0.2)
*Determined by extrapolation of Fig. 9–19 plotted points to a recovery value of 0.4.

PROBLEM 9.7 Solution: Calculating Individual Strata Injectivity Curves

$$M = (k_{rw}/k_{ro})(\mu_o/\mu_w) = (0.55/0.98)(4.15/0.7) = 3.33$$

At gas fillup:

$$(W_i/V_p) = 0.2 = S_{gi}$$

At water BT:

$$\text{DVI} = 0.55 \text{ at } 1/M \text{ of } 0.3 \text{ from Fig. 9–12}$$

$$1.0 \text{ displacement volume} = (1.0 - S_{wc} - S_{or})V_p$$
$$= (1.0 - 0.15 - 0.25)V_p = 0.6\,V_p$$

From problem 9.6 solution:

(1) DVI	(2) (W_i/V_p)	(3) E_p	(4) f_w	(5) i_r
0	0	0	0.0	1.0
—	0.2	—	0.0	4.330
0.55 (BT)	0.33	0.55	0.0 (critical)	4.330
0.75	0.45	0.68	0.50	3.165
0.90	0.55	0.74	0.67	2.770
1.00	0.60	0.77	0.73	2.630
1.10	0.66	0.80	0.78	2.510
1.20	0.72	0.82	0.82	2.420
1.30	0.78	0.83	0.83	2.400
1.40	0.84	0.85	0.86	2.330
1.50	0.90	0.86	0.88	2.280
1.75	1.05	0.88	0.90	2.230
2.00	1.20	0.90	0.925	2.180
2.25	1.35	0.91	0.940	2.140
2.50	1.50	0.92	0.950	2.110
	2.68	1.00	1.000	2.000

(4) Read from Fig. 9–11 at sweep efficiency of column 3 and 1/M = 0.3
(5) Until gas fillup $i_r = 1.0$; after gas fillup $i_r = 4.33 - 2.33$ (column 4) according to Eq. 9.48

PROBLEM 9.8 Solution: Predicting Total Flood Recovery by Analogy

For the old flood:

$$\text{Total recovery, stb/acre-ft} = \frac{\text{stb of primary production}}{\text{acre-ft of primary drainage}} + \frac{\text{stb of waterflood production}}{\text{acre-ft of gross swept volume}}$$

$$= \frac{2{,}410{,}000}{10{,}000} + \frac{1{,}570{,}000}{8{,}000} \tag{9.50}$$

$$= 437\ \frac{\text{stb}}{\text{acre-ft}} = 437 \times 1.01$$

$$= 441 \text{ res bbl/acre-ft}$$

$$\text{Fractional recovery} = \frac{441}{0.18(7{,}758)} = 0.316$$

Now, to determine the total recovery of the new flood, assume fractional:

$$\text{total recovery}_{old} = \text{fractional total recovery}_{new} = 0.316$$

For the new flood:

$$0.316 = \frac{(\text{total recovery/acre-ft})\, B_{os}}{\text{pore vol/acre-ft}} = \frac{(\text{total recovery/acre-ft})\ 1.0}{7{,}758\ (0.20)} \tag{9.49}$$

$$\text{Total recovery/acre-ft} = 490\ \frac{\text{stb}}{\text{acre-ft}}$$

$$490\ \frac{\text{stb}}{\text{acre-ft}} = \frac{\text{primary total production}}{\text{primary acre-ft}} + \frac{\text{secondary total production}}{\text{secondary acre-ft}}$$

$$\frac{\text{secondary total prod.}}{14{,}000} = 490\ \frac{\text{stb}}{\text{acre-ft}} - \frac{3{,}645{,}000}{15{,}000}\ \frac{\text{stb}}{\text{acre-ft}} \tag{9.50}$$

$$\text{Secondary total production} = 3{,}458{,}000\ \text{stb}$$

PROBLEM 9.9 Solution: Predicting Rate versus Time by Analogy

We first determine the effective injection rate for the new flood:

$$p_{wi} = 0.433\ \gamma D + p_{wh} \tag{9.52}$$

$$p_{wi(old)} = (0.433)(1.0)(3{,}000) + 600 = 1{,}899\ \text{psia}$$

$$p_{wi(proposed)} = (0.433)(1.0)(2{,}500) + 800 = 1{,}883\ \text{psia}$$

$$i_{proposed} = i_{old}\ \frac{(kh/\mu)_{proposed}\,(p_{wi} - p_{wp})_{proposed}\left(\ell n\ \frac{d}{r_w} - 0.619\right)_{old}}{(kh/\mu)_{old}\,(p_{wi} - p_{wp})_{old}\left(\ell n\ \frac{d}{r_w} - 0.619\right)_{proposed}} \tag{9.51}$$

The geometry terms cancel and:

$$i_{proposed} = 1{,}700\ \frac{(0.075 \times 780/1.0)}{(0.1 \times 401/1.0)}\ \frac{(1{,}883 - 15)}{(1{,}899 - 15)} = 2{,}458\ \text{b/d}$$

$$\text{Effective injection rate} = 0.95 \times 2{,}458 = 2{,}335\ \text{b/d}$$

The oil producing rate versus time can then be calculated in tabular form as shown in the table on p. 816. Read from data graphs of old flood.

(1) Time, years	(2) Rate, b/d	(3) Current injection rate, b/d	(4) Cumulative injection, bbl	(5) $\dfrac{b/d}{\text{current effective injection}} = \dfrac{\text{column (2)}}{0.90 \text{ column (3)}}$	(6) New rate, b/d column (5) × (2,335)	(7) $\dfrac{\text{cumulative effective injection}}{\text{sec. recovery}} = \dfrac{0.9 \text{ column (4)}}{1,570,000}$	(8) New time, years $\dfrac{\text{column (7)}}{365} \dfrac{3,458,000}{0.95} \times \dfrac{1}{2,458}$
0.50	94	1,500	191,625	0.0696	163	0.1078	0.44
1.50	1,370	3,000	1,286,625	0.507	1,184	0.718	2.91
2.25	1,170	2,200	1,888,875	0.592	1,382	1.061	4.30
3.50	580	2,000	2,801,375	0.322	752	1.572	6.38
5.25	250	1,600	3,823,375	0.174	406	2.146	8.70
5.75	195	1,800	4,151,875	0.120	280	2.340	9.49
7.50	101	1,500	5,110,000	0.075	175	2.870	11.64

PROBLEM 9.10 Solution: Calculating Injectivity Curves when using a Slug

To determine the amount of slug entering each zone, we first apply the recommended method assuming continuous injection of the pusher. Sweep efficiency at breakthrough is 0.6935. Thus, W_i/V_p at breakthrough would be:

$$0.6935 \times 0.6 = 0.416$$

Using the characteristics of zone I (the most permeable zone) to calculate t_r corresponding to breakthrough in zone I:

$$\begin{aligned} t_{rBT} &= [\Sigma i_r \Delta(W_i/V_p)](\phi/k) \qquad (9.29) \\ &= [(1 + 0.2) + (0.416 - 0.2)2.0](0.25/0.025) \\ &= (0.2 + 0.432)10 = (0.632)(10) \\ &= 6.32 \end{aligned}$$

To determine (W_i/V_p) for the other zones at zone I breakthrough, apply the previous equation to the other zones with $t_r = 6.32$ and (W_i/V_p) to be calculated. For zone 2:

$$\begin{aligned} 6.32 &= [0.2 + 2.0 \times \Delta(W_i/V_p)]20 \\ \Delta(W_i/V_p) &= 0.58 \\ (W_i/V_p) &= 0.2 + 0.058 \\ &= 0.258 \end{aligned}$$

For zone 3:

$$\begin{aligned} 6.32 &= [1.0\,\Delta(W_i/V_p)]75 \\ \Delta(W_i/V_p) &= (W_i/V_p)_3 = 0.084 \end{aligned}$$

Calculate the i_r curves for the individual zones. For zone I i_r is the same as the program to BT.

For $W_i/V_p = 0.0$–0.2:

$$i_r = 1.0$$

For $W_i/V_p = 0.2$–0.416:

$$i_r = 2.0$$

The mobility of the water before pusher injection is reduced by 13%:

$$\begin{aligned} \lambda_w &= (0.75/0.71)/1.13 = 0.9375 \\ \lambda_{slug} &= 0.75/2.37 = 0.316 \\ \lambda_o &= 0.95/3.0 = 0.317 \\ M_{slug/o} &= 0.316/0.317 = 1.0 \\ M_{w/slug} &= 0.9375/0.316 = 3.0 \end{aligned}$$

Water BT of the slug is based on the $1/M_{w/slug}$ of 0.333 when from Fig. 9–12 DVI = 0.56 or $(W_i/V_p)_w = (0.6)(0.56) = 0.336$.
Therefore, $(W_i/V_p)_{w+slug} = 0.416 + 0.336 = 0.752$ from $W_i/V_p = 0.416$–0.752.
Apply equation 9.55:

$$i_r = (1/M_{w/slug}) + f_{slug} + (1 - f_{slug})M_{slug/o}$$

Then when $M_{slug/o} = 1.0$:

$$i_r = (1/M_{w/slug}) + 1.0$$

$$i_r = 0.333 + 1.0 = 1.333$$

The injectivity ratio is only constant when slug and oil are being produced and water injected and the slug/oil mobility ratio is 1.0. After water breakthrough we base i_r on Eq. 9.56. When

$(W_i/V_p)_{total} = 1.0$:

$$(W_i/V_p)_w = 1.0 - 0.416 = 0.584$$

From Fig. 9–12 when:

$$DVI = 0.584/0.6 = 1.0$$

$$f_w = 0.73$$

$$i_r = (1/M_{w/slug})(1 + f_w) + (1 - f_w) \tag{9.56}$$

$$= (0.333)(1 + 0.73) + (1 - 0.73) = 0.846$$

This process can be repeated for other assumed values of $(W_i/V_p)_{total}$. We can estimate from Fig. 9–12 that $f_w = 1.0$ when:

$$DVI = 3.2$$

Or:

$$(W_i/V_p)_w = (0.6)(3.2) = 1.92$$

Or:

$$(W_i/V_p)_{total} = 1.92 + 0.416 = 2.336$$

Then:

$$i_r = (\lambda_w/\lambda_{slug})2 = 0.333 \times 2 = 0.666$$

Thus, the i_r curve for zone I would be as shown in Fig. C18. The i_r curves for zones I and II can be similarly calculated.

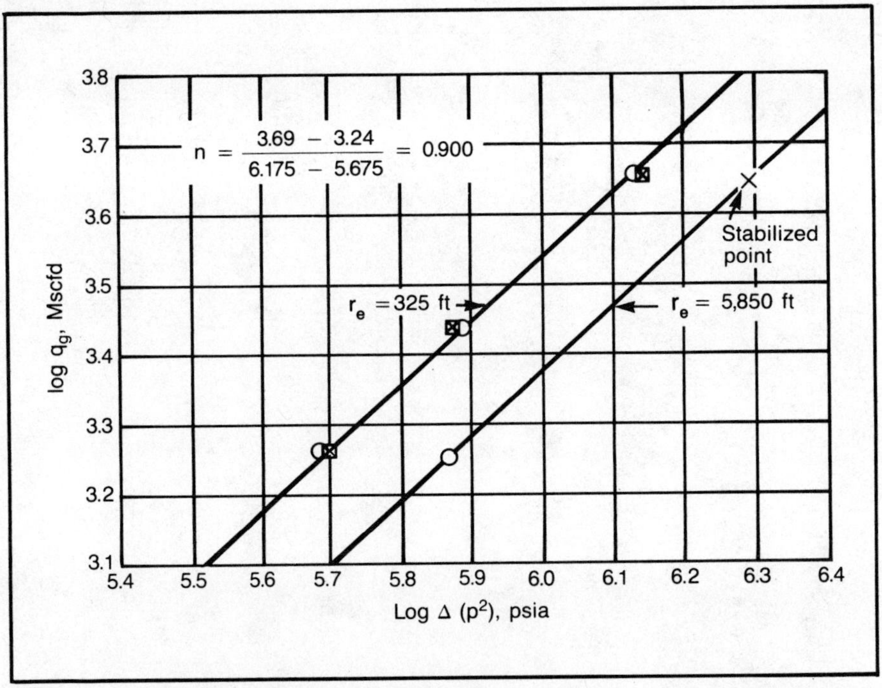

Fig. C18 Injectivity ratio curves for waterflood with slug injected until slug breakthrough, problem 9.10 solution

Index

A
abandonment (well/reservoir), xvi
absolute permeability, 14–15, 17. *See also* permeability (reservoir).
acidizing, 39, 541
Angola, xii
anomaly. *See* pressure-buildup anomaly.
aquifer, 114, 488
artesian flow, 433

B
back-pressure test, 33, 306–310
bookkeeping method (reservoir performance prediction), 580
bottom-hole pressure, 3, 5, 9, 27, 61, 137–138, 284–287, 344
boundary effect, 128–133
bounded reservoir, 33
bubble-point condition, xviii, 5, 24, 446, 468, 504, 725
Buckley-Leverett method, 384, 411–412, 417–419, 424–425, 428, 546, 548

C
Canada, xii
capillary action/effect, xvi, 15, 18–20. *See also* capillary-pressure/residual-oil relationship.
capillary-pressure/residual-oil relationship, xiii, 20–22, 384, 387–404, 414–415, 440, 442–443, 545–547, 560–563, 616–617, 628
carbon dioxide injection, xv, 371–372, 545, 655, 657–659
casing/tubing flow capacity, 347–351, 356
checklist (reservoir data), 9
coefficient of permeability variation, 597–598
Colombia, xii
combination-drive reservoir, 384–385
commingled production, xi, 7
compressibility (fluid), 11, 27, 29–30, 34–35, 72, 115–118, 270–272, 329–331, 341, 474, 694–695, 712–713, 715
compressibility (rock), 464–468, 716
computer modeling (reservoir), xi, xiii, xvii, 8, 24, 445–447, 482, 487–488, 562–563, 667–679: preparing a reservoir modeling study, 668–669; incompressible flow model, 669–672; general reservoir model, 672–677; history matching, 677–679
computer program, xvii: explicit vs. implicit, 675–677
coning, 4, 8, 383–384, 436–440, 497, 501
connate water, 19, 387, 410, 467, 547
constant-percentage decline, 517–523: use in determining reserve, 520–523
constant-pressure method, xiii, 5, 9, 315–316
constant-rate-drawdown test, 5, 135, 154–160, 329–331
constant-terminal-pressure solution, 103–115
constant-terminal-rate solution, 79–93, 317–318
core data/analysis, 2–4, 8–9, 386–388, 620–621
critical property, 262–265, 273, 718
curve-fitting test, 143
cuttings, 4–5

D
damage ratio, 43. *See also* well damage.
Darcy's law, 11–25, 33–35, 45, 52, 56–59, 70–71, 75, 147, 299–300, 305, 332, 414
data accuracy, 1–4, 5–9, 24, 50, 395–400, 472–474
data interpretation, 5–9, 24–25, 33, 47, 60, 395–400: approximation, 47–55
decline-curve analysis, xiii, 513–542: Gentry method, xiii, 534–536; Fetkovitch method, xiii, 534, 536–537, 542; decline-rate definition, 514–517; constant-percentage decline, 517–523; hyperbolic decline, 523–540; harmonic decline, 524, 540–541; log-log data plot, 536, 538–539
Deitz method, 191–192, 196
deliverability test/data, 135, 259, 306, 311–313, 344–351, 358–360, 374
density (gas), 270, 714
depletion (reservoir), 310–311, 351–355, 485

dipping formation, 58
displacement (of reservoir fluid), 383–443, 457–461, 482–486, 493: initial saturation distribution, 384–404: fluid distribution during, 405–430, 506, 509; interface tilt during, 430–440; mechanism of, 545–551. *See also* fluid distribution (reservoir) *and* saturation (reservoir).
displacing-fluid bank, 408–409
disposal-well behavior, 69, 204
distance between wells, xvi, 427. *See also* well placement/spacing.
drag zone, 407
drainage area, 4, 8, 24, 26–27, 38–39, 47–48, 64–65, 69, 188–204
drawdown analysis, 69, 79, 137, 143, 160–168, 329–331, 334
drillstem-test data/analysis, 2–4, 9, 69, 143, 220–229
drive index (calculation of), xiii

E
early well life (data), 4–5, 9
economic factor, 1–2
effective permeability, 58–61. *See also* permeability (reservoir).
Ei-function solution, 82–93, 160
electrical service (interruption of), xv
energy crisis, xv
energy demand, xv–xvi
England, xii
enhanced oil recovery, xii, 239, 383–384, 545–546, 625–664: steam and hot-water injection, 629–641; in situ combustion, 640–647; micellar and surfactant flood, 647–651; miscible displacement, 651–662; polymer flooding, 659–661; feasibility analysis of process, 661, 663–664. *See also* secondary recovery, tertiary recovery, *and* waterflooding.
equilibrium gas saturation, 21, 548. *See also* saturation (reservoir).
equipment capacity, 259, 344–351
exponential decline. *See* constant-percentage decline.

F
Fetkovitch method (of decline-curve analysis), xiii, 534, 536–537, 542
fingering, 383–384, 430, 497, 501
five-spot pattern, 34, 38, 50
flaring (of gas), 6
flow geometry, 34–38, 45, 47–52, 56–57
flow line capacity, 346–347, 350–351
flow rate, 11–13, 27, 31, 33–35, 42, 49, 60–61, 121–128, 326, 328, 344–351, 366, 730–731. *See also* equipment capacity.
flow regime (limitations), 11, 14–25: characteristics, 26–33
fluid distribution (reservoir), 383–443, 457–461, 482–486, 493; initial saturation distribution, 384–404; during displacement, 405–430, 506, 509; interface tilt during displacement, 430–440; mechanism of displacement, 545–551. *See also* displacement (of reservoir fluid) *and* saturation (reservoir).
fluid flow equation/calculation, 11–26, 34–38, 96–98, 216, 412, 473, *et passim.*
fluid flow problem, 41, 55, 61–66, 82, 85–88, 101–102, 114, 135–140, 151–154, 156–157, 297–298
fluid sample, 4–5, 9, 386
formation damage, 4–5, 39
formation thickness, 36–37
formation-volume factor, 4, 40, 60–61
fractional flow curve, 412–417, 423, 429, 441, 546, 550–552
fracturing, 39. *See also* massive hydraulic fracturing.
free gas, 468–469, 546, 548
free-water level, 394
fresh water (reservoir source), xi, 114, 433
frontal displacement (reservoir fluid), 383–443. *See also* displacement (of reservoir fluid), fluid distribution (reservoir), *and* saturation (reservoir).
fuel shortage, xv

G
gas-cap drive, xviii, 33, 36, 63, 383–384, 408, 446–447, 474–486, 502–507, 545
gas-deviation factor, xiii, 4, 44–45, 62, 258, 260–269, 286–287, 329–331, 717, 719
gas-flow equation, 44–46
gas formation volume factor, 266, 269–270, 447–448
gas in solution, 4, 462–464. *See also* solution-gas-drive reservoir.
gas injection, 504–505, 509, 545
gas liberation, 452–453, 462–464, 471
gas meter, 6, 9
gas-oil contact, 3–4, 7, 9, 383, 385–387, 430–431
gas-oil ratio, 7–8, 24, 457–461, 475–478, 481, 486, 493
gas potential, 331–334
gas reservoir, xv, 44–46, 62, 257–380, 383: properties of natural gas, 259–273; material balance, 273–298; fluid flow in, 298–344; equipment capacity limitations on deliverability, 344–351; predicting performance of, 351–362, 478–482; predicting behavior of microdarcy

reservoirs, 362–366; determining size of, 366–371
gas saturation. *See* saturation (reservoir).
gas well testing procedure, xvi
gas zone, 384–385, 405
gasoline shortage, xv
Gentry method (of decline-curve analysis), xiii, 534–535
geopressured reservoir, 118, 290–291, 293, 371
Germany, xii
gravity effect, 383–384, 416–417, 721
gross swept volume, 555–560. *See also* sweep efficiency.

H
harmonic decline, 524, 540–541
heating oil shortage, xv
hemispherical flow, xi, 34, 36–37, 46–47, 133–135, 216
histogram, 14–15
history matching, 677–679
horizontal-sweep efficiency, 563–566
Horner method/plot, xi, xvi–xvii, 3, 184–189, 195, 223, 227–228
huff-and-puff method, 630–633. *See also* steam and hot-water injection.
humping effect, 209–216
Hurst-van Everdingen constant-terminal rate solution, 79–82, 88–94, 98, 109–113
hydrodynamics (reservoir), 383–384, 430–440
hyperbolic decline, 523–540: curve-fitting method, 526, 530–532; empirical data, 695–708

I
immiscibility, 384, 408, 410, 505–545
incompressible flow, 35, 46, 51, 53: model of, 669–672
Indonesia, xii, 433
inefficiency (of production), xv–xvi
infinite-acting flow, xiii, 31, 43
in situ combustion, 640–647
initial-flow period/test, 4, 9
injection well/process, xvi, 204–205, 600–603
injectivity curve (individual strata), 591–595, 614–616
interface tilt, 430–440. *See also* hydrodynamics (reservoir).
interfacial tension, xvi, 387, 430
interference test, 69, 79, 135–136, 233
inversion (pressure-buildup analysis), 201
Iran, 13
irreducible saturation. *See* saturation (reservoir).
isochronal testing, 79, 135, 318–329, 334–335, 337, 374–375
isosaturation tilt. *See* interface tilt.

J
J-function method, 398–401, 403
Japan, xii

K
Klinkenberg effect, 58–60

L
Leighton and Higgins method, 546
liberation. *See* gas liberation.
limestone, 14
linear-diffusivity equation, 672–676
linear flow, 36, 46–47, 53, 56, 58, 383–428, 433–436, 672–675
liquefied-petroleum gas injection, 545, 651, 653–655
liquid expansion (in reservoir), 451–452
lithology, 2
log data, 2–4, 8–9, 536, 538–539, 562–563
log-inject-log method, 562
log-log data plot, 536, 538–539
logging suite, 2
low-drawdown production, xi

M
mass balance. *See* material balance.
massive hydraulic fracturing, 4, 39, 47, 366
material balance, xviii, 5–6, 8, 11, 258–259, 273–298, 419–422, 445–511: material-balance equation, 274–275, 447–453, 468–474; material-balance gas production, 275–278, 447–451, 647–651; material-balance graph, 284–289; pressure-volume-temperature relationship, 446, 453–457; behavior of produced gas-oil ratio, 457–461; with gas liberation, 462–464; pore-volume change, 464–468; modification of general equation for, 468–470; difficulty in applying equation for, 470–474; predicting gas-drive behavior, 474–486; predicting water-drive reservoir behavior, 486–500; increasing primary recovery, 500–507
material-balance equation, 274–275, 447–453, 468–474
material-balance gas production, 275–278, 447–451, 647–651
material-balance graph, 284–289
Matthews, Brons, and Hazebroek method, xi, 189–192, 196
maximum efficient rate, 406

micellar and surfactant waterflooding, 545, 616–617
micellar solution, 647. *See also* micellar and surfactant waterflooding.
microdarcy gas reservoir analysis, xiii, 362–367, 371
military supervision (of explosives), xi
miscibility, 384, 505, 545
miscible displacement, 404, 651–662
mobility-control agent, 568, 608–612
mobility (reservoir fluid), 416, 539–540, 545, 608–612, 614–616
mud filtrate, 4–5, 561
mud particle, 4–5
mud pressure, 3
multiphase flow, 11, 14–25
multiple-flow-rate capability, 3
multiple-well effect, 119–126
Muskat method, 480–482

N
natural gas property, 259–273, 371–372
natural gas supply/demand, xv, 257–258, 371
negative superposition, 143
Netherlands, 634, 636
net-pay thickness, 2
Nigeria, xii
nitrogen injection, 656
non-Darcy flow, 13, 299
nonwetting phase. *See* wettability.

O
observation well, 8–9
Odeh method, 192–194, 199
oil bank, 408, 548
oil formation volume factor, 453, 463–464, 485, 508, 727
oil reservoir, xv, 62, 370, 383, 451–452, 491–492
oil saturation. *See* saturation (reservoir).
oil zone, 384–385, 405, 463, 464, 494
oil-field unit (of measurement), 12–13, 40
open-hole completion, xi, 37

P
paraffin problem, 501
parallel flow, 53
partially quenched combustion, 644
perforation (well), xi, 36
permeability distribution. *See* permeability (reservoir).
permeability problem, 16–17, 55, 156–157, 239–240
permeability (reservoir), xi, 3–5, 7, 9, 12, 14–28, 32, 38–40, 42–43, 46, 52–55, 58, 62–63, 65, 69, 209, 214–215, 217, 247, 257, 273, 343–344, 373, 384, 394–395, 397, 403, 407–408, 411, 416, 459, 509–510, 545, 547, 597–598, 619, 691–694, 710–711
phase behavior, 259–265
piezometric-pressure drop, 431–432
pilot flooding, 598–607
piston-like displacement, 405, 407
plant shutdown, xv
plugging back, 9
polymer waterflooding, 545, 608–612, 659–663
pore (reservoir), 14–15, 18–21, 384, 389–391, 403, 408, 446, 464–468, 548
porosity (reservoir), 2–3, 9, 14–15, 18–21, 292–293, 373, 384, 397
porous-media fluid flow, 11–25, 384, 403–443
positive-displacement flow meter, 6–7, 9
Prats method (modified), 580
predrilling planning (for data acquisition), 2–4, 9
pressure buildup. *See* pressure-buildup analysis *and* pressure (reservoir).
pressure-buildup analysis, xvii, 3–5, 9, 26, 69, 79, 135, 143, 159, 168–220, 340–344
pressure-buildup anomaly, 201, 209–216. *See also* pressure-buildup analysis.
pressure-change effect (reservoir fluid withdrawal), 126–128
pressure coring, 387
pressure decline. *See* pressure (reservoir).
pressure drop, 344–351
pressure falloff test, 204–209. *See also* pressure (reservoir).
pressure maintenance. *See* pressure (reservoir).
pressure problem, 58, 82, 85–88, 92, 121, 126, 130, 135–140, 151–154, 163, 176, 178–179, 199, 204, 208–209, 219, 225–226, 245–252, 283, 309–310, 321–322, 329, 350–351, 370–371
pressure (reservoir), 3–5, 8–9, 11–12, 26–33, 39–42, 44–46, 48, 51–52, 55–62, 70–71, 115–118, 143–252, 257, 262–265, 273, 278–287, 290–291, 293, 297, 336, 370–371, 383–384, 394–404, 415, 453–457, 464–468, 470–474, 483–485
pressure-volume-temperature relationship, 453–457, 470–472, 483
primary production, xi, xv, 500–507
problem listing, xix–xxi
problem solution, 733–818
producing capacity, 344–366
producing rate, xi–xii, xv, 5, 29, 32, 41, 143, 311–313, 316, 344–366, 373–380
production allocation, 6–7, 9
production data, 352, 472–474, 508
productivity index test, xvi, 33, 43, 135,

143–154, 245, 495–496
pseudosteady-state condition/flow, xi, 8, 11, 26, 31–34, 43, 46–47, 69–140, 258, 304–306, 366: definition of, 93–103; constant-terminal-pressure solution, 103–115; effective compressibility, 115–118; superposition, 118–132
pulse-test analysis, xiii, 143, 233–244, 252
pulsed-neutron capture log, 562

R
radial-diffusivity equation, 31, 72–79, 115, 118–119, 313–315
radial flow, 28–29, 32, 34, 36–37, 39–41, 43, 45–50, 53–55, 94, 428, 430, 436–440
radius of investigation, 225
rate-change effect (reservoir fluid withdrawal), 121–126
rate of recovery, xvi, 7–8, 551–579
rate-versus-time prediction, xiii, 361–363, 546, 580–598, 603–607
real-gas function, xiii
recovery curve (individual strata), 589–590
recovery (maximizing), xvi, 7, 65–66, 500–507, 589
reduced-time curve method, 586–589
relative permeability, 14–25, 547, 549, 575–579. *See also* permeability (reservoir).
reserve estimate, 540
reservoir barrier, 4, 24, 131, 156–159, 163–168, 209, 211, 213, 246
reservoir-boundary delineation, 69, 143, 159–168, 209, 211, 213, 368, 370–371
reservoir condition, xi–xiii, 60–61, 626–629. *See also* fluid distribution (reservoir), permeability (reservoir), porosity (reservoir), pressure (reservoir), *and* saturation (reservoir).
reservoir control procedure, 500, 502
reservoir engineering data, xii, 1–9: predrilling planning for data acquisition, 2–4, early well-life data, 4–5; routine data, 509; checklist of, 9; empirical, 691–731
reservoir fluid flow, 11–66, 69–140, 298–344, 351–366, 383–443: Darcy equation, 11–33; steady state, 33–46; problems in calculating, 46–55; correcting for static pressure differences, 55–61
reservoir limit test, 159–168
reservoir model. *See* computer modeling (reservoir).
reservoir performance, 25, 33, 344, 351–366, 383: predicting, 551–607
residual-oil/capillary-pressure relationship, xiii, 20–22, 384, 387–404, 414–415, 440, 442–443, 545–547, 560–563, 616–617, 628
Reynolds' number, 13
routine reservoir data (obtaining), 5–9

S
sand production, 501
sand/sandstone reservoir, 11, 14
saturation distribution. *See* fluid distribution (reservoir) *and* saturation (reservoir).
saturation (reservoir), 2–5, 9, 18–22, 24–25, 383–443, 457–461, 482–486, 493: initial saturation distribution, 384–404; during displacement, 405–430, 506, 509; interface tilt during displacement, 430–440; undersaturated water-drive reservoir, 490–492; water-drive reservoir below saturation pressure, 492; mechanism of displacement, 545–551; oil saturation, 553–555. *See also* displacement (of reservoir fluid) *and* fluid distribution (reservoir).
Saudi Arabia, xii, 13, 37, 136–137, 147
Schilthuis method, 475–476, 480–481
secondary recovery, 33, 38, 383–384, 405, 545, 625. *See also* enhanced oil recovery *and* tertiary recovery.
segregation (of reservoir fluid), 383
semisteady-state flow. *See* pseudosteady-state condition/flow.
seven-spot pattern, 34
shutin-well pressure, 1–9, 170–220, 372
sidewall sample, 4, 9, 386
single-phase flow, 11, 14
skin damage/effect/factor, 4–5, 39–40, 42, 65, 329, 334–340. *See also* formation damage *and* well damage.
slug distribution, 613–617
solution-gas-drive reservoir, xi, xv, 25, 144–148, 384, 446, 451–453, 460, 462–464, 474–486, 492–500, 502–506, 509–510, 541, 561
spherical flow, 37, 133–135, 216–220
stabilization-time equation, 99–101
stabilized flow, 33
static pressure difference, 55–61. *See also* pressure (reservoir).
steady-state condition/flow, xi, 8, 11, 26–28, 31, 33–47, 300–304, 439–440
steam and hot-water injection, 629–641, 663
steam soak. *See* huff-and-puff method.
steam stimulation. *See* huff-and-puff method.
steamflooding, 634–641. *See also* steam and hot water injection.
Stiles prediction (of flood behavior), xvi,

567–575, 580, 619
Stone's equation, 24
strata injectivity curve (individual), 591–595, 614–616
strata recovery curve (individual), 589–590
stratification (reservoir structure/reservoir fluid flow), 24, 209, 392, 595–597
Sumatra, 433, 488
superposition, 31, 118–132
sweep efficiency (waterflooding), 563–579

T
Tarner method, 477–478, 480–481
temperature (reservoir), xi, 44–45, 61–62, 262–265, 453–457, 470–472, 483
tertiary recovery, 625. *See also* enhanced oil recovery *and* secondary recovery.
thermal recovery. *See* in situ combustion *and* steam and hot water injection.
tilt. *See* hydrodynamics (reservoir).
tortuosity, 399
tracer injection, 562
transient-pressure test, 143, 229
transition zone (reservoir), 24
tubing/casing flow capacity, 347–351, 356
tubing placement, xi
turbulence, 13–14, 45–56, 62, 299–304, 306, 317–318, 329, 334–340, 373, 720
two-rate pressure-buildup test, 199–209, 343. *See also* pressure-buildup analysis.
type-curve matching method, xiii, 229–233, 366

U
ultimate recovery, xv, 5, 7, 599–600
unsteady-state condition/flow, 26, 28–31, 33, 47, 69–140, 258, 315–318: physical description, 69–72; radial diffusivity equation, 72–79; constant-terminal-rate solution, 79–93; linear, 133; spherical, 133–135

V
vertical fluid distribution. *See* fluid distribution (reservoir).
vertical-sweep efficiency. 567–579
viscosity, 4, 12–13, 45–46, 58–62, 64, 272, 329–331, 383–384, 411, 441, 508, 548, 584, 630–633, 722–724, 728–729
Vogel IPR data, xiii, 143, 152–153
volume (reservoir). *See* gas formation volume factor *and* oil formation volume factor.

W
water bank, 409
water-drive and solution-gas-drive prediction (combined reservoir behavior), xiii, 492–500
water-drive reservoir, xi, xvi, xviii, 24, 26–27, 33, 36, 41, 46, 64–65, 69, 101, 297, 383–384, 405–406, 408, 415–416, 486–500: undersaturated, 490–492; below saturation pressure, 492; combined with solution-gas drive, 492–500, 504–506, 508, 545
water encroachment, 293–298, 373, 383, 405, 415–416, 446, 486–490, 497, 502, 510
water injection, 410, 504–505, 545
waterflood prediction (using slug), xiii
waterflooding, 38, 384, 408, 410, 443, 545–622: displacement mechanism, 546–551; predicting total recovery, 551–579; predicting rate-versus-time performance, 580–598; predicting performance by analogy or from pilot flooding, 598–607; variations, 608–617
water-oil contact, 3–4, 7, 9, 48–49, 383, 385–387, 430–431, 491
water-oil ratio, 7–8, 24, 65
water production, 6
water saturation. *See* saturation (reservoir).
water zone, 384–385, 405
Welge method, 384, 422–428
well completion, 4, 39
well control procedure, 500–502
well damage, xiii, 4–5, 38–44, 46, 52, 64
well placement/spacing, xvi, 139–140, 310–313, 355, 357–361, 368–369, 501–502. *See also* five-spot pattern *and* seven-spot pattern.
well pressure, 27, 42, 70, 137–139, 143–252, 302–303, 372: productivity-index test, 143–154; constant-rate drawdown test, 154–159; reservoir limit test, 159–168; pressure-buildup analysis, 168–220; interpreting drillstem-test data, 220–229; type-curve matching, 229–233; pulse-test interpretation, 233–244
well radius (apparent), xiii, 27, 44
well testing, 46, 69, 143–252
wet combustion, 644
wettability, 18–21, 384, 387, 403, 628–629
wetting phase. *See* wettability.
Weymouth equation, 346–347
Wyllie equation, 63

Z
zonation, 384–385, 396–397